*Principal Stress*

$$\tan 2\theta_p = \frac{\tau_{xy}}{(\sigma_x - \sigma_y)/2}$$

$$\sigma_{1,2} = \frac{\sigma_x + \sigma_y}{2} \pm \sqrt{\left(\frac{\sigma_x - \sigma_y}{2}\right)^2 + \tau_{xy}^2}$$

*Maximum In-Plane Shear Stress*

$$\tan 2\theta_s = -\frac{(\sigma_x - \sigma_y)/2}{\tau_{xy}}$$

$$\tau_{max} = \sqrt{\left(\frac{\sigma_x - \sigma_y}{2}\right)^2 + \tau_{xy}^2}$$

$$\sigma_{avg} = \frac{\sigma_x + \sigma_y}{2}$$

*Absolute Maximum Shear Stress*

$$\tau_{\substack{abs \\ max}} = \frac{\sigma_{max} - \sigma_{min}}{2}$$

$$\sigma_{avg} = \frac{\sigma_{max} + \sigma_{min}}{2}$$

## Material Property Relations

*Poisson's Ratio*

$$\nu = -\frac{\epsilon_{lat}}{\epsilon_{long}}$$

*Generalized Hooke's Law*

$$\epsilon_x = \frac{1}{E}\left[\sigma_x - \nu(\sigma_y + \sigma_z)\right]$$

$$\epsilon_y = \frac{1}{E}\left[\sigma_y - \nu(\sigma_x + \sigma_z)\right]$$

$$\epsilon_z = \frac{1}{E}\left[\sigma_z - \nu(\sigma_x + \sigma_y)\right]$$

$$\gamma_{xy} = \frac{1}{G}\tau_{xy}, \quad \gamma_{yz} = \frac{1}{G}\tau_{yz}, \quad \gamma_{zx} = \frac{1}{G}\tau_{zx}$$

*where*

$$G = \frac{E}{2(1 + \nu)}$$

## Relations Between *w, V, M*

$$\frac{dV}{dx} = -w(x), \quad \frac{dM}{dx} = V$$

## Elastic Curve

$$\frac{1}{\rho} = \frac{M}{EI}$$

$$EI\frac{d^4v}{dx^4} = -w(x)$$

$$EI\frac{d^3v}{dx^3} = V(x)$$

$$EI\frac{d^2v}{dx^2} = M(x)$$

## Buckling

*Critical Axial Load*

$$P_{cr} = \frac{\pi^2 EI}{(KL)^2}$$

*Critical Stress*

$$\sigma_{cr} = \frac{\pi^2 E}{(KL/r)^2}, \quad r = \sqrt{I/A}$$

*Secant Formula*

$$\sigma_{max} = \frac{P}{A}\left[1 + \frac{ec}{r^2}\sec\left(\frac{L}{2r}\sqrt{\frac{P}{EA}}\right)\right]$$

# Geometric Prop

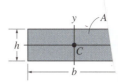

Rectangular area

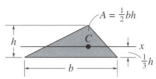

$A = \frac{1}{2}bh$

$I_x = \frac{1}{36}bh^3$

Triangular area

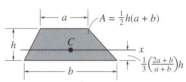

$A = \frac{1}{2}h(a + b)$

$\frac{1}{3}\left(\frac{2a + b}{a + b}\right)h$

Trapezoidal area

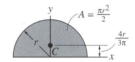

$A = \frac{\pi r^2}{2}$

$\frac{4r}{3\pi}$

$I_x = \frac{1}{8}\pi r^4$

$I_y = \frac{1}{8}\pi r^4$

Semicircular area

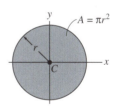

$A = \pi r^2$

$I_x = \frac{1}{4}\pi r^4$

$I_y = \frac{1}{4}\pi r^4$

Circular area

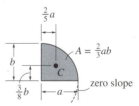

$\frac{2}{5}a$

$A = \frac{2}{3}ab$

zero slope

$\frac{3}{8}b$

Semiparabolic area

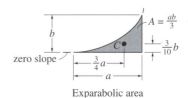

$A = \frac{ab}{3}$

$\frac{3}{10}b$

zero slope

$\frac{3}{4}a$

Exparabolic area

# STATICS AND MECHANICS OF MATERIALS

### SECOND EDITION

## TO THE STUDENT

With the hope that this work will stimulate an interest in Engineering Mechanics and Mechanics of Materials and provide an acceptable guide to its understanding.

This book represents a combined abridged version of two of the author's books, namely *Engineering Mechanics: Statics, Tenth Edition* and *Mechanics of Materials, Fifth Edition*. It is intended for those students who do not need complete coverage of these subjects. Rather, it provides a clear and thorough presentation of both the theory and application of the important fundamental topics of this material, which is often used in many engineering disciplines. Understanding is based on explaining the physical behavior of materials under load and then modeling this behavior to develop the theory. The development emphasizes the importance of satisfying equilibrium, compatibility of deformation, and material behavior requirements. The hallmark of the book, however, remains the same as the author's unabridged versions, and that is, strong emphasis is placed on drawing a free-body diagram, and the importance of selecting an appropriate coordinate system and an associated sign convention is stressed when the equations of mechanics are applied. Throughout the book, many analysis and design applications are presented, which involve mechanical elements and structural members often encountered in engineering practice.

## Organization and Approach

In order to aid both the instructor and the student, the contents of each chapter are organized into well-defined sections. Selected groups of sections contain an explanation of specific topics, followed by illustrative example problems and a set of homework problems. The topics within each section are often placed in subgroups denoted by boldface titles. The purpose of this is to present a structured method for introducing each new definition or concept and to make the book convenient for later reference and review.

As in the author's other textbooks, a "procedure for analysis" is used throughout the book, providing the student with a logical and orderly method to follow when applying the theory. The example problems are then solved using this outlined method in order to clarify its numerical application. It is to be understood, however, that once the relevant principles have been mastered and enough confidence and judgment have been acquired, the student can then develop his or her own procedures for solving problems. In most cases, it is felt that the first step in any procedure should be to draw a diagram. In doing so, the student forms the habit of tabulating the necessary data while focusing on the physical aspects of the problem and its associated geometry. If this step is correctly performed, applying the relevant equations becomes somewhat methodical, since the data can be taken directly from the diagram.

# Contents

The book is divided into two parts, and the material is covered in the traditional manner.

**Statics.**   The subject of statics is presented in 7 chapters. The text begins in Chapter 1 with an introduction to mechanics and a discussion of units. The notion of a vector and the properties of a concurrent force system are introduced in Chapter 2. Chapter 3 contains a general discussion of concentrated force systems and the methods used to simplify them. The principles of rigid-body equilibrium are developed in Chapter 4 and then applied to specific problems involving the equilibrium of trusses, frames, and machines in Chapter 5. Topics related to the center of gravity, centroid, and moment of inertia are treated in Chapter 6. Lastly, internal loadings in members is discussed in Chapter 7.

**Mechanics of Materials.**   This portion of the text is covered in 10 chapters. Chapter 8 begins with a formal definition of both normal and shear stress, and a discussion of normal stress in axially loaded members and average shear stress caused by direct shear; finally, normal and shear strain are defined. In Chapter 9 a discussion of some of the important mechanical properties of materials is given. Separate treatments of axial load, torsion, bending, and transverse shear are presented in Chapters 10, 11, 12, and 13, respectively. Chapter 14 provides a partial review of the material covered in the previous chapters, in which the state of stress resulting from combined loadings is discussed. In Chapter 15 the concepts for transforming stress and strain are presented. Chapter 16 provides a means for a further summary and review of previous material by covering design applications of beams. Also, coverage is given for various methods for computing deflections of beams and finding the reactions on these members if they are statically indeterminate. Lastly, Chapter 17 provides a discussion of column buckling.

# STATICS AND MECHANICS OF MATERIALS

SECOND EDITION

R. C. Hibbeler

Upper Saddle River, NJ 07458

**Library of Congress Cataloging-in-Publication Data on File**

Vice President and Editorial Director, ECS: *Marcia Horton*
Vice President and Director of Production and Manufacturing, ESM: *David W. Riccardi*
Associate Editor: *Dee Bernhard*
Editorial Assistant: *Brian Hoehl*
Executive Managing Editor: *Vince O'Brien*
Managing Editor: *David A. George*
Director of Creative Services: *Paul Belfanti*
Manager of Electronic Composition and Digital Content: *Jim Sullivan*
Assistant Manager of Electronic Composition and Digital Content: *Allyson Graesser*
Creative Director: *Carole Anson*
Art Director: *Jonathan Boylan*
Electronic Composition: *Clara Bartunek, Beth Gschwind, Julita Nazario, and Judith R. Wilkens*
Art Editor: *Xiaohong Zhu*
Manufacturing Manager: *Trudy Pisciotti*
Manufacturing Buyer: *Lisa McDowell*
Senior Marketing Manager: *Holly Stark*

*About the Cover*: The forces within the members of this truss bridge must be determined if they are to be properly designed. The bolts used for the connections of this steel framework are subjected to stress. Photos by R.C. Hibbeler

Sections of the book that contain more advanced material are indicated by a star (★). Time permitting, some of these topics may be included in the course. Furthermore, this material provides a suitable reference for basic principles when it is covered in other courses, and it can be used as a basis for assigning special projects.

**Alternative Method for Coverage of Mechanics of Materials.** Some instructors prefer to cover stress and strain transformations *first*, before discussing specific applications of axial load, torsion, bending, and shear. One possible method for doing this would be first to cover stress and strain and its transformations, Chapter 8 and Chapter 15. The discussion and example problems in Chapter 15 have been styled so that this is possible. Chapters 9 through 14 can then be covered with no loss in continuity.

**Problems.** Numerous problems in the book depict realistic situations encountered in engineering practice. It is hoped that this realism will both stimulate the student's interest in the subject and provide a means for developing the skill to reduce any such problem from its physical description to a model or symbolic representation to which the principles may be applied.

Throughout the text there is an approximate balance of problems using either SI or FPS units. Furthermore, in any set, an attempt has been made to arrange the problems in order of increasing difficulty. The answers to all but every fourth problem are listed in the back of the book. To alert the user to a problem without a reported answer, an asterisk (∗) is placed before the problem number. Answers are reported to three significant figures, even though the data for material properties may be known with less accuracy. Although this might appear to be poor practice, it is done simply to be consistent and to allow the student a better chance to validate his or her solution. All the problems and their solutions have been independently checked for accuracy.

**Chapter Reviews.** New chapter review sections summarize key points of the chapter, often in bulleted lists.

**Instructor's Solutions CD**—ISBN 0130408050 Provides complete solutions supported by problem statements and problem figures. All solutions appear on either one or two pages.

# Acknowledgments

Preparation of the manuscript for this book has undergone several reviews and I owe the reviewers a personal debt of gratitude. As this text is a combination of my *Statics* and *Mechanics of Materials* books, I would like to thank the reviewers who helped with these editions. Their encouragement and willingness to provide constructive criticism are very much appreciated; in particular, Patrick Kwon of Michigan State University, Cliff Lissenden of Penn State University, Dahsin Liu of Michigan State University, Ting-Wen Wu of the University of Kentucky, Javad Hashemi of Texas Tech University, and Assimina Pelegri of Rutgers—The State University of New Jersey, Paul Heyliger of Colorado State University, Kenneth Sawyers of Lehigh University, John Oyler of University of Pittsburgh, Glenn Beltz of University of California, Johannes Gessler of Colorado State University, Wilfred Nixon of University of Iowa, Jonathan Russell of U.S. Coast Guard Academy, Robert Hinks of Arizona State University, Cap. Mark Orwat of U.S. Military Academy—West Point, Cetin Cetinyaka of Clarkson University, Jack Xin of Kansas State University, Pierre Julien of Colorado State University, Stephen Bechtel of Ohio State University, W. A. Curtain of Brown University, Robert Oakberg of Montana State University, Richard Bennett of University of Tennessee.

I would also like to thank all my students who have used the manuscript and the computer tutorial as well as their revisions, and made comments to improve their contents. A particular note of thanks goes to one of my former graduate students, Kai Beng Yap, who has been a great help to me in this regard. A special note of gratitude also goes to my editors and the staff at Pearson Education, who have all been very supportive in allowing me to have a more creative and artistic license in the design and execution of this book. It has been a pleasure to work with all of them. Finally, appreciation goes to my wife, Conny, who has been a source of encouragement and has helped with the details of preparing the manuscript for publication.

RUSSELL CHARLES HIBBELER

# CONTENTS

# Statics

## 1

### General Principles

## 2

### Force Vectors

## 3

### Force System Resultants

## 4

### Equilibrium of a Rigid Body

## 5

### Structural Analysis

# Statics

Equilibrium and stability of this articulated crane boom as a function of its position can be analyzed using methods based on work and energy, which are explained in this chapter.

CHAPTER
1

# General Principles

## CHAPTER OBJECTIVES

- To provide an introduction to the basic quantities and idealizations of mechanics.
- To give a statement of Newton's Laws of Motion and Gravitation.
- To review the principles for applying the SI system of units.
- To examine the standard procedures for performing numerical calculations.
- To present a general guide for solving problems.

## 1.1   Mechanics

*Mechanics* can be defined as that branch of the physical sciences concerned with the state of rest or motion of bodies that are subjected to the action of forces. In this book we will study two very important branches of mechanics, namely, statics and mechanics of materials. These subjects form a suitable basis for the design and analysis of many types of structural, mechanical, or electrical devices encountered in engineering.

*Statics* deals with the equilibrium of bodies, that is, it is used to determine the forces acting either external to the body or within it necessary to keep the body either at rest or moving with a constant velocity. *Mechanics of materials* studies the relationships between the external loads and the intensity of internal forces acting within the body. This subject is also concerned with computing the deformations of the body, and it provides a study of the body's stability when the body is subjected to external forces. In this book we will first study the principles of statics since for the design and analysis of any structural or mechanical element, it is *first* necessary to determine the forces acting both on and within its various members. Once these internal forces are determined, the size of the members, their deflection, and their stability can then be determined using the fundamentals of mechanics of materials, which will be covered later.

## 1.2 Fundamental Concepts

Before we begin our study of engineering mechanics, it is important to understand the meaning of certain fundamental concepts and principles.

**Basic Quantities.**   The following four quantities are used throughout mechanics.

**Length.**   *Length* is needed to locate the position of a point in space and thereby describe the size of a physical system. Once a standard unit of length is defined, one can then quantitatively define distances and geometric properties of a body as multiples of the unit length.

**Time.**   *Time* is conceived as a succession of events. Although the principles of statics are time independent, this quantity does play an important role in the study of dynamics.

**Mass.**   *Mass* is a property of matter by which we can compare the action of one body with that of another. This property manifests itself as a gravitational attraction between two bodies and provides a quantitative measure of the resistance of matter to a change in velocity.

**Force.**   In general, *force* is considered as a "push" or "pull" exerted by one body on another. This interaction can occur when there is direct contact between the bodies, such as a person pushing on a wall, or it can occur through a distance when the bodies are physically separated. Examples of the latter type include gravitational, electrical, and magnetic forces. In any case, a force is completely characterized by its magnitude, direction, and point of application.

**Idealizations.** Models or idealizations are used in mechanics in order to simplify application of the theory. A few of the more important idealizations will now be defined. Others that are noteworthy will be discussed at points where they are needed.

**Particle.** A *particle* has a mass, but a size that can be neglected. For example, the size of the earth is insignificant compared to the size of its orbit, and therefore the earth can be modeled as a particle when studying its orbital motion. When a body is idealized as a particle, the principles of mechanics reduce to a rather simplified form since the geometry of the body will not be involved in the analysis of the problem.

**Rigid Body.** A *rigid body* can be considered as a combination of a large number of particles in which all the particles remain at a fixed distance from one another both before and after applying a load. As a result, the material properties of any body that is assumed to be rigid will not have to be considered when analyzing the forces acting on the body. In most cases the actual deformations occurring in structures, machines, mechanisms, and the like are relatively small, and the rigid-body assumption is suitable for analysis.

**Concentrated Force.** A *concentrated force* represents the effect of a loading which is assumed to act at a point on a body. We can represent a load by a concentrated force, provided the area over which the load is applied is very small compared to the overall size of the body. An example would be the contact force between a wheel and the ground.

**Newton's Three Laws of Motion.** The entire subject of rigid-body mechanics is formulated on the basis of Newton's three laws of motion, the validity of which is based on experimental observation. They apply to the motion of a particle as measured from a nonaccelerating reference frame. With reference to Fig. 1–1, they may be briefly stated as follows.

**First Law.** A particle originally at rest, or moving in a straight line with constant velocity, will remain in this state provided the particle is *not* subjected to an unbalanced force.

**Second Law.** A particle acted upon by an *unbalanced force* **F** experiences an acceleration that has the same direction as the force and a magnitude that is directly proportional to the force.* If **F** is applied to a particle of mass $m$, this law may be expressed mathematically as

$$\mathbf{F} = m\mathbf{a} \tag{1–1}$$

**Third Law.** The mutual forces of action and reaction between two particles are equal, opposite, and collinear.

---

*Stated another way, the unbalanced force acting on the particle is proportional to the time rate of change of the particle's linear momentum.

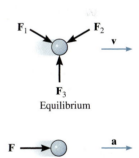

Equilibrium

Accelerated motion

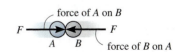

Action – reaction

**Fig. 1–1**

**Newton's Law of Gravitational Attraction.** Shortly after formulating his three laws of motion, Newton postulated a law governing the gravitational attraction between any two particles. Stated mathematically,

$$F = G\frac{m_1 m_2}{r^2} \tag{1–2}$$

where    $F$ = force of gravitation between the two particles

$G$ = universal constant of gravitation; according to experimental evidence, $G = 66.73(10^{-12})\ \text{m}^3/(\text{kg} \cdot \text{s}^2)$

$m_1, m_2$ = mass of each of the two particles

$r$ = distance between the two particles

**Weight.** According to Eq. 1–2, any two particles or bodies have a mutual attractive (gravitational) force acting between them. In the case of a particle located at or near the surface of the earth, however, the only gravitational force having any sizable magnitude is that between the earth and the particle. Consequently, this force, termed the *weight*, will be the only gravitational force considered in our study of mechanics.

From Eq. 1–2, we can develop an approximate expression for finding the weight $W$ of a particle having a mass $m_1 = m$. If we assume the earth to be a nonrotating sphere of constant density and having a mass $m_2 = M_e$, then if $r$ is the distance between the earth's center and the particle, we have

$$W = G\frac{mM_e}{r^2}$$

Letting $g = GM_e/r^2$ yields

$$\boxed{W = mg} \tag{1–3}$$

By comparison with $\mathbf{F} = m\mathbf{a}$, we term $g$ the acceleration due to gravity. Since it depends on $r$, it can be seen that the weight of a body is *not* an absolute quantity. Instead, its magnitude is determined from where the measurement was made. For most engineering calculations, however, $g$ is determined at sea level and at a latitude of 45°, which is considered the "standard location."

## 1.3 Units of Measurement

The four basic quantities—force, mass, length and time—are not all independent from one another; in fact, they are *related* by Newton's second law of motion, $\mathbf{F} = m\mathbf{a}$. Because of this, the *units* used to measure these quantities cannot *all* be selected arbitrarily. The equality $\mathbf{F} = m\mathbf{a}$ is maintained only if three of the four units, called *base units*, are *arbitrarily defined* and the fourth unit is then *derived* from the equation.

**SI Units.**   The International System of units, abbreviated SI after the French "Système International d'Unités," is a modern version of the metric system which has received worldwide recognition. As shown in Table 1–1, the SI system specifies length in meters (m), time in seconds (s), and mass in kilograms (kg). The unit of force, called a newton (N), is *derived* from $\mathbf{F} = m\mathbf{a}$. Thus, 1 newton is equal to a force required to give 1 kilogram of mass an acceleration of $1 \text{ m/s}^2$ ($\text{N} = \text{kg} \cdot \text{m/s}^2$).

If the weight of a body located at the "standard location" is to be determined in newtons, then Eq. 1–3 must be applied. Here $g = 9.806\ 65 \text{ m/s}^2$; however, for calculations, the value $g = 9.81 \text{ m/s}^2$ will be used. Thus,

$$W = mg \qquad (g = 9.81 \text{ m/s}^2) \tag{1–4}$$

Therefore, a body of mass 1 kg has a weight of 9.81 N, a 2-kg body weighs 19.62 N, and so on, Fig. 1–2a.

**U.S. Customary.**   In the U.S. Customary system of units (FPS) length is measured in feet (ft), force in pounds (lb), and time in seconds (s), Table 1–1. The unit of mass, called a *slug*, is *derived* from $\mathbf{F} = m\mathbf{a}$. Hence, 1 slug is equal to the amount of matter accelerated at $1 \text{ ft/s}^2$ when acted upon by a force of 1 lb ($\text{slug} = \text{lb} \cdot \text{s}^2/\text{ft}$).

In order to determine the mass of a body having a weight measured in pounds, we must apply Eq. 1–3. If the measurements are made at the "standard location," then $g = 32.2 \text{ ft/s}^2$ will be used for calculations. Therefore,

$$m = \frac{W}{g} \qquad (g = 32.2 \text{ ft/s}^2) \tag{1–5}$$

And so a body weighing 32.2 lb has a mass of 1 slug, a 64.4-lb body has a mass of 2 slugs, and so on, Fig. 1–2b.

Fig. 1–2

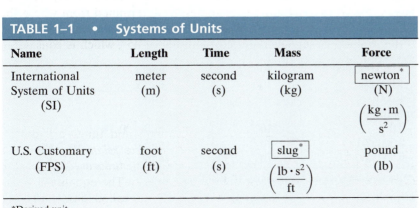

**TABLE 1–1  •  Systems of Units**

| Name | Length | Time | Mass | Force |
|---|---|---|---|---|
| International System of Units (SI) | meter (m) | second (s) | kilogram (kg) | newton* (N) $\left(\dfrac{\text{kg} \cdot \text{m}}{\text{s}^2}\right)$ |
| U.S. Customary (FPS) | foot (ft) | second (s) | slug* $\left(\dfrac{\text{lb} \cdot \text{s}^2}{\text{ft}}\right)$ | pound (lb) |

*Derived unit.

**Conversion of Units.** Table 1–2 provides a set of direct conversion factors between FPS and SI units for the basic quantities. Also, in the FPS system, recall that 1 ft = 12 in. (inches), 5280 ft = 1 mi (mile), 1000 lb = 1 kip (kilo-pound), and 2000 lb = 1 ton.

**TABLE 1–2 • Conversion Factors**

| Quantity | Unit of Measurement (FPS) | Equals | Unit of Measurement (SI) |
|---|---|---|---|
| Force | lb | | 4.448 2 N |
| Mass | slug | | 14.593 8 kg |
| Length | ft | | 0.304 8 m |

## 1.4 The International System of Units

The SI system of units is used extensively in this book since it is intended to become the worldwide standard for measurement. Consequently, the rules for its use and some of its terminology relevant to mechanics will now be presented.

**Prefixes.** When a numerical quantity is either very large or very small, the units used to define its size may be modified by using a prefix. Some of the prefixes used in the SI system are shown in Table 1–3. Each represents a multiple or submultiple of a unit which, if applied successively, moves the decimal point of a numerical quantity to every third place.*For example, 4 000 000 N = 4 000 kN (kilo-newton) = 4MN (mega-newton), or 0.005 m = 5 mm (milli-meter). Notice that the SI system does not include the multiple deca (10) or the submultiple centi (0.01), which form part of the metric system. Except for some volume and area measurements, the use of these prefixes is to be avoided in science and engineering.

**TABLE 1–3 • Prefixes**

| | Exponential Form | Prefix | SI Symbol |
|---|---|---|---|
| *Multiple* | | | |
| 1 000 000 000 | $10^9$ | giga | G |
| 1 000 000 | $10^6$ | mega | M |
| 1 000 | $10^3$ | kilo | k |
| Submultiple | | | |
| 0.001 | $10^{-3}$ | milli | m |
| 0.000 001 | $10^{-6}$ | micro | $\mu$ |
| 0.000 000 001 | $10^{-9}$ | nano | n |

*The kilogram is the only base unit that is defined with a prefix.

**Rules for Use.**   The following rules are given for the proper use of the various SI symbols:

1. A symbol is *never* written with a plural "s," since it may be confused with the unit for second (s).

2. Symbols are always written in lowercase letters, with the following exceptions: symbols for the two largest prefixes shown in Table 1–3, giga and mega, are capitalized as G and M, respectively; and symbols named after an individual are also capitalized, e.g., N.

3. Quantities defined by several units which are multiples of one another are separated by a *dot* to avoid confusion with prefix notation, as indicated by $N = kg \cdot m/s^2 = kg \cdot m \cdot s^{-2}$. Also, $m \cdot s$ (meter-second), whereas ms (milli-second).

4. The exponential power represented for a unit having a prefix refers to both the unit *and* its prefix. For example, $\mu N^2 = (\mu N)^2 = \mu N \cdot \mu N$. Likewise, $mm^2$ represents $(mm)^2 = mm \cdot mm$.

5. Physical constants or numbers having several digits on either side of the decimal point should be reported with a *space* between every three digits rather than with a comma; e.g., 73 569.213 427. In the case of four digits on either side of the decimal, the spacing is optional; e.g., 8537 or 8 537. Furthermore, always try to use decimals and avoid fractions; that is, write 15.25 rather than $15\frac{1}{4}$.

6. When performing calculations, represent the numbers in terms of their *base or derived units* by converting all prefixes to powers of 10. The final result should then be expressed using a *single prefix*. Also, after calculation, it is best to keep numerical values between 0.1 and 1000; otherwise, a suitable prefix should be chosen. For example,

$$(50 \text{ kN})(60 \text{ nm}) = [50(10^3) \text{ N}][60(10^{-9}) \text{ m}]$$
$$= 3000(10^{-6}) \text{ N} \cdot m = 3(10^{-3}) \text{ N} \cdot m = 3 \text{ mN} \cdot m$$

7. Compound prefixes should not be used; e.g., $k\mu s$ (kilo-micro-second) should be expressed as ms (milli-second) since $1 k\mu s = 1(10^3)(10^{-6}) \text{ s} = 1(10^{-3}) \text{ s} = 1 \text{ ms}$.

8. With the exception of the base unit the kilogram, in general avoid the use of a prefix in the denominator of composite units. For example, do not write N/mm, but rather kN/m; also, m/mg should be written as Mm/kg.

9. Although not expressed in multiples of 10, the minute, hour, etc., are retained for practical purposes as multiples of the second. Furthermore, plane angular measurement is made using radians (rad). In this book, however, degrees will often be used, where $180° = \pi \text{ rad}$.

## 1.5 Numerical Calculations

Numerical work in engineering practice is most often performed by using handheld calculators and computers. It is important, however, that the answers to any problem be reported with both justifiable accuracy and appropriate significant figures. In this section we will discuss these topics together with some other important aspects involved in all engineering calculations.

**Dimensional Homogeneity.** The terms of any equation used to describe a physical process must be *dimensionally homogeneous*; that is, each term must be expressed in the same units. Provided this is the case, all the terms of an equation can then be combined if numerical values are substituted for the variables. Consider, for example, the equation $s = vt + \frac{1}{2}at^2$, where, in SI units, $s$ is the position in meters, m, $t$ is time in seconds, s, $v$ is velocity in m/s, and $a$ is acceleration in m/s². Regardless of how this equation is evaluated, it maintains its dimensional homogeneity. In the form stated, each of the three terms is expressed in meters [m, (m/s̸)s̸, (m/s̸²)s̸²,] or solving for $a$, $a = 2s/t^2 - 2v/t$, the terms are each expressed in units of m/s² [m/s², m/s², (m/s)/s].

Since problems in mechanics involve the solution of dimensionally homogeneous equations, the fact that all terms of an equation are represented by a consistent set of units can be used as a partial check for algebraic manipulations of an equation.

**Significant Figures.** The accuracy of a number is specified by the number of significant figures it contains. A *significant figure* is any digit, including a zero, provided it is not used to specify the location of the decimal point for the number. For example, the numbers 5604 and 34.52 each have four significant figures. When numbers begin or end with zeros,

Computers are often used in engineering for advanced design and analysis.

however, it is difficult to tell how many significant figures are in the number. Consider the number 400. Does it have one (4), or perhaps two (40), or three (400) significant figures? In order to clarify this situation, the number should be reported using powers of 10. Using *engineering notation*, the exponent is displayed in multiples of three in order to facilitate conversion of SI units to those having an appropriate prefix. Thus, 400 expressed to one significant figure would be $0.4(10^3)$. Likewise, 2500 and 0.00546 expressed to three significant figures would be $2.50(10^3)$ and $5.46(10^{-3})$.

### Rounding Off Numbers.

For numerical calculations, the accuracy obtained from the solution of a problem generally can never be better than the accuracy of the problem data. This is what is to be expected, but often handheld calculators or computers involve more figures in the answer than the number of significant figures used for the data. For this reason, a calculated result should always be "rounded off" to an appropriate number of significant figures.

To convey appropriate accuracy, the following rules for rounding off a number to $n$ significant figures apply:

- If the $n + 1$ digit is *less than 5*, the $n + 1$ digit and others following it are dropped. For example, 2.326 and 0.451 rounded off to $n = 2$ significant figures would be 2.3 and 0.45.

- If the $n + 1$ digit is equal to 5 with zeros following it, then round off the $n$th digit to an *even number*. For example, $1.245(10^3)$ and 0.8655 rounded off to $n = 3$ significant figures become $1.24(10^3)$ and 0.866.

- If the $n + 1$ digit is *greater than* 5 or equal to 5 with any nonzero digits following it, then increase the $n$th digit by 1 and drop the $n + 1$ digit and others following it. For example, 0.723 87 and 565.500 3 rounded off to $n = 3$ significant figures become 0.724 and 566.

### Calculations.

As a general rule, to ensure accuracy of a final result when performing calculations on a pocket calculator, always retain a greater number of digits than the problem data. If possible, try to work out the computations so that numbers which are approximately equal are not subtracted since accuracy is often lost from this calculation.

In engineering we generally round off final answers to *three* significant figures since the data for geometry, loads, and other measurements are often reported with this accuracy.* Consequently, in this book the intermediate calculations for the examples are often worked out to four significant figures and the answers are generally reported to *three* significant figures.

---

*Of course, some numbers, such as $\pi$, $e$, or numbers used in derived formulas are exact and are therefore accurate to an infinite number of significant figures.

**E X A M P L E 1.1**

Convert 2 km/h to m/s. How many ft/s is this?

**Solution**
Since 1 km = 1000 m and 1 h = 3600 s, the factors of conversion are arranged in the following order, so that a cancellation of the units can be applied:

$$2 \text{ km/h} = \frac{2 \text{ km}}{h}\left(\frac{1000 \text{ m}}{\text{km}}\right)\left(\frac{1 \text{ h}}{3600 \text{ s}}\right)$$

$$= \frac{2000 \text{ m}}{3600 \text{ s}} = 0.556 \text{ m/s} \qquad Ans.$$

From Table 1–2, 1 ft = 0.3048 m. Thus

$$0.556 \text{ m/s} = \frac{0.556 \text{ m}}{\text{s}} \frac{1 \text{ ft}}{0.3048 \text{ m}}$$

$$= 1.82 \text{ ft/s} \qquad Ans.$$

**E X A M P L E 1.2**

Convert the quantities 300 lb · s and 52 slug/ft³ to appropriate SI units.

**Solution**
Using Table 1–2, 1 lb = 4.448 2 N.

$$300 \text{ lb} \cdot \text{s} = 300 \text{ lb} \cdot \text{s}\left(\frac{4.448 \ 2 \text{ N}}{\text{lb}}\right)$$

$$= 1334.5 \text{ N} \cdot \text{s} = 1.33 \text{ kN} \cdot \text{s} \qquad Ans.$$

Also, 1 slug = 14.593 8 kg and 1 ft = 0.304 8 m.

$$52 \text{ slug/ft}^3 = \frac{52 \text{ slug}}{\text{ft}^3}\left(\frac{14.593 \ 8 \text{ kg}}{1 \text{ slug}}\right)\left(\frac{1 \text{ ft}}{0.304 \ 8 \text{ m}}\right)^3$$

$$= 26.8(10^3) \text{ kg/m}^3$$

$$= 26.8 \text{ Mg/m}^3 \qquad Ans.$$

# EXAMPLE 1.3

Evaluate each of the following and express with SI units having an appropriate prefix: (a) (50 mN)(6 GN), (b) (400 mm)(0.6 MN)$^2$, (c) 45 MN$^3$/900 Gg.

## Solution

First convert each number to base units, perform the indicated operations, then choose an appropriate prefix (see Rule 6 on p. 9).

*Part (a)*

$$(50 \text{ mN})(6 \text{ GN}) = [50(10^{-3}) \text{ N}][6(10^9) \text{ N}]$$
$$= 300(10^6) \text{ N}^2$$
$$= 300(10^6) \cancel{\text{N}}^2 \left(\frac{1 \text{ kN}}{10^3 \cancel{\text{N}}}\right)\left(\frac{1 \text{ kN}}{10^3 \cancel{\text{N}}}\right)$$
$$= 300 \text{ kN}^2 \qquad \qquad \textit{Ans.}$$

Note carefully the convention kN$^2$ = (kN)$^2$ = $10^6$ N$^2$ (Rule 4 on p. 9).

*Part (b)*

$$(400 \text{ mm})(0.6 \text{ MN})^2 = [400(10^{-3}) \text{ m}][0.6(10^6) \text{ N}]^2$$
$$= [400(10^{-3}) \text{ m}][0.36(10^{12}) \text{ N}^2]$$
$$= 144(10^9) \text{ m} \cdot \text{N}^2$$
$$= 144 \text{ Gm} \cdot \text{N}^2 \qquad \qquad \textit{Ans.}$$

We can also write

$$144(10^9)\text{m} \cdot \text{N}^2 = 144(10^9) \text{ m} \cdot \cancel{\text{N}}^2 \left(\frac{1 \text{ MN}}{10^6 \cancel{\text{N}}}\right)\left(\frac{1 \text{ MN}}{10^6 \cancel{\text{N}}}\right)$$
$$= 0.144 \text{ m} \cdot \text{MN}^2$$

*Part (c)*

$$45 \text{ MN}^3/900 \text{ Gg} = \frac{45(10^6 \text{ N})^3}{900(10^6) \text{ kg}}$$
$$= 0.05(10^{12}) \text{ N}^3/\text{kg}$$
$$= 0.05(10^{12}) \cancel{\text{N}}^3 \left(\frac{1 \text{ kN}}{10^3 \cancel{\text{N}}}\right)^3 \frac{1}{\text{kg}}$$
$$= 0.05(10^3) \text{ kN}^3/\text{kg}$$
$$= 50 \text{ kN}^3/\text{kg} \qquad \qquad \textit{Ans.}$$

Here we have used Rules 4 and 8 on p. 9.

## 1.6 General Procedure for Analysis

When solving problems, do the work as neatly as possible. Being neat generally stimulates clear and orderly thinking, and vice versa.

The most effective way of learning the principles of engineering mechanics is to *solve problems*. To be successful at this, it is important to always present the work in a *logical* and *orderly manner*, as suggested by the following sequence of steps:

1. Read the problem carefully and try to correlate the actual physical situation with the theory studied.
2. Draw any necessary diagrams and tabulate the problem data.
3. Apply the relevant principles, generally in mathematical form.
4. Solve the necessary equations algebraically as far as practical, then, making sure they are dimensionally homogeneous, use a consistent set of units and complete the solution numerically. Report the answer with no more significant figures than the accuracy of the given data.
5. Study the answer with technical judgment and common sense to determine whether or not it seems reasonable.

### IMPORTANT POINTS

- Statics is the study of bodies that are at rest or move with constant velocity.
- A particle has a mass but a size that can be neglected.
- A rigid body does not deform under load.
- Concentrated forces are assumed to act at a point on a body.
- Newton's three laws of motion should be memorized.
- Mass is a property of matter that does not change from one location to another.
- Weight refers to the gravitational attraction of the earth on a body or quantity of mass. Its magnitude depends upon the elevation at which the mass is located.
- In the SI system the unit of force, the newton, is a derived unit. The meter, seond, and kilogram are base units.
- Prefixes G, M, k, m, $\mu$, n are used to represent large and small numerical quantities. Their exponential size should be known, along with the rules for using the SI units.
- Perform numerical calculations to several significant figures and then report the final answer to three significant figures.
- Algebraic manipulations of an equation can be checked in part by verifying that the equation remains dimensionally homogenous.
- Know the rules for rounding off numbers.

# PROBLEMS

**1-1.** Round off the following numbers to three significant figures: (a) 4.65735 m, (b) 55.578 s, (c) 4555 N, (d) 2768 kg.

**1-2.** Wood has a density of 4.70 slug/ft³. What is its density expressed in SI units?

**1-3.** Represent each of the following quantities in the correct SI form using an appropriate prefix: (a) 0.000431 kg, (b) 35.3(10³) N, (c) 0.00532 km.

**\*1-4.** Represent each of the following combinations of units in the correct SI form using an appropriate prefix: (a) m/ms, (b) μkm, (c) ks/mg, and (d) km · μN.

**1-5.** If a car is traveling at 55 mi/h, determine its speed in kilometers per hour and meters per second.

**1-6.** Evaluate each of the following and express with an appropriate prefix: (a) $(430 \text{ kg})^2$, (b) $(0.002 \text{ mg})^2$, and (c) $(230 \text{ m})^3$.

**1-7.** A rocket has a mass of $250(10^3)$ slugs on earth. Specify (a) its mass in SI units, and (b) its weight in SI units. If the rocket is on the moon, where the acceleration due to gravity is $g_m = 5.30$ ft/s², determine to three significant figures (c) its weight in SI units, and (d) its mass in SI units.

**\*1-8.** Represent each of the following combinations of units in the correct SI form: (a) kN/μs, (b) Mg/mN, and (c) MN/(kg · ms).

**1-9.** The *pascal* (Pa) is actually a very small unit of pressure. To show this, convert 1 Pa = 1 N/m² to lb/ft². Atmospheric pressure at sea level is 14.7 lb/in². How many pascals is this?

**1-10.** What is the weight in newtons of an object that has a mass of: (a) 10 kg, (b) 0.5 g, (c) 4.50 Mg? Express the result to three significant figures. Use an appropriate prefix.

**1-11.** Evaluate each of the following to three significant figures and express each answer in SI units using an appropriate prefix: (a) 354 mg(45 km)/(0.035 6 KN), (b) (0.004 53 Mg)(201 ms), (c) 435 MN/23.2 mm.

**\*1-12.** Convert each of the following and express the answer using an appropriate prefix: (a) 175 lb/ft³ to kN/m³, (b) 6 ft/h to mm/s, and (c) 835 lb · ft to kN · m.

**1-13.** Convert each of the following to three significant figures. (a) 20 lb · ft to N · m, (b) 450 lb/ft³ to kN/m³, and (c) 15 ft/h to mm/s.

**1-14.** If an object has a mass of 40 slugs, determine its mass in kilograms.

**1-15.** Water has a density of 1.94 slug/ft³. What is the density expressed in SI units? Express the answer to three significant figures.

**\*1-16.** Two particles have a mass of 8 kg and 12 kg, respectively. If they are 800 mm apart, determine the force of gravity acting between them. Compare this result with the weight of each particle.

**1-17.** Determine the mass of an object that has a weight of (a) 20 mN, (b) 150 kN, (c) 60 MN. Express the answer to three significant figures.

**1-18.** If a man weighs 155 lb on earth, specify (a) his mass in slugs, (b) his mass in kilograms, and (c) his weight in newtons. If the man is on the moon, where the acceleration due to gravity is $g_m = 5.30$ ft/s², determine (d) his weight in pounds, and (e) his mass in kilograms.

**1-19.** Using the base units of the SI system, show that Eq. 1-2 is a dimensionally homogeneous equation which gives $F$ in newtons. Determine to three significant figures the gravitational force acting between two spheres that are touching each other. The mass of each sphere is 200 kg and the radius is 300 mm.

**\*1-20.** Evaluate each of the following to three significant figures and express each answer in SI units using an appropriate prefix: (a) $(0.631 \text{ Mm})/(8.60 \text{ kg})^2$, (b) $(35 \text{ mm})^2 (48 \text{ kg})^3$.

This communications tower is stabilized by cables that exert forces at the points of connection. In this chapter, we will show how to determine the magnitude and direction of the resultant force at each point.

# Force Vectors

- To show how to add forces and resolve them into components using the Parallelogram Law.
- To express force and position in Cartesian vector form and explain how to determine the vector's magnitude and direction.
- To introduce the dot product in order to determine the angle between two vectors or the projection of one vector onto another.

## 2.1 Scalars and Vectors

Most of the physical quantities in mechanics can be expressed mathematically by means of scalars and vectors.

**Scalar.**    A quantity characterized by a positive or negative number is called a *scalar*. For example, mass, volume, and length are scalar quantities often used in statics. In this book, scalars are indicated by letters in italic type, such as the scalar $A$.

**Vector.**    A *vector* is a quantity that has both a magnitude and a direction. In statics the vector quantities frequently encountered are position, force, and moment. For handwritten work, a vector is generally represented by a letter with an arrow written over it, such as $\vec{A}$. The magnitude is designated $|\vec{A}|$ or simply A. In this book vectors will be symbolized in boldface type; for example, **A** is used to designate the vector "A." Its magnitude, which is always a positive quantity, is symbolized in italic type, written as $|A|$, or simply $A$ when it is understood that $A$ is a positive scalar.

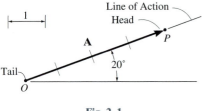

**Fig. 2–1**

A vector is represented graphically by an arrow, which is used to define its magnitude, direction, and sense. The *magnitude* of the vector is the length of the arrow, the *direction* is defined by the angle between a reference axis and the arrow's line of action, and the *sense* is indicated by the arrowhead. For example, the vector **A** shown in Fig. 2–1 has a magnitude of 4 units, a direction which is 20° measured counterclockwise from the horizontal axis, and a sense which is upward and to the right. The point $O$ is called the *tail* of the vector, the point $P$ the *tip* or *head*.

## 2.2 Vector Operations

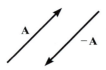

Vector **A** and its negative counterpart

**Fig. 2–2**

**Multiplication and Division of a Vector by a Scalar.** The product of vector **A** and scalar $a$, yielding $a\mathbf{A}$, is defined as a vector having a magnitude $|aA|$. The *sense* of $a\mathbf{A}$ is the *same* as **A** provided $a$ is *positive*; it is *opposite* to **A** if $a$ is *negative*. In particular, the negative of a vector is formed by multiplying the vector by the scalar $(-1)$, Fig. 2–2. Division of a vector by a scalar can be defined using the laws of multiplication, since $\mathbf{A}/a = (1/a)\mathbf{A}, a \neq 0$. Graphic examples of these operations are shown in Fig. 2–3.

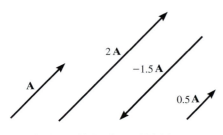

Scalar Multiplication and Division

**Fig. 2–3**

**Vector Addition.** Two vectors **A** and **B** such as force or position, Fig. 2–4a, may be added to form a "resultant" vector $\mathbf{R} = \mathbf{A} + \mathbf{B}$ by using the *parallelogram law*. To do this, **A** and **B** are joined at their tails, Fig. 2–4b. Parallel lines drawn from the head of each vector intersect at a common point, thereby forming the adjacent sides of a parallelogram. As shown, the resultant **R** is the diagonal of the parallelogram, which extends from the tails of **A** and **B** to the intersection of the lines.

We can also add **B** to **A** using a *triangle construction*, which is a special case of the parallelogram law, whereby vector **B** is added to vector **A** in a "head-to-tail" fashion, i.e., by connecting the head of **A** to the tail of **B**, Fig. 2–4c. The resultant **R** extends from the tail of **A** to the head of **B**. In a similar manner, **R** can also be obtained by adding **A** to **B**, Fig. 2–4d. By comparison, it is seen that vector addition is commutative; in other words, the vectors can be added in either order, i.e., $\mathbf{R} = \mathbf{A} + \mathbf{B} = \mathbf{B} + \mathbf{A}$.

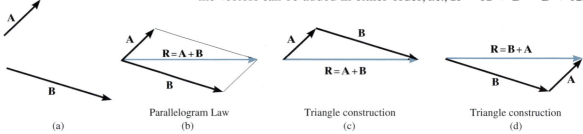

Vector Addition

**Fig. 2–4**

SECTION 2.2   Vector Operations   •   **19**

As a special case, if the two vectors **A** and **B** are *collinear*, i.e., both have the same line of action, the parallelogram law reduces to an *algebraic* or *scalar addition* $R = A + B$, as shown in Fig. 2–5.

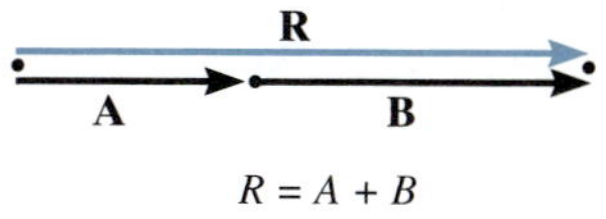

### Vector Subtraction.

The resultant *difference* between two vectors **A** and **B** of the same type may be expressed as

$$\mathbf{R'} = \mathbf{A} - \mathbf{B} = \mathbf{A} + (-\mathbf{B})$$

This vector sum is shown graphically in Fig. 2–6. Subtraction is therefore defined as a special case of addition, so the rules of vector addition also apply to vector subtraction.

$$R = A + B$$

Addition of collinear vectors

**Fig. 2–5**

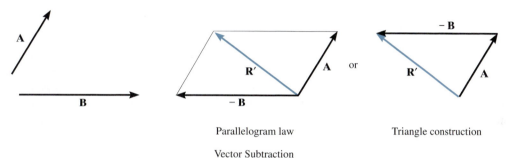

Parallelogram law

Triangle construction

Vector Subtraction

**Fig. 2–6**

### Resolution of a Vector.

A vector may be resolved into two "components" having known lines of action by using the parallelogram law. For example, if **R** in Fig. 2–7a is to be resolved into components acting along the lines *a* and *b*, one starts at the *head* of **R** and extends a line *parallel* to *a* until it intersects *b*. Likewise, a line parallel to *b* is drawn from the *head* of **R** to the point of intersection with *a*, Fig. 2–7a. The two components **A** and **B** are then drawn such that they extend from the tail of **R** to the points of intersection, as shown in Fig. 2–7b.

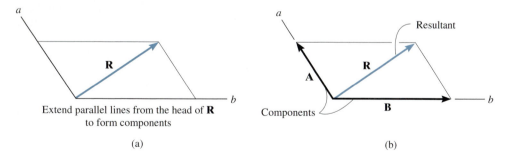

Extend parallel lines from the head of **R** to form components

(a)

Components

(b)

Resolution of a vector

**Fig. 2–7**

## 2.3 Vector Addition of Forces

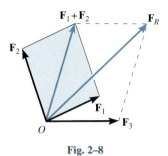

**Fig. 2–8**

Experimental evidence has shown that a force is a vector quantity since it has a specified magnitude, direction, and sense and it adds according to the parallelogram law. Two common problems in statics involve either finding the resultant force, knowing its components, or resolving a known force into two components. As described in Sec. 2.2, both of these problems require application of the parallelogram law.

If more than two forces are to be added, successive applications of the parallelogram law can be carried out in order to obtain the resultant force. For example, if three forces $\mathbf{F}_1$, $\mathbf{F}_2$, $\mathbf{F}_3$ act at a point $O$, Fig. 2–8, the resultant of any two of the forces is found—say, $\mathbf{F}_1 + \mathbf{F}_2$—and then this resultant is added to the third force, yielding the resultant of all three forces; i.e., $\mathbf{F}_R = (\mathbf{F}_1 + \mathbf{F}_2) + \mathbf{F}_3$. Using the parallelogram law to add more than two forces, as shown here, often requires extensive geometric and trigonometric calculation to determine the numerical values for the magnitude and direction of the resultant. Instead, problems of this type are easily solved by using the "rectangular-component method," which is explained in Sec. 2.4.

If we know the forces $\mathbf{F}_a$ and $\mathbf{F}_b$ that the two chains $a$ and $b$ exert on the hook, we can find their resultant force $\mathbf{F}_c$ by using the parallelogram law. This requires drawing lines parallel to $a$ and $b$ from the heads of $\mathbf{F}_a$ and $\mathbf{F}_b$ as shown thus forming a parallelogram.

In a similar manner, if the force $\mathbf{F}_c$ along chain $c$ is known, then its two components $\mathbf{F}_a$ and $\mathbf{F}_b$, that act along $a$ and $b$, can be determined from the parallelogram law. Here we must start at the head of $\mathbf{F}_c$ and construct lines parallel to $a$ and $b$, thereby forming the parallelogram.

## PROCEDURE FOR ANALYSIS

Problems that involve the addition of two forces can be solved as follows:

*Parallelogram Law.*

- Make a sketch showing the vector addition using the parallelogram law.
- Two "component" forces add according to the parallelogram law, yielding a *resultant* force that forms the diagonal of the parallelogram.
- If a force is to be resolved into *components* along two axes directed from the tail of the force, then start at the head of the force and construct lines parallel to the axes, thereby forming the parallelogram. The sides of the parallelogram represent the components.
- Label all the known and unknown force magnitudes and the angles on the sketch and identify the two unknowns.

*Trigonometry.*

- Redraw a half portion of the parallelogram to illustrate the triangular head-to-tail addition of the components.
- The magnitude of the resultant force can be determined from the law of cosines, and its direction is determined from the law of sines, Fig. 2–9.
- The magnitude of two force components are determined from the law of sines, Fig. 2–9.

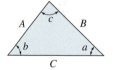

Sine law:
$$\frac{A}{\sin a} = \frac{B}{\sin b} = \frac{C}{\sin c}$$

Cosine law:
$$C = \sqrt{A^2 + B^2 - 2AB\cos c}$$

**Fig. 2–9**

## IMPORTANT POINTS

- A scalar is a positive or negative number.
- A vector is a quantity that has magnitude, direction, and sense.
- Multiplication or division of a vector by a scalar will change the magnitude of the vector. The sense of the vector will change if the scalar is negative.
- As a special case, if the vectors are collinear, the resultant is formed by an algebraic or scalar addition.

**E X A M P L E 2.1**

The screw eye in Fig. 2–10a is subjected to two forces, $\mathbf{F}_1$ and $\mathbf{F}_2$. Determine the magnitude and direction of the resultant force.

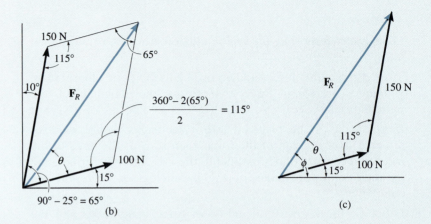

Fig. 2–10

**Solution**

*Parallelogram Law.* The parallelogram law of addition is shown in Fig. 2–10b. The two unknowns are the magnitude of $\mathbf{F}_R$ and the angle $\theta$ (theta).

*Trigonometry.* From Fig. 2–10b, the vector triangle, Fig. 2–10c, is constructed. $F_R$ is determined by using the law of cosines:

$$F_R = \sqrt{(100\text{ N})^2 + (150\text{ N})^2 - 2(100\text{ N})(150\text{ N})\cos 115°}$$
$$= \sqrt{10\,000 + 22\,500 - 30\,000(-0.4226)} = 212.6\text{ N}$$
$$= 213\text{ N}$$

The angle $\theta$ is determined by applying the law of sines, using the computed value of $F_R$.

$$\frac{150\text{ N}}{\sin \theta} = \frac{212.6\text{ N}}{\sin 115°}$$

$$\sin \theta = \frac{150\text{ N}}{212.6\text{ N}}(0.9063)$$

$$\theta = 39.8°$$

Thus, the direction $\phi$ (phi) of $\mathbf{F}_R$, measured from the horizontal, is

$$\phi = 39.8° + 15.0° = 54.8° \angle^\phi \qquad \textit{Ans.}$$

# EXAMPLE 2.2

Resolve the 200-lb force acting on the pipe, Fig. 2–11a, into components in the (a) $x$ and $y$ directions, and (b) $x'$ and $y$ directions.

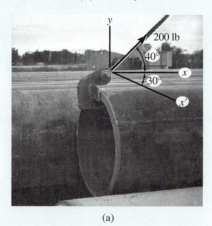

(a)

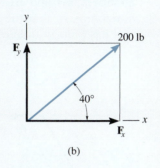

(b)

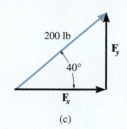

(c)

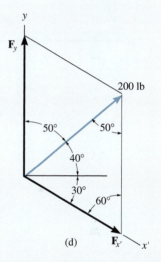

(d)

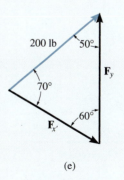

(e)

Fig. 2–11

## Solution

In each case the parallelogram law is used to resolve **F** into its two components, and then the vector triangle is constructed to determine the numerical results by trigonometry.

*Part (a).*   The vector addition $\mathbf{F} = \mathbf{F}_x + \mathbf{F}_y$ is shown in Fig. 2–11b. In particular, note that the length of the components is scaled along the $x$ and $y$ axes by first constructing lines from the tip of **F** parallel to the axes in accordance with the parallelogram law. From the vector triangle, Fig. 2–11c,

$$F_x = 200 \text{ lb} \cos 40° = 153 \text{ lb} \qquad \textit{Ans.}$$
$$F_y = 200 \text{ lb} \sin 40° = 129 \text{ lb} \qquad \textit{Ans.}$$

*Part (b).*   The vector addition $\mathbf{F} = \mathbf{F}_{x'} + \mathbf{F}_y$ is shown in Fig. 2–11d. Note carefully how the parallelogram is constructed. Applying the law of sines and using the data listed on the vector triangle, Fig. 2–11e, yields

$$\frac{F_{x'}}{\sin 50°} = \frac{200 \text{ lb}}{\sin 60°}$$

$$F_{x'} = 200 \text{ lb}\left(\frac{\sin 50°}{\sin 60°}\right) = 177 \text{ lb} \qquad \textit{Ans.}$$

$$\frac{F_y}{\sin 70°} = \frac{200 \text{ lb}}{\sin 60°}$$

$$F_y = 200 \text{ lb}\left(\frac{\sin 70°}{\sin 60°}\right) = 217 \text{ lb} \qquad \textit{Ans.}$$

The force **F** acting on the frame shown in Fig. 2–12*a* has a magnitude of 500 N and is to be resolved into two components acting along members *AB* and *AC*. Determine the angle $\theta$, measured *below* the horizontal, so that the component $\mathbf{F}_{AC}$ is directed from *A* toward *C* and has a magnitude of 400 N.

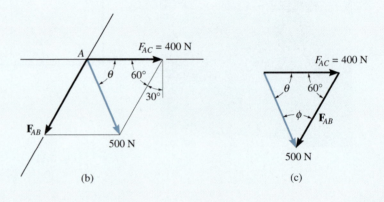

(b)          (c)

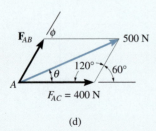

(a)

**Fig. 2–12**

**Solution**
By using the parallelogram law, the vector addition of the two components yielding the resultant is shown in Fig. 2–12*b*. Note carefully how the resultant force is resolved into the two components $\mathbf{F}_{AB}$ and $\mathbf{F}_{AC}$, which have specified lines of action. The corresponding vector triangle is shown in Fig. 2–12*c*.

The angle $\phi$ can be determined by using the law of sines:

$$\frac{400\ \text{N}}{\sin \phi} = \frac{500\ \text{N}}{\sin 60^\circ}$$

$$\sin \phi = \left(\frac{400\ \text{N}}{500\ \text{N}}\right) \sin 60^\circ = 0.6928$$

$$\phi = 43.9^\circ$$

Hence,

$$\theta = 180^\circ - 60^\circ - 43.9^\circ = 76.1^\circ \qquad \textit{Ans.}$$

Using this value for $\theta$, apply the law of cosines or the law of sines and show that $\mathbf{F}_{AB}$ has a magnitude of 561 N.

Notice that **F** can also be directed at an angle $\theta$ *above* the horizontal, as shown in Fig. 2–12*d*, and still produce the required component $\mathbf{F}_{AC}$. Show that in this case $\theta = 161.1^\circ$ and $F_{AB} = 161$ N.

(d)

# EXAMPLE 2.4

The ring shown in Fig. 2–13a is subjected to two forces, $F_1$ and $F_2$. If it is required that the resultant force have a magnitude of 1 kN and be directed vertically downward, determine (a) the magnitudes of $F_1$ and $F_2$ provided $\theta = 30°$, and (b) the magnitudes of $F_1$ and $F_2$ if $F_2$ is to be a minimum.

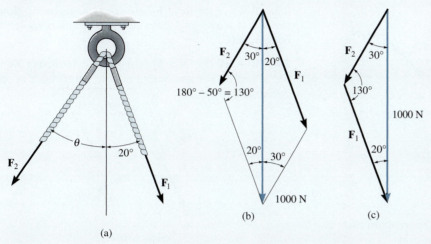

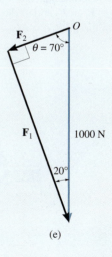

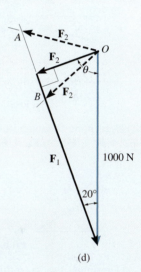

(a)

Fig. 2–13

## Solution

*Part (a).* A sketch of the vector addition according to the parallelogram law is shown in Fig. 2–13b. From the vector triangle constructed in Fig. 2–13c, the unknown magnitudes $F_1$ and $F_2$ are determined by using the law of sines:

$$\frac{F_1}{\sin 30°} = \frac{1000 \text{ N}}{\sin 130°}$$
$$F_1 = 653 \text{ N} \qquad \textit{Ans.}$$
$$\frac{F_2}{\sin 20°} = \frac{1000 \text{ N}}{\sin 130°}$$
$$F_2 = 446 \text{ N} \qquad \textit{Ans.}$$

*Part (b).* If $\theta$ is not specified, then by the vector triangle, Fig. 2–13d, $F_2$ may be added to $F_1$ in various ways to yield the resultant 1000-N force. In particular, the *minimum* length or magnitude of $F_2$ will occur when its line of action is *perpendicular to* $F_1$. Any other direction, such as *OA* or *OB*, yields a larger value for $F_2$. Hence, when $\theta = 90° - 20° = 70°$, $F_2$ is minimum. From the triangle shown in Fig. 2–13e, it is seen that

$$F_1 = 1000 \sin 70°\text{N} = 940 \text{ N} \qquad \textit{Ans.}$$
$$F_2 = 1000 \cos 70°\text{N} = 342 \text{ N} \qquad \textit{Ans.}$$

## PROBLEMS

**2-1.** Determine the magnitude of the resultant force $\mathbf{F}_R = \mathbf{F}_1 + \mathbf{F}_2$ and its direction, measured counterclockwise from the positive $x$ axis.

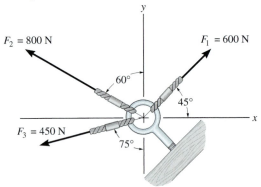

**Prob. 2–1**

**2-2.** Determine the magnitude of the resultant force if: (a) $\mathbf{F}_R = \mathbf{F}_1 + \mathbf{F}_2$; (b) $\mathbf{F}'_R = \mathbf{F}_1 - \mathbf{F}_2$.

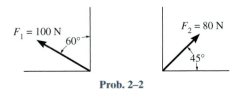

**Prob. 2–2**

**2-3.** Determine the magnitude of the resultant force $\mathbf{F}_R = \mathbf{F}_1 + \mathbf{F}_2$ and its direction, measured counterclockwise from the positive $x$ axis.

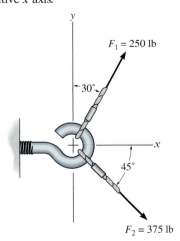

**Prob. 2–3**

**\*2-4.** Determine the magnitude of the resultant force $\mathbf{F}_R = \mathbf{F}_1 + \mathbf{F}_2$ and its direction, measured clockwise from the positive $u$ axis.

**2-5.** Resolve the force $\mathbf{F}_1$ into components acting along the $u$ and $v$ axes and determine the magnitudes of the components.

**2-6.** Resolve the force $\mathbf{F}_2$ into components acting along the $u$ and $v$ axes and determine the magnitudes of the components.

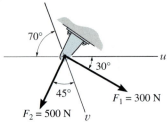

**Probs. 2–4/5/6**

**2-7.** The plate is subjected to the two forces at $A$ and $B$ as shown. If $\theta = 60°$, determine the magnitude of the resultant of these two forces and its direction measured from the horizontal.

**\*2-8.** Determine the angle $\theta$ for connecting member $A$ to the plate so that the resultant force of $F_A$ and $F_B$ is directed horizontally to the right. Also, what is the magnitude of the resultant force?

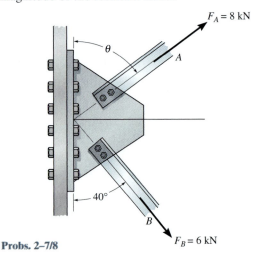

**Probs. 2–7/8**

**2-9.** The vertical force **F** acts downward at *A* on the two-membered frame. Determine the magnitudes of the two components of **F** directed along the axes of *AB* and *AC*. Set *F* = 500 N.

**2-10.** Solve Prob. 2–9 with *F* = 350 lb.

**2-13.** The 500-lb force acting on the frame is to be resolved into two components acting along the axis of the struts *AB* and *AC*. If the component of force along *AC* is required to be 300 lb, directed from *A* to *C*, determine the magnitude of force acting along *AB* and the angle $\theta$ of the 500-lb force.

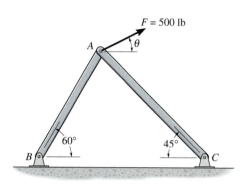

Prob. 2–13

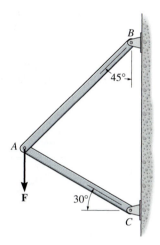

Probs. 2–9/10

**2-11.** The force acting on the gear tooth is *F* = 20 lb. Resolve this force into two components acting along the lines *aa* and *bb*.

**\*2-12.** The component of force **F** acting along line *aa* is require to be 30 lb. Determine the magnitude of **F** and its component along line *bb*.

**2-14.** The post is to be pulled out of the ground using two ropes *A* and *B*. Rope *A* is subjected to a force of 600 lb and is directed at 60° from the horizontal. If the resultant force acting on the post is to be 1200 lb, vertically upward, determine the force *T* in rope *B* and the corresponding angle $\theta$.

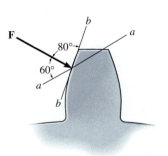

Probs. 2–11/12

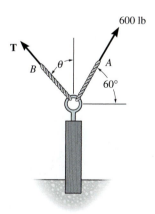

Prob. 2–14

**2-15.** Determine the design angle $\theta$ ($0° \leq \theta \leq 90°$) for strut $AB$ so that the 400-lb horizontal force has a component of 500-lb directed from $A$ towards $C$. What is the component of force acting along member $AB$? Take $\phi = 40°$.

**\*2-16.** Determine the design angle $\phi$ ($0° \leq \phi \leq 90°$) between struts $AB$ and $AC$ so that the 400-lb horizontal force has a component of 600-lb which acts up to the left, in the same direction as from $B$ towards $A$. Take $\theta = 30°$.

**2-18.** Two forces are applied at the end of a screw eye in order to remove the post. Determine the angle $\theta$ ($0° \leq \theta \leq 90°$) and the magnitude of force $\mathbf{F}$ so that the resultant force acting on the post is directed vertically upward and has a magnitude of 750 N.

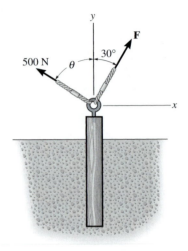

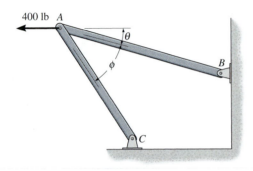

Probs. 2–15/16

Prob. 2–18

**2-17.** The chisel exerts a force of 20 lb on the wood dowel rod which is turning in a lathe. Resolve this force into components acting (a) along the $n$ and $y$ axes and (b) along the $x$ and $t$ axes.

**2-19.** If $F_1 = F_2 = 30$ lb, determine the angles $\theta$ and $\phi$ so that the resultant force is directed along the positive $x$ axis and has a magnitude of $F_R = 20$ lb.

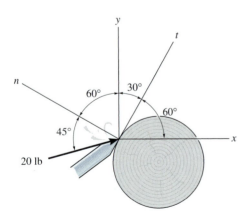

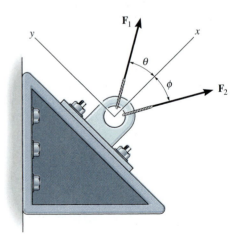

Prob. 2–17

Prob. 2–19

## 2.4 Addition of a System of Coplanar Forces

When the resultant of more than two forces has to be obtained, it is easier to find the components of each force along specified axes, add these components algebraically, and then form the resultant, rather than form the resultant of the forces by successive application of the parallelogram law as discussed in Sec. 2.3.

In this section we will resolve each force into its rectangular components $\mathbf{F}_x$ and $\mathbf{F}_y$, which lie along the $x$ and $y$ axes, respectively, Fig. 2–14a. Although the axes are horizontal and vertical, they may in general be directed at any inclination, as long as they remain perpendicular to one another, Fig. 2–14b. In either case, by the parallelogram law, we require

$$\mathbf{F} = \mathbf{F}_x + \mathbf{F}_y$$

and

$$\mathbf{F}' + \mathbf{F}'_x + \mathbf{F}'_y$$

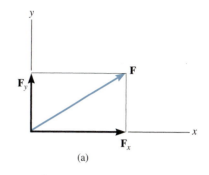

(a)

As shown in Fig. 2–14, the sense of direction of each force component is represented *graphically* by the *arrowhead*. For *analytical* work, however, we must establish a notation for representing the directional sense of the rectangular components. This can be done in one of two ways.

**Scalar Notation.** Since the $x$ and $y$ axes have designated positive and negative directions, the magnitude and directional sense of the rectangular components of a force can be expressed in terms of *algebraic scalars*. For example, the components of $\mathbf{F}$ in Fig. 2–14a can be represented by positive scalars $F_x$ and $F_y$ since their sense of direction is along the *positive x* and $y$ axes, respectively. In a similar manner, the components of $\mathbf{F}'$ in Fig. 2–14b are $F'_x$ and $-F'_y$. Here the $y$ component is negative, since $\mathbf{F}'_y$ is directed along the negative $y$ axis.

It is important to keep in mind that this scalar notation is to be used only for computational purposes, not for graphical representations in figures. Throughout the book, the *head of a vector arrow* in any figure indicates the sense of the vector *graphically*; algebraic signs are not used for this purpose. Thus, the vectors in Figs. 2–14a and 2–14b are designated by using boldface (vector) notation.* Whenever italic symbols are written near vector arrows in figures, they indicate the *magnitude* of the vector, which is *always* a *positive* quantity.

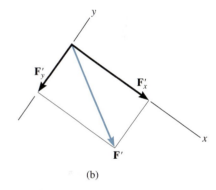

(b)

**Fig. 2–14**

---

*Negative signs are used only in figures with boldface notation when showing equal but opposite pairs of vectors as in Fig. 2–2.

**Cartesian Vector Notation.** It is also possible to represent the components of a force in terms of Cartesian unit vectors. When we do this the methods of vector algebra are easier to apply, and we will see that this becomes particularly advantageous for solving problems in three dimensions.

In two dimensions the *Cartesian unit vectors* **i** and **j** are used to designate the *directions* of the *x* and *y* axes, respectively, Fig. 2–15a.* These vectors have a dimensionless magnitude of unity, and their sense (or arrowhead) will be described analytically by a plus or minus sign, depending on whether they are pointing along the positive or negative *x* or *y* axis.

As shown in Fig. 2–15a, the *magnitude* of each component of **F** is *always a positive quantity*, which is represented by the (positive) scalars $F_x$ and $F_y$. Therefore, having established notation to represent the magnitude and the direction of each vector component, we can express **F** in Fig. 2–15a as the *Cartesian vector*,

$$\mathbf{F} = F_x\mathbf{i} + F_y\mathbf{j}$$

And in the same way, **F′** in Fig. 2–15b can be expressed as

$$\mathbf{F'} = F'_x\mathbf{i} + F'_y(-\mathbf{j})$$

or simply

$$\mathbf{F'} = F'_x\mathbf{i} - F'_y\mathbf{j}$$

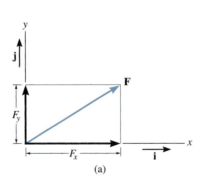

(a)

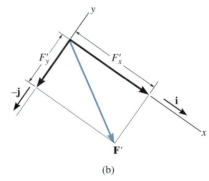

(b)

**Fig. 2–15**

*For handwritten work, unit vectors are usually indicated using a circumflex, e.g., $\hat{i}$ and $\hat{j}$.

**Coplanar Force Resultants.**   Either of the two methods just described can be used to determine the resultant of several *coplanar forces*. To do this, each force is first resolved into its $x$ and $y$ components, and then the respective components are added using *scalar algebra* since they are collinear. The resultant force is then formed by adding the resultants of the $x$ and $y$ components using the parallelogram law. For example, consider the three concurrent forces in Fig. 2–16a, which have $x$ and $y$ components as shown in Fig. 2–16b. To solve this problem using *Cartesian vector notation*, each force is first represented as a Cartesian vector, i.e.,

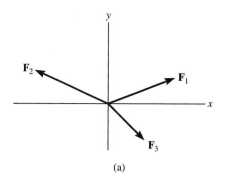

$$\mathbf{F}_1 = F_{1x}\mathbf{i} + F_{1y}\mathbf{j}$$
$$\mathbf{F}_2 = -F_{2x}\mathbf{i} + F_{2y}\mathbf{j}$$
$$\mathbf{F}_3 = F_{3x}\mathbf{i} - F_{3y}\mathbf{j}$$

(a)

The vector resultant is therefore

$$\begin{aligned}
\mathbf{F}_R &= \mathbf{F}_1 + \mathbf{F}_2 + \mathbf{F}_3 \\
&= F_{1x}\mathbf{i} + F_{1y}\mathbf{j} - F_{2x}\mathbf{i} + F_{2y}\mathbf{j} + F_{3x}\mathbf{i} - F_{3y}\mathbf{j} \\
&= (F_{1x} - F_{2x} + F_{3x})\mathbf{i} + (F_{1y} + F_{2y} - F_{3y})\mathbf{j} \\
&= (F_{Rx})\mathbf{i} + (F_{Ry})\mathbf{j}
\end{aligned}$$

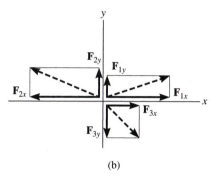

(b)

If *scalar notation* is used, then, from Fig. 2–16b, since $x$ is positive to the right and $y$ is positive upward, we have

$$(\xrightarrow{+}) \qquad\qquad F_{Rx} = F_{1x} - F_{2x} + F_{3x}$$
$$(+\uparrow) \qquad\qquad F_{Ry} = F_{1y} + F_{2y} - F_{3y}$$

These results are the *same* as the $\mathbf{i}$ and $\mathbf{j}$ components of $\mathbf{F}_R$ determined above.

In the general case, the $x$ and $y$ components of the resultant of any number of coplanar forces can be represented symbolically by the algebraic sum of the $x$ and $y$ components of all the forces, i.e.,

$$\boxed{\begin{aligned} F_{Rx} &= \Sigma F_x \\ F_{Ry} &= \Sigma F_y \end{aligned}} \qquad (2\text{–}1)$$

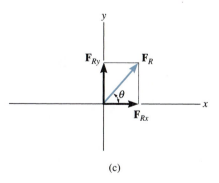

(c)

**Fig. 2–16**

When applying these equations, it is important to use the *sign convention* established for the components: components having a directional sense along the positive coordinate axes are considered positive scalars, whereas those having a directional sense along the negative coordinate axes are considered negative scalars. If this convention is followed, then the signs of the resultant components will specify the sense of these components. For example, a positive result indicates that the component has a directional sense which is in the positive coordinate direction.

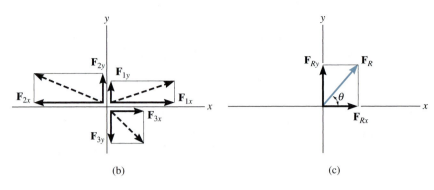

**Fig. 2–16**

Once the resultant components are determined, they may be sketched along the $x$ and $y$ axes in their proper directions, and the resultant force can be determined from vector addition, as shown in Fig. 2–16$c$. From this sketch, the magnitude of $\mathbf{F}_R$ is then found from the Pythagorean theorem; that is,

$$F_R = \sqrt{F_{Rx}^2 + F_{Ry}^2}$$

Also, the direction angle $\theta$, which specifies the orientation of the force, is determined from trigonometry:

$$\theta = \tan^{-1}\left|\frac{F_{Ry}}{F_{Rx}}\right|$$

The above concepts are illustrated numerically in the examples which follow.

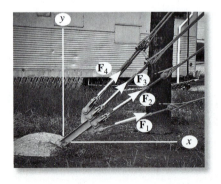

The resultant force of the four cable forces acting on the supporting bracket can be determined by adding algebraically the separate $x$ and $y$ components of each cable force. This resultant $\mathbf{F}_R$ produces the *same pulling effect* on the bracket as all four cables.

## IMPORTANT POINTS

- The resultant of several coplanar forces can easily be determined if an $x$, $y$ coordinate system is established and the forces are resolved along the axes.

- The direction of each force is specified by the angle its line of action makes with one of the axes, or by a sloped triangle.

- The orientation of the $x$ and $y$ axes is arbitrary, and their positive direction can be specified by the Cartesian unit vectors $\mathbf{i}$ and $\mathbf{j}$.

- The $x$ and $y$ components of the *resultant force* are simply the algebraic addition of the components of all the coplanar forces.

- The magnitude of the resultant force is determined from the Pythagorean theorem, and when the components are sketched on the $x$ and $y$ axes, the direction can be determined from trigonometry.

## EXAMPLE 2.5

Determine the $x$ and $y$ components of $\mathbf{F}_1$ and $\mathbf{F}_2$ acting on the boom shown in Fig. 2–17a. Express each force as a Cartesian vector.

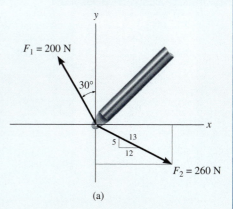

### Solution

**Scalar Notation.** By the parallelogram law, $\mathbf{F}_1$ is resolved into $x$ and $y$ components, Fig. 2–17b. The magnitude of each component is determined by trigonometry. Since $\mathbf{F}_{1x}$ acts in the $-x$ direction, and $\mathbf{F}_{1y}$ acts in the $+y$ direction, we have

$$F_{1x} = -200 \sin 30° \text{ N} = -100 \text{ N} = 100 \text{ N} \leftarrow \qquad Ans.$$
$$F_{1y} = 200 \cos 30° \text{ N} = 173 \text{ N} = 173 \text{ N} \uparrow \qquad Ans.$$

The force $\mathbf{F}_2$ is resolved into its $x$ and $y$ components as shown in Fig. 2–17c. Here the *slope* of the line of action for the force is indicated. From this "slope triangle" we could obtain the angle $\theta$, e.g., $\theta = \tan^{-1}(\frac{5}{12})$, and then proceed to determine the magnitudes of the components in the same manner as for $\mathbf{F}_1$. An easier method, however, consists of using proportional parts of similar triangles, i.e.,

$$\frac{F_{2x}}{260 \text{ N}} = \frac{12}{13} \qquad F_{2x} = 260 \text{ N}\left(\frac{12}{13}\right) = 240 \text{ N}$$

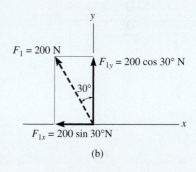

Similarly,

$$F_{2y} = 260 \text{ N}\left(\frac{5}{13}\right) = 100 \text{ N}$$

Notice that the magnitude of the *horizontal component*, $F_{2x}$, was obtained by multiplying the force magnitude by the ratio of the *horizontal leg* of the slope triangle divided by the hypotenuse; whereas the magnitude of the *vertical component*, $F_{2y}$, was obtained by multiplying the force magnitude by the ratio of the *vertical leg* divided by the hypotenuse. Hence, using scalar notation,

$$F_{2x} = 240 \text{ N} = 240 \text{ N} \rightarrow \qquad Ans.$$
$$F_{2y} = -100 \text{ N} = 100 \text{ N} \downarrow \qquad Ans.$$

**Cartesian Vector Notation.** Having determined the magnitudes and directions of the components of each force, we can express each force as a Cartesian vector.

$$\mathbf{F}_1 = \{-100\mathbf{i} + 173\mathbf{j}\} \text{ N} \qquad Ans.$$
$$\mathbf{F}_2 = \{240\mathbf{i} - 100\mathbf{j}\} \text{ N} \qquad Ans.$$

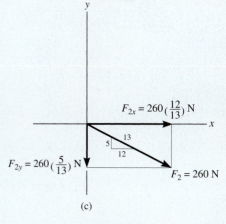

**Fig. 2–17**

E X A M P L E 2.6

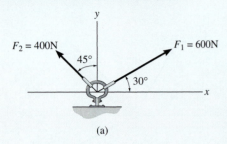

(a)

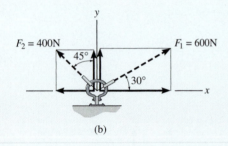

(b)

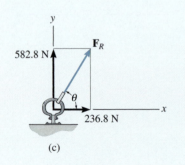

(c)

Fig. 2–18

The link in Fig. 2–18a is subjected to two forces $\mathbf{F}_1$ and $\mathbf{F}_2$. Determine the magnitude and orientation of the resultant force.

## Solution I

*Scalar Notation.* This problem can be solved by using the parallelogram law; however, here we will resolve each force into its $x$ and $y$ components, Fig. 2–18b, and sum these components algebraically. Indicating the "positive" sense of the $x$ and $y$ force components alongside each equation, we have

$$\xrightarrow{+} F_{Rx} + \Sigma F_x; \qquad F_{Rx} = 600 \cos 30° \text{ N} - 400 \sin 45° \text{ N}$$
$$= 236.8 \text{ N} \rightarrow$$

$$+\uparrow F_{Ry} = \Sigma F_y; \qquad F_{Ry} = 600 \sin 30° \text{ N} + 400 \cos 45° \text{ N}$$
$$= 582.8 \text{ N}\uparrow$$

The resultant force, shown in Fig. 2–18c, has a *magnitude* of

$$F_R = \sqrt{(236.8 \text{ N})^2 + (582.8 \text{ N})^2}$$
$$= 629 \text{ N} \qquad\qquad Ans.$$

From the vector addition, Fig. 2–18c, the direction angle $\theta$ is

$$\theta = \tan^{-1}\left(\frac{582.8\text{N}}{236.8\text{N}}\right) = 67.9° \qquad\qquad Ans.$$

## Solution II

*Cartesian Vector Notation.* From Fig. 2–18b, each force is expressed as a Cartesian vector

$$\mathbf{F}_1 = \{600 \cos 30°\mathbf{i} + 600 \sin 30°\mathbf{j}\} \text{ N}$$
$$\mathbf{F}_2 = \{-400 \sin 45°\mathbf{i} + 400 \cos 45°\mathbf{j}\} \text{ N}$$

Thus,

$$\mathbf{F}_R = \mathbf{F}_1 + \mathbf{F}_2 = (600 \cos 30° \text{ N} - 400 \sin 45° \text{ N})\mathbf{i}$$
$$+ (600 \sin 30° \text{ N} + 400 \cos 45° \text{ N})\mathbf{j}$$
$$= \{236.8\mathbf{i} + 582.8\mathbf{j}\} \text{ N}$$

The magnitude and direction of $\mathbf{F}_R$ are determined in the same manner as shown above.

Comparing the two methods of solution, note that use of scalar notation is more efficient since the components can be found *directly*, without first having to express each force as a Cartesian vector before adding the components. Later we will show that Cartesian vector analysis is very beneficial for solving three-dimensional problems.

E X A M P L E **2.7**

The end of the boom $O$ in Fig. 2–19a is subjected to three concurrent and coplanar forces. Determine the magnitude and orientation of the resultant force.

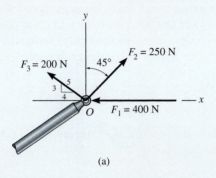

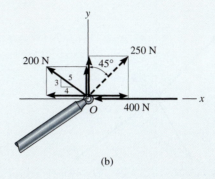

(a)                                             (b)

**Fig. 2–19**

### Solution

Each force is resolved into its $x$ and $y$ components, Fig. 2–19b. Summing the $x$ components, we have

$$\xrightarrow{+} F_{Rx} = \Sigma F_x; \qquad F_{Rx} = -400 \text{ N} + 250 \sin 45° \text{ N} - 200(\tfrac{4}{5}) \text{ N}$$
$$= -383.2 \text{ N} = 383.2 \text{ N} \leftarrow$$

The negative sign indicates that $F_{Rx}$ acts to the left, i.e., in the negative $x$ direction as noted by the small arrow. Summing the $y$ components yields

$$+\uparrow F_{Ry} = \Sigma F_y; \qquad F_{Ry} = 250 \cos 45° \text{ N} + 200(\tfrac{3}{5}) \text{ N}$$
$$= 296.8 \text{ N} \uparrow$$

The resultant force, shown in Fig. 2–19c, has a *magnitude* of

$$F_R = \sqrt{(-383.2\text{N})^2 + (296.8\text{N})^2}$$
$$= 485 \text{ N} \qquad\qquad Ans.$$

From the vector addition in Fig. 2–19c, the direction angle $\theta$ is

$$\theta = \tan^{-1}\left(\frac{296.8}{383.2}\right) = 37.8° \qquad\qquad Ans.$$

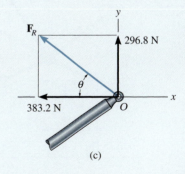

(c)

Note how convenient it is to use this method, compared to two applications of the parallelogram law.

# PROBLEMS

**\*2-20.** Determine the magnitude of the resultant force and its direction, measured counterclockwise from the positive $x$ axis.

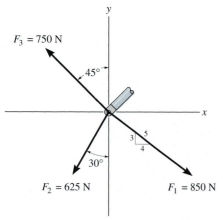

**Prob. 2–20**

**2-21.** Determine the magnitude and direction $\theta$ of $\mathbf{F}_1$ so that the resultant force is directed vertically upward and has a magnitude of 800 N.

**2-22.** Determine the magnitude and direction measured counterclockwise from the positive $x$ axis of the resultant force of the three forces acting on the ring $A$. Take $F_1 = 500$ N and $\theta = 20°$.

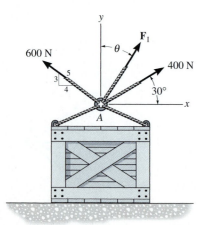

**Probs. 2–21/22**

**2-23.** Express $\mathbf{F}_1$ and $\mathbf{F}_2$ as Cartesian vectors.

**\*2-24.** Determine the magnitude of the resultant force and its direction measured counterclockwise from the positive $x$ axis.

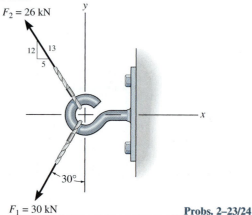

**Probs. 2–23/24**

**2-25.** Solve Prob. 2–1 by summing the rectangular or $x$, $y$ components of the forces to obtain the resultant force.

**2-26.** Solve Prob. 2–22 by summing the rectangular or $x, y$ components of the forces to obtain the resultant force.

**2-27.** Determine the magnitude and orientation $\theta$ of $\mathbf{F}_B$ so that the resultant force is directed along the positive $y$ axis and has a magnitude of 1500 N.

**\*2-28.** Determine the magnitude and orientation, measured counterclockwise from the positive $y$ axis, of the resultant force acting on the bracket, if $F_B = 600$ N and $\theta = 20°$.

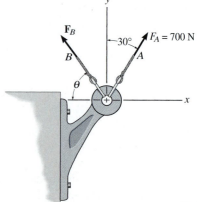

**Probs. 2–27/28**

**2-29.** Determine the x and y components of $\mathbf{F}_1$ and $\mathbf{F}_2$.

**2-30.** Determine the magnitude of the resultant force and its direction, measured counterclockwise from the positive x axis.

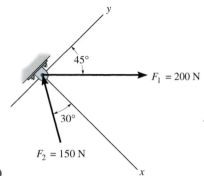

**Probs. 2–29/30**

**2-31.** Determine the x and y components of each force acting on the *gusset plate* of the bridge truss. Show that the resultant force is zero.

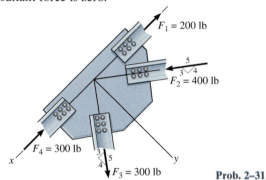

**Prob. 2–31**

**\*2-32.** If $\theta = 60°$ and $F = 20$ kN, determine the magnitude of the resultant force and its direction measured clockwise from the positive x axis.

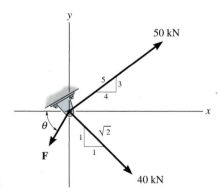

**Prob. 2–32**

**2-33.** Determine the magnitude and direction $\theta$ of $\mathbf{F}_A$ so that the resultant force is directed along the positive x axis and has a magnitude of 1250 N.

**2-34.** Determine the magnitude and direction, measured counterclockwise from the positive x axis, of the resultant force acting on the ring at $O$, if $F_A = 750$ N and $\theta = 45°$.

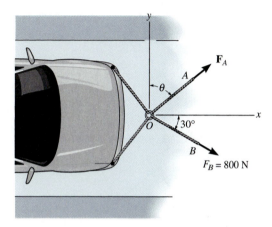

**Probs. 2–33/34**

**2-35.** Express each of the three forces acting on the column in Cartesian vector form and compute the magnitude of the resultant force.

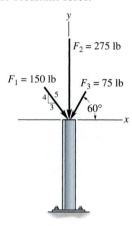

**Prob. 2–35**

**\*2-36.** Three forces act on the bracket. Determine the magnitude and orientation $\theta$ of $\mathbf{F}_2$ so that the resultant force is directed along the positive $u$ axis and has a magnitude of 50 lb.

**2-37.** If $F_2 = 150$ lb and $\theta = 55°$, determine the magnitude and orientation, measured clockwise from the positive $x$ axis, of the resultant force of the three forces acting on the bracket.

**2-38.** Determine the magnitude of force $\mathbf{F}$ so that the resultant force of the three forces is as small as possible. What is the magnitude of the resultant force?

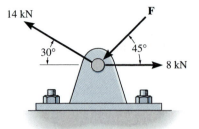

**Prob. 2–38**

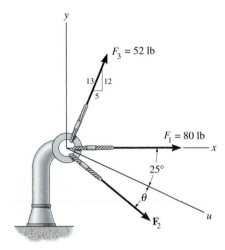

**Probs. 2–36/37**

## 2.5  Cartesian Vectors

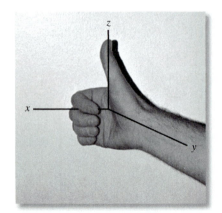

Right-handed coordinate system.

**Fig. 2–20**

The operations of vector algebra, when applied to solving problems in *three dimensions*, are greatly simplified if the vectors are first represented in Cartesian vector form. In this section we will present a general method for doing this; then in the next section we will apply this method to solving problems involving the addition of forces. Similar applications will be illustrated for the position and moment vectors given in later sections of the book.

**Right-Handed Coordinate System.**  A right-handed coordinate system will be used for developing the theory of vector algebra that follows. A rectangular or Cartesian coordinate system is said to be *right-handed* provided the thumb of the right hand points in the direction of the positive $z$ axis when the right-hand fingers are curled about this axis and directed from the positive $x$ toward the positive $y$ axis, Fig. 2–20. Furthermore, according to this rule, the $z$ axis for a two-dimensional problem as in Fig. 2–19 would be directed outward, perpendicular to the page.

Rectangular Components of a Vector. A vector **A** may have one, two, or three rectangular components along the *x, y, z* coordinate axes, depending on how the vector is oriented relative to the axes. In general, though, when **A** is directed within an octant of the *x, y, z* frame, Fig. 2–21, then by two successive applications of the parallelogram law, we may resolve the vector into components as $\mathbf{A} = \mathbf{A}' + \mathbf{A}_z$ and then $\mathbf{A}' = \mathbf{A}_x + \mathbf{A}_y$. Combining these equations, **A** is represented by the vector sum of its *three* rectangular components,

$$\mathbf{A} = \mathbf{A}_x + \mathbf{A}_y + \mathbf{A}_z \qquad (2\text{--}2)$$

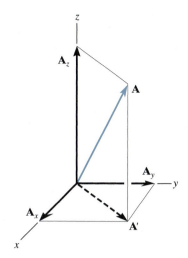

Fig. 2–21

Unit Vector. The direction of **A** can be specified using a unit vector. This vector is so named since it has a magnitude of 1. If **A** is a vector having a magnitude $A \neq 0$, then the unit vector having the *same direction* as **A** is represented by

$$\mathbf{u}_A = \frac{\mathbf{A}}{A} \qquad (2\text{--}3)$$

So that

$$\mathbf{A} = A\mathbf{u}_A \qquad (2\text{--}4)$$

Since **A** is of a certain type, e.g., a force vector, it is customary to use the proper set of units for its description. The magnitude $A$ also has this same set of units; hence, from Eq. 2–3, the *unit vector will be dimensionless* since the units will cancel out. Equation 2–4 therefore indicates that vector **A** may be expressed in terms of both its magnitude and direction *separately*; i.e., $A$ (a positive scalar) defines the *magnitude* of **A**, and $\mathbf{u}_A$ (a dimensionless vector) defines the *direction* and sense of **A**, Fig. 2–22.

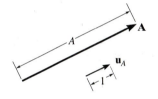

Fig. 2–22

Cartesian Unit Vectors. In three dimensions, the set of Cartesian unit vectors, **i**, **j**, **k**, is used to designate the directions of the *x, y, z* axes respectively. As stated in Sec. 2.4, the *sense* (or arrowhead) of these vectors will be described analytically by a plus or minus sign, depending on whether they are pointing along the positive or negative *x, y*, or *z* axes. The positive Cartesian unit vectors are shown in Fig. 2–23.

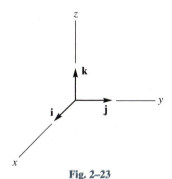

Fig. 2–23

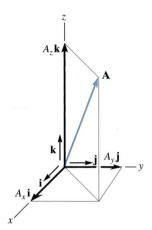

**Fig. 2–24**

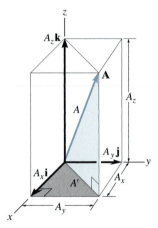

**Fig. 2–25**

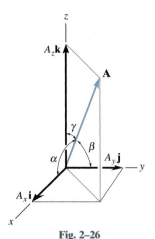

**Fig. 2–26**

**Cartesian Vector Representation.** Since the three components of **A** in Eq. 2–2 act in the positive **i**, **j**, and **k** directions, Fig. 2–24, we can write **A** in Cartesian vector form as

$$\mathbf{A} = A_x\mathbf{i} + A_y\mathbf{j} + A_z\mathbf{k} \tag{2–5}$$

There is a distinct advantage to writing vectors in this manner. Note that the *magnitude* and *direction* of each *component vector* are *separated*, and as a result this will simplify the operations of vector algebra, particularly in three dimensions.

**Magnitude of a Cartesian Vector.** It is always possible to obtain the magnitude of **A** provided it is expressed in Cartesian vector form. As shown in Fig. 2–25, from the colored right triangle, $A = \sqrt{A'^2 + A_z^2}$, and from the shaded right triangle, $A' = \sqrt{A_x^2 + A_y^2}$. Combining these equations yields

$$A = \sqrt{A_x^2 + A_y^2 + A_z^2} \tag{2–6}$$

*Hence, the magnitude of* **A** *is equal to the positive square root of the sum of the squares of its components.*

**Direction of a Cartesian Vector.** The *orientation* of **A** is defined by the *coordinate direction angles* $\alpha$ (alpha), $\beta$ (beta), and $\gamma$ (gamma), measured between the *tail* of **A** and the *positive x, y, z* axes located at the tail of **A**, Fig. 2–26. Note that regardless of where **A** is directed, each of these angles will be between 0° and 180°.

To determine $\alpha$, $\beta$, and $\gamma$, consider the projection of **A** onto the *x, y, z* axes, Fig. 2–27. Referring to the blue colored right triangles shown in each figure, we have

$$\cos\alpha = \frac{A_x}{A} \qquad \cos\beta = \frac{A_y}{A} \qquad \cos\gamma = \frac{A_z}{A} \tag{2–7}$$

These numbers are known as the *direction cosines* of **A**. Once they have been obtained, the coordinate direction angles $\alpha$, $\beta$, $\gamma$ can then be determined from the inverse cosines.

An easy way of obtaining the direction cosines of **A** is to form a unit vector in the direction of **A**, Eq. 2–3. Provided **A** is expressed in Cartesian vector form, $\mathbf{A} = A_x\mathbf{i} + A_y\mathbf{j} + A_z\mathbf{k}$ (Eq. 2–5), then

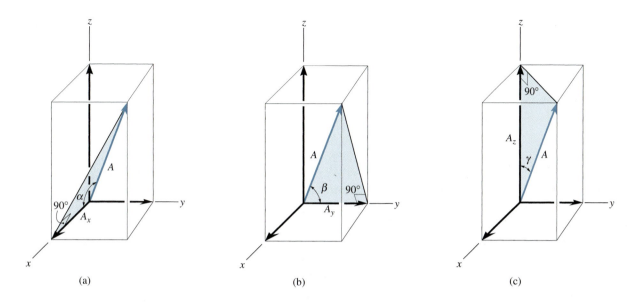

Fig. 2–27

$$\mathbf{u}_A = \frac{\mathbf{A}}{A} = \frac{A_x}{A}\mathbf{i} + \frac{A_y}{A}\mathbf{j} + \frac{A_z}{A}\mathbf{k} \qquad (2\text{--}8)$$

where $A = \sqrt{A_x^2 + A_y^2 + A_z^2}$ (Eq. 2–6). By comparison with Eqs. 2–7, it is seen that *the* $\mathbf{i}$, $\mathbf{j}$, $\mathbf{k}$ *components of* $\mathbf{u}_A$ *represent the direction cosines of* $\mathbf{A}$, i.e.,

$$\mathbf{u}_A = \cos \alpha \, \mathbf{i} + \cos \beta \, \mathbf{j} + \cos \gamma \, \mathbf{k} \qquad (2\text{--}9)$$

Since the magnitude of a vector is equal to the positive square root of the sum of the squares of the magnitudes of its components, and $\mathbf{u}_A$ has a magnitude of 1, then from Eq. 2–9 an important relation between the direction cosines can be formulated as

$$\cos^2 \alpha + \cos^2 \beta + \cos^2 \gamma = 1 \qquad (2\text{--}10)$$

Provided vector $\mathbf{A}$ lies in a known octant, this equation can be used to determine one of the coordinate direction angles if the other two are known.

Finally, if the magnitude and coordinate direction angles of $\mathbf{A}$ are given, $\mathbf{A}$ may be expressed in Cartesian vector form as

$$\begin{aligned} \mathbf{A} &= A\mathbf{u}_A \\ &= A \cos \alpha \, \mathbf{i} + A \cos \beta \, \mathbf{j} + A \cos \gamma \, \mathbf{k} \\ &= A_x\mathbf{i} + A_y\mathbf{j} + A_z\mathbf{k} \end{aligned} \qquad (2\text{--}11)$$

## 2.6 Addition and Subtraction of Cartesian Vectors

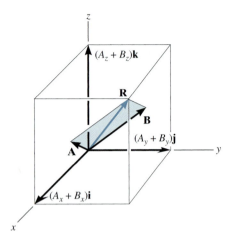

**Fig. 2–28**

The vector operations of addition and subtraction of two or more vectors are greatly simplified if the vectors are expressed in terms of their Cartesian components. For example, if $\mathbf{A} = A_x\mathbf{i} + A_y\mathbf{j} + A_z\mathbf{k}$ and $\mathbf{B} = B_x\mathbf{i} + B_y\mathbf{j} + B_z\mathbf{k}$, Fig. 2–28, then the resultant vector, $\mathbf{R}$, has components which represent the scalar sums of the $\mathbf{i}$, $\mathbf{j}$, $\mathbf{k}$ components of $\mathbf{A}$ and $\mathbf{B}$, i.e.,

$$\mathbf{R} = \mathbf{A} + \mathbf{B} = (A_x + B_x)\mathbf{i} + (A_y + B_y)\mathbf{j} + (A_z + B_z)\mathbf{k}$$

Vector subtraction, being a special case of vector addition, simply requires a scalar subtraction of the respective $\mathbf{i}$, $\mathbf{j}$, $\mathbf{k}$ components of either $\mathbf{A}$ or $\mathbf{B}$. For example,

$$\mathbf{R}' = \mathbf{A} - \mathbf{B} = (A_x - B_x)\mathbf{i} + (A_y - B_y)\mathbf{j} + (A_z - B_z)\mathbf{k}$$

**Concurrent Force Systems.**   If the above concept of vector addition is generalized and applied to a system of several concurrent forces, then the force resultant is the vector sum of all the forces in the system and can be written as

$$\mathbf{F}_R = \Sigma\mathbf{F} = \Sigma F_x\mathbf{i} + \Sigma F_y\mathbf{j} + \Sigma F_z\mathbf{k} \tag{2–12}$$

Here $\Sigma F_x$, $\Sigma F_y$, and $\Sigma F_z$ represent the algebraic sums of the respective $x$, $y$, $z$ or $\mathbf{i}$, $\mathbf{j}$, $\mathbf{k}$ components of each force in the system.

The examples which follow illustrate numerically the methods used to apply the above theory to the solution of problems involving force as a vector quantity.

The force $\mathbf{F}$ that the tie-down rope exerts on the ground support at $O$ is directed along the rope. Using the local $x, y, z$ axes, the coordinate direction angles $\alpha$, $\beta$, $\gamma$ can be measured. The cosines of their values form the components of a unit vector $\mathbf{u}$ which acts in the direction of the rope. If the force has a magnitude $F$, then the force can be written in Cartesian vector form, as $\mathbf{F} = F\mathbf{u} = F\cos\alpha\,\mathbf{i} + F\cos\beta\,\mathbf{j} + F\cos\gamma\,\mathbf{k}$.

# IMPORTANT POINTS

- Cartesian vector analysis is often used to solve problems in three dimensions.

- The positive direction of the $x$, $y$, $z$ axes are defined by the Cartesian unit vectors $\mathbf{i}$, $\mathbf{j}$, $\mathbf{k}$, respectively.

- The *magnitude* of a Cartesian vector is $A = \sqrt{A_x^2 + A_y^2 + A_z^2}$.

- The *direction* of a Cartesian vector is specified using coordinate direction angles which the tail of the vector makes with the positive $x$, $y$, $z$ axes, respectively. The components of the unit vector $\mathbf{u} = \mathbf{A}/A$ represent the direction cosines of $\alpha$, $\beta$, $\gamma$. Only two of the angles $\alpha$, $\beta$, $\gamma$ have to be specified. The third angle is determined from the relationship $\cos^2 \alpha + \cos^2 \beta + \cos^2 \gamma = 1$.

- To find the *resultant* of a concurrent force system, express each force as a Cartesian vector and add the $\mathbf{i}$, $\mathbf{j}$, $\mathbf{k}$ components of all the forces in the system.

---

## E X A M P L E  2.8

Express the force $\mathbf{F}$ shown in Fig. 2–29 as a Cartesian vector.

### Solution
Since only two coordinate direction angles are specified, the third angle $\alpha$ must be determined from Eq. 2–10; i.e.,

$$\cos^2 \alpha + \cos^2 \beta + \cos^2 \gamma = 1$$
$$\cos^2 \alpha + \cos^2 60° + \cos^2 45° = 1$$
$$\cos \alpha = \sqrt{1 - (0.5)^2 - (0.707)^2} = \pm 0.5$$

Hence, two possibilities exist, namely,

$$\alpha = \cos^{-1}(0.5) = 60° \quad \text{or} \quad \alpha = \cos^{-1}(-0.5) = 120°$$

By inspection of Fig. 2–29, it is necessary that $\alpha = 60°$, since $\mathbf{F}_x$ is in the $+x$ direction.

Using Eq. 2–11, with $F = 200$ N, we have

$$\mathbf{F} = F \cos \alpha \, \mathbf{i} + F \cos \beta \, \mathbf{j} + F \cos \gamma \, \mathbf{k}$$
$$= (200 \cos 60° \text{ N})\mathbf{i} + (200 \cos 60° \text{ N})\mathbf{j} + (200 \cos 45° \text{ N})\mathbf{k}$$
$$= \{100.0\mathbf{i} + 100.0\mathbf{j} + 141.4\mathbf{k}\} \text{ N} \qquad \qquad Ans.$$

By applying Eq. 2–6, note that indeed the magnitude of $F = 200$ N.

$$F = \sqrt{F_x^2 + F_y^2 + F_z^2}$$
$$= \sqrt{(100.0)^2 + (100.0)^2 + (141.4)^2} = 200 \text{ N}$$

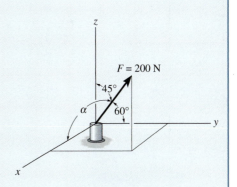

Fig. 2–29

**EXAMPLE 2.9**

Determine the magnitude and the coordinate direction angles of the resultant force acting on the ring in Fig. 2–30a.

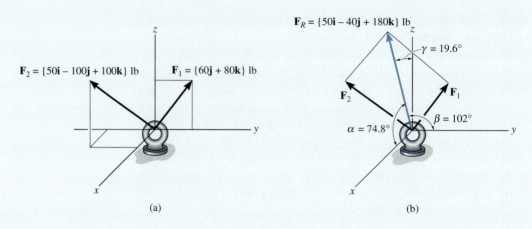

Fig. 2–30

**Solution**

Since each force is represented in Cartesian vector form, the resultant force, shown in Fig. 2–30b, is

$$\mathbf{F}_R = \Sigma\mathbf{F} = \mathbf{F}_1 + \mathbf{F}_2 = \{60\mathbf{j} + 80\mathbf{k}\}\ \text{lb} + \{50\mathbf{i} - 100\mathbf{j} + 100\mathbf{k}\}\ \text{lb}$$
$$= \{50\mathbf{i} - 40\mathbf{j} + 180\mathbf{k}\}\ \text{lb}$$

The magnitude of $\mathbf{F}_R$ is found from Eq. 2–6, i.e.,

$$F_R = \sqrt{(50)^2 + (-40)^2 + (180)^2} = 191.0$$
$$= 191\ \text{lb} \qquad\qquad Ans.$$

The coordinate direction angles $\alpha$, $\beta$, $\gamma$ are determined from the components of the unit vector acting in the direction of $\mathbf{F}_R$.

$$\mathbf{u}_{F_R} = \frac{\mathbf{F}_R}{F_R} = \frac{50}{191.0}\mathbf{i} - \frac{40}{191.0}\mathbf{j} + \frac{180}{191.0}\mathbf{k}$$
$$= 0.2617\mathbf{i} - 0.2094\mathbf{j} + 0.9422\mathbf{k}$$

so that

$$\cos\alpha = 0.2617 \qquad \alpha = 74.8° \qquad Ans.$$
$$\cos\beta = -0.2094 \qquad \beta = 102° \qquad Ans.$$
$$\cos\gamma = 0.9422 \qquad \gamma = 19.6° \qquad Ans.$$

These angles are shown in Fig. 2–30b. In particular, note that $\beta > 90°$ since the $\mathbf{j}$ component of $\mathbf{u}_{F_R}$ is negative.

# E X A M P L E  2.10

Express the force $\mathbf{F}_1$, shown in Fig. 2–31a as a Cartesian vector.

## Solution

The angles of 60° and 45° defining the direction of $\mathbf{F}_1$ are *not* coordinate direction angles. The two successive applications of the parallelogram law needed to resolve $\mathbf{F}_1$ into its $x$, $y$, $z$ components are shown in Fig. 2–31b. By trigonometry, the magnitudes of the components are

$$F_{1z} = 100 \sin 60° \text{ lb} = 86.6 \text{ lb}$$
$$F' = 100 \cos 60° \text{ lb} = 50 \text{ lb}$$
$$F_{1x} = 50 \cos 45° \text{ lb} = 35.4 \text{ lb}$$
$$F_{1y} = 50 \sin 45° \text{ lb} = 35.4 \text{ lb}$$

Realizing that $\mathbf{F}_{1y}$ has a direction defined by $-\mathbf{j}$, we have

$$\mathbf{F}_1 = \{35.4\mathbf{i} - 35.4\mathbf{j} + 86.6\mathbf{k}\} \text{ lb} \qquad \textit{Ans.}$$

To show that the magnitude of this vector is indeed 100 lb, apply Eq. 2–6,

$$F_1 = \sqrt{F_{1x}^2 + F_{1y}^2 + F_{1z}^2}$$

$$= \sqrt{(35.4)^2 + (-35.4)^2 + (86.6)^2} = 100 \text{ lb}$$

If needed, the coordinate direction angles of $\mathbf{F}_1$ can be determined from the components of the unit vector acting in the direction of $\mathbf{F}_1$. Hence,

$$\mathbf{u}_1 = \frac{\mathbf{F}_1}{F_1} = \frac{F_{1x}}{F_1}\mathbf{i} + \frac{F_{1y}}{F_1}\mathbf{j} + \frac{F_{1z}}{F_1}\mathbf{k}$$

$$= \frac{35.4}{100}\mathbf{i} - \frac{35.4}{100}\mathbf{j} + \frac{86.6}{100}\mathbf{k}$$

$$= 0.354\mathbf{i} - 0.354\mathbf{j} + 0.866\mathbf{k}$$

so that

$$\alpha_1 = \cos^{-1}(0.354) = 69.3°$$
$$\beta_1 = \cos^{-1}(-0.354) = 111°$$
$$\gamma_1 = \cos^{-1}(0.866) = 30.0°$$

These results are shown in Fig. 2–31c.

Using this same method, show that $\mathbf{F}_2$ in Fig. 2–31a can be written in Cartesian vector form as

$$\mathbf{F}_2 = \{106\mathbf{i} + 184\mathbf{j} - 212\mathbf{k}\} \text{ N} \qquad \textit{Ans.}$$

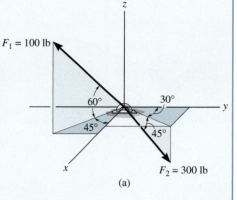

(a)

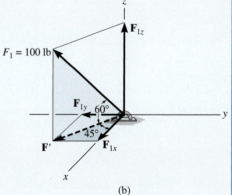

(b)

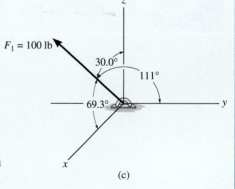

(c)

Fig. 2–31

EXAMPLE 2.11

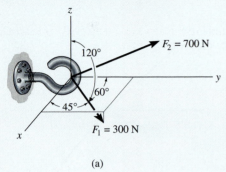

(a)

Two forces act on the hook shown in Fig. 2–32a. Specify the coordinate direction angles of $\mathbf{F}_2$ so that the resultant force $\mathbf{F}_R$ acts along the positive $y$ axis and has a magnitude of 800 N.

## Solution

To solve this problem, the resultant force $\mathbf{F}_R$ and its two components, $\mathbf{F}_1$ and $\mathbf{F}_2$, will each be expressed in Cartesian vector form. Then, as shown in Fig. 2–32b, it is necessary that $\mathbf{F}_R = \mathbf{F}_1 + \mathbf{F}_2$.

Applying Eq. 2–11,

$$\mathbf{F}_1 = F_1 \cos \alpha_1 \mathbf{i} + F_1 \cos \beta_1 \mathbf{j} + F_1 \cos \gamma_1 \mathbf{k}$$
$$= (300 \cos 45° \text{ N})\mathbf{i} + (300 \cos 60° \text{ N})\mathbf{j} + (300 \cos 120° \text{ N})\mathbf{k}$$
$$= \{212.1\mathbf{i} + 150\mathbf{j} - 150\mathbf{k}\} \text{ N}$$
$$\mathbf{F}_2 = F_{2x}\mathbf{i} + F_{2y}\mathbf{j} - F_{2z}\mathbf{k}$$

Since the resultant force $\mathbf{F}_R$ has a magnitude of 800 N and acts in the $+\mathbf{j}$ direction.

$$\mathbf{F}_R = (800 \text{ N})(+\mathbf{j}) = \{800\mathbf{j}\} \text{ N}$$

We require

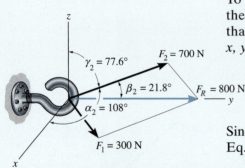

(b)

**Fig. 2–32**

$$\mathbf{F}_R = \mathbf{F}_1 + \mathbf{F}_2$$
$$800\mathbf{j} = 212.1\mathbf{i} + 150\mathbf{j} - 150\mathbf{k} + F_{2x}\mathbf{i} + F_{2y}\mathbf{j} + F_{2z}\mathbf{k}$$
$$800\mathbf{j} = (212.1 + F_{2x})\mathbf{i} + (150 + F_{2y})\mathbf{j} + (-150 + F_{2z})\mathbf{k}$$

To satisfy this equation, the corresponding $\mathbf{i}$, $\mathbf{j}$, $\mathbf{k}$ components on the left and right sides must be equal. This is equivalent to stating that the $x, y, z$ components of $\mathbf{F}_R$ must be equal to the corresponding $x, y, z$ components of $(\mathbf{F}_1 + \mathbf{F}_2)$. Hence,

$$0 = 212.1 + F_{2x} \qquad F_{2x} = -212.1 \text{ N}$$
$$800 = 150 + F_{2y} \qquad F_{2y} = 650 \text{ N}$$
$$0 = -150 + F_{2z} \qquad F_{2z} = 150 \text{ N}$$

Since the magnitudes of $\mathbf{F}_2$ and its components are known, we can use Eq. 2–11 to determine $\alpha_2, \beta_2, \gamma_2$.

$$-212.1 = 700 \cos \alpha_2; \qquad \alpha_2 = \cos^{-1}\left(\frac{-212.1}{700}\right) = 108° \qquad \textit{Ans.}$$

$$650 = 700 \cos \beta_2; \qquad \beta_2 = \cos^{-1}\left(\frac{650}{700}\right) = 21.8° \qquad \textit{Ans.}$$

$$150 = 700 \cos \gamma_2; \qquad \gamma_2 = \cos^{-1}\left(\frac{150}{700}\right) = 77.6° \qquad \textit{Ans.}$$

These results are shown in Fig. 2–32b.

# PROBLEMS

**2-39.** Determine the magnitude and coordinate direction angles of $\mathbf{F}_1 = \{60\mathbf{i} - 50\mathbf{j} + 40\mathbf{k}\}$ N and $\mathbf{F}_2 = \{-40\mathbf{i} - 85\mathbf{j} + 30\mathbf{k}\}$ N. Sketch each force on an $x, y, z$ reference.

**\*2-40.** The cable at the end of the crane boom exerts a force of 250 lb on the boom as shown. Express $\mathbf{F}$ as a Cartesian vector.

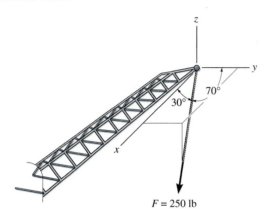

$F = 250$ lb

Prob. 2-40

**2-41.** Determine the magnitude and coordinate direction angles of the force $\mathbf{F}$ acting on the stake.

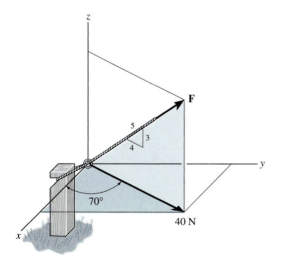

Prob. 2-41

**2-42.** Determine the magnitude and the coordinate direction angles of the resultant force.

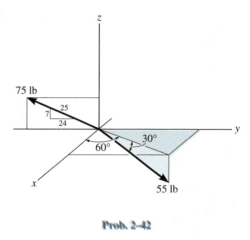

Prob. 2-42

**2-43.** The stock $S$ mounted on the lathe is subjected to a force of 60 N. Determine the coordinate direction angle $\beta$ and express the force as a Cartesian vector.

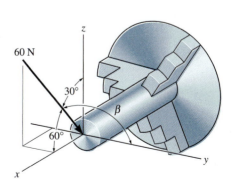

Prob. 2-43

**\*2-44.** Determine the magnitude and coordinate direction angles of the resultant force and sketch this vector on the coordinate system.

**2-45.** Specify the coordinate direction angles of $\mathbf{F}_1$ and $\mathbf{F}_2$ and express each force as a Cartesian vector.

**\*2-48.** The cables attached to the screw eye are subjected to the three forces shown. Express each force in Cartesian vector form and determine the magnitude and coordinate direction angles of the resultant force.

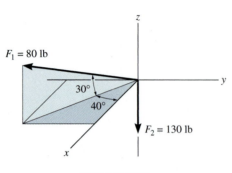

**Probs. 2–44/45**

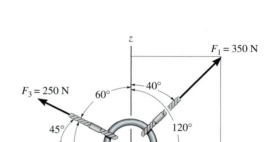

**Prob. 2–48**

**2-46.** The mast is subjected to the three forces shown. Determine the coordinate direction angles $\alpha_1$, $\beta_1$, $\gamma_1$ of $\mathbf{F}_1$ so that the resultant force acting on the mast is $\mathbf{F}_R = \{350\mathbf{i}\}$ N.

**2-47.** The mast is subjected to the three forces shown. Determine the coordinate direction angles $\alpha_1$, $\beta_1$, $\gamma_1$ of $\mathbf{F}_1$ so that the resultant force acting on the mast is zero.

**2-49.** The beam is subjected to the two forces shown. Express each force in Cartesian vector form and determine the magnitude and coordinate direction angles of the resultant force.

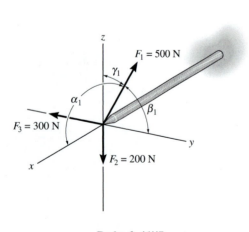

**Probs. 2–46/47**

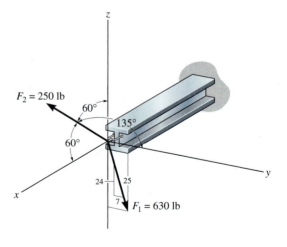

**Prob. 2–49**

**2-50.** Determine the magnitude and coordinate direction angles of the resultant force and sketch this vector on the coordinate system.

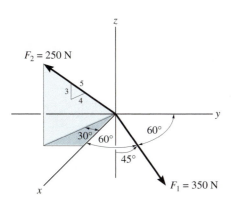

F₂ = 250 N

F₁ = 350 N

**Prob. 2–50**

**2-51.** The two forces **F**₁ and **F**₂ acting at *A* have a resultant force of $\mathbf{F}_R = \{-100\mathbf{k}\}$ lb. Determine the magnitude and coordinate direction angles of **F**₂.

*****2-52.** Determine the coordinate direction angles of the force **F**₁ and indicate them on the figure.

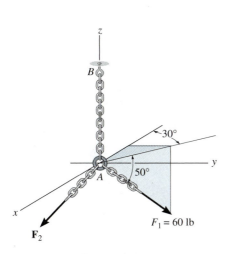

F₁ = 60 lb

F₂

**Probs. 2–51/52**

**2-53.** The bracket is subjected to the two forces shown. Express each force in Cartesian vector form and then determine the resultant force $\mathbf{F}_R$. Find the magnitude and coordinate direction angles of the resultant force.

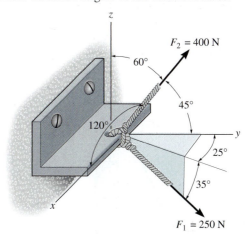

F₂ = 400 N

F₁ = 250 N

**Prob. 2–53**

**2-54.** The pole is subjected to the force **F**, which has components acting along the *x*, *y*, *z* axes as shown. If the magnitude of **F** is 3 kN, and $\beta = 30°$ and $\gamma = 75°$, determine the magnitudes of its three components.

**2-55.** The pole is subjected to the force **F** which has components $F_x = 1.5$ kN and $F_z = 1.25$ kN. If $\beta = 75°$, determine the magnitudes of **F** and **F**_y.

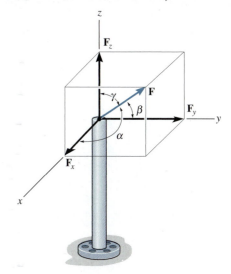

**Probs. 2–54/55**

**\*2-56.** A force **F** is applied at the top of the tower at *A*. If it acts in the direction shown such that one of its components lying in the shaded *y-z* plane has a magnitude of 80 lb, determine its magnitude *F* and coordinate direction angles $\alpha$, $\beta$, $\gamma$.

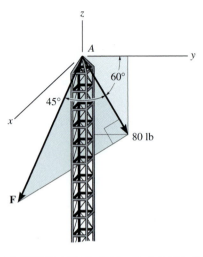

**Prob. 2–56**

**2-57.** Three forces act on the hook. If the resultant force $\mathbf{F}_R$ has a magnitude and direction as shown, determine the magnitude and the coordinate direction angles of force $\mathbf{F}_3$.

**2-58.** Determine the coordinate direction angles of $\mathbf{F}_1$ and $\mathbf{F}_R$.

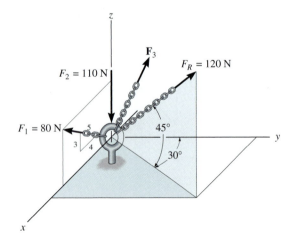

**Probs. 2–57/58**

**2-59.** The bolt is subjected to the force **F**, which has components acting along the *x, y, z* axes as shown. If the magnitude of **F** is 80 N, and $\alpha = 60°$ and $\gamma = 45°$, determine the magnitudes of its components.

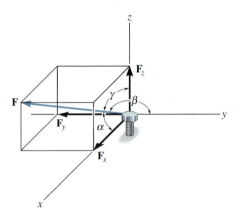

**Prob. 2–59**

**\*2-60.** Two forces $\mathbf{F}_1$ and $\mathbf{F}_2$ act on the bolt. If the resultant force $\mathbf{F}_R$ has a magnitude of 50 lb and coordinate direction angles $\alpha = 110°$ and $\beta = 80°$, as shown, determine the magnitude of $\mathbf{F}_2$ and its coordinate direction angles.

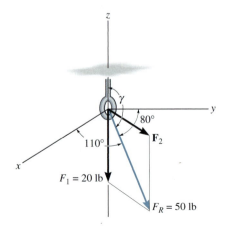

**Prob. 2–60**

## 2.7 Position Vectors

In this section we will introduce the concept of a position vector. It will be shown that this vector is of importance in formulating a Cartesian force vector directed between any two points in space. Later, in Chapter 4, we will use it for finding the moment of a force.

*x, y, z* **Coordinates.** Throughout the book we will use a *right-handed* coordinate system to reference the location of points in space. Furthermore, we will use the convention followed in many technical books, and that is to require the positive *z* axis to be directed *upward* (the zenith direction) so that it measures the height of an object or the altitude of a point. The *x, y* axes then lie in the horizontal plane, Fig. 2–33. Points in space are located relative to the origin of coordinates, *O*, by successive measurements along the *x, y, z* axes. For example, in Fig. 2–33 the coordinates of point *A* are obtained by starting at *O* and measuring $x_A = +4$ m along the *x* axis, $y_A = +2$ m along the *y* axis, and $z_A = -6$ m along the *z* axis. Thus, $A(4, 2, -6)$. In a similar manner, measurements along the *x, y, z* axes from *O* to *B* yield the coordinates of *B*, i.e., $B(0, 2, 0)$. Also notice that $C(6, -1, 4)$.

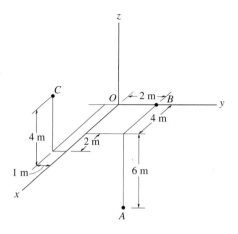

**Fig. 2–33**

**Position Vector.** The *position vector* **r** is defined as a fixed vector which locates a point in space relative to another point. For example, if **r** extends from the origin of coordinates, *O*, to point $P(x, y, z)$, Fig. 2–34*a*, then **r** can be expressed in Cartesian vector form as

$$\mathbf{r} = x\mathbf{i} + y\mathbf{j} + z\mathbf{k}$$

Note how the head-to-tail vector addition of the three components yields vector *r*, Fig. 2–34*b*. Starting at the origin *O*, one travels *x* in the +**i** direction, then *y* in the +**j** direction, and finally *z* in the +**k** direction to arrive at point $P(x, y, z)$.

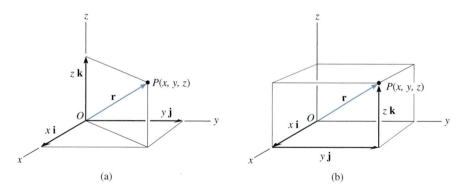

(a)                    (b)

**Fig. 2–34**

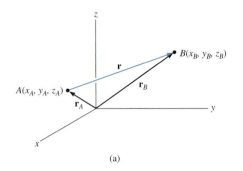

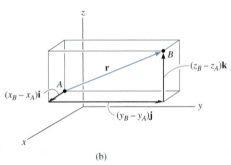

(b)

**Fig. 2–35**

In the more general case, the position vector may be directed from point $A$ to point $B$ in space, Fig. 2–35$a$. As noted, this vector is also designated by the symbol $\mathbf{r}$. As a matter of convention, however, we will *sometimes* refer to this vector with *two subscripts* to indicate from and to the point where it is directed. Thus, $\mathbf{r}$ can also be designated as $\mathbf{r}_{AB}$. Also, note that $\mathbf{r}_A$ and $\mathbf{r}_B$ in Fig. 2–35$a$ are referenced with only one subscript since they extend from the origin of coordinates.

From Fig. 2–35$a$, by the head-to-tail vector addition, we require

$$\mathbf{r}_A + \mathbf{r} = \mathbf{r}_B$$

Solving for $\mathbf{r}$ and expressing $\mathbf{r}_A$ and $\mathbf{r}_B$ in Cartesian vector form yields

$$\mathbf{r} = \mathbf{r}_B - \mathbf{r}_A = (x_B\mathbf{i} + y_B\mathbf{j} + z_B\mathbf{k}) - (x_A\mathbf{i} + y_A\mathbf{j} + z_A\mathbf{k})$$

or

$$\mathbf{r} = (x_B - x_A)\mathbf{i} + (y_B - y_A)\mathbf{j} + (z_B - z_A)\mathbf{k} \qquad (2\text{–}13)$$

*Thus, the* $\mathbf{i}$, $\mathbf{j}$, $\mathbf{k}$ *components of the position vector* $\mathbf{r}$ *may be formed by taking the coordinates of the tail of the vector,* $A(x_A, y_A, z_A)$, *and subtracting them from the corresponding coordinates of the head,* $B(x_B, y_B, z_B)$. Again note how the head-to-tail addition of these three components yields $\mathbf{r}$, i.e., going from $A$ to $B$, Fig. 2–35$b$, one first travels $(x_B - x_A)$ in the $+\mathbf{i}$ direction, then $(y_B - y_A)$ in the $+\mathbf{j}$ direction, and finally $(z_B - z_A)$ in the $+\mathbf{k}$ direction.

The length and direction of cable $AB$ used to support the stack can be determined by measuring the coordinates of points $A$ and $B$ using the $x$, $y$, $z$ axes. The position vector $\mathbf{r}$ along the cable can then be established. The magnitude $r$ represents the length of the cable, and the direction of the cable is defined by $\alpha$, $\beta$, $\gamma$, which are determined from the components of the unit vector found from the position vector, $\mathbf{u} = \mathbf{r}/r$.

# E X A M P L E   2.12

An elastic rubber band is attached to points $A$ and $B$ as shown in Fig 2–36a. Determine its length and its direction measured from $A$ toward $B$.

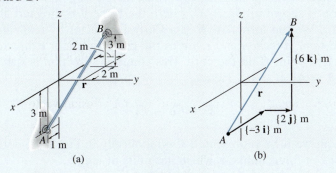

(a)          (b)

Fig. 2–36

## Solution

We first establish a position vector from $A$ to $B$, Fig. 2–36b. In accordance with Eq. 2–13, the coordinates of the tail $A(1\text{ m}, 0, -3\text{ m})$ are subtracted from the coordinates of the head $B(-2\text{ m}, 2\text{ m}, 3\text{ m})$, which yields

$$\mathbf{r} = [-2\text{ m} - 1\text{ m}]\mathbf{i} + [2\text{ m} - 0]\mathbf{j} + [3\text{ m} - (-3\text{ m})]\mathbf{k}$$
$$= \{-3\mathbf{i} + 2\mathbf{j} + 6\mathbf{k}\}\text{ m}$$

These components of $\mathbf{r}$ can also be determined *directly* by realizing from Fig. 2–36a that they represent the direction and distance one must go along each axis in order to move from $A$ to $B$, i.e., along the $x$ axis $\{-3\mathbf{i}\}$ m, along the $y$ axis $\{2\mathbf{j}\}$ m, and finally along the $z$ axis $\{6\mathbf{k}\}$ m.

The magnitude of $\mathbf{r}$ represents the length of the rubber band.

$$r = \sqrt{(-3)^2 + (2)^2 + (6)^2} = 7\text{ m} \qquad \textit{Ans.}$$

Formulating a unit vector in the direction of $\mathbf{r}$, we have

$$\mathbf{u} = \frac{\mathbf{r}}{r} = \frac{-3}{7}\mathbf{i} + \frac{2}{7}\mathbf{j} + \frac{6}{7}\mathbf{k}$$

The components of this unit vector yield the coordinate direction angles

$$\alpha = \cos^{-1}\left(\frac{-3}{7}\right) = 115° \qquad \textit{Ans.}$$

$$\beta = \cos^{-1}\left(\frac{2}{7}\right) = 73.4° \qquad \textit{Ans.}$$

$$\gamma = \cos^{-1}\left(\frac{6}{7}\right) = 31.0° \qquad \textit{Ans.}$$

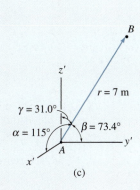

(c)

These angles are measured from the *positive axes* of a localized coordinate system placed at the tail of $\mathbf{r}$, point $A$, as shown in Fig. 2–36c.

## 2.8 Force Vector Directed Along a Line

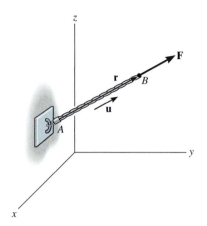

Fig. 2–37

Quite often in three-dimensional statics problems, the direction of a force is specified by two points through which its line of action passes. Such a situation is shown in Fig. 2–37, where the force **F** is directed along the cord $AB$. We can formulate **F** as a Cartesian vector by realizing that it has the *same direction* and *sense* as the position vector **r** directed from point $A$ to point $B$ on the cord. This common direction is specified by the *unit vector* $\mathbf{u} = \mathbf{r}/r$. Hence,

$$\mathbf{F} = F\mathbf{u} = F\left(\frac{\mathbf{r}}{r}\right)$$

Although we have represented **F** symbolically in Fig. 2–37, note that it has *units of force*, unlike **r**, which has units of length.

The force **F** acting along the chain can be represented as a Cartesian vector by first establishing $x$, $y$, $z$ axes and forming a position vector **r** along the length of the chain, then finding the corresponding unit vector $\mathbf{u} = \mathbf{r}/r$ that defines the direction of both the chain and the force. Finally, the magnitude of the force is combined with its direction, $\mathbf{F} = F\mathbf{u}$.

## IMPORTANT POINTS

- A position vector locates one point in space relative to another point.
- The easiest way to formulate the components of a position vector is to determine the distance and direction that must be traveled along the $x$, $y$, $z$ directions—going from the tail to the head of the vector.
- A force **F** acting in the direction of a position vector **r** can be represented in Cartesian form if the unit vector **u** of the position vector is determined and this is multiplied by the magnitude of the force, i.e., $\mathbf{F} = F\mathbf{u} = F(\mathbf{r}/r)$.

EXAMPLE 2.13

The man shown in Fig. 2–38a pulls on the cord with a force of 70 lb. Represent this force, acting on the support $A$, as a Cartesian vector and determine its direction.

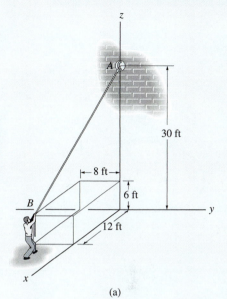

(a)

**Solution**

Force $\mathbf{F}$ is shown in Fig. 2–38b. The *direction* of this vector, $\mathbf{u}$, is determined from the position vector $\mathbf{r}$, which extends from $A$ to $B$, Fig. 2–38b. The coordinates of the end points of the cord are $A(0, 0, 30 \text{ ft})$ and $B(12 \text{ ft}, -8 \text{ ft}, 6 \text{ ft})$. Forming the position vector by subtracting the corresponding $x$, $y$, and $z$ coordinates of $A$ from those of $B$, we have

$$\mathbf{r} = (12 \text{ ft} - 0)\mathbf{i} + (-8 \text{ ft} - 0)\mathbf{j} + (6 \text{ ft} - 30 \text{ ft})\mathbf{k}$$
$$= \{12\mathbf{i} - 8\mathbf{j} - 24\mathbf{k}\} \text{ ft}$$

This result can also be determined *directly* by noting in Fig. 2–38a, that one must go from $A\{-24\mathbf{k}\}$ ft, then $\{-8\mathbf{j}\}$ ft, and finally $\{12\mathbf{i}\}$ ft to get to $B$.

The magnitude of $\mathbf{r}$, which represents the *length* of cord $AB$, is

$$r = \sqrt{(12 \text{ ft})^2 + (-8 \text{ ft})^2 + (-24 \text{ ft})^2} = 28 \text{ ft}$$

Forming the unit vector that defines the direction and sense of both $\mathbf{r}$ and $\mathbf{F}$ yields

$$\mathbf{u} = \frac{\mathbf{r}}{r} = \frac{12}{28}\mathbf{i} - \frac{8}{28}\mathbf{j} - \frac{24}{28}\mathbf{k}$$

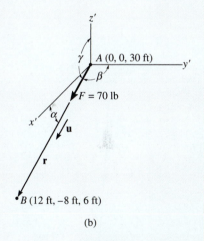

(b)

Fig. 2–38

Since $\mathbf{F}$ has a *magnitude* of 70 lb and a *direction* specified by $\mathbf{u}$, then

$$\mathbf{F} = F\mathbf{u} = 70 \text{ lb}\left(\frac{12}{28}\mathbf{i} - \frac{8}{28}\mathbf{j} - \frac{24}{28}\mathbf{k}\right)$$
$$= \{30\mathbf{i} - 20\mathbf{j} - 60\mathbf{k}\} \text{ lb} \qquad Ans.$$

The coordinate direction angles are measured between $\mathbf{r}$ (or $\mathbf{F}$) and the *positive axes* of a localized coordinate system with origin placed at $A$, Fig. 2–38b. From the components of the unit vector:

$$\alpha = \cos^{-1}\left(\frac{12}{28}\right) = 64.6° \qquad Ans.$$

$$\beta = \cos^{-1}\left(\frac{-8}{28}\right) = 107° \qquad Ans.$$

$$\gamma = \cos^{-1}\left(\frac{-24}{28}\right) = 149° \qquad Ans.$$

**E X A M P L E  2.14**

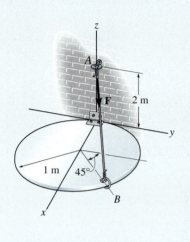

(a)

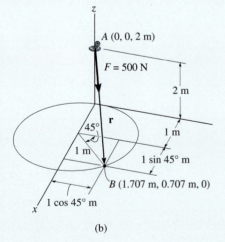

(b)

**Fig. 2–39**

The circular plate in Fig. 2–39a is partially supported by the cable AB. If the force of the cable on the hook at A is $F = 500$ N, express **F** as a Cartesian vector.

**Solution**

As shown in Fig. 2–39b, **F** has the same direction and sense as the position vector **r**, which extends from A to B. The coordinates of the end points of the cable are $A(0, 0, 2$ m$)$ and $B(1.707$ m, $0.707$ m, $0)$, as indicated in the figure. Thus,

$$\mathbf{r} = (1.707 \text{ m} - 0)\mathbf{i} + (0.707 \text{ m} - 0)\mathbf{j} + (0 - 2 \text{ m})\mathbf{k}$$
$$= \{1.707\mathbf{i} + 0.707\mathbf{j} - 2\mathbf{k}\} \text{ m}$$

Note how one can calculate these components *directly* by going from $A$, $\{-2\mathbf{k}\}$ m along the $z$ axis, then $\{1.707\mathbf{i}\}$ m along the $x$ axis, and finally $\{0.707\mathbf{j}\}$ m along the $y$ axis to get to $B$.

The magnitude of **r** is

$$r = \sqrt{(1.707)^2 + (0.707)^2 + (-2)^2} = 2.723 \text{ m}$$

Thus,

$$\mathbf{u} = \frac{\mathbf{r}}{r} = \frac{1.707}{2.723}\mathbf{i} + \frac{0.707}{2.723}\mathbf{j} - \frac{2}{2.723}\mathbf{k}$$

$$= 0.6269\mathbf{i} + 0.2597\mathbf{j} - 0.7345\mathbf{k}$$

Since $F = 500$ N and **F** has the direction **u**, we have

$$\mathbf{F} = F\mathbf{u} = 500 \text{ N}(0.6269\mathbf{i} + 0.2597\mathbf{j} - 0.7345\mathbf{k})$$
$$= \{313\mathbf{i} + 130\mathbf{j} - 367\mathbf{k}\} \text{ N} \qquad \qquad Ans.$$

Using these components, notice that indeed the magnitude of **F** is 500 N; i.e.,

$$F = \sqrt{(313)^2 + (130)^2 + (-367)^2} = 500 \text{ N}$$

Show that the coordinate direction angle $\gamma = 137°$, and indicate this angle on the figure.

EXAMPLE 2.15

The roof is supported by cables as shown in the photo. If the cables exert forces $F_{AB} = 100$ N and $F_{AC} = 120$ N on the wall hook at $A$ as shown in Fig. 2–40a, determine the magnitude of the resultant force acting at $A$.

### Solution

The resultant force $\mathbf{F}_R$ is shown graphically in Fig. 2–40b. We can express this force as a Cartesian vector by first formulating $\mathbf{F}_{AB}$ and $\mathbf{F}_{AC}$ as Cartesian vectors and then adding their components. The directions of $\mathbf{F}_{AB}$ and $\mathbf{F}_{AC}$ are specified by forming unit vectors $\mathbf{u}_{AB}$ and $\mathbf{u}_{AC}$ along the cables. These unit vectors are obtained from the associated position vectors $\mathbf{r}_{AB}$ and $\mathbf{r}_{AC}$. With reference to Fig. 2–40b, for $\mathbf{F}_{AB}$ we have

$$\mathbf{r}_{AB} = (4\text{ m} - 0)\mathbf{i} + (0 - 0)\mathbf{j} + (0 - 4\text{m})\mathbf{k}$$
$$= \{4\mathbf{i} - 4\mathbf{k}\}\text{ m}$$
$$r_{AB} = \sqrt{(4)^2 + (-4)^2} = 5.66\text{ m}$$
$$\mathbf{F}_{AB} = 100\text{ N}\left(\frac{\mathbf{r}_{AB}}{r_{AB}}\right) = 100\text{ N}\left(\frac{4}{5.66}\mathbf{i} - \frac{4}{5.66}\mathbf{k}\right)$$
$$\mathbf{F}_{AB} = \{70.7\mathbf{i} - 70.7\mathbf{k}\}\text{ N}$$

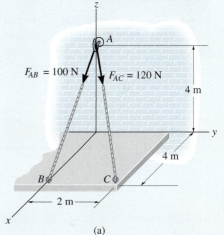

(a)

For $\mathbf{F}_{AC}$ we have

$$\mathbf{r}_{AC} = (4\text{ m} - 0)\mathbf{i} + (2\text{ m} - 0)\mathbf{j} + (0 - 4\text{ m})\mathbf{k}$$
$$= \{4\mathbf{i} + 2\mathbf{j} - 4\mathbf{k}\}\text{ m}$$
$$r_{AC} = \sqrt{(4)^2 + (2)^2 + (-4)^2} = 6\text{ m}$$
$$\mathbf{F}_{AC} = 120\text{ N}\left(\frac{\mathbf{r}_{AC}}{r_{AC}}\right) = 120\text{ N}\left(\frac{4}{6}\mathbf{i} + \frac{2}{6}\mathbf{j} - \frac{4}{6}\mathbf{k}\right)$$
$$= \{80\mathbf{i} + 40\mathbf{j} - 80\mathbf{k}\}\text{ N}$$

The resultant force is therefore

$$\mathbf{F}_R = \mathbf{F}_{AB} + \mathbf{F}_{AC} = \{70.7\mathbf{i} - 70.7\mathbf{k}\}\text{ N} + \{80\mathbf{i} + 40\mathbf{j} - 80\mathbf{k}\}\text{ N}$$
$$= \{150.7\mathbf{i} + 40\mathbf{j} - 150.7\mathbf{k}\}\text{ N}$$

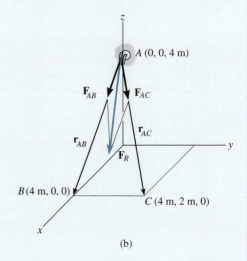

The magnitude of $\mathbf{F}_R$ is thus

$$F_R = \sqrt{(150.7)^2 + (40)^2 + (-150.7)^2}$$
$$= 217\text{ N} \qquad \textit{Ans.}$$

(b)

**Fig. 2–40**

# PROBLEMS

**2-61.** If $\mathbf{r}_1 = \{3\mathbf{i} - 4\mathbf{j} + 3\mathbf{k}\}$ m, $\mathbf{r}_2 = \{4\mathbf{i} - 5\mathbf{k}\}$ m, and $\mathbf{r}_3 = \{3\mathbf{i} - 2\mathbf{j} + 5\mathbf{k}\}$ m, determine the magnitude and direction of $\mathbf{r} = 2\mathbf{r}_1 - \mathbf{r}_2 + 3\mathbf{r}_3$.

**2-62.** Represent the position vector $\mathbf{r}$ acting from point $A(3\text{ m}, 5\text{ m}, 6\text{ m})$ to point $B(5\text{ m}, -2\text{ m}, 1\text{ m})$ in Cartesian vector form. Determine its coordinate direction angles and find the distance between points $A$ and $B$.

**2-63.** A position vector extends from the origin to point $A(2\text{ m}, 3\text{ m}, 6\text{ m})$. Determine the angles $\alpha$, $\beta$, $\gamma$ which the tail of the vector makes with the $x$, $y$, $z$ axes, respectively.

**\*2-64.** Express the position vector $\mathbf{r}$ in Cartesian vector form; then determine its magnitude and coordinate direction angles.

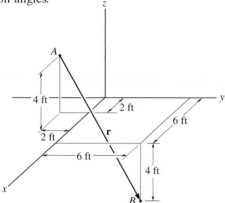

Prob. 2–64

**2-65.** Express the position vector $\mathbf{r}$ in Cartesian vector form; then determine its magnitude and coordinate direction angles.

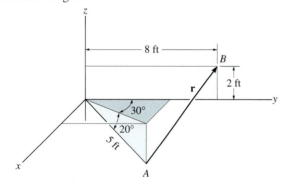

Prob. 2–65

**2-66.** Express force $\mathbf{F}$ as a Cartesian vector; then determine its coordinate direction angles.

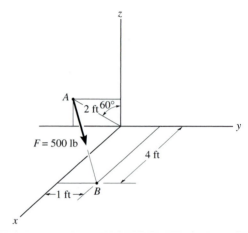

Prob. 2–66

**2-67.** Determine the length of member $AB$ of the truss by first establishing a Cartesian position vector from $A$ to $B$ and then determining its magnitude.

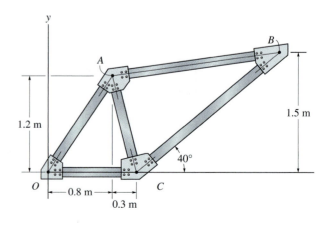

Prob. 2–67

**\*2-68.** At a given instant, the position of a plane at $A$ and a train at $B$ are measured relative to a radar antenna at $O$. Determine the distance $d$ between $A$ and $B$ at this instant. To solve the problem, formulate a position vector, directed from $A$ to $B$, and then determine its magnitude.

**2-70.** Determine the length of the crankshaft $AB$ by first formulating a Cartesian position vector from $A$ to $B$ and then determining its magnitude.

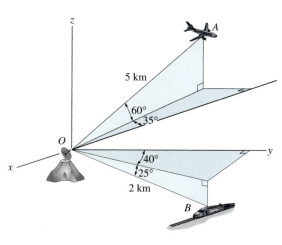

Prob. 2–68

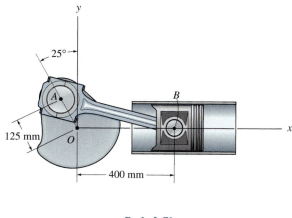

Prob. 2–70

**2-69.** The hinged plate is supported by the cord $AB$. If the force in the cord is $F = 340$ lb, express this force, directed from $A$ toward $B$, as a Cartesian vector. What is the length of the cord?

**2-71.** Determine the lengths of wires $AD$, $BD$, and $CD$. The ring at $D$ is midway between $A$ and $B$.

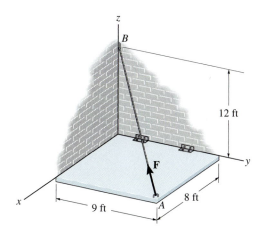

Prob. 2–69

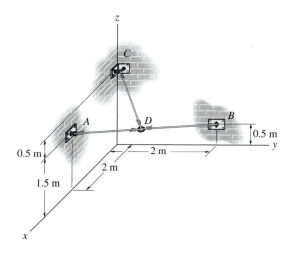

Prob. 2–71

*2-72. Express force **F** as a Cartesian vector; then determine its coordinate direction angles.

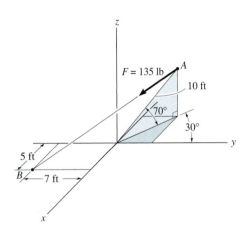

Prob. 2–72

2-73. Express force **F** as a Cartesian vector; then determine its coordinate direction angles.

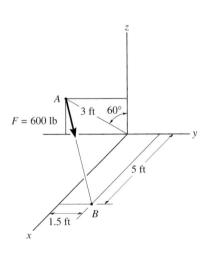

Prob. 2–73

2-74. Determine the magnitude and coordinate direction angles of the resultant force acting at point $A$.

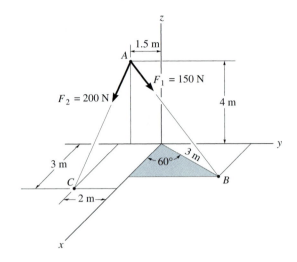

Prob. 2–74

2-75. The door is held opened by means of two chains. If the tension in $AB$ and $CD$ is $F_A = 300$ N and $F_C = 250$ N, respectively, express each of these forces in Cartesian vector form.

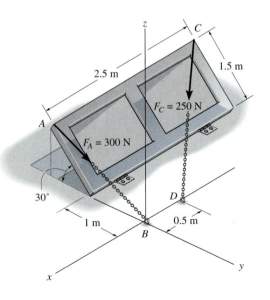

Prob. 2–75

**\*2-76.** The two mooring cables exert forces on the stern of a ship as shown. Represent each force as as Cartesian vector and determine the magnitude and direction of the resultant.

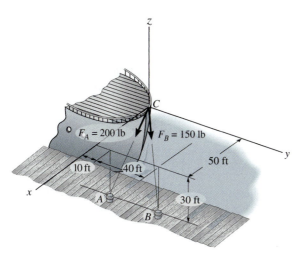

Prob. 2–76

**2-78.** The guy wires are used to support the telephone pole. Represent the force in each wire in Cartesian vector form.

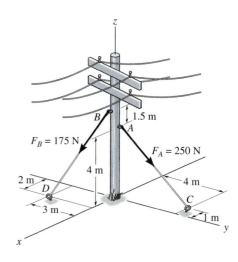

Prob. 2–78

**2-77.** Two tractors pull on the tree with the forces shown. Represent each force as a Cartesian vector and then determine the magnitude and coordinate direction angles of the resultant force.

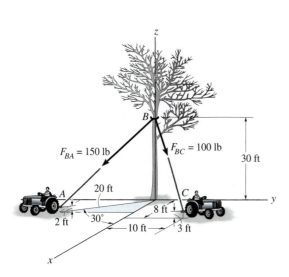

Prob. 2–77

**2-79.** Express each of the forces in Cartesian vector form and determine the magnitude and coordinate direction angles of the resultant force.

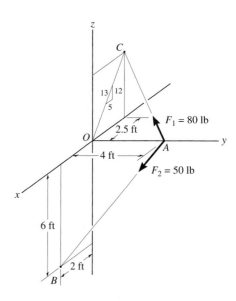

Prob. 2–79

## 2.9 Dot Product

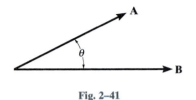

**Fig. 2–41**

Occasionally in statics one has to find the angle between two lines or the components of a force parallel and perpendicular to a line. In two dimensions, these problems can readily be solved by trigonometry since the geometry is easy to visualize. In three dimensions, however, this is often difficult, and consequently vector methods should be employed for the solution. The dot product defines a particular method for "multiplying" two vectors and is used to solve the above-mentioned problems.

The *dot product* of vectors **A** and **B**, written **A** · **B**, and read "**A** dot **B**," is defined as the product of the magnitudes of **A** and **B** and the cosine of the angle $\theta$ between their tails, Fig. 2–41. Expressed in equation form,

$$\boxed{\mathbf{A} \cdot \mathbf{B} = AB \cos \theta} \tag{2–14}$$

where $0° \leq \theta \leq 180°$. The dot product is often referred to as the *scalar product* of vectors since the result is a *scalar* and not a vector.

### Laws of Operating.

1. Commutative law:

$$\mathbf{A} \cdot \mathbf{B} = \mathbf{B} \cdot \mathbf{A}$$

2. Multiplication by a scalar:

$$a(\mathbf{A} \cdot \mathbf{B}) = (a\mathbf{A}) \cdot \mathbf{B} = \mathbf{A} \cdot (a\mathbf{B}) = (\mathbf{A} \cdot \mathbf{B})a$$

3. Distributive law:

$$\mathbf{A} \cdot (\mathbf{B} + \mathbf{D}) = (\mathbf{A} \cdot \mathbf{B}) + (\mathbf{A} \cdot \mathbf{D})$$

It is easy to prove the first and second laws by using Eq. 2–14. The proof of the distributive law is left as an exercise (see Prob. 2–109).

### Cartesian Vector Formulation.
Equation 2–14 may be used to find the dot product for each of the Cartesian unit vectors. For example, $\mathbf{i} \cdot \mathbf{i} = (1)(1) \cos 0° = 1$ and $\mathbf{i} \cdot \mathbf{j} = (1)(1) \cos 90° = 0$. In a similar manner,

$$\mathbf{i} \cdot \mathbf{i} = 1 \qquad \mathbf{j} \cdot \mathbf{j} = 1 \qquad \mathbf{k} \cdot \mathbf{k} = 1$$
$$\mathbf{i} \cdot \mathbf{j} = 0 \qquad \mathbf{i} \cdot \mathbf{k} = 0 \qquad \mathbf{k} \cdot \mathbf{j} = 0$$

These results should not be memorized; rather, it should be clearly understood how each is obtained.

Consider now the dot product of two general vectors **A** and **B** which are expressed in Cartesian vector form. We have

$$\mathbf{A} \cdot \mathbf{B} = (A_x\mathbf{i} + A_y\mathbf{j} + A_z\mathbf{k}) \cdot (B_x\mathbf{i} + B_y\mathbf{j} + B_z\mathbf{k})$$
$$= A_xB_x(\mathbf{i} \cdot \mathbf{i}) + A_xB_y(\mathbf{i} \cdot \mathbf{j}) + A_xB_z(\mathbf{i} \cdot \mathbf{k})$$
$$+ A_yB_x(\mathbf{j} \cdot \mathbf{i}) + A_yB_y(\mathbf{j} \cdot \mathbf{j}) + A_yB_z(\mathbf{j} \cdot \mathbf{k})$$
$$+ A_zB_x(\mathbf{k} \cdot \mathbf{i}) + A_zB_y(\mathbf{k} \cdot \mathbf{j}) + A_zB_z(\mathbf{k} \cdot \mathbf{k})$$

Carrying out the dot-product operations, the final result becomes

$$\mathbf{A} \cdot \mathbf{B} = A_x B_x + A_y B_y + A_z B_z \qquad (2\text{--}15)$$

*Thus, to determine the dot product of two Cartesian vectors, multiply their corresponding x, y, z components and sum their products algebraically.* Since the result is a scalar, be careful *not* to include any unit vectors in the final result.

**Applications.**   The dot product has two important applications in mechanics.

1.   *The angle formed between two vectors or intersecting lines.* The angle $\theta$ between the tails of vectors **A** and **B** in Fig. 2–41 can be determined from Eq. 2–14 and written as

$$\theta = \cos^{-1}\left(\frac{\mathbf{A} \cdot \mathbf{B}}{AB}\right) \qquad 0° \le \theta \le 180°$$

Here $\mathbf{A} \cdot \mathbf{B}$ is found from Eq. 2–15. In particular, notice that if $\mathbf{A} \cdot \mathbf{B} = 0, \theta = \cos^{-1} 0 = 90°$, so that **A** will be *perpendicular* to **B**.

2.   *The components of a vector parallel and perpendicular to a line.* The component of vector **A** parallel to or collinear with the line $aa'$ in Fig. 2–42 is defined by $\mathbf{A}_\parallel$, where $A_\parallel = A \cos\theta$. This component is sometimes referred to as the *projection* of **A** onto the line, since a right angle is formed in the construction. If the *direction* of the line is specified by the unit vector **u**, then, since $u = 1$, we can determine $A_\parallel$ directly from the dot product (Eq. 2–14); i.e.,

$$A_\parallel = A \cos\theta = \mathbf{A} \cdot \mathbf{u}$$

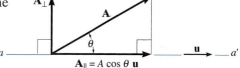

Fig. 2–42

*Hence, the scalar projection of* **A** *along a line is determined from the dot product of* **A** *and the unit vector* **u** *which defines the direction of the line.* Notice that if this result is positive, then $\mathbf{A}_\parallel$ has a directional sense which is the same as **u**, whereas if $A_\parallel$ is a negative scalar, then $\mathbf{A}_\parallel$ has the opposite sense of direction to **u**. The component $\mathbf{A}_\parallel$ represented as a *vector* is therefore

$$\mathbf{A}_\parallel = A \cos\theta\, \mathbf{u} = (\mathbf{A} \cdot \mathbf{u})\mathbf{u}$$

The component of **A** which is *perpendicular* to line $aa'$ can also be obtained, Fig. 2–42. Since $\mathbf{A} = \mathbf{A}_\parallel + \mathbf{A}_\perp$, then $\mathbf{A}_\perp = \mathbf{A} - \mathbf{A}_\parallel$. There are two possible ways of obtaining $A_\perp$. One way would be to determine $\theta$ from the dot product, $\theta = \cos^{-1}(\mathbf{A} \cdot \mathbf{u}/A)$, then $A_\perp = A \sin\theta$. Alternatively, if $A_\parallel$ is known, then by the Pythagorean theorem we can also write $A_\perp = \sqrt{A^2 - A_\parallel^2}$.

The angle $\theta$ which is made between the rope and the connecting beam $A$ can be determined by using the dot product. Simply formulate position vectors or unit vectors along the beam, $\mathbf{u}_A = \mathbf{r}_A/r_A$, and along the rope, $\mathbf{u}_r = \mathbf{r}_r/r_r$. Since $\theta$ is defined between the tails of these vectors we can solve for $\theta$ using $\theta = \cos^{-1}(\mathbf{r}_A \cdot \mathbf{r}_r/r_A r_r) = \cos^{-1} \mathbf{u}_A \cdot \mathbf{u}_r$.

If the rope exerts a force $\mathbf{F}$ on the joint, the projection of this force along beam $A$ can be determined by first defining the *direction of the beam* using the unit vector $\mathbf{u}_A = \mathbf{r}_A/r_A$ and then formulating the force as a Cartesian vector $\mathbf{F} = F(\mathbf{r}_r/r_r) = F\mathbf{u}_r$. Applying the dot product, the projection is $F_{\parallel} = \mathbf{F} \cdot \mathbf{u}_A$.

## IMPORTANT POINTS

- The dot product is used to determine the angle between two vectors or the projection of a vector in a specified direction.

- If the vectors $\mathbf{A}$ and $\mathbf{B}$ are expressed in Cartesian form, the dot product is determined by multiplying the respective $x, y, z$ scalar components together and algebraically adding the results, i.e., $\mathbf{A} \cdot \mathbf{B} = A_x B_x + A_y B_y + A_z B_z$.

- From the definition of the dot product, the angle formed between the tails of vectors $\mathbf{A}$ and $\mathbf{B}$ is $\theta = \cos^{-1}(\mathbf{A} \cdot \mathbf{B}/AB)$.

- The magnitude of the projection of vector $\mathbf{A}$ along a line whose direction is specified by $\mathbf{u}$ is determined from the dot product $A_{\parallel} = \mathbf{A} \cdot \mathbf{u}$.

## EXAMPLE 2.16

The frame shown in Fig. 2–43a is subjected to a horizontal force $\mathbf{F} = \{300\mathbf{j}\}$ N. Determine the magnitude of the components of this force parallel and perpendicular to member $AB$.

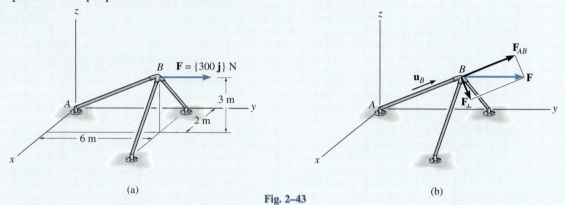

(a)

(b)

**Fig. 2–43**

### Solution

The magnitude of the component of $\mathbf{F}$ along $AB$ is equal to the dot product of $\mathbf{F}$ and the unit vector $\mathbf{u}_B$, which defines the direction of $AB$, Fig. 2–43b. Since

$$\mathbf{u}_B = \frac{\mathbf{r}_B}{r_B} = \frac{2\mathbf{i} + 6\mathbf{j} + 3\mathbf{k}}{\sqrt{(2)^2 + (6)^2 + (3)^2}} = 0.286\mathbf{i} + 0.857\mathbf{j} + 0.429\mathbf{k}$$

then

$$F_{AB} = F\cos\theta = \mathbf{F}\cdot\mathbf{u}_B = (300\mathbf{j})\cdot(0.286\mathbf{i} + 0.857\mathbf{j} + 0.429\mathbf{k})$$
$$= (0)(0.286) + (300)(0.857) + (0)(0.429)$$
$$= 257.1 \text{ N} \qquad\qquad Ans.$$

Since the result is a positive scalar, $\mathbf{F}_{AB}$ has the same sense of direction as $\mathbf{u}_B$, Fig. 2–43b.

Expressing $\mathbf{F}_{AB}$ in Cartesian vector form, we have

$$\mathbf{F}_{AB} = F_{AB}\mathbf{u}_B = (257.1 \text{ N})(0.286\mathbf{i} + 0.857\mathbf{j} + 0.429\mathbf{k})$$
$$= \{73.5\mathbf{i} + 220\mathbf{j} + 110\mathbf{k}\} \text{ N} \qquad\qquad Ans.$$

The perpendicular component, Fig. 2–43b, is therefore

$$\mathbf{F}_\perp = \mathbf{F} - \mathbf{F}_{AB} = 300\mathbf{j} - (73.5\mathbf{i} + 220\mathbf{j} + 110\mathbf{k})$$
$$= \{-73.5\mathbf{i} + 80\mathbf{j} - 110\mathbf{k}\} \text{ N}$$

Its magnitude can be determined either from this vector or from the Pythagorean theorem, Fig. 2–43b:

$$F_\perp = \sqrt{F^2 - F^2_{AB}}$$
$$= \sqrt{(300 \text{ N})^2 - (257.1 \text{ N})^2}$$
$$= 155 \text{ N} \qquad\qquad Ans.$$

**E X A M P L E 2.17**

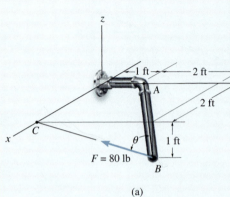

(a)

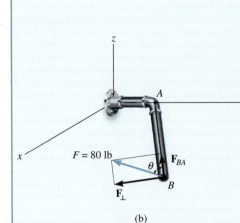

(b)

**Fig. 2–44**

The pipe in Fig. 2–44a is subjected to the force of $F = 80$ lb. Determine the angle $\theta$ between **F** and the pipe segment $BA$, and the magnitudes of the components of **F**, which are parallel and perpendicular to $BA$.

**Solution**

*Angle* $\theta$. First we will establish position vectors from $B$ to $A$ and $B$ to $C$. Then we will determine the angle $\theta$ between the tails of these two vectors.

$$\mathbf{r}_{BA} = \{-2\mathbf{i} - 2\mathbf{j} + 1\mathbf{k}\}\ \text{ft}$$
$$\mathbf{r}_{BC} = \{-3\mathbf{j} + 1\mathbf{k}\}\ \text{ft}$$

Thus,

$$\cos\theta = \frac{\mathbf{r}_{BA}\cdot\mathbf{r}_{BC}}{r_{BA}r_{BC}} = \frac{(-2)(0) + (-2)(-3) + (1)(1)}{3\sqrt{10}}$$
$$= 0.7379$$
$$\theta = 42.5° \qquad\qquad Ans.$$

***Components of F.*** The force **F** is resolved into components as shown in Fig. 2–44b. Since $F_{BA} = \mathbf{F}\cdot\mathbf{u}_{BA}$, we must first formulate the unit vector along $BA$ and force **F** as Cartesian vectors.

$$\mathbf{u}_{BA} = \frac{\mathbf{r}_{BA}}{r_{BA}} = \frac{(-2\mathbf{i} - 2\mathbf{j} + 1\mathbf{k})}{3} = -\frac{2}{3}\mathbf{i} - \frac{2}{3}\mathbf{j} + \frac{1}{3}\mathbf{k}$$

$$\mathbf{F} = 80\ \text{lb}\left(\frac{\mathbf{r}_{BC}}{r_{BC}}\right) = 80\left(\frac{-3\mathbf{j} + 1\mathbf{k}}{\sqrt{10}}\right) = -75.89\mathbf{j} + 25.30\mathbf{k}$$

Thus,

$$F_{BA} = \mathbf{F}\cdot\mathbf{u}_{BA} = (-75.89\mathbf{j} + 25.30\mathbf{k})\cdot\left(-\frac{2}{3}\mathbf{i} - \frac{2}{3}\mathbf{j} + \frac{1}{3}\mathbf{k}\right)$$
$$= 0 + 50.60 + 8.43$$
$$= 59.01\ \text{lb} \qquad\qquad Ans.$$

Since $\theta$ was calculated in Fig. 2–44b, this same result can also be obtained directly from trigonometry.

$$F_{BA} = 80\cos 42.5°\ \text{lb} = 59.0\ \text{lb} \qquad\qquad Ans.$$

The perpendicular component can be obtained by trigonometry,

$$F_\perp = F\sin\theta$$
$$= 80\sin 42.5°\ \text{lb}$$
$$= 54.0\ \text{lb} \qquad\qquad Ans.$$

Or, by the Pythagorean theorem,

$$F_\perp = \sqrt{F^2 - F_{BA}^2} = \sqrt{(80)^2 - (59.0)^2}$$
$$= 54.0\ \text{lb} \qquad\qquad Ans.$$

# PROBLEMS

**\*2-80.** Given the three vectors **A**, **B**, and **D**, show that
**A** · (**B** + **D**) = (**A** · **B**) + (**A** · **D**).

**2-81.** Determine the angle $\theta$ between the tails of the two vectors.

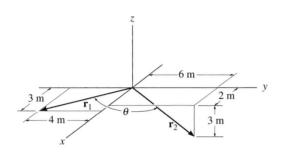

**Prob. 2–81**

**2-82.** Determine the angle $\theta$ between the tails of the two vectors.

**2-83.** Determine the magnitude of the projected component of $\mathbf{r}_1$ along $\mathbf{r}_2$, and the projection of $\mathbf{r}_2$ along $\mathbf{r}_1$.

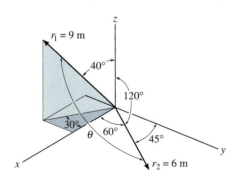

**Probs. 2–82/83**

**\*2-84.** Determine the angle $\theta$ between the y axis of the pole and the wire $AB$.

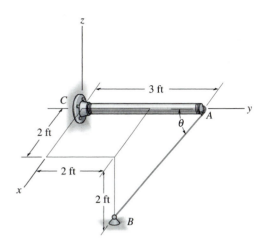

**Prob. 2–84**

**2-85.** The force **F** = {25**i** − 50**j** + 10**k**} N acts at the end $A$ of the pipe assembly. Determine the magnitude of the components $\mathbf{F}_1$ and $\mathbf{F}_2$ which act along the axis of $AB$ and perpendicular to it.

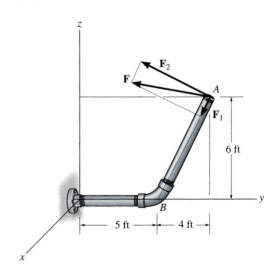

**Prob. 2–85**

**2-86.** Determine the angle $\theta$ between the sides of the triangular plate.

**2-87.** Determine the length of side $BC$ of the triangular plate. Solve the problem by finding the magnitude of $\mathbf{r}_{BC}$; then check the result by first finding $\theta$, $r_{AB}$, and $r_{AC}$ and then use the cosine law.

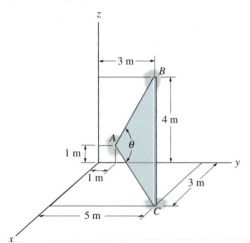

**Probs. 2–86/87**

**\*2-88.** Determine the components of $\mathbf{F}$ that act along rod $AC$ and perpendicular to it. Point $B$ is located at the midpoint of the rod.

**2-89.** Determine the components of $\mathbf{F}$ that act along rod $AC$ and perpendicular to it. Point $B$ is located 3 m along the rod from end $C$.

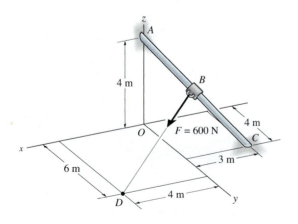

**Probs. 2–88/89**

**2-90.** The clamp is used on a jig. If the vertical force acting on the bolt is $\mathbf{F} = \{-500\mathbf{k}\}$ N, determine the magnitudes of the components $\mathbf{F}_1$ and $\mathbf{F}_2$ which act along the $OA$ axis and perpendicular to it.

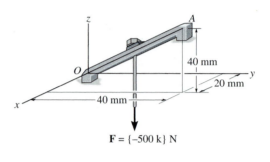

$\mathbf{F} = \{-500\ \mathbf{k}\}$ N

**Prob. 2–90**

**2-91.** Determine the projection of the force $\mathbf{F}$ along the pole.

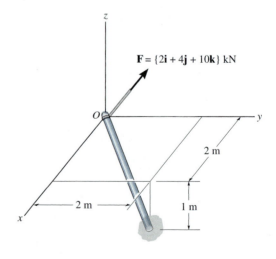

$\mathbf{F} = \{2\mathbf{i} + 4\mathbf{j} + 10\mathbf{k}\}$ kN

**Prob. 2–91**

*2-92.   Determine the projected component of the 80-N force acting along the axis $AB$ of the pipe.

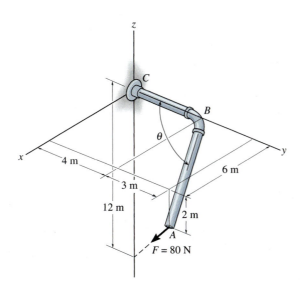

Prob. 2–92

2-93.   Cable $OA$ is used to support column $OB$. Determine the angle $\theta$ it makes with beam $OC$.

2-94.   Cable $OA$ is used to support column $OB$. Determine the angle $\phi$ it makes with beam $OD$.

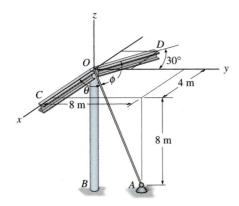

Probs. 2–93/94

**2-95.** The force **F** acts at the end $A$ of the pipe assembly. Determine the magnitudes of the components $\mathbf{F}_1$ and $\mathbf{F}_2$ which act along the axis of $AB$ and perpendicular to it.

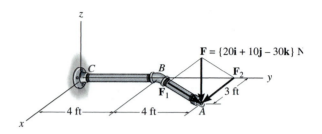

**Prob. 2–95**

*2-96.** Two cables exert forces on the pipe. Determine the magnitude of the projected component of $\mathbf{F}_1$ along the line of action of $\mathbf{F}_2$.

**2-97.** Determine the angle $\theta$ between the two cables attached to the pipe.

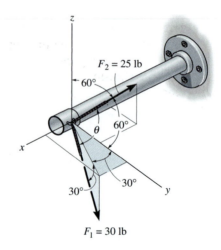

**Probs. 2–96/97**

# CHAPTER REVIEW

- *Parallelogram Law.* Two vectors add according to the parallelogram law. The *components* form the sides of the parallelogram and the *resultant* is the diagonal. To obtain the components or the resultant, show how the vectors add by the tip-to-tail addition using the triangle rule, and then use the law of sines and the law of cosines to calculate their values.

- *Cartesian Vectors.* A vector can be resolved into its Cartesian components along the *x, y, z* axes so that $\mathbf{F} = F_x\mathbf{i} + F_y\mathbf{j} + F_z\mathbf{k}$.

  The magnitude of $\mathbf{F}$ is determined from $F = \sqrt{F_x^2 + F_y^2 + F_z^2}$ and the coordinate direction angles $\alpha, \beta, \gamma$ are determined by formulating a unit vector in the direction of $\mathbf{F}$, that is $\mathbf{u} = (F_x/F)\mathbf{i} + (F_y/F)\mathbf{j} + (F_z/F)\mathbf{k}$. The components of $\mathbf{u}$ represent $\cos\alpha, \cos\beta, \cos\gamma$. These three angles are related by $\cos^2\alpha + \cos^2\beta + \cos^2\gamma = 1$, so that only two of the three angles are independent of one another.

- *Force and Position Vectors.* A position vector is directed between two points. It can be formulated by finding the distance and the direction one has to travel along the *x, y, z* axes from one point (the tail) to the other point (the tip). If the line of action of a force passes through these two points, then it acts in the same direction $\mathbf{u}$ as the position vector. The force can be expressed as a Cartesian vector using $\mathbf{F} = F\mathbf{u} = F(\mathbf{r}/r)$.

- *Dot Product.* The dot product between two vectors $\mathbf{A}$ and $\mathbf{B}$ is defined by $\mathbf{A} \cdot \mathbf{B} = AB\cos\theta$. If $\mathbf{A}$ and $\mathbf{B}$ are expressed as Cartesian vectors, then $\mathbf{A} \cdot \mathbf{B} = A_xB_x + A_yB_y + A_zB_z$. In statics, the dot product is used to determine the angle between the tails of the vectors, $\theta = \cos^{-1}(\mathbf{A} \cdot \mathbf{B}/AB)$. It is also used to determine the projected component of a vector $\mathbf{A}$ onto an axis defined by its unit vector $\mathbf{u}$, so that $A = A\cos\theta = \mathbf{A} \cdot \mathbf{u}$.

## REVIEW PROBLEMS

**2-98.** The force **F** has a magnitude of 80 lb and acts at the midpoint $C$ of the thin rod. Express the force as a Cartesian vector.

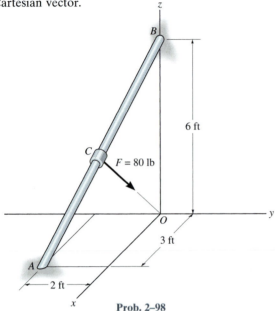

**Prob. 2–98**

**2-99.** Express each force acting on the screw eye as a Cartesian vector, and then determine the magnitude and orientation of the resultant force.

**2-100.** Use the parallelogram law to determine the magnitude of the resultant force acting on the screw eye. Determine its orientation, measured counterclockwise from the positive $x$ axis.

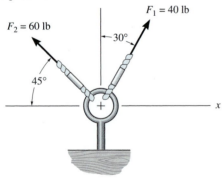

**Probs. 2–99/100**

**2-101.** The backbone exerts a vertical force of 110 lb on the lumbosacral joint of a person who is standing erect. Determine the magnitudes of the components of this force directed perpendicular to the surface of the sacrum, $v$ axis, and parallel to it, $u$ axis.

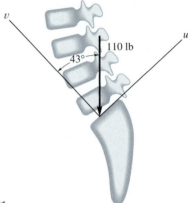

**Prob. 2–101**

**2-102.** The cord $AB$ exerts a force of $\mathbf{F} = \{8\mathbf{i} + 12\mathbf{j} - 9\mathbf{k}\}$ lb on hook $A$. If the cord is 8.5 ft long, determine the location $x$, $y$ of the point of attachment $B$, and the height $z$ of the hook.

**2-103.** The cord exerts a force of $F = 60$ N on hook $A$. If the cord is 10 ft long, $z = 5$ ft, and the $x$ component of the force is $F_x = 50$ N, determine the location $x$, $y$ of the point of attachment $B$ of the cord to the ground.

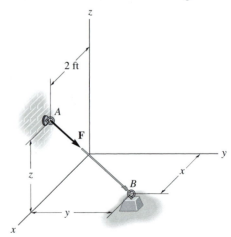

**Probs. 2–102/103**

*2-104. Determine the magnitude of the projected component of force **F** along the *Oa* axis.

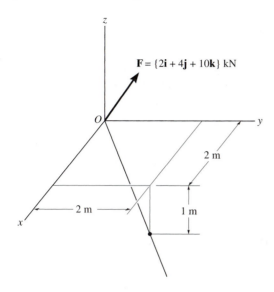

**Prob. 2–104**

2-105. Express each of the three forces acting on the column in Cartesian vector form and determine the magnitude of the resultant force.

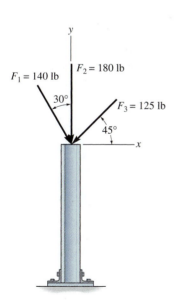

**Prob. 2–105**

2-106. Determine the angle $\theta$ ($0° \leq \theta \leq 90°$) for cable $AB$ so that the 500-N vertical force has a component of 200 N directed along cable $AB$ from $A$ toward $B$. What is the corresponding component of force acting along cable $CB$? Set $\phi = 60°$.

2-107. Determine the angle $\phi$ ($0° \leq \phi \leq 90°$) between cables $BA$ and $BC$ so that the 500-N vertical force has a component of 250 N which acts along cable $CB$, from $C$ to $B$. Set $\theta = 60°$.

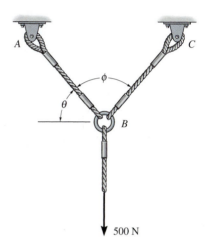

**Probs. 2–106/107**

This utility pole is subjected to many forces, caused by the cables and the weight of the transformer. In some cases, it is important to be able to simplify this system to a single resultant force and specify where this resultant acts on the pole.

# Force System Resultants

- To discuss the concept of the moment of a force and show how to calculate it in two and three dimensions.
- To provide a method for finding the moment of a force about a specified axis.
- To define the moment of a couple.
- To present methods for determining the resultants of nonconcurrent force systems.
- To indicate how to reduce a simple distributed loading to a resultant force having a specified location.

## 3.1 Moment of a Force—Scalar Formulation

The *moment* of a force about a point or axis provides a measure of the tendency of the force to cause a body to rotate about the point or axis. For example, consider the horizontal force $\mathbf{F}_x$, which acts perpendicular to the handle of the wrench and is located a distance $d_y$ from point $O$, Fig. 3–1a. It is seen that this force tends to cause the pipe to turn about the $z$ axis. The larger the force or the distance $d_y$, the greater the turning effect. This tendency for rotation caused by $\mathbf{F}_x$ is sometimes called a *torque*, but most often it is called the *moment of a force* or simply the *moment* $(\mathbf{M}_O)_z$. Note that the *moment axis* ($z$) is perpendicular to the shaded plane ($x$–$y$) which contains both $\mathbf{F}_x$ and $d_y$ and that this axis intersects the plane at point $O$.

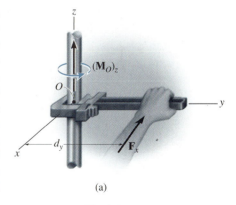

(a)

Fig. 3–1

75

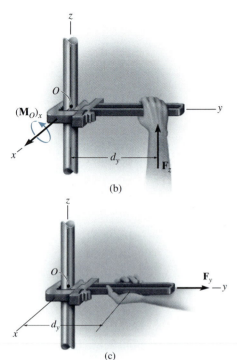

(b)

(c)

**Fig. 3–1**

Moment axis

Sense of rotation

(a)

(b)

**Fig. 3–2**

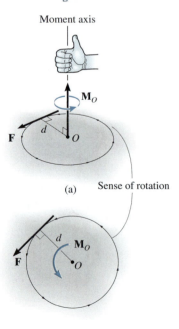

Now consider applying the force $\mathbf{F}_z$ to the wrench, Fig. 3–1b. This force will *not* rotate the pipe about the $z$ axis. Instead, it tends to rotate it about the $x$ axis. Keep in mind that although it may not be possible to actually "rotate" or turn the pipe in this manner, $\mathbf{F}_z$ still creates the *tendency* for rotation and so the moment $(\mathbf{M}_O)_x$ is produced. As before, the force and distance $d_y$ lie in the shaded plane ($y$–$z$) which is perpendicular to the moment axis ($x$). Lastly, if a force $\mathbf{F}_y$ is applied to the wrench, Fig. 3–1c, no moment is produced about point $O$. This results in a lack of turning since the line of action of the force passes through $O$ and therefore no tendency for rotation is possible.

We will now generalize the above discussion and consider the force $\mathbf{F}$ and point $O$ which lie in a shaded plane as shown in Fig. 3–2a. The moment $\mathbf{M}_O$ about point $O$, or about an axis passing through $O$ and perpendicular to the plane, is a *vector quantity* since it has a specified magnitude and direction.

**Magnitude.**   The magnitude of $M_O$ is

$$M_O = Fd \qquad (3\text{–}1)$$

where $d$ is referred to as the *moment arm* or perpendicular distance from the axis at point $O$ to the line of action of the force. Units of moment magnitude consist of force times distance, e.g., $\mathrm{N \cdot m}$ or $\mathrm{lb \cdot ft}$.

**Direction.**   The direction of $\mathbf{M}_O$ will be specified by using the "right-hand rule." To do this, the fingers of the right hand are curled such that they follow the sense of rotation, which would occur if the force could rotate about point $O$, Fig. 3–2a. The *thumb* then *points* along the *moment axis* so that it gives the direction and sense of the moment vector, which is *upward* and *perpendicular* to the shaded plane containing $\mathbf{F}$ and $d$.

In three dimensions, $\mathbf{M}_O$ is illustrated by a vector arrow with a curl on it to *distinguish* it from a force vector, Fig. 3–2a. Many problems in mechanics, however, involve coplanar force systems that may be conveniently viewed in two dimensions. For example, a two-dimensional view of Fig. 3–2a is given in Fig. 3–2b. Here $\mathbf{M}_O$ is simply represented by the (counterclockwise) curl, which indicates the action of $\mathbf{F}$. The arrowhead on this curl is used to show the *sense of rotation* caused by $\mathbf{F}$. Using the right-hand rule, however, realize that the direction and sense of the moment vector in Fig. 3–2b are specified by the thumb, which points *out* of the page since the fingers follow the curl. In particular, notice that *this curl or sense of rotation can always be determined by observing in which direction the force would "orbit" about point $O$* (counterclockwise in Fig. 3–2b). In two dimensions we will often refer to finding the moment of a force "about a point" ($O$). Keep in mind, however, that the moment *always acts about an axis* which is perpendicular to the plane containing $\mathbf{F}$ and $d$, and this axis intersects the plane at the point ($O$), Fig. 3–2a.

**Resultant Moment of a System of Coplanar Forces.**    If a system of forces lies in an $x$–$y$ plane, then the moment produced by each force about point $O$ will be directed along the $z$ axis, Fig. 3–3. Consequently, the resultant moment $\mathbf{M}_{R_O}$ of the system can be determined by simply adding the moments of all forces *algebraically* since all the moment vectors are collinear. We can write this vector sum symbolically as

$$\zeta + M_{R_O} = \Sigma Fd \qquad (3\text{–}2)$$

Here the counterclockwise curl written alongside the equation indicates that, by the scalar sign convention, the moment of any force will be positive if it is directed along the $+z$ axis, whereas a negative moment is directed along the $-z$ axis.

The following examples illustrate numerical application of Eqs. 3–1 and 3–2.

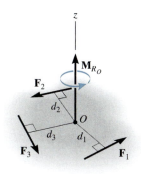

Fig. 3–3

By pushing down on the pry bar the load on the ground at $A$ can be lifted. The turning effect, caused by the applied force, is due to the moment about $A$. To produce this moment with minimum effort we instinctively know that the force should be applied to the *end* of the bar; however, the *direction* in which this force is applied is also important. This is because moment is the product of the force and the moment arm. Notice that when the force is at an angle $\theta < 90°$, then the moment arm distance is *shorter* than when the force is applied perpendicular to the bar $\theta = 90°$, i.e., $d' < d$. Hence the greatest moment is produced when the force is farthest from point $A$ and applied perpendicular to the axis of the bar so as to maximize the moment arm.

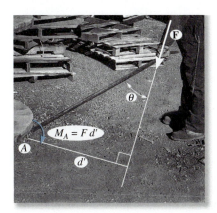

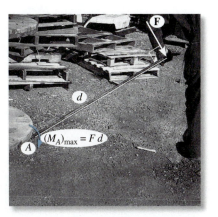

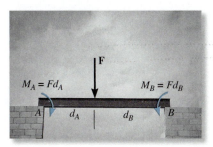

The moment of a force does not always cause a rotation. For example, the force $\mathbf{F}$ tends to rotate the beam clockwise about its support at $A$ with a moment $M_A = Fd_A$. The actual rotation would occur if the support at $B$ were removed. In the same manner, $\mathbf{F}$ creates a tendency to rotate the beam counterclockwise about $B$ with a moment $M_B = Fd_B$. Here the support at $A$ prevents the rotation.

**E X A M P L E 3.1**

For each case illustrated in Fig. 3–4, determine the moment of the force about point $O$.

**Solution (*Scalar Analysis*)**
The line of action of each force is extended as a dashed line in order to establish the moment arm $d$. Also illustrated is the tendency of rotation of the member as caused by the force. Furthermore, the orbit of the force is shown as a colored curl. Thus,

| | | |
|---|---|---|
| Fig. 3–4$a$ | $M_O = (100 \text{ N})(2 \text{ m}) = 200 \text{ N} \cdot \text{m} \downarrow$ | *Ans.* |
| Fig. 3–4$b$ | $M_O = (50 \text{ N})(0.75 \text{ m}) = 37.5 \text{ N} \cdot \text{m} \downarrow$ | *Ans.* |
| Fig. 3–4$c$ | $M_O = (40 \text{ lb})(4 \text{ ft} + 2 \cos 30° \text{ ft}) = 229 \text{ lb} \cdot \text{ft} \downarrow$ | *Ans.* |
| Fig. 3–4$d$ | $M_O = (60 \text{ lb})(1 \sin 45° \text{ ft}) = 42.4 \text{ lb} \cdot \text{ft} \uparrow$ | *Ans.* |
| Fig. 3–4$e$ | $M_O = (7 \text{ kN})(4 \text{ m} - 1 \text{ m}) = 21.0 \text{ kN} \cdot \text{m} \uparrow$ | *Ans.* |

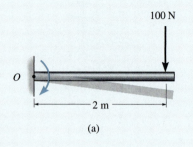

(a)

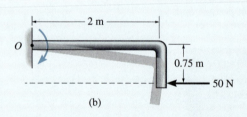

(b)

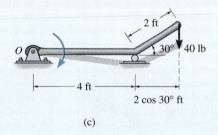

(c)

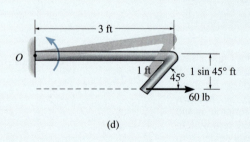

(d)

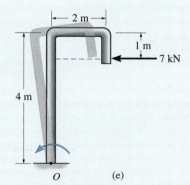

(e)

**Fig. 3–4**

E X A M P L E   **3.2**

Determine the moments of the 800-N force acting on the frame in Fig. 3–5 about points $A$, $B$, $C$, and $D$.

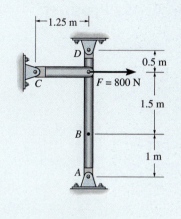

**Solution (*Scalar Analysis*)**

In general, $M = Fd$, where $d$ is the moment arm or *perpendicular distance* from the *point* on the moment axis to the *line of action* of the force. Hence,

$M_A = 800\text{ N}(2.5\text{ m}) = 2000\text{ N}\cdot\text{m}\downarrow$      *Ans.*

$M_B = 800\text{ N}(1.5\text{ m}) = 1200\text{ N}\cdot\text{m}\downarrow$      *Ans.*

$M_C = 800\text{ N}(0) = 0$    (line of action of **F** passes through $C$)     *Ans.*

$M_D = 800\text{ N}(0.5\text{ m}) = 400\text{ N}\cdot\text{m}\uparrow$      *Ans.*

**Fig. 3–5**

The curls indicate the sense of rotation of the moment, which is defined by the direction the force orbits about each point.

E X A M P L E   **3.3**

Determine the resultant moment of the four forces acting on the rod shown in Fig. 3–6 about point $O$.

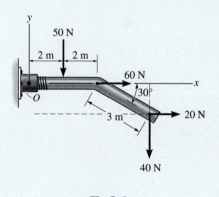

**Solution**

Assuming that positive moments act in the $+\mathbf{k}$ direction, i.e., counter-clockwise, we have

$\downarrow + M_{R_O} = \Sigma Fd;$

$\quad M_{R_O} = -50\text{ N}(2\text{m}) + 60\text{ N}(0) + 20\text{ N}(3\sin 30°\text{ m})$

$\quad\quad\quad -40\text{ N}(4\text{ m} + 3\cos 30°\text{ m})$

$\quad M_{R_O} = -334\text{ N}\cdot\text{m} = 334\text{ N}\cdot\text{m}\downarrow$      *Ans.*

**Fig. 3–6**

For this calculation, note how the moment-arm distances for the 20-N and 40-N forces are established from the extended (dashed) lines of action of each of these forces.

## 3.2 Cross Product

The concept of the *moment* of a force will be developed using vectors in the next two section's. Before doing this, however, it is first necessary to expand our knowledge of vector algebra and introduce the cross-product method of vector multiplication.

The *cross product* of two vectors **A** and **B** yields the vector **C**, which is written

$$\mathbf{C} = \mathbf{A} \times \mathbf{B}$$

and is read "**C** equals **A** cross **B**."

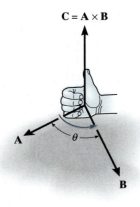

C = A × B

**Magnitude.**   The *magnitude* of **C** is defined as the product of the magnitudes of **A** and **B** and the sine of the angle $\theta$ between their tails $(0° \leq \theta \leq 180°)$. Thus, $C = AB \sin \theta$.

**Direction.**   Vector **C** has a *direction* that is perpendicular to the plane containing **A** and **B** such that **C** is specified by the right-hand rule; i.e., curling the fingers of the right hand from vector **A** (cross) to vector **B**, the thumb then points in the direction of **C**, as shown in Fig. 3–7.

Knowing both the magnitude and direction of **C**, we can write

$$\mathbf{C} = \mathbf{A} \times \mathbf{B} = (AB \sin \theta)\mathbf{u}_C \qquad (3\text{–}3)$$

where the scalar $AB \sin \theta$ defines the *magnitude* of **C** and the unit vector $\mathbf{u}_C$ defines the *direction* of **C**. The terms of Eq. 3–3 are illustrated graphically in Fig. 3–8.

Fig. 3–7

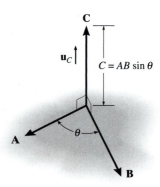

$C = AB \sin \theta$

Fig. 3–8

## Laws of Operation.

1. The commutative law is *not* valid; i.e.,

$$\mathbf{A} \times \mathbf{B} \neq \mathbf{B} \times \mathbf{A}$$

Rather,

$$\mathbf{A} \times \mathbf{B} = -\mathbf{B} \times \mathbf{A}$$

This is shown in Fig. 3–9 by using the right-hand rule. The cross product $\mathbf{B} \times \mathbf{A}$ yields a vector that acts in the opposite direction to $\mathbf{C}$; i.e., $\mathbf{B} \times \mathbf{A} = -\mathbf{C}$.

2. Multiplication by a scalar:

$$a(\mathbf{A} \times \mathbf{B}) = (a\mathbf{A}) \times \mathbf{B} = \mathbf{A} \times (a\mathbf{B}) = (\mathbf{A} \times \mathbf{B})a$$

This property is easily shown since the magnitude of the resultant vector ($|a|AB \sin \theta$) and its direction are the same in each case.

3. The distributive law:

$$\mathbf{A} \times (\mathbf{B} + \mathbf{D}) = (\mathbf{A} \times \mathbf{B}) + (\mathbf{A} \times \mathbf{D})$$

The proof of this identity is left as an exercise (see Prob. 3–1). It is important to note that *proper order* of the cross products must be maintained, since they are not commutative.

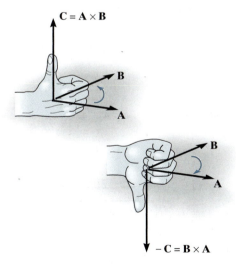

**Fig. 3–9**

**Cartesian Vector Formulation.** Equation 3–3 may be used to find the cross product of a pair of Cartesian unit vectors. For example, to find $\mathbf{i} \times \mathbf{j}$, the *magnitude* of the resultant vector is $(i)(j)(\sin 90°) = (1)(1)(1) = 1$, and its *direction* is determined using the right-hand rule. As shown in Fig. 3–10, the resultant vector points in the $+\mathbf{k}$ direction. Thus, $\mathbf{i} \times \mathbf{j} = (1)\mathbf{k}$. In a similar manner,

$$\mathbf{i} \times \mathbf{j} = \mathbf{k} \quad \mathbf{i} \times \mathbf{k} = -\mathbf{j} \quad \mathbf{i} \times \mathbf{i} = \mathbf{0}$$
$$\mathbf{j} \times \mathbf{k} = \mathbf{i} \quad \mathbf{j} \times \mathbf{i} = -\mathbf{k} \quad \mathbf{j} \times \mathbf{j} = \mathbf{0}$$
$$\mathbf{k} \times \mathbf{i} = \mathbf{j} \quad \mathbf{k} \times \mathbf{j} = -\mathbf{i} \quad \mathbf{k} \times \mathbf{k} = \mathbf{0}$$

These results should *not* be memorized; rather, it should be clearly understood how each is obtained by using the right-hand rule and the definition of the cross product. A simple scheme shown in Fig. 3–11 is helpful for obtaining the same results when the need arises. If the circle is constructed as shown, then "crossing" two unit vectors in a *counterclockwise* fashion around the circle yields the *positive* third unit vector; e.g., $\mathbf{k} \times \mathbf{i} = \mathbf{j}$. Moving *clockwise*, a *negative* unit vector is obtained; e.g., $\mathbf{i} \times \mathbf{k} = -\mathbf{j}$.

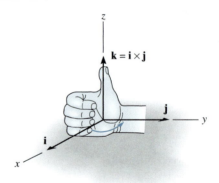

Fig. 3–10

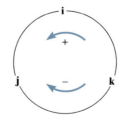

Fig. 3–11

Consider now the cross product of two general vectors **A** and **B** which are expressed in Cartesian vector form. We have

$$\mathbf{A} \times \mathbf{B} = (A_x\mathbf{i} + A_y\mathbf{j} + A_z\mathbf{k}) \times (B_x\mathbf{i} + B_y\mathbf{j} + B_z\mathbf{k})$$
$$= A_xB_x(\mathbf{i} \times \mathbf{i}) + A_xB_y(\mathbf{i} \times \mathbf{j}) + A_xB_z(\mathbf{i} \times \mathbf{k})$$
$$+ A_yB_x(\mathbf{j} \times \mathbf{i}) + A_yB_y(\mathbf{j} \times \mathbf{j}) + A_yB_z(\mathbf{j} \times \mathbf{k})$$
$$+ A_zB_x(\mathbf{k} \times \mathbf{i}) + A_zB_y(\mathbf{k} \times \mathbf{j}) + A_zB_z(\mathbf{k} \times \mathbf{k})$$

Carrying out the cross-product operations and combining terms yields

$$\mathbf{A} \times \mathbf{B} = (A_yB_z - A_zB_y)\mathbf{i} - (A_xB_z - A_zB_x)\mathbf{j} + (A_xB_y - A_yB_x)\mathbf{k} \quad (3\text{--}4)$$

This equation may also be written in a more compact determinant form as

$$\mathbf{A} \times \mathbf{B} = \begin{vmatrix} \mathbf{i} & \mathbf{j} & \mathbf{k} \\ A_x & A_y & A_z \\ B_x & B_y & B_z \end{vmatrix} \quad (3\text{--}5)$$

Thus, to find the cross product of any two Cartesian vectors **A** and **B**, it is necessary to expand a determinant whose first row of elements consists of the unit vectors **i**, **j**, and **k** and whose second and third rows represent the *x, y, z* components of the two vectors **A** and **B**, respectively.*

---

*A determinant having three rows and three columns can be expanded using three minors, each of which is multiplied by one of the three terms in the first row. There are four elements in each minor, e.g.,

$$\begin{vmatrix} A_{11} & A_{12} \\ A_{21} & A_{22} \end{vmatrix}$$

By *definition*, this notation represents the terms $(A_{11}A_{22} - A_{12}A_{21})$, which is simply the product of the two elements of the arrow slanting downward to the right $(A_{11}A_{22})$ *minus* the product of the two elements intersected by the arrow slanting downward to the left $(A_{12}A_{21})$. For a $3 \times 3$ determinant, such as Eq. 3–5, the three minors can be generated in accordance with the following scheme:

For element **i**: $\begin{vmatrix} \mathbf{i} & \mathbf{j} & \mathbf{k} \\ A_x & A_y & A_z \\ B_x & B_y & B_z \end{vmatrix} = \mathbf{i}(A_yB_z - A_zB_y)$

For element **j**: $\begin{vmatrix} \mathbf{i} & \mathbf{j} & \mathbf{k} \\ A_x & A_y & A_z \\ B_x & B_y & B_z \end{vmatrix} = -\mathbf{j}(A_xB_z - A_zB_x)$

For element **k**: $\begin{vmatrix} \mathbf{i} & \mathbf{j} & \mathbf{k} \\ A_x & A_y & A_z \\ B_x & B_y & B_z \end{vmatrix} = \mathbf{k}(A_xB_y - A_yB_x)$

Adding the results and noting that the **j** element *must include the minus sign* yields the expanded form of **A** $\times$ **B** given by Eq. 3–4.

## 3.3 Moment of a Force—Vector Formulation

The moment of a force $\mathbf{F}$ about point $O$, or actually about the moment axis passing through $O$ and perpendicular to the plane containing $O$ and $\mathbf{F}$, Fig. 3–12a, can be expressed using the vector cross product, namely,

$$\mathbf{M}_O = \mathbf{r} \times \mathbf{F} \qquad (3\text{–}6)$$

Here $\mathbf{r}$ represents a position vector drawn *from O* to *any point* lying on the line of action of $\mathbf{F}$. We will now show that indeed the moment $\mathbf{M}_O$, when determined by this cross product, has the proper magnitude and direction.

**Magnitude.**  The magnitude of the cross product is defined from Eq. 3–1 as $M_O = rF \sin \theta$, where the angle $\theta$ is measured between the *tails* of $\mathbf{r}$ and $\mathbf{F}$. To establish this angle, $\mathbf{r}$ must be treated as a sliding vector so that $\theta$ can be constructed properly, Fig. 3–12b. Since the moment arm $d = r \sin \theta$, then

$$M_O = rF \sin \theta = F(r \sin \theta) = Fd$$

which agrees with Eq. 3–4.

**Direction.**  The direction and sense of $\mathbf{M}_O$ in Eq. 3–6 are determined by the right-hand rule as it applies to the cross product. Thus, extending $\mathbf{r}$ to the dashed position and curling the right-hand fingers from $\mathbf{r}$ toward $\mathbf{F}$, "$\mathbf{r}$ cross $\mathbf{F}$," the thumb is directed upward or perpendicular to the plane containing $\mathbf{r}$ and $\mathbf{F}$ and this is in the *same direction* as $\mathbf{M}_O$, the moment of the force about point $O$, Fig. 3–12b. Note that the "curl" of the fingers, like the curl around the moment vector, indicates the sense of rotation caused by the force. Since the cross product is not commutative, it is important that the *proper order* of $\mathbf{r}$ and $\mathbf{F}$ be maintained in Eq. 3–6.

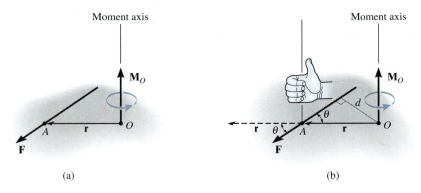

(a)                                         (b)

**Fig. 3–12**

**Principle of Transmissibility.** Consider the force **F** applied at point $A$ in Fig. 3–13. The moment created by **F** about $O$ is $\mathbf{M}_O = \mathbf{r}_A \times \mathbf{F}$; however, it was shown that "**r**" can extend from $O$ to *any point* on the line of action of **F**. Consequently, **F** may be applied at point $B$ or $C$, and the same moment $\mathbf{M}_O = \mathbf{r}_B \times \mathbf{F} = \mathbf{r}_C \times \mathbf{F}$ will be computed. As a result, **F** has the properties of a *sliding vector* and can therefore act at *any point along its line of action and still create the same moment about point $O$*. We refer to this as the *principle of transmissibility*, and we will discuss this property further in Sec. 3.7.

**Cartesian Vector Formulation.** If we establish $x, y, z$ coordinate axes, then the position vector **r** and force **F** can be expressed as Cartesian vectors, Fig. 3–14. Applying Eq. 3–3 we have

$$\mathbf{M}_O = \mathbf{r} \times \mathbf{F} = \begin{vmatrix} \mathbf{i} & \mathbf{j} & \mathbf{k} \\ r_x & r_y & r_z \\ F_x & F_y & F_z \end{vmatrix} \quad (3\text{–}7)$$

where

$r_x, r_y, r_z$ represent the $x, y, z$ components of the position vector drawn from point $O$ to *any point* on the line of action of the force

$F_x, F_y, F_z$ represent the $x, y, z$ components of the force vector

If the determinant is expanded, then like Eq. 3–2 we have

$$\mathbf{M}_O = (r_y F_z - r_z F_y)\mathbf{i} - (r_x F_z - r_z F_x)\mathbf{j} + (r_x F_y - r_y F_x)\mathbf{k} \quad (3\text{–}8)$$

The physical meaning of these three moment components becomes evident by studying Fig. 3–14a. For example, the **i** component of $\mathbf{M}_O$ is

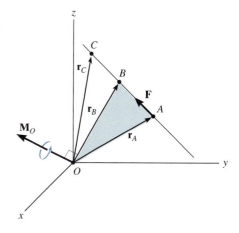

Fig. 3–13

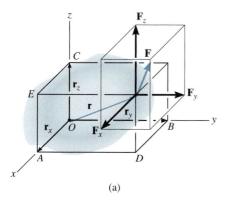

(a)

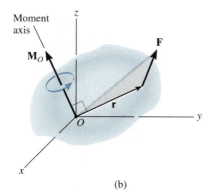

(b)

Fig. 3–14

determined from the moments of $\mathbf{F}_x$, $\mathbf{F}_y$, and $\mathbf{F}_z$ about the $x$ axis. In particular, note that $\mathbf{F}_x$ does *not* create a moment or tendency to cause turning about the $x$ axis since this force is *parallel* to the $x$ axis. The line of action of $\mathbf{F}_y$ passes through point $E$, and so the magnitude of the moment of $\mathbf{F}_y$ about point $A$ on the $x$ axis is $r_z F_y$. By the right-hand rule this component acts in the negative $\mathbf{i}$ direction. Likewise, $\mathbf{F}_z$ contributes a moment component of $r_y F_z \mathbf{i}$. Thus, $(M_O)_x = (r_y F_z - r_z F_y)$ as shown in Eq. 3–8. As an exercise, establish the $\mathbf{j}$ and $\mathbf{k}$ components of $\mathbf{M}_O$ in this manner and show that indeed the expanded form of the determinant, Eq. 3–8, represents the moment of $\mathbf{F}$ about point $O$. Once $\mathbf{M}_O$ is determined, realize that it will always be *perpendicular* to the shaded plane containing vectors $\mathbf{r}$ and $\mathbf{F}$, Fig. 3–14b.

It will be shown in Example 3.4 that the computation of the moment using the cross product has a distinct advantage over the scalar formulation when solving problems in *three dimensions*. This is because it is generally easier to establish the position vector $\mathbf{r}$ to the force, rather than determining the moment-arm distance $d$ that must be directed *perpendicular* to the line of action of the force.

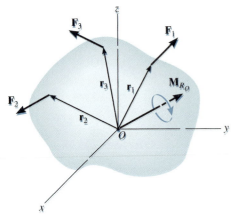

**Fig. 3–15**

**Resultant Moment of a System of Forces.**   If a body is acted upon by a system of forces, Fig. 3–15, the resultant moment of the forces about point $O$ can be determined by vector addition resulting from successive applications of Eq. 3–6. This resultant can be written symbolically as

$$\mathbf{M}_{R_O} = \Sigma(\mathbf{r} \times \mathbf{F}) \qquad (3\text{–}9)$$

and is shown in Fig. 3–15.

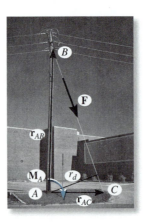

If we pull on cable $BC$ with a force $\mathbf{F}$ at *any point along the cable*, the moment of this force about the base of the utility pole at $A$ will always be the *same*. This is a consequence of the principle of transmissibility. Note that the moment arm, or perpendicular distance from $A$ to the cable, is $r_d$, and so $M_A = r_d F$. In three dimensions this distance is often difficult to determine, and so we can use the vector cross product to obtain the moment in a more direct manner. For example, $\mathbf{M}_A = \mathbf{r}_{AB} \times \mathbf{F} = \mathbf{r}_{AC} \times \mathbf{F}$. As required, both of these vectors are directed from point $A$ to a point on the line of action of the force.

# EXAMPLE 3.4

The pole in Fig. 3–16a is subjected to a 60-N force that is directed from C to B. Determine the magnitude of the moment created by this force about the support at A.

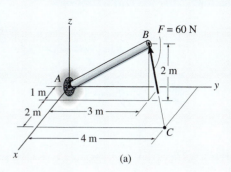

(a)

## Solution (*Vector Analysis*)

As shown in Fig. 3–16b, either one of two position vectors can be used for the solution, since $\mathbf{M}_A = \mathbf{r}_B \times \mathbf{F}$ or $\mathbf{M}_A = \mathbf{r}_C \times \mathbf{F}$. The position vectors are represented as

$$\mathbf{r}_B = \{1\mathbf{i} + 3\mathbf{j} + 2\mathbf{k}\}\ \text{m} \quad \text{and} \quad \mathbf{r}_C = \{3\mathbf{i} + 4\mathbf{j}\}\ \text{m}$$

The force has a magnitude of 60 N and a direction specified by the unit vector $\mathbf{u}_F$, directed from C to B. Thus,

$$\mathbf{F} = (60\ \text{N})\mathbf{u}_F = (60\ \text{N})\left[\frac{(1-3)\mathbf{i} + (3-4)\mathbf{j} + (2-0)\mathbf{k}}{\sqrt{(-2)^2 + (-1)^2 + (2)^2}}\right]$$
$$= \{-40\mathbf{i} - 20\mathbf{j} + 40\mathbf{k}\}\ \text{N}$$

Substituting into the determinant formulation, Eq. 3–7, and following the scheme for determinant expansion as stated in the footnote on page 84, we have

$$\mathbf{M}_A = \mathbf{r}_B \times \mathbf{F} = \begin{vmatrix} \mathbf{i} & \mathbf{j} & \mathbf{k} \\ 1 & 3 & 2 \\ -40 & -20 & 40 \end{vmatrix}$$
$$= [3(40) - 2(-20)]\mathbf{i} - [1(40) - 2(-40)]\mathbf{j} + [1(-20) - 3(-40)]\mathbf{k}$$

or

$$\mathbf{M}_A = \mathbf{r}_C \times \mathbf{F} = \begin{vmatrix} \mathbf{i} & \mathbf{j} & \mathbf{k} \\ 3 & 4 & 0 \\ -40 & -20 & 40 \end{vmatrix}$$
$$= [4(40) - 0(-20)]\mathbf{i} - [3(40) - 0(-40)]\mathbf{j} + [3(-20) - 4(-40)]\mathbf{k}$$

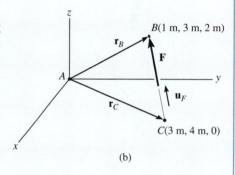

(b)

In both cases,

$$\mathbf{M}_A = \{160\mathbf{i} - 120\mathbf{j} + 100\mathbf{k}\}\ \text{N} \cdot \text{m}$$

The *magnitude* of $\mathbf{M}_A$ is therefore

$$M_A = \sqrt{(160)^2 + (-120)^2 + (100)^2} = 224\ \text{N} \cdot \text{m} \qquad Ans.$$

As expected, $\mathbf{M}_A$ acts perpendicular to the shaded plane containing vectors $\mathbf{F}$, $\mathbf{r}_B$, and $\mathbf{r}_C$, Fig. 3–16c. (How would you find its coordinate direction angles $\alpha = 44.3°$, $\beta = 122°$, $\gamma = 63.4°$?) Had this problem been worked using a scalar approach, where $M_A = Fd$, notice the difficulty that can arise in obtaining the moment arm $d$.

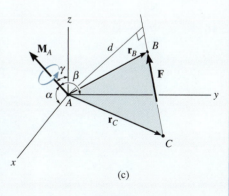

(c)

Fig. 3–16

**E X A M P L E  3.5**

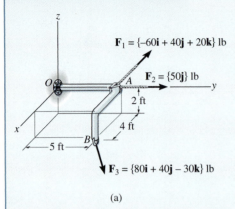

(a)

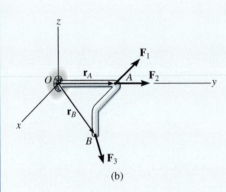

(b)

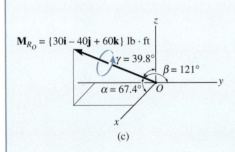

(c)

Fig. 3–17

Three forces act on the rod shown in Fig. 3–17a. Determine the resultant moment they create about the flange at $O$ and determine the coordinate direction angles of the moment axis.

**Solution**

Position vectors are directed from point $O$ to each force as shown in Fig. 3–17b. These vectors are

$$\mathbf{r}_A = \{5\mathbf{j}\} \text{ ft}$$
$$\mathbf{r}_B = \{4\mathbf{i} + 5\mathbf{j} - 2\mathbf{k}\} \text{ ft}$$

The resultant moment about $O$ is therefore

$$\mathbf{M}_{R_O} = \Sigma(\mathbf{r} \times \mathbf{F})$$
$$= \mathbf{r}_A \times \mathbf{F}_1 + \mathbf{r}_A \times \mathbf{F}_2 + \mathbf{r}_B \times \mathbf{F}_3$$
$$= \begin{vmatrix} \mathbf{i} & \mathbf{j} & \mathbf{k} \\ 0 & 5 & 0 \\ -60 & 40 & 20 \end{vmatrix} + \begin{vmatrix} \mathbf{i} & \mathbf{j} & \mathbf{k} \\ 0 & 5 & 0 \\ 0 & 50 & 0 \end{vmatrix} + \begin{vmatrix} \mathbf{i} & \mathbf{j} & \mathbf{k} \\ 4 & 5 & -2 \\ 80 & 40 & -30 \end{vmatrix}$$
$$= [5(20) - 40(0)]\mathbf{i} - [0\mathbf{j}] + [0(40) - (-60)(5)]\mathbf{k} + [0\mathbf{i} - 0\mathbf{j} + 0\mathbf{k}]$$
$$+ [5(-30) - (40)(-2)]\mathbf{i} - [4(-30) - 80(-2)]\mathbf{j} + [4(40) - 80(5)]\mathbf{k}$$
$$= \{30\mathbf{i} - 40\mathbf{j} + 60\mathbf{k}\} \text{ lb} \cdot \text{ft} \qquad \qquad Ans.$$

The moment axis is directed along the line of action of $\mathbf{M}_{R_O}$. Since the magnitude of this moment is

$$M_{R_O} = \sqrt{(30)^2 + (-40)^2 + (60)^2} = 78.10 \text{ lb} \cdot \text{ft}$$

the unit vector which defines the direction of the moment axis is

$$\mathbf{u} = \frac{\mathbf{M}_{R_O}}{M_{R_O}} = \frac{30\mathbf{i} - 40\mathbf{j} + 60\mathbf{k}}{78.10} = 0.3841\mathbf{i} - 0.5121\mathbf{j} + 0.7682\mathbf{k}$$

Therefore, the coordinate direction angles of the moment axis are

$$\cos \alpha = 0.3841; \qquad \alpha = 67.4° \qquad \qquad Ans.$$
$$\cos \beta = -0.5121; \qquad \beta = 121° \qquad \qquad Ans.$$
$$\cos \gamma = 0.7682; \qquad \gamma = 39.8° \qquad \qquad Ans.$$

These results are shown in Fig. 3–17c. Realize that the three forces tend to cause the rod to rotate about this axis in the manner shown by the curl indicated on the moment vector.

## 3.4  Principle of Moments

A concept often used in mechanics is the *principle of moments*, which is sometimes referred to as *Varignon's theorem* since it was originally developed by the French mathematician Varignon (1654–1722). It states that *the moment of a force about a point is equal to the sum of the moments of the force's components about the point*. The proof follows directly from the distributive law of the vector cross product. To show this, consider the force $\mathbf{F}$ and two of its rectangular components, where $\mathbf{F} = \mathbf{F}_1 + \mathbf{F}_2$, Fig. 3–18. We have

$$\mathbf{M}_O = \mathbf{r} \times \mathbf{F}_1 + \mathbf{r} \times \mathbf{F}_2 = \mathbf{r} \times (\mathbf{F}_1 + \mathbf{F}_2) = \mathbf{r} \times \mathbf{F}$$

This concept has important applications to the solution of problems and proofs of theorems that follow, since it is often easier to determine the moments of a force's components rather than the moment of the force itself.

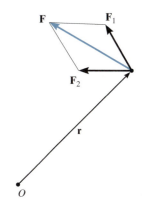

Fig. 3–18

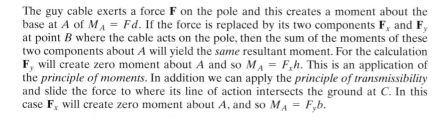

The guy cable exerts a force $\mathbf{F}$ on the pole and this creates a moment about the base at $A$ of $M_A = Fd$. If the force is replaced by its two components $\mathbf{F}_x$ and $\mathbf{F}_y$ at point $B$ where the cable acts on the pole, then the sum of the moments of these two components about $A$ will yield the *same* resultant moment. For the calculation $\mathbf{F}_y$ will create zero moment about $A$ and so $M_A = F_x h$. This is an application of the *principle of moments*. In addition we can apply the *principle of transmissibility* and slide the force to where its line of action intersects the ground at $C$. In this case $\mathbf{F}_x$ will create zero moment about $A$, and so $M_A = F_y b$.

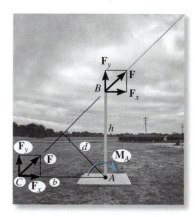

## IMPORTANT POINTS

- The moment of a force indicates the tendency of a body to turn about an axis passing through a specific point $O$.
- Using the right-hand rule, the sense of rotation is indicated by the fingers, and the thumb is directed along the moment axis, or line of action of the moment.
- The magnitude of the moment is determined from $M_O = Fd$, where $d$ is the perpendicular or shortest distance from point $O$ to the line of action of the force $\mathbf{F}$.
- In three dimensions use the vector cross product to determine the moment, i.e., $\mathbf{M}_O = \mathbf{r} \times \mathbf{F}$. Remember that $\mathbf{r}$ is directed *from* point $O$ *to any point* on the line of action of $\mathbf{F}$.
- The principle of moments states that the moment of a force about a point is equal to the sum of the moments of the force's components about the point. This is a very convenient method to use in two dimensions.

E X A M P L E  **3.6**

A 200-N force acts on the bracket shown in Fig. 3–19*a*. Determine the moment of the force about point *A*.

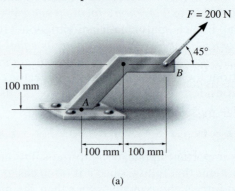

(a)

**Fig. 3–19**

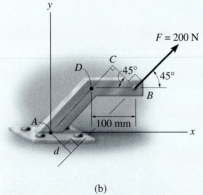

(b)

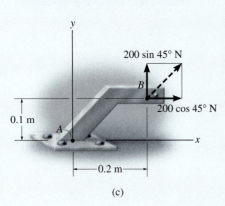

(c)

### Solution I

The moment arm *d* can be found by trigonometry, using the construction shown in Fig. 3–19*b*. From the right triangle *BCD*,

$$CB = d = 100 \cos 45° = 70.71 \text{ mm} = 0.070\ 71 \text{ m}$$

Thus,

$$M_A = Fd = 200 \text{ N}(0.070\ 71 \text{ m}) = 14.1 \text{ N} \cdot \text{m} \curvearrowleft$$

According to the right-hand rule, $\mathbf{M}_A$ is directed in the $+\mathbf{k}$ direction since the force tends to rotate or orbit *counterclockwise* about point *A*. Hence, reporting the moment as a Cartesian vector, we have

$$\mathbf{M}_A = \{14.1\mathbf{k}\} \text{ N} \cdot \text{m} \qquad \textit{Ans.}$$

### Solution II

The 200-N force may be resolved into *x* and *y* components, as shown in Fig. 3–19*c*. In accordance with the principle of moments, the moment of **F** computed about point *A* is equivalent to the sum of the moments produced by the two force components. Assuming counterclockwise rotation as positive, i.e., in the $+\mathbf{k}$ direction, we can apply Eq. 3–5 ($M_A = \Sigma Fd$), in which case

$$\curvearrowleft + M_A = (200 \sin 45°\text{N})(0.20 \text{ m}) - (200 \cos 45°\text{N})(0.10 \text{ m})$$
$$= 14.1 \text{ N} \cdot \text{m} \curvearrowleft$$

Thus

$$\mathbf{M}_A = \{14.1\mathbf{k}\} \text{ N} \cdot \text{m} \qquad \textit{Ans.}$$

By comparison, it is seen that Solution II provides a more *convenient method* for analysis than Solution I since the moment arm for each component force is easier to establish.

E X A M P L E   **3.7**

The force **F** acts at the end of the angle bracket shown in Fig. 3–20*a*. Determine the moment of the force about point *O*.

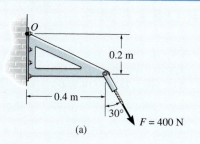

(a)

**Solution I (*Scalar Analysis*)**
The force is resolved into its *x* and *y* components as shown in Fig. 3–20*b*, and the moments of the components are computed about point *O*. Taking positive moments as counterclockwise, i.e., in the +**k** direction, we have

$$\downarrow + M_O = 400 \sin 30°\text{N}(0.2 \text{ m}) - 400 \cos 30°\text{N}(0.4 \text{ m})$$
$$= -98.6 \text{ N}\cdot\text{m} = 98.6 \text{ N}\cdot\text{m} \downarrow$$

or

$$M_O = \{-98.6\mathbf{k}\} \text{ N}\cdot\text{m} \qquad \textit{Ans.}$$

**Solution II (*Vector Analysis*)**
Using a Cartesian vector approach, the force and position vectors shown in Fig. 3–20*c* can be represented as

$$\mathbf{r} = \{0.4\mathbf{i} - 0.2\mathbf{j}\} \text{ m}$$
$$\mathbf{F} = \{400 \sin 30°\mathbf{i} - 400 \cos 30°\mathbf{j}\} \text{ N}$$
$$= \{200.0\mathbf{i} - 346.4\mathbf{j}\} \text{ N}$$

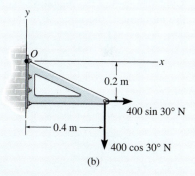

(b)

The moment is therefore

$$\mathbf{M}_O = \mathbf{r} \times \mathbf{F} = \begin{vmatrix} \mathbf{i} & \mathbf{j} & \mathbf{k} \\ 0.4 & -0.2 & 0 \\ 200.0 & -346.4 & 0 \end{vmatrix}$$

$$= 0\mathbf{i} - 0\mathbf{j} + [0.4(-346.4) - (-0.2)(200.0)]\mathbf{k}$$
$$= \{-98.6\mathbf{k}\} \text{ N}\cdot\text{m} \qquad \textit{Ans.}$$

By comparison, it is seen that the scalar analysis (Solution I) provides a more *convenient method* for analysis than Solution II since the direction of the moment and the moment arm for each component force are easy to establish. Hence, this method is generally recommended for solving problems displayed in two dimensions. On the other hand, Cartesian vector analysis is generally recommended only for solving three-dimensional problems, where the moment arms and force components are often more difficult to determine.

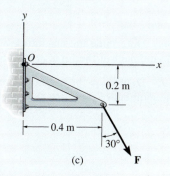

(c)

**Fig. 3–20**

# PROBLEMS

**3-1.** If **A**, **B**, and **D** are given vectors, prove the distributive law for the vector cross product, i.e., $\mathbf{A} \times (\mathbf{B} + \mathbf{D}) = (\mathbf{A} \times \mathbf{B}) + (\mathbf{A} \times \mathbf{D})$.

**3-2.** Prove the triple scalar product identity $\mathbf{A} \cdot \mathbf{B} \times \mathbf{C} = \mathbf{A} \times \mathbf{B} \cdot \mathbf{C}$.

**3-3.** Given the three nonzero vectors **A**, **B**, and **C**, show that if $\mathbf{A} \cdot (\mathbf{B} \times \mathbf{C}) = 0$, the three vectors *must* lie in the same plane.

**\*3-4.** Determine the magnitude and directional sense of the moment of the force at A about point O.

**3-5.** Determine the magnitude and directional sense of the moment of the force at A about point P.

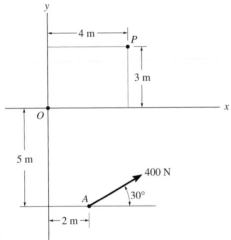

Probs. 3–4/5

**3-6.** Determine the magnitude and directional sense of the moment of the force at A about point O.

**3-7.** Determine the magnitude and directional sense of the moment of the force at A about point P.

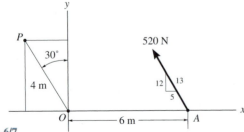

Probs. 3–6/7

**\*3-8.** Determine the magnitude and directional sense of the resultant moment of the forces about point O.

**3-9.** Determine the magnitude and directional sense of the resultant moment of the forces about point P.

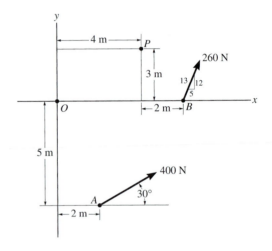

Probs. 3–8/9

**3-10.** The wrench is used to loosen the bolt. Determine the moment of each force about the bolt's axis passing through point O.

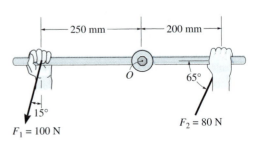

Prob. 3–10

**3-11.** Determine the magnitude and directional sense of the resultant moment of the forces about point $O$.

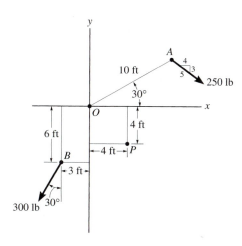

Prob, 3-11

**3-14.** Determine the moment of each force about the bolt located at $A$. Take $F_B = 40$ lb, $F_C = 50$ lb.

**3-15.** If $F_B = 30$ lb and $F_C = 45$ lb, determine the resultant moment about the bolt located at $A$.

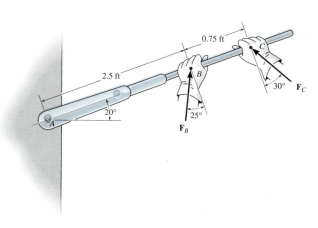

Probs, 3-14/15

**\*3-12.** Determine the moment about point $A$ of each of the three forces acting on the beam.

**3-13.** Determine the moment about point $B$ of each of the three forces acting on the beam.

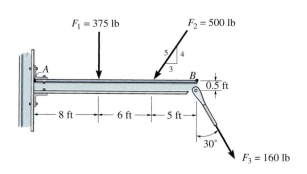

Probs, 3-12/13

**\*3-16.** The power pole supports the three lines, each line exerting a vertical force on the pole due to its weight as shown. Determine the resultant moment at the base $D$ due to all of these forces. If it is possible for wind or ice to snap the lines, determine which line(s) when removed create(s) a condition for the greatest moment about the base. What is this resultant moment?

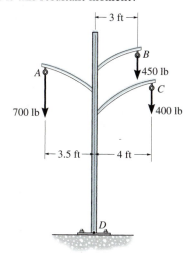

Prob, 3-16

**3-17.** A force of 80 N acts on the handle of the paper cutter at $A$. Determine the moment created by this force about the hinge at $O$, if $\theta = 60°$. At what angle $\theta$ should the force be applied so that the moment it creates about point $O$ is a maximum (clockwise)? What is this maximum moment?

**3-19.** The hub of the wheel can be attached to the axle either with negative offset (left) or with positive offset (right). If the tire is subjected to both a normal and radial load as shown, determine the resultant moment of these loads about the axle, point $O$ for both cases.

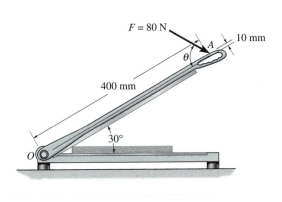

Prob. 3–17

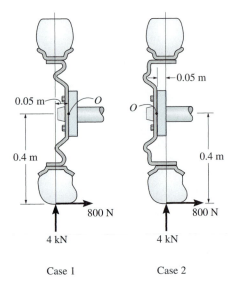

Case 1    Case 2

Prob. 3–19

**3-18.** Determine the direction $\theta$ ($0° \leq \theta \leq 180°$) of the force $F = 40$ lb so that it produces (a) the maximum moment about point $A$ and (b) the minimum moment about point $A$. Compute the moment in each case.

**\*3-20.** The boom has a length of 30 ft, a weight of 800 lb, and mass center at $G$. If the maximum moment that can be developed by the motor at $A$ is $M = 20(10^3)$ lb · ft, determine the maximum load $W$, having a mass center at $G'$, that can be lifted. Take $\theta = 30°$.

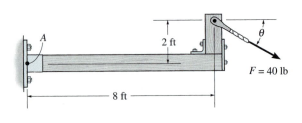

Prob. 3–18

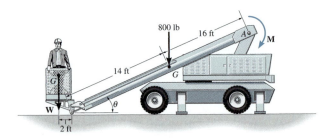

Prob. 3–20

**3-21.** The tool at *A* is used to hold a power lawnmower blade stationary while the nut is being loosened with the wrench. If a force of 50 N is applied to the wrench at *B* in the direction shown, determine the moment it creates about the nut at *C*. What is the magnitude of force **F** at *A* so that it creates the opposite moment about *C*?

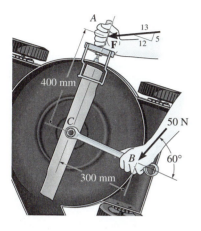

Prob. 3–21

**3-22.** Determine the moment of each of the three forces about point *A*. Solve the problem first by using each force as a whole, and then by using the principle of moments.

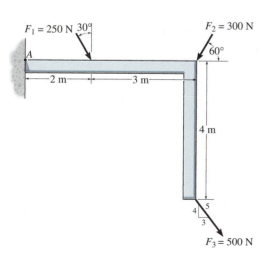

$F_1 = 250$ N  $30°$   $F_2 = 300$ N

$60°$

*A*

2 m    3 m

4 m

4  5
3

$F_3 = 500$ N

Prob. 3–22

**3-23.** As part of an acrobatic stunt, a man supports a girl who has a weight of 120 lb and is seated on a chair on top of the pole. If her center of gravity is at *G*, and if the maximum counterclockwise moment the man can exert on the pole at *A* is 250 lb · ft, determine the maximum angle of tilt, $\theta$, which will not allow the girl to fall, i.e., so her clockwise moment about *A* does not exceed 250 lb · ft.

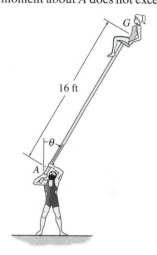

Prob. 3–23

**\*3-24.** The two boys push on the gate with forces of $F_A = 30$ lb and $F_B = 50$ lb as shown. Determine the moment of each force about *C*. Which way will the gate rotate, clockwise or counterclockwise? Neglect the thickness of the gate.

**3-25.** Two boys push on the gate as shown. If the boy at *B* exerts a force of $F_B = 30$ lb, determine the magnitude of the force $F_A$ the boy at *A* must exert in order to prevent the gate from turning. Neglect the thickness of the gate.

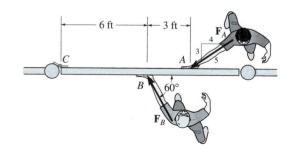

Probs. 3–24/25

**3-26.** Determine the angle $\theta$ at which the 500-N force must act at $A$ so that the moment of this force about point $B$ is equal to zero.

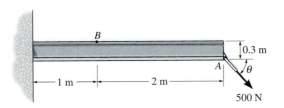

Prob. 3-26

**3-27.** Segments of drill pipe $D$ for an oil well are tightened a prescribed amount by using a set of tongs $T$, which grip the pipe, and a hydraulic cylinder (not shown) to regulate the force $\mathbf{F}$ applied to the tongs. This force acts along the cable which passes around the small pulley $P$. If the cable is originally perpendicular to the tongs as shown, determine the magnitude of force $\mathbf{F}$ which must be applied so that the moment about the pipe is $M = 2000$ lb · ft. In order to maintain this same moment what magnitude of $\mathbf{F}$ is required when the tongs rotate $30°$ to the dashed position? *Note:* The angle $DAP$ is not $90°$ in this position.

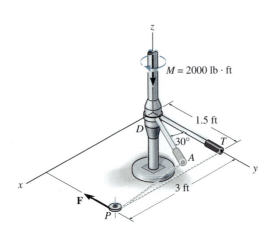

Prob. 3-27

*3-28.** Determine the moment of the force at $A$ about point $O$. Express the result as a Cartesian vector.

**3-29.** Determine the moment of the force at $A$ about point $P$. Express the result as a Cartesian vector.

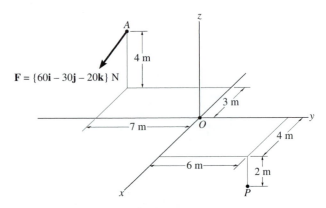

Probs. 3-28/29

**3-30.** Determine the moment of the force $\mathbf{F}$ at $A$ about point $O$. Express the result as a Cartesian vector.

**3-31.** Determine the moment of the force $\mathbf{F}$ at $A$ about point $P$. Express the result as a Cartesian vector.

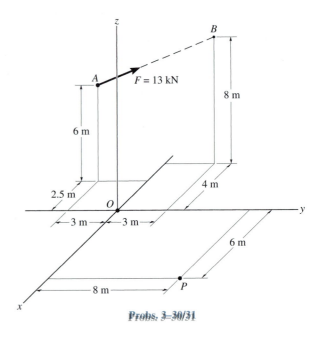

Probs. 3-30/31

**\*3-32.** The curved rod lies in the x–y plane and has a radius of 3 m. If a force of $F = 80$ N acts at its end as shown, determine the moment of this force about point O.

**3-33.** The curved rod lies in the x–y plane and has a radius of 3 m. If a force of $F = 80$ N acts at its end as shown, determine the moment of this force about point B.

**3-35.** The curved rod has a radius of 5 ft. If a force of 60 lb acts at its end as shown, determine the moment of this force about point C.

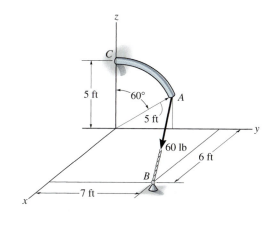

Prob. 3–35

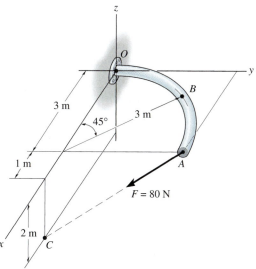

Probs. 3–32/33

**3-34.** The force $\mathbf{F} = \{600\mathbf{i} + 300\mathbf{j} - 600\mathbf{k}\}$ N acts at the end of the beam. Determine the moment of the force about point A.

**\*3-36.** A force **F** having a magnitude of $F = 100$ N acts along the diagonal of the parallelepiped. Determine the moment of **F** about point A, using $\mathbf{M}_A = \mathbf{r}_B \times \mathbf{F}$ and $\mathbf{M}_A = \mathbf{r}_C \times \mathbf{F}$.

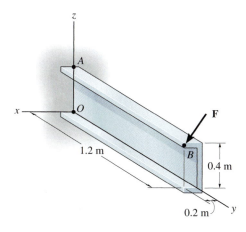

Prob. 3–34

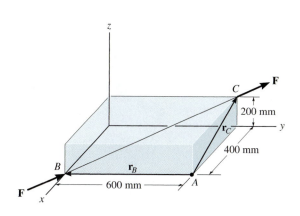

Prob. 3–36

**3-37.** Determine the smallest force $F$ that must be applied along the rope in order to cause the curved rod, which has a radius of 5 ft, to fail at the support $C$. This requires a moment of $M = 80$ lb · ft to be developed at $C$.

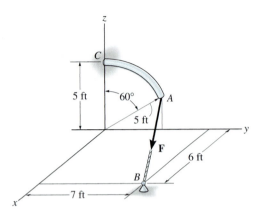

**Prob. 3-37**

**3-38.** The pipe assembly is subjected to the 80-N force. Determine the moment of this force about point $A$.

**3-39.** The pipe assembly is subjected to the 80-N force. Determine the moment of this force about point $B$.

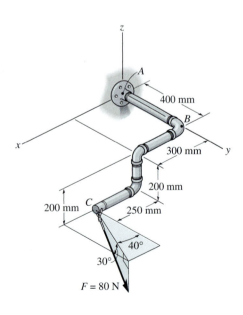

**Probs. 3-38/39**

**\*3-40.** Strut $AB$ of the 1-m-diameter hatch door exerts a force of 450 N on point $B$. Determine the moment of this force about point $O$.

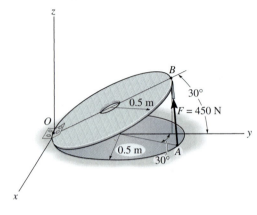

**Prob. 3-40**

**3-41.** Using Cartesian vector analysis, determine the resultant moment of the three forces about the base of the column at $A$. Take $\mathbf{F}_1 = \{400\mathbf{i} + 300\mathbf{j} + 120\mathbf{k}\}$ N.

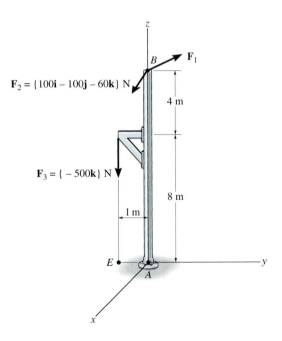

**Prob. 3-41**

## 3.5  Moment of a Force about a Specified Axis

Recall that when the moment of a force is computed about a point, the moment and its axis are *always* perpendicular to the plane containing the force and the moment arm. In some problems it is important to find the *component* of this moment along a *specified axis* that passes through the point. To solve this problem either a scalar or vector analysis can be used.

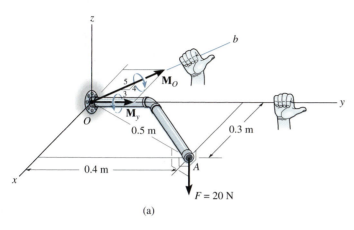

(a)

**Fig. 3–21**

**Scalar Analysis.**  As a numerical example of this problem, consider the pipe assembly shown in Fig. 3–21a, which lies in the horizontal plane and is subjected to the vertical force of $F = 20$ N applied at point $A$. The moment of this force about point $O$ has a *magnitude* of $M_O = (20 \text{ N})(0.5 \text{ m}) = 10 \text{ N} \cdot \text{m}$, and a *direction* defined by the right-hand rule, as shown in Fig. 3–21a. This moment tends to turn the pipe about the $Ob$ axis. For practical reasons, however, it may be necessary to determine the *component* of $\mathbf{M}_O$ about the $y$ axis, $\mathbf{M}_y$, since this component tends to unscrew the pipe from the flange at $O$. From Fig. 3–21a, $\mathbf{M}_y$ has a magnitude of $M_y = \frac{3}{5}(10 \text{ N} \cdot \text{m}) = 6 \text{ N} \cdot \text{m}$ and a sense of direction shown by the vector resolution. Rather than performing this *two-step* process of first finding the moment of the force about point $O$ and then resolving the moment along the $y$ axis, it is also possible to solve this problem *directly*. To do so, it is necessary to determine the perpendicular or moment-arm distance from the line of action of $\mathbf{F}$ to the $y$ axis. From Fig. 3–21a this distance is 0.3 m. Thus the *magnitude* of the moment of the force about the $y$ axis is again $M_y = 0.3(20 \text{ N}) = 6 \text{ N} \cdot \text{m}$, and the *direction* is determined by the right-hand rule as shown.

In general, then, *if the line of action of a force* $\mathbf{F}$ *is perpendicular to any specified axis aa*, the magnitude of the moment of $\mathbf{F}$ about the axis can be determined from the equation

$$M_a = F d_a \qquad\qquad (3\text{-}10)$$

Here $d_a$ is the *perpendicular or shortest distance* from the force line of action to the axis. The direction is determined from the thumb of the right hand when the fingers are curled in accordance with the direction of rotation as produced by the force. In particular, realize that a *force will not contribute a moment about a specified axis if the force line of action is parallel to the axis or its line of action passes through the axis*.

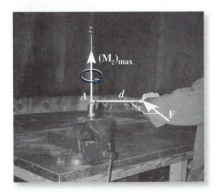

If a horizontal force **F** is applied to the handle of the flex-headed wrench, it tends to turn the socket at $A$ about the $z$ axis. This effect is caused by the moment of **F** about the $z$ axis. The *maximum moment* is determined when the wrench is in the horizontal plane so that full leverage from the handle can be achieved, i.e., $(M_z)_{max} = Fd$. If the handle is not in the horizontal position, then the moment about the $z$ axis is determined from $M_z = Fd'$, where $d'$ is the perpendicular distance from the force line of action to the axis. We can also determine this moment by first finding the moment of **F** about $A$, $M_A = Fd$, then finding the projection or component of this moment along $z$, i.e., $M_z = M_A \cos \theta$.

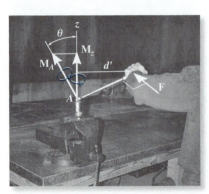

**Vector Analysis.** The previous two-step solution of first finding the moment of the force about a point on the axis and then finding the projected component of the moment about the axis can also be performed using a vector analysis, Fig. 3–21$b$. Here the moment about point $O$ is first determined from $\mathbf{M}_O = \mathbf{r}_A \times \mathbf{F} = (0.3\mathbf{i} + 0.4\mathbf{j}) \times (-20\mathbf{k}) = \{-8\mathbf{i} + 6\mathbf{j}\}\ \text{N} \cdot \text{m}$. The component or projection of this moment along the $y$ axis is then determined from the dot product (Sec. 2.9). Since the unit vector for this axis (or line) is $\mathbf{u}_a = \mathbf{j}$, then $M_y = \mathbf{M}_O \cdot \mathbf{u}_a = (-8\mathbf{i} + 6\mathbf{j}) \cdot \mathbf{j} = 6\ \text{N} \cdot \text{m}$. This result, of course, is to be expected, since it represents the **j** component of $\mathbf{M}_O$.

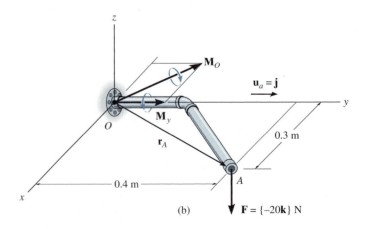

(b)

**Fig. 3–21**

A vector analysis such as this is particularly advantageous for finding the moment of a force about an axis when the force components or the appropriate moment arms are difficult to determine. For this reason, the above two-step process will now be generalized and applied to a body of arbitrary shape. To do so, consider the body in Fig. 3–22, which is subjected to the force **F** acting at point $A$. Here we wish to determine the effect of **F** in tending to rotate the body about the $aa'$ axis. This tendency for rotation is measured by the moment component $\mathbf{M}_a$. To determine $\mathbf{M}_a$ we first compute the moment of **F** about any *arbitrary point $O$* that lies on the $aa'$ axis. In this case, $\mathbf{M}_O$ is expressed by the cross product $\mathbf{M}_O = \mathbf{r} \times \mathbf{F}$, where **r** is directed from $O$ to $A$. Here $\mathbf{M}_O$ acts along the moment axis $bb'$, and so the component or projection of $\mathbf{M}_O$ onto the $aa'$ axis is then $\mathbf{M}_a$. The *magnitude* of $\mathbf{M}_a$ is determined by the dot product, $M_a = M_O \cos \theta = \mathbf{M}_O \cdot \mathbf{u}_a$ where $\mathbf{u}_a$ is a unit vector that defines the direction of the $aa'$ axis. Combining these two steps as a general expression, we have $M_a = (\mathbf{r} \times \mathbf{F}) \cdot \mathbf{u}_a$. Since the dot product is commutative, we can also write

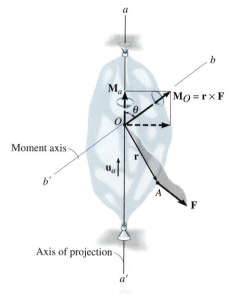

Fig. 3–22

$$M_a = \mathbf{u}_a \cdot (\mathbf{r} \times \mathbf{F})$$

In vector algebra, this combination of dot and cross product yielding the scalar $M_a$ is called the *triple scalar product*. Provided $x, y, z$ axes are established and the Cartesian components of each of the vectors can be determined, then the triple scalar product may be written in determinant form as

$$M_a = (u_{a_x}\mathbf{i} + u_{a_y}\mathbf{j} + u_{a_z}\mathbf{k}) \cdot \begin{vmatrix} \mathbf{i} & \mathbf{j} & \mathbf{k} \\ r_x & r_y & r_z \\ F_x & F_y & F_z \end{vmatrix}$$

or simply

$$M_a = \mathbf{u}_a \cdot (\mathbf{r} \times \mathbf{F}) = \begin{vmatrix} u_{a_x} & u_{a_y} & u_{a_z} \\ r_x & r_y & r_z \\ F_x & F_y & F_z \end{vmatrix} \qquad (3\text{–}11)$$

where

| | |
|---|---|
| $u_{a_x}, u_{a_y}, u_{a_z}$ | represent the $x, y, z$ components of the unit vector defining the direction of the $aa'$ axis |
| $r_x, r_y, r_z$ | represent the $x, y, z$ components of the position vector drawn from *any point $O$* on the $aa'$ axis to *any point $A$* on the line of action of the force |
| $F_x, F_y, F_z$ | represent the $x, y, z$ components of the force vector. |

When $M_a$ is evaluated from Eq. 3–11, it will yield a positive or negative scalar. The sign of this scalar indicates the sense of direction of $\mathbf{M}_a$ along the $aa'$ axis. If it is positive, then $\mathbf{M}_a$ will have the same sense as $\mathbf{u}_a$, whereas if it is negative, then $\mathbf{M}_a$ will act opposite to $\mathbf{u}_a$.

Once $M_a$ is determined, we can then express $\mathbf{M}_a$ as a Cartesian vector, namely,

$$\mathbf{M}_a = M_a \mathbf{u}_a = [\mathbf{u}_a \cdot (\mathbf{r} \times \mathbf{F})]\mathbf{u}_a \qquad (3\text{--}12)$$

Finally, if the resultant moment of a series of forces is to be computed about the $aa'$ axis, then the moment components of each force are added together *algebraically*, since each component lies along the same axis. Thus the magnitude of $\mathbf{M}_a$ is

$$M_a = \Sigma[\mathbf{u}_a \cdot (\mathbf{r} \times \mathbf{F})] = \mathbf{u}_a \cdot \Sigma(\mathbf{r} \times \mathbf{F})$$

The examples which follow illustrate a numerical application of the above concepts.

Wind blowing on the face of this traffic sign creates a resultant force $\mathbf{F}$ that tends to tip the sign over due to the moment $\mathbf{M}_A$ created about the $a−a$ axis. The moment of $\mathbf{F}$ about a point $A$ that lies on the axis is $\mathbf{M}_A = \mathbf{r} \times \mathbf{F}$. The projection of this moment along the axis, whose direction is defined by the unit vector $\mathbf{u}_a$, is $M_a = \mathbf{u}_a \cdot (\mathbf{r} \times \mathbf{F})$. Had this moment been calculated using scalar methods, then the perpendicular distance from the force line of action to the $a−a$ axis would have to be determined, which in this case would be a more difficult task.

## IMPORTANT POINTS

- The moment of a force about a specified axis can be determined provided the perpendicular distance $d_a$ from *both* the force line of action and the axis can be determined. $M_a = F d_a$.

- If vector analysis is used, $M_a = \mathbf{u}_a \cdot (\mathbf{r} \times \mathbf{F})$, where $\mathbf{u}_a$ defines the direction of the axis and $\mathbf{r}$ is directed from *any point* on the axis to *any point* on the line of action of the force.

- If $M_a$ is calculated as a negative scalar, then the sense of direction of $\mathbf{M}_a$ is opposite to $\mathbf{u}_a$.

- The moment $\mathbf{M}_a$ expressed as a Cartesian vector is determined from $\mathbf{M}_a = M_a \mathbf{u}_a$.

EXAMPLE **3.8**

The force $\mathbf{F} = \{-40\mathbf{i} + 20\mathbf{j} + 10\mathbf{k}\}$ N acts at point $A$ shown in Fig. 3–23a. Determine the moments of this force about the $x$ and $a$ axes.

### Solution I (*Vector Analysis*)

We can solve this problem by using the position vector $\mathbf{r}_A$. Why? Since $\mathbf{r}_A = \{-3\mathbf{i} + 4\mathbf{j} + 6\mathbf{k}\}$ m and $\mathbf{u}_x = \mathbf{i}$, then applying Eq. 3–11,

$$M_x = \mathbf{i} \cdot (\mathbf{r}_A \times \mathbf{F}) = \begin{vmatrix} 1 & 0 & 0 \\ -3 & 4 & 6 \\ -40 & 20 & 10 \end{vmatrix}$$

$$= 1[4(10) - 6(20)] - 0[(-3)(10) - 6(-40)] + 0[(-3)(20) - 4(-40)]$$

$$= -80 \text{ N} \cdot \text{m} \qquad\qquad Ans.$$

The negative sign indicates that the sense of $\mathbf{M}_x$ is opposite to $\mathbf{i}$.

   We can compute $M_a$ also using $\mathbf{r}_A$ because $\mathbf{r}_A$ extends from a point on the $a$ axis to the force. Also, $\mathbf{u}_a = -\frac{3}{5}\mathbf{i} + \frac{4}{5}\mathbf{j}$. Thus,

$$M_a = \mathbf{u}_a \cdot (\mathbf{r}_A \times \mathbf{F}) = \begin{vmatrix} -\frac{3}{5} & \frac{4}{5} & 0 \\ -3 & 4 & 6 \\ -40 & 20 & 10 \end{vmatrix}$$

$$= -\tfrac{3}{5}[4(10) - 6(20)] - \tfrac{4}{5}[(-3)(10) - 6(-40)] + 0[(-3)(20) - 4(-40)]$$

$$= -120 \text{ N} \cdot \text{m} \qquad\qquad Ans.$$

What does the negative sign indicate?

   The moment components are shown in Fig. 3–23b.

### Solution II (*Scalar Analysis*)

Since the force components and moment arms are easy to determine for computing $M_x$, a scalar analysis can be used to solve this problem. Referring to Fig. 3–23c, only the 10-N and 20-N forces contribute moments about the $x$ axis. (The line of action of the 40-N force is parallel to this axis and hence its moment about the $x$ axis is zero.) Using the right-hand rule, the algebraic sum of the moment components about the $x$ axis is therefore

$$M_x = (10 \text{ N})(4 \text{ m}) - (20 \text{ N})(6 \text{ m}) = -80 \text{ N} \cdot \text{m} \qquad Ans.$$

Although not required here, note also that

$$M_y = (10 \text{ N})(3 \text{ m}) - (40 \text{ N})(6 \text{ m}) = -210 \text{ N} \cdot \text{m}$$

$$M_z = (40 \text{ N})(4 \text{ m}) - (20 \text{ N})(3 \text{ m}) = 100 \text{ N} \cdot \text{m}$$

   If we were to determine $M_a$ by this scalar method, it would require much more effort since the force components of 40 N and 20 N are *not perpendicular* to the direction of the $a$ axis. The vector analysis yields a more direct solution.

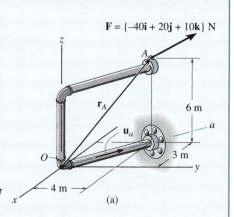

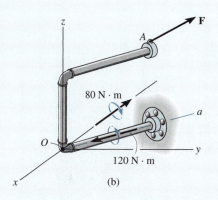

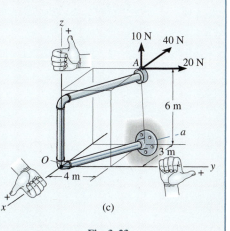

Fig. 3–23

EXAMPLE 3.9

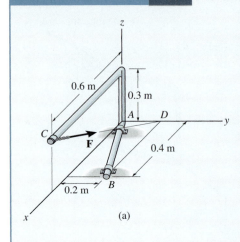

(a)

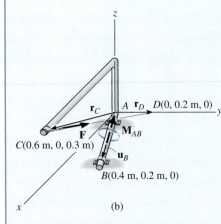

(b)

**Fig. 3–24**

The rod shown in Fig. 3–24a is supported by two brackets at $A$ and $B$. Determine the moment $\mathbf{M}_{AB}$ produced by $\mathbf{F} = \{-600\mathbf{i} + 200\mathbf{j} - 300\mathbf{k}\}\,\text{N}$, which tends to rotate the rod about the $AB$ axis.

**Solution**

A vector analysis using $M_{AB} = \mathbf{u}_B \cdot (\mathbf{r} \times \mathbf{F})$ will be considered for the solution since the moment arm or perpendicular distance from the line of action of $\mathbf{F}$ to the $AB$ axis is difficult to determine. Each of the terms in the equation will now be identified.

Unit vector $\mathbf{u}_B$ defines the direction of the $AB$ axis of the rod, Fig. 3–24b, where

$$\mathbf{u}_B = \frac{\mathbf{r}_B}{r_B} = \frac{0.4\mathbf{i} + 0.2\mathbf{j}}{\sqrt{(0.4)^2 + (0.2)^2}} = 0.894\mathbf{i} + 0.447\mathbf{j}$$

Vector $\mathbf{r}$ is directed from *any point* on the $AB$ axis to *any point* on the line of action of the force. For example, position vectors $\mathbf{r}_C$ and $\mathbf{r}_D$ are suitable, Fig. 3–24b. (Although not shown, $\mathbf{r}_{BC}$ or $\mathbf{r}_{BD}$ can also be used.) For simplicity, we choose $\mathbf{r}_D$, where

$$\mathbf{r}_D = \{0.2\mathbf{j}\}\,\text{m}$$

The force is

$$\mathbf{F} = \{-600\mathbf{i} + 200\mathbf{j} - 300\mathbf{k}\}\,\text{N}$$

Substituting these vectors into the determinant form and expanding, we have

$$M_{AB} = \mathbf{u}_B \cdot (\mathbf{r}_D \times \mathbf{F}) = \begin{vmatrix} 0.894 & 0.447 & 0 \\ 0 & 0.2 & 0 \\ -600 & 200 & -300 \end{vmatrix}$$

$$= 0.894[0.2(-300) - 0(200)] - 0.447[0(-300) - 0(-600)] + 0[0(200) - 0.2(-600)]$$

$$= -53.67\,\text{N} \cdot \text{m}$$

The negative sign indicates that the sense of $\mathbf{M}_{AB}$ is opposite to that of $\mathbf{u}_B$.

Expressing $\mathbf{M}_{AB}$ as a Cartesian vector yields

$$\mathbf{M}_{AB} = M_{AB}\mathbf{u}_B = (-53.67\,\text{N} \cdot \text{m})(0.894\mathbf{i} + 0.447\mathbf{j})$$

$$= \{-48.0\mathbf{i} - 24.0\mathbf{j}\}\,\text{N} \cdot \text{m} \qquad \textit{Ans.}$$

The result is shown in Fig. 3–24b.

Note that if axis $AB$ is defined using a unit vector directed from $B$ toward $A$, then in the above formulation $-\mathbf{u}_B$ would have to be used. This would lead to $M_{AB} = +53.67\,\text{N} \cdot \text{m}$. Consequently, $\mathbf{M}_{AB} = M_{AB}(-\mathbf{u}_B)$, and the above result would again be determined.

# PROBLEMS

**3-42.** Determine the moment of the force **F** about the *Oa* axis. Express the result as a Cartesian vector.

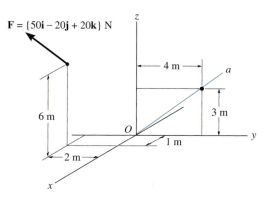

Prob. 3–42

**3-43.** Determine the moment of the force **F** about the *a–a* axis. Express the result as a Cartesian vector.

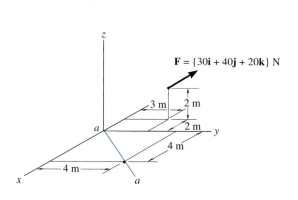

Prob. 3–43

*3-44.** Determine the resultant moment of the two forces about the *Oa* axis. Express the result as a Cartesian vector.

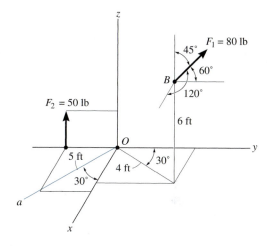

Prob. 3–44

**3-45.** Determine the magnitude of the moment of each of the three forces about the axis *AB*. Solve the problem (a) using a Cartesian vector approach and (b) using a scalar approach.

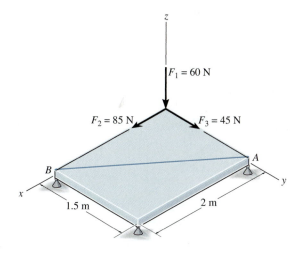

Prob. 3–45

**3-46.** The chain $AB$ exerts a force of 20 lb on the door at $B$. Determine the magnitude of the moment of this force along the hinged axis $x$ of the door.

**\*3-48.** The cutting tool on the lathe exerts a force $\mathbf{F}$ on the shaft in the direction shown. Determine the moment of this force about the $y$ axis of the shaft.

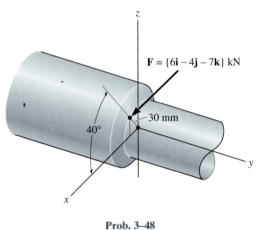

$\mathbf{F} = \{6\mathbf{i} - 4\mathbf{j} - 7\mathbf{k}\}$ kN

**Prob. 3–48**

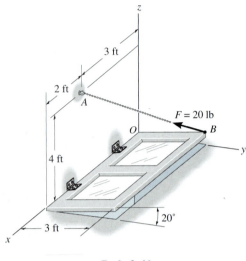

**Prob. 3–46**

**3-47.** The force of $F = 30$ N acts on the bracket as shown. Determine the moment of the force about the $a$–$a$ axis of the pipe. Also, determine the coordinate direction angles of $F$ in order to produce the maximum moment about the $a$–$a$ axis. What is this moment?

**3-49.** The hood of the automobile is supported by the strut $AB$, which exerts a force of $F = 24$ lb on the hood. Determine the moment of this force about the hinged axis $y$.

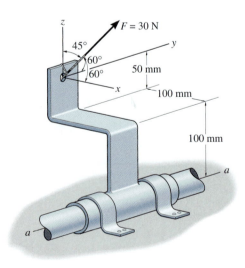

**Prob. 3–47**

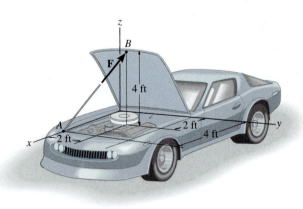

**Prob. 3–49**

**3-50.** Determine the magnitude of the moments of the force **F** about the *x*, *y*, and *z* axes. Solve the problem (a) using a Cartesian vector approach and (b) using a scalar approach.

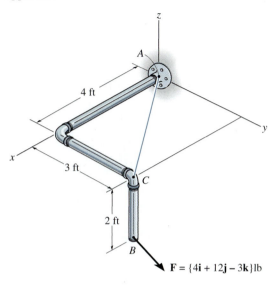

**Prob. 3–50**

**3-51.** Determine the moment of the force **F** about an axis extending between *A* and *C*. Express the result as a Cartesian vector.

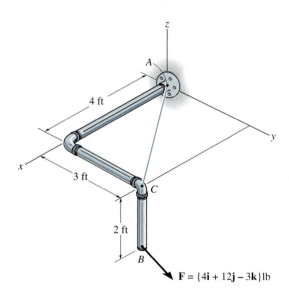

**Probs. 3–51**

**\*3-52.** The lug and box wrenches are used in combination to remove the lug nut from the wheel hub. If the applied force on the end of the box wrench is **F** = {4**i** − 12**j** + 2**k**} N, determine the magnitude of the moment of this force about the *x* axis which is effective in unscrewing the lug nut.

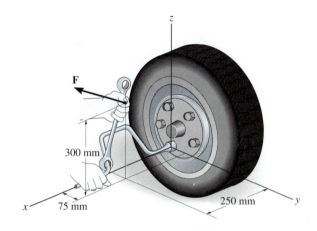

**Prob. 3–52**

**3-53.** A 70-lb force acts vertically on the "Z" bracket. Determine the magnitude of the moment of this force about the bolt axis (z axis).

**3-54.** Determine the magnitude of the moment of the force $\mathbf{F} = \{50\mathbf{i} - 20\mathbf{j} - 80\mathbf{k}\}$ N about the base line $CA$ of the tripod.

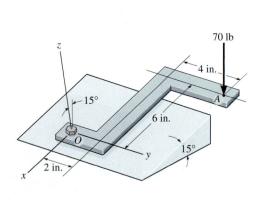

**Prob. 3–53**

**Prob. 3–54**

## 3.6 Moment of a Couple

A *couple* is defined as two parallel forces that have the same magnitude, have opposite directions, and are separated by a perpendicular distance $d$, Fig. 3–25. Since the resultant force is zero, the only effect of a couple is to produce a rotation or tendency of rotation in a specified direction.

The moment produced by a couple is called a *couple moment*. We can determine its value by finding the sum of the moments of both couple forces about *any* arbitrary point. For example, in Fig. 3–26, position vectors $\mathbf{r}_A$ and $\mathbf{r}_B$ are directed from point $O$ to points $A$ and $B$ lying on the line of action of $-\mathbf{F}$ and $\mathbf{F}$. The couple moment computed about $O$ is therefore

$$\mathbf{M} = \mathbf{r}_A \times (-\mathbf{F}) + \mathbf{r}_B \times \mathbf{F}$$

Rather than sum the moments of both forces to determine the couple moment, it is simpler to take moments about a point lying on the line of action of one of the forces. If point $A$ is chosen, then the moment of $-\mathbf{F}$ about $A$ is zero, and we have

$$\mathbf{M} = \mathbf{r} \times \mathbf{F} \qquad (3\text{–}13)$$

The fact that we obtain the *same result* in both cases can be demonstrated by noting that in the first case we can write $\mathbf{M} = (\mathbf{r}_B - \mathbf{r}_A) \times \mathbf{F}$; and by the triangle rule of vector addition, $\mathbf{r}_A + \mathbf{r} = \mathbf{r}_B$ or $\mathbf{r} = \mathbf{r}_B - \mathbf{r}_A$, so that upon substitution we obtain Eq. 3–13. This result indicates that a couple moment is a *free vector*, i.e., it can act at *any point* since $\mathbf{M}$ depends *only* upon the position vector $\mathbf{r}$ directed *between* the forces and *not* the position vectors $\mathbf{r}_A$ and $\mathbf{r}_B$, directed from the arbitrary point $O$ to the forces. This concept is therefore unlike the moment of a force, which requires a definite point (or axis) about which moments are determined.

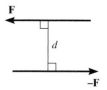

Fig. 3–25

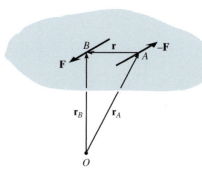

Fig. 3–26

### Scalar Formulation.

The moment of a couple, $\mathbf{M}$, Fig. 3–27, is defined as having a *magnitude* of

$$\boxed{M = Fd} \qquad (3\text{–}14)$$

where $F$ is the magnitude of one of the forces and $d$ is the perpendicular distance or moment arm between the forces. The *direction* and sense of the couple moment are determined by the right-hand rule, where the thumb indicates the direction when the fingers are curled with the sense of rotation caused by the two forces. In all cases, $\mathbf{M}$ acts perpendicular to the plane containing these forces.

### Vector Formulation.

The moment of a couple can also be expressed by the vector cross product using Eq. 3–13, i.e.,

$$\boxed{\mathbf{M} = \mathbf{r} \times \mathbf{F}} \qquad (3\text{–}15)$$

Application of this equation is easily remembered if one thinks of taking the moments of both forces about a point lying on the line of action of one of the forces. For example, if moments are taken about point $A$ in Fig. 3–26, the moment of $-\mathbf{F}$ is *zero* about this point, and the moment

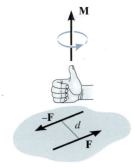

Fig. 3–27

(a)

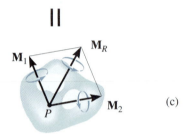

(b)

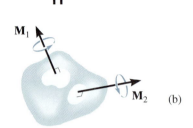

(c)

**Fig. 3–28**

or **F** is defined from Eq. 3–15. Therefore, in the formulation **r** is crossed with the force **F** to which it is directed.

**Equivalent Couples.**   Two couples are said to be equivalent if they produce the same moment. Since the moment produced by a couple is always perpendicular to the plane containing the couple forces, it is therefore necessary that the forces of equal couples lie either in the same plane or in planes that are *parallel* to one another. In this way, the direction of each couple moment will be the same, that is, perpendicular to the parallel planes.

**Resultant Couple Moment.**   Since couple moments are free vectors, they may be applied at any point $P$ on a body and added vectorially. For example, the two couples acting on different planes of the body in Fig. 3–28a may be replaced by their corresponding couple moments $\mathbf{M}_1$ and $\mathbf{M}_2$, Fig. 3–28b, and then these free vectors may be moved to the *arbitrary point P* and added to obtain the resultant couple moment $\mathbf{M}_R = \mathbf{M}_1 + \mathbf{M}_2$, shown in Fig. 3–28c.

If more than two couple moments act on the body, we may generalize this concept and write the vector resultant as

$$\mathbf{M}_R = \Sigma(\mathbf{r} \times \mathbf{F}) \qquad (3\text{–}16)$$

These concepts are illustrated numerically in the examples which follow. In general, problems projected in two dimensions should be solved using a scalar analysis since the moment arms and force components are easy to compute.

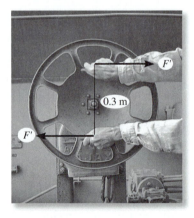

The frictional forces of the floor on the blades of the concrete finishing machine create a couple moment $\mathbf{M}_c$ on the machine that tends to turn it. An equal but opposite couple moment must be applied by the hands of the operator to prevent the turning. Here the couple moment, $M_c = Fd$, is applied on the handle, although it could be applied at any other point on the machine.

A moment of 12 N · m is needed to turn the shaft connected to the center of the wheel. To do this it is efficient to apply a couple since this effect produces a pure rotation. The couple forces can be made as small as possible by placing the hands on the *rim* of the wheel, where the spacing is 0.4 m. In this case 12 N · m = $F(0.4$ m$)$, $F = 30$ N. An equivalent couple moment of 12 N · m can be produced if one grips the wheel within the inner hub, although here much larger forces are needed. If the distance between the hands becomes 0.3 m, then 12 N · m = $F'(0.3)$, $F' = 40$ N. Also, realize that if the wheel was connected to the shaft at a point other than at its center, the wheel would still turn when the forces are applied since the 12-N · m couple moment is a *free vector*.

## IMPORTANT POINTS

- A couple moment is produced by two noncollinear forces that are equal but opposite. Its effect is to produce pure rotation, or tendency for rotation in a specified direction.

- A couple moment is a free vector, and as a result it causes the same effect of rotation on a body regardless of where the couple moment is applied to the body.

- The moment of the two couple forces can be computed about *any point*. For convenience, this point is often chosen on the line of action of one of the forces in order to eliminate the moment of this force about the point.

- In three dimensions the couple moment is often determined using the vector formulation, $\mathbf{M} = \mathbf{r} \times \mathbf{F}$, where $\mathbf{r}$ is directed from *any point* on the line of action of one of the forces to *any point* on the line of action of the other force $\mathbf{F}$.

- A resultant couple moment is simply the vector sum of all the couple moments of the system.

## E X A M P L E 3.10

A couple acts on the gear teeth as shown in Fig. 3–29a. Replace it by an equivalent couple having a pair of forces that act through points $A$ and $B$.

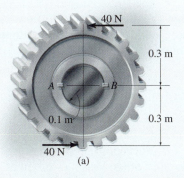

(a)

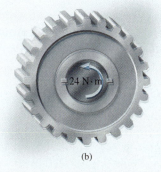

(b)

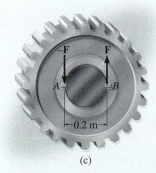

(c)

Fig. 3–29

### Solution (*Scalar Analysis*)

The couple has a magnitude of $M = Fd = 40(0.6) = 24\ \text{N}\cdot\text{m}$ and a direction that is out of the page since the forces tend to rotate counterclockwise. $\mathbf{M}$ is a free vector, and so it can be placed at any point on the gear, Fig. 3–29b. To preserve the counterclockwise rotation of $\mathbf{M}$, *vertical* forces acting through points $A$ and $B$ must be directed as shown in Fig. 3–29c. The magnitude of each force is

$$M = Fd \qquad 24\ \text{N}\cdot\text{m} = F(0.2\ \text{m})$$

$$F = 120\ \text{N} \qquad\qquad\qquad\qquad \textit{Ans.}$$

**E X A M P L E   3.11**

Determine the moment of the couple acting on the member shown in Fig. 3–30a.

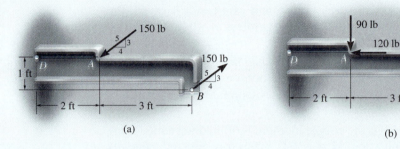

(a)                    (b)

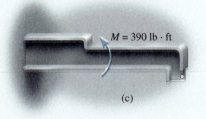

(c)

Fig. 3–30

**Solution (*Scalar Analysis*)**

Here it is somewhat difficult to determine the perpendicular distance between the forces and compute the couple moment as $M = Fd$. Instead, we can resolve each force into its horizontal and vertical components, $F_x = \frac{4}{5}(150 \text{ lb}) = 120 \text{ lb}$ and $F_y = \frac{3}{5}(150 \text{ lb}) = 90 \text{ lb}$, Fig. 3–30b, and then use the principle of moments. The couple moment can be determined about *any point*. For example, if point $D$ is chosen, we have for all four forces,

$$\zeta + M = 120 \text{ lb}(0 \text{ ft}) - 90 \text{ lb}(2 \text{ ft}) + 90 \text{ lb}(5 \text{ ft}) + 120 \text{ lb}(1 \text{ ft})$$
$$= 390 \text{ lb} \cdot \text{ft} \uparrow \qquad\qquad\qquad Ans.$$

It is easier, however, to determine the moments about point $A$ or $B$ in order to *eliminate* the moment of the forces acting at the moment point. For point $A$, Fig. 3–30b, we have

$$\zeta + M = 90 \text{ lb}(3 \text{ ft}) + 120 \text{ lb}(1 \text{ ft})$$
$$= 390 \text{ lb} \cdot \text{ft} \uparrow \qquad\qquad\qquad Ans.$$

Show that one obtains this same result if moments are summed about point $B$. Notice also that the couple in Fig. 3–30a can be replaced by *two* couples in Fig. 3–30b. Using $M = Fd$, one couple has a moment of $M_1 = 90 \text{ lb}(3 \text{ ft}) = 270 \text{ lb} \cdot \text{ft}$ and the other has a moment of $M_2 = 120 \text{ lb}(1 \text{ ft}) = 120 \text{ lb} \cdot \text{ft}$. By the right-hand rule, both couple moments are counterclockwise and are therefore directed out of the page. Since these couples are free vectors, they can be moved to any point and added, which yields $M = 270 \text{ lb} \cdot \text{ft} + 120 \text{ lb} \cdot \text{ft} = 390 \text{ lb} \cdot \text{ft} \uparrow$, the same result determined above. **M** is a free vector and can therefore act at any point on the member, Fig. 3–30c. Also, realize that the external effect, such as the support reactions on the member, will be the *same* if the member supports the couple, Fig. 3–30a, or the couple moment, Fig. 3–30c.

# EXAMPLE 3.12

Determine the couple moment acting on the pipe shown in Fig. 3–31a. Segment $AB$ is directed 30° below the $x$–$y$ plane.

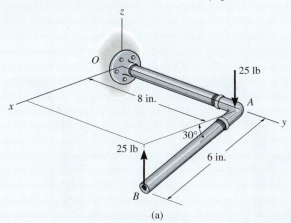

(a)

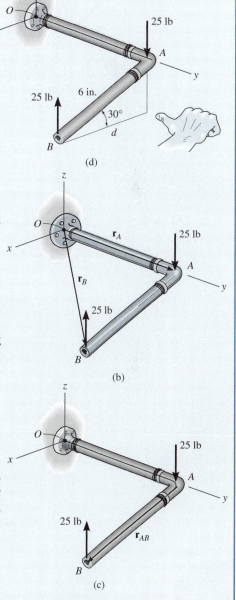

## Solution I (*Vector Analysis*)

The moment of the two couple forces can be found about *any point*. If point $O$ is considered, Fig. 3–31b, we have

$$\mathbf{M} = \mathbf{r}_A \times (-25\mathbf{k}) + \mathbf{r}_B \times (25\mathbf{k})$$
$$= (8\mathbf{j}) \times (-25\mathbf{k}) + (6\cos 30°\mathbf{i} + 8\mathbf{j} - 6\sin 30°\mathbf{k}) \times (25\mathbf{k})$$
$$= -200\mathbf{i} - 129.9\mathbf{j} + 200\mathbf{i}$$
$$= \{-130\mathbf{j}\}\ \text{lb}\cdot\text{in.} \qquad\qquad Ans.$$

It is *easier* to take moments of the couple forces about a point lying on the line of action of one of the forces, e.g., point $A$, Fig. 3–31c. In this case the moment of the force $A$ is zero, so that

$$\mathbf{M} = \mathbf{r}_{AB} \times (25\mathbf{k})$$
$$= (6\cos 30°\mathbf{i} - 6\sin 30°\mathbf{k}) \times (25\mathbf{k})$$
$$= \{-130\mathbf{j}\}\ \text{lb}\cdot\text{in.} \qquad\qquad Ans.$$

## Solution II (*Scalar Analysis*)

Although this problem is shown in three dimensions, the geometry is simple enough to use the scalar equation $M = Fd$. The perpendicular distance between the lines of action of the forces is $d = 6\cos 30° = 5.20$ in., Fig. 3–31d. Hence, taking moments of the forces about either point $A$ or $B$ yields

$$M = Fd = 25\ \text{lb}(5.20\ \text{in.}) = 129.9\ \text{lb}\cdot\text{in.}$$

Applying the right-hand rule, $\mathbf{M}$ acts in the $-\mathbf{j}$ direction. Thus,

$$\mathbf{M} = \{-130\mathbf{j}\}\ \text{lb}\cdot\text{in.} \qquad\qquad Ans.$$

**Fig. 3–31**

E X A M P L E  3.13

Replace the two couples acting on the pipe column in Fig. 3–32a by a resultant couple moment.

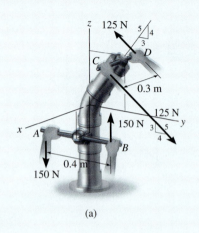

(a)

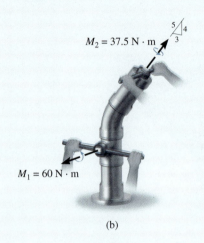

(b)

(c)

Fig. 3–32

### Solution (*Vector Analysis*)

The couple moment $M_1$, developed by the forces at $A$ and $B$, can easily be determined from a scalar formulation.

$$M_1 = Fd = 150 \text{ N}(0.4 \text{ m}) = 60 \text{ N} \cdot \text{m}$$

By the right-hand rule, $M_1$ acts in the $+\mathbf{i}$ direction, Fig. 3–32b. Hence,

$$M_1 = \{60\mathbf{i}\} \text{ N} \cdot \text{m}$$

Vector analysis will be used to determine $M_2$, caused by forces at $C$ and $D$. If moments are computed about point $D$, Fig. 3–32a, $M_2 = \mathbf{r}_{DC} \times \mathbf{F}_C$, then

$$M_2 = \mathbf{r}_{DC} \times \mathbf{F}_C = (0.3\mathbf{i}) \times [125(\tfrac{4}{5})\mathbf{j} - 125(\tfrac{3}{5})\mathbf{k}]$$
$$= (0.3\mathbf{i}) \times [100\mathbf{j} - 75\mathbf{k}] = 30(\mathbf{i} \times \mathbf{j}) - 22.5(\mathbf{i} \times \mathbf{k})$$
$$= \{22.5\mathbf{j} + 30\mathbf{k}\} \text{ N} \cdot \text{m}$$

Try to establish $M_2$ by using a scalar formulation, Fig. 3–32b.

Since $M_1$ and $M_2$ are free vectors, they may be moved to some arbitrary point $P$ and added vectorially, Fig. 3–32c. The resultant couple moment becomes

$$M_R = M_1 + M_2 = \{60\mathbf{i} + 22.5\mathbf{j} + 30\mathbf{k}\} \text{ N} \cdot \text{m} \qquad \textit{Ans.}$$

# PROBLEMS

**3-55.** Determine the magnitude and sense of the couple moment.

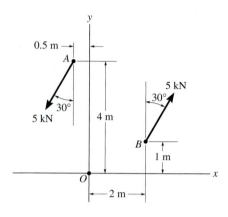

Prob. 3–55

**3-57.** Determine the magnitude and sense of the couple moment. Each force has a magnitude of $F = 8$ kN.

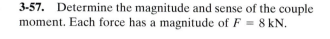
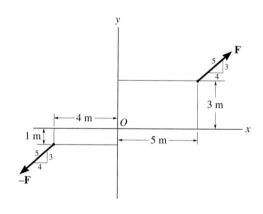

Prob. 3–57

**\*3-56.** If the couple moment has a magnitude of 250 N · m, determine the magnitude $F$ of the couple forces.

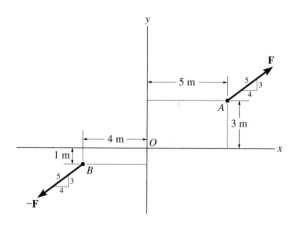

Prob. 3–56

**3-58.** If the couple moment has a magnitude of 300 lb · ft, determine the magnitude $F$ of the couple forces.

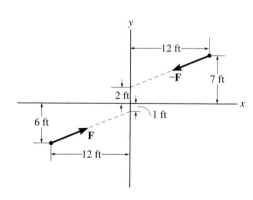

Prob. 3–58

**3-59.** A twist of 4 N·m is applied to the handle of the screwdriver. Resolve this couple moment into a pair of couple forces **F** exerted on the handle and **P** exerted on the blade.

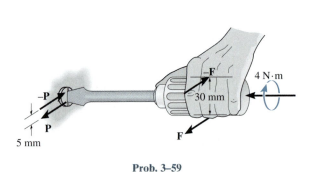

**Prob. 3–59**

**3-61.** A device called a rolamite is used in various ways to replace slipping motion with rolling motion. If the belt, which wraps between the rollers, is subjected to a tension of 15 N, determine the reactive forces *N* of the top and bottom plates on the rollers so that the resultant couple acting on the rollers is equal to zero.

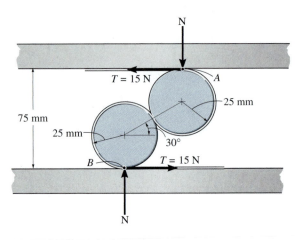

**Prob. 3–61**

*\*3-60.** The resultant couple moment created by the two couples acting on the disk is $\mathbf{M}_R = \{10\mathbf{k}\}$ kip·in. Determine the magnitude of force **T**.

**3-62.** The caster wheel is subjected to the two couples. Determine the forces **F** that the bearings create on the shaft so that the resultant couple moment on the caster is zero.

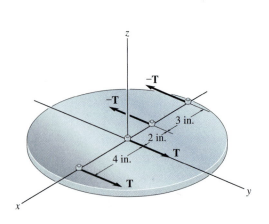

**Prob. 3–60**

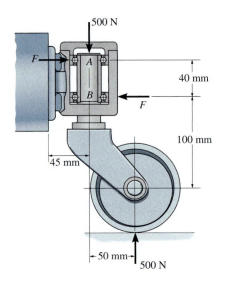

**Prob. 3–62**

**3-63.** Two couples act on the beam as shown. Determine the magnitude of **F** so that the resultant couple moment is 300 lb·ft counterclockwise. Where on the beam does the resultant couple act?

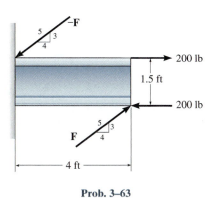

**Prob. 3–63**

**\*3-64.** Two couples act on the frame. If the resultant couple moment is to be zero, determine the distance *d* between the 80-lb couple forces.

**\*3-65.** Two couples act on the frame. If *d* = 4 ft, determine the resultant couple moment. Compute the result by resolving each force into *x* and *y* components and (a) finding the moment of each couple (Eq. 3–13) and (b) summing the moments of all the force components about point *A*.

**3-66.** Two couples act on the frame. If *d* = 4 ft, determine the resultant couple moment. Compute the result by resolving each force into *x* and *y* components and (a) finding the moment of each couple (Eq. 3–13) and (b) summing the moments of all the force components about point *B*.

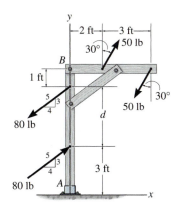

**Probs. 3–64/65/66**

**3-67.** Determine the couple moment. Express the result as a Cartesian vector.

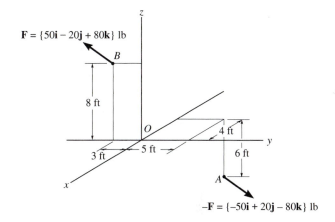

**Prob. 3–67**

**\*3-68.** Determine the couple moment. Express the result as a Cartesian vector. Each force has a magnitude of *F* = 120 lb.

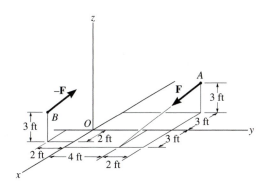

**Prob. 3–68**

**3-69.** The gear reducer is subject to the couple moments shown. Determine the resultant couple moment and specify its magnitude and coordinate direction angles.

**3-71.** A couple acts on each of the handles of the minidual valve. Determine the magnitude and coordinate direction angles of the resultant couple moment.

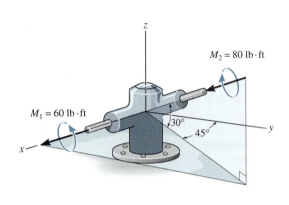

**Prob. 3–69**

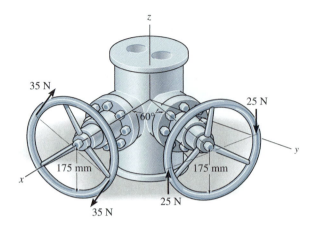

**Prob. 3–71**

**3-70.** The meshed gears are subjected to the couple moments shown. Determine the magnitude of the resultant couple moment and specify its coordinate direction angles.

**\*3-72.** Determine the resultant couple moment of the two couples that act on the pipe assembly. The distance from $A$ to $B$ is $d = 400$ mm. Express the result as a Cartesian vector.

**3-73.** Determine the distance $d$ between $A$ and $B$ so that the resultant couple moment has a magnitude of $M_R = 20$ N · m.

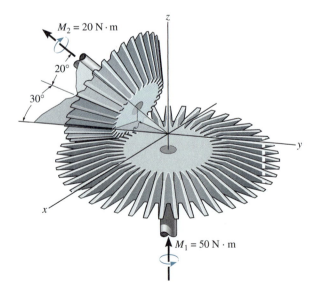

**Prob. 3–70**

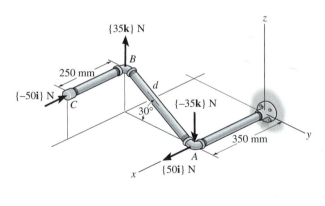

**Probs. 3–72/73**

## 3.7  Equivalent System

A force has the effect of both translating and rotating a body, and the amount by which it does so depends upon where and how the force is applied. In the next section we will discuss the method used to *simplify* a system of forces and couple moments acting on a body to a single resultant force and couple moment acting at a specified point $O$. To do this, however, it is necessary that the force and couple moment system produce the *same* "external" effects of translation and rotation of the body as their resultants. When this occurs these two sets of loadings are said to be *equivalent*.

In this section we wish to show how to maintain this equivalency when a single force is applied to a specific point on a body and when it is located at another point $O$. Two cases for the location of point $O$ will now be considered.

### Point O Is On the Line of Action of the Force.

Consider the body shown in Fig. 3–33$a$, which is subjected to the force $\mathbf{F}$ applied to point $A$. In order to apply the force to point $O$ without altering the external effects on the body, we will first apply equal but opposite forces $\mathbf{F}$ and $-\mathbf{F}$ at $O$, as shown in Fig. 3–33$b$. The two forces indicated by the slash across them can be canceled, leaving the force at point $O$ as required, Fig. 3–33$c$. By using this construction procedure, an *equivalent system* has been maintained between each of the diagrams, as shown by the equal signs. Note, however, that the force has simply been "transmitted" along its line of action, from point $A$, Fig. 3–33$a$, to point $O$, Fig. 3–33$c$. In other words, the force can be considered as a *sliding vector* since it can act at any point $O$ along its line of action. In Sec. 3.3 we referred to this concept as the *principle of transmissibility*. It is important to realize that only the *external effects*, such as the body's motion or the forces needed to support the body if it is stationary, remain *unchanged* after $\mathbf{F}$ is moved. Certainly the *internal effects* depend on where $\mathbf{F}$ is located. For example, when $\mathbf{F}$ acts at $A$, the internal forces in the body have a high intensity around $A$; whereas movement of $\mathbf{F}$ away from this point will cause these internal forces to decrease.

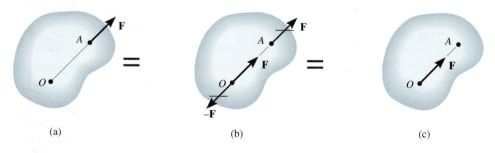

(a)   (b)   (c)

Fig. 3–33

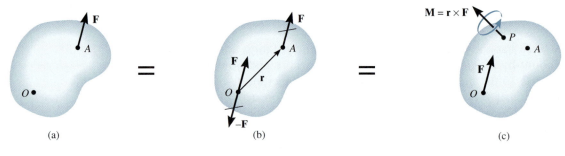

(a)  (b)  (c)

$M = r \times F$

**Fig. 3–34**

### Point *O* Is Not On the Line of Action of the Force.

This case is shown in Fig. 3–34*a*, where **F** is to be moved to point *O* without altering the external effects on the body. Following the same procedure as before, we first apply equal but opposite forces **F** and −**F** at point *O*, Fig. 3–34*b*. Here the two forces indicated by a slash across them form a couple which has a moment that is perpendicular to **F** and is defined by the cross product **M** = **r** × **F**. Since the couple moment is a *free vector*, it may be applied at *any point P* on the body as shown in Fig. 3–34*c*. In addition to this couple moment, **F** now acts at point *O* as required.

To summarize these concepts, when the point on the body is *on the line of action of the force*, simply transmit or slide the force along its line of action to the point. When the point is not on the line of action of the force, then move the force to the point and add a couple moment anywhere to the body. This couple moment is found by taking the moment of the force about the point. When these rules are carried out, equivalent external effects will be produced.

Consider the effects on the hand when a stick of negligible weight supports a force **F** at its end. When the force is applied horizontally, the same force is felt at the grip, regardless of where it is applied along its line of action. This is a consequence of the principle of transmissibility.

When the force is applied vertically it causes both a downward force **F** to be felt at the grip and a clockwise couple moment or twist of $M = Fd$. These same effects are felt if **F** is applied at the grip and **M** is applied anywhere on the stick. In both cases the systems are equivalent.

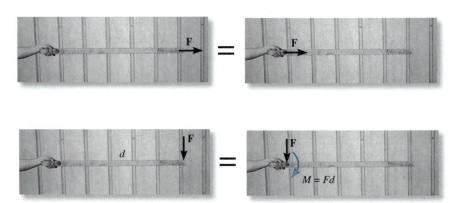

## 3.8  Resultants of a Force and Couple System

When a rigid body is subjected to a *system* of forces and couple moments, it is often simpler to study the external effects on the body by *replacing* the system by an equivalent single resultant force acting at a specified point $O$ and a resultant couple moment. To show how to determine these resultants we will consider the rigid body in Fig. 3–35a and use the concepts discussed in the previous section. Since point $O$ is not on the line of action of the forces, an equivalent effect is produced if the forces are moved to point $O$ *and* the corresponding couple moments $\mathbf{M}_1 = \mathbf{r}_1 \times \mathbf{F}_1$ and $\mathbf{M}_2 = \mathbf{r}_2 \times \mathbf{F}_2$ are applied to the body. Furthermore, the couple moment $\mathbf{M}_c$ is simply moved to point $O$ since it is a free vector. These results are shown in Fig. 3–35b. By vector addition, the resultant force is $\mathbf{F}_R = \mathbf{F}_1 + \mathbf{F}_2$, and the resultant couple moment is $\mathbf{M}_{R_O} = \mathbf{M}_c + \mathbf{M}_1 + \mathbf{M}_2$, Fig. 3–35c. Since equivalency is maintained between the diagrams in Fig. 3–35, each force and couple system will cause the *same external effects*, i.e., the same translation and rotation of the body. Note that both the magnitude and direction of $\mathbf{F}_R$ are independent of the location of point $O$; however, $\mathbf{M}_{R_O}$ depends upon this location since the moments $\mathbf{M}_1$ and $\mathbf{M}_2$ are determined using the position vectors $\mathbf{r}_1$ and $\mathbf{r}_2$. Also note that $\mathbf{M}_{R_O}$ is a free vector and can act at *any point* on the body, although point $O$ is generally chosen as its point of application.

The above method of simplifying any force and couple moment system to a resultant force acting at point $O$ and a resultant couple moment can be generalized and represented by application of the following two equations.

$$\boxed{\begin{aligned} \mathbf{F}_R &= \Sigma\mathbf{F} \\ \mathbf{M}_{R_O} &= \Sigma\mathbf{M}_c + \Sigma\mathbf{M}_O \end{aligned}} \qquad (3\text{--}17)$$

The first equation states that the resultant force of the system is equivalent to the sum of all the forces; and the second equation states that the resultant couple moment of the system is equivalent to the sum of all the couple moments $\Sigma\mathbf{M}_c$, plus the moments about point $O$ of all the forces $\Sigma\mathbf{M}_O$. If the force system lies in the $x$–$y$ plane and any couple moments are perpendicular to this plane, that is along the $z$ axis, then the above equations reduce to the following three scalar equations.

$$\boxed{\begin{aligned} F_{R_x} &= \Sigma F_x \\ F_{R_y} &= \Sigma F_y \\ M_{R_O} &= \Sigma M_c + \Sigma M_O \end{aligned}} \qquad (3\text{--}18)$$

Note that the resultant force $\mathbf{F}_R$ is equivalent to the vector sum of its two components $\mathbf{F}_{R_x}$ and $\mathbf{F}_{R_y}$.

(a)

(b)

(c)

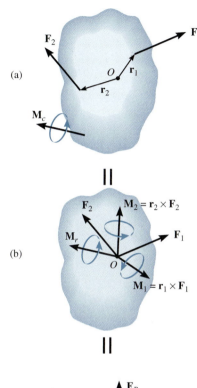

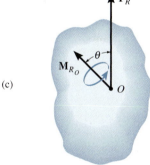

**Fig. 3–35**

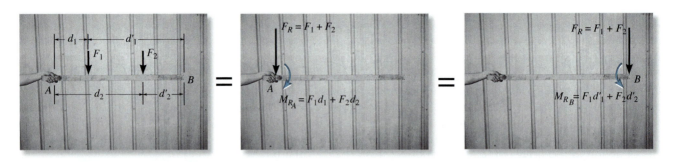

If the two forces acting on the stick are replaced by an equivalent resultant force and couple moment at point $A$, or by the equivalent resultant force and couple moment at point $B$, then in each case the hand must provide the same resistance to translation and rotation in order to keep the stick in the horizontal position. In other words, the external effects on the stick are the *same* in each case.

## PROCEDURE FOR ANALYSIS

The following points should be kept in mind when applying Eqs. 3–17 or 3–18.

- Establish the coordinate axes with the origin located at the point $O$ and the axes having a selected orientation.

*Force Summation.*

- If the force system is *coplanar*, resolve each force into its $x$ and $y$ components. If a component is directed along the positive $x$ or $y$ axis, it represents a positive scalar; whereas if it is directed along the negative $x$ or $y$ axis, it is a negative scalar.

- In three dimensions, represent each force as a Cartesian vector before summing the forces.

*Moment Summation.*

- When determining the moments of a *coplanar* force system about point $O$, it is generally advantageous to use the principle of moments, i.e., determine the moments of the components of each force rather than the moment of the force itself.

- In three dimensions use the vector cross product to determine the moment of each force about the point. Here the position vectors extend from point $O$ to any point on the line of action of each force.

EXAMPLE **3.14**

Replace the forces acting on the brace shown in Fig. 3–36a by an equivalent resultant force and couple moment acting at point A.

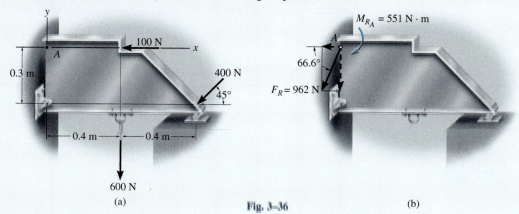

(a)                                                                                    (b)

**Fig. 3–36**

### Solution (*Scalar Analysis*)

The principle of moments will be applied to the 400-N force, whereby the moments of its two rectangular components will be considered.

***Force Summation.*** The resultant force has $x$ and $y$ components of

$$\xrightarrow{+} F_{R_x} = \Sigma F_x; \quad F_{R_x} = -100\ \text{N} - 400\cos 45°\text{N} = -382.8\ \text{N} = 382.8\ \text{N} \leftarrow$$

$$+\uparrow F_{R_y} = \Sigma F_y; \quad F_{R_y} = -600\ \text{N} - 400\sin 45°\text{N} = -882.8\ \text{N} = 882.8\ \text{N}\downarrow$$

As shown in Fig. 3–36b, $\mathbf{F}_R$ has a magnitude of

$$F_R = \sqrt{(F_{R_x})^2 + (F_{R_y})^2} = \sqrt{(382.8)^2 + (882.8)^2} = 962\ \text{N} \quad \textit{Ans.}$$

and a direction of

$$\theta = \tan^{-1}\left(\frac{F_{R_y}}{F_{R_x}}\right) = \tan^{-1}\left(\frac{882.8}{382.8}\right) = 66.6° \quad \theta \nearrow \quad \textit{Ans.}$$

***Moment Summation.*** The resultant couple moment $\mathbf{M}_{R_A}$ is determined by summing the moments of the forces about point A. Assuming that positive moments act counterclockwise, i.e., in the $+\mathbf{k}$ direction, we have

$$\zeta + M_{R_A} = \Sigma M_A;$$

$$M_{R_A} = 100\ \text{N}(0) - 600\ \text{N}(0.4\ \text{m}) - (400\sin 45°\text{N})(0.8\ \text{m})$$

$$\qquad\qquad\qquad - (400\cos 45°\ \text{N})(0.3\ \text{m})$$

$$= -551\ \text{N}\cdot\text{m} = 551\ \text{N}\cdot\text{m} \downarrow \quad \textit{Ans.}$$

In conclusion, when $\mathbf{M}_{R_A}$ and $\mathbf{F}_R$ act on the brace at point $A$, Fig. 3–36b, they will produce the *same* external effect or reactions at the supports as that produced by the force system in Fig. 3–36a.

EXAMPLE **3.15**

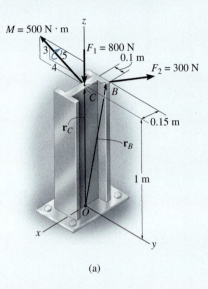

(a)

Fig. 3–37

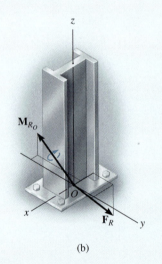

(b)

A structural member is subjected to a couple moment $\mathbf{M}$ and forces $\mathbf{F}_1$ and $\mathbf{F}_2$ as shown in Fig. 3–37a. Replace this system by an equivalent resultant force and couple moment acting at its base, point $O$.

### Solution (*Vector Analysis*)

The three-dimensional aspects of the problem can be simplified by using a Cartesian vector analysis. Expressing the forces and couple moment as Cartesian vectors, we have

$$\mathbf{F}_1 = \{-800\mathbf{k}\} \text{ N}$$

$$\mathbf{F}_2 = (300 \text{ N})\mathbf{u}_{CB} = (300 \text{ N})\left(\frac{r_{CB}}{r_{CB}}\right)$$

$$= 300\left[\frac{-0.15\mathbf{i} + 0.1\mathbf{j}}{\sqrt{(-0.15)^2 + (0.1)^2}}\right] = \{-249.6\mathbf{i} + 166.4\mathbf{j}\} \text{ N}$$

$$\mathbf{M} = -500(\tfrac{4}{5})\mathbf{j} + 500(\tfrac{3}{5})\mathbf{k} = \{-400\mathbf{j} + 300\mathbf{k}\} \text{ N} \cdot \text{m}$$

### *Force Summation.*

$$\mathbf{F}_R = \Sigma\mathbf{F}; \qquad \mathbf{F}_R = \mathbf{F}_1 + \mathbf{F}_2 = -800\mathbf{k} - 249.6\mathbf{i} + 166.4\mathbf{j}$$

$$= \{-249.6\mathbf{i} + 166.4\mathbf{j} - 800\mathbf{k}\} \text{ N} \qquad\qquad Ans.$$

### *Moment Summation.*

$$\mathbf{M}_{R_O} = \Sigma\mathbf{M}_C + \Sigma\mathbf{M}_O$$

$$\mathbf{M}_{R_O} = \mathbf{M} + \mathbf{r}_C \times \mathbf{F}_1 + \mathbf{r}_B \times \mathbf{F}_2$$

$$\mathbf{M}_{R_O} = (-400\mathbf{j} + 300\mathbf{k}) + (1\mathbf{k}) \times (-800\mathbf{k}) + \begin{vmatrix} \mathbf{i} & \mathbf{j} & \mathbf{k} \\ -0.15 & 0.1 & 1 \\ -249.6 & 166.4 & 0 \end{vmatrix}$$

$$= (-400\mathbf{j} + 300\mathbf{k}) + (\mathbf{0}) + (-166.4\mathbf{i} - 249.6\mathbf{j})$$

$$= \{-166\mathbf{i} - 650\mathbf{j} + 300\mathbf{k}\} \text{ N} \cdot \text{m} \qquad\qquad Ans.$$

The results are shown in Fig. 3–37b.

## 3.9  Further Reduction of a Force and Couple System

**Simplification to a Single Resultant Force.** Consider now a special case for which the system of forces and couple moments acting on a rigid body, Fig. 3–38a, reduces at point $O$ to a resultant force $\mathbf{F}_R = \Sigma\mathbf{F}$ and resultant couple moment $\mathbf{M}_{R_O} = \Sigma\mathbf{M}_O$, which are *perpendicular* to one another, Fig. 3–38b. Whenever this occurs, we can further simplify the force and couple moment system by moving $\mathbf{F}_R$ to another point $P$, located either on or off the body so that no resultant couple moment has to be applied to the body, Fig. 3–38c. In other words, if the force and couple moment system in Fig. 3–38a is reduced to a resultant system at point $P$, only the force resultant will have to be applied to the body, Fig. 3–38c.

The location of point $P$, measured from point $O$, can always be determined provided $\mathbf{F}_R$ and $\mathbf{M}_{R_O}$ are known, Fig. 3–38b. As shown in Fig. 3–38c, $P$ must lie on the $bb$ axis, which is perpendicular to both the line of action of $\mathbf{F}_R$ and the $aa$ axis. This point is chosen such that the distance $d$ satisfies the scalar equation $M_{R_O} = F_R d$ or $d = M_{R_O}/F_R$. With $\mathbf{F}_R$ so located, it will produce the same external effects on the body as the force and couple moment system in Fig. 3–38a, or the force and couple moment resultants in Fig. 3–38b.

If a system of forces is either concurrent, coplanar, or parallel, it can always be reduced, as in the above case, to a single resultant force $\mathbf{F}_R$. This is because in each of these cases $\mathbf{F}_R$ and $\mathbf{M}_{R_O}$ will always be perpendicular to each other when the force system is simplified at *any* point $O$.

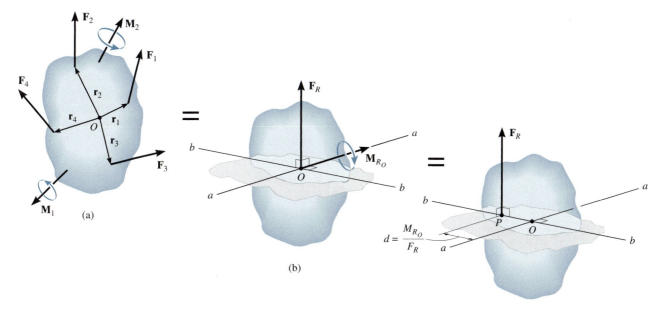

(a)

(b)

(c)

**Fig. 3–38**

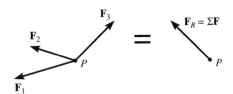

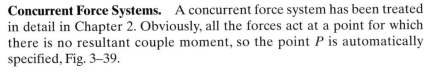

**Fig. 3–39**

**Concurrent Force Systems.** A concurrent force system has been treated in detail in Chapter 2. Obviously, all the forces act at a point for which there is no resultant couple moment, so the point $P$ is automatically specified, Fig. 3–39.

**Coplanar Force Systems.** Coplanar force systems, which may include couple moments directed perpendicular to the plane of the forces as shown in Fig. 3–40a, can be reduced to a single resultant force, because when each force in the system is moved to any point $O$ in the $x$–$y$ plane, it produces a couple moment that is *perpendicular* to the plane, i.e., in the $\pm\mathbf{k}$ direction. The resultant moment $\mathbf{M}_{R_O} = \Sigma\mathbf{M} + \Sigma(\mathbf{r} \times \mathbf{F})$ is thus perpendicular to the resultant force $\mathbf{F}_R$. Fig. 3–40b; and so $\mathbf{F}_R$ can be positioned a distance $d$ from $O$ so as to create this same moment $\mathbf{M}_{R_O}$ about $O$, Fig. 3–40c.

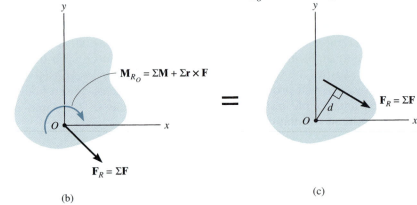

**Fig. 3–40**

**Parallel Force Systems.** Parallel force systems, which can include couple moments that are perpendicular to the forces, as shown in Fig. 3–41a, can be reduced to a single resultant force because when each force is moved to any point $O$ in the $x$–$y$ plane, it produces a couple moment that has components only about the $x$ and $y$ axes. The resultant moment $\mathbf{M}_{R_O} = \Sigma\mathbf{M}_O + \Sigma(\mathbf{r} \times \mathbf{F})$ is thus perpendicular to the resultant force $\mathbf{F}_R$, Fig. 3–41b; and so $\mathbf{F}_R$ can be moved to a point a distance $d$ away so that it produces the same moment about $O$.

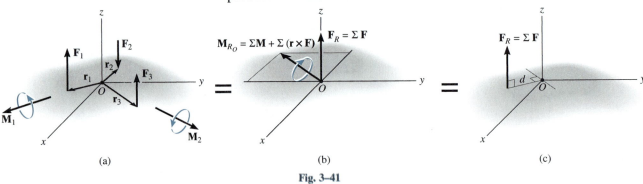

**Fig. 3–41**

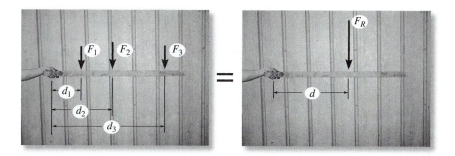

The three parallel forces acting on the stick can be replaced by a single resultant force $F_R$ acting at a distance $d$ from the grip. To be equivalent we require the resultant force to equal the sum of the forces, $F_R = F_1 + F_2 + F_3$, and to find the distance $d$ the moment of the resultant force about the grip must be equal to the moment of all the forces about the grip, $F_R d = F_1 d_1 + F_2 d_2 + F_3 d_3$.

## PROCEDURE FOR ANALYSIS

The technique used to reduce a coplanar or parallel force system to a single resultant force follows a similar procedure outlined in the previous section.

- Establish the $x$, $y$, $z$, axes and locate the resultant force $\mathbf{F}_R$ an arbitrary distance away from the origin of the coordinates.

*Force Summation.*

- The resultant force is equal to the sum of all the forces in the system.

- For a coplanar force system, resolve each force into its $x$ and $y$ components. Positive components are directed along the positive $x$ and $y$ axes, and negative components are directed along the negative $x$ and $y$ axes.

*Moment Summation.*

- The moment of the resultant force about point $O$ is equal to the sum of all the couple moments in the system plus the moments about point $O$ of all the forces in the system.

- This moment condition is used to find the location of the resultant force from point $O$.

**E X A M P L E  3.16**

The beam $AE$ in Fig. 3–42$a$ is subjected to a system of coplanar forces. Determine the magnitude, direction, and location on the beam of a resultant force which is equivalent to the given system of forces measured from $E$.

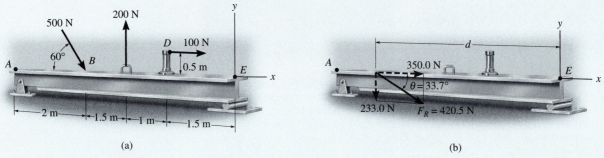

(a)

**Fig. 3–42**

(b)

**Solution**

The origin of coordinates is located at point $E$ as shown in Fig. 3–42$a$.

*Force Summation.* Resolving the 500-N force into $x$ and $y$ components and summing the force components yields

$$\xrightarrow{+} F_{R_x} = \Sigma F_x; \quad F_{R_x} = 500\cos 60°\text{N} + 100\text{ N} = 350.0\text{ N} \rightarrow$$
$$+\uparrow F_{R_y} = \Sigma F_y; \quad F_{R_y} = -500\sin 60°\text{N} + 200\text{ N} = -233.0\text{ N}$$
$$= 233.0\text{ N}\downarrow$$

The magnitude and direction of the resultant force are established from the vector addition shown in Fig. 3–42$b$. We have

$$F_R = \sqrt{(350.0)^2 + (233.0)^2} = 420.5\text{ N} \qquad Ans.$$

$$\theta = \tan^{-1}\left(\frac{233.0}{350.0}\right) = 33.7° \,\,\searrow_\theta \qquad Ans.$$

*Moment Summation.* Moments will be summed about point $E$. Hence, from Figs. 3–42$a$ and 3–42$b$, we require the moments of the components of $\mathbf{F}_R$ (or the moment of $\mathbf{F}_R$) about point $E$ to equal the moments of the force system about E. Assuming positive moments are counterclockwise, we have

$$\zeta+M_{R_E} = \Sigma M_E$$
$$233.0\text{ N}(d) + 350.0\text{ N}(0) = (500\sin 60°\text{N})(4\text{ m}) + (500\cos 60°\text{N})(0)$$
$$-(100\text{ N})(0.5\text{ m}) - (200\text{ N})(2.5\text{ m})$$

$$d = \frac{1182.1}{233.0} = 5.07\text{ m} \qquad Ans.$$

Note that using a clockwise sign convention would yield this same result. Since $d$ is *positive*, $\mathbf{F}_R$ acts to the left of $E$ as shown. Try to solve this problem by summing moments about point $A$ and show $d' = 0.927$ m, measured to the right of $A$.

# EXAMPLE 3.17

The jib crane shown in Fig. 3–43a is subjected to three coplanar forces. Replace this loading by an equivalent resultant force and specify where the resultant's line of action intersects the column $AB$ and boom $BC$.

## Solution

*Force Summation.* Resolving the 250-lb force into $x$ and $y$ components and summing the force components yields

$$\xrightarrow{+} F_{R_x} = \Sigma F_x; \quad F_{R_x} = -250 \text{ lb}(\tfrac{3}{5}) - 175 \text{ lb} = -325 \text{ lb} = 325 \text{ lb} \leftarrow$$

$$+\uparrow F_{R_y} = \Sigma F_y; \quad F_{R_y} = -250 \text{ lb}(\tfrac{4}{5}) - 60 \text{ lb} = -260 \text{ lb} = 260 \text{ lb} \downarrow$$

As shown by the vector addition in Fig. 3–44b,

$$F_R = \sqrt{(325)^2 + (260)^2} = 416 \text{ lb} \qquad \textit{Ans.}$$

$$\theta = \tan^{-1}\left(\frac{260}{325}\right) = 38.7° \; \theta \overleftarrow{\phantom{x}} \qquad \textit{Ans.}$$

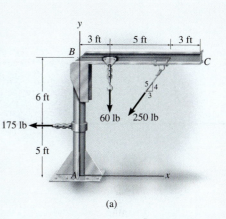

(a)

*Moment Summation.* Moments will be summed about the arbitrary point $A$. Assuming the line of action of $\mathbf{F}_R$ intersects $AB$, Fig. 3–43b, we require the moment of the components of $\mathbf{F}_R$ in Fig. 3–43b about $A$ to equal the moments of the force system in Fig. 3–43a about $A$; i.e.,

$$\downarrow + M_{R_A} = \Sigma M_A; \quad 325 \text{ lb}(y) + 260 \text{ lb}(0)$$

$$= 175 \text{ lb}(5 \text{ ft}) - 60 \text{ lb}(3 \text{ ft}) + 250 \text{ lb}(\tfrac{3}{5})(11 \text{ ft}) - 250 \text{ lb}(\tfrac{4}{5})(8 \text{ ft})$$

$$y = 2.29 \text{ ft} \qquad \textit{Ans.}$$

By the principle of transmissibility, $\mathbf{F}_R$ can also be treated as intersecting $BC$, Fig. 3–44b, in which case we have

$$\downarrow + M_{R_A} = \Sigma M_A; \quad 325 \text{ lb}(11 \text{ ft}) - 260 \text{ lb}(x)$$

$$= 175 \text{ lb}(5 \text{ ft}) - 60 \text{ lb}(3 \text{ ft}) + 250 \text{ lb}(\tfrac{3}{5})(11 \text{ ft}) - 250 \text{ lb}(\tfrac{4}{5})(8 \text{ ft})$$

$$x = 10.9 \text{ ft} \qquad \textit{Ans.}$$

We can also solve for these positions by assuming $\mathbf{F}_R$ acts at the arbitrary point $(x, y)$ on its line of action, Fig. 3–43b. Summing moments about point $A$ yields

$$\downarrow + M_{R_A} = \Sigma M_A; \quad 325 \text{ lb}(y) - 260 \text{ lb}(x)$$

$$= 175 \text{ lb}(5 \text{ ft}) - 60 \text{ lb}(3 \text{ ft}) + 250 \text{ lb}(\tfrac{3}{5})(11 \text{ ft}) - 250 \text{ lb}(\tfrac{4}{5})(8 \text{ ft})$$

$$325y - 260x = 745$$

which is the equation of the colored dashed line in Fig. 3–43b. To find the points of intersection with the crane along $AB$, set $x = 0$, then $y = 2.29$ ft, and along $BC$ set $y = 11$ ft, then $x = 10.9$ ft.

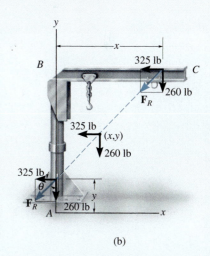

(b)

**Fig. 3–43**

EXAMPLE 3.18

The slab in Fig. 3–44a is subjected to four parallel forces. Determine the magnitude and direction of a resultant force equivalent to the given force system and locate its point of application on the slab.

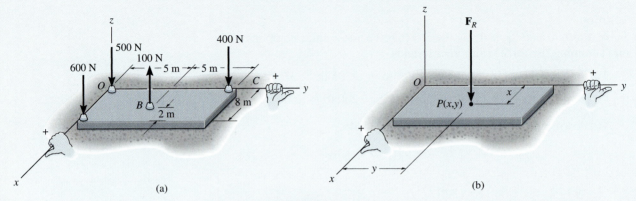

Fig. 3–44

**Solution (*Scalar Analysis*)**
***Force Summation.*** From Fig. 3–44a, the resultant force is

$$+\uparrow F_R = \Sigma F; \quad F_R = -600\ \text{N} + 100\ \text{N} - 400\ \text{N} - 500\ \text{N}$$
$$= -1400\ \text{N} = 1400\ \text{N}\downarrow \qquad \qquad Ans.$$

***Moment Summation.*** We require the moment about the $x$ axis of the resultant force, Fig. 3–44b, to be equal to the sum of the moments about the $x$ axis of all the forces in the system, Fig. 3–44a. The moment arms are determined from the $y$ coordinates since these coordinates represent the perpendicular distances from the $x$ axis to the lines of action of the forces. Using the right-hand rule, where positive moments act in the $+\mathbf{i}$ direction, we have

$$M_{R_x} = \Sigma M_x;$$
$$-(1400\ \text{N})y = 600\ \text{N}(0) + 100\ \text{N}(5\ \text{m}) - 400\ \text{N}(10\ \text{m}) + 500\ \text{N}(0)$$
$$-1400y = -3500 \qquad y = 2.50\ \text{m} \qquad \qquad Ans.$$

In a similar manner, assuming that positive moments act in the $+\mathbf{j}$ direction, a moment equation can be written about the $y$ axis using moment arms defined by the $x$ coordinates of each force.

$$M_{R_y} = \Sigma M_y;$$
$$(1400\ \text{N})x = 600\ \text{N}(8\ \text{m}) - 100\ \text{N}(6\ \text{m}) + 400\ \text{N}(0) + 500\ \text{N}(0)$$
$$1400x = 4200 \qquad x = 3.00\ \text{m} \qquad \qquad Ans.$$

Hence, a force of $F_R = 1400\ \text{N}$ placed at point $P(3.00\ \text{m}, 2.50\ \text{m})$ on the slab, Fig. 3–44b, is equivalent to the parallel force system acting on the slab in Fig. 3–44a.

# EXAMPLE 3.19

Three parallel bolting forces act on the rim of the circular cover plate in Fig. 3–45a. Determine the magnitude and direction of a resultant force equivalent to the given force system and locate its point of application, $P$, on the cover plate.

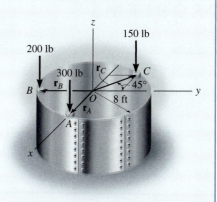

(a)

### Solution (*Vector Analysis*)

*Force Summation.* From Fig. 3–45a, the force resultant $\mathbf{F}_R$ is

$$\mathbf{F}_R = \Sigma\mathbf{F}; \quad \mathbf{F}_R = -300\mathbf{k} - 200\mathbf{k} - 150\mathbf{k}$$
$$= \{-650\mathbf{k}\} \text{ lb} \qquad \qquad \textit{Ans.}$$

*Moment Summation.* Choosing point $O$ as a reference for computing moments and assuming that $\mathbf{F}_R$ acts at a point $P(x, y)$, Fig. 3–45b, we require

$$\mathbf{M}_{R_O} = \Sigma\mathbf{M}_O;$$
$$\mathbf{r} \times \mathbf{F}_R = \mathbf{r}_A \times (-300\mathbf{k}) + \mathbf{r}_B \times (-200\mathbf{k}) + \mathbf{r}_C \times (-150\mathbf{k})$$
$$(x\mathbf{i} + y\mathbf{j}) \times (-650\mathbf{k}) = (8\mathbf{i}) \times (-300\mathbf{k}) + (-8\mathbf{j}) \times (-200\mathbf{k})$$
$$+(-8\sin 45°\mathbf{i} + 8\cos 45°\mathbf{j}) \times (-150\mathbf{k})$$
$$650x\mathbf{j} - 650y\mathbf{i} = 2400\mathbf{j} + 1600\mathbf{i} - 848.5\mathbf{j} - 848.5\mathbf{i}$$

Equating the corresponding $\mathbf{j}$ and $\mathbf{i}$ components yields

$$650x = 2400 - 848.5 \qquad (1)$$
$$-650y = 1600 - 848.5 \qquad (2)$$

Solving these equations, we obtain the coordinates of point $P$,

$$x = 2.39 \text{ ft} \qquad y = -1.16 \text{ ft} \qquad \textit{Ans.}$$

The negative sign indicates that it was wrong to have assumed a $+y$ position for $\mathbf{F}_R$ as shown in Fig. 3–45b.

It is also possible to establish Eqs. 1 and 2 directly by summing moments about the $y$ and $x$ axes. Using the right-hand rule we have

$$M_{R_y} = \Sigma M_y; \qquad 650x = 300 \text{ lb } (8 \text{ ft}) - 150 \text{ lb } (8\sin 45° \text{ ft})$$
$$M_{R_x} = \Sigma M_x; \qquad -650y = 200 \text{ lb } (8 \text{ ft}) - 150 \text{ lb } (8\cos 45° \text{ ft})$$

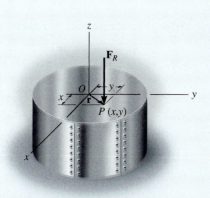

(b)

Fig. 3–45

# PROBLEMS

**3-74.** Replace the force at $A$ by an equivalent force and couple moment at point $O$.

**3-75.** Replace the force at $A$ by an equivalent force and couple moment at point $P$.

**3-78.** Replace the force system by an equivalent force and couple moment at point $O$.

**3-79.** Replace the force system by an equivalent force and couple moment at point $P$.

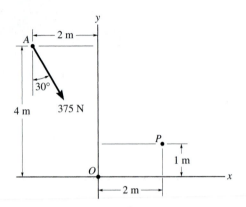

Probs. 3–74/75

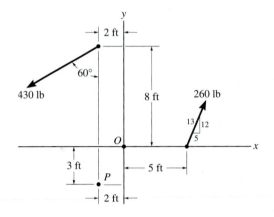

Probs. 3–78/79

**\*3-76.** Replace the force and couple moment system by an equivalent force and couple moment acting at point $O$.

**3-77.** Replace the force and couple moment system by an equivalent force and couple moment acting at point $P$.

**\*3-80.** Replace the force and couple system by an equivalent force and couple moment acting at point $O$.

**3-81.** Replace the force and couple system by an equivalent force and couple moment acting at point $P$.

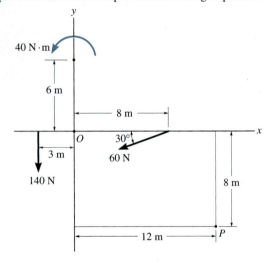

Probs. 3–76/77

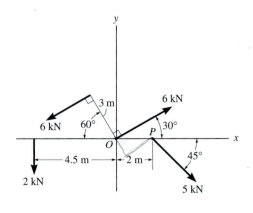

Probs. 3–80/81

**3-82.** Replace the force and couple system by an equivalent force and couple moment at point *O*.

**3-83.** Replace the force and couple system by an equivalent force and couple moment at point *P*.

**3-86.** Replace the force system acting on the beam by an equivalent force and couple moment at point *A*.

**3-87.** Replace the force system acting on the beam by an equivalent force and couple moment at point *B*.

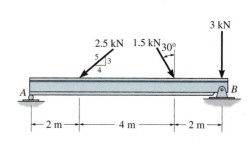

Probs. 3–86/87

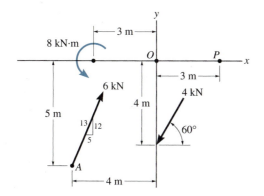

Probs. 3–82/83

**\*3-84.** Replace the force system by a single force resultant and specify its point of application, measured along the *x* axis from point *O*.

**3-85.** Replace the force system by a single force resultant and specify its point of application, measured along the *x* axis from point *P*.

**\*3-88.** Replace the three forces acting on the shaft by a single resultant force. Specify where the force acts, measured from end *A*.

**3-89.** Replace the three forces acting on the shaft by a single resultant force. Specify where the force acts, measured from end *B*.

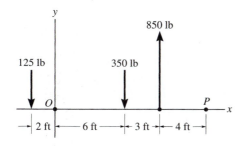

Probs. 3–84/85

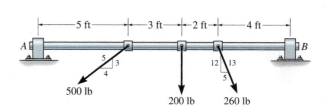

Probs. 3–88/89

**3-90.** Replace the loading on the frame by a single resultant force. Specify where its line of action intersects member $AB$, measured from $A$.

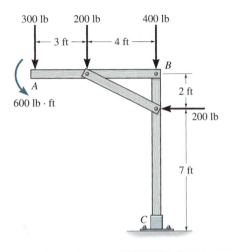

**Prob. 3–90**

**3-91.** Replace the loading acting on the beam by a single resultant force. Specify where the force acts, measured from end $A$.

**\*3-92.** Replace the loading acting on the beam by a single resultant force. Specify where the force acts, measured from $B$.

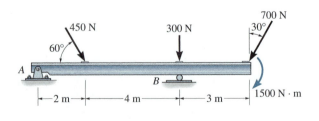

**Probs. 3–91/92**

**3-93.** Determine the magnitudes of $\mathbf{F}_1$ and $\mathbf{F}_2$ and the direction of $\mathbf{F}_1$ so that the loading creates a zero resultant force and couple moment on the wheel.

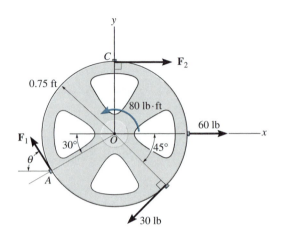

**Prob. 3–93**

**3-94.** The weights of the various components of the truck are shown. Replace this system of forces by an equivalent resultant force and couple moment acting at point $A$.

**3-95.** The weights of the various components of the truck are shown. Replace this system of forces by an equivalent resultant force and specify its location measured from point $A$.

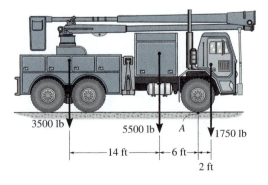

**Probs. 3–94/95**

**\*3-96.** Replace the force system by an equivalent force and couple moment at point $A$.

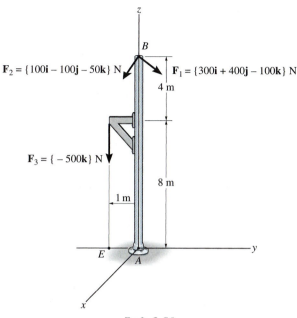

$F_2 = \{100\mathbf{i} - 100\mathbf{j} - 50\mathbf{k}\}$ N

$F_1 = \{300\mathbf{i} + 400\mathbf{j} - 100\mathbf{k}\}$ N

$F_3 = \{-500\mathbf{k}\}$ N

4 m

8 m

1 m

**Prob. 3–96**

**3-97.** The slab is to be hoisted using the three slings shown. Replace the system of forces acting on slings by an equivalent force and couple moment at point $O$. The force $\mathbf{F}_1$, is vertical.

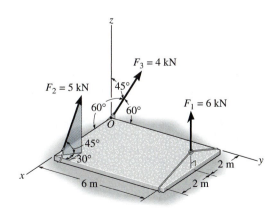

$F_3 = 4$ kN

$F_2 = 5$ kN

$45°$

$60°$

$60°$

$F_1 = 6$ kN

$45°$

$30°$

6 m

2 m

2 m

**Prob. 3–97**

**3-98.** A biomechanical model of the lumbar region of the human trunk is shown. The forces acting in the four muscle groups consist of $F_R = 35$ N for the rectus, $F_O = 45$ N for the oblique, $F_L = 23$ N for the lumbar latissimus dorsi, and $F_E = 32$ N for the erector spinae. These loadings are symmetric with respect to the $y$–$z$ plane. Replace this system of parallel forces by an equivalent force and couple moment acting at the spine, point $O$. Express the results in Cartesian vector form.

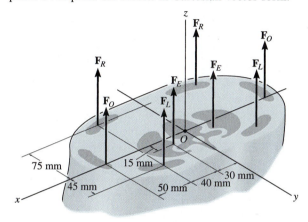

$F_R$ $F_O$

$F_R$ $F_E$ $F_L$

$F_E$

$F_O$ $F_L$

$O$

75 mm $\quad$ 15 mm

45 mm $\quad$ 50 mm $\quad$ 40 mm $\quad$ 30 mm

**Prob. 3–98**

**3-99.** The building slab is subjected to four parallel column loadings. Determine the equivalent resultant force and specify its location $(x, y)$ on the slab. Take $F_1 = 30$ kN, $F_2 = 40$ kN.

**\*3-100.** The building slab is subjected to four parallel column loadings. Determine the equivalent resultant force and specify its location $(x, y)$ on the slab. Take $F_1 = 20$ kN, $F_2 = 50$ kN.

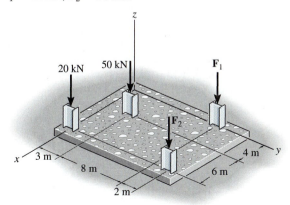

20 kN $\quad$ 50 kN $\quad$ $F_1$

$F_2$

3 m

8 m $\quad$ 4 m

6 m

2 m

**Probs. 3–99/100**

## CHAPTER REVIEW

- *Moment of a Force.* A force produces a turning effect about a point $O$ that does not lie on its line of action. In scalar form, the moment *magnitude* is $M_O = Fd$, where $d$ is the moment arm or perpendicular distance from point $O$ to the line of action of the force. The *direction* of the moment is defined using the right-hand rule. Rather than finding $d$, it is normally easier to resolve the force into its $x$ and $y$ components, determine the moment of each component about the point, and then sum the results. Since three-dimensional geometry is generally more difficult to visualize, the vector cross product can be used to determine the moment, $\mathbf{M}_O = \mathbf{r} \times \mathbf{F}$, where $\mathbf{r}$ is a position vector that extends from point $O$ to any point on the line of action of $\mathbf{F}$.

- *Moment about a Specified Axis.* If the moment of a force is to be determined about an arbitrary axis, then the projection of the moment onto the axis must be obtained. Provided the distance $d_a$ that is perpendicular to *both* the line of action of the force and the axis can be determined, then the moment of the force about the axis is simply $M_a = Fd_a$. If this distance $d_a$ cannot be found, then the vector triple product should be used, where $M_a = \mathbf{u}_a \cdot \mathbf{r} \times \mathbf{F}$. Here $\mathbf{u}_a$ is the unit vector that specifies the direction of the axis and $\mathbf{r}$ is a position vector that is directed from any point on the axis to any point on the line of action of the force.

- *Couple Moment.* A couple consists of two equal but opposite forces that act a perpendicular distance $d$ apart. Couples tend to produce a rotation without translation. The moment of the couple is determined from $M = Fd$, and its direction is established using the right-hand rule. If the vector cross product is used to determine the moment of the couple then $\mathbf{M} = \mathbf{r} \times \mathbf{F}$. Here $\mathbf{r}$ extends from any point on the line of action of one of the forces to any point on the line of action of the force $\mathbf{F}$ used in the cross product.

- *Reduction of a Force and Couple System.* Any system of forces and couples can be reduced to a single resultant force and resultant couple moment acting at a point. The resultant force is the sum of all the forces in the system, and the resultant couple moment is equal to the sum of all the forces and couple moments about the point. Further simplification to a single resultant force is possible provided the force system is *concurrent, coplanar,* or *parallel.* For this case, to find the location of the resultant force from a point, it is necessary to equate the moment of the resultant force about the point to the moment of the forces and couples in the system about the same point.

## REVIEW PROBLEMS

**3-101.** Determine the moment about point $A$ of each of the three forces acting on the beam. What is the resultant moment of all the forces about point $A$?

**3-102.** Determine the moment about point $B$ of each of the three forces acting on the beam. What is the resultant moment of all the forces about point $B$?

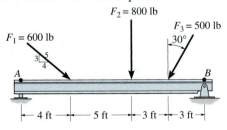

**Probs. 3–101/102**

**3-103.** Determine the moment of the force $\mathbf{F}$ which is directed along $ED$ of the block about the axis $BA$. Solve the problem by using two different position vectors $\mathbf{r}$. Express the result as a Cartesian vector.

**\*3-104.** Determine the moment of the force $\mathbf{F}$ which is directed along $ED$ of the block about the axis $OC$. Solve the problem by using two different position vectors $\mathbf{r}$. Express the result as a Cartesian vector.

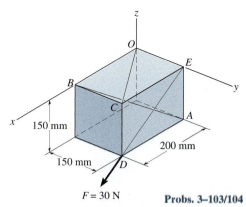

F = 30 N

**Probs. 3–103/104**

**3-105.** The main beam along the wing of an airplane is swept back at an angle of 25°. From load calculations it is determined that the beam is subjected to couple moments $M_x = 23$ kN·m and $M_y = 34$ kN·m. Determine the resultant couple moments created about the $x'$ and $y'$ axes. The axes all lie in the same horizontal plane.

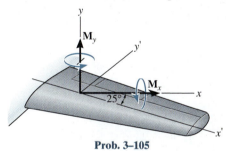

**Prob. 3–105**

**3-106.** Determine the moment of the force $\mathbf{F}_c$ about the door hinge at $A$. Express the result as a Cartesian vector.

**3-107.** Determine the magnitude of the moment of the force $\mathbf{F}_c$ about the hinged axis $aa$ of the door.

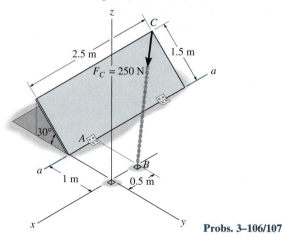

**Probs. 3–106/107**

*3-108.** The belt passing over the pulley is subjected to forces $\mathbf{F}_1$ and $\mathbf{F}_2$, each having a magnitude of 53 N. $\mathbf{F}_1$ acts in the $-\mathbf{k}$ direction. Replace these forces by an equivalent force and couple moment at point $A$. Express the result in Cartesian vector form. Set $\theta = 0°$ so that $\mathbf{F}_2$ acts in the $-\mathbf{j}$ direction.

**3-109.** The belt passing over the pulley is subjected to two forces $\mathbf{F}_1$ and $\mathbf{F}_2$, each having a magnitude of 53 N. $\mathbf{F}_1$ acts in the $-\mathbf{k}$ direction. Replace these forces by an equivalent force and couple moment at point $A$. Express the result in Cartesian vector form. Take $\theta = 30°$.

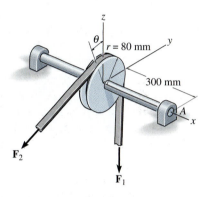

**Probs. 3–108/109**

**3-110.** The forces and couple moments that are exerted on the toe and heel plates of a snow ski are $\mathbf{F}_t = \{-50\mathbf{i} + 80\mathbf{j} - 158\mathbf{k}\}$ N, $\mathbf{M}_t = \{-6\mathbf{i} + 4\mathbf{j} + 2\mathbf{k}\}$ N·m, and $\mathbf{F}_h = \{-20\mathbf{i} + 60\mathbf{j} - 250\mathbf{k}\}$ N, $\mathbf{M}_h = \{-20\mathbf{i} + 8\mathbf{j} + 3\mathbf{k}\}$ N·m, respectively. Replace this system by an equivalent force and couple moment acting at point $P$. Express the results in Cartesian vector form.

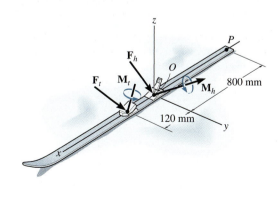

**Prob. 3–110**

The tower crane is subjected to its weight and the load it supports. In order to calculate the support reactions for the crane, it is necessary to apply the principles of equilibrium.

# CHAPTER 4

# Equilibrium of a Rigid Body

## CHAPTER OBJECTIVES

- To develop the equations of equilibrium for a rigid body.
- To introduce the concept of the free-body diagram for a rigid body.
- To show how to solve rigid body equilibrium problems using the equations of equilibrium.
- To develop the concepts of static and kinetic friction with applications to equilibrium problems.

## 4.1 Conditions for Equilibrium

In this section we will develop both the necessary and sufficient conditions required for equilibrium of a rigid body. To do this, consider the rigid body in Fig. 4–1a, which is fixed in the *x, y, z* reference and is either at rest or moves with the reference at constant velocity. A free-body diagram of the arbitrary *i*th particle of the body is shown in Fig. 4–1b. There are two types of forces which act on it. The resultant *internal force*, $\mathbf{f}_i$, is caused by interactions with adjacent particles. The resultant *external force* $\mathbf{F}_i$ represents, for example, the effects of gravitational, electrical, magnetic, or contact forces between the *i*th particle and adjacent bodies or particles *not* included within the body. If the particle is in equilibrium, then applying Newton's first law we have

$$\mathbf{F}_i + \mathbf{f}_i = \mathbf{0}$$

When the equation of equilibrium is applied to each of the other particles of the body, similar equations will result. If all these equations are added together *vectorially*, we obtain

$$\Sigma \mathbf{F}_i + \Sigma \mathbf{f}_i = \mathbf{0}$$

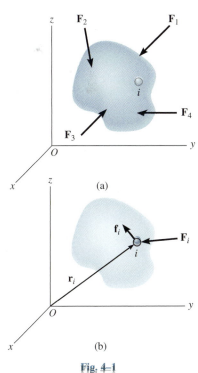

(a)

(b)

Fig. 4–1

139

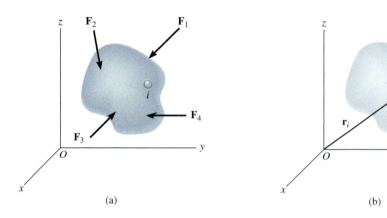

(a)                    (b)

Fig. 4–1

The summation of the internal forces will equal zero since the internal forces between particles within the body will occur in equal but opposite collinear pairs, Newton's third law. Consequently, only the sum of the *external forces* will remain; and therefore, letting $\Sigma \mathbf{F}_i = \Sigma \mathbf{F}$, the above equation can be written as

$$\Sigma \mathbf{F} = \mathbf{0}$$

Let us now consider the moments of the forces acting on the $i$th particle about the arbitrary point $O$, Fig. 4–1$b$. Using the above particle equilibrium equation and the distributive law of the vector cross product we have

$$\mathbf{r}_i \times (\mathbf{F}_i + \mathbf{f}_i) = \mathbf{r}_i \times \mathbf{F}_i + \mathbf{r}_i \times \mathbf{f}_i = \mathbf{0}$$

Similar equations can be written for the other particles of the body, and adding them together vectorially, we obtain

$$\Sigma(\mathbf{r}_i \times \mathbf{F}_i) + \Sigma(\mathbf{r}_i \times \mathbf{f}_i) = \mathbf{0}$$

The second term is zero since, as stated above, the internal forces occur in equal but opposite collinear pairs, and therefore the resultant moment of each pair of forces about point $O$ is zero. Hence, using the notation $\Sigma \mathbf{M}_O = \Sigma(\mathbf{r}_i \times \mathbf{F}_i)$ we have

$$\Sigma \mathbf{M}_O = \mathbf{0}$$

Hence the two *equations of equilibrium* for a rigid body can be summarized as follows:

$$\begin{array}{c} \Sigma \mathbf{F} = \mathbf{0} \\ \Sigma \mathbf{M}_O = \mathbf{0} \end{array} \qquad (4\text{–}1)$$

These equations require that a rigid body will remain in equilibrium provided the sum of all the *external forces* acting on the body is equal to zero and the sum of the moments of the external forces about a point is equal to zero. The fact that these conditions are *necessary* for equilibrium has now been proven. They are also *sufficient* for maintaining equilibrium. To show this, let us assume that the body is in equilibrium and the force system acting on the body satisfies Eqs. 4–1. Suppose that an *additional force* $\mathbf{F}'$ is applied to the body. As a result, the equilibrium equations become

$$\Sigma \mathbf{F} + \mathbf{F}' = \mathbf{0}$$

$$\Sigma \mathbf{M}_O + \mathbf{M}'_O = \mathbf{0}$$

where $\mathbf{M}'_O$ is the moment of $\mathbf{F}'$ about $O$. Since $\Sigma \mathbf{F} = \mathbf{0}$ and $\Sigma \mathbf{M}_O = \mathbf{0}$, then we require $\mathbf{F}' = \mathbf{0}$ (also $\mathbf{M}'_O = \mathbf{0}$). Consequently, the additional force $\mathbf{F}'$ is not required, and indeed Eqs. 4–1 are also sufficient conditions for maintaining equilibrium.

Many types of engineering problems involve symmetric loadings and can be solved by projecting all the forces acting on a body onto a single plane. Hence, in the next section, the equilibrium of a body subjected to a *coplanar* or *two-dimensional force system* will be considered. Ordinarily the geometry of such problems is not very complex, so a scalar solution is suitable for analysis. The more general discussion of rigid bodies subjected to *three-dimensional force systems* is given in the latter part of this chapter. It will be seen that many of these types of problems can best be solved by using vector analysis.

# Equilibrium in Two Dimensions

## 4.2 Free-Body Diagrams

Successful application of the equations of equilibrium requires a complete specification of *all* the known and unknown external forces that act *on* the body. The best way to account for these forces is to draw the body's free-body diagram. This diagram is a sketch of the outlined shape of the body, which represents it as being *isolated* or "free" from its surroundings, i.e., a "free body." On this sketch it is necessary to show *all* the forces and couple moments that the surroundings exert *on the body* so that these effects can be accounted for when the equations of equilibrium are applied. For this reason, *a thorough understanding of how to draw a free-body diagram is of primary importance for solving problems in mechanics.*

## TABLE 4–1 • Supports for Rigid Bodies Subjected to Two-Dimensional Force Systems

| Types of Connection | Reaction | Number of Unknowns |
|---|---|---|
| (1) <br> cable | | One unknown. The reaction is a tension force which acts away from the member in the direction of the cable. |
| (2) <br> weightless link | or | One unknown. The reaction is a force which acts along the axis of the link. |
| (3) <br> roller | | One unknown. The reaction is a force which acts perpendicular to the surface at the point of contact. |
| (4) <br> roller or pin in confined smooth slot | or | One unknown. The reaction is a force which acts perpendicular to the slot. |
| (5) <br> rocker | | One unknown. The reaction is a force which acts perpendicular to the surface at the point of contact. |
| (6) <br> smooth contacting surface | | One unknown. The reaction is a force which acts perpendicular to the surface at the point of contact. |
| (7) <br> member pin connected to collar on smooth rod | or | One unknown. The reaction is a force which acts perpendicular to the rod. |

continued

**TABLE 4–1 • Continued**

| Types of Connection | Reaction | Number of Unknowns |
|---|---|---|
| (8) 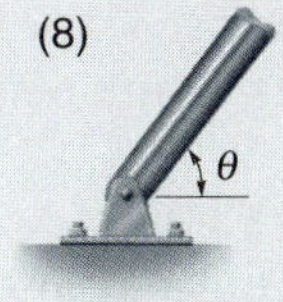 smooth pin or hinge | 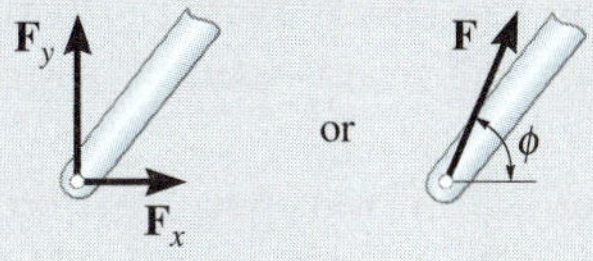 $F_y$, $F_x$ or $F$, $\phi$ | Two unknowns. The reactions are two components of force, or the magnitude and direction $\phi$ of the resultant force. Note that $\phi$ and $\theta$ are not necessarily equal [usually not, unless the rod shown is a link as in (2)]. |
| (9) 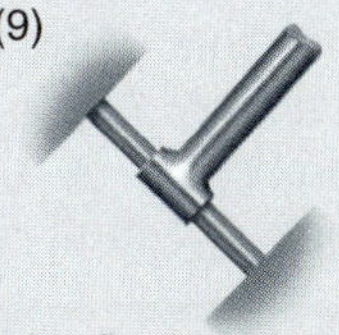 member fixed connected to collar on smooth rod | 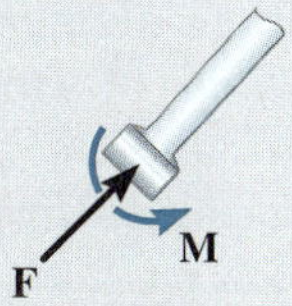 $F$, $M$ | Two unknowns. The reactions are the couple moment and the force which acts perpendicular to the rod. |
| (10) 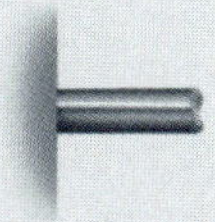 fixed support | 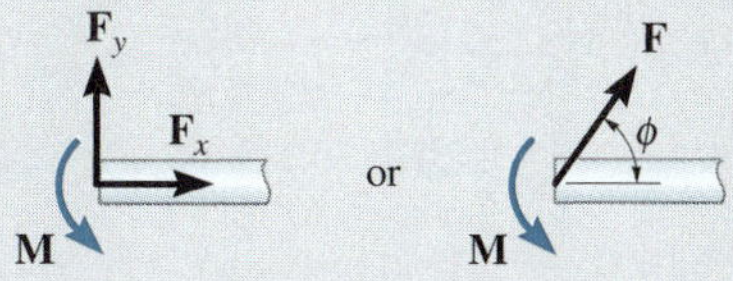 $F_y$, $F_x$, $M$ or $F$, $\phi$, $M$ | Three unknowns. The reactions are the couple moment and the two force components, or the couple moment and the magnitude and direction $\phi$ of the resultant force. |

**Support Reactions.**  Before presenting a formal procedure as to how to draw a free-body diagram, we will first consider the various types of reactions that occur at supports and points of support between bodies subjected to coplanar force systems. *As a general rule, if a support prevents the translation of a body in a given direction, then a force is developed on the body in that direction. Likewise, if rotation is prevented, a couple moment is exerted on the body.*

For example, let us consider three ways in which a horizontal member, such as a beam, is supported at its end. One method consists of a *roller* or cylinder, Fig. 4–2a. Since this support only prevents the beam from *translating* in the vertical direction, the roller can only exert a *force* on the beam in this direction, Fig. 4–2b.

The beam can be supported in a more restrictive manner by using a *pin* as shown in Fig. 4–3a. The pin passes through a hole in the beam and two leaves which are fixed to the ground. Here the pin can prevent *translation* of the beam in *any direction* $\phi$, Fig. 4–3b, and so the pin must exert a *force* **F** on the beam in this direction. For purposes of analysis, it is generally easier to represent this resultant force **F** by its two components **F**$_x$ and **F**$_y$, Fig. 4–3c. If $F_x$ and $F_y$ are known, then $F$ and $\phi$ can be calculated.

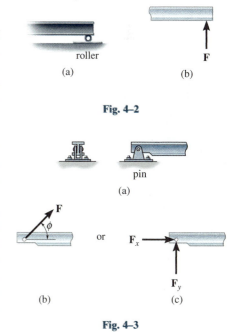

roller

(a)              (b)

**Fig. 4–2**

pin

(a)

$F$, $\phi$  or  $F_x$, $F_y$

(b)              (c)

**Fig. 4–3**

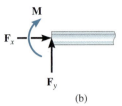

fixed support

(a)

M

$\mathbf{F}_x$

$\mathbf{F}_y$

(b)

**Fig. 4–4**

The most restrictive way to support the beam would be to use a *fixed support* as shown in Fig. 4–4a. This support will prevent both *translation and rotation* of the beam, and so to do this a *force and couple moment* must be developed on the beam at its point of connection, Fig. 4–4b. As in the case of the pin, the force is usually represented by its components $\mathbf{F}_x$ and $\mathbf{F}_y$.

Table 4–1 lists other common types of supports for bodies subjected to coplanar force systems. (In all cases the angle $\theta$ is assumed to be known.) Carefully study each of the symbols used to represent these supports and the types of reactions they exert on their contacting members. Although concentrated forces and couple moments are shown in this table, they actually represent the *resultants* of small *distributed surface loads* that exist between each support and its contacting member. It is these *resultants* which will be determined from the equations of equilibrium.

Typical examples of actual supports that are referenced to Table 4–1 are shown in the following sequence of photos.

The cable exerts a force on the bracket in the direction of the cable. (1)

The rocker support for this bridge girder allows horizontal movement so the bridge is free to expand and contract due to temperature. (5)

This concrete girder rests on the ledge that is assumed to act as a smooth contacting surface. (6)

This utility building is pin supported at the top of the column. (8)

The floor beams of this building are welded together and thus form fixed connections. (10)

**Springs.**   If a *linear elastic spring* is used to support a body, the length of the spring will change in direct proportion to the force acting on it. A characteristic that defines the "elasticity" of a spring is the *spring constant* or *stiffness k*. Specifically, the magnitude of force developed by a linear elastic spring which has a stiffness $k$, and is deformed (elongated or compressed) a distance $s$ measured from its unloaded position, is

$$F = ks$$                                                    (4–2)

Note that $s$ is determined from the difference in the spring's deformed length $l$ and its undeformed length $l_0$, i.e., $s = l - l_0$. Thus, if $s$ is positive, $\mathbf{F}$ "pulls" on the spring; whereas if $s$ is negative, $\mathbf{F}$ must "push" on it. For example, the spring shown in Fig. 4–5 has an undeformed length $l_0 = 0.4$ m and stiffness $k = 500\ \text{N/m}$. To stretch it so that $l = 0.6$ m, a force $F = ks = (500\ \text{N/m})(0.6\ \text{m} - 0.4\ \text{m}) = 100\ \text{N}$ is needed. Likewise, to compress it to a length $l = 0.2$ m, a force $F = ks = (500\ \text{N/m})(0.2\ \text{m} - 0.4\ \text{m}) = -100\ \text{N}$ is required, Fig. 4–5.

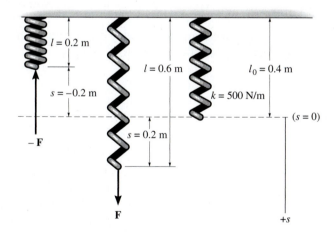

**Fig. 4–5**

**External and Internal Forces.** Since a rigid body is a composition of particles, both *external* and *internal* loadings may act on it. It is important to realize, however, that if the free-body diagram for the body is drawn, the forces that are *internal* to the body are *not represented* on the free-body diagram. As discussed in Sec. 4.1, these forces always occur in equal but opposite collinear pairs, and therefore their *net effect* on the body is zero.

In some problems, a free-body diagram for a "system" of connected bodies may be used for an analysis. An example would be the free-body diagram of an entire automobile (system) composed of its many parts. Obviously, the connecting forces between its parts would represent *internal forces* which would *not* be included on the free-body diagram of the automobile. To summarize, internal forces act between particles which are contained within the boundary of the free-body diagram. Particles or bodies outside this boundary exert external forces on the system, and these alone must be shown on the free-body diagram.

**Weight and the Center of Gravity.** When a body is subjected to a gravitational field, then each of its particles has a specified weight. For the entire body it is appropriate to consider these gravitational forces to be represented as a *system of parallel forces* acting on all the particles contained within the boundary of the body. It was shown in Sec. 3.9 that such a system can be reduced to a single resultant force acting through a specified point. We refer to this force resultant as the *weight* **W** of the body and to the location of its point of application as the *center of gravity*. The methods used for its calculation will be developed in Chapter 6.

In the examples and problems that follow, if the weight of the body is important for the analysis, this force will then be reported in the problem statement. Also, when the body is *uniform* or made of homogeneous material, the center of gravity will be located at the body's *geometric center* or *centroid*; however, if the body is nonhomogeneous or has an unusual shape, then the location of its center of gravity will be given.

**Idealized Models.** In order to perform a correct force analysis of any object, it is important to consider a corresponding analytical or idealized model that gives results that approximate as closely as possible the actual situation. To do this, careful choices have to be made so that selection of the type of supports, the material behavior, and the object's dimensions can be justified. This way the engineer can feel confident that any design or analysis will yield results which can be trusted. In complex cases this process may require developing several different models of the object that must be analyzed, but in any case, this selection process requires both skill and experience.

To illustrate what is required to develop a proper model, we will now consider a few cases. As shown in Fig. 4–6a, the steel beam is to be used to support the roof joists of a building. For a force analysis it is reasonable to assume the material is rigid since only very small deflections will occur when the beam is loaded. A bolted connection at A will allow for any slight rotation that occurs when the load is applied, and so a *pin* can be considered for this support. At B a *roller* can be considered since the support offers no resistance to horizontal movement here. Building code requirements are used to specify the roof loading which results in a calculation of the joist loads **F**. These forces will be larger than any actual loading on the beam since they account for extreme loading cases and for dynamic or vibrational effects. The weight of the beam is generally neglected when it is small compared to the load the beam supports. The idealized model of the beam is shown with average dimensions a, b, c, and d in Fig. 4–6b.

As a second case, consider the lift boom in Fig. 4–7a. By inspection, it is supported by a pin at A and by the hydraulic cylinder BC, which can be approximated as a weightless link. The material can be assumed rigid, and with its density known, the weight of the boom and the location of its center of gravity G are determined. When a design loading **P** is specified, the idealized model shown in Fig. 4–7b can be used for a force analysis. Average dimensions (not shown) are used to specify the location of the loads and the supports.

Idealized models of specific objects will be given in some of the examples throughout the text. It should be realized, however, that each case represents the reduction of a practical situation using simplifying assumptions like the ones illustrated here.

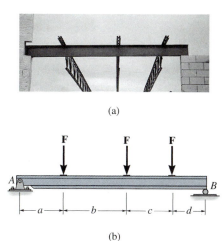

(a)

(b)

**Fig. 4–6**

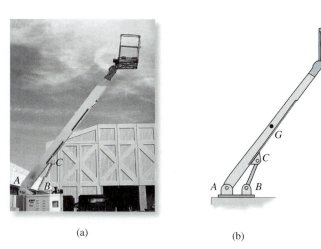

(a)          (b)

**Fig. 4–7**

## PROCEDURE FOR DRAWING A FREE-BODY DIAGRAM

To construct a free-body diagram for a rigid body or group of bodies considered as a single system, the following steps should be performed:

*Draw Outlined Shape.* Imagine the body to be *isolated* or cut "free" from its constraints and connections and draw (sketch) its outlined shape.

*Show All Forces and Couple Moments.* Identify all the external forces and couple moments that act on the body. Those generally encountered are due to (1) applied loadings, (2) reactions occurring at the supports or at points of contact with other bodies (see Table 4–1), and (3) the weight of the body. To account for all these effects, it may help to trace over the boundary, carefully noting each force or couple moment acting on it.

*Identify Each Loading and Give Dimensions.* The forces and couple moments that are known should be labeled with their proper magnitudes and directions. Letters are used to represent the magnitudes and direction angles of forces and couple moments that are *unknown*. Establish an *x, y* coordinate system so that these unknowns, $A_x$, $B_y$, etc., can be identified. Indicate the dimensions of the body necessary for calculating the moments of forces.

## IMPORTANT POINTS

- No equilibrium problem should be solved without *first* drawing the free-body diagram, so as to account for all the forces and couple moments that act on the body.

- If a support *prevents translation* of a body in a particular direction, then the support exerts a *force* on the body in that direction.

- If *rotation is prevented*, then the support exerts a *couple moment* on the body.

- Study Table 4–1.

- Internal forces are never shown on the free-body diagram, since they occur in equal but opposite collinear pairs and therefore cancel out.

- The weight of a body is an external force, and its effect is shown as a single resultant force acting through the body's center of gravity *G*.

- *Couple moments* can be placed anywhere on the free-body diagram, since they are *free vectors*. *Forces* can act at any point along their lines of action, since they are *sliding vectors*.

# EXAMPLE 4.1

Draw the free-body diagram of the uniform beam shown in Fig. 4–8a. The beam has a mass of 100 kg.

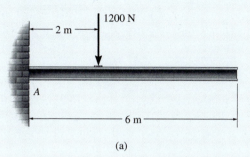

(a)

## Solution

The free-body diagram of the beam is shown in Fig. 4–8b. Since the support at $A$ is a fixed wall, there are three reactions acting *on the beam* at $A$, denoted as $\mathbf{A}_x$, $\mathbf{A}_y$, and $\mathbf{M}_A$ drawn in an arbitrary direction. The magnitudes of these vectors are *unknown*, and their sense has been *assumed*. The weight of the beam, $W = 100(9.81) = 981$ N, acts through the beam's center of gravity $G$, which is 3 m from $A$ since the beam is uniform.

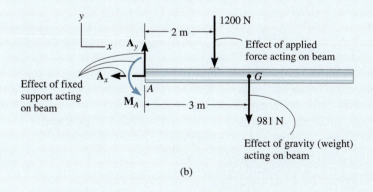

(b)

**Fig. 4–8**

**E X A M P L E 4.2**

Draw the free-body diagram of the foot lever shown in Fig. 4–9a. The operator applies a vertical force to the pedal so that the spring is stretched 1.5 in and the force in the short link at $B$ is 20 lb.

(a)

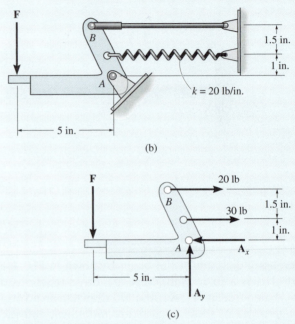

(b)

$k = 20$ lb/in.

1.5 in.

1 in.

F

B

A

5 in.

20 lb

30 lb

1.5 in.

1 in.

$\mathbf{A}_x$

$\mathbf{A}_y$

F

B

A

5 in.

(c)

Fig. 4–9

**Solution**

By inspection, the lever is loosely bolted to the frame at $A$. The rod at $B$ is pinned at its ends and acts as a "short link." After making the proper measurements, the idealized model of the lever is shown in Fig. 4–9b. From this the free-body diagram must be drawn. As shown in Fig. 4–9c, the pin support at $A$ exerts force components $\mathbf{A}_x$ and $\mathbf{A}_y$ on the lever, where each component has a known line of action but unknown magnitude. The link at $B$ exerts a force of 20 lb, acting in the direction of the link. In addition the spring also exerts a horizontal force on the lever. If the stiffness is measured and found to be $k = 20$ lb/in., then since the stretch $s = 1.5$ in., using Eq. 4–2, $F_s = ks = (20 \text{ lb/in.})(1.5 \text{ in.}) = 30$ lb. Finally, the operator's shoe applies a vertical force of $\mathbf{F}$ on the pedal. The dimensions of the lever are also shown on the free-body diagram, since this information will be useful when computing the moments of the forces. As usual, the senses of the unknown forces at $A$ have been assumed. The correct senses will become apparent after solving the equilibrium equations.

# EXAMPLE 4.3

Two smooth pipes, each having a mass of 300 kg, are supported by the forks of the tractor in Fig. 4–10a. Draw the free-body diagrams for each pipe and both pipes together.

(a)

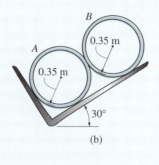

(b)

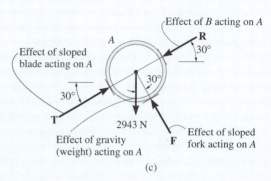

(c)

**Fig. 4–10**

## Solution

The idealized model from which we must draw the free-body diagrams is shown in Fig. 4–10b. Here the pipes are identified, the dimensions have been added, and the physical situation reduced to its simplest form.

The free-body diagram for pipe A is shown in Fig. 4–10c. Its weight is $W = 300(9.81) = 2943$ N. Assuming all contacting surfaces are *smooth*, the reactive forces **T**, **F**, **R** act in a direction *normal* to the tangent at their surfaces of contact.

The free-body diagram of pipe B is shown in Fig. 4–10d. Can you identify each of the three forces acting *on this pipe*? In particular, note that **R**, representing the force of A on B, Fig. 4–10d, is equal and opposite to **R** representing the force of B on A, Fig. 4–10c. This is a consequence of Newton's third law of motion.

The free-body diagram of both pipes combined ("system") is shown in Fig. 4–10e. Here the contact force **R**, which acts between A and B, is considered as an *internal* force and hence is not shown on the free-body diagram. That is, it represents a pair of equal but opposite collinear forces which cancel each other.

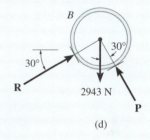

(d)

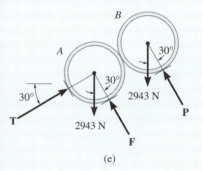

(e)

# 4.3 Equations of Equilibrium

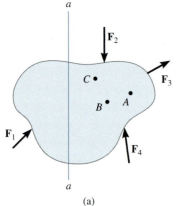

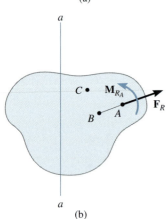

(a)

In Sec. 4.1, we developed the two equations which are both necessary and sufficient for the equilibrium of a rigid body, namely, $\Sigma \mathbf{F} = \mathbf{0}$ and $\Sigma \mathbf{M}_O = \mathbf{0}$. When the body is subjected to a system of forces, which all lie in the $x$–$y$ plane, then the forces can be resolved into their $x$ and $y$ components. Consequently, the conditions for equilibrium in two dimensions are

$$
\boxed{
\begin{aligned}
\Sigma F_x &= 0 \\
\Sigma F_y &= 0 \\
\Sigma M_O &= 0
\end{aligned}
}
\tag{4–3}
$$

Here $\Sigma F_x$ and $\Sigma F_y$ represent, respectively, the algebraic sums of the $x$ and $y$ components of all the forces acting on the body, and $\Sigma M_O$ represents the algebraic sum of the couple moments and the moments of all the force components about an axis perpendicular to the $x$–$y$ plane and passing through the arbitrary point $O$, which may lie either on or off the body.

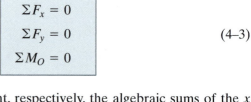

(b)

**Alternative Sets of Equilibrium Equations.** Although Eqs. 4–3 are *most often* used for solving coplanar equilibrium problems, two *alternative* sets of three independent equilibrium equations may also be used. One such set is

$$
\boxed{
\begin{aligned}
\Sigma F_a &= 0 \\
\Sigma M_A &= 0 \\
\Sigma M_B &= 0
\end{aligned}
}
\tag{4–4}
$$

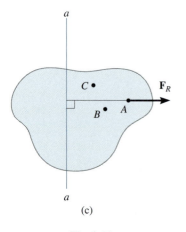

(c)

**Fig. 4–11**

When using these equations, it is required that a line passing through points $A$ and $B$ is *not perpendicular* to the $a$ axis. To prove that Eqs. 4–4 provide the *conditions* for equilibrium, consider the free-body diagram of an arbitrarily shaped body shown in Fig. 4–11a. Using the methods of Sec. 3.8, all the forces on the free-body diagram may be replaced by an equivalent resultant force $\mathbf{F}_R = \Sigma \mathbf{F}$, acting at point $A$, and a resultant couple moment $\mathbf{M}_{R_A} = \Sigma \mathbf{M}_A$, Fig. 4–11b. If $\Sigma M_A = 0$ is satisfied, it is necessary that $\mathbf{M}_{R_A} = \mathbf{0}$. Furthermore, in order that $\mathbf{F}_R$ satisfy $\Sigma F_a = 0$, it must have *no component* along the $a$ axis, and therefore its line of action must be perpendicular to the $a$ axis, Fig. 4–11c. Finally, if it is required that $\Sigma M_B = 0$, where $B$ does not lie on the line of action of $\mathbf{F}_R$, then $\mathbf{F}_R = \mathbf{0}$. Since $\Sigma \mathbf{F} = \mathbf{0}$ and $\Sigma \mathbf{M}_A = \mathbf{0}$, indeed the body in Fig. 4–11a must be in equilibrium.

A second alternative set of equilibrium equations is

$$\Sigma M_A = 0$$
$$\Sigma M_B = 0 \qquad (4\text{–}5)$$
$$\Sigma M_C = 0$$

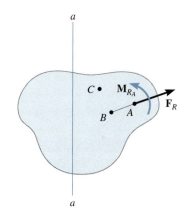

Fig. 4–12

Here it is necessary that points $A$, $B$, and $C$ do not lie on the same line. To prove that these equations, when satisfied, ensure equilibrium, consider the free-body diagram in Fig. 4–12. If $\Sigma M_A = 0$ is to be satisfied, then $\mathbf{M}_{R_A} = \mathbf{0}$. $\Sigma M_B = 0$ is satisfied if the line of action of $\mathbf{F}_R$ passes through point $B$ as shown. Finally, if we require $\Sigma M_C = 0$, where $C$ does not lie on line $AB$, it is necessary that $\mathbf{F}_R = \mathbf{0}$, and the body in Fig. 4–11$a$ must then be in equilibrium.

## PROCEDURE FOR ANALYSIS

Coplanar force equilibrium problems for a rigid body can be solved using the following procedure.

### Free-Body Diagram.

- Establish the $x, y$ coordinate axes in any suitable orientation.
- Draw an outlined shape of the body.
- Show all the forces and couple moments acting on the body.
- Label all the loadings and specify their directions relative to the $x, y$ axes. The sense of a force or couple moment having an *unknown* magnitude but known line of action can be *assumed*.
- Indicate the dimensions of the body necessary for computing the moments of forces.

### Equations of Equilibrium.

- Apply the moment equation of equilibrium, $\Sigma M_O = 0$, about a point ($O$) that lies at the intersection of the lines of action of two unknown forces. In this way, the moments of these unknowns are zero about $O$, and a *direct solution* for the third unknown can be determined.
- When applying the force equilibrium equations, $\Sigma F_x = 0$ and $\Sigma F_y = 0$, orient the $x$ and $y$ axes along lines that will provide the simplest resolution of the forces into their $x$ and $y$ components.
- If the solution of the equilibrium equations yields a negative scalar for a force or couple moment magnitude, this indicates that the sense is opposite to that which was assumed on the free-body diagram.

**E X A M P L E   4.4**

Determine the horizontal and vertical components of reaction for the beam loaded as shown in Fig. 4–13a. Neglect the weight of the beam in the calculations.

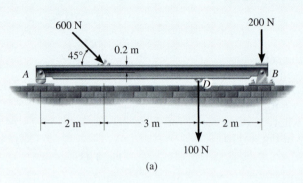

(a)

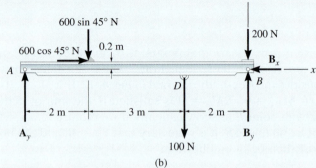

(b)

**Fig. 4–13**

## Solution

*Free-Body Diagram.*   Can you identify each of the forces shown on the free-body diagram of the beam, Fig. 4–13b? For simplicity, the 600-N force is represented by its $x$ and $y$ components as shown. Also, note that a 200-N force acts on the beam at $B$ and is independent of the force components $\mathbf{B}_x$ and $\mathbf{B}_y$, which represent the effect of the pin on the beam.

*Equations of Equilibrium.*   Summing forces in the $x$ direction yields

$$\xrightarrow{+} \Sigma F_x = 0; \qquad 600 \cos 45° \text{ N} - B_x = 0$$
$$B_x = 424 \text{ N} \qquad\qquad Ans.$$

A direct solution for $\mathbf{A}_y$ can be obtained by applying the moment equation $\Sigma M_B = 0$ about point $B$. For the calculation, it should be apparent that forces 200 N, $\mathbf{B}_x$, and $\mathbf{B}_y$ all create zero moment about $B$. Assuming counterclockwise rotation about $B$ to be positive (in the $+\mathbf{k}$ direction), Fig. 4–13b, we have

$$\zeta + \Sigma M_B = 0; \qquad 100 \text{ N}(2 \text{ m}) + (600 \sin 45° \text{ N})(5 \text{ m})$$
$$- (600 \cos 45° \text{ N})(0.2 \text{ m}) - A_y(7 \text{ m}) = 0$$
$$A_y = 319 \text{ N} \qquad\qquad Ans.$$

Summing forces in the $y$ direction, using this result, gives

$$+\uparrow \Sigma F_y = 0; \qquad 319 \text{ N} - 600 \sin 45° \text{ N} - 100 \text{ N} - 200 \text{ N} + B_y = 0$$
$$B_y = 405 \text{ N} \qquad\qquad Ans.$$

We can check this result by summing moments about point $A$.

$$\zeta + \Sigma M_A = 0; \qquad -(600 \sin 45° \text{ N})(2 \text{ m}) - (600 \cos 45° \text{ N})(0.2 \text{ m})$$
$$- (100 \text{ N})(5 \text{ m}) - (200 \text{ N})(7 \text{ m}) + B_y(7 \text{ m}) = 0$$
$$B_y = 405 \text{ N} \qquad\qquad Ans.$$

# E X A M P L E 4.5

The cord shown in Fig. 4–14a supports a force of 100 lb and wraps over the frictionless pulley. Determine the tension in the cord at $C$ and the horizontal and vertical components of reaction at pin $A$.

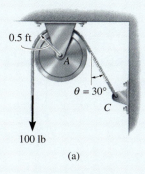

0.5 ft

$A$

$\theta = 30°$

$C$

100 lb

(a)

**Fig. 4–14**

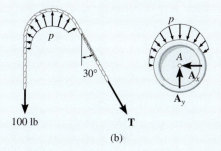

$p$

$p$

$A$

$A_x$

$A_y$

$30°$

100 lb     T

(b)

## Solution

*Free-Body Diagrams.* The free-body diagrams of the cord and pulley are shown in Fig. 4–14b. Note that the principle of action, equal but opposite reaction must be carefully observed when drawing each of these diagrams: the cord exerts an unknown load distribution $p$ along part of the pulley's surface, whereas the pulley exerts an equal but opposite effect on the cord. For the solution, however, it is simpler to *combine* the free-body diagrams of the pulley and the contacting portion of the cord, so that the distributed load becomes *internal* to the system and is therefore eliminated from the analysis, Fig. 4–14c.

*Equations of Equilibrium.* Summing moments about point $A$ to eliminate $\mathbf{A}_x$ and $\mathbf{A}_y$, Fig. 4–14c, we have

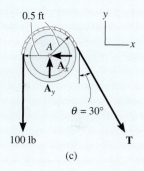

0.5 ft    $y$

$A$

$A_x$

$x$

$A_y$

$\theta = 30°$

100 lb     T

(c)

$\zeta + \Sigma M_A = 0;$    $100 \text{ lb}(0.5 \text{ ft}) - T(0.5 \text{ ft}) = 0$

$$T = 100 \text{ lb} \qquad Ans.$$

It is seen that the tension remains *constant* as the cord passes over the pulley. (This of course is true for *any angle* $\theta$ at which the cord is directed and for *any radius* $r$ of the pulley.) Using the result for $T$, a force summation is applied to determine the components of reaction at pin $A$.

$\xrightarrow{+} \Sigma F_x = 0;$    $-A_x + 100 \sin 30° \text{ lb} = 0$

$$A_x = 50.0 \text{ lb} \qquad Ans.$$

$+\uparrow \Sigma F_y = 0;$    $A_y - 100 \text{ lb} - 100 \cos 30° \text{ lb} = 0$

$$A_y = 187 \text{ lb} \qquad Ans.$$

**EXAMPLE 4.6**

The link shown in Fig. 4–15a is pin-connected at $A$ and rests against a smooth support at $B$. Compute the horizontal and vertical components of reaction at the pin $A$.

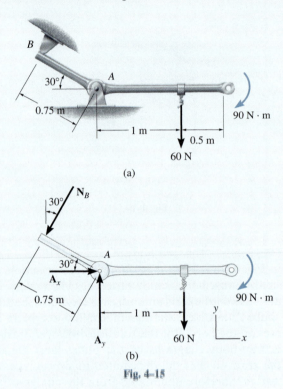

(a)

(b)

**Fig. 4–15**

### Solution

*Free-Body Diagram.* As shown in Fig. 4–15b, the reaction $\mathbf{N}_B$ is perpendicular to the link at $B$. Also, horizontal and vertical components of reaction are represented at $A$.

*Equations of Equilibrium.* Summing moments about $A$, we obtain a direct solution for $N_B$,

$$\zeta + \Sigma M_A = 0; \qquad -90 \text{ N} \cdot \text{m} - 60 \text{ N}(1 \text{ m}) + N_B(0.75 \text{ m}) = 0$$
$$N_B = 200 \text{ N}$$

Using this result,

$$\xrightarrow{+} \Sigma F_x = 0; \qquad A_x - 200 \sin 30° \text{ N} = 0$$
$$A_x = 100 \text{ N} \qquad \qquad Ans.$$

$$+\uparrow \Sigma F_y = 0; \qquad A_y - 200 \cos 30° \text{ N} - 60 \text{ N} = 0$$
$$A_y = 233 \text{ N} \qquad \qquad Ans.$$

# EXAMPLE 4.7

The box wrench in Fig. 4–16a is used to tighten the bolt at $A$. If the wrench does not turn when the load is applied to the handle, determine the torque or moment applied to the bolt and the force of the wrench on the bolt.

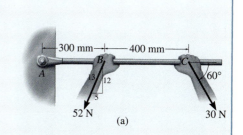

(a)

## Solution

*Free-Body Diagram.*   The free-body diagram for the wrench is shown in Fig. 4–16b. Since the bolt acts as a "fixed support," it exerts force components $\mathbf{A}_x$ and $\mathbf{A}_y$ and a torque $\mathbf{M}_A$ on the wrench at $A$.

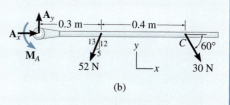

(b)

**Fig. 4–16**

*Equations of Equilibrium.*

$\xrightarrow{+} \Sigma F_x = 0;$   $A_x - 52\left(\frac{5}{13}\right)\text{N} + 30\cos 60° \text{ N} = 0$

$\qquad\qquad\qquad A_x = 5.00 \text{ N}$   *Ans.*

$+\uparrow \Sigma F_y = 0;$   $A_y - 52\left(\frac{12}{13}\right)\text{N} - 30\sin 60° \text{ N} = 0\ A_y = 74.0 \text{ N}$

$\qquad\qquad\qquad A_y = 74.0 \text{ N}$   *Ans.*

$\zeta+\Sigma M_A = 0;$   $M_A - 52\left(\frac{12}{13}\right)\text{N }(0.3 \text{ m}) - (30\sin 60° \text{ N})(0.7 \text{ m}) = 0$

$\qquad\qquad\qquad M_A = 32.6 \text{ N}\cdot\text{m}$   *Ans.*

Point $A$ was chosen for summing moments because the lines of action of the *unknown* forces $\mathbf{A}_x$ and $\mathbf{A}_y$ pass through this point, and therefore these forces were not included in the moment summation. Realize, however, that $\mathbf{M}_A$ must be *included* in this moment summation. This couple moment is a free vector and represents the twisting resistance of the bolt on the wrench. By Newton's third law, the wrench exerts an equal but opposite moment or torque on the bolt. Furthermore, the resultant force on the wrench is

$$F_A = \sqrt{(5.00)^2 + (74.0)^2} = 74.1 \text{ N}$$   *Ans.*

Because the force components $A_x$ and $A_y$ were calculated as positive quantities, their directional sense is shown correctly on the free-body diagram in Fig. 4–16b. Hence

$$\theta = \tan^{-1}\left(\frac{74.0 \text{ N}}{5.00 \text{ N}}\right) = 86.1° \measuredangle$$

Realize that $\mathbf{F}_A$ acts in the opposite direction on the bolt. Why?

Although only *three* independent equilibrium equations can be written for a rigid body, it is a good practice to *check* the calculations using a fourth equilibrium equation. For example, the above computations may be verified in part by summing moments about point $C$:

$\zeta+\Sigma M_C = 0;\ 52\left(\frac{12}{13}\right)\text{N }(0.4 \text{ m}) + 32.6 \text{ N}\cdot\text{m} - 74.0 \text{ N }(0.7 \text{ m}) = 0$

$\qquad\qquad 19.2 \text{ N}\cdot\text{m} + 32.6 \text{ N}\cdot\text{m} - 51.8 \text{ N}\cdot\text{m} = 0$

**EXAMPLE** **4.8**

(a)

Placement of concrete from the truck is accomplished using the chute shown in the photos, Fig. 4–17a. Determine the force that the hydraulic cylinder and the truck frame exert on the chute to hold it in the position shown. The chute and wet concrete contained along its length have a uniform weight of 35 lb/ft.

**Solution**

The idealized model of the chute is shown in Fig. 4–17b. Here the dimensions are given, and it is assumed the chute is pin connected to the frame at $A$ and the hydraulic cylinder $BC$ acts as a short link.

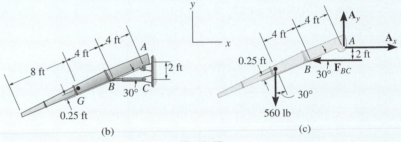

(b)                                     (c)

**Fig. 4–17**

**Free-Body Diagram.** Since the chute has a length of 16 ft, the total supported weight is $(35 \text{ lb/ft})(16 \text{ ft}) = 560$ lb, which is assumed to act at its midpoint, $G$. The hydraulic cylinder exerts a horizontal force $\mathbf{F}_{BC}$ on the chute, Fig. 4–17c.

**Equations of Equilibrium.** A direct solution for $\mathbf{F}_{BC}$ is possible by summing moments about the pin at $A$. To do this we will use the principle of moments and resolve the weight into components parallel and perpendicular to the chute. We have,

$\zeta + \Sigma M_A = 0;$

$\quad -F_{BC}(2 \text{ ft}) + 560 \cos 30° \text{ lb}(8 \text{ ft}) + 560 \sin 30° \text{ lb}(0.25 \text{ ft}) = 0$

$\quad\quad\quad\quad\quad F_{BC} = 1975 \text{ lb}$  *Ans.*

Summing forces to obtain $A_x$ and $A_y$, we obtain

$\xrightarrow{+} \Sigma F_x = 0; \quad -A_x + 1975 \text{ lb} = 0$

$\quad\quad\quad\quad\quad A_x = 1975 \text{ lb}$  *Ans.*

$+\uparrow \Sigma F_y = 0; \quad A_y - 560 \text{ lb} = 0$

$\quad\quad\quad\quad\quad A_y = 560 \text{ lb}$  *Ans.*

To verify this solution we can sum moments about point $B$.

$\zeta + \Sigma M_B = 0; \quad -1975 \text{ lb}(2 \text{ ft}) + 560 \text{ lb}(4 \cos 30° \text{ ft}) +$

$\quad\quad\quad\quad 560 \cos 30° \text{ lb}(4 \text{ ft}) + 560 \sin 30° \text{ lb}(0.25 \text{ ft}) = 0$

## 4.4  Two- and Three-Force Members

The solution to some equilibrium problems can be simplified if one is able to recognize members that are subjected to only two or three forces.

**Two-Force Members.**   When a member is subject to *no couple moments* and forces are applied at only two points on a member, the member is called a *two-force member*. An example is shown in Fig. 4–18a. The forces at $A$ and $B$ are summed to obtain their respective *resultants* $\mathbf{F}_A$ and $\mathbf{F}_B$, Fig. 4–18b. These two forces will maintain *translational or force equilibrium* ($\Sigma\mathbf{F} = \mathbf{0}$) provided $\mathbf{F}_A$ is of equal magnitude and opposite direction to $\mathbf{F}_B$. Furthermore, *rotational or moment equilibrium* ($\Sigma\mathbf{M}_O = \mathbf{0}$) is satisfied if $\mathbf{F}_A$ is *collinear* with $\mathbf{F}_B$. As a result, the line of action of both forces is known since it always passes through $A$ and $B$. Hence, only the force magnitude must be determined or stated. Other examples of two-force members held in equilibrium are shown in Fig. 4–19.

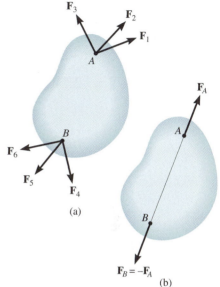

(a)

(b)

Two-force member

**Fig. 4–18**

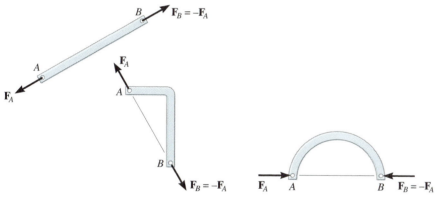

Two-force members

**Fig. 4–19**

**Three-Force Members.**   If a member is subjected to only three forces, then it is necessary that the forces be either *concurrent* or *parallel* for the member to be in equilibrium. To show the concurrency requirement, consider the body in Fig. 4–20a and suppose that any two of the three forces acting on the body have lines of action that intersect at point $O$. To satisfy moment equilibrium about $O$, i.e., $\Sigma M_O = 0$, the third force must also pass through $O$, which then makes the force system *concurrent*. If two of the three forces are parallel, Fig. 4–20b, the point of concurrency, $O$, is considered to be at "infinity," and the third force must be parallel to the other two forces to intersect at this "point."

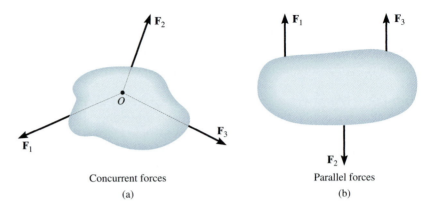

Concurrent forces

(a)

Parallel forces

(b)

Three-force members

**Fig. 4–20**

---

Many mechanical elements act as two- or three-force members, and the ability to recognize them in a problem will considerably simplify an equilibrium analysis.

- The bucket link $AB$ on the back-hoe is a typical example of a two-force member since it is pin connected at its ends and, provided its weight is neglected, no other force acts on this member.

- The hydraulic cylinder $BC$ is pin connected at its ends. It is a two-force member. The boom $ABD$ is subjected to the weight of the suspended motor at $D$, the force of the hydraulic cylinder at $B$, and the force of the pin at $A$. If the boom's weight is neglected, it is a three-force member.

- The dump bed of the truck operates by extending the telescopic hydraulic cylinder $AB$. If the weight of $AB$ is neglected, we can classify it as a two-force member since it is pin connected at its end points.

# EXAMPLE 4.9

The lever $ABC$ is pin-supported at $A$ and connected to a short link $BD$ as shown in Fig. 4–21a. If the weight of the members is negligible, determine the force of the pin on the lever at $A$.

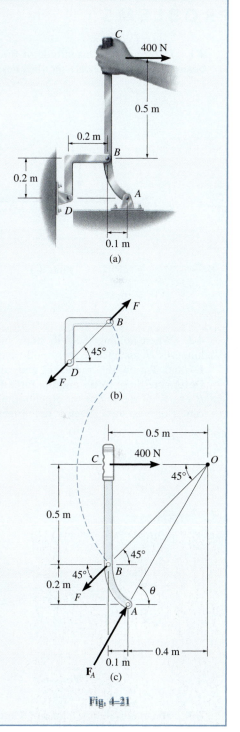

(a)

(b)

(c)

Fig. 4–21

### Solution

*Free-Body Diagrams.* As shown by the free-body diagram, Fig. 4–21b, the short link $BD$ is a *two-force member*, so the *resultant forces* at pins $D$ and $B$ must be equal, opposite, and collinear. Although the magnitude of the force is unknown, the line of action is known since it passes through $B$ and $D$.

Lever $ABC$ is a *three-force member*, and therefore, in order to satisfy moment equilibrium, the three nonparallel forces acting on it must be concurrent at $O$, Fig. 4–21c. In particular, note that the force $F$ on the lever at $B$ is equal but opposite to the force $F$ acting at $B$ on the link. Why? The distance $CO$ must be 0.5 m since the lines of action of $\mathbf{F}$ and the 400-N force are known.

*Equations of Equilibrium.* By requiring the force system to be concurrent at $O$, since $\Sigma M_O = 0$, the angle $\theta$ which defines the line of action of $\mathbf{F}_A$ can be determined from trigonometry,

$$\theta = \tan^{-1}\left(\frac{0.7}{0.4}\right) = 60.3° \; \angle^\theta \qquad Ans.$$

Using the $x$, $y$ axes and applying the force equilibrium equations, we can obtain $F_A$ and $F$.

$$\xrightarrow{+} \Sigma F_x = 0; \qquad F_A \cos 60.3° - F \cos 45° + 400 \text{ N} = 0$$
$$+\uparrow \Sigma F_y = 0; \qquad F_A \sin 60.3° - F \sin 45° = 0$$

Solving, we get

$$F_A = 1.07 \text{ kN} \qquad Ans.$$
$$F = 1.32 \text{ kN}$$

*Note:* We can also solve this problem by representing the force at $A$ by its two components $\mathbf{A}_x$ and $\mathbf{A}_y$ and applying $\Sigma M_A = 0$, $\Sigma F_x = 0$, $\Sigma F_y = 0$ to the lever. Once $A_x$ and $A_y$ are determined, how would you find $F_A$ and $\theta$?

# PROBLEMS

**4-1.** Determine the magnitude of the reactions on the beam at $A$ and $B$. Neglect the thickness of the beam.

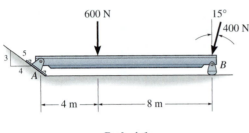

Prob. 4–1

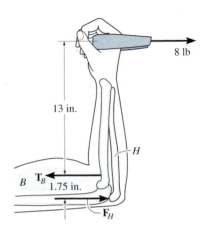

Prob. 4–3

**4-2.** When holding the 5-lb stone in equilibrium, the humerus $H$, assumed to be smooth, exerts normal forces $\mathbf{F}_C$ and $\mathbf{F}_A$ on the radius $C$ and ulna $A$ as shown. Determine these forces and the force $\mathbf{F}_B$ that the biceps $B$ exerts on the radius for equilibrium. The stone has a center of mass at $G$. Neglect the weight of the arm.

*\*4-4.** The ramp of a ship has a weight of 200 lb and a center of gravity at $G$. Determine the cable force in $CD$ needed to just start lifting the ramp, (i.e., so the reaction at $B$ becomes zero). Also, determine the horizontal and vertical components of force at the hinge (pin) at $A$.

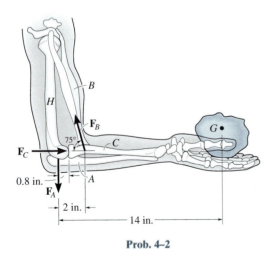

Prob. 4–2

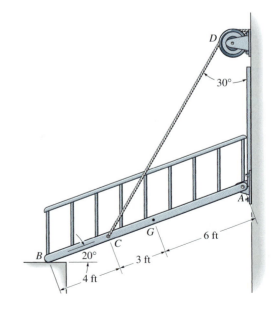

Prob. 4–4

**4-3.** The man is pulling a load of 8 lb with one arm held as shown. Determine the force $\mathbf{F}_H$ this exerts on the humerus bone $H$, and the tension developed in the biceps muscle $B$. Neglect the weight of the man's arm.

**4-5.** Determine the magnitude of force at the pin $A$ and in the cable $BC$ needed to support the 500-lb load. Neglect the weight of the boom $AB$.

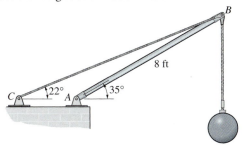

**Prob. 4–5**

**4-6.** Compare the force exerted on the toe and heel of a 120-lb woman when she is wearing regular shoes and stiletto heels. Assume all her weight is placed on one foot and the reactions occur at points $A$ and $B$ as shown.

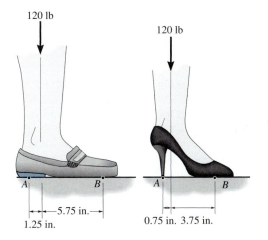

**Prob. 4–6**

**4-7.** Determine the reactions at the pins $A$ and $B$. The spring has an unstretched length of 80 mm.

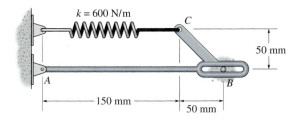

**Prob. 4–7**

**\*4-8.** The platform assembly has a weight of 250 lb and center of gravity at $G_1$. If it is intended to support a maximum load of 400 lb placed at point $G_2$, determine the smallest counterweight $W$ that should be placed at $B$ in order to prevent the platform from tipping over.

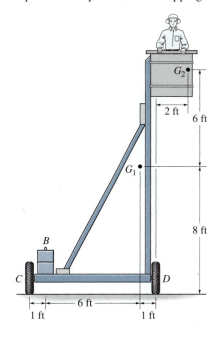

**Prob. 4–8**

**4-9.** Determine the tension in the cable and the horizontal and vertical components of reaction of the pin $A$. The pulley at $D$ is frictionless and the cylinder weighs 80 lb.

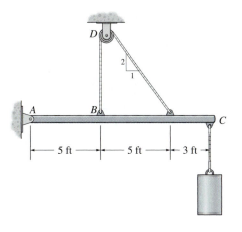

**Prob. 4–9**

**4-10.** The device is used to hold an elevator door open. If the spring has a stiffness of $k = 40$ N/m and it is compressed 0.2 m, determine the horizontal and vertical components of reaction at the pin $A$ and the resultant force at the wheel bearing $B$.

**\*4-12.** The cantilevered jib crane is used to support the load of 780 lb. If the trolley $T$ can be placed anywhere between $1.5$ ft $\le x \le 7.5$ ft, determine the maximum magnitude of reaction at the supports $A$ and $B$. Note that the supports are collars that allow the crane to rotate freely about the vertical axis. The collar at $B$ supports a force in the vertical direction, whereas the one at $A$ does not.

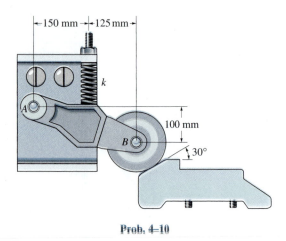

**Prob. 4-10**

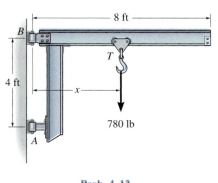

**Prob. 4-12**

**4-11.** The cutter is subjected to a horizontal force of 580 lb and a normal force of 350 lb. Determine the horizontal and vertical components of force acting on the pin $A$ and the force along the hydraulic cylinder $BC$ (a two-force member).

**4-13.** The sports car has a mass of 1.5 Mg and mass center at $G$. If the front two springs each have a stiffness of $k_A = 58$ kN/m and the rear two springs each have a stiffness of $k_B = 65$ kN/m, determine their compression when the car is parked on the 30° incline. Also, what friction force $\mathbf{F}_B$ must be applied to each of the rear wheels to hold the car in equilibrium? *Hint:* First determine the normal force at $A$ and $B$, then determine the compression in the springs.

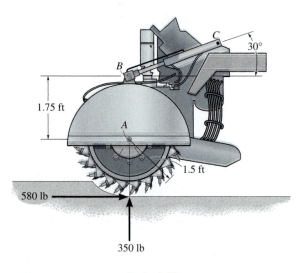

**Prob. 4-11**

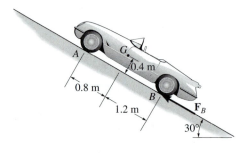

**Prob. 4-13**

**4-14.** The power pole supports the three lines, each line exerting a vertical force on the pole due to its weight as shown. Determine the reactions at the fixed support D. If it is possible for wind or ice to snap the lines, determine which line(s) when removed create(s) a condition for the greatest moment reaction at D.

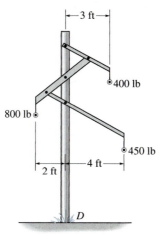

Prob. 4–14

**4-15.** The jib crane is pin-connected at A and supported by a smooth collar at B. Determine the roller placement x of the 5000-lb load so that it gives the maximum and minimum reactions at the supports. Calculate these reactions in each case. Neglect the weight of the crane. Require 4 ft ≤ x ≤ 10 ft.

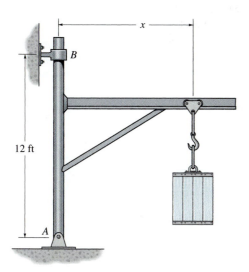

Prob. 4–15

**\*4-16.** If the wheelbarrow and its contents have a mass of 60 kg and center of mass at G, determine the magnitude of the resultant force which the man must exert on *each* of the two handles in order to hold the wheelbarrow in equilibrium.

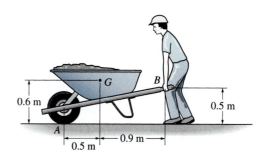

Prob. 4–16

**4-17.** The telephone pole of negligible thickness is subjected to the force of 80 lb directed as shown. It is supported by the cable BCD and can be assumed pinned at its base A. In order to provide clearance for a sidewalk right of way, where D is located, the strut CE is attached at C, as shown by the dashed lines (cable segment CD is removed). If the tension in CD' is to be twice the tension in BCD, determine the height h for placement of the strut CE.

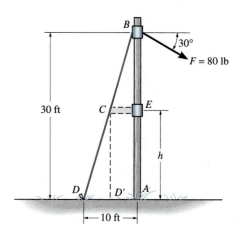

Prob. 4–17

**4-18.** The worker uses the hand truck to move material down the ramp. If the truck and its contents are held in the position shown and have a weight of 100 lb with center of gravity at $G$, determine the resultant normal force of both wheels on the ground $A$ and the magnitude of the force required at the grip $B$.

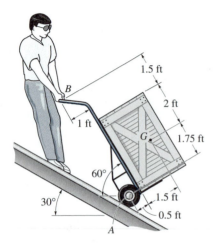

**Prob. 4-18**

**4-19.** The shelf supports the electric motor which has a mass of 15 kg and mass center at $G_m$. The platform upon which it rests has a mass of 4 kg and mass center at $G_p$. Assuming that a single bolt $B$ holds the shelf up and the bracket bears against the smooth wall at $A$, determine this normal force at $A$ and the horizontal and vertical components of reaction of the bolt on the bracket.

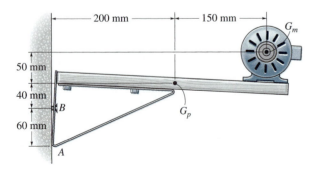

**Prob. 4-19**

**\*4-20.** The upper portion of the crane boom consists of the jib $AB$, which is supported by the pin at $A$, the guy line $BC$, and the backstay $CD$, each cable being separately attached to the mast at $C$. If the 5-kN load is supported by the hoist line, which passes over the pulley at $B$, determine the magnitude of the resultant force the pin exerts on the jib at $A$ for equilibrium, the tension in the guy line $BC$, and the tension $T$ in the hoist line. Neglect the weight of the jib. The pulley at $B$ has a radius of 0.1 m.

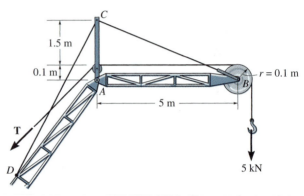

**Prob. 4-20**

**4-21.** The mobile crane has a weight of 120,000 lb and center of gravity at $G_1$; the boom has a weight of 30,000 lb and center of gravity at $G_2$. Determine the smallest angle of tilt $\theta$ of the boom, without causing the crane to overturn if the suspended load is $W = 40,000$ lb. Neglect the thickness of the tracks at $A$ and $B$.

**4-22.** The mobile crane has a weight of 120,000 lb and center of gravity at $G_1$; the boom has a weight of 30,000 lb and center of gravity at $G_2$. If the suspended load has a weight of $W = 16,000$ lb, determine the normal reactions at the tracks $A$ and $B$. For the calculation, neglect the thickness of the tracks and take $\theta = 30°$.

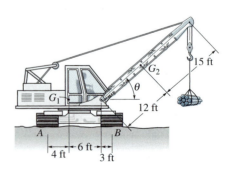

**Probs. 4-21/22**

**4-23.** The winch consists of a drum of radius 4 in., which is pin-connected at its center *C*. At its outer rim is a ratchet gear having a mean radius of 6 in. The pawl *AB* serves as a two-force member (short link) and holds the drum from rotating. If the suspended load is 500 lb, determine the horizontal and vertical components of reaction at the pin *C*.

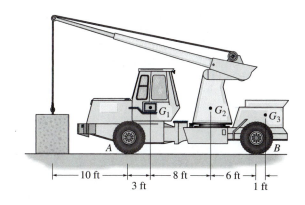

**Prob. 4–24**

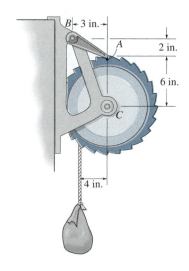

**Prob. 4–23**

**4-25.** The boom supports the two vertical loads. Neglect the size of the collars at *D* and *B* and the thickness of the boom, and compute the horizontal and vertical components of force at the pin *A* and the force in cable *CB*. Set $F_1 = 800$ N and $F_2 = 350$ N.

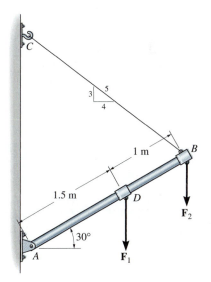

**Prob. 4–25**

*4-24.** The crane consists of three parts, which have weights of $W_1 = 3500$ lb, $W_2 = 900$ lb, $W_3 = 1500$ lb and centers of gravity at $G_1$, $G_2$, and $G_3$, respectively. Neglecting the weight of the boom, determine (a) the reactions on each of the four tires if the load is hoisted at constant velocity and has a weight of 800 lb, and (b), with the boom held in the position shown, the maximum load the crane can lift without tipping over.

**4-26.** The boom is intended to support two vertical loads, $F_1$ and $F_2$. If the cable $CB$ can sustain a maximum load of 1500 lb before it fails, determine the critical loads if $F_1 = 2F_2$. Also, what is the magnitude of the maximum reaction at pin $A$?

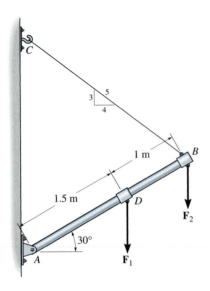

**Prob. 4–26**

**4-27.** Three uniform books, each having a weight $W$ and length $a$, are stacked as shown. Determine the maximum distance $d$ that the top book can extend out from the bottom one so the stack does not topple over.

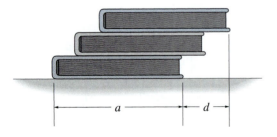

**Prob. 4–27**

**\*4-28.** The toggle switch consists of a cocking lever that is pinned to a fixed frame at $A$ and held in place by the spring which has an unstretched length of 200 mm. Determine the magnitude of the resultant force at $A$ and the normal force on the peg at $B$ when the lever is in the position shown.

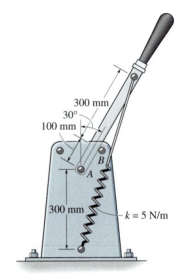

**Prob. 4–28**

**4-29.** The rigid beam of negligible weight is supported horizontally by two springs and a pin. If the springs are uncompressed when the load is removed, determine the force in each spring when the load $P$ is applied. Also, compute the vertical deflection of end $C$. Assume the spring stiffness $k$ is large enough so that only small deflections occur. *Hint:* The beam rotates about $A$ so the deflections in the springs can be related.

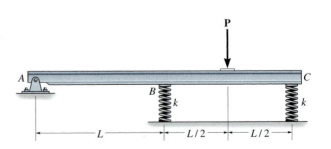

**Prob. 4–29**

**4-30.** The uniform rod $AB$ has a weight of 15 lb and the spring is unstretched when $\theta = 0°$. If $\theta = 30°$, determine the stiffness $k$ of the spring, so that the rod is in equilibrium.

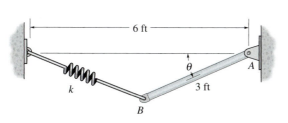

Prob. 4-30

**4-31.** The smooth pipe rests against the wall at the points of contact $A$, $B$, and $C$. Determine the reactions at these points needed to support the vertical force of 45 lb. Neglect the pipe's thickness in the calculation.

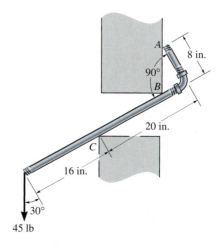

Prob. 4-31

# Equilibrium in Three Dimensions

## 4.5  Free-Body Diagrams

The first step in solving three-dimensional equilibrium problems, as in the case of two dimensions, is to draw a free-body diagram of the body (or group of bodies considered as a system). Before we show this, however, it is necessary to discuss the types of reactions that can occur at the supports.

Support Reactions.   The reactive forces and couple moments acting at various types of supports and connections, when the members are viewed in three dimensions, are listed in Table 4–2. It is important to recognize the symbols used to represent each of these supports and to understand clearly how the forces and couple moments are developed by each support. As in the two-dimensional case, *a force is developed by a support that restricts the translation of the attached member, whereas a couple moment is developed when rotation of the attached member is prevented.* For example, in Table 4–2, the ball-and-socket joint (4) prevents any translation of the connecting member; therefore, a force must act on the member at the point of connection. This force has three components having unknown magnitudes, $F_x$, $F_y$, $F_z$. Provided these components are known, one can obtain the magnitude of force. $F = \sqrt{F_x^2 + F_y^2 + F_z^2}$, and the force's orientation defined by the coordinate direction angles $\alpha$, $\beta$, $\gamma$, Eqs. 2–7.* Since the connecting member is allowed to rotate freely about *any* axis, no couple moment is resisted by a ball-and-socket joint.

It should be noted that the *single* bearing supports (5) and (7), the *single* pin (8), and the *single* hinge (9) are shown to support both force and couple-moment components. If, however, these supports are used in conjunction with *other* bearings, pins, or hinges to hold a rigid body in equilibrium and the supports are *properly aligned* when connected to the body, then the *force reactions* at these supports *alone* may be adequate for supporting the body. In other words, the couple moments become redundant and are not shown on the free-body diagram. The reason for this should become clear after studying the examples which follow.

---

*The three unknowns may also be represented as an unknown force magnitude $F$ and two unknown coordinate direction angles. The third direction angle is obtained using the identity $\cos^2 \alpha + \cos^2 \beta + \cos^2 \gamma = 1$, Eq. 2–10.

**TABLE 4–2  •  Supports for Rigid Bodies Subjected to Three-Dimensional Force Systems**

| Types of Connection | Reaction | Number of Unknowns |
|---|---|---|
| (1) <br><br> cable | | One unknown. The reaction is a force which acts away from the member in the known direction of the cable. |
| (2) <br><br> smooth surface support | | One unknown. The reaction is a force which acts perpendicular to the surface at the point of contact. |
| (3) <br><br> roller | | One unknown. The reaction is a force which acts perpendicular to the surface at the point of contact. |
| (4) <br><br> ball and socket | | Three unknowns. The reactions are three rectangular force components. |
| (5) <br><br> single journal bearing | | Four unknowns. The reactions are two force and two couple-moment components which act perpendicular to the shaft. |

*continued*

**TABLE 4–2   •   Continued**

| Types of Connection | Reaction | Number of Unknown |
|---|---|---|
| (6) single journal bearing with square shaft | | Five unknowns. The reactions are two force and three couple-moment components. |
| (7) single thrust bearing | | Five unknowns. The reactions are three force and two couple-moment components. |
| (8) single smooth pin | | Five unknowns. The reactions are three force and two couple-moment components. |
| (9) single hinge | | Five unknowns. The reactions are three force and two couple-moment components. |
| (10) fixed support | | Six unknowns. The reactions are three force and three couple-moment components. |

Typical examples of actual supports that are referenced to Table 4–2 are shown in the following sequence of photos.

This ball-and-socket joint provides a connection for the housing of an earth grader to its frame. (4)

This journal bearing supports the end of the shaft. (5)

This thrust bearing is used to support the drive shaft on a machine. (7)

This pin is used to support the end of the strut used on a tractor. (8)

**Free-Body Diagrams.**   The general procedure for establishing the free-body diagram of a rigid body has been outlined in Sec. 4.2. Essentially it requires first "isolating" the body by drawing its outlined shape. This is followed by a careful *labeling* of *all* the forces and couple moments in reference to an established *x, y, z* coordinate system. As a general rule, *components of reaction* having an *unknown magnitude* are shown acting on the free-body diagram in the *positive sense*. In this way, if any negative values are obtained, they will indicate that the components act in the negative coordinate directions.

(a)         (b)

It is a mistake to support a door using a single hinge since the hinge must develop a force $C_y$ to support the weight $\mathbf{W}$ of the door and a couple moment $\mathbf{M}$ to support the moment of $\mathbf{W}$, i.e., $M = Wd$. If instead two properly aligned hinges are used, then the weight is carried by both hinges, $A_y + B_y = W$, and the moment of the door is resisted by the two hinge forces $\mathbf{F}_x$ and $-\mathbf{F}_x$. These forces form a couple, such that $F_x d' = Wd$. In other words, no couple moments are produced by the hinges on the door provided they are in *proper alignment*. Instead, the forces $\mathbf{F}_x$ and $-\mathbf{F}_x$ resist the rotation caused by $\mathbf{W}$.

**E X A M P L E  4.10**

Several examples of objects along with their associated free-body diagrams are shown in Fig. 4–22. In all cases, the x, y, z axes are established and the unknown reaction components are indicated in the positive sense. The weight of the objects is neglected.

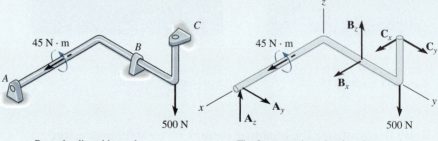

Properly aligned journal bearings at A, B, C.

The force reactions developed by the bearings are sufficient for equilibrium since they prevent the shaft from rotating about each of the coordinate axes.

(a)

**Fig. 4–22**

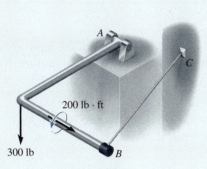

Pin at $A$ and cable $BC$.

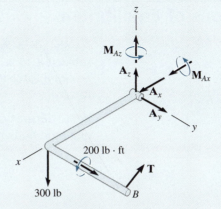

Moment components are developed
by the pin on the rod to prevent
rotation about the $x$ and $z$ axes.

(b)

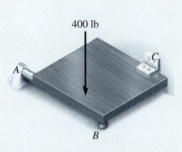

Properly aligned journal bearing
at $A$ and hinge at $C$. Roller at $B$.

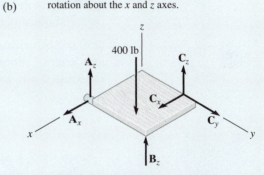

Only force reactions are developed by
the bearing and hinge on the plate to
prevent rotation about each coordinate axis.
No moments at the hinge are developed.

(c)

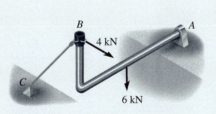

Thrust bearing at $A$ and
cable $BC$

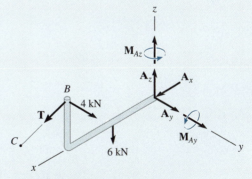

Moment components are developed
by the bearing on the rod in order to
prevent rotation about the $y$ and $z$ axes.

(d)

**Fig. 4–22**

## 4.6 Equations of Equilibrium

As stated in Sec. 4.1, the conditions for equilibrium of a rigid body subjected to a three-dimensional force system require that both the *resultant* force and *resultant* couple moment acting on the body be equal to *zero*.

**Vector Equations of Equilibrium.** The two conditions for equilibrium of a rigid body may be expressed mathematically in vector form as

$$\Sigma \mathbf{F} = \mathbf{0}$$
$$\Sigma \mathbf{M}_O = \mathbf{0}$$

(4–6)

where $\Sigma \mathbf{F}$ is the vector sum of all the external forces acting on the body and $\Sigma \mathbf{M}_O$ is the sum of the couple moments and the moments of all the forces about any point $O$ located either on or off the body.

**Scalar Equations of Equilibrium.** If all the applied external forces and couple moments are expressed in Cartesian vector form and substituted into Eqs. 4–6, we have

$$\Sigma \mathbf{F} = \Sigma F_x \mathbf{i} + \Sigma F_y \mathbf{j} + \Sigma F_z \mathbf{k} = \mathbf{0}$$
$$\Sigma \mathbf{M}_O = \Sigma M_x \mathbf{i} + \Sigma M_y \mathbf{j} + \Sigma M_z \mathbf{k} = \mathbf{0}$$

Since the $\mathbf{i}, \mathbf{j}$, and $\mathbf{k}$ components are independent from one another, the above equations are satisfied provided

$$\Sigma F_x = 0$$
$$\Sigma F_y = 0$$
$$\Sigma F_z = 0$$

(4–7a)

and

$$\Sigma M_x = 0$$
$$\Sigma M_y = 0$$
$$\Sigma M_z = 0$$

(4–7b)

These *six scalar equilibrium equations* may be used to solve for at most six unknowns shown on the free-body diagram. Equations 4–7a express the fact that the sum of the external force components acting in the $x$, $y$, and $z$ directions must be zero, and Eqs. 4–7b require the sum of the moment components about the $x$, $y$, and $z$ axes to be zero.

## PROCEDURE FOR ANALYSIS

The following procedure provides a method for solving three-dimensional equilibrium problems.

*Free-Body Diagram.* Construct the free-body diagram for the body. Be sure to include *all* the forces and couple moments that act *on* the body. These interactions are commonly caused by the externally applied loadings, contact forces exerted by adjacent bodies, support reactions, and the weight of the body if it is significant compared to the magnitudes of the other applied forces. Establish the origin of the *x, y, z* axes at a convenient point, and orient the axes so that they are parallel to as many of the external forces and moments as possible. Identify the unknowns, and in general show all the unknown components having a positive sense if the sense cannot be determined. Dimensions of the body, necessary for calculating the moments of forces, are also included on the free-body diagram.

*Equations of Equilibrium.* Apply the equations of equilibrium. In many cases, problems can be solved by *direct application* of the six scalar equations $\Sigma F_x = 0$, $\Sigma F_y = 0$, $\Sigma F_z = 0$, $\Sigma M_x = 0$, $\Sigma M_y = 0$, $\Sigma M_z = 0$, Eqs. 4–7; however, if the force components or moment arms seem difficult to determine, it is recommended that the solution be obtained by using vector equations: $\Sigma \mathbf{F} = \mathbf{0}$, $\Sigma \mathbf{M}_O = \mathbf{0}$, Eqs. 4–6. In any case, it is *not necessary* that the set of axes chosen for force summation *coincide* with the set of axes chosen for moment summation. Instead, it is recommended that one *choose the direction of an axis for moment summation such that it intersects the lines of action of as many unknown forces as possible.* The moments of forces passing through points on this axis or forces which are parallel to the axis will then be zero. Furthermore, *any set of three nonorthogonal axes* may be chosen for either the force or moment summations. By the proper choice of axes, it may be possible to solve directly for an unknown quantity, or at least reduce the need for solving a large number of simultaneous equations for the unknowns.

E X A M P L E  **4.11**

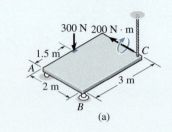

300 N  200 N · m

1.5 m

*A*

2 m    3 m

*C*

*B*

(a)

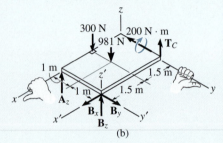

*z*

300 N   200 N · m

981 N   $T_C$

1 m

*z'*   1.5 m

1 m   1.5 m

*x*   $A_z$   *y*

*x'*   $B_x$  $B_y$   *y'*

$B_z$

(b)

**Fig. 4–23**

The homogeneous plate shown in Fig. 4–23*a* has a mass of 100 kg and is subjected to a force and couple moment along its edges. If it is supported in the horizontal plane by means of a roller at *A*, a ball-and-socket joint at *B*, and a cord at *C*, determine the components of reaction at the supports.

**Solution (*Scalar Analysis*)**

*Free-Body Diagram.*   There are five unknown reactions acting on the plate, as shown in Fig. 4–23*b*. Each of these reactions is assumed to act in a positive coordinate direction.

*Equations of Equilibrium.*   Since the three-dimensional geometry is rather simple, a *scalar analysis* provides a *direct solution* to this problem. A force summation along each axis yields

$$\Sigma F_x = 0; \qquad\qquad B_x = 0 \qquad\qquad\qquad Ans.$$
$$\Sigma F_y = 0; \qquad\qquad B_y = 0 \qquad\qquad\qquad Ans.$$
$$\Sigma F_z = 0; \qquad\qquad A_z + B_z + T_C - 300\text{ N} - 981\text{ N} = 0 \qquad (1)$$

Recall that the moment of a force about an axis is equal to the product of the force magnitude and the perpendicular distance (moment arm) from the line of action of the force to the axis. The sense of the moment is determined by the right-hand rule. Also, forces that are parallel to an axis or pass through it create no moment about the axis. Hence, summing moments of the forces on the free-body diagram, with positive moments acting along the positive *x* or *y* axis, we have

$$\Sigma M_x = 0; \qquad T_C(2\text{ m}) - 981\text{ N}(1\text{ m}) + B_z(2\text{ m}) = 0 \qquad (2)$$
$$\Sigma M_y = 0;$$
$$300\text{ N}(1.5\text{ m}) + 981\text{ N}(1.5\text{ m}) - B_z(3\text{ m}) - A_z(3\text{ m}) - 200\text{ N}\cdot\text{m} = 0 \qquad (3)$$

The components of force at *B* can be eliminated if the *x'*, *y'*, *z'* axes are used. We obtain

$$\Sigma M_{x'} = 0; \qquad 981\text{ N}(1\text{ m}) + 300\text{ N}(2\text{ m}) - A_z(2\text{ m}) = 0 \qquad (4)$$
$$\Sigma M_{y'} = 0;$$
$$-300\text{ N}(1.5\text{ m}) - 981\text{ N}(1.5\text{ m}) - 200\text{ N}\cdot\text{m} + T_C(3\text{ m}) = 0 \qquad (5)$$

Solving Eqs. 1 through 3 or the more convenient Eqs. 1, 4, and 5 yields

$$A_z = 790\text{ N} \qquad B_z = -217\text{ N} \qquad T_C = 707\text{ N} \qquad Ans.$$

The negative sign indicates that $\mathbf{B}_z$ acts downward.

Note that the solution of this problem does not require the use of a summation of moments about the *z* axis. The plate is partially constrained since the supports cannot prevent it from turning about the *z* axis if a force is applied to it in the *x–y* plane.

# EXAMPLE 4.12

The windlass shown in Fig. 4–24a is supported by a thrust bearing at A and a smooth journal bearing at B, which are properly aligned on the shaft. Determine the magnitude of the vertical force **P** that must be applied to the handle to maintain equilibrium of the 100-kg bucket. Also calculate the reactions at the bearings.

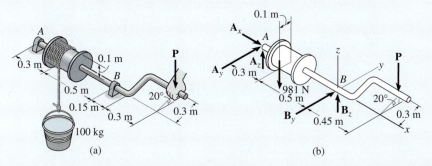

**Fig. 4–24**

## Solution (*Scalar Analysis*)

*Free-Body Diagram.*  Since the bearings at A and B are aligned correctly, *only* force reactions occur at these supports, Fig. 4–24b. Why are there no moment reactions?

*Equations of Equilibrium.*  Summing moments about the x axis yields a direct solution for **P**. Why? For a scalar moment summation, it is necessary to determine the moment of each force as the product of the force magnitude and the *perpendicular distance* from the x axis to the line of action of the force. Using the right-hand rule and assuming positive moments act in the $+\mathbf{i}$ direction, we have

$$\Sigma M_x = 0; \qquad 981 \text{ N}(0.1 \text{ m}) - P(0.3 \cos 20° \text{ m}) = 0$$
$$P = 348.0 \text{ N} \qquad \textit{Ans.}$$

Using this result and summing moments about the y and z axes yields

$$\Sigma M_y = 0;$$
$$-981 \text{ N}(0.5 \text{ m}) + A_z(0.8 \text{ m}) + (348.0 \text{ N})(0.45 \text{ m}) = 0$$
$$A_z = 417.4 \text{ N} \qquad \textit{Ans.}$$
$$\Sigma M_z = 0; \qquad -A_y(0.8 \text{ m}) = 0 \quad A_y = 0 \qquad \textit{Ans.}$$

The reactions at B are determined by a force summation, using the results obtained above.

$$\Sigma F_x = 0; \qquad\qquad A_x = 0 \qquad \textit{Ans.}$$
$$\Sigma F_y = 0; \qquad\qquad 0 + B_y = 0 \quad B_y = 0 \qquad \textit{Ans.}$$
$$\Sigma F_z = 0; \qquad 417.4 - 981 - B_z - 348.0 = 0 \quad B_z = 911.6 \text{ N} \qquad \textit{Ans.}$$

EXAMPLE 4.13

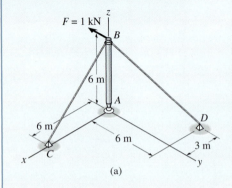

(a)

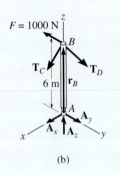

(b)

Fig. 4–25

Determine the tension in cables $BC$ and $BD$ and the reactions at the ball-and-socket joint $A$ for the mast shown in Fig. 4–25a.

**Solution (Vector Analysis)**

*Free-Body Diagram.* There are five unknown force magnitudes shown on the free-body diagram, Fig. 4–25b.

*Equations of Equilibrium.* Expressing each force in Cartesian vector form, we have

$$\mathbf{F} = \{-1000\mathbf{j}\}\,\text{N}$$
$$\mathbf{F}_A = A_x\mathbf{i} + A_y\mathbf{j} + A_z\mathbf{k}$$
$$\mathbf{T}_C = 0.707T_C\mathbf{i} - 0.707T_C\mathbf{k}$$
$$\mathbf{T}_D = T_D\left(\frac{\mathbf{r}_{BD}}{r_{BD}}\right) = -0.333T_D\mathbf{i} + 0.667T_D\mathbf{j} - 0.667T_D\mathbf{k}$$

Applying the force equation of equilibrium gives

$$\Sigma\mathbf{F} = \mathbf{0}; \qquad \mathbf{F} + \mathbf{F}_A + \mathbf{T}_C + \mathbf{T}_D = \mathbf{0}$$
$$(A_x + 0.707T_C - 0.333T_D)\mathbf{i} + (-1000 + A_y + 0.667T_D)\mathbf{j}$$
$$+ (A_z - 0.707T_C - 0.667T_D)\mathbf{k} = \mathbf{0}$$

$$\Sigma F_x = 0; \qquad\qquad A_x + 0.707T_C - 0.333T_D = 0 \qquad (1)$$
$$\Sigma F_y = 0; \qquad\qquad A_y + 0.667T_D - 1000 = 0 \qquad (2)$$
$$\Sigma F_z = 0; \qquad\qquad A_z - 0.707T_C - 0.667T_D = 0 \qquad (3)$$

Summing moments about point $A$, we have

$$\Sigma\mathbf{M}_A = \mathbf{0}; \qquad \mathbf{r}_B \times (\mathbf{F} + \mathbf{T}_C + \mathbf{T}_D) = \mathbf{0}$$
$$6\mathbf{k} \times (-1000\mathbf{j} + 0.707T_C\mathbf{i} - 0.707T_C\mathbf{k}$$
$$- 0.333T_D\mathbf{i} + 0.667T_D\mathbf{j} - 0.667T_D\mathbf{k}) = \mathbf{0}$$

Evaluating the cross product and combining terms yields

$$(-4T_D + 6000)\mathbf{i} + (4.24T_C - 2T_D)\mathbf{j} = \mathbf{0}$$
$$\Sigma M_x = 0; \qquad\qquad -4T_D + 6000 = 0 \qquad (4)$$
$$\Sigma M_y = 0; \qquad\qquad 4.24T_C - 2T_D = 0 \qquad (5)$$

The moment equation about the $z$ axis, $\Sigma M_z = 0$, is automatically satisfied. Why? Solving Eqs. 1 through 5 we have

$$T_C = 707\,\text{N} \qquad T_D = 1500\,\text{N} \qquad\qquad \textit{Ans.}$$
$$A_x = 0\,\text{N} \qquad A_y = 0\,\text{N} \qquad A_z = 1500\,\text{N} \qquad \textit{Ans.}$$

Since the mast is a two-force member, note that the value $A_x = A_y = 0$ could have been determined *by inspection.*

# PROBLEMS

**\*4-32.** Determine the $x, y, z$ components of reaction at the fixed wall $A$.

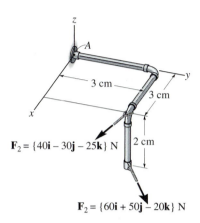

$$F_2 = \{40i - 30j - 25k\} \text{ N}$$

$$F_2 = \{60i + 50j - 20k\} \text{ N}$$

**Prob. 4–32**

**4-33.** The nonhomogeneous door of a large pressure vessel has a weight of 125 lb and a center of gravity at $G$. Determine the magnitudes of the resultant force and resultant couple moment, developed at the hinge $A$, needed to support the door in any open position.

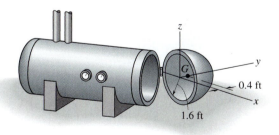

**Prob. 4–33**

**4-34.** The power drill is subjected to the forces shown acting on the grips. Determine the $x, y, z$ components of force and the $y$ and $z$ components of moment reaction acting on the drill bit at $A$.

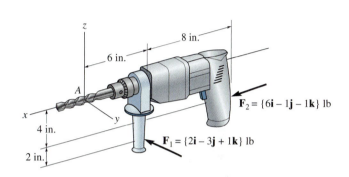

$$F_2 = \{6i - 1j - 1k\} \text{ lb}$$

$$F_1 = \{2i - 3j + 1k\} \text{ lb}$$

**Prob. 4–34**

**4-35.** Determine the floor reaction on each wheel of the engine stand. The engine weighs 750 lb and has a center of gravity at $G$.

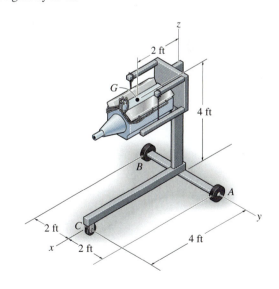

**Prob. 4–35**

*4-36. The shaft is supported by a thrust bearing at $A$ and a journal bearing at $B$. Determine the $x, y, z$ components of reaction at these supports and the magnitude of force acting on the gear at $C$ necessary to hold the shaft in equilibrium. The bearings are in proper alignment and exert only force reactions on the shaft.

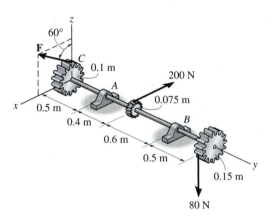

**Prob. 4–36**

4-37. The windlass is subjected to a load of 150 lb. Determine the horizontal force **P** needed to hold the handle in the position shown, and the $x, y, z$ components of reaction at the ball-and-socket $A$ and the smooth journal bearing $B$. The bearing is in proper alignment and exerts only force reactions on the shaft.

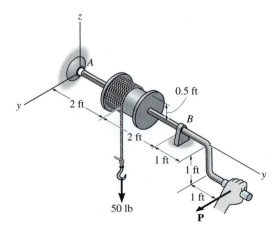

**Prob. 4–37**

**4-38.** The shaft is supported by journal bearings at $A$ and $B$. A key is inserted into the bearing at $B$ in order to prevent the shaft from rotating about and translating along its axis. Determine the $x$, $y$, $z$ components of reaction at the bearings when the 600-N force is applied to the arm. The bearings are in proper alignment and exert only force reactions on the shaft.

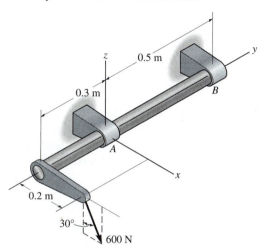

**Prob. 4–38**

**4-39.** The bent rod is supported at $A$, $B$, and $C$ by journal bearings. Determine the $x$, $y$, $z$ reaction components at the bearings if the rod is subjected to a 200-lb vertical force and a 30-lb · ft couple moment as shown. The bearings are in proper alignment and exert only force reactions on the rod.

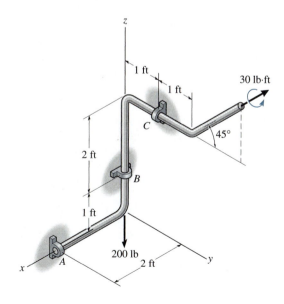

**Prob. 4–39**

## 4.7 Friction

*Friction* may be defined as a force of resistance acting on a body which prevents or retards slipping of the body relative to a second body or surface with which it is in contact with. This force always acts *tangent* to the surface at points of contact with other bodies and is directed so as to oppose the possible or existing motion of the body relative to these points.

In general, two types of friction can occur between surfaces. *Fluid friction* exists when the contacting surfaces are separated by a film of fluid (gas or liquid). The nature of fluid friction is studied in fluid mechanics since it depends upon knowledge of the velocity of the fluid and the fluid's ability to resist shear force. In this book only the effects of *dry friction* will be presented. This type of friction is often called *Coulomb friction* since its characteristics were studied extensively by C.A. Coulomb in 1781. Specifically, dry friction occurs between the contacting surfaces of bodies in the absence of a lubricating fluid.

### Theory of Dry Friction.

The theory of dry friction can best be explained by considering what effects are caused by pulling horizontally on a block of uniform weight **W** which is resting on a rough horizontal surface, Fig. 4–26a. To properly develop a full understanding of the nature of friction, it is necessary to consider the surfaces of contact to be *nonrigid or deformable*. The other portion of the block, however, will be considered rigid. As shown on the free-body diagram of the block, Fig. 4–26b, the floor exerts a *distribution* of both *normal force* $\Delta \mathbf{N}_n$ and *frictional force* $\Delta \mathbf{F}_n$ along the contacting surface. For equilibrium, the normal forces must act *upward* to balance the block's weight **W**, and the frictional forces act to the left to prevent the applied force **P** from moving the block to the right. Close examination of the contacting surfaces between the floor and block reveals how these frictional and normal forces develop, Fig. 4–26c. It can be seen that many microscopic irregularities exist between the two surfaces and, as a result, reactive forces $\Delta \mathbf{R}_n$ are developed at each of the protuberances.* These forces act at all points of contact, and, as shown, each reactive force contributes both a frictional component $\Delta \mathbf{F}_n$ and a normal component $\Delta \mathbf{N}_n$.

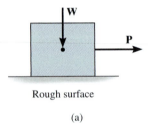

Rough surface

(a)

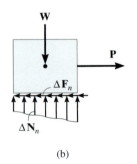

(b)

**Fig. 4–26**

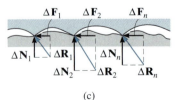

(c)

**Equilibrium.**  For simplicity in the following analysis, the effect of the *distributed* normal and frictional loadings will be indicated by their *resultants* **N** and **F**, which are represented on the free-body diagram as shown in Fig. 4–26*d*. Clearly, the distribution of $\Delta \mathbf{F}_n$ in Fig. 4–26*b* indicates that **F** always acts *tangent to the contacting surface, opposite* to the direction of **P**. On the other hand, the normal force **N** is determined from the distribution of $\Delta \mathbf{N}_n$ in Fig. 4–26*b* and is directed upward to balance the block's weight **W**. Notice that **N** acts a distance $x$ to the right of the line of action of **W**, Fig. 4–26*d*. This location, which coincides with the centroid or geometric center of the loading diagram in Fig. 4–26*b*, is necessary in order to balance the "tipping effect" caused by **P**. For example, if **P** is applied at a height $h$ from the surface, Fig. 4–26*d*, then moment equilibrium about point $O$ is satisfied if $Wx = Ph$ or $x = Ph/W$. In particular, the block will be on the verge of *tipping* if $N$ acts at the right corner of the block, $x = a/2$.

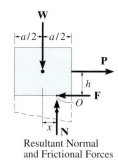

Resultant Normal
and Frictional Forces

(d)

**Fig. 4–26**

The heat generated by the abrasive action of friction can be noticed when using this grinder to sharpen a metal blade

*Besides mechanical interactions as explained here, which is referred to as a classical approach, a detailed treatment of the nature of frictional forces must also include the effects of temperature, density, cleanliness, and atomic or molecular attraction between the contacting surfaces. See J. Krim, *Scientific American*, October, 1996.

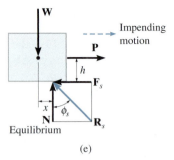

Fig. 4–26

**Impending Motion.** In cases where $h$ is small or the surfaces of contact are rather "slippery," the frictional force $\mathbf{F}$ may *not* be great enough to balance $\mathbf{P}$, and consequently the block will tend to slip *before* it can tip. In other words, as $P$ is slowly increased, $F$ correspondingly increases until it attains a certain *maximum value* $F_s$, called the *limiting static frictional force*, Fig. 4–26e. When this value is reached, the block is in *unstable equilibrium* since any further increase in $P$ will cause deformations and fractures at the points of surface contact, and consequently the block will begin to move. Experimentally, it has been determined that the limiting static frictional force $F_s$ is *directly proportional* to the resultant normal force $N$. This may be expressed mathematically as

$$F_s = \mu_s N \qquad (4\text{–}8)$$

where the constant of proportionality, $\mu_s$ (mu "sub" s), is called the *coefficient of static friction.*

Thus, when the block is on the *verge of sliding*, the normal force $\mathbf{N}$ and frictional force $\mathbf{F}_s$ combine to create a resultant $\mathbf{R}_s$, Fig. 4–26e. The angle $\phi_s$ that $\mathbf{R}_s$ makes with $\mathbf{N}$ is called the *angle of static friction*. From the figure,

$$\phi_s = \tan^{-1}\left(\frac{F_s}{N}\right) = \tan^{-1}\left(\frac{\mu_s N}{N}\right) = \tan^{-1}\mu_s$$

**Tabular Values of $\mu_s$.** Typical values for $\mu_s$, found in many engineering handbooks, are given in Table 4–3. Although this coefficient is generally less than 1, be aware that in some cases it is possible, as in the case of aluminum on aluminum, for $\mu_s$ to be greater than 1. Physically this means, of course, that in this case the frictional force is greater than the corresponding normal force. Furthermore, it should be noted that $\mu_s$ is dimensionless and depends only on the characteristics of the two surfaces in contact. A wide range of values is given for each value of $\mu_s$ since experimental testing was done under variable conditions of roughness and cleanliness of the contacting surfaces. For applications, therefore, it is important that both caution and judgment be exercised when selecting a coefficient of friction for a given set of conditions. When a more accurate calculation of $F_s$ is required, the coefficient of friction should be determined directly by an experiment that involves the two materials to be used.

| TABLE 4–3 • Typical Values for $\mu_s$ | |
| --- | --- |
| **Contact Materials** | **Coefficient of Static Friction ($\mu_s$)** |
| Metal on ice | 0.03–0.05 |
| Wood on wood | 0.30–0.70 |
| Leather on wood | 0.20–0.50 |
| Leather on metal | 0.30–0.60 |
| Aluminum on aluminum | 1.10–1.70 |

**Motion.** If the magnitude of **P** acting on the block is increased so that it becomes greater than $F_s$, the frictional force at the contacting surfaces drops slightly to a smaller value $F_k$, called the *kinetic frictional force*. The block will *not* be held in equilibrium ($P > F_k$); instead, it will begin to slide with increasing speed, Fig. 4–27a. The drop made in the frictional force magnitude, from $F_s$ (static) to $F_k$ (kinetic), can be explained by again examining the surfaces of contact, Fig. 4–27b. Here it is seen that when $P > F_s$, then $P$ has the capacity to shear off the peaks at the contact surfaces and cause the block to "lift" somewhat out of its settled position and "ride" on top of these peaks. Once the block begins to slide, high local temperatures at the points of contact cause momentary adhesion (welding) of these points. The continued shearing of these welds is the dominant mechanism creating friction. Since the resultant contact forces $\Delta \mathbf{R}_n$ are aligned slightly more in the vertical direction than before, they thereby contribute *smaller* frictional components, $\Delta \mathbf{F}_n$, than when the irregularities are meshed.

Experiments with sliding blocks indicate that the magnitude of the resultant frictional force $\mathbf{F}_k$ is directly proportional to the magnitude of the resultant normal force **N**. This may be expressed mathematically as

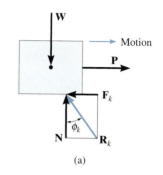

(a)

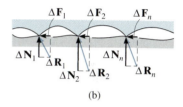

(b)

**Fig. 4–27**

$$F_k = \mu_k N \qquad (4\text{–}9)$$

Here the constant of proportionality, $\mu_k$, is called the *coefficient of kinetic friction*. Typical values for $\mu_k$ are approximately 25 percent *smaller* than those listed in Table 4–3 for $\mu_s$.

As shown in Fig. 4–27a, in this case, the resultant $\mathbf{R}_k$ has a line of action defined by $\phi_k$. This angle is referred to as the *angle of kinetic friction*, where

$$\phi_k = \tan^{-1}\left(\frac{F_k}{N}\right) = \tan^{-1}\left(\frac{\mu_k N}{N}\right) = \tan^{-1}\mu_k$$

By comparison, $\phi_s \geq \phi_k$.

The above effects regarding friction can be summarized by reference to the graph in Fig. 4–28, which shows the variation of the frictional force $F$ versus the applied load $P$. Here the frictional force is categorized in three different ways: namely, $F$ is a *static-frictional force* if equilibrium is maintained; $F$ is a *limiting static-frictional force* $F_s$ when it reaches a maximum value needed to maintain equilibrium; and finally, $F$ is termed a *kinetic-frictional force* $F_k$ when sliding occurs at the contacting surface. Notice also from the graph that for very large values of $P$ or for high speeds, because of aerodynamic effects, $F_k$ and likewise $\mu_k$ begin to decrease.

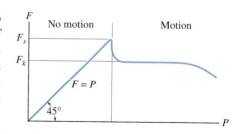

**Fig. 4–28**

**Characteristics of Dry Friction.** As a result of *experiments* that pertain to the foregoing discussion, the following rules which apply to bodies subjected to dry friction may be stated.

1. The frictional force acts *tangent* to the contacting surfaces in a direction *opposed* to the *relative motion* or tendency for motion of one surface against another.

2. The magnitude of the maximum static frictional force $\mathbf{F}_s$ that can be developed is independent of the area of contact, provided the normal pressure is not very low nor great enough to severely deform or crush the contacting surfaces of the bodies.

3. The magnitude of the maximum static frictional force is generally greater than the magnitude of the kinetic frictional force for any two surfaces of contact. However, if one of the bodies is moving with a *very low velocity* over the surface of another, $F_k$ becomes approximately equal to $F_s$, i.e., $\mu_s \approx \mu_k$.

4. When *slipping* at the surface of contact is *about to occur*, the magnitude of the maximum static frictional force is proportional to the magnitude of the normal force, such that $F_s = \mu_s N$, Eq. 4–8.

5. When *slipping* at the surface of contact is *occurring*, the magnitude of the kinetic frictional force is proportional to the magnitude of the normal force, such that $F_k = \mu_k N$, Eq. 4–9.

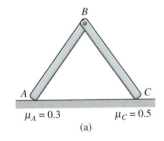

$\mu_A = 0.3$     $\mu_C = 0.5$

(a)

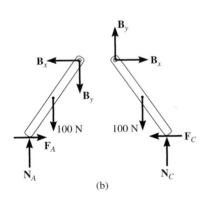

(b)

**Fig. 4–29**

**Types of Friction Problems.** If a body is in equilibrium when it is subjected to a system of forces that includes the effect of friction, the force system must satisfy not only the equations of equilibrium but *also* the laws that govern the frictional forces. In general, there are three types of mechanics problems involving dry friction. They can easily be classified once the free-body diagrams are drawn and the total number of unknowns are identified and compared with the total number of available equilibrium equations. Each type of problem will now be explained and illustrated graphically by examples. In all these cases the geometry and dimensions for the problem are assumed to be known.

**Equilibrium.** Problems in this category are strictly equilibrium problems which require *the total number of unknowns to be equal to the total number of available equilibrium equations*. Once the frictional forces are determined from the solution, however, their numerical values must be checked to be sure they satisfy the inequality $F \le \mu_s N$; otherwise, slipping will occur and the body will not remain in equilibrium. A problem of this type is shown in Fig. 4–29a. Here we must determine the frictional forces at $A$ and $C$ to check if the equilibrium position of the bars can be maintained. If the bars are uniform and have known weights of 100 N each, then the free-body diagrams are as shown in Fig. 4–29b. There are six unknown force components which can be determined *strictly* from the six equilibrium equations (three for each member). Once $\mathbf{F}_A$, $\mathbf{N}_A$, $\mathbf{F}_C$, and $N_C$ are determined, then the bars will remain in equilibrium provided $F_A \leq 0.3N_A$ and $F_C \leq 0.5N_C$ are satisfied.

**Impending Motion at All Points.** In this case *the total number of unknowns will equal the total number of available equilibrium equations plus the total number of available frictional equations,* $F = \mu N$. In particular, if *motion is impending* at the points of contact, then $F_s = \mu_s N$; whereas if the body is *slipping*, then $F_k = \mu_k N$. For example, consider the problem of finding the smallest angle $\theta$ at which the 100-N bar in Fig. 4–30*a* can be placed against the wall without slipping. The free-body diagram is shown in Fig. 4–30*b*. Here there are *five* unknowns: $F_A$, $N_A$, $F_B$, $N_B$, $\theta$. For the solution there are *three* equilibrium equations and *two* static frictional equations which apply at *both* points of contact, so that $F_A = 0.3N_A$ and $F_B = 0.4N_B$. (It should also be noted that the bar will not be in a state where motion impends *unless* the bar slips at *both* points $A$ and $B$ simultaneously.)

**Tipping or Impending Motion at Some Points.** Here *the total number of unknowns will be less than the number of available equilibrium equations plus the total number of frictional equations or conditional equations for tipping.* As a result, several possibilities for motion or impending motion will exist and the problem will involve a determination of the kind of motion which actually occurs. For example, consider the two-member frame shown in Fig. 4–31*a*. In this problem we wish to determine the horizontal force **P** needed to cause movement of the frame. If each member has a weight of 100 N, then the free-body diagrams are as shown in Fig. 4–31*b*. There are *seven* unknowns: $N_A$, $F_A$, $N_C$, $F_C$, $B_x$, $B_y$, $P$. For a unique solution we must satisfy the *six* equilibrium equations (three for each member) and only *one* of two possible static frictional equations. This means that as $P$ increases its magnitude it will either cause slipping at $A$ and no slipping at $C$, so that $F_A = 0.3N_A$ and $F_C \leq 0.5N_C$; or slipping occurs at $C$ and no slipping at $A$, in which case $F_C = 0.5N_C$ and $F_A \leq 0.3N_A$. The actual situation can be determined by calculating $P$ for each case and then choosing the case for which $P$ is *smallest.* If in both cases the *same value* for $P$ is calculated, which in practice would be highly improbable, then slipping at both points occurs simultaneously; i.e., the *seven unknowns* will satisfy *eight equations.*

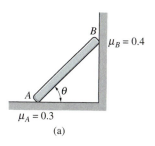

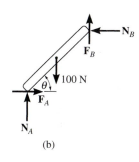

Fig. 4–30

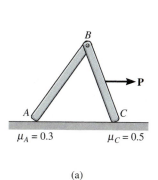

(a)

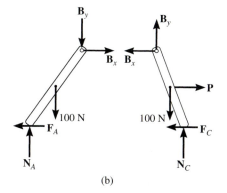

(b)

Fig. 4–31

As a second example, consider a block having a width $b$, height $h$, and weight $W$ which is resting on a rough surface, Fig. 4–32a. The force **P** needed to cause motion is to be determined. Inspection of the free-body diagram, Fig. 4–32b, indicates that there are *four unknowns*, namely, $P$, $F$, $N$, and $x$. For a unique solution, however, we must satisfy the *three* equilibrium equations and either *one* static friction equation or *one* conditional equation which requires the block not to tip. Hence two possibilities of motion exist. Either the block will *slip*, Fig. 4–32b, in which case $F = \mu_s N$ and the value obtained for $x$ must satisfy $0 \leq x \leq b/2$; or the block will *tip*, Fig. 4–32c, in which case $x = b/2$ and the frictional force will satisfy the inequality $F \leq \mu_s N$. The solution yielding the *smallest* value of $P$ will define the type of motion the block undergoes. If it happens that the same value of $P$ is calculated for both cases, although this would be very improbable, then slipping and tipping will occur simultaneously; i.e., the *four unknowns* will satisfy *five equations*.

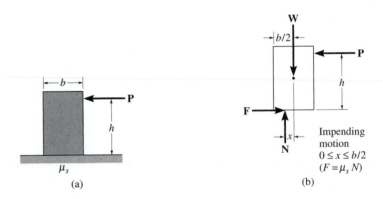

(a)

(b)

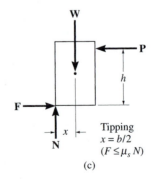

(c)

**Fig. 4–32**

**Equilibrium Versus Frictional Equations.** It was stated earlier that the frictional force *always* acts so as to either oppose the relative motion or impede the motion of a body over its contacting surface. Realize, however, that we can *assume* the sense of the frictional force in problems which require $F$ to be an "equilibrium force" and satisfy the inequality $F < \mu_s N$. The correct sense is made known *after* solving the equations of equilibrium for $F$. For example, if $F$ is a negative scalar, the sense of $\mathbf{F}$ is the reverse of that which was assumed. This convenience of *assuming* the sense of $\mathbf{F}$ is possible because the equilibrium equations equate to zero the *components of vectors* acting in the *same direction*. In cases where the frictional equation $F = \mu N$ is used in the solution of a problem, however, the convenience of *assuming* the sense of $\mathbf{F}$ is *lost*, since the frictional equation relates only the *magnitudes* of two perpendicular vectors. Consequently, $\mathbf{F}$ *must always* be shown acting with its *correct sense* on the free-body diagram whenever the frictional equation is used for the solution of a problem.

## PROCEDURE FOR ANALYSIS

The following procedure provides a method for solving equilibrium problems involving dry friction.

*Free-Body Diagrams.* Draw the necessary free-body diagrams and determine the number of unknowns or equations required for a complete solution. Unless stated in the problem, *always* show the frictional forces as *unknowns;* i.e., *do not assume that $F = \mu N$.* Recall that only three equations of coplanar equilibrium can be written for each body. Consequently, if there are more unknowns than equations of equilibrium, it will be necessary to apply the frictional equation at some, if not all, points of contact to obtain the extra equations needed for a complete solution.

*Equations of Friction and Equilibrium.* Apply the equations of equilibrium and the necessary frictional equations (or conditional equations if tipping is involved) and solve for the unknowns. If the problem involves a three-dimensional force system such that it becomes difficult to obtain the force components or the necessary moment arms, apply the equations of equilibrium using Cartesian vectors.

E X A M P L E **4.14**

The uniform crate shown in Fig. 4–33a has a mass of 20 kg. If a force $P = 80$ N is applied to the crate, determine if it remains in equilibrium. The coefficient of static friction is $\mu_s = 0.3$.

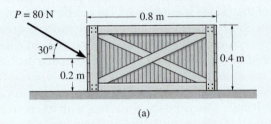

(a)

**Fig. 4–33**

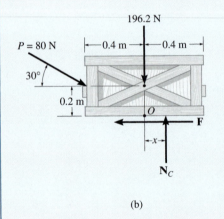

(b)

### Solution

*Free-Body Diagram.* As shown in Fig. 4–33b, the *resultant* normal force $\mathbf{N}_C$ must act a distance $x$ from the crate's center line in order to counteract the tipping effect caused by **P**. There are *three unknowns*, $F$, $N_C$, and $x$, which can be determined strictly from the *three* equations of equilibrium.

*Equations of Equilibrium.*

$$\xrightarrow{+} \Sigma F_x = 0; \qquad\qquad 80 \cos 30° - F = 0$$
$$+\uparrow \Sigma F_y = 0; \qquad\qquad -80 \sin 30° + N_C - 196.2 = 0$$
$$\downarrow + \Sigma M_O = 0; \quad 80 \sin 30°(0.4) - 80 \cos 30°(0.2) + N_C(x) = 0$$

Solving,

$$F = 69.3 \text{ N}$$
$$N_C = 236 \text{ N}$$
$$x = -0.00908 \text{ m} = -9.08 \text{ mm}$$

Since $x$ is negative it indicates that the *resultant* normal force acts (slightly) to the *left* of the crate's center line. No tipping will occur since $x \leq 0.4$ m. Also, the *maximum* frictional force which can be developed at the surface of contact is $F_{max} = \mu_s N_C = 0.3(236 \text{ N}) = 70.8$ N. Since $F = 69.3$ N $< 70.8$ N, the crate will *not slip*, although it is very close to doing so.

# E X A M P L E  4.15

It is observed that when the bed of the dump truck is raised to an angle
of $\theta = 25°$ the vending machines begin to slide off the bed, Fig. 4–34a.
Determine the static coefficient of friction between them and the
surface of the truck.

(a)

## Solution

An idealized model of a vending machine resting on the bed of the
truck is shown in Fig. 4–34b. The dimensions have been measured and
the center of gravity has been located. We will assume that the
machine weighs $W$.

*Free-Body Diagram.*  As shown in Fig. 4–34c, the dimension $x$ is used
to locate the position of the resultant normal force $\mathbf{N}$. There are four
unknowns, $N$, $F$, $\mu_s$, and $x$.

*Equations of Equilibrium.*

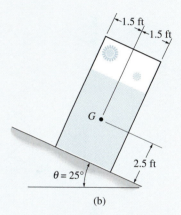

(b)

$$+\searrow\Sigma F_x = 0; \qquad W \sin 25° - F = 0 \qquad (1)$$
$$+\nearrow\Sigma F_y = 0; \qquad N - W \cos 25° = 0 \qquad (2)$$
$$\lfloor+\Sigma M_O = 0; \quad -W \sin \theta\,(2.5\text{ ft}) + W \cos \theta\,(x) = 0 \qquad (3)$$

Since slipping impends at $\theta = 25°$, using the first two equations, we
have

$$F_s = \mu_s N; \qquad W \sin 25° = \mu_s (W \cos 25°)$$
$$\mu_s = \tan 25° = 0.466 \qquad \textit{Ans.}$$

The angle of $\theta = 25°$ is referred to as the *angle of repose*, and by
comparison, it is equal to the angle of static friction $\theta = \phi_s$. Notice
from the calculation that $\theta$ is independent of the weight of the vending
machine, and so knowing $\theta$ provides a convenient method for
determining the coefficient of static friction.

From Eq. 3, with $\theta = 25°$, we find $x = 1.17$ ft. Since 1.17 ft $<$ 1.5 ft,
indeed the vending machine will slip before it can tip as observed in
Fig. 4–34a.

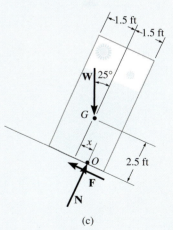

(c)

Fig. 4–34

## E X A M P L E  4.16

The uniform rod having a weight $W$ and length $l$ is supported at its ends against the surface at $A$ and $B$ in Fig. 4–35a. If the rod is on the verge of slipping when $\theta = 30°$, determine the coefficient of static friction $\mu_s$ at $A$ and $B$. Neglect the thickness of the rod for the calculation.

### Solution

*Free-Body Diagram.* As shown in Fig. 4–35b, there are *five* unknowns: $F_A$, $N_A$, $F_B$, $N_B$, and $\mu_s$. These can be determined from the *three* equilibrium equations and *two* frictional equations applied at points $A$ and $B$. The frictional forces must be drawn with their correct sense so that they oppose the tendency for motion of the rod. Why?

*Equations of Friction and Equilibrium.* Writing the frictional equations,

$$F = \mu_s N; \qquad\qquad F_A = \mu_s N_A$$
$$F_B = \mu_s N_B$$

Using these results and applying the equations of equilibrium yields

$$\xrightarrow{+} \Sigma F_x = 0; \quad \mu_s N_A + \mu_s N_B \cos 30° - N_B \sin 30° = 0 \tag{1}$$
$$+\uparrow \Sigma F_y = 0; \quad N_A - W + N_B \cos 30° + \mu_s N_B \sin 30° = 0 \tag{2}$$
$$\curvearrowright + \Sigma M_A = 0; \quad N_B l - W\left(\frac{l}{2}\right)\cos 30° = 0 \tag{3}$$

$$N_B = 0.4330\,W$$

From Eqs. 1 and 2,

$$\mu_s N_A = 0.2165\,W - (0.3750\,W)\mu_s$$
$$N_A = 0.6250\,W - (0.2165\,W)\mu_s$$

By division,

$$0.6250\mu_s - 0.2165\mu_s^2 = 0.2165 - 0.375\mu_s$$

or,

$$\mu_s^2 - 4.619\,\mu_s + 1 = 0$$

Solving for the smallest root,

$$\mu_s = 0.228 \qquad\qquad\qquad Ans.$$

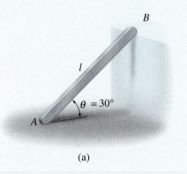

(a)

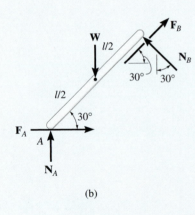

(b)

Fig. 4–35

# EXAMPLE 4.17

The concrete pipes are stacked in the yard as shown in Fig. 4–36a. Determine the minimum coefficient of static friction at each point of contact so that the pile does not collapse.

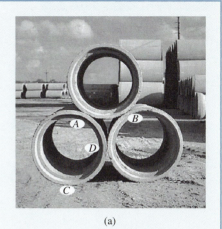

(a)

### Solution

*Free-body Diagrams*   Recognize that the coefficient of static friction between two pipes, at $A$ and $B$, and between a pipe and the ground, at $C$, will be different since the contacting surfaces are different. We will assume each pipe has an outer radius $r$ and weight $W$. The free-body diagrams for two of the pipes are shown in Fig. 4–36b. There are six unknowns, $N_A$, $F_A$, $N_B$, $F_B$, $N_C$, $F_C$. (Note that when collapse is about to occur the normal force at $D$ is zero.) Since only the six equations of equilibrium are necessary to obtain the unknowns, the sense of direction of the frictional forces can be verified from the solution.

*Equations of Equilibrium.*   For the top pipe we have

$$\zeta + \Sigma M_O = 0; \quad -F_A(r) + F_B(r) = 0; \quad F_A = F_B = F$$
$$\overset{+}{\rightarrow} \Sigma F_x = 0; \quad N_A \sin 30° - F \cos 30° - N_B \sin 30° + F \cos 30° = 0$$
$$N_A = N_B = N$$
$$+\uparrow \Sigma F_y = 0; \quad 2N \cos 30° + 2F \sin 30° - W = 0 \qquad (1)$$

For the bottom pipe, using $F_A = F$ and $N_A = N$, we have,

$$\zeta + \Sigma M_{O'} = 0; \quad F_C(r) - F(r) = 0; \quad F_C = F$$
$$\overset{+}{\rightarrow} \Sigma F_x = 0; \quad -N \sin 30° + F \cos 30° + F = 0 \qquad (2)$$
$$+\uparrow \Sigma F_y = 0; \quad N_C - W - N \cos 30° - F \sin 30° = 0 \qquad (3)$$

From Eq. 2, $F = 0.2679 \, N$, so that between the pipes

$$(\mu_s)_{min} = \frac{F}{N} = 0.268 \qquad \qquad Ans.$$

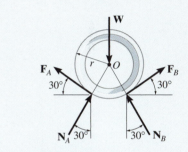

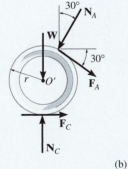

(b)

Fig. 4–36

Using this result in Eq. 1.

$$N = 0.5 \, W$$

From Eq. 3,

$$N_C - W - (0.5 \, W) \cos 30° - 0.2679 \, (0.5 \, W) \sin 30° = 0$$
$$N_C = 1.5 \, W$$

At the ground, the smallest required coefficient of static friction would be

$$(\mu_s')_{min} = \frac{F}{N_C} = \frac{0.2679(0.5 \, W)}{1.5 \, W} = 0.0893 \qquad Ans.$$

Hence a greater coefficient of static friction is required between the pipes than that required at the ground; and so it is likely that if slipping would occur between the pipes the bottom two pipes would roll away from one another without slipping as the top pipe falls downward.

**EXAMPLE 4.18**

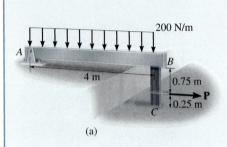

(a)

Beam $AB$ is subjected to a uniform load of 200 N/m and is supported at $B$ by post $BC$, Fig. 4–37a. If the coefficients of static friction at $B$ and $C$ are $\mu_B = 0.2$ and $\mu_C = 0.5$, determine the force **P** needed to pull the post out from under the beam. Neglect the weight of the members and the thickness of the post.

**Solution**

*Free-Body Diagrams.* The free-body diagram of beam $AB$ is shown in Fig. 4–37b. Applying $\Sigma M_A = 0$, we obtain $N_B = 400$ N. This result is shown on the free-body diagram of the post, Fig. 4–37c. Referring to this member, the *four* unknowns $F_B$, $P$, $F_C$, and $N_C$ are determined from the *three* equations of equilibrium and *one* frictional equation applied either at $B$ or $C$.

*Equations of Equilibrium and Friction.*

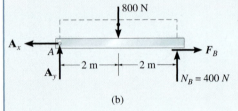

(b)

$$\xrightarrow{+} \Sigma F_x = 0; \qquad\qquad P - F_B - F_C = 0 \qquad\qquad (1)$$

$$+\uparrow \Sigma F_y = 0; \qquad\qquad N_C - 400\ \text{N} = 0 \qquad\qquad (2)$$

$$\zeta+\Sigma M_C = 0; \qquad -P(0.25\ \text{m}) + F_B(1\ \text{m}) = 0 \qquad (3)$$

*(Post Slips Only at B)* This requires $F_C \leq \mu_C N_C$ and

$$F_B = \mu_B N_B; \qquad\qquad F_B = 0.2(400\ \text{N}) = 80\ \text{N}$$

Using this result and solving Eqs. 1 through 3, we obtain

$$P = 320\ \text{N}$$
$$F_C = 240\ \text{N}$$
$$N_C = 400\ \text{N}$$

Since $F_C = 240$ N $> \mu_C N_C = 0.5(400$ N$) = 200$ N, the other case of movement must be investigated.

*(Post Slips Only at C.)* Here $F_B \leq \mu_B N_B$ and

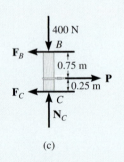

(c)

Fig. 4–37

$$F_C = \mu_C N_C; \qquad\qquad F_C = 0.5\ N_C \qquad\qquad (4)$$

Solving Eqs. 1 through 4 yields

$$P = 267\ \text{N} \qquad\qquad\qquad Ans.$$
$$N_C = 400\ \text{N}$$
$$F_C = 200\ \text{N}$$
$$F_B = 66.7\ \text{N}$$

Obviously, this case occurs first since it requires a *smaller* value for $P$.

# E X A M P L E  **4.19**

Determine the normal force $P$ that must be exerted on the rack to begin pushing the 100-kg pipe shown in Fig. 4–38a up the 20° incline. The coefficients of static friction at the points of contact are $(\mu_s)_A = 0.15$, and $(\mu_s)_B = 0.4$.

### Solution

***Free-Body Diagram.***   As shown in Fig. 4–38b, the rack must exert a force $P$ on the pipe due to force equilibrium in the $x$ direction. There are four unknowns $P$, $F_A$, $N_A$, and $F_B$ acting on the pipe Fig. 4–38c. These can be determined from the *three* equations of equilibrium and *one* frictional equation, which apply either at $A$ or $B$. If slipping begins to occur only at $B$, the pipe will begin to roll up the incline; whereas if slipping occurs only at $A$, the pipe will begin to *slide* up the incline. Here we must find $N_B$.

***Equations of Equilibrium and Friction (for Fig. 4–38c)***

$$+\nearrow\Sigma F_x = 0; \qquad -F_A + P - 981 \sin 20° \text{ N} = 0 \qquad (1)$$

$$+\nwarrow\Sigma F_y = 0; \qquad N_A - F_B - 981 \cos 20° \text{ N} = 0 \qquad (2)$$

$$\downarrow{+}\Sigma M_O = 0; \qquad F_B(400 \text{ mm}) - F_A(400 \text{ mm}) = 0 \qquad (3)$$

***(Pipe Rolls up Incline.)***   In this case $F_A \le 0.15N_A$ and

$$(F_s)_B = (\mu_s)_B N_B; \qquad F_B = 0.4P \qquad (4)$$

The direction of the frictional force at $B$ must be specified correctly. Why? Since the spool is being forced up the incline, $\mathbf{F}_B$ acts downward to prevent any clockwise rolling motion of the pipe, Fig. 4–38c. Solving Eqs. 1 through 4, we have

$$N_A = 1146 \text{ N} \quad F_A = 224 \text{ N} \quad F_B = 224 \text{ N} \quad P = 559 \text{ N}$$

The assumption regarding no slipping at $A$ should be checked.

$$F_A \overset{?}{\le} (\mu_s)_A N_A; \quad 224 \text{ N} \overset{?}{\le} 0.15(1146 \text{ N}) = 172 \text{ N}$$

The inequality does *not apply*, and therefore slipping occurs at $A$ and not at $B$. Hence, the other case of motion will occur.

***(Pipe Slides up Incline.)***   In this case, $P \le 0.4N_B$ and

$$(F_s)_A = (\mu_s)_A N_A; \qquad F_A = 0.15N_A \qquad (5)$$

Solving Eqs. 1 through 3 and 5 yields

$$N_A = 1085 \text{ N} \quad F_A = 163 \text{ N} \quad F_B = 163 \text{ N} \quad P = 498 \text{ N} \qquad \textit{Ans.}$$

The validity of the solution ($P = 498$ N) can be checked by testing the assumption that indeed no slipping occurs at $B$.

$$F_B \le (\mu_s)_B P; \quad 163 \text{ N} < 0.4(498 \text{ N}) = 199 \text{ N} \qquad \text{(check)}$$

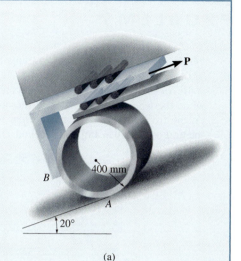

(a)

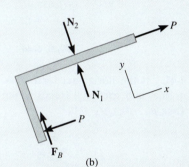

(b)

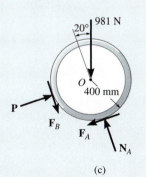

(c)

**Fig. 4–38**

# PROBLEMS

*4-40. The mine car and its contents have a total mass of 6 Mg and a center of gravity at $G$. If the coefficient of static friction between the wheels and the tracks is $\mu_s = 0.4$ when the wheels are locked, find the normal force acting on the front wheels at $B$ and the rear wheels at $A$ when the brakes at both $A$ and $B$ are locked. Does the car move?

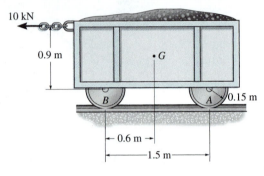

Prob. 4-40

4-41. If the horizontal force $P = 80$ lb, determine the normal and frictional forces acting on the 300-lb crate. Take $\mu_s = 0.3$, $\mu_k = 0.2$.

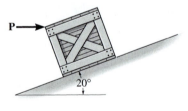

Prob. 4-41

4-42. The uniform pole has a weight of 30 lb and a length of 26 ft. If it is placed against the smooth wall and on the rough floor in the position $d = 10$ ft, will it remain in this position when it is released? The coefficient of static friction is $\mu_s = 0.3$.

4-43. The uniform pole has a weight of 30 lb and a length of 26 ft. Determine the maximum distance $d$ it can be placed from the smooth wall and not slip. The coefficient of static friction between the floor and the pole is $\mu_s = 0.3$.

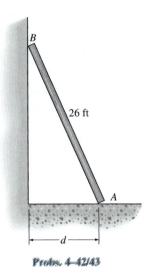

Probs. 4-42/43

*4-44. The uniform 20-lb ladder rests on the rough floor for which the coefficient of static friction is $\mu_s = 0.8$ and against the smooth wall at $B$. Determine the horizontal force $P$ the man must exert on the ladder in order to cause it to move.

4-45. The uniform 20-lb ladder rests on the rough floor for which the coefficient of static friction is $\mu_s = 0.4$ and against the smooth wall at $B$. Determine the horizontal force $P$ the man must exert on the ladder in order to cause it to move.

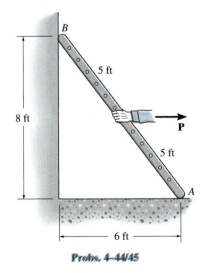

Probs. 4-44/45

**4-46.** An axial force of $T = 800$ lb is applied to the bar. If the coefficient of static friction at the jaws $C$ and $D$ is $\mu_s = 0.5$, determine the smallest normal force that the screw at $A$ must exert on the smooth surface of the links at $B$ and $C$ in order to hold the bar stationary. The links are pin-connected at $F$ and $G$.

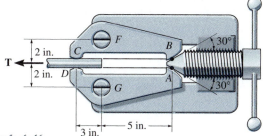

**Prob. 4-46**

**4-47.** The winch on the truck is used to hoist the garbage bin onto the bed of the truck. If the loaded bin has a weight of 8500 lb and center of gravity at $G$, determine the force in the cable needed to begin the lift. The coefficients of static friction at $A$ and $B$ are $\mu_A = 0.3$ and $\mu_B = 0.2$, respectively. Neglect the height of the support at $A$.

**Prob. 4-47**

**\*4-48.** The 15-ft ladder has a uniform weight of 80 lb and rests against the smooth wall at $B$. If the coefficient of static friction at $A$ is $\mu_A = 0.4$, determine if the ladder will slip. Take $\theta = 60°$.

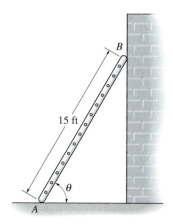

**Prob. 4-48**

**4-49.** The block brake is used to stop the wheel from rotating when the wheel is subjected to a couple moment $\mathbf{M}_0$. If the coefficient of static friction between the wheel and the block is $\mu_s$, determine the smallest force $P$ that should be applied.

**4-50.** Show that the brake in Prob. 4-49 is self locking, i.e., $P \leq 0$, provided $b/c \leq \mu_s$.

**4-51.** Solve Prob. 4-49 if the couple moment $\mathbf{M}_0$ is applied counterclockwise.

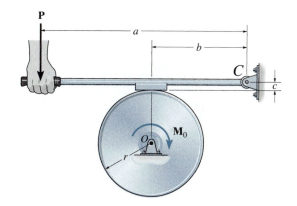

**Probs. 4-49/50/51**

**\*4-52.** The block brake consists of a pin-connected lever and friction block at $B$. The coefficient of static friction between the wheel and the lever is $\mu_s = 0.3$, and a torque of 5 N·m is applied to the wheel. Determine if the brake can hold the wheel stationary when the force applied to the lever is (a) $P = 30$ N, (b) $P = 70$ N.

**4-53.** Solve Prob. 4-52 if the 5-N·m torque is applied counter-clockwise.

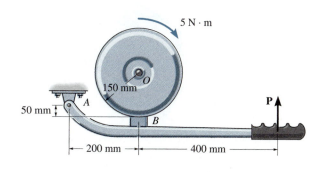

**Probs. 4-52/53**

**4-54.** The tractor has a weight of 4500 lb with center of gravity at $G$. The driving traction is developed at the rear wheels $B$, while the front wheels at $A$ are free to roll. If the coefficient of static friction between the wheels at $B$ and the ground is $\mu_s = 0.5$, determine if it is possible to pull at $P = 1200$ lb without causing the wheels at $B$ to slip or the front wheels at $A$ to lift off the ground.

**\*4-56.** The drum has a weight of 100 lb and rests on the floor for which the coefficient of static friction is $\mu_s = 0.6$. If $a = 2$ ft and $b = 3$ ft, determine the smallest magnitude of the force $P$ that will cause impending motion of the drum.

**4-57.** The drum has a weight of 100 lb and rests on the floor for which the coefficient of static friction is $\mu_s = 0.5$. If $a = 3$ ft and $b = 4$ ft, determine the smallest magnitude of the force $P$ that will cause impending motion of the drum.

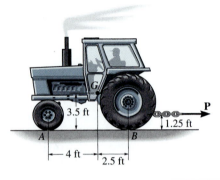

Prob. 4–54

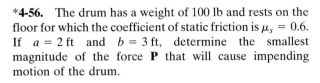

Probs. 4–56/57

**4-55.** The car has a mass of 1.6 Mg and center of mass at $G$. If the coefficient of static friction between the shoulder of the road and the tires is $\mu_s = 0.4$, determine the greatest slope $\theta$ the shoulder can have without causing the car to slip or tip over if the car travels along the shoulder at constant velocity.

**4-58.** The coefficient of static friction between the shoes at $A$ and $B$ of the tongs and the pallet is $\mu'_s = 0.5$, and between the pallet and the floor $\mu_s = 0.4$. If a horizontal towing force of $P = 300$ N is applied to the tongs, determine the largest mass that can be towed.

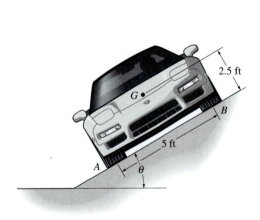

Prob. 4–55

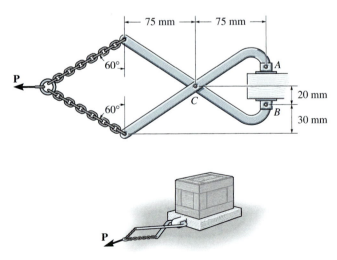

Prob. 4–58

**4-59.** The pipe is hoisted using the tongs. If the coefficient of static friction at $A$ and $B$ is $\mu_s$, determine the smallest dimension $b$ so that any pipe of inner diameter $d$ can be lifted.

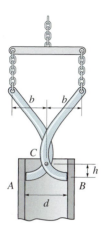

**Prob. 4–59**

***4-60.** Determine the maximum weight $W$ the man can lift with constant velocity using the pulley system, without and then with the "leading block" or pulley at $A$. The man has a weight of 200 lb and the coefficient of static friction between his feet and the ground is $\mu_s = 0.6$.

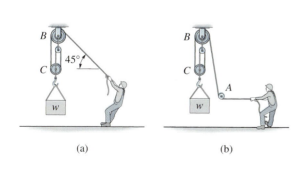

(a)                    (b)

**Prob. 4–60**

**4-61.** The uniform dresser has a weight of 90 lb and rests on a tile floor for which $\mu_s = 0.25$. If the man pushes on it in the horizontal direction $\theta = 0°$, determine the smallest magnitude of force **F** needed to move the dresser. Also, if the man has a weight of 150 lb, determine the smallest coefficient of static friction between his shoes and the floor so that he does not slip.

**4-62.** The uniform dresser has a weight of 90 lb and rests on a tile floor for which $\mu_s = 0.25$. If the man pushes on it in the direction $\theta = 30°$, determine the smallest magnitude of force **F** needed to move the dresser. Also, if the man has a weight of 150 lb, determine the smallest coefficient of static friction between his shoes and the floor so that he does not slip.

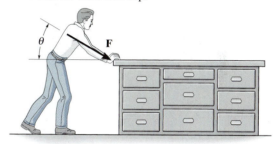

**Probs. 4–61/62**

**4-63.** The 5-kg cylinder is suspended from two equal-length cords. The end of each cord is attached to a ring of negligible mass, which passes along a horizontal shaft. If the coefficient of static friction between each ring and the shaft is $\mu_s = 0.5$, determine the greatest distance $d$ by which the rings can be separated and still support the cylinder.

**Prob. 4–63**

*4-64. The board can be adjusted vertically by tilting it up and sliding the smooth pin A along the vertical guide G. When placed horizontally, the bottom C then bears along the edge of the guide, where $\mu_s = 0.4$. Determine the largest dimension d which will support any applied force **F** without causing the board to slip downward.

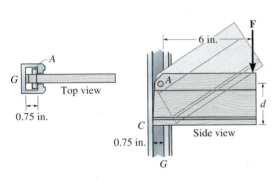

**Prob. 4–64**

4-65. The homogeneous semicylinder has a mass m and mass center at G. Determine the largest angle $\theta$ of the inclined plane upon which it rests so that it does not slip down the plane. The coefficient of static friction between the plane and the cylinder is $\mu_s = 0.3$. Also, what is the angle $\phi$ for this case?

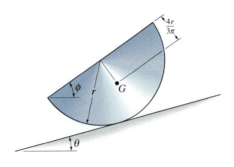

**Prob. 4–65**

4-66. Car A has a mass of 1.4 Mg and mass center at G. If car B exerts a horizontal force on A of 2 kN, determine if this force is great enough to move car A. The coefficients of static and kinetic friction between the tires and the road are $\mu_s = 0.5$ and $\mu_k = 0.35$. Assume B's bumper is smooth.

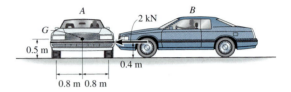

**Prob. 4–66**

4-67. A 35-kg disk rests on an inclined surface for which $\mu_s = 0.2$. Determine the maximum vertical force **P** that may be applied to link AB without causing the disk to slip at C.

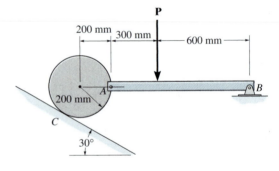

**Prob. 4–67**

*4-68. The crate has a weight W and the coefficient of static friction at the surface is $\mu_s = 0.3$. Determine the orientation of the cord and the smallest possible force **P** that has to be applied to the cord so that the crate is on the verge of moving.

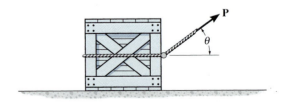

**Prob. 4–68**

**4-69.** The 800-lb concrete pipe is being lowered from the truck bed when it is in the position shown. If the coefficient of static friction at the points of support $A$ and $B$ is $\mu_s = 0.4$, determine where it begins to slip first: at $A$ or $B$, or both at $A$ and $B$.

**4-71.** The semicylinder of mass $m$ and radius $r$ lies on the rough inclined plane for which $\phi = 10°$ and the coefficient of static friction is $\mu_s = 0.3$. Determine if the semicylinder slides down the plane, and if not, find the angle of tip $\theta$ of its base $AB$.

**\*4-72.** The semicylinder of mass $m$ and radius $r$ lies on the rough inclined plane. If the inclination $\phi = 15°$, determine the smallest coefficient of static friction which will prevent the semicylinder from slipping.

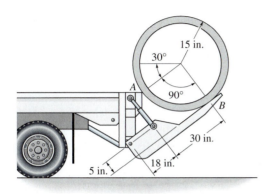

Prob. 4–69

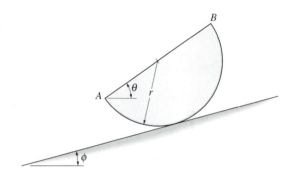

Probs. 4–71/72

**4-70.** The friction pawl is pinned at $A$ and rests against the wheel at $B$. It allows freedom of movement when the wheel is rotating counterclockwise about $C$. Clockwise rotation is prevented due to friction of the pawl which tends to bind the wheel. If $(\mu_s)_B = 0.6$, determine the design angle $\theta$ which will prevent clockwise motion for any value of applied moment $M$. *Hint:* Neglect the weight of the pawl so that it becomes a two-force member.

**4-73.** The door brace $AB$ is to be designed to prevent opening the door. If the brace forms a pin connection under the doorknob and the coefficient of static friction with the floor is $\mu_s = 0.5$, determine the largest length $L$ the brace can have to prevent the door from being opened. Neglect the weight of the brace.

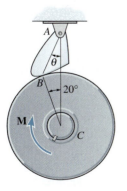

Prob. 4–70

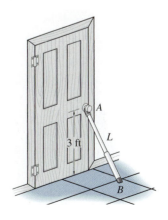

Prob. 4–73

# CHAPTER REVIEW

- *Free-Body Diagram.*  Before analyzing any equilibrium problem it is first necessary to draw a free-body diagram. This is an outlined shape of the body, which shows all the forces and couple moments that act on the body. Remember that a support will exert a *force* on the body in a particular direction if it prevents *translation* of the body in that direction, and it will exert a *couple moment* on the body if it prevents *rotation*. Angles used to resolve forces, and dimensions used to take moments of the forces, should also be shown on the free-body diagram.

- *Two Dimensions.*  Normally the three scalar equations of equilibrium, $\Sigma F_x = 0$, $\Sigma F_y = 0$, $\Sigma M_o = 0$, can be applied when solving problems in two dimensions, since the geometry is easy to visualize. For the most direct solution, try to sum forces along an axis that will eliminate as many unknown forces as possible. Sum moments about a point $O$ that passes through the line of action of as many unknown forces as possible.

- *Three Dimensions.*  In three dimensions, it is often advantageous to use a Cartesian vector analysis when applying the equations of equilibrium. To do this, first express each known and unknown force and couple moment shown on the free-body diagram as a Cartesian vector. Then set the force summation equal to zero, $\Sigma \mathbf{F} = \mathbf{0}$. Take moments about a point $O$ that lies on the line of action of as many unknown force components as possible. From point $O$ direct position vectors to each force, and then use the cross product to determine the moment of each force. Require $\Sigma \mathbf{M}_o = \Sigma (\mathbf{r} \times \mathbf{F}) = \mathbf{0}$. The six scalar equations of equilibrium are established by setting the respective $\mathbf{i}, \mathbf{j}$, and $\mathbf{k}$ components of these force and moment sums equal to zero.

- *Dry Friction.*  Frictional forces exist at rough surfaces of contact. They act on a body so as to oppose the motion or tendency of motion of the body. A static friction force approaches a maximum value of $F_s = \mu_s N$, where $\mu_s$ is the *coefficient of static friction*. In this case motion between the contacting surfaces is about to impend. If slipping occurs, then the friction force remains essentially constant and equal to a value of $F_k = \mu_k N$. Here $\mu_k$ is the *coefficient of kinetic friction*. The solution of a problem involving friction requires first drawing the free-body diagram of the body. If the unknowns cannot be determined strictly from the equations of equilibrium, and the possibility of slipping can occur, then the friction equation should be applied at the appropriate points of contact in order to complete the solution. It may also be possible for slender objects to tip over, and this situation should also be investigated.

# REVIEW PROBLEMS

**4-74.** The horizontal beam is supported by springs at its ends. Each spring has a stiffness of $k = 5$ kN/m and is originally unstretched so that the beam is in the horizontal position. Determine the angle of tilt of the beam if a load of 800 N is applied at point $C$ as shown.

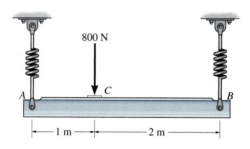

**Prob. 4-74**

**4-75.** The horizontal beam is supported by springs at its ends. If the stiffness of the spring at $A$ is $k_A = 5$ kN/m, determine the required stiffness of the spring at $B$ so that if the beam is loaded with the 800-N force it remains in the horizontal position both before and after loading.

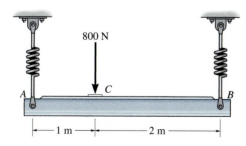

**Prob. 4-75**

**\*4-76.**  A vertical force of 80 lb acts on the crankshaft. Determine the horizontal equilibrium force **P** that must be applied to the handle and the x, y, z components of force at the smooth journal bearing A and the thrust bearing B. The bearings are properly aligned and exert only force reactions on the shaft.

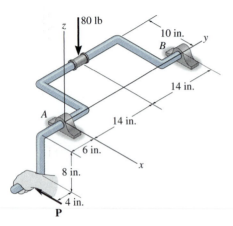

Prob. 4–76

**4-77.**  The spring has a stiffness of $k = 80$ N/m and an unstretched length of 2 m. Determine the force in cables BC and BD when the spring is held in the position shown.

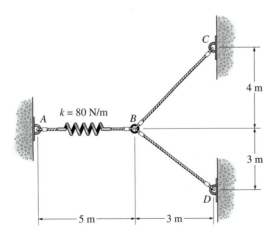

Prob. 4–77

**4-78.**  A skeletal diagram of the lower leg is shown in the lower figure. Here it can be noted that this portion of the leg is lifted by the quadriceps muscle attached to the hip at A and to the patella bone at B. This bone slides freely over cartilage at the knee joint. The quadriceps is further extended and attached to the tibia at C. Using the mechanical system shown in the upper figure to model the lower leg, determine the tension **T** in the quadriceps and the magnitude of the resultant force at the femur (pin), D, in order to hold the lower leg in the position shown. The lower leg has a mass of 3.2 kg and a mass center at $G_1$, the foot has a mass of 1.6 kg and a mass center at $G_2$.

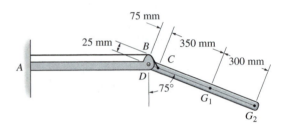

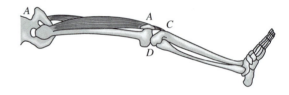

Prob. 4–78

**4-79.** The man has a mass of 40 kg. He plans to scale the vertical crevice using the method shown. If the coefficient of static friction between his shoes and the rock is $\mu_s = 0.4$ and between his backside and the rock, $\mu'_s = 0.3$, determine the smallest horizontal force his body must exert on the rock in order to do this.

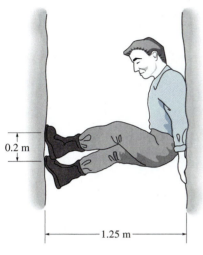

Prob. 4–79

**4-81.** The stiff-leg derrick used on ships is supported by a ball-and-socket joint at $D$ and two cables $BA$ and $BC$. The cables are attached to a smooth collar ring at $B$, which allows rotation of the derrick about the $z$ axis. If the derrick supports a crate having a mass of 100 kg, determine the tension in the supporting cables and the $x$, $y$, $z$ components of reaction at $D$.

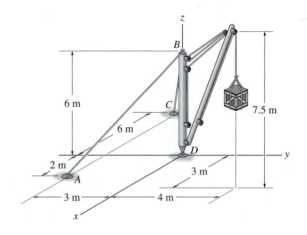

Prob. 4–81

**\*4-80.** Determine the reactions at the supports $A$ and $B$ of the frame.

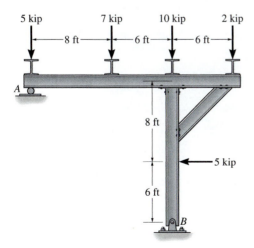

Prob. 4–80

The forces within the members of this truss bridge must be determined if they are to be properly designed.

# CHAPTER 5

# Structural Analysis

## CHAPTER OBJECTIVES

- To show how to determine the forces in the members of a truss using the method of joints and the method of sections.
- To analyze the forces acting on the members of frames and machines composed of pin-connected members.

## 5.1 Simple Trusses

A *truss* is a structure composed of slender members joined together at their end points. The members commonly used in construction consist of wooden struts or metal bars. The joint connections are usually formed by bolting or welding the ends of the members to a common plate, called a *gusset plate*, as shown in Fig. 5–1a, or by simply passing a large bolt or pin through each of the members, Fig. 5–1b.

(a)

(b)

Fig. 5–1

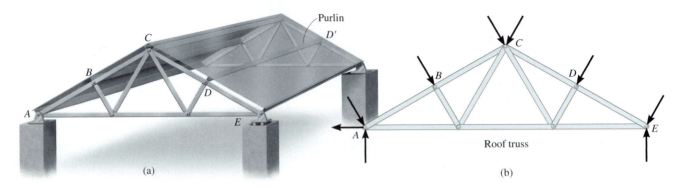

Fig. 5–2

**Planar Trusses.** *Planar* trusses lie in a single plane and are often used to support roofs and bridges. The truss *ABCDE*, shown in Fig. 5–2a, is an example of a typical roof-supporting truss. In this figure, the roof load is transmitted to the truss *at the joints* by means of a series of *purlins*, such as *DD'*. Since the imposed loading acts in the same plane as the truss, Fig. 5–2b, the analysis of the forces developed in the truss members is two-dimensional.

In the case of a bridge, such as shown in Fig. 5–3a, the load on the deck is first transmitted to *stringers*, then to *floor beams*, and finally to the *joints B, C*, and *D* of the two supporting side trusses. Like the roof truss, the bridge truss loading is also coplanar, Fig. 5–3b.

When bridge or roof trusses extend over large distances, a rocker or roller is commonly used for supporting one end, e.g., joint *E* in Figs. 5–2a and 5–3a. This type of support allows freedom for expansion or contraction of the members due to temperature or application of loads.

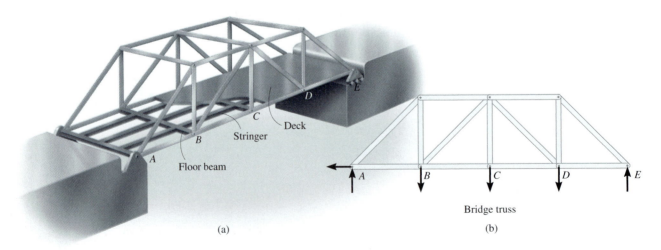

Fig. 5–3

**Assumptions for Design.** To design both the members and the connections of a truss, it is first necessary to determine the *force* developed in each member when the truss is subjected to a given loading. In this regard, two important assumptions will be made:

1. *All loadings are applied at the joints.* In most situations, such as for bridge and roof trusses, this assumption is true. Frequently in the force analysis the weight of the members is neglected since the forces supported by the members are usually large in comparison with their weight. If the member's weight is to be included in the analysis, it is generally satisfactory to apply it as a vertical force, half of its magnitude applied at each end of the member.

2. *The members are joined together by smooth pins.* In cases where bolted or welded joint connections are used, this assumption is satisfactory provided the center lines of the joining members are *concurrent*, as in Fig. 5–1a.

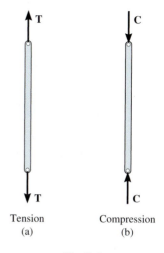

Tension
(a)

Compression
(b)

**Fig. 5–4**

Because of these two assumptions, *each truss member acts as a two-force member*, and therefore the forces at the ends of the member must be directed along the axis of the member. If the force tends to *elongate* the member, it is a *tensile force* (T), Fig. 5–4a; whereas if it tends to *shorten* the member, it is a *compressive force* (C), Fig. 5–4b. In the actual design of a truss it is important to state whether the nature of the force is tensile or compressive. Often, compression members must be made *thicker* than tension members because of the buckling or column effect that occurs when a member is in compression.

**Simple Truss.** To prevent collapse, the form of a truss must be rigid. Obviously, the four-bar shape *ABCD* in Fig. 5–5 will collapse unless a diagonal member, such as *AC*, is added for support. The simplest form that is rigid or stable is a *triangle*. Consequently, a *simple truss* is constructed by *starting* with a basic triangular element, such as *ABC* in Fig. 5–6, and connecting two members (*AD* and *BD*) to form an additional element. As each additional element consisting of two members and a joint is placed on the truss, it is possible to construct a simple truss.

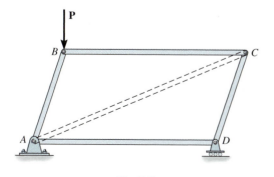

**Fig. 5–5**

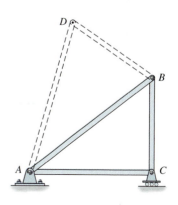

**Fig. 5–6**

## 5.2  The Method of Joints

These Howe trusses are used to support the roof of the metal building. Note how the members come together at a common point on the gusset plate and how the roof purlins transmit the load to the joints.

In order to analyze or design a truss, we must obtain the force in each of its members. If we were to consider a free-body diagram of the entire truss, then the forces in the members would be *internal forces,* and they could not be obtained from an equilibrium analysis. Instead, if we consider the equilibrium of a joint of the truss then a member force becomes an *external force* on the joint's free-body diagram, and the equations of equilibrium can be applied to obtain its magnitude. This forms the basis for the *method of joints.*

Because the truss members are all straight two-force members lying in the same plane, the force system acting at each joint is *coplanar and concurrent.* Consequently, rotational or moment equilibrium is automatically satisfied at the joint (or pin), and it is only necessary to satisfy $\Sigma F_x = 0$ and $\Sigma F_y = 0$ to ensure equilibrium.

When using the method of joints, it is *first* necessary to draw the joint's free-body diagram before applying the equilibrium equations. To do this, recall that the *line of action* of each member force acting on the joint is *specified* from the geometry of the truss since the force in a member passes along the axis of the member. As an example, consider the pin at joint $B$ of the truss in Fig. 5–7a. Three forces act on the pin, namely, the 500-N force and the forces exerted by members $BA$ and $BC$. The free-body diagram is shown in Fig. 5–7b. As shown, $\mathbf{F}_{BA}$ is "pulling" on the pin, which means that member $BA$ is in *tension;* whereas $\mathbf{F}_{BC}$ is "pushing" on the pin, and consequently member $BC$ is in *compression.* These effects are clearly demonstrated by isolating the joint with small segments of the member connected to the pin, Fig. 5–7c. The pushing or pulling on these small segments indicates the effect of the member being either in compression or tension.

In all cases, the analysis should start at a joint having at least one known force and at most two unknown forces, as in Fig. 5–7b. In this way, application of $\Sigma F_x = 0$ and $\Sigma F_y = 0$ yields two algebraic equations which can be solved for the two unknowns. When applying these equations, the correct sense of an unknown member force can be determined using one of two possible methods:

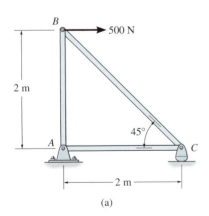

(a)

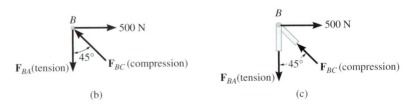

(b)                    (c)

Fig. 5–7

- *Always assume* the *unknown member forces* acting on the joint's free-body diagram to be in *tension*, i.e., "pulling" on the pin. If this is done, then numerical solution of the equilibrium equations will yield *positive scalars for members in tension and negative scalars for members in compression.* Once an unknown member force is found, use its *correct* magnitude and sense (T or C) on subsequent joint free-body diagrams.
- The *correct* sense of direction of an unknown member force can, in many cases, be determined "by inspection." For example, $F_{BC}$ in Fig. 5–7b must push on the pin (compression) since its horizontal component, $F_{BC} \sin 45°$, must balance the 500-N force ($\Sigma F_x = 0$). Likewise, $\mathbf{F}_{BA}$ is a tensile force since it balances the vertical component, $F_{BC} \cos 45°$ ($\Sigma F_y = 0$). In more complicated cases, the sense of an unknown member force can be *assumed;* then, after applying the equilibrium equations, the assumed sense can be verified from the numerical results. A *positive* answer indicates that the sense is *correct,* whereas a *negative* answer indicates that the sense shown on the free-body diagram must be *reversed.* This is the method we will use in the example problems which follow.

## PROCEDURE FOR ANALYSIS

The following procedure provides a typical means for analyzing a truss using the method of joints.

- Draw the free-body diagram of a joint having at least one known force and at most two unknown forces. (If this joint is at one of the supports, then it may be necessary to know the external reactions at the truss support.)
- Use one of the two methods described above for establishing the sense of an unknown force.
- Orient the $x$ and $y$ axes such that the forces on the free-body diagram can be easily resolved into their $x$ and $y$ components and then apply the two force equilibrium equations $\Sigma F_x = 0$ and $\Sigma F_y = 0$. Solve for the two unknown member forces and verify their correct sense.
- Continue to analyze each of the other joints, where again it is necessary to choose a joint having at most two unknowns and at least one known force.
- Once the force in a member is found from the analysis of a joint at one of its ends, the result can be used to analyze the forces acting on the joint at its other end. Remember that a member in *compression* "pushes" on the joint and a member in *tension* "pulls" on the joint.

EXAMPLE 5.1

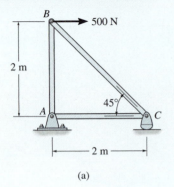

2 m

B

500 N

45°

A

C

2 m

(a)

B

500 N

45°

$F_{BA}$   $F_{BC}$

(b)

707.1 N

45°

$F_{CA}$   C

$C_y$

(c)

$F_{BA}$ = 500 N

$A_x$   A   $F_{CA}$ = 500 N

$A_y$

(d)

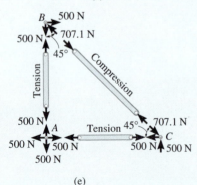

B   500 N

500 N   707.1 N

45°

Tension

Compression

500 N   A   Tension   45°   707.1 N

500 N   500 N   500 N   C   500 N

500 N

(e)

**Fig. 5–8**

Determine the force in each member of the truss shown in Fig. 5–8a and indicate whether the members are in tension or compression.

### Solution

By inspection of Fig. 5–8a, there are two unknown member forces at joint B, two unknown member forces and an unknown reaction force at joint C, and two unknown member forces and two unknown reaction forces at joint A. Since we should have no more than two unknowns at the joint and at least one known force acting there, we will begin the analysis at joint B.

*Joint B.* The free-body diagram of the pin at B is shown in Fig. 5–8b. Applying the equations of joint equilibrium, we have

$$\xrightarrow{+} \Sigma F_x = 0; \quad 500 \text{ N} - F_{BC} \sin 45° = 0 \qquad F_{BC} = 707.1 \text{ N (C)} \quad Ans.$$
$$+\uparrow \Sigma F_y = 0; \quad F_{BC} \cos 45° - F_{BA} = 0 \qquad F_{BA} = 503 \text{ N} \quad \text{(T)} \quad Ans.$$

Since the force in member BC has been calculated, we can proceed to analyze joint C in order to determine the force in member CA and the support reaction at the rocker.

*Joint C.* From the free-body diagram of joint C, Fig. 5–8c, we have

$$\xrightarrow{+} \Sigma F_x = 0; \quad -F_{CA} + 707.1 \cos 45° \text{N} = 0 \quad F_{CA} = 500 \text{ N (T)} \quad Ans.$$
$$+\uparrow \Sigma F_y = 0; \quad C_y - 707.1 \sin 45° \text{N} = 0 \quad C_y = 500 \text{ N} \quad Ans.$$

*Joint A.* Although it is not necessary, we can determine the support reactions at joint A using the results of $F_{CA}$ = 500 N and $F_{BA}$ = 500 N. From the free-body diagram, Fig. 5–8d, we have

$$\xrightarrow{+} \Sigma F_x = 0; \quad 500 \text{ N} - A_x = 0 \quad A_x = 500 \text{ N}$$
$$+\uparrow \Sigma F_y = 0; \quad 500 \text{ N} - A_y = 0 \quad A_y = 500 \text{ N}$$

The results of the analysis are summarized in Fig. 5–8e. Note that the free-body diagram of each pin shows the effects of all the connected members and external forces applied to the pin, whereas the free-body diagram of each member shows only the effects of the end pins on the member.

# E X A M P L E   5.2

Determine the forces acting in all the members of the truss shown in Fig. 5–9*a*.

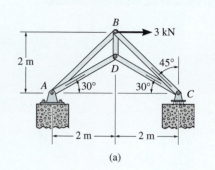

(a)

## Solution

By inspection, there are more than two unknowns at each joint. Consequently, the support reactions on the truss must first be determined. Show that they have been correctly calculated on the free-body diagram in Fig. 5–9*b*. We can now begin the analysis at joint *C*. Why?

*Joint C.* From the free-body diagram, Fig. 5–9*c*,

$$\xrightarrow{+} \Sigma F_x = 0; \qquad -F_{CD} \cos 30° + F_{CB} \sin 45° = 0$$

$$+\uparrow \Sigma F_y = 0; \quad 1.5 \text{ kN} + F_{CD} \sin 30° - F_{CB} \cos 45° = 0$$

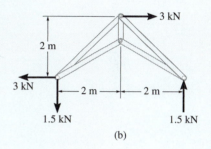

(b)

These two equations must be solved *simultaneously* for each of the two unknowns. Note, however, that a *direct solution* for one of the unknown forces may be obtained by applying a force summation along an axis that is *perpendicular* to the direction of the other unknown force. For example, summing forces along the *y′* axis, which is perpendicular to the direction of $\mathbf{F}_{CD}$, Fig. 5–9*d*, yields a direct solution for $F_{CB}$.

$$+\nearrow \Sigma F_{y'} = 0;$$

$$1.5 \cos 30° \text{kN} - F_{CB} \sin 15° = 0 \qquad F_{CB} = 5.02 \text{ kN} \quad \text{(C)} \qquad Ans.$$

In a similar fashion, summing forces along the *y″* axis, Fig. 5–9*e*, yields a direct solution for $F_{CD}$.

$$+\nearrow \Sigma F_{y'} = 0;$$

$$1.5 \cos 45° \text{kN} - F_{CD} \sin 15° = 0 \qquad F_{CD} = 4.10 \text{ kN} \quad \text{(T)} \qquad Ans.$$

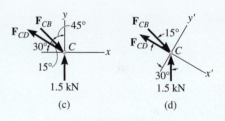

(c)  (d)

*Joint D.* We can now proceed to analyze joint *D*. The free-body diagram is shown in Fig. 5–9*f*.

$$\xrightarrow{+} \Sigma F_x = 0; \quad -F_{DA} \cos 30° + 4.10 \cos 30° \text{kN} = 0$$

$$F_{DA} = 4.10 \text{ kN} \quad \text{(T)} \qquad \qquad Ans.$$

$$+\uparrow \Sigma F_y = 0; \qquad F_{DB} - 2(4.10 \sin 30° \text{kN}) = 0$$

$$F_{DB} = 4.10 \text{ kN} \quad \text{(T)} \qquad \qquad Ans.$$

The force in the last member, *BA*, can be obtained from joint *B* or joint *A*. As an exercise, draw the free-body diagram of joint *B*, sum the forces in the horizontal direction, and show that $F_{BA} = 0.776 \text{ kN}$ (C).

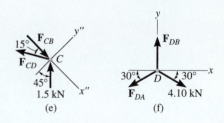

(e)  (f)

**Fig. 5–9**

**E X A M P L E   5.3**

Determine the force in each member of the truss shown in Fig. 5–10a. Indicate whether the members are in tension or compression.

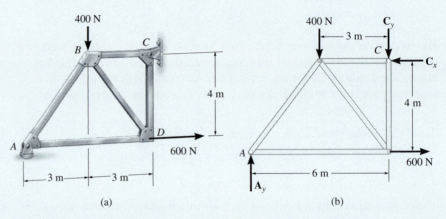

(a)                                    (b)

**Fig. 5–10**

**Solution**

***Support Reactions.*** No joint can be analyzed until the support reactions are determined. Why? A free-body diagram of the entire truss is given in Fig. 5–10b. Applying the equations of equilibrium, we have

$$\xrightarrow{+} \Sigma F_x = 0; \quad 600\text{ N} - C_x = 0 \quad C_x = 600\text{ N}$$
$$\zeta + \Sigma M_C = 0; \quad -A_y(6\text{ m}) + 400\text{ N}(3\text{ m}) + 600\text{ N}(4\text{ m}) = 0$$
$$A_y = 600\text{ N}$$
$$+\uparrow \Sigma F_y = 0; \quad 600\text{ N} - 400\text{ N} - C_y = 0 \quad C_y = 200\text{ N}$$

The analysis can now start at either joint A or C. The choice is arbitrary since there are one known and two unknown member forces acting on the pin at each of these joints.

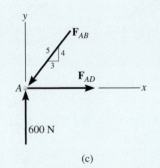

(c)

***Joint A (Fig. 5–10c).*** As shown on the free-body diagram, there are three forces that act on the pin at joint A. The inclination of $\mathbf{F}_{AB}$ is determined from the geometry of the truss. By inspection, can you see why this force is assumed to be compressive and $\mathbf{F}_{AD}$ tensile? Applying the equations of equilibrium, we have

$$+\uparrow \Sigma F_y = 0; \quad 600\text{ N} - \tfrac{4}{5}F_{AB} = 0 \quad F_{AB} = 750\text{ N} \;\;(\text{C}) \qquad Ans.$$
$$\xrightarrow{+} \Sigma F_x = 0; \quad F_{AD} - \tfrac{3}{5}(750\text{ N}) = 0 \quad F_{AD} = 450\text{ N} \;\;(\text{T}) \qquad Ans.$$

*Joint D (Fig. 5–10d).* The pin at this joint is chosen next since, by inspection of Fig. 5–10*a*, the force in *AD* is known and the unknown forces in *DB* and *DC* can be determined. Summing forces in the horizontal direction, Fig. 5–10*d*, we have

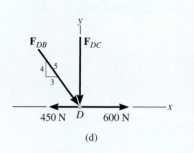

(d)

$$\xrightarrow{+} \Sigma F_x = 0; \quad -450\text{ N} + \tfrac{3}{5}F_{DB} + 600\text{ N} = 0 \qquad F_{DB} = -250\text{ N}$$

The negative sign indicates that $\mathbf{F}_{DB}$ acts in the *opposite sense* to that shown in Fig. 5–10*d**. Hence,

$$F_{DB} = 250\text{ N} \quad (\text{T}) \qquad\qquad Ans.$$

To determine $\mathbf{F}_{DC}$, we can either correct the sense of $\mathbf{F}_{DB}$ and then apply $\Sigma F_y = 0$, or apply this equation and retain the negative sign for $F_{DB}$, i.e.,

$$+\uparrow \Sigma F_y = 0; \quad -F_{DC} - \tfrac{4}{5}(-250\text{ N}) = 0 \qquad F_{DC} = 200\text{ N} \quad (\text{C}) \qquad Ans.$$

*Joint C (Fig. 6–10e).*

$$\xrightarrow{+} \Sigma F_x = 0; \quad F_{CB} - 600\text{ N} = 0 \qquad F_{CB} = 600\text{ N} \quad (\text{C}) \qquad Ans.$$
$$+\uparrow \Sigma F_y = 0; \quad 200\text{ N} - 200\text{ N} = 0 \quad (\text{check})$$

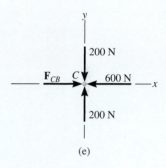

(e)

The analysis is summarized in Fig. 5–10*f*, which shows the correct free-body diagram for each pin and member.

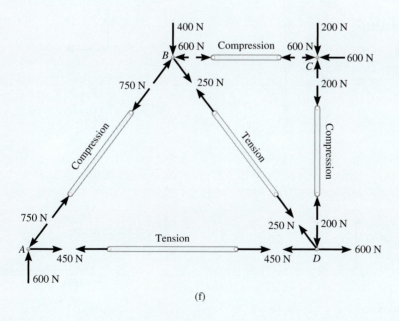

(f)

*The proper sense could have been determined by inspection, prior to applying $\Sigma F_x = 0$.

## 5.3 Zero-Force Members

Truss analysis using the method of joints is greatly simplified if one is first able to determine those members which support *no loading*. These *zero-force members* are used to increase the stability of the truss during construction and to provide support if the applied loading is changed.

The zero-force members of a truss can generally be determined *by inspection* of each of its joints. For example, consider the truss shown in Fig. 5–11*a*. If a free-body diagram of the pin at joint *A* is drawn, Fig. 5–11*b*, it is seen that members *AB* and *AF* are zero-force members. On the other hand, notice that we could not have come to this conclusion if we had considered the free-body diagrams of joints *F* or *B* simply because there are five unknowns at each of these joints. In a similar manner, consider the free-body diagram of joint *D*, Fig. 5–11*c*. Here again it is seen that *DC* and *DE* are zero-force members. As a general rule, *if only two members form a truss joint and no external load or support reaction is applied to the joint, the members must be zero-force members*. The load on the truss in Fig. 5–11*a* is therefore supported by only five members as shown in Fig. 5–11*d*.

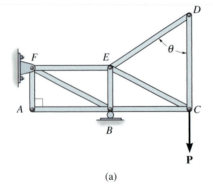

(a)

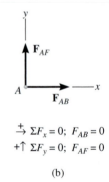

$$\overset{+}{\rightarrow} \Sigma F_x = 0; \quad F_{AB} = 0$$
$$+\uparrow \Sigma F_y = 0; \quad F_{AF} = 0$$

(b)

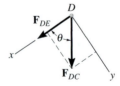

$$+ \searrow \Sigma F_y = 0; \ F_{DC} \sin \theta = 0; \quad F_{DC} = 0 \text{ since } \sin \theta \neq 0$$
$$+ \swarrow \Sigma F_x = 0; \ F_{DE} + 0 = 0; \quad F_{DE} = 0$$

(c)

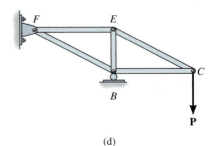

(d)

**Fig. 5–11**

Now consider the truss shown in Fig. 5–12a. The free-body diagram of the pin at joint $D$ is shown in Fig. 5–12b. By orienting the $y$ axis along members $DC$ and $DE$ and the $x$ axis along member $DA$, it is seen that $DA$ is a zero-force member. Note that this is also the case for member $CA$, Fig. 5–12c. In general, *if three members form a truss joint for which two of the members are collinear, the third member is a zero-force member provided no external force or support reaction is applied to the joint.* The truss shown in Fig. 5–12d is therefore suitable for supporting the load **P**.

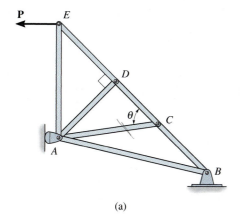

(a)

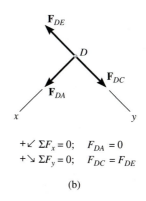

$$+\swarrow \Sigma F_x = 0; \quad F_{DA} = 0$$
$$+\searrow \Sigma F_y = 0; \quad F_{DC} = F_{DE}$$

(b)

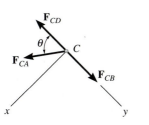

$$+\swarrow \Sigma F_x = 0; \quad F_{CA}\sin\theta = 0; \quad F_{CA} = 0 \text{ since } \sin\theta \neq 0;$$
$$+\searrow \Sigma F_y = 0; \quad F_{CB} = F_{CD}$$

(c)

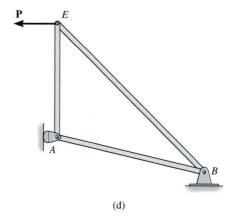

(d)

**Fig. 5–12**

## E X A M P L E  5.4

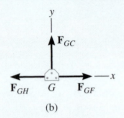

(b)

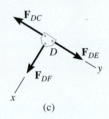

(c)

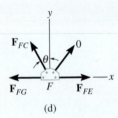

(d)

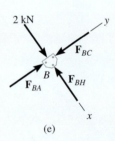

(e)

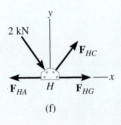

(f)

Using the method of joints, determine all the zero-force members of the *Fink roof truss* shown in Fig. 5–13*a*. Assume all joints are pin connected.

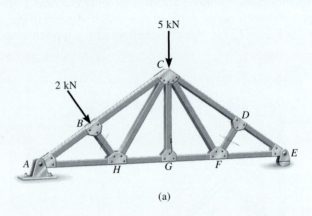

(a)

**Fig. 5–13**

### Solution

Look for joint geometries that have three members for which two are collinear. We have

**Joint G (Fig. 5–13b).**

$$+\uparrow \Sigma F_y = 0; \qquad F_{GC} = 0 \qquad\qquad\qquad Ans.$$

Realize that we could not conclude that *GC* is a zero-force member by considering joint *C*, where there are five unknowns. The fact that *GC* is a zero-force member means that the 5-kN load at *C* must be supported by members *CB, CH, CF,* and *CD.*

**Joint D (Fig. 5–13c).**

$$+\nearrow \Sigma F_x = 0; \qquad F_{DF} = 0 \qquad\qquad\qquad Ans.$$

**Joint F (Fig. 5–13d).**

$$+\uparrow \Sigma F_y = 0; \qquad F_{FC}\cos\theta = 0 \quad \text{Since } \theta \neq 90°, \qquad F_{FC} = 0 \quad Ans.$$

Note that if joint *B* is analyzed, Fig. 5–13*e*,

$$+\searrow \Sigma F_x = 0; \quad 2\text{ kN} - F_{BH} = 0 \quad F_{BH} = 2\text{ kN} \qquad (C)$$

Note that $F_{HC}$ must satisfy $\Sigma F_y = 0$, Fig. 5–13*f*, and therefore *HC* is *not* a zero-force member.

# PROBLEMS

**5-1.** Determine the force in each member of the truss and state if the members are in tension or compression. Set $P_1 = 800$ lb and $P_2 = 400$ lb.

**5-2.** Determine the force on each member of the truss and state if the members are in tension or compression. Set $P_1 = 500$ lb and $P_2 = 100$ lb.

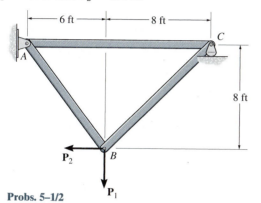

**Probs. 5–1/2**

**5-3.** The truss, used to support a balcony, is subjected to the loading shown. Approximate each joint as a pin and determine the force in each member. State whether the members are in tension or compression. Set $P_1 = 600$ lb, $P_2 = 400$ lb.

**\*5-4.** The truss, used to support a balcony, is subjected to the loading shown. Approximate each joint as a pin and determine the force in each member. State whether the members are in tension or compression. Set $P_1 = 800$ lb, $P_2 = 0$.

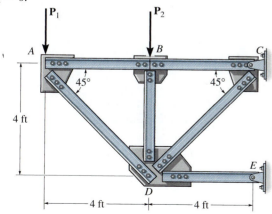

**Probs. 5–3/4**

**5-5.** Determine the force in each member of the truss and state if the members are in tension or compression. Assume each joint as a pin. Set $P = 4$ kN.

**5-6.** Assume that each member of the truss is made of steel having a mass per length of 4 kg/m. Set $P = 0$, determine the force in each member, and indicate if the members are in tension or compression. Neglect the weight of the gusset plates and assume each joint is a pin. Solve the problem by assuming the weight of each member can be represented as a vertical force, half of which is applied at the end of each member.

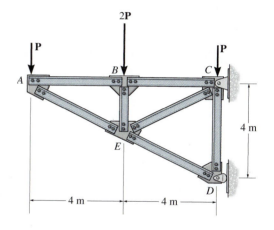

**Probs. 5–5/6**

**5-7.** Determine the force in each member of the truss and state if the members are in tension or compression.

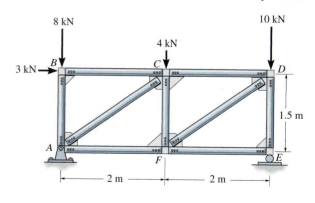

**Prob. 5–7**

*5-8. Determine the force in each member of the truss and state if the members are in tension or compression. Set $P_1 = 2$ kN and $P_2 = 1.5$ kN.

5-9. Determine the force in each member of the truss and state if the members are in tension or compression. Set $P_1 = P_2 = 4$ kN.

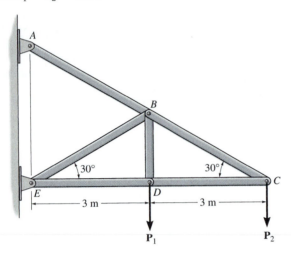

Probs. 5–8/9

5-10. Determine the force in each member of the truss and state if the members are in tension or compression. Set $P_1 = 0$, $P_2 = 1000$ lb.

5-11. Determine the force in each member of the truss and state if the members are in tension or compression. Set $P_1 = 500$ lb, $P_2 = 1500$ lb.

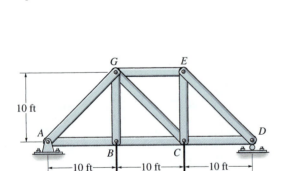

Probs. 5–10/11

*5-12. Determine the force in each member of the truss and state if the members are in tension or compression. Set $P_1 = 10$ kN, $P_2 = 15$ kN.

5-13. Determine the force in each member of the truss and state if the members are in tension or compression. Set $P_1 = 0$, $P_2 = 20$ kN.

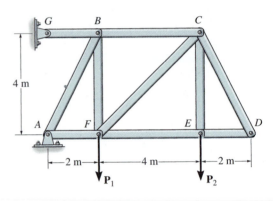

Probs. 5–12/13

5-14. Determine the force in each member of the truss and state if the members are in tension or compression. Set $P_1 = 100$ lb, $P_2 = 200$ lb, $P_3 = 300$ lb.

5-15. Determine the force in each member of the truss and state if the members are in tension or compression. Set $P_1 = 400$ lb, $P_2 = 400$ lb, $P_3 = 0$.

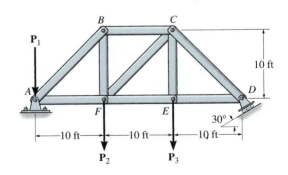

Probs. 5–14/15

## 5.4 The Method of Sections

The *method of sections* is used to determine the loadings acting within a body. It is based on the principle that if a body is in equilibrium then any part of the body is also in equilibrium. For example, consider the two truss members shown on the left in Fig. 5–14. If the forces within the members are to be determined, then an imaginary section indicated by the blue line, can be used to cut each member into two parts and thereby "expose" each internal force as "external" to the free-body diagrams shown on the right. Clearly, it can be seen that equilibrium requires that the member in tension (T) be subjected to a "pull," whereas the member in compression (C) is subjected to a "push."

The method of sections can also be used to "cut" or section the members of an entire truss. If the section passes through the truss and the free-body diagram of either of its two parts is drawn, we can then apply the equations of equilibrium to that part to determine the member forces at the "cut section." Since only *three* independent equilibrium equations ($\Sigma F_x = 0$, $\Sigma F_y = 0$, $\Sigma M_O = 0$) can be applied to the isolated part of the truss, try to select a section that, in general, passes through not more than *three* members in which the forces are unknown. For example, consider the truss in Fig. 5–15a. If the force in member *GC* is to be determined, section *aa* would be appropriate. The free-body diagrams of the two parts are shown in Figs. 5–15b and 5–15c. In particular, note that the line of action of each member force is specified from the *geometry* of the truss, since the force in a member passes along its axis. Also, the member forces acting on one part of the truss are equal but opposite to those acting on the other part—Newton's third law. As noted above, members assumed to be in *tension* (*BC* and *GC*) are subjected to a "pull," whereas the member in *compression* (*GF*) is subjected to a "push."

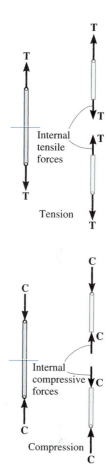

**Fig. 5–14**

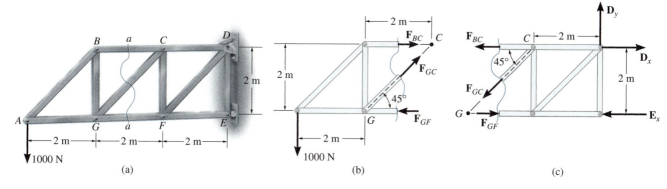

**Fig. 5–15**

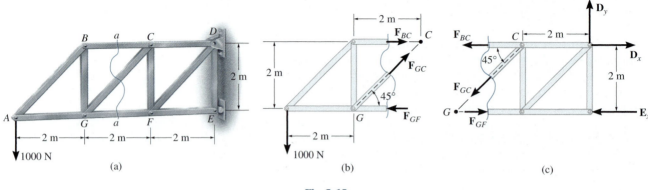

Fig. 5–15

Two Pratt trusses are used to construct this pedestrian bridge.

The three unknown member forces $\mathbf{F}_{BC}$, $\mathbf{F}_{GC}$, and $\mathbf{F}_{GF}$ can be obtained by applying the three equilibrium equations to the free-body diagram in Fig. 5–15b. If, however, the free-body diagram in Fig. 5–15c is considered, the three support reactions $\mathbf{D}_x$, $\mathbf{D}_y$ and $\mathbf{E}_x$ will have to be determined *first*. Why? (This, of course, is done in the usual manner by considering a free-body diagram of the *entire truss*.)

When applying the equilibrium equations, one should consider ways of writing the equations so as to yield a *direct solution* for each of the unknowns, rather than having to solve simultaneous equations. For example, summing moments about $C$ in Fig. 5–15b would yield a direct solution for $\mathbf{F}_{GF}$ since $\mathbf{F}_{BC}$ and $\mathbf{F}_{GC}$ create zero moment about $C$. Likewise, $\mathbf{F}_{BC}$ can be directly obtained by summing moments about $G$. Finally, $\mathbf{F}_{GC}$ can be found directly from a force summation in the vertical direction since $\mathbf{F}_{GF}$ and $\mathbf{F}_{BC}$ have no vertical components. This ability to *determine directly* the force in a particular truss member is one of the main advantages of using the method of sections.*

---

*By comparison, if the method of joints were used to determine, say, the force in member $GC$, it would be necessary to analyze joints $A$, $B$, and $G$ in sequence.

As in the method of joints, there are two ways in which one can determine the correct sense of an unknown member force:

- *Always assume* that the unknown member forces at the cut section are in *tension,* i.e., "pulling" on the member. By doing this, the numerical solution of the equilibrium equations will yield *positive scalars for members in tension and negative scalars for members in compression.*

- The correct sense of an unknown member force can in many cases be determined "by inspection." For example, $\mathbf{F}_{BC}$ is a tensile force as represented in Fig. 5–15*b* since moment equilibrium about *G* requires that $\mathbf{F}_{BC}$ create a moment opposite to that of the 1000-N force. Also, $\mathbf{F}_{GC}$ is tensile since its vertical component must balance the 1000-N force which acts downward. In more complicated cases, the sense of an unknown member force may be *assumed.* If the solution yields a *negative* scalar, it indicates that the force's sense is *opposite* to that shown on the free-body diagram. This is the method we will use in the example problems which follow.

## PROCEDURE FOR ANALYSIS

The forces in the members of a truss may be determined by the method of sections using the following procedure.

### *Free-Body Diagram.*

- Make a decision as to how to "cut" or section the truss through the members where forces are to be determined.

- Before isolating the appropriate section, it may first be necessary to determine the truss's *external* reactions. Then three equilibrium equations are available to solve for member forces at the cut section.

- Draw the free-body diagram of that part of the sectioned truss which has the least number of forces acting on it.

- Use one of the two methods described above for establishing the sense of an unknown member force.

### *Equations of Equilibrium.*

- Moments should be summed about a point that lies at the intersection of the lines of action of two unknown forces, so that the third unknown force is determined directly from the moment equation.

- If two of the unknown forces are *parallel,* forces may be summed *perpendicular* to the direction of these unknowns to determine *directly* the third unknown force.

EXAMPLE 5.5

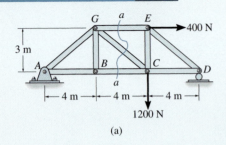

(a)

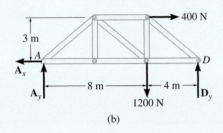

(b)

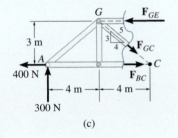

(c)

Fig. 5–16

Determine the force in members $GE$, $GC$, and $BC$ of the truss shown in Fig. 5–16a. Indicate whether the members are in tension or compression.

### Solution

Section $aa$ in Fig. 5–16a has been chosen since it cuts through the *three* members whose forces are to be determined. In order to use the method of sections, however, it is *first* necessary to determine the external reactions at $A$ or $D$. Why? A free-body diagram of the entire truss is shown in Fig. 5–16b. Applying the equations of equilibrium, we have

$$\xrightarrow{+} \Sigma F_x = 0; \quad 400 \text{ N} - A_x = 0 \quad A_x = 400 \text{ N}$$

$$\zeta + \Sigma M_A = 0; \quad -1200 \text{ N}(8 \text{ m}) - 400 \text{ N}(3 \text{ m}) + D_y(12 \text{ m}) = 0$$

$$D_y = 900 \text{ N}$$

$$+\uparrow \Sigma F_y = 0; \quad A_y - 1200 \text{ N} + 900 \text{ N} = 0 \quad A_y = 300 \text{ N}$$

*Free-Body Diagram.* The free-body diagram of the left portion of the sectioned truss is shown in Fig. 5–16c. For the analysis this diagram will be used since it involves the least number of forces.

*Equations of Equilibrium.* Summing moments about point $G$ eliminates $\mathbf{F}_{GE}$ and $\mathbf{F}_{GC}$ and yields a direct solution for $F_{BC}$.

$$\zeta + \Sigma M_G = 0; \quad -300 \text{ N}(4 \text{ m}) - 400 \text{ N}(3 \text{ m}) + F_{BC}(3 \text{ m}) = 0$$

$$F_{BC} = 800 \text{ N} \quad (\text{T}) \qquad \textit{Ans.}$$

In the same manner, by summing moments about point $C$ we obtain a direct solution for $F_{GE}$.

$$\zeta + \Sigma M_C = 0; \quad -300 \text{ N}(8 \text{ m}) + F_{GE}(3 \text{ m}) = 0$$

$$F_{GE} = 800 \text{ N} \quad (\text{C}) \qquad \textit{Ans.}$$

Since $\mathbf{F}_{BC}$ and $\mathbf{F}_{GE}$ have no vertical components, summing forces in the $y$ direction directly yields $F_{GC}$, i.e.,

$$+\uparrow \Sigma F_y = 0 \quad 300 \text{ N} - \tfrac{3}{5} F_{GC} = 0$$

$$F_{GC} = 500 \text{ N} \quad (\text{T}) \qquad \textit{Ans.}$$

As an exercise, obtain these results by applying the equations of equilibrium to the free-body diagram of the right portion of the sectioned truss.

# EXAMPLE 5.6

Determine the force in member *CF* of the bridge truss shown in Fig. 5–17a. Indicate whether the member is in tension or compression. Assume each member is pin-connected.

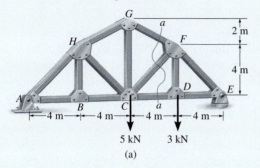
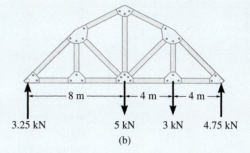

5 kN     3 kN

(a)

3.25 kN     5 kN     3 kN     4.75 kN

(b)

**Fig. 5–17**

## Solution

*Free-Body Diagram.* Section *aa* in Fig. 5–17a will be used since this section will "expose" the internal force in member *CF* as "external" on the free-body diagram of either the right or left portion of the truss. It is first necessary, however, to determine the external reactions on either the left or right side. Verify the results shown on the free-body diagram in Fig. 5–17b.

The free-body diagram of the right portion of the truss, which is the easiest to analyze, is shown in Fig. 5–17c. There are three unknowns, $F_{FG}$, $F_{CF}$, and $F_{CD}$.

*Equations of Equilibrium.* The most direct method for solving this problem requires application of the moment equation about a point that eliminates two of the unknown forces. Hence, to obtain $\mathbf{F}_{CF}$, we will eliminate $\mathbf{F}_{FG}$ and $\mathbf{F}_{CD}$ by summing moments about point *O*, Fig. 5–17c. Note that the location of point *O* measured from *E* is determined from proportional triangles, i.e., $4/(4 + x) = 6/(8 + x)$, $x = 4$ m. Or, stated in another manner, the slope of member *GF* has a drop of 2 m to a horizontal distance of 4 m. Since *FD* is 4 m, Fig. 5–17c, then from *D* to *O* the distance must be 8 m.

An easy way to determine the moment of $\mathbf{F}_{CF}$ about point *O* is to use the principle of transmissibility and move $\mathbf{F}_{CF}$ to point *C*, and then resolve $\mathbf{F}_{CF}$ into its two rectangular components. We have

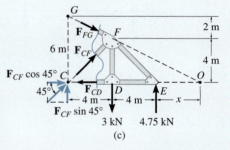

(c)

$\zeta + \Sigma M_O = 0;$

$$-F_{CF} \sin 45°(12 \text{ m}) + (3 \text{ kN})(8 \text{ m}) - (4.75 \text{ kN})(4 \text{ m}) = 0$$

$$F_{CF} = 0.589 \text{ kN} \quad \text{(C)} \qquad Ans.$$

EXAMPLE 5.7

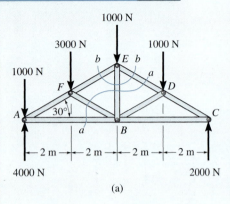

(a)

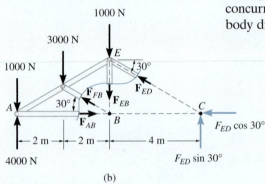

(b)

Fig. 5–18

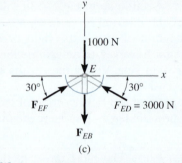

(c)

Determine the force in member *EB* of the roof truss shown in Fig. 5–18*a*. Indicate whether the member is in tension or compression.

**Solution**

*Free-Body Diagrams.* By the method of sections, any imaginary vertical section that cuts through *EB*, Fig. 5–18*a*, will also have to cut through three other members for which the forces are unknown. For example, section *aa* cuts through *ED*, *EB*, *FB*, and *AB*. If the components of reaction at *A* are calculated first ($A_x = 0$, $A_y = 4000$ N) and a free-body diagram of the left side of this section is considered, Fig. 5–18*b*, it is possible to obtain $\mathbf{F}_{ED}$ by summing moments about *B* to eliminate the other three unknowns; however, $\mathbf{F}_{EB}$ cannot be determined from the remaining two equilibrium equations. One possible way of obtaining $\mathbf{F}_{EB}$ is first to determine $\mathbf{F}_{ED}$ from section *aa*, then use this result on section *bb*, Fig. 5–18*a*, which is shown in Fig. 5–18*c*. Here the force system is concurrent and our sectioned free-body diagram is the same as the free-body diagram for the pin at *E* (method of joints).

*Equations of Equilibrium.* In order to determine the moment of $\mathbf{F}_{ED}$ about point *B*, Fig. 5–18*b*, we will resolve the force into its rectangular components and, by the principle of transmissibility, extend it to point *C* as shown. The moments of 1000 N, $F_{AB}$, $F_{FB}$, $F_{EB}$, and $F_{ED} \cos 30°$ are all zero about *B*. Therefore,

$$\zeta + \Sigma M_B = 0; \quad 1000\text{ N}(4\text{ m}) + 3000\text{ N}(2\text{ m}) - 4000\text{ N}(4\text{ m}) + F_{ED}\sin 30°(4) = 0$$
$$F_{ED} = 3000\text{ N} \quad (C)$$

Considering now the free-body diagram of section *bb*, Fig. 5–18*c*, we have

$$\xrightarrow{+} \Sigma F_x = 0; \quad F_{EF}\cos 30° - 3000\cos 30°\text{N} = 0$$
$$F_{EF} = 3000\text{ N} \quad (C)$$
$$+\uparrow\Sigma F_y = 0; \quad 2(3000\sin 30°\text{N}) - 1000\text{ N} - F_{EB} = 0$$
$$F_{EB} = 2000\text{ N} \quad (T) \qquad Ans.$$

# PROBLEMS

**\*5-16.** Determine the force in members *BC*, *HC*, and *HG* of the bridge truss, and indicate whether the members are in tension or compression.

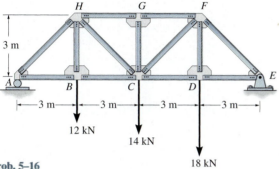

**Prob. 5–16**

**5-17.** Determine the force in members *GF*, *CF*, and *CD* of the bridge truss, and indicate whether the members are in tension or compression

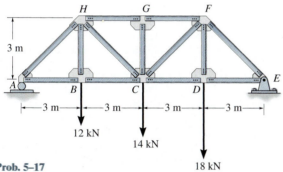

**Prob. 5–17**

**5-18.** Determine the force in members *DE*, *DF*, and *GF* of the cantilevered truss and state if the members are in tension or compression.

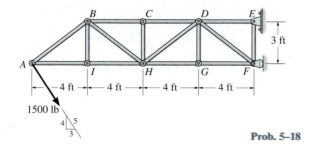

**Prob. 5–18**

**5-19.** The roof truss supports the vertical loading shown. Determine the force in members *BC*, *CK*, and *KJ* and state if these members are in tension or compression.

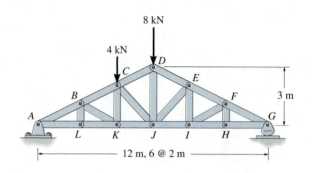

**Prob. 5–19**

**\*5-20.** Determine the force in members *CD*, *CJ*, *KJ*, and *DJ* of the truss which serves to support the deck of a bridge. State if these members are in tension or compression.

**5-21.** Determine the force in members *EI* and *JI* of the truss which serves to support the deck of a bridge. State if these members are in tension or compression.

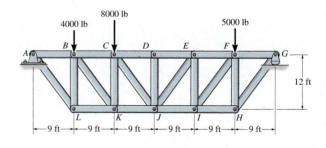

**Probs. 5–20/21**

**5-22.** Determine the force in members *BC, CG*, and *GF* of the *Warren truss*. Indicate if the members are in tension or compression.

**5-23.** Determine the force in members *CD, CF*, and *FG* of the *Warren truss*. Indicate if the members are in tension or compression.

**■5-25.** The truss supports the vertical load of 600 N. Determine the force in members *BC, BG*, and *HG* as the dimension *L* varies. Plot the results of *F* (ordinate with tension as positive) versus *L* (abscissa) for $0 \leq L \leq 3$ m.

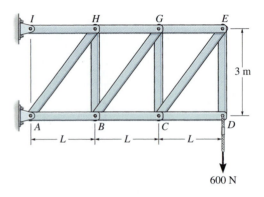

**Prob. 5-25**

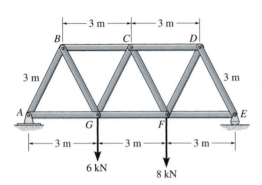

**Probs. 5-22/23**

**5-26.** Determine the force in members *IC* and *CG* of the truss and state if these members are in tension or compression. Also, indicate all zero-force members.

*__**5-24.**__ Determine the force developed in members *GB* and *GF* of the bridge truss and state if these members are in tension or compression.

**5-27.** Determine the force in members *JE* and *GF* of the truss and state if these members are in tension or compression. Also, indicate all zero-force members.

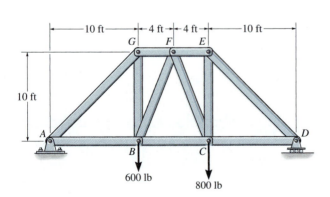

**Prob. 5-24**

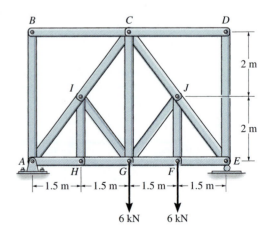

**Probs. 5-26/27**

*5-28. Determine the force in members $BC$, $HC$, and $HG$. After the truss is sectioned use a single equation of equilibrium for the calculation of each force. State if these members are in tension or compression.

5-29. Determine the force in members $CD$, $CF$, and $CG$ and state if these members are in tension or compression.

5-31. Determine the force in member $GJ$ of the truss and state if this member is in tension or compression.

*5-32. Determine the force in member $GC$ of the truss and state if this member is in tension or compression.

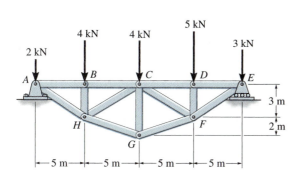

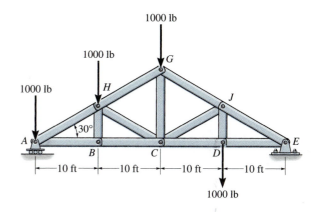

Probs. 5-28/29

Probs. 5-31/32

5-30. Determine the force in members $GF$, $FB$, and $BC$ of the *Fink truss* and state if the members are in tension or compression.

5-33. Determine the force in members $GF$, $CF$, and $CD$ of the roof truss and indicate if the members are in tension or compression.

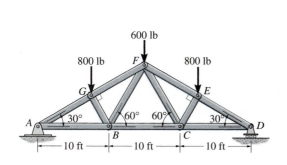

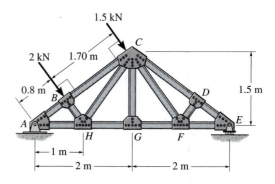

Prob. 5-30

Prob. 5-33

# 5.5  Frames and Machines

Frames and machines are two common types of structures which are often composed of pin-connected *multiforce members*, i.e., members that are subjected to more than two forces. *Frames* are generally stationary and are used to support loads, whereas *machines* contain moving parts and are designed to transmit and alter the effect of forces. Provided a frame or machine is properly constrained and contains no more supports or members than are necessary to prevent collapse, the forces acting at the joints and supports can be determined by applying the equations of equilibrium to each member. Once the forces at the joints are obtained, it is then possible to *design* the size of the members, connections, and supports using the theory of mechanics of materials and an appropriate engineering design code.

**Free-Body Diagrams.**  In order to determine the forces acting at the joints and supports of a frame or machine, the structure must be disassembled and the free-body diagrams of its parts must be drawn. The following important points *must* be observed:

- Isolate each part by drawing its *outlined shape*. Then show all the forces and/or couple moments that act on the part. Make sure to *label* or *identify* each known and unknown force and couple moment with reference to an established *x, y* coordinate system. Also, indicate any dimensions used for taking moments. Most often the equations of equilibrium are easier to apply if the forces are represented by their rectangular components. As usual, the sense of an unknown force or couple moment can be assumed.

- Identify all the two-force members in the structure and represent their free-body diagrams as having two equal but opposite collinear forces acting at their points of application. (See Sec. 4.4.) By recognizing the two-force members, we can avoid solving an unnecessary number of equilibrium equations.

- Forces common to any two *contacting* members act with equal magnitudes but opposite sense on the respective members. If the two members are treated as a *"system" of connected members*, then these forces are *"internal"* and are *not shown* on the *free-body diagram of the system*; however, if the free-body diagram of *each member* is drawn, the forces are *"external"* and *must* be shown on each of the free-body diagrams.

The following examples graphically illustrate application of these points in drawing the free-body diagrams of a dismembered frame or machine. In all cases, the weight of the members is neglected.

# EXAMPLE 5.8

For the frame shown in Fig. 5–19a, draw the free-body diagram of (a) each member, (b) the pin at B, and (c) the two members connected together.

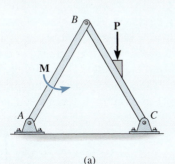

(a)

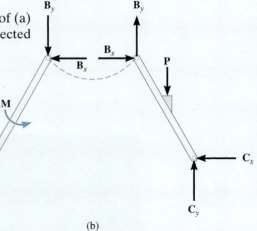

(b)

## Solution

*Part (a).* By inspection, members *BA* and *BC* are *not* two-force members. Instead, as shown on the free-body diagrams, Fig. 5–19b, *BC* is subjected to *not* five but *three forces*, namely, the resultant force from pins *B* and *C* and the external force **P**. Likewise, *AB* is subjected to the *resultant* forces from the pins at *A* and *B* and the external couple moment **M**.

*Part (b).* It can be seen in Fig. 5–19a that the pin at *B* is subjected to only *two forces*, i.e., the force of member *BC* on the pin and the force of member *AB* on the pin. For *equilibrium* these forces and therefore their respective components must be equal but opposite, Fig. 5–19c. Notice carefully how Newton's third law is applied between the pin and its contacting members, i.e., the effect of the pin on the two members, Fig. 5–19b, and the equal but opposite effect of the two members on the pin, Fig. 5–19c. Also note that $\mathbf{B}_x$ and $\mathbf{B}_y$, shown equal but opposite in Fig. 5–19b on members *AB* and *BC*, is *not* the effect of Newton's third law; instead, this results from the *equilibrium* analysis of the pin, Fig. 5–19c.

*Part (c).* The free-body diagram of both members connected together, yet removed from the supporting pins at *A* and *C*, is shown in Fig. 5–19d. The force components $\mathbf{B}_x$ and $\mathbf{B}_y$ are *not shown* on this diagram since they form equal but opposite collinear pairs of *internal* forces (Fig. 5–19b) and therefore cancel out. Also, to be consistent when later applying the equilibrium equations, the unknown force components at *A* and *C* must act in the *same sense* as those shown in Fig. 5–19b. Here the couple moment **M** can be applied at any point on the frame in order to determine the reactions at *A* and *C*. Note, however, that it must act on member *AB* in Fig. 5–19b and *not* on member *BC*.

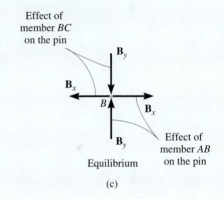

Effect of member *BC* on the pin

Equilibrium

Effect of member *AB* on the pin

(c)

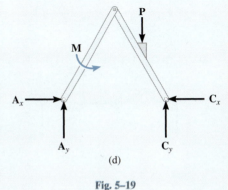

(d)

Fig. 5–19

**E X A M P L E** **5.9**

A constant tension in the conveyor belt is maintained by using the device shown in Fig. 5–20a. Draw the free-body diagrams of the frame and the cylinder which supports the belt. The suspended block has a weight of W.

(a)

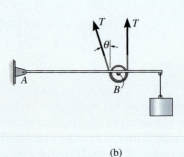

(b)

Fig. 5–20

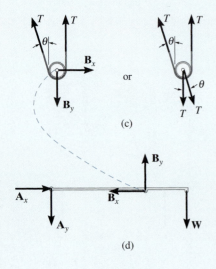

(c)

(d)

**Solution**

The idealized model of the device is shown in Fig. 5–20b. Here the angle $\theta$ is assumed to be known. Notice that the tension in the belt is the same on each side of the cylinder, since the cylinder is free to turn. From this model, the free-body diagrams of the frame and cylinder are shown in Figs. 5–20c and 5–20d, respectively. Note that the force that the pin at B exerts on the cylinder can be represented by either its horizontal and vertical components $\mathbf{B}_x$ and $\mathbf{B}_y$, which can be determined by using the force equations of equilibrium applied to the cylinder, or by the two components $T$, which provide equal but opposite couple moments on the cylinder and thus keep it from turning. Also, realize that once the pin reactions at A have been determined, half of their values act on each side of the frame since pin connections occur on each side, Fig. 5–20a.

EXAMPLE 5.10

Draw the free-body diagram of each part of the smooth piston and link mechanism used to crush recycled cans, which is shown in Fig. 5–21a.

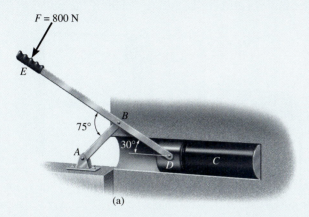

(a)

### Solution

By inspection, member $AB$ is a two-force member. The free-body diagrams of the parts are shown in Fig. 5–21b. Since the pins at $B$ and $D$ *connect only two parts together,* the forces there are shown as equal but opposite on the separate free-body diagrams of their connected members. In particular, four components of force act on the piston: $\mathbf{D}_x$ and $\mathbf{D}_y$ represent the effect of the pin (or lever $EBD$), $\mathbf{N}_w$ is the *resultant force* of the floor, and $\mathbf{P}$ is the resultant compressive force caused by the can $C$.

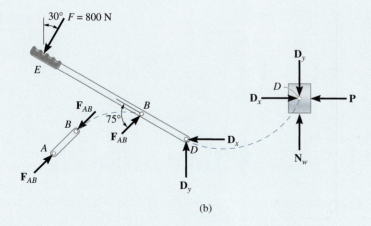

(b)

**Fig. 5–21**

**EXAMPLE** **5.11**

For the frame shown in Fig. 5–22a, draw the free-body diagrams of (a) the entire frame including the pulleys and cords, (b) the frame without the pulleys and cords, and (c) each of the pulleys.

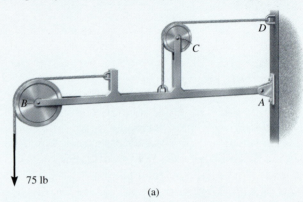

(a)

**Solution**

*Part (a).* When the entire frame including the pulleys and cords is considered, the interactions at the points where the pulleys and cords are connected to the frame become pairs of *internal forces* which cancel each other and therefore are not shown on the free-body diagram, Fig. 5–22b.

*Part (b).* When the cords and pulleys are removed, their effect *on the frame* must be shown, Fig. 5–22c.

*Part (c).* The force components $B_x$, $B_y$, $C_x$, $C_y$ of the pins on the pulleys, Fig. 5–22d, are equal but opposite to the force components exerted by the pins on the frame, Fig. 5–22c. Why?

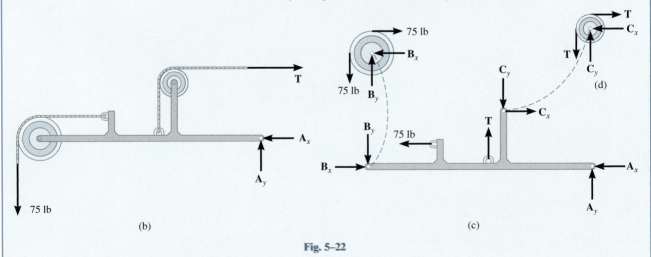

(b)

(c)

(d)

**Fig. 5–22**

*Before proceeding, it is recommended to cover the solutions to the previous examples and attempt to draw the requested free-body diagrams. When doing so, make sure the work is neat and that all the forces and couple moments are properly labeled.*

**Equations of Equilibrium.** Provided the structure (frame or machine) is properly supported and contains no more supports or members than are necessary to prevent its collapse, then the unknown forces at the supports and connections can be determined from the equations of equilibrium. If the structure lies in the x–y plane, then for *each* free-body diagram drawn the loading must satisfy $\Sigma F_x = 0$, $\Sigma F_y = 0$, and $\Sigma M_O = 0$. The selection of the free-body diagrams used for the analysis is *completely arbitrary*. They may represent each of the members of the structure, a portion of the structure, or its entirety. For example, consider finding the six components of the pin reactions at *A, B,* and *C* for the frame shown in Fig. 5–23a. If the frame is dismembered, as it is in Fig. 5–23b, these unknowns can be determined by applying the three equations of equilibrium to each of the two members (total of six equations). The free-body diagram of the *entire frame* can also be used for part of the analysis, Fig. 5–23c. Hence, if so desired, all six unknowns can be determined by applying the three equilibrium equations to the entire frame, Fig. 5–23c, and also to either one of its members. Furthermore, the answers can be checked in part by applying the three equations of equilibrium to the remaining "second" member. In general, then, this problem can be solved by writing *at most* six equilibrium equations using free-body diagrams of the members and/or the combination of connected members. Any more than six equations written would *not* be unique from the original six and would only serve to check the results.

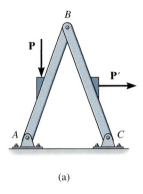

(a)

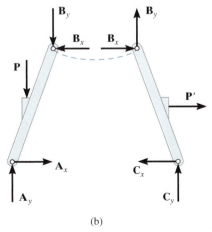

(b)

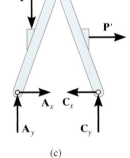

(c)

**Fig. 5–23**

## PROCEDURE FOR ANALYSIS

The joint reactions on frames or machines (structures) composed of multiforce members can be determined using the following procedure.

### *Free-Body Diagram.*

- Draw the free-body diagram of the entire structure, a portion of the structure, or each of its members. The choice should be made so that it leads to the most direct solution of the problem.

- When the free-body diagram of a group of members of a structure is drawn, the forces at the connected parts of this group are internal forces and are not shown on the free-body diagram of the group.

- Forces common to two members which are in contact act with equal magnitude but opposite sense on the respective free-body diagrams of the members.

- Two-force members, regardless of their shape, have equal but opposite collinear forces acting at the ends of the member.

- In many cases it is possible to tell by inspection the proper sense of the unknown forces acting on a member; however, if this seems difficult, the sense can be assumed.

- A couple moment is a free vector and can act at any point on the free-body diagram. Also, a force is a sliding vector and can act at any point along its line of action.

### *Equations of Equilibrium.*

- Count the number of unknowns and compare it to the total number of equilibrium equations that are available. In two dimensions, there are three equilibrium equations that can be written for each member.

- Sum moments about a point that lies at the intersection of the lines of action of as many unknown forces as possible.

- If the solution of a force or couple moment magnitude is found to be negative, it means the sense of the force is the reverse of that shown on the free-body diagrams.

# EXAMPLE 5.12

Determine the horizontal and vertical components of force which the pin at $C$ exerts on member $CB$ of the frame in Fig. 5–24a.

**Solution I**

*Free-Body Diagrams.*  By inspection it can be seen that $AB$ is a two-force member. The free-body diagrams are shown in Fig. 5–24b.

*Equations of Equilibrium.*  The *three unknowns*, $C_x$, $C_y$, and $F_{AB}$, can be determined by applying the three equations of equilibrium to member $CB$.

$\zeta + \Sigma M_C = 0;$  $2000\text{ N}(2\text{ m}) - (F_{AB}\sin 60°)(4\text{ m}) = 0$  $F_{AB} = 1154.7\text{ N}$

$\overset{+}{\rightarrow} \Sigma F_x = 0;$  $1154.7\cos 60°\text{N} - C_x = 0$  $C_x = 577\text{ N}$  *Ans.*

$+\uparrow \Sigma F_y = 0;$  $1154.7\sin 60°\text{N} - 2000\text{ N} + C_y = 0$  $C_y = 1000\text{ N}$ *Ans.*

**Solution II**

*Free-Body Diagrams.*  If one does not recognize that $AB$ is a two-force member, then more work is involved in solving this problem. The free-body diagrams are shown in Fig. 5–24c.

*Equations of Equilibrium.*  The *six unknowns*, $A_x$, $A_y$, $B_x$, $B_y$, $C_x$, $C_y$, are determined by applying the three equations of equilibrium to each member.

 *Member AB*

$\zeta + \Sigma M_A = 0;$  $B_x(3\sin 60°\text{ m}) - B_y(3\cos 60°\text{ m}) = 0$  (1)

$\overset{+}{\rightarrow} \Sigma F_x = 0;$  $A_x - B_x = 0$  (2)

$+\uparrow \Sigma F_y = 0;$  $A_y - B_y = 0$  (3)

 *Member BC*

$\zeta + \Sigma M_C = 0;$  $2000\text{ N}(2\text{ m}) - B_y(4\text{ m}) = 0$  (4)

$\overset{+}{\rightarrow} \Sigma F_x = 0;$  $B_x - C_x = 0$  (5)

$+\uparrow \Sigma F_y = 0;$  $B_y - 2000\text{ N} + C_y = 0$  (6)

 The results for $C_x$ and $C_y$ can be determined by solving these equations in the following sequence: 4, 1, 5, then 6. The results are

$$B_y = 1000\text{ N}$$
$$B_x = 577\text{ N}$$
$$C_x = 577\text{ N} \qquad Ans.$$
$$C_y = 1000\text{ N} \qquad Ans.$$

By comparison, Solution I is simpler since the requirement that $F_{AB}$ in Fig. 5–24b be equal, opposite, and collinear at the ends of member $AB$ automatically satisfies Eqs. 1, 2, and 3 above and therefore eliminates the need to write these equations. *As a result, always identify the two-force members before starting the analysis!*

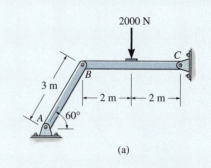

(a)

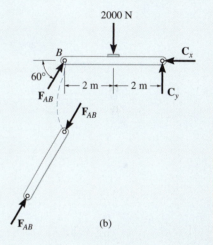

(b)

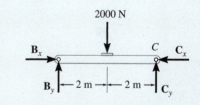

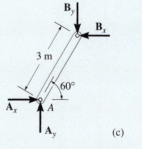

(c)

**Fig. 5–24**

**E X A M P L E 5.13**

The compound beam shown in Fig. 5–25a is pin connected at B. Determine the reactions at its supports. Neglect its weight and thickness.

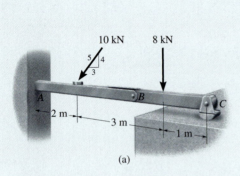

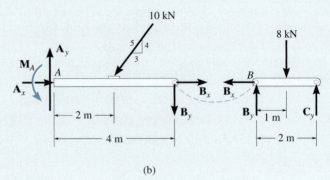

(a)

(b)

**Fig. 5–25**

**Solution**

*Free-Body Diagrams.* By inspection, if we consider a free-body diagram of the entire beam $ABC$, there will be three unknown reactions at $A$ and one at $C$. These four unknowns cannot all be obtained from the three equations of equilibrium, and so it will become necessary to dismember the beam into its two segments as shown in Fig. 5–25b.

*Equations of Equilibrium.* The six unknowns are determined as follows:

*Segment BC*

$$\xrightarrow{+} \Sigma F_x = 0; \qquad\qquad\qquad B_x = 0$$

$$\zeta + \Sigma M_B = 0; \qquad -8 \text{ kN}(1 \text{ m}) + C_y(2 \text{ m}) = 0$$

$$+\uparrow \Sigma F_y = 0; \qquad\qquad B_y - 8 \text{ kN} + C_y = 0$$

*Segment AB*

$$\xrightarrow{+} \Sigma F_x = 0; \qquad\qquad A_x - (10 \text{ kN})(\tfrac{3}{5}) + B_x = 0$$

$$\zeta + \Sigma M_A = 0; \qquad M_A - (10 \text{ kN})(\tfrac{4}{5})(2 \text{ m}) - B_y(4 \text{ m}) = 0$$

$$+\uparrow \Sigma F_y = 0; \qquad\qquad A_y - (10 \text{ kN})(\tfrac{4}{5}) - B_y = 0$$

Solving each of these equations successively, using previously calculated results, we obtain

$$A_x = 6 \text{ kN} \qquad A_y = 12 \text{ kN} \qquad M_A = 32 \text{ kN} \cdot \text{m} \qquad \textit{Ans.}$$
$$B_x = 0 \qquad\qquad B_y = 4 \text{ kN}$$
$$C_y = 4 \text{ kN} \qquad\qquad\qquad\qquad\qquad\qquad\qquad\quad \textit{Ans.}$$

# E X A M P L E  5.14

The smooth disk shown in Fig. 5–26a is pinned at $D$ and has a weight of 20 lb. Neglecting the weights of the other members, determine the horizontal and vertical components of reaction at pins $B$ and $D$.

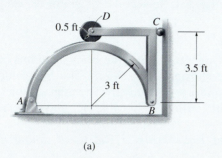

(a)

## Solution

**Free-Body Diagrams.**  By inspection, the three components of reaction at the supports can be determined from a free-body diagram of the entire frame, Fig. 5–26b. Also, free-body diagrams of the members are shown in Fig. 5–26c.

**Equations of Equilibrium.**  The eight unknowns can of course be obtained by applying the eight equilibrium equations to each member—three to member $AB$, three to member $BCD$, and two to the disk. (Moment equilibrium is automatically satisfied for the disk.) If this is done, however, all the results can be obtained only from a simultaneous solution of some of the equations. (Try it and find out.) To avoid this situation, it is best to first determine the three support reactions on the *entire* frame; then, using these results, the remaining five equilibrium equations can be applied to two other parts in order to solve successively for the other unknowns.

*Entire Frame*

$$\zeta + \Sigma M_A = 0; \quad -20 \text{ lb}(3 \text{ ft}) + C_x(3.5 \text{ ft}) = 0 \quad C_x = 17.1 \text{ lb}$$
$$\stackrel{+}{\rightarrow} \Sigma F_x = 0; \qquad A_x - 17.1 \text{ lb} = 0 \qquad A_x = 17.1 \text{ lb}$$
$$+\uparrow \Sigma F_y = 0; \qquad A_y - 20 \text{ lb} = 0 \qquad A_y = 20 \text{ lb}$$

*Member AB*

$$\stackrel{+}{\rightarrow} \Sigma F_x = 0; \qquad 17.1 \text{ lb} - B_x = 0 \qquad B_x = 17.1 \text{ lb} \qquad \textit{Ans.}$$
$$\zeta + \Sigma M_B = 0; \quad -20 \text{ lb}(6 \text{ ft}) + N_D(3 \text{ ft}) = 0 \qquad N_D = 40 \text{ lb}$$
$$+\uparrow \Sigma F_y = 0; \qquad 20 \text{ lb} - 40 \text{ lb} + B_y = 0 \qquad B_y = 20 \text{ lb} \qquad \textit{Ans.}$$

*Disk*

$$\stackrel{+}{\rightarrow} \Sigma F_x = 0; \qquad D_x = 0 \qquad \qquad \textit{Ans.}$$
$$+\uparrow \Sigma F_y = 0; \qquad 40 \text{ lb} - 20 \text{ lb} - D_y = 0 \qquad D_y = 20 \text{ lb} \qquad \textit{Ans.}$$

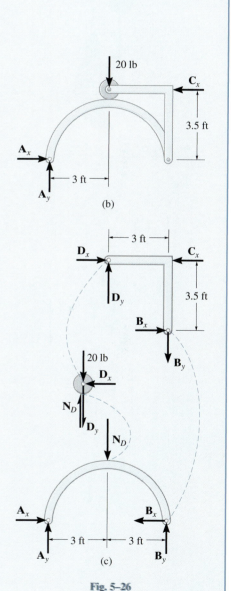

(b)

(c)

**Fig. 5–26**

**E X A M P L E  5.15**

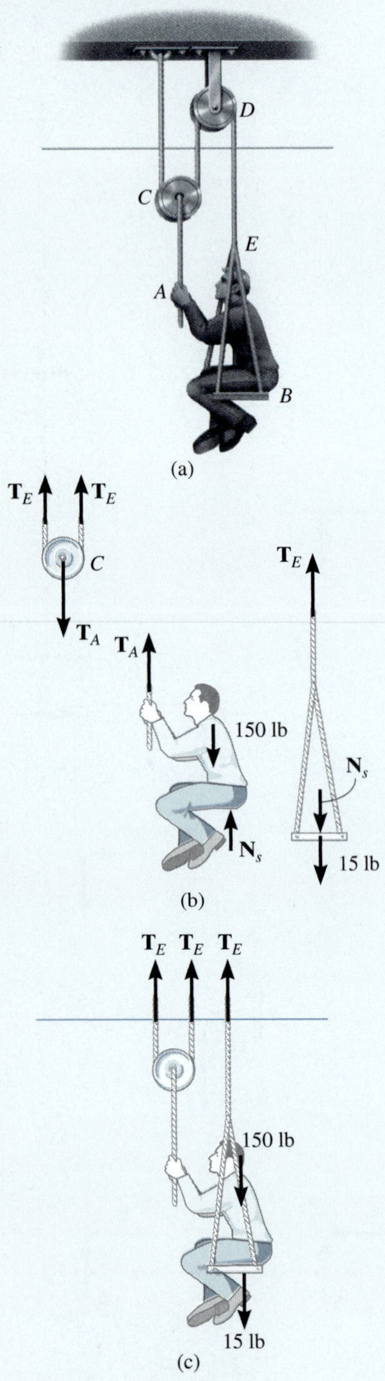

(a)

(b)

(c)

**Fig. 5–27**

A man having a weight of 150 lb supports himself by means of the cable and pulley system shown in Fig. 5–27a. If the seat has a weight of 15 lb, determine the force that he must exert on the cable at $A$ and the force he exerts on the seat. Neglect the weight of the cables and pulleys.

**Solution I**

*Free-Body Diagrams.* The free-body diagrams of the man, seat, and pulley $C$ are shown in Fig. 5–27b. The *two* cables are subjected to tensions $\mathbf{T}_A$ and $\mathbf{T}_E$, respectively. The man is subjected to three forces: his weight, the tension $\mathbf{T}_A$ of cable $AC$, and the reaction $\mathbf{N}_s$ of the seat.

*Equations of Equilibrium.* The three unknowns are obtained as follows:

*Man*

$$+\uparrow\Sigma F_y = 0; \qquad\qquad T_A + N_s - 150 \text{ lb} = 0 \qquad\qquad (1)$$

*Seat*

$$+\uparrow\Sigma F_y = 0; \qquad\qquad T_E - N_s - 15 \text{ lb} = 0 \qquad\qquad (2)$$

*Pulley C*

$$+\uparrow\Sigma F_y = 0; \qquad\qquad 2T_E - T_A = 0 \qquad\qquad (3)$$

Here $T_E$ can be determined by adding Eqs. 1 and 2 to eliminate $N_s$ and then using Eq. 3. The other unknowns are then obtained by resubstitution of $T_E$.

$$T_A = 110 \text{ lb} \qquad\qquad Ans.$$
$$T_E = 55 \text{ lb}$$
$$N_s = 40 \text{ lb} \qquad\qquad Ans.$$

**Solution II**

*Free-Body Diagrams.* By using the blue section shown in Fig. 5–27a, the man, pulley, and seat can be considered as a *single system*, Fig. 5–27c. Here $\mathbf{N}_s$ and $\mathbf{T}_A$ are *internal* forces and hence are not included on this "combined" free-body diagram.

*Equations of Equilibrium.* Applying $\Sigma F_y = 0$ yields a *direct* solution for $T_E$.

$$+\uparrow\Sigma F_y = 0; \quad 3T_E - 15 \text{ lb} - 150 \text{ lb} = 0 \quad T_E = 55 \text{ lb}$$

The other unknowns can be obtained from Eqs. 2 and 3.

The hand exerts a force of 8 lb on the grip of the spring compressor shown in Fig. 5–28a. Determine the force in the spring needed to maintain equilibrium of the mechanism.

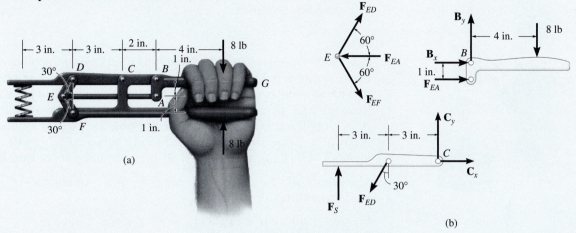

**Fig. 5–28**

## Solution

*Free-Body Diagrams.* By inspection, members *EA, ED*, and *EF* are all two-force members. The free-body diagrams for parts *DC* and *ABG* are shown in Fig. 5–28b. The pin at *E* has also been included here since *three* force interactions occur on this pin. They represent the effects of members *ED, EA*, and *EF*. Note carefully how equal and opposite force reactions occur between each of the parts.

*Equations of Equilibrium.* By studying the free-body diagrams, the most direct way to obtain the spring force is to apply the equations of equilibrium in the following sequence:

  *Lever ABG*

$\zeta+\Sigma M_B = 0$;   $F_{EA}(1 \text{ in.}) - 8 \text{ lb}(4 \text{ in.}) = 0$   $F_{EA} = 32 \text{ lb}$

  *Pin E*

$+\uparrow\Sigma F_y = 0$;   $F_{ED} \sin 60° - F_{EF} \sin 60° = 0$   $F_{ED} = F_{EF} = F$

$\xrightarrow{+}\Sigma F_x = 0$;       $2F \cos 60° - 32 \text{ lb} = 0$       $F = 32 \text{ lb}$

  *Arm DC*

$\zeta+\Sigma M_C = 0$;   $-F_s(6 \text{ in.}) + 32 \cos 30° \text{lb}(3 \text{ in.}) = 0$

                    $F_s = 13.9 \text{ lb}$                    *Ans.*

# EXAMPLE 5.17

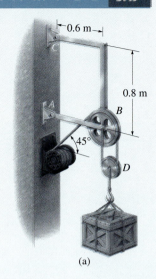

(a)

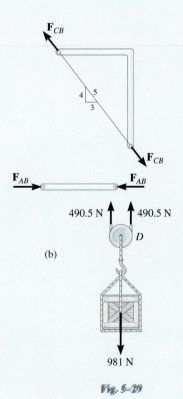

490.5 N ↑ ↑ 490.5 N

D

(b)

981 N

Fig. 5–29

The 100-kg block is held in equilibrium by means of the pulley and continuous cable system shown in Fig. 5–29a. If the cable is attached to the pin at $B$, compute the forces which this pin exerts on each of its connecting members.

## Solution

*Free-Body Diagrams.* A free-body diagram of each member of the frame is shown in Fig. 5–29b. By inspection, members $AB$ and $CB$ are two-force members. Furthermore, the cable must be subjected to a force of 490.5 N in order to hold pulley $D$ and the block in equilibrium. A free-body diagram of the pin at $B$ is needed since *four interactions* occur at this pin. These are caused by the attached cable (490.5 N), member $AB$ ($\mathbf{F}_{AB}$), member $CB$ ($\mathbf{F}_{CB}$), and pulley $B$ ($\mathbf{B}_x$ and $\mathbf{B}_y$).

*Equations of Equilibrium.* Applying the equations of force equilibrium to pulley $B$, we have

$$\xrightarrow{+}\Sigma F_x = 0; \quad B_x - 490.5\cos 45°\text{N} = 0 \quad B_x = 346.8\text{ N} \quad \textit{Ans.}$$

$$+\uparrow\Sigma F_y = 0; \quad B_y - 490.5\sin 45°\text{N} - 490.5\text{ N} = 0$$

$$B_y = 837.3\text{ N} \quad \textit{Ans.}$$

Using these results, equilibrium of the pin requires that

$$+\uparrow\Sigma F_y = 0; \quad \tfrac{4}{5}F_{CB} - 837.3\text{ N} - 490.5\text{ N} \quad F_{CB} = 1660\text{ N} \quad \textit{Ans.}$$

$$\xrightarrow{+}\Sigma F_x = 0; \quad F_{AB} - \tfrac{3}{5}(1660\text{ N}) - 346.8\text{ N} = 0 \quad F_{AB} = 1343\text{ N} \quad \textit{Ans.}$$

It may be noted that the two-force member $CB$ is subjected to bending as caused by the force $\mathbf{F}_{CB}$. From the standpoint of design, it would be better to make this member *straight* (from $C$ to $B$) so that the force $\mathbf{F}_{CB}$ would create only tension in the member.

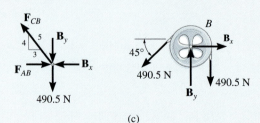

(c)

Before solving the following problems, it is suggested that a brief review be made of all the previous examples. This may be done by covering each solution, trying to locate the two-force members, drawing the free-body diagrams, and conceiving ways of applying the equations of equilibrium to obtain the solution.

## PROBLEMS

**5-34.** In each case, determine the force **P** required to maintain equilibrium. The block weighs 100 lb.

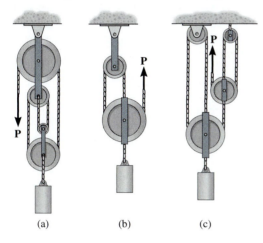

(a)      (b)      (c)

**Prob. 5–34**

**5-35.** The eye hook has a positive locking latch when it supports the load because its two parts are pin-connected at *A* and they bear against one another along the smooth surface at *B*. Determine the resultant force at the pin and the normal force at *B* when the eye hook supports a load of 800 lb.

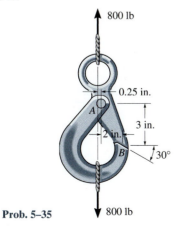

**Prob. 5–35**

**\*5-36.** Determine the force **P** needed to support the 100-lb weight. Each pulley has a weight of 10 lb. Also, what are the cord reactions at *A* and *B*?

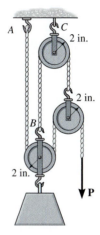

**Prob. 5–36**

**5-37.** The link is used to hold the rod in place. Determine the required axial force on the screw at *E* if the largest force to be exerted on the rod at *B, C* or *D* is to be 100 lb. Also, find the magnitude of the force reaction at pin *A*. Assume all surfaces of contact are smooth.

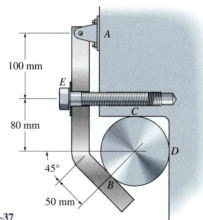

**Prob. 5–37**

**5-38.** The principles of a *differential chain block* are indicated schematically in the figure. Determine the magnitude of force **P** needed to support the 800-N force. Also, find the distance *x* where the cable must be attached to bar *AB* so the bar remains horizontal. All pulleys have a radius of 60 mm.

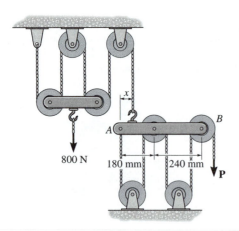

800 N   180 mm   240 mm

**Prob. 5–38**

**5-39.** Determine the force *P* needed to support the 20-kg mass using the *Spanish Burton rig*. Also, what are the reactions at the supporting hooks *A*, *B*, and *C*?

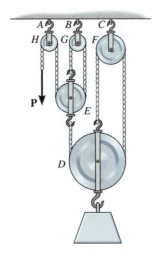

**Prob. 5–39**

***5-40.** The compound beam is fixed at *A* and supported by a rocker at *B* and *C*. There are hinges (pins) at *D* and *E*. Determine the reactions at the supports.

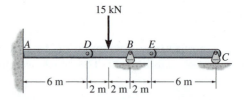

15 kN

6 m   2 m 2 m 2 m   6 m

**Prob. 5–40**

**5-41.** The compound beam is pin-supported at *C* and supported by a roller at *A* and *B*. There is a hinge (pin) at *D*. Determine the reactions at the supports. Neglect the thickness of the beam.

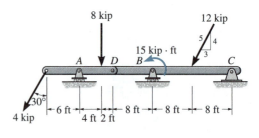

8 kip   12 kip

15 kip · ft

30°   6 ft   8 ft   8 ft   8 ft

4 kip   4 ft 2 ft

**Prob. 5–41**

**5-42.** Determine the greatest force $P$ that can be applied to the frame if the largest force resultant acting at $A$ can have a magnitude of 2 kN.

**\*5-44.** The three-hinged arch supports the loads $F_1 = 8$ kN and $F_2 = 5$ kN. Determine the horizontal and vertical components of reaction at the pin supports $A$ and $B$. Take $h = 2$ m.

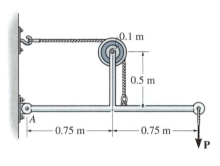

**Prob. 5–42**

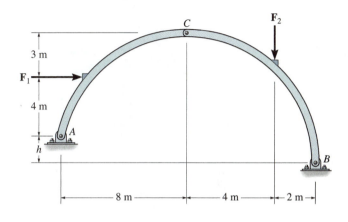

**Prob. 5–44**

**5-43.** Determine the horizontal and vertical components force at pins $A$ and $C$ of the two-member frame.

**5-45.** Determine the horizontal and vertical components of force at pins $A$, $B$, and $C$, and the reactions to the fixed support $D$ of the three-member frame.

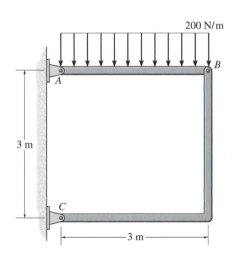

**Prob. 5–43**

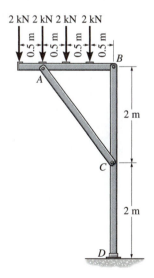

**Prob. 5–45**

**5-46.** Determine the horizontal and vertical components of force at $C$ which member $ABC$ exerts on member $CEF$.

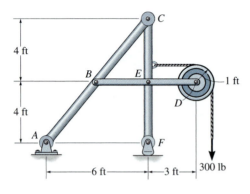

**Prob. 5-46**

***5-48.** The hoist supports the 125-kg engine. Determine the force the load creates in member $DB$ and in member $FB$, which contains the hydraulic cylinder $H$.

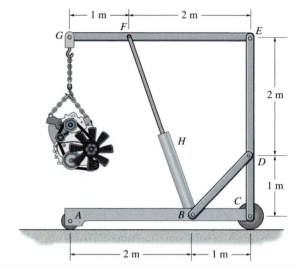

**Prob. 5-48**

**5-47.** Determine the horizontal and vertical components of force that the pins at $A$, $B$, and $C$ exert on their connecting members.

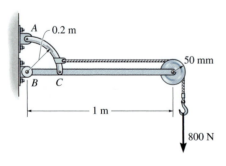

**Prob. 5-47**

**5-49.** Determine the force $P$ on the cord, and the angle $\theta$ that the pulley-supporting link $AB$ makes with the vertical. Neglect the mass of the pulleys and the link. The block has a weight of 200 lb and the cord is attached to the pin at $B$. The pulleys have radii of $r_1 = 2$ in. and $r_2 = 1$ in.

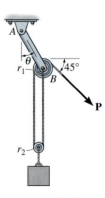

**Prob. 5-49**

**5-50.** The front of the car is to be lifted using a smooth, rigid 10-ft long board. The car has a weight of 3500 lb and a center of gravity at $G$. Determine the position $x$ of the fulcrum so that an applied force of 100 lb at $E$ will lift the front wheels of the car.

**\*5-52.** Determine the force that the smooth roller $C$ exerts on beam $AB$. Also, what are the horizontal and vertical components of reaction at pin $A$? Neglect the weight of the frame and roller.

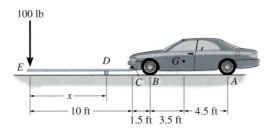

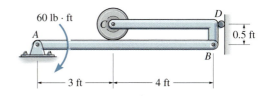

Prob. 5-52

Prob. 5-50

**5-51.** The wall crane supports a load of 700 lb. Determine the horizontal and vertical components of reaction at the pins $A$ and $D$. Also, what is the force in the cable at the winch $W$?

**5-53.** Determine the horizontal and vertical components of force which the pins exert on member $ABC$.

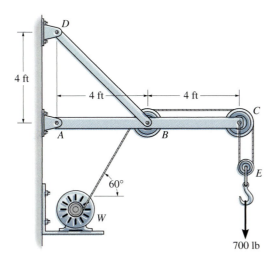

Prob. 5-51

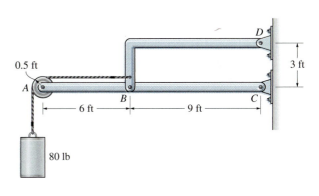

Prob. 5-53

**5-54.** The engine hoist is used to support the 200-kg engine. Determine the force acting in the hydraulic cylinder $AB$, the horizontal and vertical components of force at the pin $C$, and the reactions at the fixed support $D$.

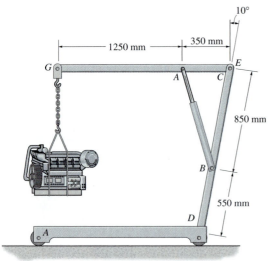

Prob. 5–54

**5-55.** Determine the horizontal and vertical components of force at pins $B$ and $C$.

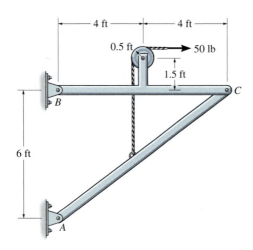

Prob. 5–55

*****5-56.** The pipe cutter is clamped around the pipe $P$. If the wheel at $A$ exerts a normal force of $F_A = 80$ N on the pipe, determine the normal forces of wheels $B$ and $C$ on the pipe. Also compute the pin reaction on the wheel at $C$. The three wheels each have a radius of 7 mm and the pipe has an outer radius of 10 mm.

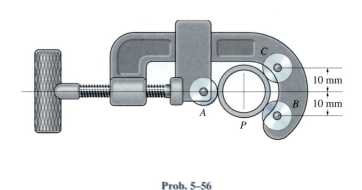

Prob. 5–56

**5-57.** Determine the horizontal and vertical components of force at each pin. The suspended cylinder has a weight of 80 lb.

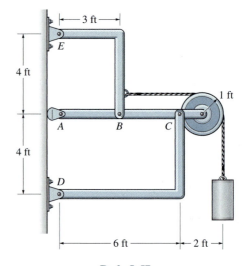

Prob. 5–57

**5-58.** The toggle clamp is subjected to a force **F** at the handle. Determine the vertical clamping force acting at *E*.

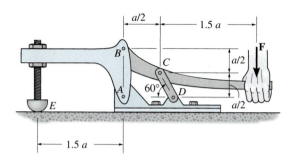

**Prob. 5–58**

**5-59.** Determine the horizontal and vertical components of force which the pins at *A, B,* and *C* exert on member *ABC* of the frame.

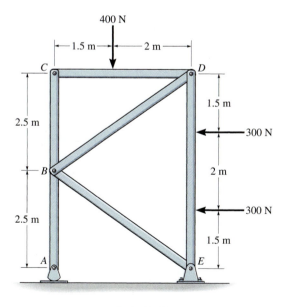

**Prob. 5–59**

**\*5-60.** The derrick is pin-connected to the pivot at *A*. Determine the largest mass that can be supported by the derrick if the maximum force that can be sustained by the pin at *A* is 18 kN.

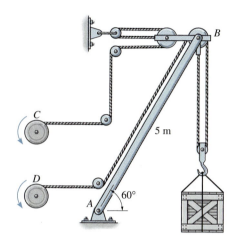

**Prob. 5–60**

**5-61.** Determine the required mass of the suspended cylinder if the tension in the chain wrapped around the freely turning gear is to be 2 kN. Also, what is the magnitude of the resultant force on pin *A*?

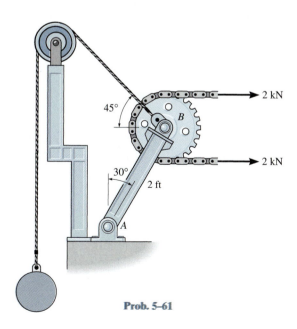

**Prob. 5–61**

**5-62.** The pumping unit is used to recover oil. When the walking beam $ABC$ is horizontal, the force acting in the wireline at the well head is 250 lb. Determine the torque **M** which must be exerted by the motor in order to overcome this load. The horse-head $C$ weighs 60 lb and has a center of gravity at $G_C$. The walking beam $ABC$ has a weight of 130 lb and a center of gravity at $G_B$, and the counterweight has a weight of 200 lb and a center of gravity at $G_W$. The pitman, $AD$, is pin-connected at its ends and has negligible weight.

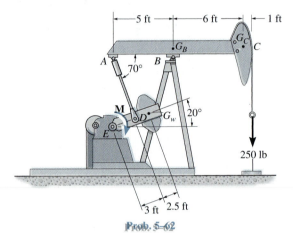

Prob. 5-62

**5-63.** Determine the force $P$ on the cable if the spring is compressed 0.5 in. when the mechanism is in the position shown. The spring has a stiffness of $k = 800$ lb/ft.

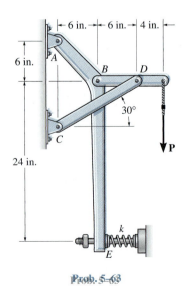

Prob. 5-63

**\*5-64.** Determine the force that the jaws $J$ of the metal cutters exert on the smooth cable $C$ if 100-N forces are applied to the handles. The jaws are pinned at $E$ and $A$, and $D$ and $B$. There is also a pin at $F$.

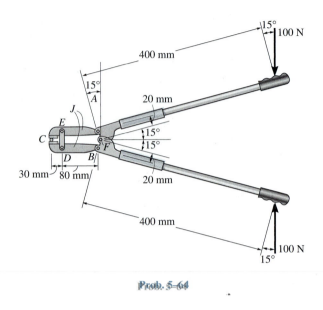

Prob. 5-64

**5-65.** The compound arrangement of the pan scale is shown. If the mass on the pan is 4 kg, determine the horizontal and vertical components at pins $A$, $B$, and $C$ and the distance $x$ of the 25-g mass to keep the scale in balance.

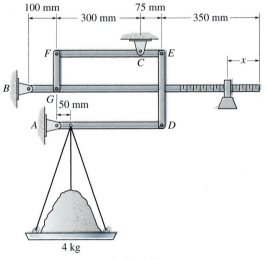

Prob. 5-65

**5-66.** The scissors lift consists of *two* sets of cross members and *two* hydraulic cylinders, *DE*, symmetrically located on *each side* of the platform. The platform has a uniform mass of 60 kg, with a center of gravity at $G_1$. The load of 85 kg, with center of gravity at $G_2$, is centrally located between each side of the platform. Determine the force in each of the hydraulic cylinders for equilibrium. Rollers are located at *B* and *D*.

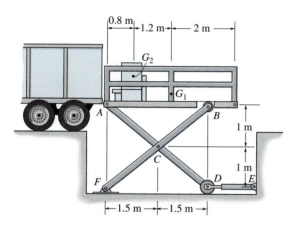

**Prob. 5–66**

**5-67.** Determine the horizontal and vertical components of force that the pins at *A*, *B*, and *C* exert on the frame. The cylinder has a mass of 80 kg. The pulley has a radius of 0.1 m.

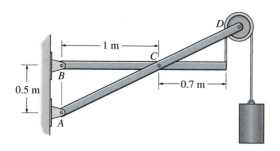

**Prob. 5–67**

**\*5-68.** By squeezing on the hand brake of the bicycle, the rider subjects the brake cable to a tension of 50 lb. If the caliper mechanism is pin-connected to the bicycle frame at *B*, determine the normal force each brake pad exerts on the rim of the wheel. Is this the force that stops the wheel from turning? Explain.

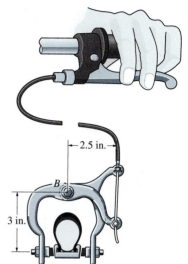

**Prob. 5–68**

**5-69.** If a force of *P* = 6 lb is applied perpendicular to the handle of the mechanism, determine the magnitude of force **F** for equilibrium. The members are pin-connected at *A*, *B*, *C*, and *D*.

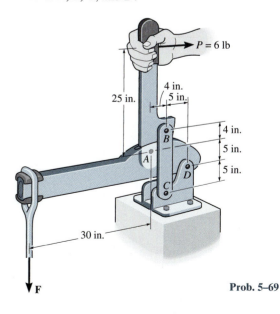

**Prob. 5–69**

**5-70.** The bucket of the backhoe and its contents have a weight of 1200 lb and a center of gravity at $G$. Determine the forces of the hydraulic cylinder $AB$ and in links $AC$ and $AD$ in order to hold the load in the position shown. The bucket is pinned at $E$.

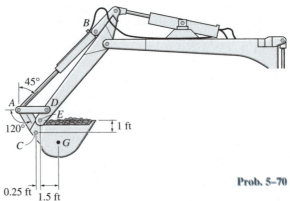

**Prob. 5–70**

## CHAPTER REVIEW

- *Truss Analysis.* A simple truss consists of triangular elements connected together by pin joints. The forces within it members can be determined by assuming the members are all two-force members, connected concurrently at each joint.

- *Method of Joints.* If a truss is in equilibrium, then each of its joints is also in equilibrium. For a coplanar truss, the concurrent force system at each joint must satisfy force equilibrium, $\Sigma F_x = 0$, $\Sigma F_y = 0$. To obtain a numerical solution for the forces in the members, select a joint that has a free-body diagram with at most two unknown forces and one known force. (This may require first finding the reactions at the supports.) Once a member force is determined, use its value and apply it to an adjacent joint. Remember that forces that are found to *pull* on the joint are in *tension*, and those that *push* on the joint are in *compression*. To avoid a simultaneous solution of two equations, try to sum forces in a direction that is perpendicular to one of the unknowns. This will allow a direct solution for the other unknown. To further simplify the analysis, first identify all the zero-force members.

- *Method of Sections.* If a truss is in equilibrium, then each section of the truss is also in equilibrium. Pass a section through the member whose force is to be determined. Then draw the free-body diagram of the sectioned part having the least number of forces on it.

Sectioned members subjected to *pulling* are in *tension*, and those that are subjected to *pushing* are in *compression*. If the force system is coplanar, then three equations of equilibrium are available to determine the unknowns. If possible, sum forces in a direction that is perpendicular to two of the three unknown forces. This will yield a direct solution for the third force. Likewise, sum moments about a point that passes through the line of action of two of the three unknown forces, so that the third unknown force can be determined directly.

• *Frames and Machines.* The forces acting at the joints of a frame or machine can be determined by drawing the free-body diagrams of each of its members or parts. The principle of action-reaction should be carefully observed when drawing these forces on each adjacent member or pin. For a coplanar force system, there are three equilibrium equations available for each member.

# REVIEW PROBLEMS

**5-71.** Determine the reactions at the supports of the compound beam. There is a short vertical link at $C$.

**\*5-72.** The two-bar mechanism consists of a lever arm $AB$ and smooth link $CD$, which has a fixed collar at its end $C$ and a roller at the other end $D$. Determine the force $\mathbf{P}$ needed to bold the lever in the position $\theta$. The spring has a stiffness $k$ and unstretched length $2L$. The roller contacts either the top or bottom portion of the horizontal guide.

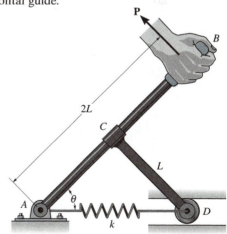

Prob. 5–71

Prob. 5–72

**5-73.** Determine the force in each member of the truss and indicate whether the members are in tension or compression.

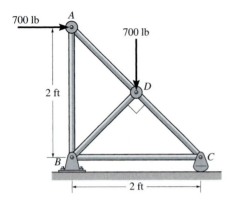

**Prob. 5–73**

**5-75.** The *Howe bridge truss* is subjected to the loading shown. Determine the force in members. *HI, HB,* and *BC,* and indicate whether the members are in tension or compression.

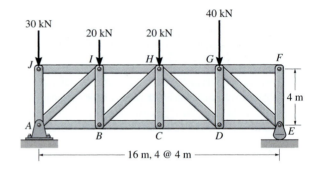

**Prob. 5–75**

**5-74.** The *Howe bridge truss* is subjected to the loading shown. Determine the force in members *HD, CD,* and *GD,* and indicate whether the members are in tension or compression.

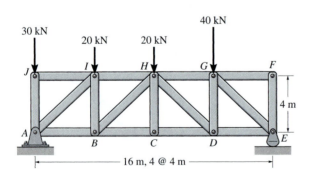

**Prob. 5–74**

***5-76.** Determine the horizontal and vertical components of force at pins *A, B,* and *C* of the two-member frame.

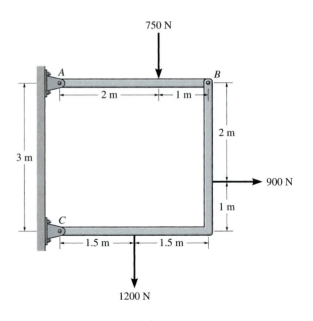

**Prob. 5–76**

**5-77.** The compound beam is supported by a rocker at *B* and fixed to the wall at *A*. If it is hinged (pinned) together at *C*, determine the reactions at the supports.

**5-78.** Determine the horizontal and vertical components of reaction at *A* and *B*. The pin at *C* is fixed to member *AE* and fits through a smooth slot in member *BD*.

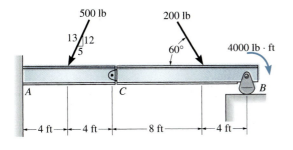

**Prob. 5-77**

**Prob. 5-78**

When a pressure vessel is designed, it is important to be able to determine the center of gravity of its component parts, calculate its volume and surface area, and reduce three-dimensional distributed loadings to their resultants. These topics are discussed in this chapter.

# CHAPTER 6

# Geometric Properties and Distributed Loadings

## CHAPTER OBJECTIVES

- To discuss the concept of the center of gravity, center of mass, and the centroid.
- To show how to determine the location of the center of gravity and centroid for a system of discrete particles and a body of arbitrary shape.
- To present a method for finding the resultant of a general distributed loading.
- To develop a method for determining the moment of inertia for an area.

## 6.1 Center of Gravity and Center of Mass for a System of Particles

**Center of Gravity.** The *center of gravity* $G$ is a point which locates the resultant weight of a system of particles. To show how to determine this point consider the system of $n$ particles fixed within a region of space as shown in Fig. 6–1a. The weights of the particles comprise a system of parallel forces[*] which can be replaced by a single (equivalent) resultant weight having the defined point $G$ of application. To find the $\bar{x}, \bar{y}, \bar{z}$ coordinates of $G$, we must use the principles outlined in Sec. 3.9.

---

[*]This is not true in the exact sense, since the weights are not parallel to each other; rather they are all *concurrent* at the earth's center. Furthermore, the acceleration of gravity $g$ is actually different for each particle since it depends on the distance from the earth's center to the particle. For all practical purposes, however, both of these effects can generally be neglected.

259

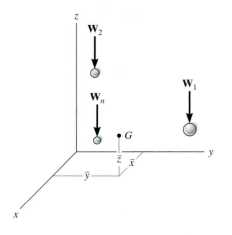

(a)

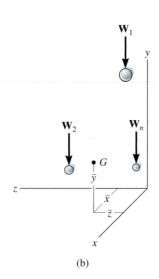

(b)

Fig. 6–1

This requires that the resultant weight be equal to the total weight of all $n$ particles; that is,

$$W_R = \Sigma W$$

The sum of the moments of the weights of all the particles about the $x$, $y$, and $z$ axes is then equal to the moment of the resultant weight about these axes. Thus, to determine the $\bar{x}$ coordinate of $G$, we can sum moments about the $y$ axis. This yields

$$\bar{x}W_R = \tilde{x}_1 W_1 + \tilde{x}_2 W_2 + \cdots + \tilde{x}_n W_n$$

Likewise, summing moments about the $x$ axis, we can obtain the $\bar{y}$ coordinate; i.e.,

$$\bar{y}W_R = \tilde{y}_1 W_1 + \tilde{y}_2 W_2 + \cdots + \tilde{y}_n W_n$$

Although the weights do not produce a moment about the $z$ axis, we can obtain the $\bar{z}$ coordinate of $G$ by imagining the coordinate system, with the particles fixed in it, as being rotated 90° about the $x$ (or $y$) axis, Fig. 6–1$b$. Summing moments about the $x$ axis, we have

$$\bar{z}W_R = \tilde{z}_1 W_1 + \tilde{z}_2 W_2 + \cdots + \tilde{z}_n W_n$$

We can generalize these formulas, and write them symbolically in the form

$$\bar{x} = \frac{\Sigma \tilde{x} W}{\Sigma W} \qquad \bar{y} = \frac{\Sigma \tilde{y} W}{\Sigma W} \qquad \bar{z} = \frac{\Sigma \tilde{z} W}{\Sigma W} \qquad (6–1)$$

Here

$\bar{x}, \bar{y}, \bar{z}$   represent the coordinates of the center of gravity $G$ of the system of particles.

$\tilde{x}, \tilde{y}, \tilde{z}$   represent the coordinates of each particle in the system.

$\Sigma W$ is the resultant sum of the weights of all the particles in the system.

These equations are easily remembered if it is kept in mind that they simply represent a balance between the sum of the moments of the weights of each particle of the system and the moment of the *resultant* weight for the system.

**Center of Mass.**   To study problems concerning the motion of *matter* under the influence of force, i.e., dynamics, it is necessary to locate a point called the *center of mass*. Provided the acceleration due to gravity $g$ for every particle is constant, then $W = mg$. Substituting into Eqs. 6–1 and canceling $g$ from both the numerator and denominator yields

$$\bar{x} = \frac{\Sigma \tilde{x} m}{\Sigma m} \qquad \bar{y} = \frac{\Sigma \tilde{y} m}{\Sigma m} \qquad \bar{z} = \frac{\Sigma \tilde{z} m}{\Sigma m} \qquad (6–2)$$

By comparison, then, the location of the center of gravity *coincides* with that of the center of mass. *Recall, however, that particles have "weight" only when under the influence of a gravitational attraction, whereas the center of mass is independent of gravity. For example, it would be meaningless to define the center of gravity of a system of particles representing the planets of our solar system, while the center of mass of this system is important.

## 6.2  Center of Gravity and Centroid for a Body

**Center of Gravity.**  A rigid body is composed of an infinite number of particles, and so if the principles used to determine Eqs. 6–1 are applied to the system of particles composing a rigid body, it becomes necessary to use integration rather than a discrete summation of the terms. Considering the arbitrary particle located at $(\tilde{x}, \tilde{y}, \tilde{z})$ and having a weight $dW$, Fig. 6–2, the resulting equations are

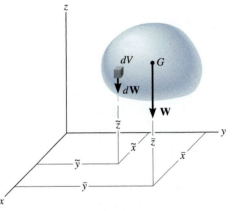

Fig. 6–2

$$\bar{x} = \frac{\displaystyle\int \tilde{x}\,dW}{\displaystyle\int dW} \qquad \bar{y} = \frac{\displaystyle\int \tilde{y}\,dW}{\displaystyle\int dW} \qquad \bar{z} = \frac{\displaystyle\int \tilde{z}\,dW}{\displaystyle\int dW} \qquad (6\text{–}3)$$

In order to apply these equations properly, the differential weight $dW$ must be expressed in terms of its associated volume $dV$. If $\gamma$ represents the *specific weight* of the body, measured as a weight per unit volume, then $dW = \gamma\,dV$ and therefore

$$\bar{x} = \frac{\displaystyle\int_V \tilde{x}\gamma\,dV}{\displaystyle\int_V \gamma\,dV} \qquad \bar{y} = \frac{\displaystyle\int_V \tilde{y}\gamma\,dV}{\displaystyle\int_V \gamma\,dV} \qquad \bar{z} = \frac{\displaystyle\int_V \tilde{z}\gamma\,dV}{\displaystyle\int_V \gamma\,dV} \qquad (6\text{–}4)$$

Here integration must be performed throughout the entire volume of the body.

**Center of Mass.**  The *density* $\rho$, or mass per unit volume, is related to $\gamma$ by the equation $\gamma = \rho g$, where $g$ is the acceleration due to gravity. Substituting this relationship into Eqs. 6–4 and canceling $g$ from both the numerators and denominators yields similar equations (with $\rho$ replacing $\gamma$) that can be used to determine the body's *center of mass*.

---

*This is true as long as the gravity field is assumed to have the same magnitude and direction everywhere. That assumption is appropriate for most engineering applications, since gravity does not vary appreciably between, for instance, the bottom and the top of a building.

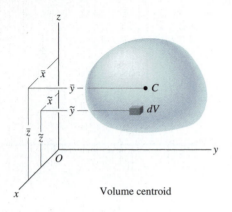

Volume centroid

**Fig. 6–3**

**Centroid.** The *centroid C* is a point which defines the *geometric center* of an object. Its location can be determined from formulas similar to those used to determine the body's center of gravity or center of mass. In particular, if the material composing a body is uniform or *homogeneous*, the *density or specific weight* will be *constant* throughout the body, and therefore this term will factor out of the integrals and *cancel* from both the numerators and denominators of Eqs. 6–4. The resulting formulas define the centroid of the body since they are independent of the body's weight and instead depend only on the body's geometry. Three specific cases will be considered.

**Volume.** If an object is subdivided into volume elements $dV$, Fig. 6–3, the location of the centroid $C(\bar{x}, \bar{y}, \bar{z})$ for the volume of the object can be determined by computing the "moments" of the elements about each of the coordinate axes. The resulting formulas are

$$\bar{x} = \frac{\int_V \tilde{x}\, dV}{\int_V dV} \qquad \bar{y} = \frac{\int_V \tilde{y}\, dV}{\int_V dV} \qquad \bar{z} = \frac{\int_V \tilde{z}\, dV}{\int_V dV} \qquad (6\text{–}5)$$

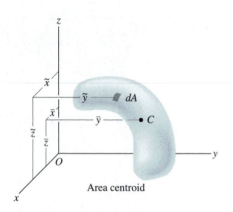

Area centroid

**Fig. 6–4**

**Area.** In a similar manner, the centroid for the surface area of an object, such as a plate or shell, Fig. 6–4, can be found by subdividing the area into differential elements $dA$ and computing the "moments" of these area elements about each of the coordinate axes, namely,

$$\bar{x} = \frac{\int_A \tilde{x}\, dA}{\int_A dA} \qquad \bar{y} = \frac{\int_A \tilde{y}\, dA}{\int_A dA} \qquad \bar{z} = \frac{\int_A \tilde{z}\, dA}{\int_A dA} \qquad (6\text{–}6)$$

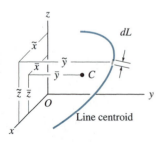

Line centroid

**Fig. 6–5**

**Line.** If the geometry of the object, such as a thin rod or wire, takes the form of a line, Fig. 6–5, the balance of moments of the differential elements $dL$ about each of the coordinate axes yields

$$\bar{x} = \frac{\int_L \tilde{x}\, dL}{\int_L dL} \qquad \bar{y} = \frac{\int_L \tilde{y}\, dL}{\int_L dL} \qquad \bar{z} = \frac{\int_L \tilde{z}\, dL}{\int_L dL} \qquad (6\text{–}7)$$

Remember that when applying Eqs. 6–4 through 6–7 it is best to choose a coordinate system that simplifies as much as possible the equation used to describe the object's boundary. For example, polar coordinates are generally appropriate for areas having circular boundaries. Also, the terms $\tilde{x}, \tilde{y}, \tilde{z}$ in the equations refer to the "moment arms" or coordinates of the *center of gravity or centroid for the differential element* used. If possible, this differential element should be chosen such that it has a differential size or thickness in only *one direction*. When this is done, only a single integration is required to cover the entire region.

**Symmetry.** The *centroids* of some shapes may be partially or completely specified by using conditions of *symmetry*. In cases where the shape has an axis of symmetry, the centroid of the shape will lie along that axis. For example, the centroid $C$ for the line shown in Fig. 6–6 must lie along the $y$ axis since for every elemental length $dL$ at a distance $+\tilde{x}$ to the right of the $y$ axis there is an identical element at a distance $-\tilde{x}$ to the left. The total moment for all the elements about the axis of symmetry will therefore cancel; i.e., $\int \tilde{x} \, dL = 0$ (Eq. 6–7), so that $\bar{x} = 0$. In cases where a shape has two or three axes of symmetry, it follows that the centroid lies at the intersection of these axes, Fig. 6–7 and Fig. 6–8.

Integration must be used to determine the location of the center of gravity of this goal post.

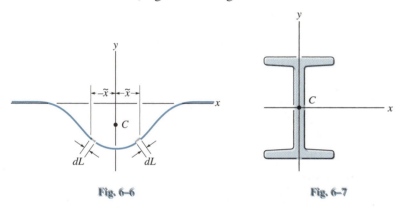

**Fig. 6–6**

**Fig. 6–7**

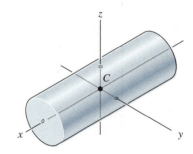

**Fig. 6–8**

## IMPORTANT POINTS

- The centroid represents the geometric center of a body. This point coincides with the center of mass or the center of gravity only if the material composing the body is uniform or homogeneous.

- Formulas used to locate the center of gravity or the centroid simply represent a balance between the sum of moments of all the parts of the system and the moment of the "resultant" for the system.

- In some cases the centroid is located at a point that is not on the object, as in the case of a ring, where the centroid is at its center. Also, this point will lie on any axis of symmetry for the body.

## PROCEDURE FOR ANALYSIS

The center of gravity or centroid of an object or shape can be determined by single integrations using the following procedure.

### Differential Element.

- Select an appropriate coordinate system, specify the coordinate axes, and then choose a differential element for integration.
- For lines the element $dL$ is represented as a differential line segment.
- For areas the element $dA$ is generally a rectangle having a finite length and differential width.
- For volumes the element $dV$ is either a circular disk having a finite radius and differential thickness, or a shell having a finite length and radius and a differential thickness.
- Locate the element at an arbitrary point $(x, y, z)$ on the curve that defines the shape.

### Size and Moment Arms.

- Express the length $dL$, area $dA$, or volume $dV$ of the element in terms of the coordinates of the curve used to define the geometric shape.
- Determine the coordinates or moment arms $\tilde{x}$, $\tilde{y}$, $\tilde{z}$ for the centroid or center of gravity of the element.

### Integrations.

- Substitute the formulations for $\tilde{x}$, $\tilde{y}$, $\tilde{z}$ and $dL$, $dA$, or $dV$ into the appropriate equations (Eqs. 6–4 through 6–7) and perform the integrations.[*]
- Express the function in the integrand in terms of the *same variable as the differential thickness of the element* in order to perform the integration.
- The limits of the integral are defined from the two extreme locations of the element's differential thickness, so that when the elements are "summed" or the integration performed, the entire region is covered.

[*]Formulas for integration are given in Appendix A.

# E X A M P L E   6.1

Determine the distance $\overline{y}$ from the $x$ axis to the centroid of the area of the triangle shown in Fig. 6–9.

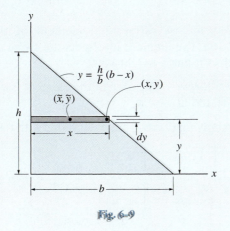

**Fig. 6–9**

## Solution

*Differential Element.*  Consider a rectangular element having thickness $dy$ which intersects the boundary at $(x, y)$, Fig. 6–9.

*Area and Moment Arms.*  The area of the element is $dA = x\, dy = \dfrac{b}{h}(h - y)\, dy$, and its centroid is located a distance $\widetilde{y} = y$ from the $x$ axis.

*Integrations.*  Applying the second of Eqs. 6–6 and integrating with respect to $y$ yields

$$\overline{y} = \frac{\displaystyle\int_A \widetilde{y}\, dA}{\displaystyle\int_A dA} = \frac{\displaystyle\int_0^h y\frac{b}{h}(h - y)\, dy}{\displaystyle\int_0^h \frac{b}{h}(h - y)\, dy} = \frac{\frac{1}{6}bh^2}{\frac{1}{2}bh}$$

$$= \frac{h}{3} \qquad\qquad\qquad \textit{Ans.}$$

## EXAMPLE 6.2

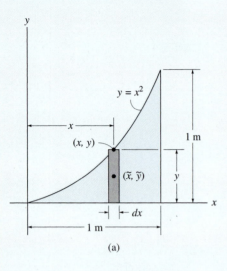

(a)

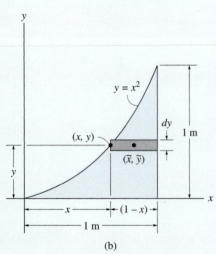

(b)

Fig. 6–10

Locate the centroid of the area shown in Fig. 6–10a.

### Solution I

*Differential Element.* A differential element of thickness $dx$ is shown in Fig. 6–10a. The element intersects the curve at the *arbitrary point* $(x, y)$, and so it has a height $y$.

*Area and Moment Arms.* The area of the element is $dA = y\,dx$, and its centroid is located at $\tilde{x} = x$, $\tilde{y} = y/2$.

*Integrations.* Applying Eqs. 6–6 and integrating with respect to $x$ yields

$$\bar{x} = \frac{\displaystyle\int_A \tilde{x}\,dA}{\displaystyle\int_A dA} = \frac{\displaystyle\int_0^1 xy\,dx}{\displaystyle\int_0^1 y\,dx} = \frac{\displaystyle\int_0^1 x^3\,dx}{\displaystyle\int_0^1 x^2\,dx} = \frac{0.250}{0.333} = 0.75 \text{ m} \qquad \textit{Ans.}$$

$$\bar{y} = \frac{\displaystyle\int_A \tilde{y}\,dA}{\displaystyle\int_A dA} = \frac{\displaystyle\int_0^1 (y/2)y\,dx}{\displaystyle\int_0^1 y\,dx} = \frac{\displaystyle\int_0^1 (x^2/2)x^2\,dx}{\displaystyle\int_0^1 x^2\,dx} = \frac{0.100}{0.333} = 0.3 \text{ m} \qquad \textit{Ans.}$$

### Solution II

*Differential Element.* The differential element of thickness $dy$ is shown in Fig. 6–10b. The element intersects the curve at the *arbitrary point* $(x, y)$, and so it has a length $(1 - x)$.

*Area and Moment Arms.* The area of the element is $dA = (1 - x)\,dy$, and its centroid is located at

$$\tilde{x} = x + \left(\frac{1 - x}{2}\right) = \frac{1 + x}{2}, \quad \tilde{y} = y$$

*Integrations.* Applying Eqs. 6–6 and integrating with respect to $y$, we obtain

$$\bar{x} = \frac{\displaystyle\int_A \tilde{x}\,dA}{\displaystyle\int_A dA} = \frac{\displaystyle\int_0^1 [(1 + x)/2](1 - x)\,dy}{\displaystyle\int_0^1 (1 - x)\,dy} = \frac{\dfrac{1}{2}\displaystyle\int_0^1 (1 - y)\,dy}{\displaystyle\int_0^1 (1 - \sqrt{y})\,dy} = \frac{0.250}{0.333} = 0.75 \text{ m} \quad \textit{Ans.}$$

$$\bar{y} = \frac{\displaystyle\int_A \tilde{y}\,dA}{\displaystyle\int_A dA} = \frac{\displaystyle\int_0^1 y(1 - x)\,dy}{\displaystyle\int_0^1 (1 - x)\,dy} = \frac{\displaystyle\int_0^1 (y - y^{3/2})\,dy}{\displaystyle\int_0^1 (1 - \sqrt{y})\,dy} = \frac{0.100}{0.333} = 0.3 \text{ m} \qquad \textit{Ans.}$$

# EXAMPLE 6.3

Locate the $\bar{y}$ centroid for the paraboloid of revolution, which is generated by revolving the shaded area shown in Fig. 6–11a about the y axis.

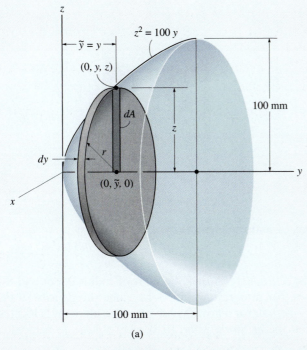

(a)

**Fig. 6–11**

## Solution

*Differential Element.* An element having the shape of a *thin disk* is chosen, Fig. 6–11a. This element has a thickness $dy$. In this "disk" method of analysis, the element of planar area, $dA$, is always taken *perpendicular* to the axis of revolution. Here the element intersects the generating curve at the *arbitrary point* $(0, y, z)$, and so its radius is $r = z$.

*Area and Moment Arm.* The volume of the element is $dV = (\pi z^2)\, dy$, and its centroid is located at $\tilde{y} = y$.

*Integration.* Applying the second of Eqs. 6–5 and integrating with respect to y yields

$$\bar{y} = \frac{\displaystyle\int_V \tilde{y}\, dV}{\displaystyle\int_V dV} = \frac{\displaystyle\int_0^{100} y(\pi z^2)\, dy}{\displaystyle\int_0^{100} (\pi z^2)\, dy} = \frac{100\pi \displaystyle\int_0^{100} y^2\, dy}{100\pi \displaystyle\int_0^{100} y\, dy} = 66.7 \text{ mm} \qquad \textit{Ans.}$$

**E X A M P L E 6.4**

Determine the location of the center of mass of the cylinder shown in Fig. 6–12 if its density varies directly with its distance from the base, i.e., $\rho = 200z$ kg/m³.

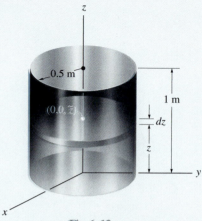

Fig. 6–12

**Solution**

For reasons of material symmetry,

$$\bar{x} = \bar{y} = 0 \qquad \qquad Ans.$$

*Differential Element.* A disk element of radius 0.5 m and thickness $dz$ is chosen for integration, Fig. 6–12, since the *density of the entire element is constant* for a given value of $z$. The element is located along the $z$ axis at the *arbitrary point* $(0, 0, z)$.

*Volume and Moment Arm.* The volume of the element is $dV = \pi(0.5)^2\, dz$, and its centroid is located at $\tilde{z} = z$.

*Integrations.* Using an equation similar to the third of Eqs. 6–4 and integrating with respect to $z$, noting that $\rho = 200z$, we have

$$\bar{z} = \frac{\displaystyle\int_V \tilde{z}\rho\, dV}{\displaystyle\int_V \rho\, dV} = \frac{\displaystyle\int_0^1 z(200z)\pi(0.5)^2\, dz}{\displaystyle\int_0^1 (200z)\pi(0.5)^2\, dz}$$

$$= \frac{\displaystyle\int_0^1 z^2\, dz}{\displaystyle\int_0^1 z\, dz} = 0.667 \text{ m} \qquad \qquad Ans.$$

# PROBLEMS

**6-1.** Locate the centroid $(\bar{x}, \bar{y})$ of the shaded area.

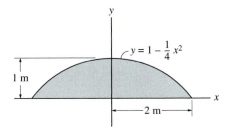

Probs. 6–1

**6-3.** Locate the centroid $\bar{y}$ of the shaded area.

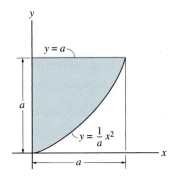

Prob. 6–3

**6-2.** The plate has a thickness of 0.25 ft and a specific weight of $\gamma = 180 \, \text{lb/ft}^3$. Determine the location of its center of gravity. Also, find the tension in each of the cords used to support it.

*6-4.** Locate the centroid of the shaded area bounded by the parabola and the line $y = a$.

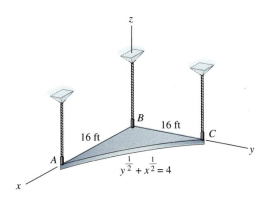

Prob. 6–2

Prob. 6–4

**6-5.** Locate the centroid of the quarter elliptical area.

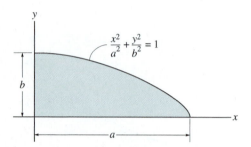

**Prob. 6–5**

**6-7.** Locate the centroid of the shaded area.

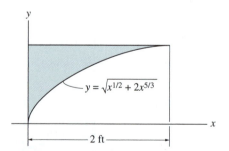

**Prob. 6–7**

**6-6.** Locate the centroid $(\bar{x}, \bar{y})$ of the shaded area.

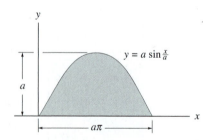

**Prob. 6–6**

**■*6-8.** Locate the centroid $\bar{x}$ of the shaded area. Solve the problem by evaluating the integrals using Simpson's rule.

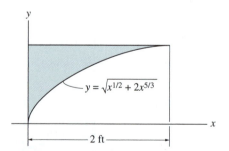

**Prob. 6–8**

**6-9.** Locate the center of gravity of the volume. The material is homogeneous.

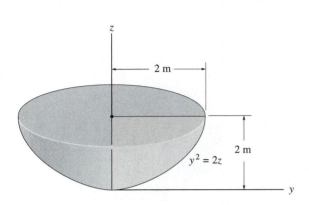

Prob. 6–9

**6-11.** Locate the centroid of the solid.

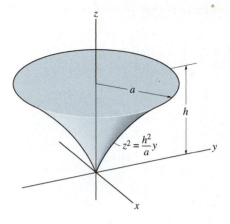

Prob. 6–11

**6-10.** Locate the centroid $\bar{z}$ of the hemisphere.

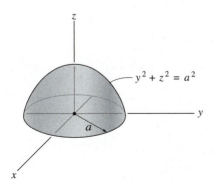

Prob. 6–10

**\*6-12.** Locate the centroid of the quarter-cone.

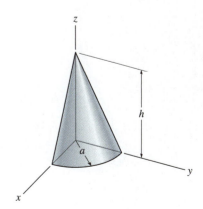

Prob. 6–12

**6-13.** Locate the center of mass $\bar{x}$ of the hemisphere. The density of the material varies linearly from zero at the origin $O$ to $\rho_0$ at the surface. *Suggestion:* Choose a hemispherical shell element for integration.

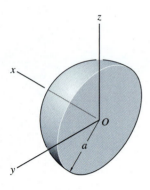

**Prob. 6–13**

**6-15.** Locate the centroid $\bar{y}$ of the paraboloid.

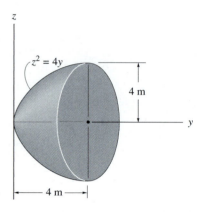

**Prob. 6–15**

**6-14.** Locate the centroid $\bar{z}$ of the right-elliptical cone.

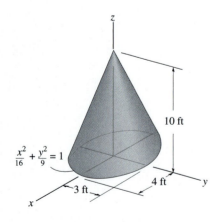

**Prob. 6–14**

*6-16.** The king's chamber of the Great Pyramid of Giza is located at its centroid. Assuming the pyramid to be a solid, prove that this point is at $\bar{z} = \frac{1}{4}h$. *Suggestion:* Use a rectangular differential plate element having a thickness $dz$ and area $(2x)(2y)$.

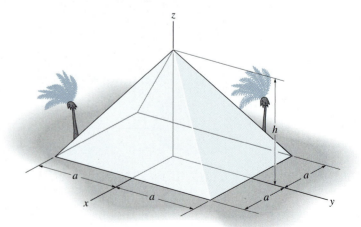

**Prob. 6–16**

## 6.3 Composite Bodies

A *composite body* consists of a series of connected "simpler" shaped bodies, which may be rectangular, triangular, semicircular, etc. Such a body can often be sectioned or divided into its composite parts and, provided the *weight* and location of the center of gravity of each of these parts are known, we can eliminate the need for integration to determine the center of gravity for the entire body. The method for doing this requires treating each composite part like a particle and following the procedure outlined in Sec. 6.1. Formulas analogous to Eqs. 6–1 result since we must account for a finite number of weights. Rewriting these formulas, we have

$$\bar{x} = \frac{\Sigma \tilde{x} W}{\Sigma W} \qquad \bar{y} = \frac{\Sigma \tilde{y} W}{\Sigma W} \qquad \bar{z} = \frac{\Sigma \tilde{z} W}{\Sigma W} \qquad (6\text{--}8)$$

Here

$\bar{x}, \bar{y}, \bar{z}$    represent the coordinates of the center of gravity $G$ of the composite body.

$\tilde{x}, \tilde{y}, \tilde{z}$    represent the coordinates of the center of gravity of each composite part of the body.

$\Sigma W$    is the sum of the weights of all the composite parts of the body, or simply the total weight of the body.

In order to determine the force required to tip over this concrete barrier it is first necessary to determine the location of its center of gravity.

When the body has a *constant density or specific weight*, the center of gravity *coincides* with the centroid of the body. The centroid for composite lines, areas, and volumes can be found using relations analogous to Eqs. 6–8; however, the $W$'s are replaced by $L$'s, $A$'s, and $V$'s, respectively. Centroids for common shapes of lines, areas, shells, and volumes that often make up a composite body are given in the table in Appendix C.

## PROCEDURE FOR ANALYSIS

The location of the center of gravity of a body or the centroid of a composite geometrical object represented by a line, area, or volume can be determined using the following procedure.

*Composite Parts.*

• Using a sketch, divide the body or object into a finite number of composite parts that have simpler shapes.

• If a composite part has a *hole*, or a geometric region having no material, then consider the composite part without the hole and consider the hole as an *additional* composite part having *negative* weight or size.

*Moment Arms.*

• Establish the coordinate axes on the sketch and determine the coordinates $\tilde{x}, \tilde{y}, \tilde{z}$ of the center of gravity or centroid of each part.

*Summations.*

• Determine $\bar{x}, \bar{y}, \bar{z}$ by applying the center of gravity equations, Eqs. 6–8, or the analogous centroid equations.

• If an object is *symmetrical* about an axis, the centroid of the object lies on this axis.

If desired, the calculations can be arranged in tabular form, as indicated in the following three examples.

EXAMPLE 6.5

Locate the centroid of the plate area shown in Fig. 6–13a.

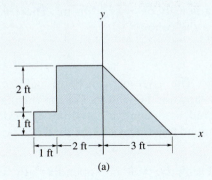

(a)

Fig. 6–13

**Solution**

*Composite Parts.* The plate is divided into three segments as shown in Fig. 6–13b. Here the area of the small rectangle ③ is considered "negative" since it must be subtracted from the larger one ②.

*Moment Arms.* The centroid of each segment is located as indicated in the figure. Note that the $\tilde{x}$ coordinates of ② and ③ are *negative*.

*Summations.* Taking the data from Fig. 6–13b, the calculations are tabulated as follows:

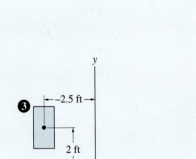

(b)

| Segment | $A$ (ft$^2$) | $\tilde{x}$ (ft) | $\tilde{y}$ (ft) | $\tilde{x}A$ (ft$^3$) | $\tilde{y}A$ (ft$^3$) |
|---------|-----------|------------|------------|--------------|--------------|
| 1 | $\frac{1}{2}(3)(3) =$ 4.5 | 1 | 1 | 4.5 | 4.5 |
| 2 | $(3)(3) =$ 9 | −1.5 | 1.5 | −13.5 | 13.5 |
| 3 | $-(2)(1) =$ −2 | −2.5 | 2 | 5 | −4 |
|   | $\Sigma A$ = 11.5 | | | $\Sigma\tilde{x}A$ = −4 | $\Sigma\tilde{y}A$ = 14 |

Thus,

$$\bar{x} = \frac{\Sigma\tilde{x}A}{\Sigma A} = \frac{-4}{11.5} = -0.348 \text{ ft} \qquad Ans.$$

$$\bar{y} = \frac{\Sigma\tilde{y}A}{\Sigma A} = \frac{14}{11.5} = 1.22 \text{ ft} \qquad Ans.$$

# EXAMPLE 6.6

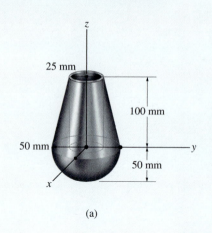

(a)

Fig. 6–14

Locate the center of mass of the composite assembly shown in Fig. 6–14a. The conical frustum has a density of $\rho_c = 8\ \text{Mg/m}^3$, and the hemisphere has a density of $\rho_h = 4\ \text{Mg/m}^3$. There is a 25-mm radius cylindrical hole in the center.

## Solution

*Composite Parts.* The assembly can be thought of as consisting of four segments as shown in Fig. 6–14b. For the calculations, ③ and ④ must be considered as "negative" volumes in order that the four segments, when added together, yield the total composite shape shown in Fig. 6–14a.

*Moment Arm.* Using the table in Appendix C, the computations for the centroid $\tilde{z}$ of each piece are shown in the figure.

*Summations.* Because of *symmetry*, note that

$$\bar{x} = \bar{y} = 0 \qquad Ans.$$

Since $W = mg$ and $g$ is constant, the third of Eqs. 6–8 becomes $\bar{z} = \Sigma\tilde{z}m/\Sigma m$. The mass of each piece can be computed from $m = \rho V$ and used for the calculations. Also, $1\ \text{Mg/m}^3 = 10^{-6}\ \text{kg/mm}^3$, so that

| Segment | $m$ (kg) | $\tilde{z}$ (mm) | $\tilde{z}m$ (kg·mm) |
|---------|----------|------------------|----------------------|
| 1 | $8(10^{-6})(\frac{1}{3})\pi(50)^2(200) = 4.189$ | 50 | 209.440 |
| 2 | $4(10^{-6})(\frac{2}{3})\pi(50)^3 = 1.047$ | $-18.75$ | $-19.635$ |
| 3 | $-8(10^{-6})(\frac{1}{3})\pi(25)^2(100) = -0.524$ | $100 + 25 = 125$ | $-65.450$ |
| 4 | $-8(10^{-6})\pi(25)^2(100) = -1.571$ | 50 | $-78.540$ |
| | $\Sigma m = 3.141$ | | $\Sigma\tilde{z}m = 45.815$ |

Thus,

$$\tilde{z} = \frac{\Sigma\tilde{z}m}{\Sigma m} = \frac{45.815}{3.141} = 14.6\ \text{mm} \qquad Ans.$$

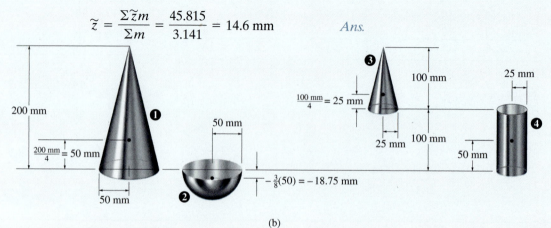

(b)

# PROBLEMS

**6–17.** Determine the location $\bar{y}$ of the centroidal axis $\bar{x},\bar{x}$ of the beam's cross-sectional area. Neglect the size of the corner welds at $A$ and $B$ for the calculation.

**6-19.** Determine the location $\bar{x}$ of the centroid $C$ of the shaded area which is part of a circle having a radius $r$.

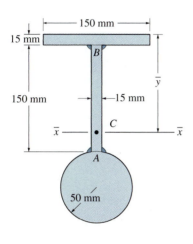

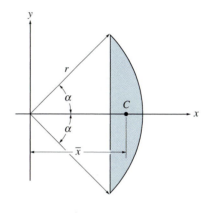

Prob. 6–19

Prob. 6–17

**6-18.** Locate the centroid $\bar{y}$ of the cross-sectional area of the beam.

**\*6-20.** Locate the centroid $\bar{y}$ of the channel's cross-sectional area.

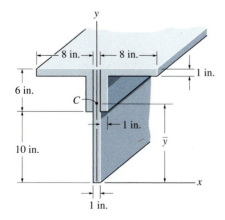

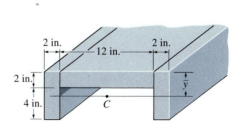

Prob. 6–18

Prob. 6–20

**6-21.** Locate the centroid $\bar{y}$ of the cross-sectional area of the beam constructed from a channel and a plate. Assume all corners are square and neglect the size of the weld at $A$.

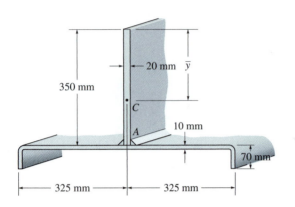

350 mm

20 mm  $\bar{y}$

$C$

$A$

10 mm

70 mm

325 mm

325 mm

**Prob. 6–21**

**6-22.** Locate the centroid $(\bar{x}, \bar{y})$ of the member's cross-sectional area.

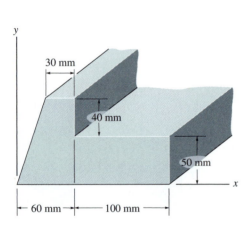

$y$

30 mm

40 mm

50 mm

60 mm

100 mm

$x$

**Prob. 6–22**

**6-23.** Locate the centroid $\bar{y}$ of the concrete beam having the tapered cross section shown.

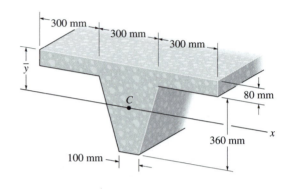

300 mm

300 mm

300 mm

$\bar{y}$

80 mm

$C$

$x$

360 mm

100 mm

**Prob. 6–23**

**\*6-24.** Locate the centroid $\bar{y}$ of the beam's cross-section built up from a channel and a wide-flange beam.

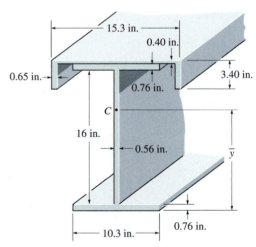

15.3 in.

0.40 in.

0.65 in.

3.40 in.

0.76 in.

$C$

16 in.

0.56 in.

$\bar{y}$

10.3 in.

0.76 in.

**Prob. 6–24**

**6-25.** Locate the centroid $\bar{y}$ of the bulb-tee cross section.

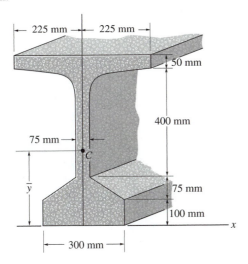

**Prob. 6–25**

**6-26.** Determine the distance $h$ to which a 100-mm-diameter hole must be bored into the base of the cone so that the center of mass of the resulting shape is located at $\bar{z} = 115$ mm. The material has a density of 8 Mg/m³.

**6-27.** Determine the distance $\bar{z}$ to the centroid of the shape which consists of a cone with a hole of height $h = 50$ mm bored into its base.

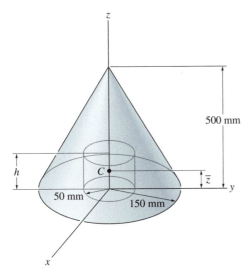

**Probs. 6–26/27**

**\*6-28.** The sheet metal part has the dimensions shown. Determine the location $(\bar{x}, \bar{y}, \bar{z})$ of its centroid.

**6-29.** The sheet metal part has a weight per unit area of 2 lb/ft² and is supported by the smooth rod and at $C$. If the cord is cut, the part will rotate about the $y$ axis until it reaches equilibrium. Determine the equilibrium angle of tilt, measured downward from the negative $x$ axis, that $AD$ makes with the $-x$ axis.

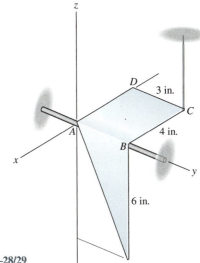

**Probs. 6–28/29**

**6-30.** Determine the location $(\bar{x}, \bar{y})$ of the centroid $C$ of the cross-sectional area for the structural member constructed from two equal-sized channels welded together as shown. Assume all corners are square. Neglect the size of the welds.

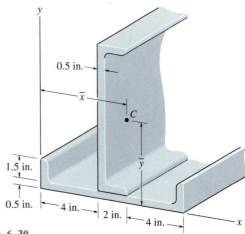

**Prob. 6–30**

**6-31.** Locate the centroid $\bar{z}$ of the top made from a hemisphere and a cone.

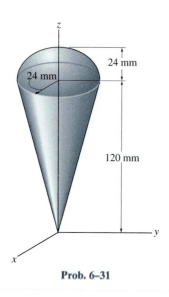

**Prob. 6–31**

**6-33.** Locate the center of gravity of the two-block assembly. The specific weights of the materials $A$ and $B$ are $\gamma_A = 150$ lb/ft$^3$ and $\gamma_B = 400$ lb/ft$^3$, respectively.

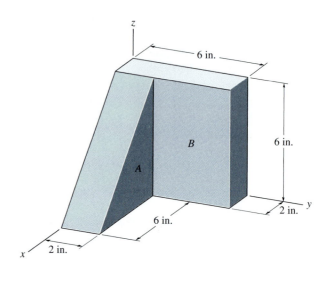

**Prob. 6–33**

**\*6-32.** Locate the center of mass $\bar{z}$ of the assembly. The material has a density of $\rho = 3$ Mg/m$^3$. There is a 30-mm diameter hole bored through the center.

**Prob. 6–32**

**6-34.** The buoy is made from two homogeneous cones each having a radius of 1.5 ft. If $h = 1.2$ ft, find the distance $\bar{z}$ to the buoy's center of gravity $G$.

**6-35.** The buoy is made from two homogeneous cones each having a radius of 1.5 ft. If it is required that the buoy's center of gravity $G$ be located at $\bar{z} = 0.5$ ft, determine the height $h$ of the top cone.

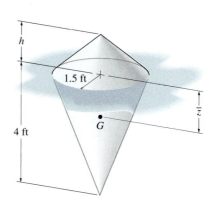

**Probs. 6–34/35**

## 6.4 Resultant of a Distributed Force System

In Chapter 4 we considered ways of simplifying a system of concentrated forces which act on a body. In many practical situations, however, the body may be subjected to loadings distributed over its surface. We have already encountered this situation in Sec. 4.7, while studying frictional and normal forces acting on the bottom of a block resting on a flat surface. Other examples of distributed loadings result from wind and hydrostatic pressure. The effects of these loadings can be studied in a simple manner if we replace them by their resultants. Here we will use the methods of Sec. 3.8 and show how to compute the resultant force of a distributed loading and specify its line of action. As a specific application, in Sec. 6.5 we will find the resultant loading acting on the surface of a body that is submerged in a fluid.

**Pressure Distribution over a Surface.**   Consider the flat plate shown in Fig. 6–15a, which is subjected to the loading function $p = p(x, y)$ Pa, where Pa (pascal) $= 1$ N/m². Knowing this function, we can determine the force $dF$ acting on the differential area $dA$ m² of the plate, located at the arbitrary point $(x, y)$. This force magnitude is simply $dF = [p(x, y)$ N/m²$](dA$ m²$) = [p(x, y) \, dA]$ N. The entire loading on the plate is therefore represented as a system of *parallel forces* infinite in number and each acting on a separate differential area $dA$. This system will now be simplified to a single resultant force $\mathbf{F}_R$ acting through a unique point $(\overline{x}, \overline{y})$ on the plate, Fig. 6–15b.

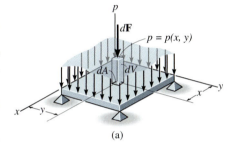

(a)

**Magnitude of Resultant Force.**   To determine the *magnitude* of $\mathbf{F}_R$, it is necessary to sum each of the differential forces $dF$ acting over the plate's *entire surface area A*. This sum may be expressed mathematically as an integral:

$$F_R = \Sigma F; \qquad \boxed{F_R = \int_A p(x, y) \, dA = \int_V dV} \qquad (6\text{--}9)$$

Here $p(x, y) \, dA = dV$, the colored differential *volume element* shown in Fig. 6–15a. Therefore, the result indicates that the *magnitude of the resultant force is equal to the total volume under the distributed-loading diagram*.

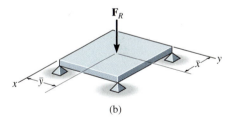

(b)

**Fig. 6–15**

**Location of Resultant Force.**   The location $(\overline{x}, \overline{y})$ of $\mathbf{F}_R$ is determined by setting the moments of $\mathbf{F}_R$ equal to the moments of all the forces $dF$ about the respective $y$ and $x$ axes: From Figs. 6–15a and 6–15b, using Eq. 6–9, this results in

$$\boxed{\overline{x} = \frac{\displaystyle\int_A x p(x, y) \, dA}{\displaystyle\int_A p(x, y) \, dA} = \frac{\displaystyle\int_V x \, dV}{\displaystyle\int_V dV} \qquad \overline{y} = \frac{\displaystyle\int_A y p(x, y) \, dA}{\displaystyle\int_A p(x, y) \, dA} = \frac{\displaystyle\int_V y \, dV}{\displaystyle\int_V dV}} \qquad (6\text{--}10)$$

Hence, it can be seen that the *line of action of the resultant force passes through the geometric center or centroid of the volume under the distributed loading diagram*.

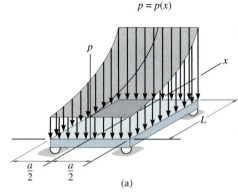

(a)

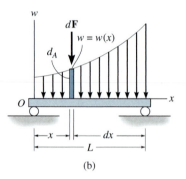

(b)

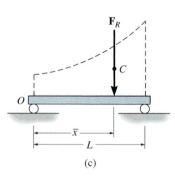

(c)

**Fig. 6–16**

**Linear Distribution of Load Along a Staight Line.**  In many cases a pressure distribution acting on a flat surface will be symmetric about an axis of the surface upon which it acts. An example of such a loading is shown in Fig. 6–16a. Here the loading function $p = p(x)$ Pa is only a function of $x$ since the pressure is uniform along the $y$ axis. If we multiply $p = p(x)$ by the width of $a$ m of the plate, we obtain $w = [p(x) \text{ N/m}^2](a \text{ m}) = w(x)$ N/m. This loading function, shown in Fig. 6–16b, is a measure of load distribution along the line $y = 0$, which is in the plane of symmetry of the loading, Fig. 6–16c. As noted, it is measured as a force per unit length, rather than a force per unit area. Consequently, the load intensity diagram for $w = w(x)$ represents a system of *coplanar* parallel forces, shown in two dimensions in Fig. 6–16b. Each infinitesimal force $d\mathbf{F}$ acts over an element of length $dx$ such that at any point $x$, $dF = (w(x) \text{ N/m})(dx \text{ m}) = w(x)\, dx$ N. Using the same method as above, we will now simplify the loading to a single resultant force and specify its location $\bar{x}$.

**Magnitude of Resultant Force.**  The magnitude of $\mathbf{F}_R$ is determined from the sum

$$F_R = \int_L w(x)\, dx \qquad (6\text{–}11)$$

Note that $w(x)\, dx = dA$, where $dA$ is the colored differential area shown in Fig. 6–11. Thus we can state *the magnitude of the resultant force is equal to the total area under the distributed-loading diagram.*

**Location of the Resultant Force.**  The location $\bar{x}$ of the line of action of $\mathbf{F}_R$ is determined by equating the moments of the force resultant and the force distribution about point $O$, Fig. 6–16b. This yields

$$\bar{x} = \frac{\displaystyle\int_L x w(x)\, dx}{\displaystyle\int_L w(x)\, dx} \qquad (6\text{–}12)$$

Setting $dA = w(x)\, dx$, we can also write

$$\bar{x} = \frac{\displaystyle\int_A x\, dA}{\displaystyle\int_A dA} \qquad (6\text{–}13)$$

This equation locates the $\bar{x}$ coordinate for the geometric center or centroid for the area. Hence, *the line of action of the resultant force passes through the geometric center or centroid of the area under the distributed-loading diagram.* Once $\bar{x}$ is determined, by symmetry, $\mathbf{F}_R$ passes through point $(\bar{x}, a/2)$ on the surface of the plate, Fig. 6–16c.

# E X A M P L E  6.7

In each case, determine the magnitude and location of the resultant of the distributed load acting on the beams in Fig. 6–17.

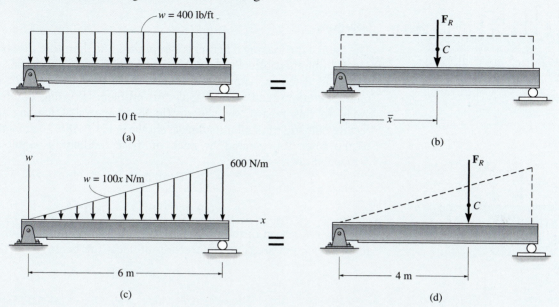

(a)

(b)

(c)

(d)

Fig. 6–17

## Solution

*Uniform Loading.*  As indicated $w = 400$ lb/ft, which is constant over the entire beam, Fig. 6–17a. This loading forms a rectangle, the area of which is equal to the resultant force, Fig. 6–17b; i.e.,

$$F_R = (400 \text{ lb/ft})(10 \text{ ft}) = 4000 \text{ lb} \qquad \textit{Ans.}$$

The location of $\mathbf{F}_R$ passes through the geometric center or centroid $C$ of this rectangular area, so

$$\bar{x} = 5 \text{ ft} \qquad \textit{Ans.}$$

*Triangular Loading.*  Here the loading varies uniformly in intensity from 0 to 600 N/m, Fig. 6–17c. These values can be verified by substitution of $x = 0$ and $x = 6$ m into the loading function $w = 100x$ N/m. The area of this triangular loading is equal to $\mathbf{F}_R$, Fig. 6–17d. From Appendix C, $A = \frac{1}{2}bh$, so that

$$F_R = \frac{1}{2}(6 \text{ m})(600 \text{ N/m}) = 1800 \text{ N} \qquad \textit{Ans.}$$

The line of action of $\mathbf{F}_R$ passes through the centroid $C$ of the triangle. Using Appendix B, this point lies at a distance of one third the length of the beam, measured from the right side. Hence,

$$\bar{x} = 6 \text{ m} - \frac{1}{3}(6 \text{ m}) = 4 \text{ m} \qquad \textit{Ans.}$$

## EXAMPLE 6.8

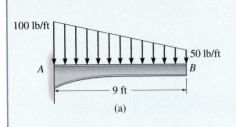

(a)

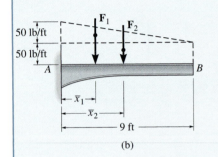

(b)

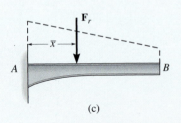

(c)

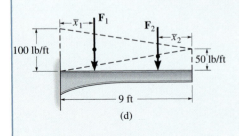

(d)

**Fig. 6-18**

Determine the magnitude and location of the resultant of the distributed load acting on the beam shown in Fig. 6–18a.

### Solution

The area of the loading diagram is a *trapezoid*, and therefore the solution can be obtained directly from the area and centroid formulas for a trapezoid listed in Appendix C. Since these formulas are not easily remembered, instead we will solve this problem by using "composite" areas. In this regard, we can divide the trapezoidal loading into a rectangular and triangular loading as shown in Fig. 6–18b. The magnitude of the force represented by each of these loadings is equal to its associated *area*,

$$F_1 = \tfrac{1}{2}(9 \text{ ft})(50 \text{ lb/ft}) = 225 \text{ lb}$$

$$F_2 = (9 \text{ ft})(50 \text{ lb/ft}) = 450 \text{ lb}$$

The lines of action of these parallel forces act through the *centroid* of their associated areas and therefore intersect the beam at

$$\bar{x}_1 = \tfrac{1}{3}(9 \text{ ft}) = 3 \text{ ft}$$

$$\bar{x}_2 = \tfrac{1}{2}(9 \text{ ft}) = 4.5 \text{ ft}$$

The two parallel forces $\mathbf{F}_1$ and $\mathbf{F}_2$ can be reduced to a single resultant $\mathbf{F}_R$. The magnitude of $\mathbf{F}_R$ is

$$+\downarrow F_R = \Sigma F; \qquad F_R = 225 + 450 = 675 \text{ lb} \qquad \textit{Ans.}$$

With reference to point $A$, Fig. 6–18b and c, we can define the location of $\mathbf{F}_R$. We require

$$\curvearrowright +M_{R_A} = \Sigma M_A; \qquad \bar{x}(675) = 3(225) + 4.5(450)$$

$$\bar{x} = 4 \text{ ft} \qquad \textit{Ans.}$$

*Note:* The trapezoidal area in Fig. 6–18a can also be divided into two triangular areas as shown in Fig. 6–18d. In this case

$$F_1 = \tfrac{1}{2}(9 \text{ ft})(100 \text{ lb/ft}) = 450 \text{ lb}$$

$$F_2 = \tfrac{1}{2}(9 \text{ ft})(50 \text{ lb/ft}) = 225 \text{ lb}$$

and

$$\bar{x}_1 = \tfrac{1}{3}(9 \text{ ft}) = 3 \text{ ft}$$

$$\bar{x}_2 = \tfrac{1}{3}(9 \text{ ft}) = 3 \text{ ft}$$

Using these results, show that again $F_R = 675$ lb and $\bar{x} = 4$ ft.

# EXAMPLE 6.9

Determine the magnitude and location of the resultant force acting on the beam in Fig. 6–19a.

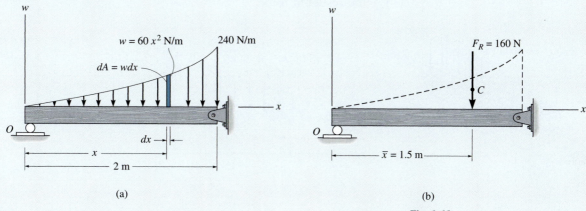

(a)

(b)

Fig. 6–19

## Solution

Since $w = w(x)$ is given, this problem will be solved by integration. the colored differential area element $dA = w\,dx = 60x^2\,dx$. Applying Eq. 6–11, by summing these elements from $x = 0$ to $x = 2$ m, we obtain the resultant force $\mathbf{F}_R$,

$$F_R = \Sigma F;$$

$$F_R = \int_L w(x)\,dx = \int_0^2 60x^2\,dx = 60\left[\frac{x^3}{3}\right] = 60\left[\frac{2^3}{3} - \frac{0^3}{3}\right]$$

$$= 160 \text{ N} \qquad \qquad \textit{Ans.}$$

Since the element of area $dA$ is located an arbitrary distance $x$ from $O$, the location $\bar{x}$ of $\mathbf{F}_R$ measured from $O$, Fig. 6–19b, is determined from Eq. 6–12.

$$\bar{x} = \frac{\int_L x\,w(x)\,dx}{\int_L w(x)\,dx} = \frac{\int_0^2 x(60x^2)\,dx}{160} = \frac{60\left[\frac{x^4}{4}\right]_0^2}{160} = \frac{60\left[\frac{2^4}{4} - \frac{0^4}{4}\right]}{160}$$

$$= 1.5 \text{ m} \qquad \qquad \textit{Ans.}$$

These results may be checked by using Appendix C, where it is shown that for an exparabolic area of length $a$, height $b$, and shape shown in Fig. 6–19a,

$$A = \frac{ab}{3} = \frac{2(240)}{3} = 160 \text{ N} \quad \text{and} \quad \bar{x} = \frac{3}{4}a = \frac{3}{4}(2) = 1.5 \text{ m}$$

E X A M P L E  **6.10**

A distributed loading of $p = 800x$ Pa acts over the top surface of the beam shown in Fig. 6–20a. Determine the magnitude and location of the resultant force.

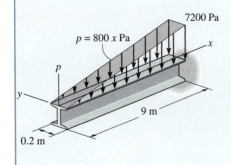

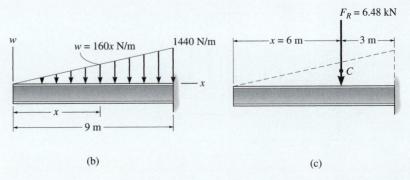

(a)                (b)                (c)

Fig. 6–20

## Solution

The loading function $p = 800x$ Pa indicates that the load intensity varies uniformly from $p = 0$ at $x = 0$ to $p = 7200$ Pa at $x = 9$ m. Since the intensity is uniform along the width of the beam (the $y$ axis), the loading may be viewed in two dimensions as shown in Fig. 6–20b. Here

$$w = (800x \text{ N/m}^2)(0.2 \text{ m})$$

$$= (160x) \text{ N/m}$$

At $x = 9$ m, note that $w = 1440$ N/m. Although we may again apply Eqs. 6–11 and 6–12 as in Example 6–9, it is simpler to use Appendix C.

The magnitude of the resultant force is

$$F_R = \tfrac{1}{2}(9 \text{ m})(1440 \text{ N/m}) = 6480 \text{ N} = 6.48 \text{ kN} \qquad \textit{Ans.}$$

The line of action of $\mathbf{F}_R$ passes through the *centroid C* of the triangle. Hence,

$$\bar{x} = 9 \text{ m} - \tfrac{1}{3}(9 \text{ m}) = 6\text{m} \qquad \textit{Ans.}$$

The results are shown in Fig. 6–20c.

We may also view the resultant $\mathbf{F}_R$ as acting through the *centroid* of the *volume* of the loading diagram $p = p(x)$ in Fig. 6–20a. Hence $\mathbf{F}_R$ intersects the $x$–$y$ plane at the point (6 m, 0). Furthermore, the *magnitude* of $\mathbf{F}_R$ is equal to the *volume* under the loading diagram; i.e.,

$$F_R = V = \tfrac{1}{2}(7200 \text{ N/m}^2)(9 \text{ m})(0.2 \text{ m}) = 6.48 \text{ kN} \qquad \textit{Ans.}$$

# PROBLEMS

**\*6-36.** Determine the resultant moment of both the 100-lb force and the triangular distributed load about point $O$.

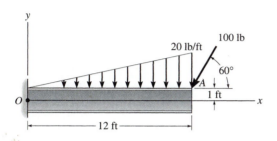

**Prob. 6–36**

**6-37.** Replace the loading by an equivalent resultant force and specify the location of the force on the beam, measured from point $B$.

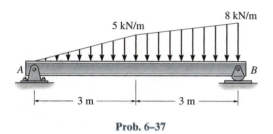

**Prob. 6–37**

**6-38.** Replace the distributed loading by an equivalent resultant force, and specify its location on the beam measured from the pin at $C$.

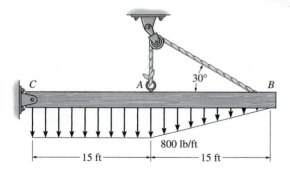

**Prob. 6–38**

**6-39.** Determine the magnitude of the equivalent resultant force of the distributed loading and specify its location on the beam, measured from point $A$.

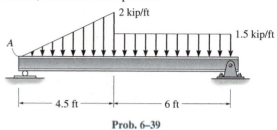

**Prob. 6–39**

**\*6-40.** The column is used to support the floor which exerts a force of 3000 lb on the top of the column. The effect of soil pressure along the side of the column is distributed as shown. Replace this loading by an equivalent resultant force and specify where it acts along the column measured from its base $A$.

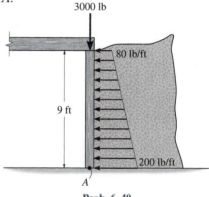

**Prob. 6–40**

**6-41.** The beam is subjected to the distributed loading. Determine the length $b$ of the uniform load and its position $a$ on the beam such that the resultant force and couple moment acting on the beam are zero.

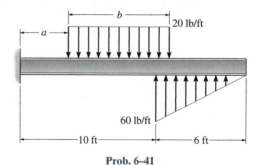

**Prob. 6–41**

**6-42.** Wet concrete exerts a pressure distribution along the wall of the form. Determine the resultant force of this distribution and specify the height $h$ where the bracing strut should be placed so that it lies through the line of action of the resultant force. The wall has a width of 5 m.

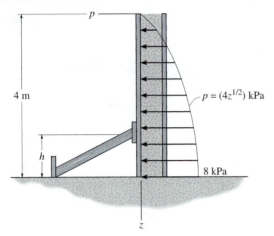

**Prob. 6-42**

**6-43.** Determine the magnitude of the resultant force of the loading acting on the beam and specify where it acts from point $O$.

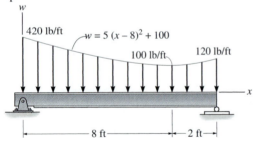

**Prob. 6-43**

***6-44.** The load over the plate varies linearly along the sides of the plate such that $p = \frac{2}{3}[x(4-y)]$ kPa. Determine the resultant force and its position $(\bar{x}, \bar{y})$ on the plate.

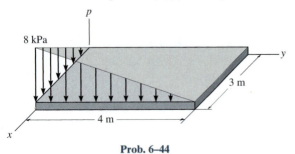

**Prob. 6-44**

**6-45.** Determine the magnitude and location of the resultant force of the parabolic loading acting on the plate.

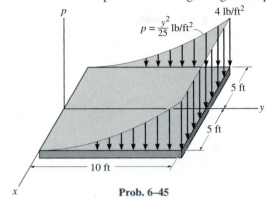

**Prob. 6-45**

**6-46.** The loading acting on a square plate is represented by a parabolic pressure distribution. Determine the magnitude and location of the resultant force. Also, what are the reactions at the rollers $B$ and $C$ and the ball-and-socket joint $A$? Neglect the weight of the plate.

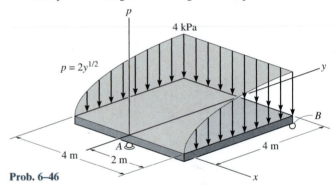

**Prob. 6-46**

**6-47.** The pressure loading on the plate is described by the function $p = [-240/(x+1) + 340]$ Pa. Determine the magnitude of the resultant force and coordinates of the point where the line of action of the force intersects the plate.

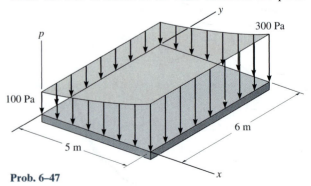

**Prob. 6-47**

# 6.5   Moments of Inertia for Areas

In the first few sections of this chapter, we determined the centroid for an area by considering the first moment of the area about an axis; that is, for the computation we had to evaluate an integral of the form $\int x\, dA$. An integral of the second moment of an area, such as $\int x^2\, dA$, is referred to as the *moment of inertia* for the area. The terminology "moment of inertia" as used here is actually a misnomer; however, it has been adopted because of the similarity with integrals of the same form related to mass.

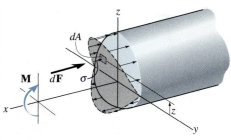

The moment of inertia of an area orginates whenever one relates the normal stress $\sigma$ (sigma), or force per unit area, acting on the transverse cross section of an elastic beam, to the applied external moment **M**, which causes bending of the beam. From the theory of mechanics of materials, it can be shown that the stress within the beam varies linearly with its distance from an axis passing through the centroid $C$ of the beam's cross-sectional area; i.e., $\sigma = kz$, Fig. 6–21. The magnitude of force acting on the area element $dA$, shown in the figure, is therefore $dF = \sigma\, dA = kz\, dA$. Since this force is located a distance $z$ from the $y$ axis, the moment of $d\mathbf{F}$ about the $y$ axis is $dM = dFz = kz^2\, dA$. The resulting moment of the entire stress distribution is equal to the applied moment **M**; hence, $M = k \int z^2\, dA$. Here the integral represents the moment of inertia of the area about the $y$ axis. Since integrals of this form often arise in formulas used in mechanics of materials, structural mechanics, fluid mechanics, and machine design, the engineer should become familiar with the methods used for their computation.

**Fig. 6–21**

**Moment of Inertia.**   Consider the area $A$, shown in Fig. 6–22, which lies in the $x$–$y$ plane. By definition, the moments of inertia of the differential planar area $dA$ about the $x$ and $y$ axes are $dI_x = y^2\, dA$ and $dI_y = x^2\, dA$, respectively. For the entire area the *moments of inertia* are determined by integration; i.e.,

$$I_x = \int_A y^2\, dA$$

$$I_y = \int_A x^2\, dA$$

(6–14)

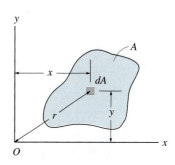

**Fig. 6–22**

We can also formulate the second moment of $dA$ about the pole $O$ or $z$ axis, Fig. 6–22. This is referred to as the polar moment of inertia, $dJ_O = r^2\, dA$. Here $r$ is the perpendicular distance from the pole ($z$ axis) to the element $dA$. For the entire area the *polar moment of inertia* is

$$J_O = \int_A r^2\, dA = I_x + I_y \qquad (6\text{--}15)$$

The relationship between $J_O$ and $I_x$, $I_y$ is possible since $r^2 = x^2 + y^2$, Fig. 6–22.

From the above formulations it is seen that $I_x$, $I_y$, and $J_O$ will *always* be *positive* since they involve the product of distance squared and area. Furthermore, the units for moment of inertia involve length raised to the fourth power, e.g., $m^4$, $mm^4$, or $ft^4$, $in^4$.

## 6.6 Parallel-Axis Theorem

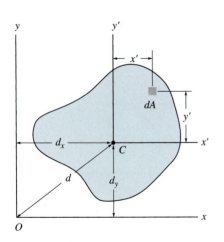

Fig. 6–23

If the moment of inertia for an area is known about an axis passing through its centroid, which is often the case, it is convenient to determine the moment of inertia of the area about a corresponding parallel axis using the *parallel-axis theorem*. To derive this theorem, consider finding the moment of inertia of the shaded area shown in Fig. 6–23 about the $x$ axis. In this case, a differential element $dA$ is located at an arbitrary distance $y'$ from the *centroidal* $x'$ axis, whereas the *fixed distance* between the parallel $x$ and $x'$ axes is defined as $d_y$. Since the moment of inertia of $dA$ about the $x$ axis is $dI_x = (y' + d_y)^2\, dA$, then for the entire area,

$$I_x = \int_A (y' + d_y)^2\, dA$$

$$= \int_A y'^2\, dA + 2d_y \int_A y'\, dA + d_y^2 \int_A dA$$

The first integral represents the moment of inertia of the area about the centroidal axis, $\bar{I}_{x'}$. The second integral is zero since the $x'$ axis passes through the area's centroid $C$; i.e., $\int y' dA = \bar{y} \int dA = 0$ since $\bar{y} = 0$. Realizing that the third integral represents the total area $A$, the final result is therefore

$$I_x = \bar{I}_{x'} + Ad_y^2 \qquad\qquad (6\text{--}16)$$

A similar expression can be written for $I_y$; i.e.,

$$I_y = \bar{I}_{y'} + Ad_x^2 \qquad\qquad (6\text{--}17)$$

And finally, for the polar moment of inertia about an axis perpendicular to the $x$–$y$ plane and passing through the pole $O$ ($z$ axis), Fig. 6–23, we have

$$J_O = \bar{J}_C + Ad^2 \qquad\qquad (6\text{--}18)$$

The form of each of these three equations states that *the moment of inertia of an area about an axis is equal to the moment of inertia of the area about a parallel axis passing through the area's centroid plus the product of the area and the square of the perpendicular distance between the axes.*

## 6.7  Moments of Inertia for an Area by Integration

When the boundaries for a planar area are expressed by mathematical functions, Eqs. 6–14 may be integrated to determine the moments of inertia for the area. If the element of area chosen for integration has a differential size in two directions as shown in Fig. 6–22, a double integration must be performed to evaluate the moment of inertia. Most often, however, it is easier to perform only a single integration by choosing an element having a differential size or thickness in only one direction.

## PROCEDURE FOR ANALYSIS

- If a single integration is performed to determine the moment of inertia of an area about an axis, it will be necessary to specify the differential element $dA$.

- Most often this element will be rectangular, such that it will have a finite length and differential width.

- The element should be located so that it intersects the boundary of the area at the *arbitrary point* $(x, y)$. There are two possible ways to orient the element with respect to the axis about which the moment of inertia is to be determined.

*Case 1*

- The *length* of the element can be oriented *parallel* to the axis. This situation occurs when the rectangular element shown in Fig. 6–24 is used to determine $I_y$ for the area. Direct application of Eq. 6–14, i.e., $I_y = \int x^2\, dA$, can be made in this case since the element has an infinitesimal thickness $dx$ and therefore *all parts* of the element lie at the *same* moment-arm distance $x$ from the $y$ axis.*

*Case 2*

- The *length* of the element can be oriented *perpendicular* to the axis. Here Eq. 6–14 *does not apply* since all parts of the element will *not* lie at the same moment-arm distance from the axis. For example, if the rectangular element in Fig. 6–24 is used for determining $I_x$ for the area, it will first be necessary to calculate the moment of inertia of the *element* about a horizontal axis passing through the element's centroid and then determine the moment of inertia of the *element* about the $x$ axis by using the parallel-axis theorem. Integration of this result will yield $I_x$.

*In the case of the element $dA = dx\,dy$, Fig. 6–21, the moment arms $y$ and $x$ are appropriate for the formulation of $I_x$ and $I_y$ (Eq. 6–14) since the *entire* element, because of its infinitesimal size, lies at the specified $y$ and $x$ perpendicular distances from the $x$ and $y$ axes.

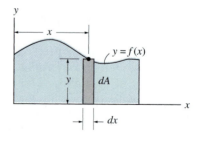

**Fig. 6–24**

EXAMPLE 6.11

Determine the moment of inertia for the rectangular area shown in Fig. 6–25 with respect to (a) the centroidal $x'$ axis, (b) the axis $x_b$ passing through the base of the rectangle, and (c) the pole or $z'$ axis perpendicular to the $x'-y'$ plane and passing through the centroid $C$.

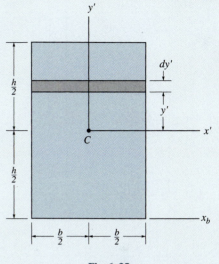

### Solution (Case 1)

**Part (a).** The differential element shown in Fig. 6–25 is chosen for integration. Because of its location and orientation, the *entire element* is at a distance $y'$ from the $x'$ axis. Here it is necessary to integrate from $y' = -h/2$ to $y' = h/2$. Since $dA = b\,dy'$, then

$$\bar{I}_{x'} = \int_A y'^2 \, dA = \int_{-h/2}^{h/2} y'^2 (b\,dy') = b\int_{-h/2}^{h/2} y'^2 \, dy$$

$$= \frac{1}{12}bh^3 \qquad\qquad Ans.$$

**Fig. 6–25**

**Part (b).** The moment of inertia about an axis passing through the base of the rectangle can be obtained by using the result of part (a) and applying the parallel-axis theorem, Eq. 6–16.

$$I_{x_b} = \bar{I}_{x'} + A d_y^2$$

$$= \frac{1}{12}bh^3 + bh\left(\frac{h}{2}\right)^2 = \frac{1}{3}bh^3 \qquad Ans.$$

**Part (c).** To obtain the polar moment of inertia about point $C$, we must first obtain $\bar{I}_{y'}$, which may be found by interchanging the dimensions $b$ and $h$ in the result of part (a), i.e.,

$$\bar{I}_{y'} = \frac{1}{12}hb^3$$

Using Eq. 6–15, the polar moment of inertia about $C$ is therefore

$$\bar{J}_C = \bar{I}_{x'} + \bar{I}_{y'} = \frac{1}{12}bh(h^2 + b^2) \qquad Ans.$$

E X A M P L E 6.12

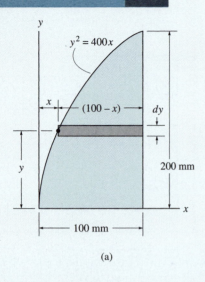

(a)

Determine the moment of inertia of the shaded area shown in Fig. 6–26a about the x axis.

### Solution I (Case 1)

A differential element of area that is *parallel* to the x axis, as shown in Fig. 6–24a, is chosen for integration. Since the element has a thickness $dy$ and intersects the curve at the *arbitrary point* $(x, y)$, the area is $dA = (100 - x)\,dy$. Furthermore, all parts of the element lie at the same distance $y$ from the x axis. Hence, integrating with respect to $y$, from $y = 0$ to $y = 200$ mm, yields

$$I_x = \int_A y^2\,dA = \int_A y^2(100 - x)\,dy$$

$$= \int_0^{200} y^2 \left(100 - \frac{y^2}{400}\right) dy = 100 \int_0^{200} y^2\,dy - \frac{1}{400}\int_0^{200} y^4\,dy$$

$$= 107(10^6)\ \text{mm}^4 \qquad\qquad Ans.$$

### Solution II (Case 2)

A differential element *parallel* to the y axis, as shown in Fig. 6–26b, is chosen for integration. It intersects the curve at the *arbitrary point* $(x, y)$. In this case, all parts of the element do *not* lie at the same distance from the x axis, and therefore the parallel-axis theorem must be used to determine the *moment of inertia of the element* with respect to this axis. For a rectangle having a base $b$ and height $h$, the moment of inertia about its centroidal axis has been determined in part (a) of Example 6.11. There it was found that $\bar{I}_{x'} = \frac{1}{12}bh^3$. For the differential element shown in Fig. 6–26b, $b = dx$ and $h = y$, and thus $d\bar{I}_{x'} = \frac{1}{12}dx\,y^3$. Since the centroid of the element is at $\tilde{y} = y/2$ from the x axis, the moment of inertia of the element about this axis is

$$dI_x = d\bar{I}_{x'} + dA\,\tilde{y}^2 = \frac{1}{12}\,dx\,y^3 + y\,dx\left(\frac{y}{2}\right)^2 = \frac{1}{3}y^3\,dx$$

This result can also be concluded from part (b) of Example 6.11. Integrating with respect to x, from $x = 0$ to $x = 100$ mm, yields

$$I_x = \int dI_x = \int_A \frac{1}{3}y^3\,dx = \int_0^{100} \frac{1}{3}(400x)^{3/2}\,dx$$

$$= 107(10^6)\ \text{mm}^4 \qquad\qquad Ans.$$

Fig. 6–26

# EXAMPLE 6.13

Determine the moment of inertia with respect to the $x$ axis of the circular area shown in Fig. 6–27a.

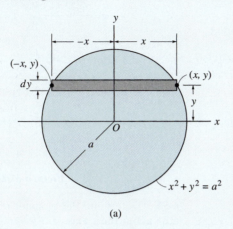

(a)

**Fig. 6–27**

## Solution I *(Case 1)*

Using the differential element shown in Fig. 6–27a, since $dA = 2x\,dy$, we have

$$I_x = \int_A y^2\,dA = \int_A y^2(2x)\,dy$$

$$= \int_{-a}^{a} y^2(2\sqrt{a^2 - y^2})\,dy = \frac{\pi a^4}{4} \qquad Ans.$$

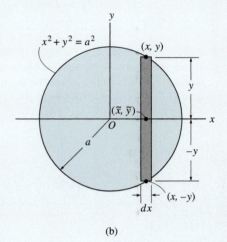

## Solution II *(Case 2)*

When the differential element is chosen as shown in Fig. 6–27a, the centroid for the element happens to lie on the $x$ axis, and so, applying Eq. 6–16, noting that $d_y = 0$ and for a rectangle $\bar{I}_{x'} = \frac{1}{12}bh^3$, we have

$$dI_x = \frac{1}{12}dx(2y)^3$$

$$= \frac{2}{3}y^3\,dx$$

(b)

Integrating with respect to $x$ yields

$$I_x = \int_{-a}^{a} \frac{2}{3}(a^2 - x^2)^{3/2}\,dx = \frac{\pi a^4}{4} \qquad Ans.$$

## PROBLEMS

*6-48. Determine the moment of inertia of the shaded area about the $x$ axis.

6-49. Determine the moment of inertia of the shaded area about the $y$ axis.

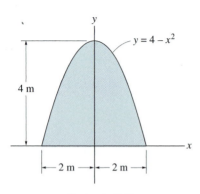

$y = 4 - x^2$

4 m

2 m    2 m

**Probs. 6–48/49**

6-50. Determine the moment of inertia of the area about the $x$ axis. Solve the problem in two ways, using rectangular differential elements: (a) having a thickness $dx$ and (b) having a thickness of $dy$.

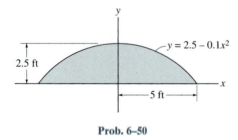

$y = 2.5 - 0.1x^2$

2.5 ft

5 ft

**Prob. 6–50**

6-51. Determine the moment of inertia of the area about the $x$ axis. Solve the problem in two ways, using rectangular differential elements: (a) having a thickness of $dx$, and (b) having a thickness of $dy$.

*6-52. Determine the moment of inertia of the area about the $y$ axis. Solve the problem in two ways, using rectangular differential elements: (a) having a thickness of $dx$, and (b) having a thickness of $dy$.

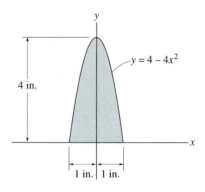

$y = 4 - 4x^2$

4 in.

1 in.   1 in.

**Probs. 6–51/52**

6-53. Determine the moment of inertia of the shaded area about the $x$ axis.

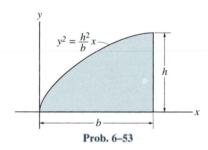

$y^2 = \dfrac{h^2}{b} x$

h

b

**Prob. 6–53**

6-54. Determine the moment of inertia of the shaded area about the $x$ axis.

6-55. Determine the moment of inertia of the shaded area about the $y$ axis.

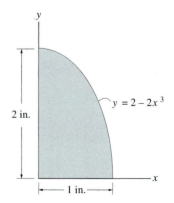

$y = 2 - 2x^3$

2 in.

1 in.

**Probs. 6–54/55**

**\*6-56.** Determine the moment of inertia of the shaded area about the $x$ axis.

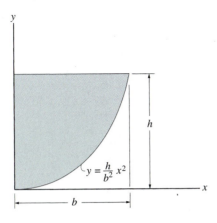

Prob. 6–56

**6-57.** Determine the moment of inertia of the shaded area about the $y$ axis.

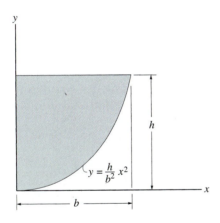

Prob. 6–57

**6-58.** Determine the moment of inertia of the shaded area about the $x$ axis.

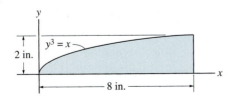

Prob. 6–58

**\*6-59.** Determine the moment of inertia of the shaded area about the $x$ axis.

**6-60.** Determine the moment of inertia of the shaded area about the $y$ axis.

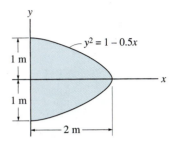

Probs. 6–59/60

**6-61.** Determine the moment of inertia of the shaded area about the $x$ axis.

**6-62.** Determine the moment of inertia of the shaded area about the $y$ axis.

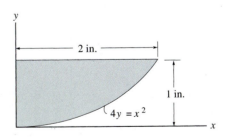

Probs. 6–61/62

## 6.8 Moments of Inertia for Composite Areas

Structural members have various cross-sectional shapes, and it is necessary to calculate their moments of inertia in order to determine the stress in these members.

A composite area consists of a series of connected "simpler" parts or shapes, such as semicircles, rectangles, and triangles. Provided the moment of inertia of each of these parts is known or can be determined about a common axis, then the moment of inertia of the composite area equals the *algebraic sum* of the moments of inertia of all its parts.

### PROCEDURE FOR ANALYSIS

The moment of inertia of a composite area about a reference axis can be determined using the following procedure.

*Composite Parts.*

*   Using a sketch, divide the area into its composite parts and indicate the perpendicular distance from the *centroid* of each part to the reference axis.

*Parallel-Axis Theorem.*

*   The moment of inertia of each part should be determined about its centroidal axis, which is parallel to the reference axis. For the calculation use the table given in Appendix C.
*   If the centroidal axis does not coincide with the reference axis, the parallel-axis theorem, $I = \bar{I} + Ad^2$, should be used to determine the moment of inertia of the part about the reference axis.

*Summation.*

*   The moment of inertia of the entire area about the reference axis is determined by summing the results of its composite parts.
*   If a composite part has a "hole," its moment of inertia is found by "subtracting" the moment of inertia for the hole from the moment of inertia of the entire part including the hole.

E X A M P L E   6.14

Compute the moment of inertia of the composite area shown in Fig. 6–28a about the $x$ axis.

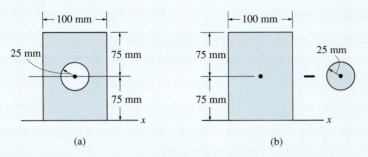

(a)                                    (b)

Fig. 6–28

**Solution**

*Composite Parts.*   The composite area is obtained by *subtracting* the circle from the rectangle as shown in Fig. 6–28b. The centroid of each area is located in the figure.

*Parallel-Axis Theorem.*   The moments of inertia about the $x$ axis are determined using the parallel-axis theorem and the data in the table in Appendix C.

*Circle*

$$I_x = \overline{I}_{x'} + Ad_y^2$$

$$= \frac{1}{4}\pi(25)^4 + \pi(25)^2(75)^2 = 11.4(10^6) \text{ mm}^4$$

*Rectangle*

$$I_x = \overline{I}_{x'} + Ad_y^2$$

$$= \frac{1}{12}(100)(150)^3 + (100)(150)(75)^2 = 112.5(10^6) \text{ mm}^4$$

*Summation.*   The moment of inertia for the composite area is thus

$$I_x = -11.4(10^6) + 112.5(10^6)$$

$$= 101(10^6) \text{ mm}^4 \qquad\qquad Ans.$$

**E X A M P L E   6.15**

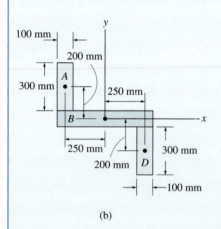

(a)

(b)

**Fig. 6–29**

Determine the moments of inertia of the beam's cross-sectional area shown in Fig. 6–29a about the $x$ and $y$ centroidal axes.

**Solution**

*Composite Parts.* The cross section can be considered as three composite rectangular areas $A$, $B$, and $D$ shown in Fig. 6–29b. For the calculation, the centroid of each of these rectangles is located in the figure.

*Parallel-Axis Theorem.* From the table in Appendix C, or Example 6.11, the moment of inertia of a rectangle about its centroidal axis is $\bar{I} = \frac{1}{12}bh^3$. Hence, using the parallel-axis theorem for rectangles $A$ and $D$, the calculations are as follows:

*Rectangle A*

$$I_x = \bar{I}_{x'} + Ad_y^2 = \frac{1}{12}(100)(300)^3 + (100)(300)(200)^2$$

$$= 1.425(10^9) \text{ mm}^4$$

$$I_y = \bar{I}_{y'} + Ad_x^2 = \frac{1}{12}(300)(100)^3 + (100)(300)(250)^2$$

$$= 1.90(10^9) \text{ mm}^4$$

*Rectangle B*

$$I_x = \frac{1}{12}(600)(100)^3 = 0.05(10^9) \text{ mm}^4$$

$$I_y = \frac{1}{12}(100)(600)^3 = 1.80(10^9) \text{ mm}^4$$

*Rectangle D*

$$I_x = \bar{I}_{x'} + Ad_y^2 = \frac{1}{12}(100)(300)^3 + (100)(300)(200)^2$$

$$= 1.425(10^9) \text{ mm}^4$$

$$I_y = \bar{I}_{y'} + Ad_x^2 = \frac{1}{12}(300)(100)^3 + (100)(300)(250)^2$$

$$= 1.90(10^9) \text{ mm}^4$$

*Summation.* The moments of inertia for the entire cross section are thus

$$I_x = 1.425(10^9) + 0.05(10^9) + 1.425(10^9)$$

$$= 2.90(10^9) \text{ mm}^4 \qquad\qquad Ans.$$

$$I_y = 1.90(10^9) + 1.80(10^9) + 1.90(10^9)$$

$$= 5.60(10^9) \text{ mm}^4 \qquad\qquad Ans.$$

# PROBLEMS

**6-63.** Locate the centroid $\bar{y}$ of the cross-sectional area for the angle. Then find the moment of inertia $\bar{I}_{x'}$ about the $x'$ centroidal axis.

**\*6-64.** Locate the centroid $\bar{x}$ of the cross-sectional area for the angle. Then find the moment of inertia $\bar{I}_{y'}$ about the $y'$ centroidal axis.

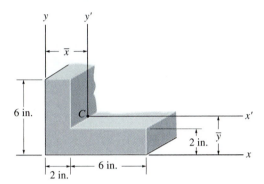

Probs. 6–63/64

**6-65.** Determine the distance $\bar{x}$ to the centroid of the beam's cross-sectional area: then find the moment of inertia about the $y'$ axis.

**6-66.** Determine the moment of inertia of the beam's cross-sectional area about the $x'$ axis.

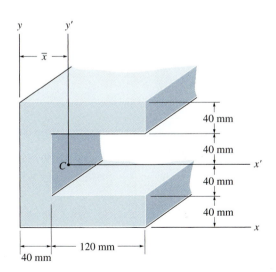

Probs. 6–65/66

**6-67.** Determine the moments of inertia of the shaded area about the $x$ and $y$ axes.

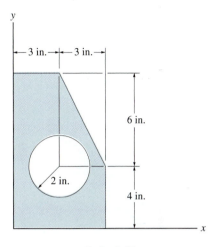

Prob. 6–67

**\*6-68.** Determine the moment of inertia of the beam's cross-sectional area about the $x'$ axis. Neglect the size of the corner welds at $A$ and $B$ for the calculation, $\bar{y} = 154.4$ mm.

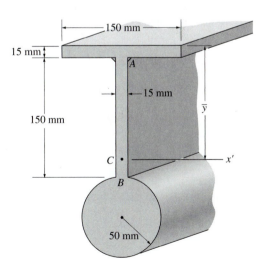

Prob. 6–68

**6-69.** Compute the moments of inertia $I_x$ and $I_y$ for the beam's cross-sectional area about the $x$ and $y$ axes.

**6-70.** Determine the distance $\bar{y}$ to the centroid $C$ of the beam's cross-sectional area and then compute the moment of inertia $\bar{I}_{x'}$ about the $x'$ axis.

**6-71.** Determine the distance $\bar{x}$ to the centroid $C$ of the beam's cross-sectional area and then compute the moment of inertia $\bar{I}_{y'}$ about the $y'$ axis.

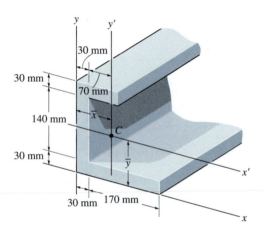

**Probs. 6–69/70/71**

*\*6-72.** Locate the centroid $\bar{y}$ of the cross section and determine the moment of inertia of the section about the $x'$ axis.

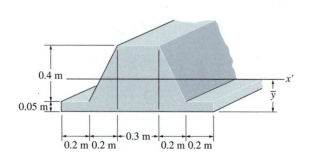

**Prob. 6–72**

**6-73.** Determine $\bar{y}$, which locates the centroidal axis $x'$ for the cross-sectional area of the T-beam, and then find the moments of inertia $\bar{I}_{x'}$ and $\bar{I}_{y'}$.

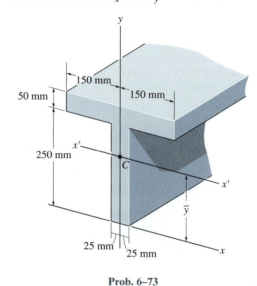

**Prob. 6–73**

**6-74.** Determine the distance $\bar{y}$ to the centroid for the beam's cross-sectional area; then determine the moment of inertia about the $x'$ axis.

**6-75.** Determine the moment of inertia of the beam's cross-sectional area about the $y$ axis.

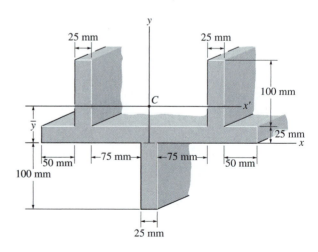

**Probs. 6–74/75**

**\*6-76.** Determine the moment of inertia $I_x$ of the shaded area about the $x$ axis.

**6-77.** Determine the moment of inertia $I_y$ of the shaded area about the $y$ axis.

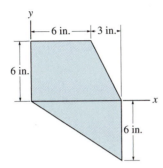

Probs. 6–76/77

**6-78.** Locate the centroid $\bar{y}$ of the channel's cross-sectional area, and then determine the moment of inertia with respect to the $x'$ axis passing through the centroid.

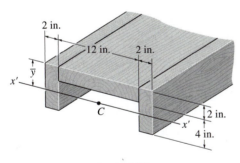

Prob. 6–78

**6-79.** Determine the moments of inertia $I_x$ and $I_y$ of the shaded area.

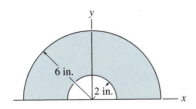

Prob. 6–79

**\*6-80.** Determine the moment of inertia of the parallelogram about the $x'$ axis, which passes through the centroid $C$ of the area.

**6-81.** Determine the moment of inertia of the parallelogram about the $y'$ axis, which passes through the centroid $C$ of the area.

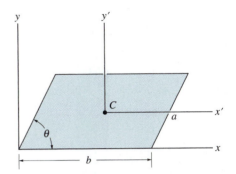

Probs. 6–80/81

**6-82.** An aluminum strut has a cross section referred to as a deep hat. Determine the location $\bar{y}$ of the centroid of its area and the moment of inertia of the area about the $x'$ axis. Each segment has a thickness of 10 mm.

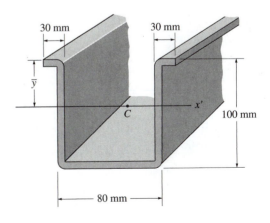

Prob. 6–82

**6-83.** Determine the moment of inertia of the beam's cross-sectional area with respect to the $x'$ axis passing through the centroid $C$ of the cross section. Neglect the size of the corner welds at $A$ and $B$ for the calculation, $\bar{y} = 104.3$ mm.

**6-85.** Determine the moments of inertia of the triangular area about the $x'$ and $y'$ axes, which pass through the centroid $C$ of the area.

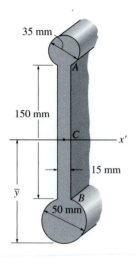

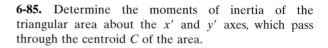

**Prob. 6–83**

**Prob. 6–85**

***6-84.** Determine the location $\bar{y}$ of the centroid of the channel's cross-sectional area and then calculate the moment of inertia of the area about this axis.

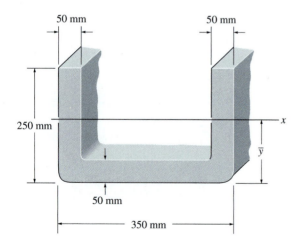

**Prob. 6–84**

# CHAPTER REVIEW

- *Center of Gravity and Centroid.* The *center of gravity* represents a point where the weight of the body can be considered concentrated. The distance $\bar{s}$ to this point can be determined from a balance of moments. This requires that the moment of the weight of all the particles of the body about some point must equal the moment of the entire body about the point, $\bar{s}W = \Sigma \tilde{s}W$. The *centroid* is the location of the geometric center for the body. It is determined in a similar manner, using a moment balance of geometric elements such as line, area, or volume segments. For bodies having a continuous shape, moments are summed (integrated) using differential elements. If the body is a composite of several shapes, each having a known location for its center of gravity or centroid, then the location is determined from a discrete summation using its composite parts.

- *Distributed Force System.* A distributed load can be replaced by resultant force that is equivalent to the area or volume under the loading diagram. This resultant has a line of action that passes through the centroid or geometric center of the area or volume in the diagram.

- *Area Moment of Inertia.* The *area moment of inertia* represents the second moment of the area about an axis, $I = \int r^2 \, dA$. It is frequently used in formulas related to strength and stability of structural members or mechanical elements. If the area shape is irregular, then a differential element must be selected and integration over the entire area must be performed. Tabular values of the moment of inertia of common shapes about their *centroidal axis* are available. To determine the moment of inertia of these shapes about some *other axis*, the parallel-axis theorem must be used, $I = \bar{I} + Ad^2$. If an area is a composite of these shapes, then its moment of inertia is equal to the sum of the moments of inertia of each of its parts.

# REVIEW PROBLEMS

**6-86.** Determine the distance $\bar{y}$ to the centroidal axis $\bar{x}-\bar{x}$ of the beam's cross-sectional area. Neglect the size of the corner welds at $A$ and $B$ for the calculation.

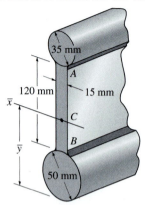

**Prob. 6–86**

**6-87.** Determine the distance $\bar{y}$ to the centroid of the plate area.

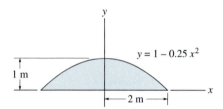

$y = 1 - 0.25 x^2$

**Prob. 6–87**

**\*6-88.** Determine the location $(\bar{x}, \bar{y})$ of the centroid of the parabolic area.

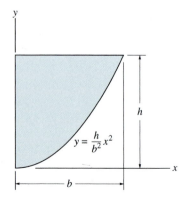

$y = \dfrac{h}{b^2} x^2$

**Prob. 6–88**

**6-89.** Determine the distance $\bar{z}$ to the center of gravity for the frustum of the paraboloid. The material is homogeneous.

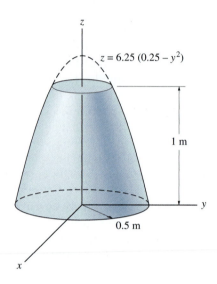

$z = 6.25\,(0.25 - y^2)$

**Prob. 6–89**

**6-90.** Determine the moments of inertia of the "Z" section with respect to the $x$ and $y$ centroidal axes.

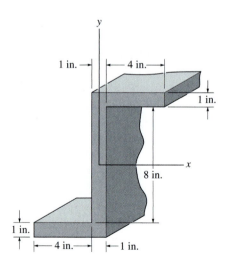

**Prob. 6–90**

**6-91.** The form is used to cast concrete columns. Determine the resultant force that wet concrete exerts along the plate $A$, $0.5\,m \le z \le 3\,m$, if the pressure due to the concrete varies as shown. Specify the location of the resultant force, measured from the top of the column.

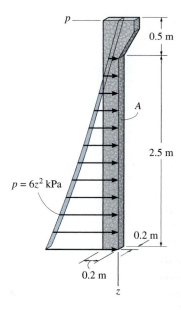

Prob. 6–91

**\*6-92.** Determine the moment of inertia of the beam's cross-sectional area about the $x$ axis.

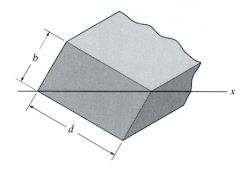

Prob. 6–92

**6-93.** Determine the location $(\bar{x}, \bar{y})$ of the centroid of the area.

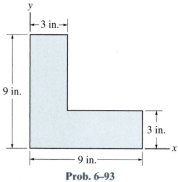

Prob. 6–93

**6-94.** Determine the location $(\bar{x}, \bar{y})$ of the centroid of the homogeneous plate.

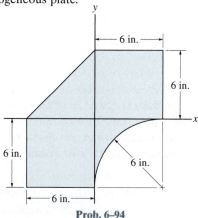

Prob. 6–94

**6-95.** Determine the moment of inertia of the beam's cross-sectional area about the $x$ axis which passes through the centroid $C$.

**\*6-96.** Determine the moment of inertia of the beam's cross-sectional area about the $y$ axis, which passes through the centroid $C$.

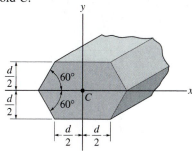

Probs. 6–95/96

The design and analysis of any structural member requires knowledge of the internal loadings acting within it, not only when it is in place and subjected to service loads, but also when it is being hoisted as shown here. In this chapter, we will discuss how engineers determine these loadings.

# CHAPTER

# 7

# Internal Loadings

## CHAPTER OBJECTIVES

- To show how to use the method of sections for determining the internal loadings in a member.
- To generalize this procedure by formulating equations that can be plotted so that they describe the internal shear and moment throughout a member.

## 7.1 Internal Loadings Developed in Structural Members

The design of any structural or mechanical member requires an investigation of the loading acting within the member in order to be sure the material can resist this loading. These internal loadings can be determined by using the *method of sections*. To illustrate the procedure, consider the "simply supported" beam shown in Fig. 7–1a, which is subjected to the forces $\mathbf{F}_1$ and $\mathbf{F}_2$ and the *support reactions* $\mathbf{A}_x$, $\mathbf{A}_y$, and $\mathbf{B}_y$, Fig. 7–1b. If the *internal loadings* acting on the cross section at $C$ are to be determined, then an imaginary section is passed through the beam, cutting it into two segments. By doing this the internal loadings at the section become *external* on the free-body diagram of each segment,

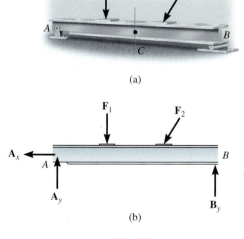

(a)

(b)

**Fig. 7–1**

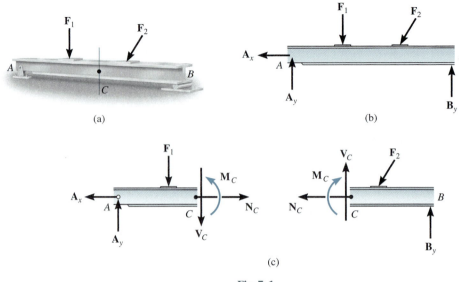

(a)

(b)

(c)

**Fig. 7–1**

Fig. 7–1c. Since both segments (*AC* and *CB*) were in equilibrium *before* the beam was sectioned, equilibrium of each segment is maintained provided rectangular force components $\mathbf{N}_C$ and $\mathbf{V}_C$ and a resultant couple moment $\mathbf{M}_C$ are developed at the section. Note that these loadings must be equal in magnitude and opposite in direction on each of the segments (Newton's third law). The magnitude of each of these loadings can now be determined by applying the three equations of equilibrium to either segment *AC* or *CB*. A *direct solution* for $\mathbf{N}_C$ is obtained by applying $\Sigma F_x = 0$; $\mathbf{V}_C$ is obtained directly from $\Sigma F_y = 0$; and $\mathbf{M}_C$ is determined by summing moments about point *C*, $\Sigma M_C = 0$, in order to eliminate the moments of the unknowns $\mathbf{N}_C$ and $\mathbf{V}_C$.

To save on material the beams used to support the roof of this shelter were tapered since the roof loading will produce a larger internal moment at the beams' centers than at their ends.

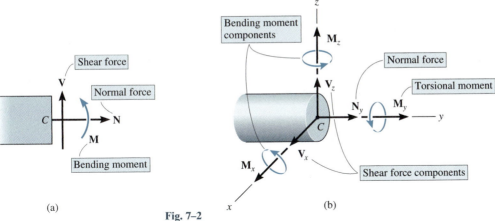

(a)

(b)

**Fig. 7–2**

In mechanics, the force components **N**, acting normal to the beam at the cut section, and **V**, acting tangent to the section, are termed the *normal or axial force* and the *shear force*, respectively. The couple moment **M** is referred to as the *bending moment*, Fig. 7–2a. In three dimensions, a general internal force and couple moment resultant will act at the section. The $x, y, z$ components of these loadings are shown in Fig. 7–2b. Here $\mathbf{N}_y$ is the *normal force*, and $\mathbf{V}_x$ and $\mathbf{V}_z$ are *shear force* components. $\mathbf{M}_y$ is a *torsional or twisting moment*, and $\mathbf{M}_x$ and $\mathbf{M}_z$ are *bending moment components*. For most applications, these *resultant loadings* will act at the geometric center or centroid ($C$) of the section's cross-sectional area. Although the magnitude for each loading generally will be different at various points along the axis of the member, the method of sections can always be used to determine their values.

**Free-Body Diagrams.**  Since frames and machines are composed of *multiforce members*, each of these members will generally be subjected to internal normal, shear, and bending loadings. For example, consider the frame shown in Fig. 7–3a. If the blue section is passed through the frame to determine the internal loadings at points $H$, $G$, and $F$, the resulting free-body diagram of the top portion of this section is shown in Fig. 7–3b. At each point where a member is sectioned there is an unknown normal force, shear force, and bending moment. As a result, we cannot apply the *three* equations of equilibrium to this section in order to obtain these *nine unknowns*.* Instead, to solve this problem we must *first dismember* the frame and determine the reactions at the connections of the members using the techniques of Sec. 6.6. Once this is done, *each member* may then be sectioned at its appropriate point, and the three equations of equilibrium can be applied to determine **N**, **V**, and **M**. For example, the free-body diagram of segment $DG$, Fig. 7–3c, can be used to determine the internal loadings at $G$ provided the reactions of the pin, $\mathbf{D}_x$ and $\mathbf{D}_y$, are known.

*Recall that this method of analysis worked well for trusses since truss members are *straight two-force members* which support only an axial or normal load.

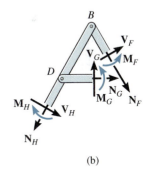

(a)

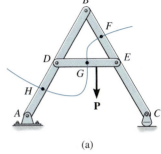

(b)

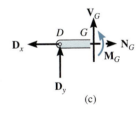

(c)

**Fig. 7–3**

In each case, the link on the backhoe is a two-force member. In the top photo it is subjected to both bending and axial load at its center. By making the member straight, as in the bottom photo, then only an axial force acts within the member.

## PROCEDURE FOR ANALYSIS

The method of sections can be used to determine the internal loadings at a specific location in a member using the following procedure.

### Support Reactions.

* Before the member is "cut" or sectioned, it may first be necessary to determine the member's support reactions, so that the equilibrium equations are used only to solve for the internal loadings when the member is sectioned.

* If the member is part of a frame or machine, the reactions at its connections are determined using the methods of Sec. 5.5.

### Free-Body Diagram.

* Keep all distributed loadings, couple moments, and forces acting on the member in their *exact locations*, then pass an imaginary section through the member, perpendicular to its axis at the point where the internal loading is to be determined.

* After the section is made, draw a free-body diagram of the segment that has the least number of loads on it, and indicate the $x, y, z$ components of the force and couple moment resultants at the section.

* If the member is subjected to a *coplanar* system of forces, the internal loading at the section reduces to normal force $\mathbf{N}$, shear $\mathbf{V}$, and bending moment $\mathbf{M}$.

* In many cases it may be possible to tell by inspection the proper sense of the unknown loadings; however, if this seems difficult, the sense can be assumed.

### Equations of Equilibrium.

* Moments should be summed at the section about axes passing through the *centroid* or geometric center of the member's cross-sectional area in order to eliminate the unknown normal and shear forces and thereby obtain direct solutions for the moment components.

* If the solution of the equilibrium equations yields a negative scalar, the assumed sense of the quantity is opposite to that shown on the free-body diagram.

**E X A M P L E   7.1**

Determine the resultant internal loadings acting on the cross section at *C* of the beam shown in Fig. 7–4a.

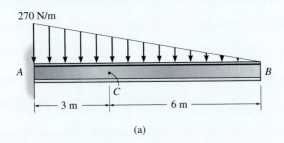

(a)

**Fig. 7–4**

## Solution

*Support Reactions.*   This problem can be solved in the most direct manner by considering segment *CB* of the beam, since then the support reactions at *A* do not have to be computed.

*Free-Body Diagram.*   Passing an imaginary section perpendicular to the longitudinal axis of the beam yields the free-body diagram of segment *CB* shown in Fig. 7–4b. It is important to keep the distributed loading exactly where it is on the segment until *after* the section is made. Only then should this loading be replaced by a single resultant force. Notice that the intensity of the distributed loading at *C* is found by proportion, i.e., from Fig. 7–4a, $w/6\,\text{m} = (270\,\text{N/m})/9\text{m}, w = 180\,\text{N/m}$. The magnitude of the resultant of the distributed load is equal to the area under the loading curve (triangle) and acts through the centroid of this area. Thus, $F = \frac{1}{2}(180\,\text{N/m})(6\,\text{m}) = 540\,\text{N}$, which acts $1/3(6\,\text{m}) = 2\,\text{m}$ from *C* as shown in Fig. 7–4b.

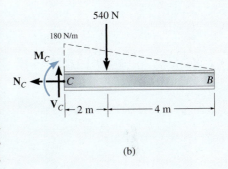

(b)

*Equations of Equilibrium.*   Applying the equations of equilibrium we have

$$\xrightarrow{+}\ \Sigma F_x = 0; \qquad\qquad -N_C = 0$$
$$N_C = 0 \qquad\qquad Ans.$$
$$+\uparrow \Sigma F_y = 0; \qquad V_C - 540\,\text{N} = 0$$
$$V_C = 540\,\text{N} \qquad\qquad Ans.$$
$$\zeta+\ \Sigma M_C = 0; \qquad -M_C - 540\,\text{N}(2\,\text{m}) = 0$$
$$M_C = -1080\,\text{N}\cdot\text{m} \qquad\qquad Ans.$$

The negative sign indicates that $\mathbf{M}_C$ acts in the opposite direction to that shown on the free-body diagram. Try solving this problem using segment *AC*, by first obtaining the support reactions at *A*, which are given in Fig. 7–4c.

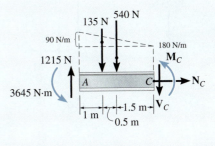

(c)

**E X A M P L E  7.2**

Determine the resultant internal loadings acting on the cross section at $C$ of the machine shaft shown in Fig. 7–5a. The shaft is supported by bearings at $A$ and $B$, which exert only vertical forces on the shaft.

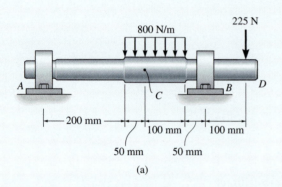

(a)

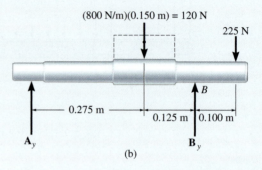

(b)

**Fig. 7–5**

**Solution**

We will solve this problem using segment $AC$ of the shaft.

*Support Reactions.* A free-body diagram of the entire shaft is shown in Fig. 7–5b. Since segment $AC$ is to be considered, only the reaction at $A$ has to be determined. Why?

$$\stackrel{+}{\curvearrowleft} \Sigma M_B = 0; \quad -A_y(0.400 \text{ m}) +120 \text{ N}(0.125 \text{ m}) - 225 \text{ N}(0.100 \text{ m}) = 0$$
$$A_y = -18.75 \text{ N}$$

The negative sign for $A_y$ indicates that $\mathbf{A}_y$ acts in the *opposite sense* to that shown on the free-body diagram.

*Free-Body Diagram.* Passing an imaginary section perpendicular to the axis of the shaft through $C$ yields the free-body diagram of segment $AC$ shown in Fig. 7–5c.

*Equations of Equilibrium.*

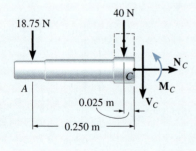

(c)

$$\stackrel{+}{\rightarrow} \Sigma F_x = 0; \quad\quad\quad N_C = 0 \quad\quad\quad Ans.$$
$$+\uparrow \Sigma F_y = 0; \quad\quad -18.75 \text{ N} - 40 \text{ N} - V_C = 0$$
$$V_C = -58.8 \text{ N} \quad\quad Ans.$$
$$\stackrel{+}{\curvearrowleft} \Sigma M_C = 0; \quad M_C +40 \text{ N}(0.025 \text{ m}) +18.75 \text{ N}(0.250 \text{ m}) = 0$$
$$M_C = -5.69 \text{ N} \cdot \text{m} \quad\quad Ans.$$

What do the negative signs for $V_C$ and $M_C$ indicate? As an exercise, calculate the reaction at $B$ and try to obtain the same results using segment $CBD$ of the shaft.

# E X A M P L E 7.3

The hoist in Fig. 7–6a consists of the beam $AB$ and attached pulleys, the cable, and the motor. Determine the resultant internal loadings acting on the cross section at $C$ if the motor is lifting the 500-lb load $W$ with constant velocity. Neglect the weight of the pulleys and beam.

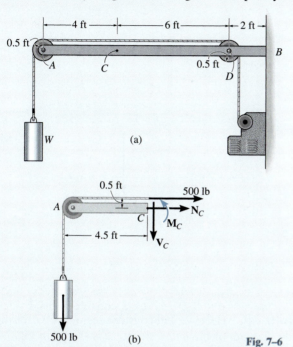

(a)

(b)

**Fig. 7–6**

## Solution

The most direct way to solve this problem is to section both the cable and the beam at $C$ and then consider the entire left segment.

*Free-Body Diagram.*   See Fig. 7–6b.

*Equations of Equilibrium.*

$\xrightarrow{+} \Sigma F_x = 0;$     500 lb + $N_C$ = 0     $N_C$ = −500 lb     *Ans.*

$+\uparrow \Sigma F_y = 0;$     −500 lb − $V_C$ = 0     $V_C$ = −500 lb     *Ans.*

$\zeta + \Sigma M_C = 0;$     500 lb(4.5 ft) − 500 lb(0.5 ft) + $M_C$ = 0

$M_C$ = −2000 lb · ft     *Ans.*

As an exercise, try obtaining these same results by considering just the beam segment $AC$, i.e., remove the pulley at $A$ from the beam and show the 500-lb force components of the pulley acting on the beam segment $AC$. Also, this problem can be worked by first finding the reactions at $B$, $(B_x = 0, B_y = 1000$ lb, $M_B = 7000$ lb·ft) and then considering segment $CB$.

E X A M P L E  **7.4**

Determine the resultant internal loadings acting on the cross section at *G* of the wooden beam shown in Fig. 7–7a. Assume the joints at *A*, *B*, *C*, *D*, and *E* are pin connected.

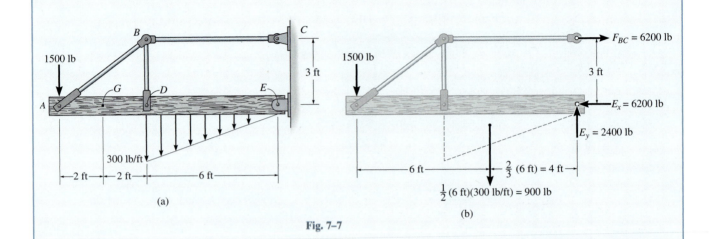

(a)

(b)

**Fig. 7–7**

### Solution

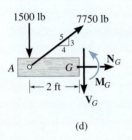

(c)

(d)

*Support Reactions.* Here we will consider segment *AG* for the analysis. A free-body diagram of the *entire* structure is shown in Fig. 7–7b. Verify the computed reactions at *E* and *C*. In particular, note that *BC* is a *two-force member* since only two forces act on it. For this reason the reaction at *C* must be horizontal as shown.

Since *BA* and *BD* are also two-force members, the free-body diagram of joint *B* is shown in Fig. 7–7c. Again, verify the magnitudes of the computed forces $\mathbf{F}_{BA}$ and $\mathbf{F}_{BD}$.

*Free-Body Diagram.* Using the result for $\mathbf{F}_{BA}$, the left section *AG* of the beam is shown in Fig. 7–7d.

*Equations of Equilibrium.* Applying the equations of equilibrium to segment *AG*, we have

$\xrightarrow{+} \Sigma F_x = 0;$   $7750 \text{ lb}(\tfrac{4}{5}) + N_G = 0$   $N_G = -6200 \text{ lb}$    *Ans.*

$+\uparrow \Sigma F_y = 0;$   $-1500 \text{ lb} + 7750 \text{ lb}(\tfrac{3}{5}) - V_G = 0$

$$V_G = 3150 \text{ lb} \qquad Ans.$$

$\stackrel{+}{\downarrow} \Sigma M_G = 0;$   $M_G - (7750 \text{ lb})(\tfrac{3}{5})(2 \text{ ft}) + 1500 \text{ lb}(2 \text{ ft}) = 0$

$$M_G = 6300 \text{ lb} \cdot \text{ft} \qquad Ans.$$

As an exercise, compute these same results using segment *GE*.

**EXAMPLE  7.5**

Determine the resultant internal loadings acting on the cross section at $B$ of the pipe shown in Fig. 7–8a. The pipe has a mass of 2 kg/m and is subjected to both a vertical force of 50 N and a couple moment of 70 N · m at its end $A$. It is fixed to the wall at $C$.

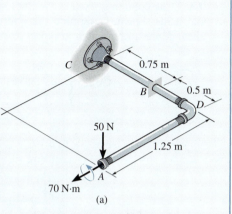

(a)

**Solution**

The problem can be solved by considering segment $AB$, which does not involve the support reactions at $C$.

*Free-Body Diagram.* The $x$, $y$, $z$ axes are established at $B$ and the free-body diagram of segment $AB$ is shown in Fig. 7–8b. The resultant force and moment components at the section are assumed to act in the positive coordinate directions and to pass through the *centroid* of the cross-sectional area at $B$. The weight of each segment of pipe is calculated as follows:

$$W_{BD} = (2 \text{ kg/m})(0.5 \text{ m})(9.81 \text{ N/kg}) = 9.81 \text{ N}$$
$$W_{AD} = (2 \text{ kg/m})(1.25 \text{ m})(9.81 \text{ N/kg}) = 24.525 \text{ N}$$

These forces act through the center of gravity of each segment.

*Equations of Equilibrium.* Applying the six scalar equations of equilibrium, we have*

$$\Sigma F_x = 0; \qquad\qquad (F_B)_y = 0 \qquad\qquad Ans.$$
$$\Sigma F_y = 0; \qquad\qquad (F_B)_y = 0 \qquad\qquad Ans.$$
$$\Sigma F_z = 0; \quad (F_B)_z - 9.81 \text{ N} - 24.525 \text{ N} - 50 \text{ N} = 0$$
$$(F_B)_z = 84.3 \text{ N} \qquad\qquad Ans.$$
$$\Sigma (M_B)_x = 0; \quad (M_B)_x + 70 \text{ N} \cdot \text{m} - 50 \text{ N}(0.5 \text{ m}) - 24.525 \text{ N}(0.5 \text{ m})$$
$$- 9.81 \text{ N}(0.25 \text{ m}) = 0$$
$$(M_B)_x = -30.3 \text{ N} \cdot \text{m} \qquad\qquad Ans.$$
$$\Sigma (M_B)_y = 0; \quad (M_B)_y + 24.525 \text{ N}(0.625 \text{ m}) + 50 \text{ N}(1.25 \text{ m}) = 0$$
$$(M_B)_y = -77.8 \text{ N} \cdot \text{m} \qquad\qquad Ans.$$
$$\Sigma (M_B)_z = 0; \qquad\qquad (M_B)_z = 0 \qquad\qquad Ans.$$

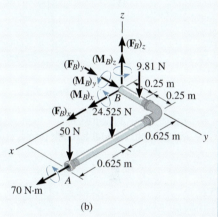

(b)

**Fig. 7–8**

What do the negative signs for $(M_B)_x$ and $(M_B)_y$ indicate? Note that the normal force $N_B = (F_B)_y = 0$, whereas the shear force is $V_B = \sqrt{(0)^2 + (84.3)^2} = 84.3 \text{ N}$. Also, the torsional moment is $T_B = (M_B)_y = 77.8 \text{ N} \cdot \text{m}$ and the bending moment is $M_B = \sqrt{(30.3)^2 + (0)^2} = 30.3 \text{ N} \cdot \text{m}$.

---

*The *magnitude* of each moment about an axis is equal to the magnitude of each force times the perpendicular distance from the axis to the line of action of the force. The *direction* of each moment is determined using the right-hand rule, with positive moments (thumb) directed along the positive coordinate axes.

# PROBLEMS

**7-1.** The column is fixed to the floor and is subjected to the loads shown. Determine the internal normal force, shear force, and moment at points $A$ and $B$.

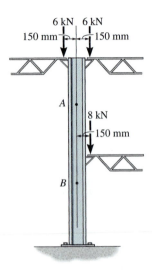

**Prob. 7–1**

**7-2.** The rod is subjected to the forces shown. Determine the internal normal force at points $A$, $B$, and $C$.

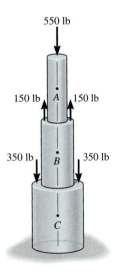

**Prob. 7–2**

**7-3.** The forces act on the shaft shown. Determine the internal normal force at points $A$, $B$, and $C$.

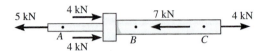

**Prob. 7–3**

***7-4.** The shaft is supported by the two smooth bearings $A$ and $B$. The four pulleys attached to the shaft are used to transmit power to adjacent machinery. If the torques applied to the pulleys are as shown, determine the internal torques at points $C$, $D$, and $E$.

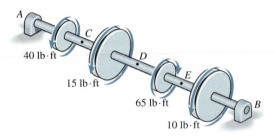

**Prob. 7–4**

**7-5.** The shaft is supported by a journal bearing at $A$ and a thrust bearing at $B$. Determine the normal force, shear force, and moment at a section passing through (a) point $C$, which is just to the right of the bearing at $A$, and (b) point $D$, which is just to the left of the 3000-lb force.

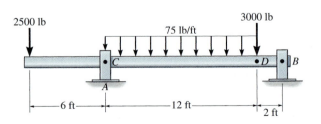

**Prob. 7–5**

**7-6.** Determine the internal normal force and shear force, and the bending moment in the beam at points $C$ and $D$. Assume the support at $B$ is a roller. Point $C$ is located just to the right of the 8-kip load.

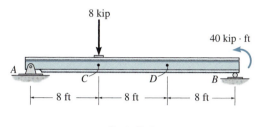

8 kip

40 kip · ft

$A$

$C$

$D$

$B$

8 ft — 8 ft — 8 ft

**Prob. 7–6**

**7-7.** Determine the shear force and moment at points $C$ and $D$.

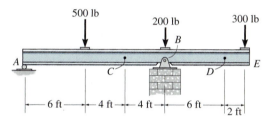

500 lb

200 lb

$B$

300 lb

$A$

$C$

$D$

$E$

6 ft — 4 ft — 4 ft — 6 ft

2 ft

**Prob. 7–7**

*__7-8.__ Determine the normal force, shear force, and moment at a section passing through point $C$. Assume the support at $A$ can be approximated by a pin and $B$ as a roller.

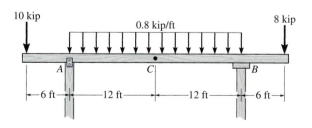

10 kip

0.8 kip/ft

8 kip

$A$

$C$

$B$

6 ft — 12 ft — 12 ft — 6 ft

**Prob. 7–8**

**7-9.** Determine the normal force, shear force, and moment at a section passing through point $D$. Take $w = 150$ N/m.

**7-10.** The beam $AB$ will fail if the maximum internal moment at $D$ reaches 800 N · m or the normal force in member $BC$ becomes 1500 N. Determine the largest load $w$ it can support.

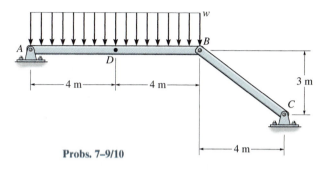

$w$

$A$

$D$

$B$

$C$

4 m — 4 m

3 m

4 m

**Probs. 7–9/10**

**7-11.** Determine the shear force and moment acting at a section passing through point $C$ in the beam.

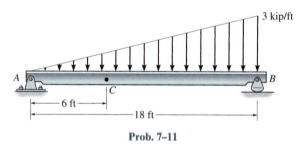

3 kip/ft

$A$

$C$

$B$

6 ft

18 ft

**Prob. 7–11**

*__7-12.__ The boom $DF$ of the jib crane and the column $DE$ have a uniform weight of 50 lb/ft. If the hoist and load weigh 300 lb, determine the normal force, shear force, and moment in the crane at sections passing through points $A$, $B$, and $C$.

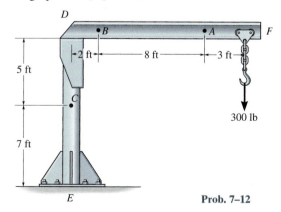

$D$

$B$

$A$

$F$

2 ft — 8 ft — 3 ft

5 ft

$C$

300 lb

7 ft

$E$

**Prob. 7–12**

**7-13.** Determine the internal normal force, shear force, and moment acting at point $C$ and at point $D$, which is located just to the right of the roller support at $B$.

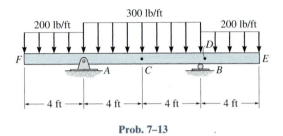

**Prob. 7–13**

**7-14.** Determine the resultant internal normal and shear force in the member at (*a*) section *a–a* and (*b*) section *b–b*, each of which passes through point $A$. The 500-lb load is applied along the centroidal axis of the member.

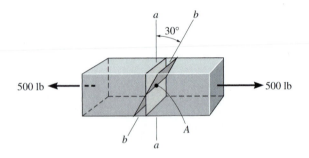

**Prob. 7–14**

**7-15.** Determine the normal force, shear force, and moment at a section passing through point $E$ of the two-member frame.

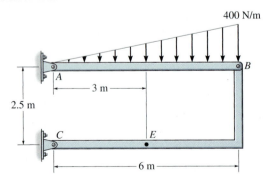

**Prob. 7–15**

**\*7-16.** The strongback or lifting beam is used for materials handling. If the suspended load has a weight of 2 kN and a center of gravity of $G$, determine the placement $d$ of the padeyes on the top of the beam so that there is no moment developed within the length $AB$ of the beam. The lifting bridle has two legs that are positioned at 45°, as shown.

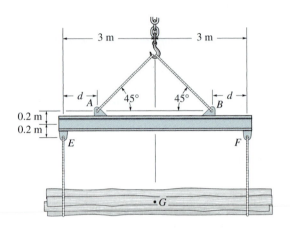

**Prob. 7–16**

**7-17.** Determine the normal force, shear force, and moment acting at a section passing through point $C$.

**7-18.** Determine the normal force, shear force, and moment acting at a section passing through point $D$.

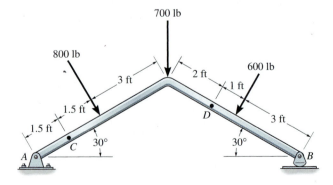

**Probs. 7–17/18**

**7-19.** Determine the normal force, shear force, and moment at a section passing through point C. Take P = 8 kN.

***7-20.** The cable will fail when subjected to a tension of 2 kN. Determine the largest vertical load P the frame will support and calculate the internal normal force, shear force, and moment at a section passing through point C for this loading.

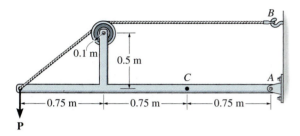

Probs. 7–19/20

**7-21.** Determine the internal normal force, shear force, and bending moment in the beam at point B.

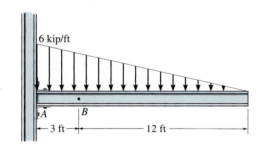

Prob. 7–21

**7-22.** Determine the ratio of a/b for which the shear force will be zero at the midpoint C of the beam.

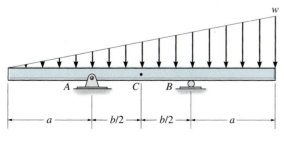

Prob. 7–22

**7-23.** Determine the internal normal force, shear force, and bending moment at point C.

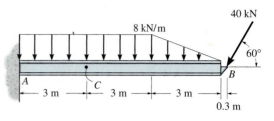

Prob. 7–23

***7-24.** The jack AB is used to straighten the bent beam DE using the arrangement shown. If the axial compressive force in the jack is 5000 lb, determine the internal moment developed at point C of the top beam. Neglect the weight of the beams.

**7-25.** Solve Prob. 7–24 assuming that each beam has a uniform weight of 150 lb/ft.

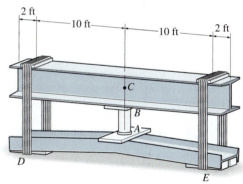

Probs. 7–24/25

**7-26.** Determine the normal force, shear force, and moment in the beam at sections passing through points D and E. Point E is just to the right of the 3-kip load.

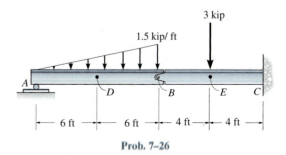

Prob. 7–26

**7-27.** Determine the internal normal force, shear force, and bending moment in the beam at points $D$ and $E$. Point $E$ is just to the right of the 4-kip load. Assume $A$ is a roller support, the splice at $B$ is a pin, and $C$ is a fixed support.

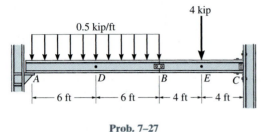

**Prob. 7–27**

**\*7-28.** Determine the internal normal force, shear force, and bending moment at points $E$ and $F$ of the frame.

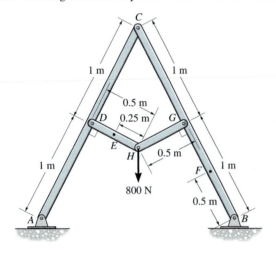

**Prob. 7–28**

**7-29.** The semicircular arch is subjected to a uniform distributed load along its axis of $w_0$ per unit length. Determine the internal normal force, shear force, and moment in the arch at $\theta = 45°$.

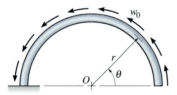

**Probs. 7–29**

**7-30.** Determine the $x$, $y$, $z$ components of internal loading at a section passing through point $C$ in the pipe assembly. Neglect the weight of the pipe. Take $\mathbf{F}_1 = \{-80\mathbf{i} + 200\mathbf{j} - 300\mathbf{k}\}$ lb and $\mathbf{F}_2 = \{250\mathbf{i} - 150\mathbf{j} - 200\mathbf{k}\}$ lb.

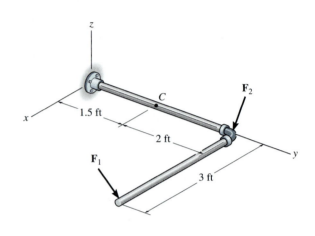

**Prob. 7–30**

**7-31.** Determine the $x$, $y$, $z$ components of force and moment at point $C$ in the pipe assembly. Neglect the weight of the pipe. The load acting at $(0, 3.5 \text{ ft}, 3 \text{ ft})$ is $\mathbf{F}_1 = \{-24\mathbf{i} - 10\mathbf{k}\}$ lb and $\mathbf{M} = \{-30\mathbf{k}\}$ lb · ft and at point $(0, 3.5 \text{ ft}, 0)$ $\mathbf{F}_2 = \{-80\mathbf{i}\}$ lb.

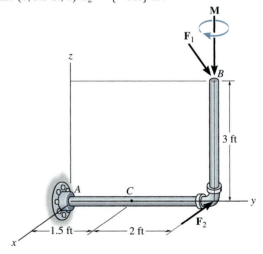

**Prob. 7–31**

## 7.2  Shear and Moment Equations and Diagrams

*Beams* are structural members which are designed to support loadings applied perpendicular to their axes. In general, beams are long, straight bars having a constant cross-sectional area. Often they are classified as to how they are supported. For example, a *simply supported beam* is pinned at one end and roller-supported at the other, Fig. 7–9, whereas a *cantilevered beam* is fixed at one end and free at the other. The actual design of a beam requires a detailed knowledge of the *variation* of the internal shear force $V$ and bending moment $M$ acting at *each point* along the axis of the beam. After this force and bending-moment analysis is complete, one can then use the theory of mechanics of materials and an appropriate engineering design code to determine the beam's required cross-sectional area.

The *variations* of $V$ and $M$ as functions of the position $x$ along the beam's axis can be obtained by using the method of sections discussed in Sec. 7.1. Here, however, it is necessary to section the beam at an arbitrary distance $x$ from one end rather than at a specified point. If the results are plotted, the graphical variations of $V$ and $M$ as functions of $x$ are termed the *shear diagram* and *bending-moment diagram*, respectively.

In general, the internal shear and bending-moment functions generally will be discontinuous, or their slopes will be discontinuous at points where a distributed load changes or where concentrated forces or couple moments are applied. Because of this, these functions must be determined for *each segment* of the beam located between any two discontinuities of loading. For example, sections located at $x_1$, $x_2$, and $x_3$ will have to be used to describe the variation of $V$ and $M$ throughout the length of the beam in Fig. 7–9. These functions will be valid *only* within regions from $O$ to $a$ for $x_1$, from $a$ to $b$ for $x_2$, and from $b$ to $L$ for $x_3$.

The internal normal force will not be considered in the following discussion for two reasons. In most cases, the loads applied to a beam act perpendicular to the beam's axis and hence produce only an internal shear force and bending moment. For design purposes, the beam's resistance to shear, and particularly to bending, is more important than its ability to resist a normal force.

The designer of this shop crane realized the need for additional reinforcement around the joint in order to prevent severe internal bending of the joint when a large load is suspended from the chain hoist.

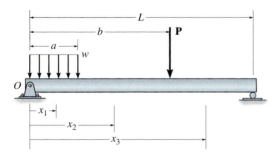

**Fig. 7–9**

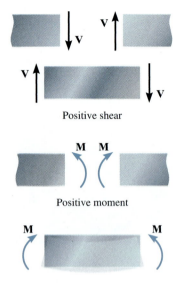

Positive shear

Positive moment

Beam sign convention

**Fig. 7–10**

**Sign Convention.** Before presenting a method for determining the shear and bending moment as functions of $x$ and later plotting these functions (shear and bending-moment diagrams), it is first necessary to establish a *sign convention* so as to define a "positive" and "negative" shear force and bending moment acting in the beam. [This is analogous to assigning coordinate directions $x$ positive to the right and $y$ positive upward when plotting a function $y = f(x)$.] Although the choice of a sign convention is arbitrary, here we will choose the one used for the majority of engineering applications. It is illustrated in Fig. 7–10. Here the positive directions are denoted by an internal *shear force* that causes *clockwise rotation* of the member on which it acts, and by an internal *moment* that causes *compression or pushing on the upper part* of the member. Also, positive moment would tend to bend the member if it were elastic, concave upward. Loadings that are opposite to these are considered negative.

## PROCEDURE FOR ANALYSIS

The shear and bending-moment diagrams for a beam can be constructed using the following procedure.

*Support Reactions.*

- Determine all the reactive forces and couple moments acting on the beam and resolve all the forces into components acting perpendicular and parallel to the beam's axis.

*Shear and Moment Functions.*

- Specify separate coordinates $x$ having an origin at the beam's *left end* and extending to regions of the beam *between* concentrated forces and/or couple moments, or where there is no discontinuity of distributed loading.
- Section the beam perpendicular to its axis at each distance $x$ and draw the free-body diagram of one of the segments. Be sure **V** and **M** are shown acting in their *positive sense*, in accordance with the sign convention given in Fig. 7–10.
- The shear $V$ is obtained by summing forces perpendicular to the beam's axis.
- The moment $M$ is obtained by summing moments about the sectioned end of the segment.

*Shear and Moment Diagrams.*

- Plot the shear diagram ($V$ versus $x$) and the moment diagram ($M$ versus $x$). If computed values of the functions describing $V$ and $M$ are *positive*, the values are plotted above the $x$ axis, whereas *negative* values are plotted below the $x$ axis.
- Generally, it is convenient to plot the shear and bending-moment diagrams directly below the free-body diagram of the beam.

# E X A M P L E  7.6

Draw the shear and moment diagrams for the beam shown in Fig. 7-11a.

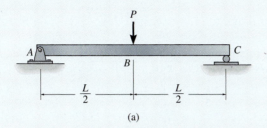

(a)

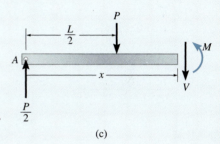

(b)

## Solution

***Support Reactions.*** The support reactions have been determined,
Fig. 7–11d.

***Shear and Moment Functions.*** The beam is sectioned at an arbitrary
distance $x$ from the support $A$, extending within region $AB$, and the
free-body diagram of the left segment is shown in Fig. 7–11b. The
unknowns $V$ and $M$ are indicated acting in the *positive sense* on the
right-hand face of the segment according to the established sign
convention. Applying the equilibrium equations yields

$$+\uparrow \Sigma F_y = 0; \qquad\qquad V = \frac{P}{2} \qquad\qquad (1)$$

$$\zeta + \Sigma M = 0; \qquad\qquad M = \frac{P}{2}x \qquad\qquad (2)$$

A free-body diagram for a left segment of the beam extending
a distance $x$ within region $BC$ is shown in Fig. 7–11c. As always, $V$ and
$M$ are shown acting in the positive sense. Hence,

$$+\uparrow \Sigma F_y = 0; \qquad \frac{P}{2} - P - V = 0$$

$$V = -\frac{P}{2} \qquad\qquad (3)$$

$$\zeta + \Sigma M = 0; \qquad M + P\left(x - \frac{L}{2}\right) - \frac{P}{2}x = 0$$

$$M = \frac{P}{2}(L - x) \qquad\qquad (4)$$

The shear diagram represents a plot of Eqs. 1 and 3, and the
moment diagram represents a plot of Eqs. 2 and 4, Fig. 7–11d. These
equations can be checked in part by noting that $dV/dx = -w$ and
$dM/dx = V$ in each case.

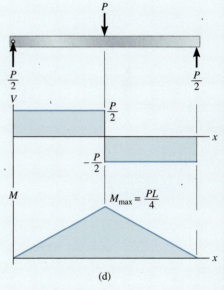

(c)

(d)

**Fig. 7–11**

Draw the shear and moment diagrams for the beam shown in Fig. 7–12a.

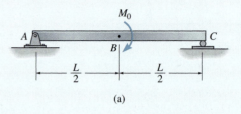

(a)

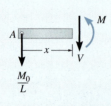

(b)

**Solution**

*Support Reactions.* The support reactions have been determined in Fig. 7–12d.

*Shear and Moment Functions.* This problem is similar to the previous example, where two $x$ coordinates must be used to express the shear and moment in the beam throughout its length. For the segment within region $AB$, Fig. 7–12b, we have

$$+\uparrow \Sigma F_y = 0; \qquad V = -\frac{M_0}{L}$$

$$\curvearrowleft + \Sigma M = 0; \qquad M = -\frac{M_0}{L}x$$

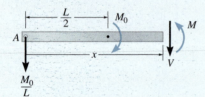

(c)

And for the segment within region $BC$, Fig. 7–12c,

$$+\uparrow \Sigma F_y = 0; \qquad V = -\frac{M_0}{L}$$

$$\curvearrowleft + \Sigma M = 0; \qquad M = M_0 - \frac{M_0}{L}x$$

$$M = M_0\left(1 - \frac{x}{L}\right)$$

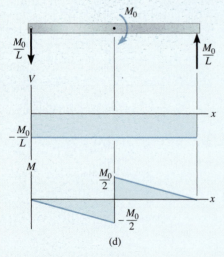

(d)

**Fig. 7–12**

*Shear and Moment Diagrams.* When the above functions are plotted, the shear and moment diagrams shown in Fig. 7–12d are obtained. In this case, notice that the shear is constant over the entire length of the beam; i.e., it is not affected by the couple moment $\mathbf{M}_0$ acting at the center of the beam. Just as a force creates a jump in the shear diagram, Example 7.6, a couple moment creates a jump in the moment diagram.

# EXAMPLE 7.8

Draw the shear and moment diagrams for the beam shown in Fig. 7–13.

## Solution

***Support Reactions.***   The support reactions have been computed in Fig. 7–13c.

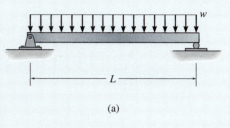

(a)

***Shear and Moment Functions.***   A free-body diagram of the left segment of the beam is shown in Fig. 7–13b. The distributed loading on this segment is represented by its resultant force only *after* the segment is isolated as a free-body diagram. Since the segment has a length $x$, the *magnitude* of the *resultant force* is $wx$. This force acts through the centroid of the area comprising the distributed loading, a distance of $x/2$ from the right end. Applying the two equations of equilibrium yields

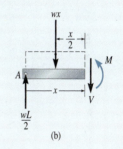

$$+\uparrow \Sigma F_y = 0; \qquad \frac{wL}{2} - wx - V = 0$$

$$V = w\left(\frac{L}{2} - x\right) \qquad (1)$$

$$\zeta + \Sigma M = 0; \qquad -\left(\frac{wL}{2}\right)x + (wx)\left(\frac{x}{2}\right) + M = 0$$

$$M = \frac{w}{2}(Lx - x^2) \qquad (2)$$

(b)

These results for $V$ and $M$ can be checked by noting that $dV/dx = -w$. This is indeed correct, since positive $w$ acts downward. Also, notice that $dM/dx = V$.

***Shear and Moment Diagrams.***   The shear and moment diagrams shown in Fig. 7–13c are obtained by plotting Eqs. 1 and 2. The point of *zero shear* can be found from Eq. 1:

$$V = w\left(\frac{L}{2} - x\right) = 0$$

$$x = \frac{L}{2}$$

From the moment diagram, this value of $x$ happens to represent the point on the beam where the *maximum moment* occurs, since the slope $V = 0 = dM/dx$. From Eq. 2, we have

$$M_{max} = \frac{w}{2}\left[L\left(\frac{L}{2}\right) - \left(\frac{L}{2}\right)^2\right]$$

$$= \frac{wL^2}{8}$$

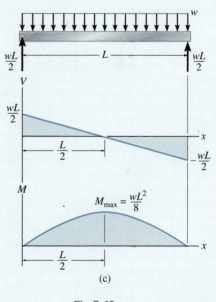

(c)

**Fig. 7–13**

Draw the shear and moment diagrams for the beam shown in Fig. 7–14a.

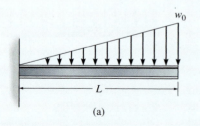

(a)

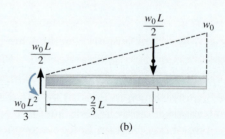

(b)

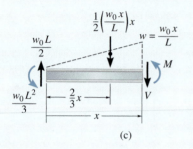

(c)

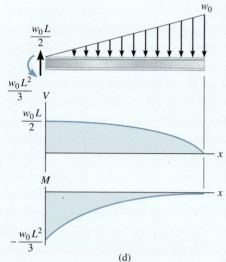

(d)

**Fig. 7–14**

### Solution

***Support Reactions.***   The distributed load is replaced by its resultant force and the reactions have been determined as shown in Fig. 7–14.

***Shear and Moment Functions.***   A free-body diagram of a beam segment of length $x$ is shown in Fig. 7–14c. Note that the intensity of the triangular load at the section is found by proportion, that is, $w/x = w_0/L$ or $w = w_0x/L$. With the load intensity known, the resultant of the distributed loading is determined from the area under the diagram, Fig. 7–14c. Thus,

$$+\uparrow \Sigma F_y = 0; \qquad \frac{w_0L}{2} - \frac{1}{2}\left(\frac{w_0x}{L}\right)x - V = 0$$

$$V = \frac{w_0}{2L}(L^2 - x^2) \qquad (1)$$

$$\zeta+ \Sigma M = 0; \quad \frac{w_0L^2}{3} - \frac{w_0L}{2}(x) + \frac{1}{2}\left(\frac{w_0x}{L}\right)x\left(\frac{1}{3}x\right) + M = 0$$

$$M = \frac{w_0}{6L}(-2L^3 + 3L^2x - x^3) \qquad (2)$$

These results can be checked as

$$w = -\frac{dV}{dx} = -\frac{w_0}{2L}(0 - 2x) = \frac{w_0x}{L} \qquad \text{OK}$$

$$V = \frac{dM}{dx} = \frac{w_0}{6L}(-0 + 3L^2 - 3x^2) = \frac{w_0}{2L}(L^2 - x^2) \qquad \text{OK}$$

***Shear and Moment Diagrams.***   The graphs of Eqs. 1 and 2 are shown in Fig. 7–14d.

## PROBLEMS

For each of the following problems, establish the *x* axis with the origin at the left side of the beam, and obtain the internal shear and moment as a function of *x*. Use these results to plot the shear and moment diagrams.

**\*7-32.** Draw the shear and moment diagrams for the beam (a) in terms of the parameters shown; (b) set $P = 600$ lb, $a = 5$ ft, $b = 7$ ft.

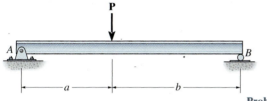

**Prob. 7–32**

**7-33.** Draw the shear and moment diagrams for the cantilevered beam.

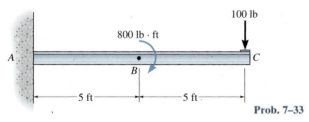

**Prob. 7–33**

**7-34.** The suspender bar supports the 600-lb engine. Draw the shear and moment diagrams for the bar.

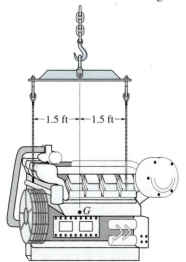

**Prob. 7–34**

**7-35.** Draw the shear and moment diagrams for the beam (a) in terms of the parameters shown; (b) set $M_0 = 500$ N·m, $L = 8$ m.

**\*7-36.** If $L = 9$ m, the beam will fail when the maximum shear force is $V_{max} = 5$ kN or the maximum bending moment is $M_{max} = 2$ kN·m. Determine the magnitude $M_0$ of the largest couple moments it will support.

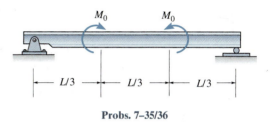

**Probs. 7–35/36**

**7-37.** The shaft is supported by a thrust bearing at $A$ and a journal bearing at $B$. If $L = 10$ ft, the shaft will fail when the maximum moment is $M_{max} = 5$ kip·ft. Determine the largest uniform distributed load $w$ the shaft will support.

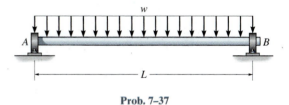

**Prob. 7–37**

**7-38.** Draw the shear and moment diagrams for the beam.

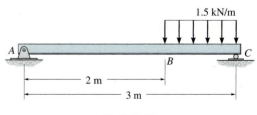

**Prob. 7–38**

**7-39.** Draw the shear and moment diagrams for the beam.

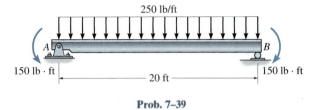

250 lb/ft

$A$       $B$

150 lb · ft      — 20 ft —      150 lb · ft

**Prob. 7–39**

**\*7-40.** Draw the shear and moment diagrams for the beam.

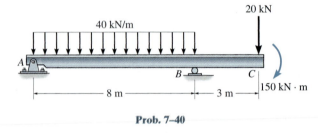

20 kN

40 kN/m

$A$    $B$    $C$

     — 8 m —    — 3 m —    150 kN · m

**Prob. 7–40**

**7-41.** Draw the shear and bending-moment diagrams for each of the two segments of the compound beam.

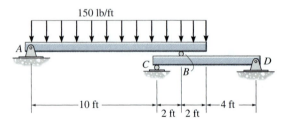

150 lb/ft

$A$       $C$   $D$    $B$

— 10 ft —   2 ft   2 ft   — 4 ft —

**Prob. 7–41**

**7-42.** Draw the shear and bending-moment diagrams for beam $ABC$. Note that there is a pin at $B$.

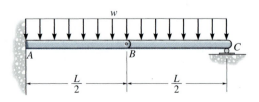

$w$

$A$    $B$    $C$

$\dfrac{L}{2}$      $\dfrac{L}{2}$

**Prob. 7–42**

**7-43.** Draw the shear and moment diagrams for the compound beam. The beam is pin-connected at $E$ and $F$.

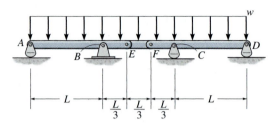

$w$

$A$    $B$   $E$   $F$   $C$    $D$

— $L$ —   $\dfrac{L}{3}$   $\dfrac{L}{3}$   $\dfrac{L}{3}$   — $L$ —

**Prob. 7–43**

**\*7-44.** Draw the shear and moment diagrams for the beam.

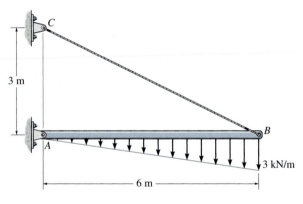

$C$

3 m

$B$

$A$      3 kN/m

— 6 m —

**Prob. 7–44**

**7-45.** If $L = 18$ ft, the beam will fail when the maximum shear force is $V_{max} = 800$ lb, or the maximum moment is $M_{max} = 1200$ lb · ft. Determine the largest intensity $w$ of the distributed loading it will support.

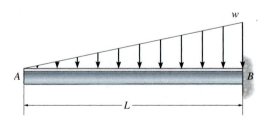

$w$

$A$       $B$

— $L$ —

**Prob. 7–45**

**7-46.** The beam will fail when the maximum internal moment is $M_{max}$. Determine the position $x$ of the concentrated force **P** and its smallest magnitude that will cause failure.

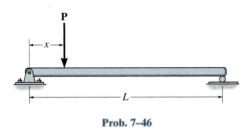

**Prob. 7–46**

**7-47.** Draw the shear and bending-moment diagrams for the beam.

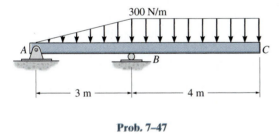

**Prob. 7–47**

***7-48.** Draw the shear and moment diagrams for the beam (a) in terms of the parameters shown; (b) set $p = 800$ lb, $a = 5$ ft, $L = 12$ ft.

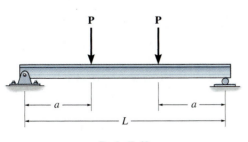

**Prob. 7–48**

**7-49.** Express the $x$, $y$, $z$ components of internal loading in the rod as a function of $y$, where $0 \le y \le 4$ ft.

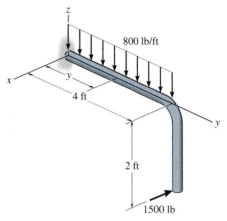

**Prob. 7–49**

**7-50.** Determine the normal force, shear force, and moment in the curved rod as a function of $\theta$.

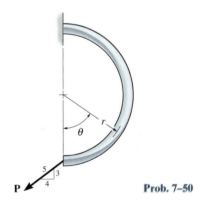

**Prob. 7–50**

**7-51.** Express the internal shear and moment components acting in the rod as a function of $y$, where $0 \le y \le 4$ ft.

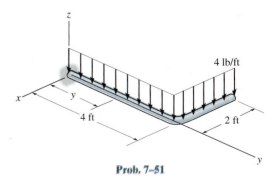

**Prob. 7–51**

## 7.3 Relations between Distributed Load, Shear, and Moment

In cases where a beam is subjected to several concentrated forces, couple moments, and distributed loads, the method of constructing the shear and bending-moment diagrams discussed in Sec. 7.2 may become quite tedious. In this section a simpler method for constructing these diagrams is discussed—a method based on differential relations that exist between the load, shear, and bending moment.

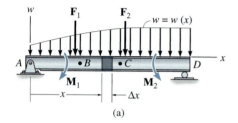

(a)

**Distributed Load.** Consider the beam $AD$ shown in Fig. 7–15a, which is subjected to an arbitrary load $w = w(x)$ and a series of concentrated forces and couple moments. In the following discussion, the *distributed load* will be considered *positive* when the *loading acts downward* as shown. A free-body diagram for a small segment of the beam having a length $\Delta x$ is chosen at a point $x$ along the beam which is *not* subjected to a concentrated force or couple moment, Fig. 7–15b. Hence any results obtained will not apply at points of concentrated loading. The internal shear force and bending moment shown on the free-body diagram are assumed to act in the *positive sense* according to the established sign convention. Note that both the shear force and moment acting on the right-hand face must be increased by a small, finite amount in order to keep the segment in equilibrium. The distributed loading has been replaced by a resultant force $\Delta F = w(x)\,\Delta x$ that acts at a fractional distance $k\,(\Delta x)$ from the right end, where $0 < k < 1$ [for example, if $w(x)$ is *uniform*, $k = \frac{1}{2}$]. Applying the equations of equilibrium, we have

$$+\uparrow\Sigma F_y = 0; \qquad V - w(x)\,\Delta x - (V + \Delta V) = 0$$
$$\Delta V = -w(x)\,\Delta x$$
$$\zeta + \Sigma M_O = 0; \quad -V\Delta x - M + w(x)\,\Delta x[k(\Delta x)] + (M + \Delta M) = 0$$
$$\Delta M = V\Delta x - w(x)k(\Delta x)^2$$

Dividing by $\Delta x$ and taking the limit as $\Delta x \rightarrow 0$, these two equations become

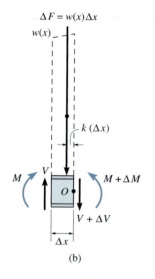

$$\Delta F = w(x)\Delta x$$

$$k\,(\Delta x)$$

$$M + \Delta M$$

$$V + \Delta V$$

(b)

**Fig. 7–15**

$$\frac{dV}{dx} = -w(x)$$

| Slope of shear diagram | = | Negative of distributed load intensity |
|---|---|---|

(7–1)

$$\frac{dM}{dx} = V$$

| Slope of moment diagram | = | Shear |
|---|---|---|

(7–2)

These two equations provide a convenient means for plotting the shear and moment diagrams for a beam. At a specific point in a beam, Eq. 7–1 states that the *slope of the shear diagram is equal to the negative of the intensity of the distributed load*, while Eq. 7–2 states that the *slope of the moment diagram is equal to the shear*. In particular, if the shear is equal to zero, $dM/dx = 0$, and therefore *a point of zero shear corresponds to a point of maximum (or possibly minimum) moment.*

Equations 7–1 and 7–2 may also be rewritten in the form $dV = -w(x)\,dx$ and $dM = V\,dx$. Noting that $w(x)\,dx$ and $V\,dx$ represent differential areas under the distributed-loading and shear diagrams, respectively, we can integrate these areas between two points $B$ and $C$ along the beam, Fig. 7–15a, and write

This concrete beam is used to support the roof. Its size and the placement of steel reinforcement within it can be determined once the shear and moment diagrams have been established.

$$\Delta V_{BC} = -\int w(x)\,dx$$

$$\frac{\text{Change}}{\text{in shear}} = \frac{\text{Negative of area under}}{\text{loading curve}} \qquad (7\text{–}3)$$

and

$$\Delta M_{BC} = \int V\,dx$$

$$\frac{\text{Change}}{\text{in moment}} = \frac{\text{Area under}}{\text{shear diagram}} \qquad (7\text{–}4)$$

Equation 7–3 states that the *change in shear between points B and C is equal to the negative of the area under the distributed-loading curve between these points*. Similarly, from Eq. 7–4, the *change in moment between B and C is equal to the area under the shear diagram within region BC*. Because two integrations are involved, first to determine the change in shear, Eq. 7–3, then to determine the change in moment, Eq. 7–4, we can state that if the loading curve $w = w(x)$ is a polynomial of degree $n$, then $V = V(x)$ will be a curve of degree $n + 1$, and $M = M(x)$ will be a curve of degree $n + 2$.

As stated previously, the above equations do not apply at points where a *concentrated* force or couple moment acts. These two special cases create *discontinuities* in the shear and moment diagrams, and as a result, each deserves separate treatment.

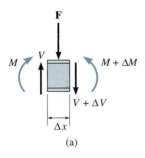

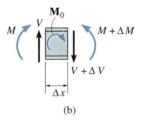

(a)

(b)

**Fig. 7–16**

**Force.** A free-body diagram of a small segment of the beam in Fig. 7–15a, taken from under one of the forces, is shown in Fig. 7–16a. Here it can be seen that force equilibrium requires

$$+\uparrow \Sigma F_y = 0; \qquad\qquad \Delta V = -F \qquad\qquad (7–5)$$

Thus, the *change in shear is negative*, so that on the shear diagram the shear will "jump" *downward when* **F** *acts downward* on the beam. Likewise, the jump in shear ($\Delta V$) is upward when **F** acts upward.

**Couple Moment.** If we remove a segment of the beam in Fig. 7–15a that is located at the couple moment, the free-body diagram shown in Fig. 7–16b results. In this case letting $\Delta x \to 0$, moment equilibrium requires

$$\zeta + \Sigma M = 0; \qquad\qquad \Delta M = M_0 \qquad\qquad (7–6)$$

Thus, the *change in moment is positive*, or the moment diagram will "jump" *upward if* **M**$_0$ *is clockwise*. Likewise, the jump $\Delta M$ is downward when **M**$_0$ is counterclockwise.

The examples which follow illustrate application of the above equations for the construction of the shear and moment diagrams. After working through these examples, it is recommended that Examples 7.6 and 7.7 be solved using this method.

Each outrigger such as *AB* supporting this crane acts as a beam which is fixed to the frame of the crane at one end and subjected to a force **F** on the footing at its other end. A proper design requires that the outrigger is able to resist its maximum internal shear and moment. The shear and moment diagrams indicate that the shear will be constant throughout its length and the maximum moment occurs at the support *A*.

## IMPORTANT POINTS

- The slope of the shear diagram is equal to the negative of the intensity of the distributed loading, where positive distributed loading is downward, i.e., $dV/dx = -w(x)$.

- If a concentrated force acts downward on the beam, the shear will jump downward by the amount of the force.

- The change in the shear $\Delta V$ between two points is equal to *the negative of the area* under the distributed-loading curve between the points.

- The slope of the moment diagram is equal to the shear, i.e., $dM/dx = V$.

- The change in the moment $\Delta M$ between two points is equal to the *area* under the shear diagram between the two points.

- If a *clockwise* couple moment acts on the beam, the shear will not be affected, however, the moment diagram will jump *upward* by the amount of the moment.

- Points of *zero shear* represent points of *maximum or minimum moment* since $dM/dx = 0$.

# E X A M P L E 7.10

Draw the shear and moment diagrams for the beam shown in Fig. 7–17a.

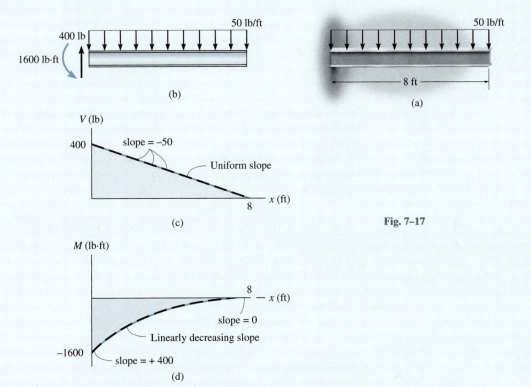

Fig. 7–17

## Solution

*Support Reactions.* The reactions at the fixed support have been calculated and are shown on the free-body diagram of the beam, Fig. 7–17b.

*Shear Diagram.* The shear at the end points is plotted first, Fig. 7–17c. From the sign convention, Fig. 7–10, $V = +400$ at $x = 0$ and $V = 0$ at $x = 8$. Since $dV/dx = -w = -50$, a straight, *negative* sloping line connects the end points.

*Moment Diagram.* From our sign convention, Fig. 7–10, the moments at the beam's end points, $M = -1600$ at $x = 0$ and $M = 0$ at $x = 8$, are plotted first, Fig. 7–17d. Successive values of shear taken from the shear diagram, Fig. 7–17c, indicate that the *slope* $dM/dx = V$ of the moment diagram, Fig. 7–17d, is always positive yet *linearly decreasing* from $dM/dx = 400$ at $x = 0$ to $dM/dx = 0$ at $x = 8$. Thus, due to the integrations, $w$ a constant yields $V$ a sloping line (first-degree curve) and $M$ a parabola (second-degree curve).

E X A M P L E 7.11

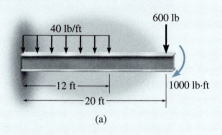

40 lb/ft

600 lb

12 ft

20 ft

1000 lb·ft

(a)

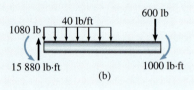

1080 lb

40 lb/ft

600 lb

15 880 lb·ft

1000 lb·ft

(b)

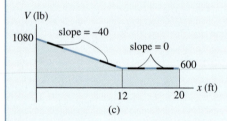

V (lb)

1080

slope = –40

slope = 0

600

12      20      x (ft)

(c)

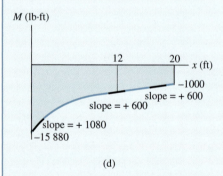

M (lb·ft)

12      20      x (ft)

–1000

slope = + 600

slope = + 600

slope = + 1080

–15 880

(d)

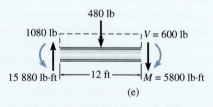

480 lb

1080 lb

V = 600 lb

15 880 lb·ft      12 ft      M = 5800 lb·ft

(e)

**Fig. 7–18**

Draw the shear and moment diagrams for the cantilevered beam shown in Fig. 7–18a.

### Solution

*Support Reactions.* The reactions at the fixed support have been calculated and are shown on the free-body diagram of the beam, Fig. 7–18b.

*Shear Diagram.* Using the established sign convention, Fig. 7–10, the shear at the ends of the beam is plotted first; i.e., $x = 0$, $V = +1080$; $x = 20$, $V = +600$, Fig. 7–18c.

Since the uniform distributed load is downward and *constant*, the slope of the shear diagram is $dV/dx = -w = -40$ for $0 \le x < 12$ as indicated.

The magnitude of shear at $x = 12$ is $V = +600$. This can be determined by first finding the area under the load diagram between $x = 0$ and $x = 12$. This represents the change in shear. That is, $\Delta V = -\int w(x)\, dx = -40(12) = -480$. Thus $V|_{x=12} = V|_{x=0} + (-480) = 1080 - 480 = 600$. Also, we can obtain this value by using the method of sections, Fig. 7–18e, where for equilibrium $V = +600$.

Since the load between $12 < x \le 20$ is $w = 0$, the slope $dV/dx = 0$ as indicated. This brings the shear to the required value of $V = +600$ at $x = 20$.

*Moment Diagram.* Again, using the established sign convention, the moments at the ends of the beam are plotted first; i.e., $x = 0$, $M = -15\,880$; $x = 20$, $M = -1000$, Fig. 7–18d.

Each value of shear gives the slope of the moment diagram since $dM/dx = V$. As indicated, at $x = 0$, $dM/dx = +1080$; and at $x = 12$, $dM/dx = +600$. For $0 \le x < 12$, specific values of the shear diagram are positive but linearly decreasing. Hence, the moment diagram is parabolic with a linearly decreasing positive slope.

The magnitude of moment at $x = 12$ is $-5800$. This can be found by first determining the trapezoidal area under the shear diagram, which represents the change in moment, $\Delta M = \int V\, dx = 600(12) + \frac{1}{2}(1080 - 600)(12) = +10080$. Thus, $M|_{x=12} = M|_{x=0} + 10080 = -15880 + 10080 = -5800$. The more "basic" method of sections can also be used, where equilibrium at $x = 12$ requires $M = -5800$, Fig. 7–18e.

The moment diagram has a constant slope for $12 < x \le 20$ since, from the shear diagram, $dM/dx = V = +600$. This brings the value of $M = -1000$ at $x = 20$, as required.

# E X A M P L E  **7.12**

Draw the shear and moment diagrams for the shaft in Fig. 7–19*a*. The support at *A* is a thrust bearing and the support at *B* is a journal bearing.

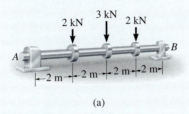

(a)

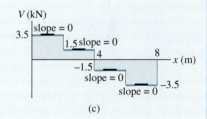

(b)

## Solution

*Support Reactions.*  The reactions at the supports are shown on the free-body diagram in Fig. 7–19*b*.

*Shear Diagram.*  The end points $x = 0$, $V = +3.5$ and $x = 8$, $V = -3.5$ are plotted first, as shown in Fig. 7–19*c*.

Since there is no distributed load on the shaft, the slope of the shear diagram throughout the shaft's length is zero; i.e., $dV/dx = -w = 0$. There is a discontinuity or "jump" of the shear diagram, however, at each concentrated force. From Eq. 7–5, $\Delta V = -F$, the change in shear is negative when the force acts downward and positive when the force acts upward. Stated another way, the "jump" follows the force, i.e., a downward force causes a downward jump, and vice versa. Thus, the 2-kN force at $x = 2$ m changes the shear from 3.5 kN to 1.5 kN; the 3-kN force at $x = 4$ m changes the shear from 1.5 kN to −1.5 kN, etc. We can *also* obtain numerical values for the shear at a specified point in the shaft by using the method of sections, as for example, $x = 2^+$ m, $V = 1.5$ kN in Fig. 7–19*e*.

*Moment Diagram.*  The end points $x = 0$, $M = 0$ and $x = 8$, $M = 0$ are plotted first, as shown in Fig. 7–19*d*.

Since the shear is constant in each region of the shaft, the moment diagram has a corresponding constant positive or negative slope as indicated on the diagram. Numerical values for the change in moment at any point can be computed from the *area* under the shear diagram. For example, at $x = 2$ m, $\Delta M = \int V\, dx = 3.5(2) = 7$. Thus, $M|_{x=2} = M|_{x=0} + 7 = 0 + 7 = 7$. Also, by the method of sections, we can determine the moment at a specified point, as for example, $x = 2^+$ m, $M = 7$ kN · m, Fig. 7–19*e*.

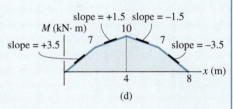

(c)

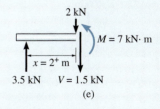

(d)

(e)

**Fig. 7–19**

## E X A M P L E  7.13

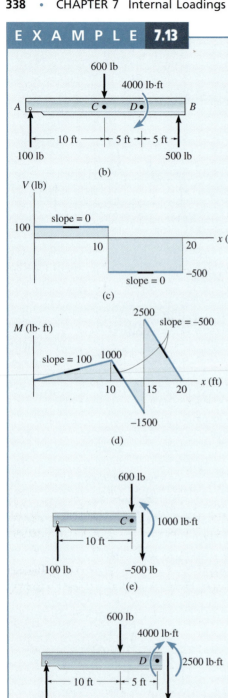

(b)

(c)

(d)

(e)

(f)

**Fig. 7–20**

Sketch the shear and moment diagrams for the beam shown in Fig. 7–20a.

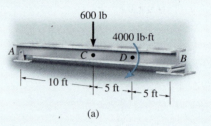

(a)

### Solution

*Support Reactions.* The reactions are calculated and indicated on the free-body diagram, Fig. 7–20b.

*Shear Diagram.* As in Example 7.12, the shear diagram can be constructed by "following the load" on the free-body diagram. In this regard, beginning at $A$, the reaction is up so $V_A = +100$ lb, Fig. 7–20c. No load acts between $A$ and $C$, so the shear remains constant; i.e., $dV/dx = -w(x) = 0$. At $C$ the 600-lb force acts downward, so the shear jumps down 600 lb, from 100 lb to $-500$ lb. Again the shear is constant (no load) and ends at $-500$ lb, point $B$. Notice that no jump or discontinuity in shear occurs at $D$, the point where the 4000-lb · ft couple moment is applied, Fig. 7–20b. This is because, for force equilibrium, $\Delta V = 0$ in Fig. 7–16b.

*Moment Diagram.* The moment at each end of the beam is zero. These two points are plotted first, Fig. 7–20d. The slope of the moment diagram from $A$ to $C$ is constant since $dM/dx = V = +100$. The value of the moment at $C$ can be determined by the method of sections, Fig. 7–20e where $M_C = +1000$ lb · ft; or by first computing the rectangular area under the shear diagram between $A$ and $C$ to obtain the change in moment $\Delta M_{AC} = (100 \text{ lb})(10 \text{ ft}) = 1000$ lb · ft. Since $M_A = 0$, then $M_C = 0 + 1000$ lb · ft $= 1000$ lb · ft. From $C$ to $D$ the slope of the moment diagram is $dM/dx = V = -500$, Fig. 7–20c. The area under the shear diagram between points $C$ and $D$ is $\Delta M_{CD} = (-500 \text{ lb})(5 \text{ ft}) = -2500$ lb · ft, so that $M_D = M_C + \Delta M_{CD} = 1000 - 2500 = -1500$ lb · ft. A jump in the moment diagram occurs at point $D$, which is caused by the concentrated couple moment of 4000 lb · ft. From Eq. 7–6 the jump is *positive* since the couple moment is *clockwise*. Thus, at $x = 15^+$ ft, the moment is $M_D = -1500 + 4000 = 2500$ lb · ft. This value can *also* be determined by the method of sections, Fig. 7–20f. From point $D$ the slope of $dM/dx = -500$ is maintained until the diagram closes to zero at $B$, Fig. 7–20d.

# PROBLEMS

**\*7-52.** Draw the shear and moment diagrams for the beam.

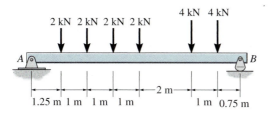

Prob. 7–52

**7-53.** Draw the shear and moment diagrams for the beam *ABCDE*. All pulleys have a radius of 1 ft. Neglect the weight of the beam and pulley arrangement. The load weighs 500 lb.

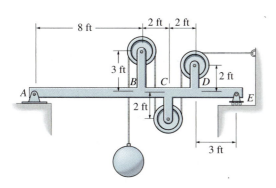

Prob. 7–53

**7-54.** Draw the shear and moment diagrams for the beam.

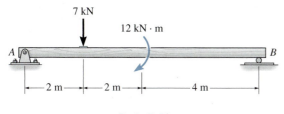

Prob. 7–54

**7-55.** Draw the shear and moment diagrams for the beam.

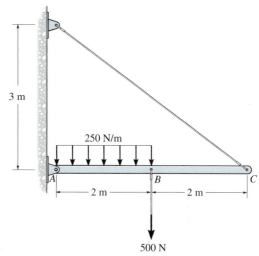

Prob. 7–55

**\*7-56.** Draw the shear and moment diagrams for the beam.

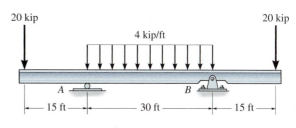

Prob. 7–56

**7-57.** Draw the shear and moment diagrams for the beam.

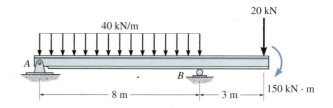

Prob. 7–57

**7-58.** Draw the shear and moment diagrams for the shaft. The support at *A* is a journal bearing and at *B* it is a thrust bearing.

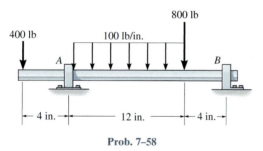

**Prob. 7–58**

**7-59.** Draw the shear and moment diagrams for the beam.

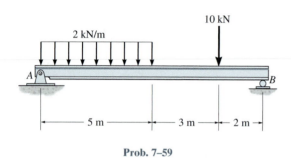

**Prob. 7–59**

**\*7-60.** Draw the shear and moment diagrams for the shaft. The support at *A* is a journal bearing and at *B* it is a thrust bearing.

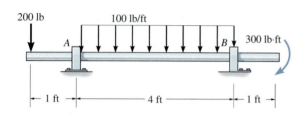

**Prob. 7–60**

**7-61.** Draw the shear and moment diagrams for the beam.

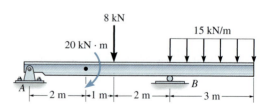

**Prob. 7–61**

**7-62.** Draw the shear and moment diagrams for the shaft. The support at *A* is a thrust bearing and at *B* it is a journal bearing.

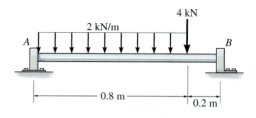

**Prob. 7–62**

**7-63.** Draw the shear and moment diagrams for the beam.

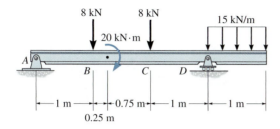

**Prob. 7–63**

**7-65.** The beam consists of two segments pin connected at $B$. Draw the shear and moment diagrams for the beam.

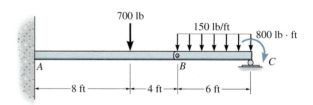

**Prob. 7–65**

***7-64.** The beam will fail when the maximum moment is $M_{max} = 30 \text{ kip} \cdot \text{ft}$ or the maximum shear is $V_{max} = 8 \text{ kip}$. Determine the largest distributed load $w$ the beam will support.

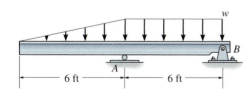

**Prob. 7–64**

**7-66.** Draw the shear and moment diagrams for the beam.

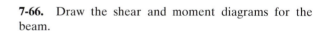
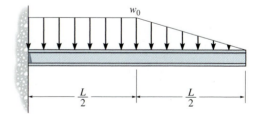

**Prob. 7–66**

**7-67.** Draw the shear and moment diagrams for the beam.

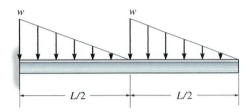

Prob. 7–67

**7-69.** Draw the shear and moment diagrams for the beam.

3 kip/ft    3 kip/ft

A    B

6 ft    6 ft

Prob. 7–69

*7-68.** Draw the shear and moment diagrams for the beam.

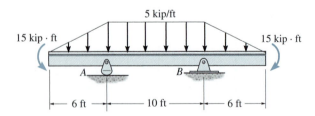

Prob. 7–68

# CHAPTER REVIEW

- *Internal Loadings.* If a coplanar force system acts on a member, then in general a resultant internal *normal force N, shear force V,* and *bending moment M* will act at any cross section along the member. These resultants are determined using the method of sections. To find them, the member is sectioned at the point where the internal loadings are to be determined. A free-body diagram of one of the sectioned parts is then drawn. The normal force is determined by summing forces normal to the cross section. The shear force is found by summing forces tangent to the cross section, and the bending moment is found by summing moments about the centroid of the cross-sectional area. If the member is subjected to a three-dimensional loading, then, in general, a *torsional loading* will also act on the cross section. It can be determined by summing moments about an axis that is perpendicular to the cross section and passes through its centroid.

- *Shear and Moment Diagrams as Functions of x.* To construct the shear and moment diagrams for a member, it is necessary to section the member at an arbitrary point, located a distance $x$ from one end. The unknown shear and moment are indicated on the cross section in the positive direction according to the established sign convention. Application of the equilibrium equations will give these loadings as a function of $x$, which can then be plotted. If the external loading consists of changes in the distributed load, or a series of concentrated forces and couple moments act on the member, then different expressions for $V$ and $M$ must be determined within regions between these different loadings.

- *Graphical Methods for Establishing Shear and Moment Diagrams.* It is possible to plot the shear and moment diagrams quickly by using differential relationships that exist between the distributed loading $w$ and $V$ and $M$. The slope of the shear diagram is equal to the distributed loading at any point, $dV/dx = -w$; and the slope of the moment diagram is equal to the shear at any point, $V = dM/dx$. Also, the change in shear between any two points is equal to the area under the distributed loading between the points, $\Delta V = \int w\, dx$, and the change in the moment is equal to the area under the shear diagram between the points, $\Delta M = \int V\, dx$.

# REVIEW PROBLEMS

**7-70.** Draw the shear and moment diagrams for the beam.

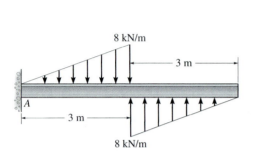

**Prob. 7-70**

**7-71.** Determine the normal force, shear force, and moment at point *C*. Assume the support at *A* is a roller and the support at *B* is a pin.

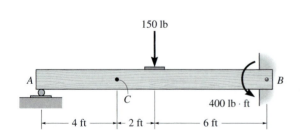

**Prob. 7-71**

**\*7-72.** Draw the shear and moment diagrams for the beam.

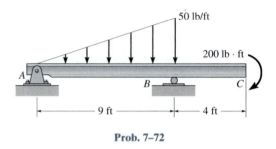

**Prob. 7-72**

**7-73.** Draw the shear and moment diagrams for the beam.

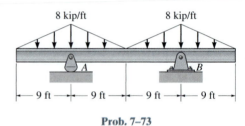

**Prob. 7-73**

**7-74.** The beam is supported by a pin at *C* and a rod *AB*. Determine the normal force, shear force, and moment at point *D*.

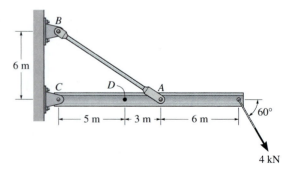

**Prob. 7-74**

**7-75.** Draw the shear and moment diagrams for the shaft. The supports at *A* and *B* are journal bearings.

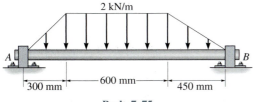

Prob. 7–75

**\*7-76.** Determine the normal force, shear force, and moment at points *D* and *E* of the frame.

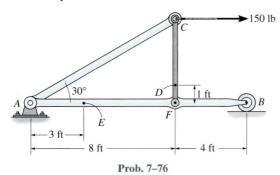

Prob. 7–76

**7-77.** The sheave on a traveling block supports a load of 15 kip. Draw the moment diagram for the supporting pin *AB* if the force is (a) assumed concentrated as a single force acting at the *center E* of the pin, and (b) uniformly distributed over the sheave contact area along *CD*.

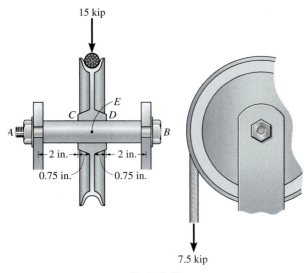

Prob. 7–77

**7-78.** Draw the shear and moment diagrams for the beam.

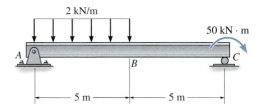

Prob. 7–78

**7-79.** Each of the street lights has a weight of 20 lb, and the supporting arm weighs 5 lb/ft. Draw the shear and moment diagrams for the arm.

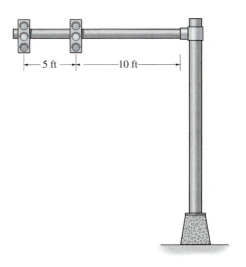

Prob. 7–79

# Mechanics
# of
# Materials

The bolts used for the connections of this steel framework are subjected to stress. In this chapter we will discuss how engineers design these connections and their fasteners.

# 8

# Stress and Strain

- To introduce the concepts of normal and shear stress, and to use them in the analysis and design of members subject to axial load and direct shear.
- To define normal and shear strain, and show how they can be determined for various types of problems.

## 8.1 Introduction

*Mechanics of materials* is a branch of mechanics that studies the relationships between the *external* loads applied to a deformable body and the intensity of *internal* forces acting within the body. This subject also involves computing the *deformations* of the body, and it provides a study of the body's *stability* when the body is subjected to external forces.

In the design of any structure or machine, it is *first* necessary to use the principles of statics to determine the forces acting both on and within its various members. The size of the members, their deflection, and their stability depend not only on the internal loadings, but also on the type of material from which the members are made. Consequently, an accurate determination and fundamental understanding of *material behavior* will be of vital importance for developing the necessary equations used in mechanics of materials. Realize that many formulas and rules for design, as defined in engineering codes and used in practice, are based on the fundamentals of mechanics of materials, and for this reason an understanding of the principles of this subject is very important.

## 8.2 Stress

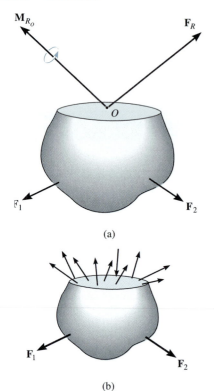

Fig. 8–1

Consider an arbitrary body subject to both external and internal forces. The body is sectioned across the middle; Fig. 8–1a is the free-body diagram of the lower part. Force and moment shown acting at $O$ represent the resultant effects of the actual *distribution of force* acting over the sectioned area, Fig. 8–1b. Obtaining this *distribution* of internal loading is of primary importance in mechanics of materials. To solve this problem it is necessary to establish the concept of stress.

Consider the sectioned area to be subdivided into small areas, such as $\Delta A$ shown dark shaded in Fig. 8–2a. As we reduce $\Delta A$ to a smaller and smaller size, we must make two assumptions regarding the properties of the material. We will consider the material to be **continuous**, that is, to consist of a *continuum* or uniform distribution of matter having no voids, rather than being composed of a finite number of distinct atoms or molecules. Furthermore, the material must be **cohesive**, meaning that all portions of it are connected together, rather than having breaks, cracks, or separations. A typical finite yet very small force $\Delta\mathbf{F}$, acting on its associated area $\Delta A$, is shown in Fig. 8–2a. This force, like all the others, will have a unique direction, but for further discussion we will replace it by its *three components*, namely, $\Delta\mathbf{F}_x$, $\Delta\mathbf{F}_y$, and $\Delta\mathbf{F}_z$, which are taken tangent and normal to the area, respectively. As the area $\Delta A$ approaches zero, so do the force $\Delta\mathbf{F}$ and its components; however, the quotient of the force and area will, in general, approach a finite limit. This quotient is called *stress*, and as noted, it describes the *intensity of the internal force* on a *specific plane* (area) passing through a point.

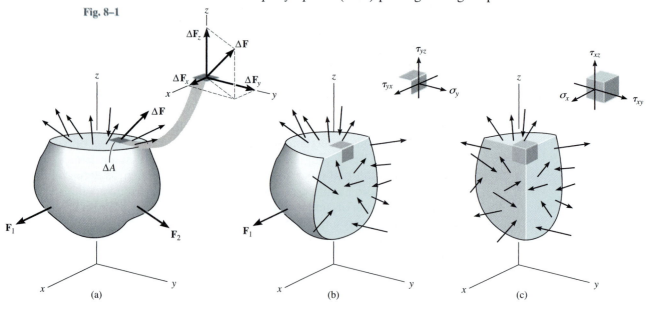

Fig. 8–2

## Normal Stress.

The *intensity* of force, or force per unit area, acting normal to $\Delta A$ is defined as the **normal stress**, $\sigma$ (sigma). Since $\Delta F_z$ is normal to the area then

$$\sigma_z = \lim_{\Delta A \to 0} \frac{\Delta F_z}{\Delta A} \qquad (8\text{–}1)$$

If the normal force or stress "pulls" on the area element $\Delta A$ as shown in Fig. 8–2a, it is referred to as *tensile stress*, whereas if it "pushes" on $\Delta A$ it is called *compressive stress*.

## Shear Stress.

The intensity of force, or force per unit area, acting tangent to $\Delta A$ is called the **shear stress**, $\tau$ (tau). Here we have shear stress components,

$$\tau_{zx} = \lim_{\Delta A \to 0} \frac{\Delta F_x}{\Delta A}$$

$$\tau_{zy} = \lim_{\Delta A \to 0} \frac{\Delta F_y}{\Delta A} \qquad (8\text{–}2)$$

Note that the subscript notation $z$ in $\sigma_z$ is used to reference the *direction* of the outward normal line, which specifies the orientation of the area $\Delta A$, Fig. 8–3. Two subscripts are used for the shear-stress components, $\tau_{zx}$ and $\tau_{zy}$. The $z$ axis specifies the orientation of the area, and $x$ and $y$ refer to the direction lines for the shear stresses.

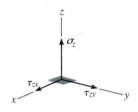

Fig. 8–3

## General State of Stress.

If the body is further sectioned by planes parallel to the $x$–$z$ plane, Fig. 8–2b, and the $y$–$z$ plane, Fig. 8–2c, we can then "cut out" a cubic volume element of material that represents the **state of stress** acting around the chosen point in the body, Fig. 8–4. This state of stress is then characterized by three components acting on each face of the element. These stress components describe the state of stress at the point only for the element orientated along the $x$, $y$, $z$ axes. Had the body been sectioned into a cube having some other orientation, then the state of stress would be defined using a different set of stress components.

Fig. 8–4

## Units.

In the International Standard or SI system, the magnitudes of both normal and shear stress are specified in the basic units of newtons per square meter ($N/m^2$). This unit, called a *pascal* ($1\ Pa = 1\ N/m^2$) is rather small, and in engineering work prefixes such as kilo- ($10^3$), symbolized by k, mega- ($10^6$), symbolized by M, or giga- ($10^9$), symbolized by G, are used to represent larger, more realistic values of stress.* Likewise, in the U.S. Customary or Foot-Pound-Second system of units, engineers usually express stress in pounds per square inch (psi) or kilopounds per square inch (ksi), where 1 kilopound (kip) = 1000 lb.

---

*Sometimes stress is expressed in units of $N/mm^2$, where $1\ mm = 10^{-3}\ m$. However, in the SI system, prefixes are not allowed in the denominator of a fraction and therefore it is better to use the equivalent $1\ N/mm^2 = 1\ MN/m^2 = 1\ MPa$.

## 8.3 Average Normal Stress in an Axially Loaded Bar

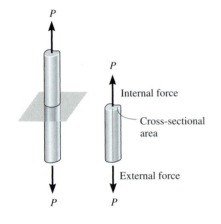

(a)    (b)

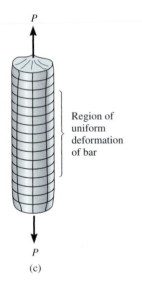

Region of uniform deformation of bar

(c)

**Fig. 8–5**

Frequently structural or mechanical members are made long and slender. Also, they are subjected to axial loads that are usually applied to the ends of the member. Truss members, hangers, and bolts are typical examples. In this section we will determine the average stress distribution acting on the cross section of an axially loaded bar, such as the one having the general form shown in Fig. 8–5a. This section defines the ***cross-sectional area*** of the bar, and since all such cross sections are the same, the bar is referred to as being ***prismatic***. If we neglect the weight of the bar and section it as indicated, then, for equilibrium of the bottom segment, Fig. 8–5b, the internal resultant force acting on the cross-sectional area must be equal in magnitude, opposite in direction, and collinear to the external force acting at the bottom of the bar.

**Assumptions.**   Before we determine the average stress distribution acting over the bar's cross-sectional area, it is necessary to make two simplifying assumptions concerning the material description and the specific application of the load.

1.  It is necessary that the bar remains straight both before and after the load is applied, and also, the cross section should remain flat or plane during the deformation, that is, during the time the bar changes its volume and shape. If this occurs, then horizontal and vertical grid lines inscribed on the bar will *deform uniformly* when the bar is subjected to the load, Fig. 8–5c. Here we will not consider regions of the bar near its ends, where application of the external loads can cause *localized distortions*. Instead we will focus only on the stress distribution within the bar's midsection.

2.  In order for *the bar to undergo uniform deformation*, it is necessary that **P** be applied along the *centroidal axis* of the cross section, and the material must be homogeneous and isotropic. ***Homogeneous material*** has the same physical and mechanical properties throughout its volume, and ***isotropic material*** has these same properties in all directions. Many engineering materials may be approximated as being both homogeneous and isotropic as assumed here. Steel, for example, contains thousands of randomly oriented crystals in each cubic millimeter of its volume, and since most problems involving this material have a physical size that is much larger than a single crystal, the above assumption regarding its material composition is quite realistic. It should be mentioned, however, that steel can be made anisotropic by cold-rolling, i.e., rolling or forging it at subcritical temperatures. ***Anisotropic materials*** have different properties in different directions, and although this is the case, if the anisotropy is oriented along the bar's axis, then the bar will also deform uniformly when subjected to an axial load. For example, timber, due to its grains or fibers of wood, is an engineering material that is homogeneous and anisotropic and is therefore suited for the following analysis.

**Average Normal Stress Distribution.** Provided the bar is subjected to a constant uniform deformation as noted, then this deformation is the result of a *constant* normal stress $\sigma$, Fig. 8–6. As a result, each area $\Delta A$ on the cross section is subjected to a force $\Delta F = \sigma \Delta A$, and the *sum* of these forces acting over the entire cross-sectional area must be equivalent to the internal resultant force **P** at the section. If we let $\Delta A \to dA$ and therefore $\Delta F \to dF$, then, recognizing $\sigma$ is *constant*, we have

$$+\uparrow F_{Rz} = \Sigma F_z; \qquad \int dF = \int_A \sigma \, dA$$

$$P = \sigma A$$

$$\boxed{\sigma = \frac{P}{A}} \qquad (8\text{--}3)$$

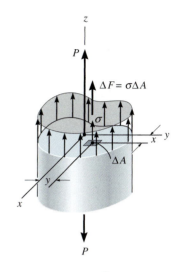

Fig. 8–6

Here

$\sigma$ = average normal stress at any point on the cross-sectional area

$P$ = internal resultant normal force, which is applied through the centroid of the cross-sectional area. $P$ is determined using the method of sections and the equations of equilibrium.

$A$ = cross-sectional area of the bar

The internal load $P$ must pass through the centroid of the crosssection since the uniform stress distribution will produce zero moments about any $x$ and $y$ axes passing through this point, Fig. 8–6. When this occurs,

$$(M_R)_x = \Sigma M_x; \qquad 0 = \int_A y \, dF = \int_A y\sigma \, dA = \sigma \int_A y \, dA$$

$$(M_R)_y = \Sigma M_y; \qquad 0 = -\int_A x \, dF = -\int_A x\sigma \, dA = -\sigma \int_A x \, dA$$

These equations are indeed satisfied, since by definition of the centroid, $\int y \, dA = 0$ and $\int x \, dA = 0$.

**Equilibrium.** It should be apparent that only a normal stress exists on any volume element of material located at each point on the cross section of an axially loaded bar. If we consider vertical equilibrium of the element, Fig. 8–7, then applying the equation of force equilibrium,

$$\Sigma F_z = 0; \qquad \sigma(\Delta A) - \sigma'(\Delta A) = 0$$

$$\sigma = \sigma'$$

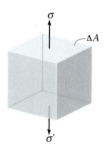

Fig. 8–7

In other words, the two normal stress components on the element must be equal in magnitude but opposite in direction. This is referred to as *uniaxial stress*.

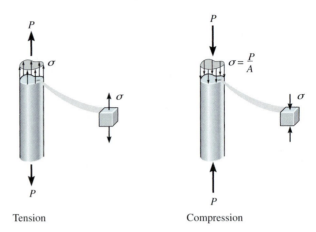

Tension                    Compression

**Fig. 8–8**

The previous analysis applies to members subjected to either tension or compression, as shown in Fig. 8–8. As a graphical interpretation, the *magnitude* of the internal resultant force **P** is *equivalent* to the *volume* under the stress diagram; that is, $P = \sigma A$ (volume = height × base). Furthermore, as a consequence of the balance of moments, ***this resultant passes through the centroid of this volume***.

Although we have developed this analysis for *prismatic* bars, this assumption can be relaxed somewhat to include bars that have a *slight taper*. For example, it can be shown, using the more exact analysis of the theory of elasticity, that for a tapered bar of rectangular cross section, for which the angle between two adjacent sides is 15°, the average normal stress, as calculated by $\sigma = P/A$, is only 2.2% *less* than its value found from the theory of elasticity.

**Maximum Average Normal Stress.**   In our analysis both the internal force $P$ and the cross-sectional area $A$ were *constant* along the longitudinal axis of the bar, and as a result the normal stress $\sigma = P/A$ is also *constant* throughout the bar's length. Occasionally, however, the bar may be subjected to *several* external loads along its axis, or a change in its cross-sectional area may occur. As a result, the normal stress within the bar could be different from one section to the next, and, if the *maximum* average normal stress is to be determined, then it becomes important to find the location where the ratio $P/A$ is a *maximum*. To do this it is necessary to determine the internal force $P$ at various sections along the bar. Here it may be helpful to show this variation by drawing an ***axial or normal force diagram***. Specifically, this diagram is a plot of the normal force $P$ versus its position $x$ along the bar's length. As a sign convention, $P$ will be positive if it causes tension in the member, and negative if it causes compression. Once the internal loading throughout the bar is known, the maximum ratio of $P/A$ can then be identified.

This steel tie rod is used to suspend a portion of a staircase, and as a result it is subjected to tensile stress.

## IMPORTANT POINTS

- When a body that is subjected to an external load is sectioned, there is a distribution of force acting over the sectioned area which holds each segment of the body in equilibrium. The intensity of this internal force at a point in the body is referred to as *stress*.

- Stress is the limiting value of force per unit area, as the area approaches zero. For this definition, the material at the point is considered to be continuous and cohesive.

- In general, there are six independent components of stress at each point in the body, consisting of *normal stress*, $\sigma_x$, $\sigma_y$, $\sigma_z$, and *shear stress*, $\tau_{xy}$, $\tau_{yz}$, $\tau_{xz}$.

- The magnitude of these components depends upon the type of loading acting on the body, and the orientation of the element at the point.

- When a prismatic bar is made from homogeneous and isotropic material, and is subjected to axial force acting through the centroid of the cross-sectioned area, then the material within the bar is subjected *only to normal stress*. This stress is assumed to be uniform or *averaged* over the cross-sectional area.

## PROCEDURE FOR ANALYSIS

The equation $\sigma = P/A$ gives the *average* normal stress on the cross-sectional area of a member when the section is subjected to an internal resultant normal force **P**. For axially loaded members, application of this equation requires the following steps.

*Internal Loading.*

- Section the member *perpendicular* to its longitudinal axis at the point where the normal stress is to be determined and use the necessary free-body diagram and equation of force equilibrium to obtain the internal axial force **P** at the section.

*Average Normal Stress.*

- Determine the member's cross-sectional area at the section and compute the average normal stress $\sigma = P/A$.

- It is suggested that $\sigma$ be shown acting on a small volume element of the material located at a point on the section where stress is calculated. To do this, first draw $\sigma$ on the face of the element coincident with the sectioned area A. Here $\sigma$ acts in the *same direction* as the internal force **P** since all the normal stresses on the cross section act in this direction to develop this resultant. The normal stress $\sigma$ acting on the opposite face of the element can be drawn in its appropriate direction.

E X A M P L E  8.1

The bar in Fig. 8–9a has a constant width of 35 mm and a thickness of 10 mm. Determine the maximum average normal stress in the bar when it is subjected to the loading shown.

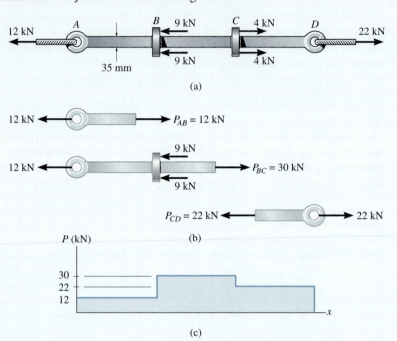

(a)

(b)

$P$ (kN)

(c)

## Solution

*Internal Loading.*   By inspection, the internal axial forces in regions $AB$, $BC$, and $CD$ are all constant yet have different magnitudes. Using the method of sections, these loadings are determined in Fig. 8–9b; and the normal force diagram which represents these results graphically is shown in Fig. 8–9c. By inspection, the largest loading is in region $BC$, where $P_{BC} = 30$ kN. Since the cross-sectional area of the bar is *constant*, the largest average normal stress also occurs within this region of the bar.

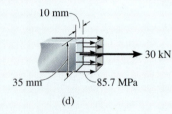

(d)

**Fig. 8–9**

*Average Normal Stress.*   Applying Eq. 8–3, we have

$$\sigma_{BC} = \frac{P_{BC}}{A} = \frac{30(10^3)\text{ N}}{(0.035\text{ m})(0.010\text{ m})} = 85.7\text{ MPa} \qquad \textit{Ans.}$$

The stress distribution acting on an arbitrary cross section of the bar within region $BC$ is shown in Fig. 8–9d. Graphically the *volume* (or "block") represented by this distribution of stress is equivalent to the load of 30 kN; that is, 30 kN = (85.7 MPa)(35 mm)(10 mm).

EXAMPLE 8.2

The 80-kg lamp is supported by two rods $AB$ and $BC$ as shown in Fig. 8–10a. If $AB$ has a diameter of 10 mm and $BC$ has a diameter of 8 mm, determine the average normal stress in each rod.

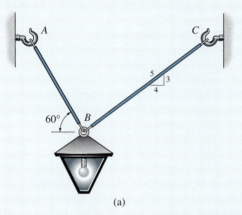

(a)

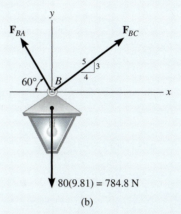

(b)

**Fig. 8–10**

## Solution

*Internal Loading.* We must first determine the axial force in each rod. A free-body diagram of the lamp is shown in Fig. 8–10b. Applying the equations of force equilibrium yields

$$\xrightarrow{+} \Sigma F_x = 0; \qquad\qquad F_{BC}(\tfrac{4}{5}) - F_{BA} \cos 60° = 0$$

$$+\uparrow \Sigma F_y = 0; \qquad F_{BC}(\tfrac{3}{5}) + F_{BA} \sin 60° - 784.8 \text{ N} = 0$$

$$F_{BC} = 395.2 \text{ N}, \qquad F_{BA} = 632.4 \text{ N}$$

By Newton's third law of action, equal but opposite reaction, these forces subject the rods to tension throughout their length.

*Average Normal Stress.* Applying Eq. 8–3, we have

$$\sigma_{BC} = \frac{F_{BC}}{A_{BC}} = \frac{395.2 \text{ N}}{\pi(0.004 \text{ m})^2} = 7.86 \text{ MPa} \qquad \textit{Ans.}$$

$$\sigma_{BA} = \frac{F_{BA}}{A_{BA}} = \frac{632.4 \text{ N}}{\pi(0.005 \text{ m})^2} = 8.05 \text{ MPa} \qquad \textit{Ans.}$$

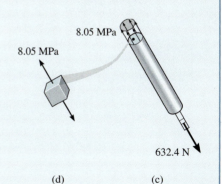

(d)                    (c)

The average normal stress distribution acting over a cross section of rod $AB$ is shown in Fig. 8–10c, and at a point on this cross section, an element of material is stressed as shown in Fig. 8–10d.

E X A M P L E   **8.3**

The casting shown in Fig. 8–11a is made of steel having a specific weight of $\gamma_{st} = 490$ lb/ft³. Determine the average compressive stress acting at points $A$ and $B$.

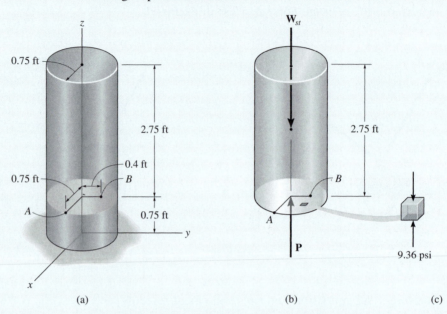

Fig. 8–11    (a)    (b)    (c)

### Solution

*Internal Loading.* A free-body diagram of the top segment of the casting where the section passes through points $A$ and $B$ is shown in Fig. 8–11b. The weight of this segment is determined from $W_{st} = \gamma_{st}V_{st}$. Thus the internal axial force $P$ at the section is

$$+\uparrow \Sigma F_z = 0; \qquad P - W_{st} = 0$$
$$P - (490 \text{ lb/ft}^3)(2.75 \text{ ft})\pi(0.75 \text{ ft})^2 = 0$$
$$P = 2381 \text{ lb}$$

*Average Compressive Stress.* The cross-sectional area at the section is $A = \pi(0.75 \text{ ft})^2$, and so the average compressive stress becomes

$$\sigma = \frac{P}{A} = \frac{2381 \text{ lb}}{\pi(0.75 \text{ ft})^2}$$
$$= 1347.5 \text{ lb/ft}^2 = 1347.5 \text{ lb/ft}^2 \,(1 \text{ ft}^2/144 \text{ in}^2)$$
$$= 9.36 \text{ psi} \qquad\qquad\qquad Ans.$$

The stress shown on the volume element of material in Fig. 8–11c is representative of the conditions at either point $A$ or $B$. Notice that this stress acts *upward* on the bottom or shaded face of the element since this face forms part of the bottom surface area of the cut section, and on this surface, the resultant internal force **P** is pushing upward.

E X A M P L E  **8.4**

Member $AC$ shown in Fig. 8–12$a$ is subjected to a vertical force of 3 kN. Determine the position $x$ of this force so that the average compressive stress at the smooth support $C$ is equal to the average tensile stress in the tie rod $AB$. The rod has a cross-sectional area of 400 mm$^2$ and the contact area at $C$ is 650 mm$^2$.

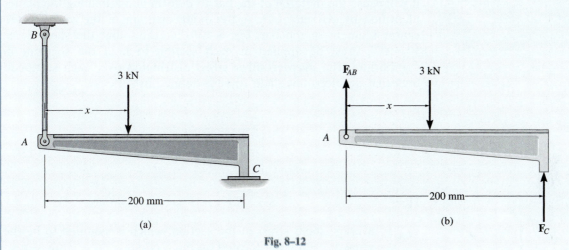

(a)                                                            (b)

**Fig. 8–12**

**Solution**

*Internal Loading.*    The forces at $A$ and $C$ can be related by considering the free-body diagram for member $AC$, Fig. 8–12$b$. There are three unknowns, namely, $F_{AB}$, $F_C$, and $x$. To solve this problem we will work in units of newtons and millimeters.

$$+\uparrow \Sigma F_y = 0; \qquad\qquad F_{AB} + F_C - 3000 \text{ N} = 0 \qquad\qquad (1)$$
$$\zeta + \Sigma M_A = 0; \qquad -3000 \text{ N}(x) + F_C(200 \text{ mm}) = 0 \qquad\qquad (2)$$

*Average Normal Stress.*    A necessary third equation can be written that requires the tensile stress in the bar $AB$ and the compressive stress at $C$ to be equivalent, i.e.,

$$\sigma = \frac{F_{AB}}{400 \text{ mm}^2} = \frac{F_C}{650 \text{ mm}^2}$$
$$F_C = 1.625 F_{AB}$$

Substituting this into Eq. 1, solving for $F_{AB}$, then solving for $F_C$, we obtain

$$F_{AB} = 1143 \text{ N}$$
$$F_C = 1857 \text{ N}$$

The position of the applied load is determined from Eq. 2,

$$x = 124 \text{ mm} \qquad\qquad\qquad Ans.$$

Note that $0 < x < 200$ mm, as required.

## 8.4 Average Shear Stress

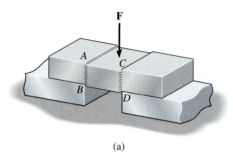

(a)

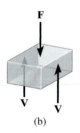

(b)

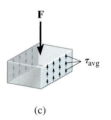

(c)

**Fig. 8–13**

Shear stress has been defined in Sec. 8–2 as the stress component that acts *in the plane* of the sectioned area. In order to show how this stress can develop, we will consider the effect of applying a force **F** to the bar in Fig. 8–13*a*. If the supports are considered rigid, and **F** is large enough, it will cause the material of the bar to deform and fail along the planes identified by *AB* and *CD*. A free-body diagram of the unsupported center segment of the bar, Fig. 8–13*b*, indicates that the shear force $V = F/2$ must be applied at each section to hold the segment in equilibrium. The ***average shear stress*** distributed over each sectioned area that develops this shear force is defined by

$$\tau_{avg} = \frac{V}{A} \tag{8–4}$$

Here

$\tau_{avg}$ = average shear stress at the section, which is assumed to be the same at each point located on the section

$V$ = internal resultant shear force at the section determined from the equations of equilibrium

$A$ = area at the section

The distribution of average shear stress is shown acting over the sections in Fig. 8–13*c*. Notice that $\tau_{avg}$ is in the *same direction* as **V**, since the shear stress must create associated forces all of which contribute to the internal resultant force **V** at the section.

The loading case discussed in Fig. 8–13 is an example of ***simple or direct shear***, since the shear is caused by the *direct action* of the applied load **F**. This type of shear often occurs in various types of simple connections that use bolts, pins, welding material, etc. In all these cases, however, application of Eq. 8–4 is *only approximate*. A more precise investigation of the shear-stress distribution over the critical section often reveals that much larger shear stresses occur in the material than those predicted by this equation. Although this may be the case, application of Eq. 8–4 is generally acceptable for many problems in engineering design and analysis. For example, engineering codes allow its use when considering design sizes for fasteners such as bolts and for obtaining the bonding strength of joints subjected to shear loadings. In this regard, two types of shear frequently occur in practice, which deserve separate treatment.

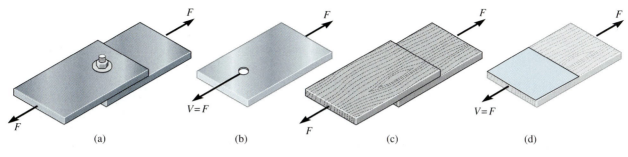

(a)  (b)  (c)  (d)

**Fig. 8–14**

**Single Shear.** The steel and wood joints shown in Fig. 8–14a and 8–14c, respectively, are examples of ***single-shear connections*** and are often referred to as *lap joints*. Here we will assume that the members are thin and that the nut in Fig. 8–14a is not tightened to any great extent so friction between the members can be neglected. Passing a section between the members yields the free-body diagrams shown in Fig. 8–14b and 8–14d. Since the members are thin, we can neglect the moment created by the force $F$. Hence for equilibrium the cross-sectional area of the bolt in Fig. 8–14d and the bonding surface between the members in Fig. 8–14d are subjected only to a *single shear force* $V = F$. This force is used in Eq. 8–4 to determine the average shear stress acting on the colored section of Fig. 8–14d.

**Double Shear.** When the joint is constructed as shown in Fig. 8–15a or 8–15c, two shear surfaces must be considered. These types of connections are often called *double lap joints*. If we pass a section between each of the members, the free-body diagrams of the center member are shown in Fig. 8–15b and 8–15d. Here we have a condition of **double shear**. Consequently, $V = F/2$ acts on *each* sectioned area and this shear must be considered when applying $\tau_{avg} = V/A$.

The pin on this tractor is subjected to double shear.

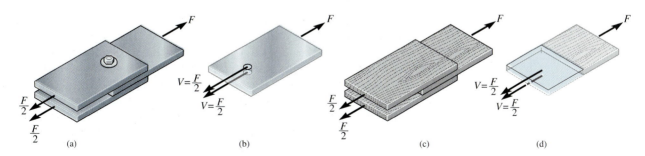

(a)  (b)  (c)  (d)

**Fig. 8–15**

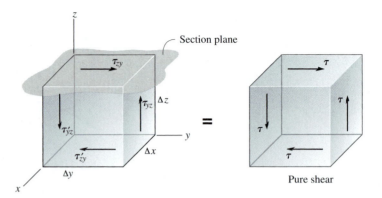

**Fig. 8–16**

**Equilibrium.**  Consider a volume element of material taken at a point located on the surface of any sectioned area on which the average shear stress acts, Fig. 8–16a. If we consider force equilibrium in the y direction, then

$$\Sigma F_y = 0; \qquad \tau_{zy}(\Delta x \Delta y) - \tau'_{zy}\Delta x \Delta y = 0$$

$$\tau_{zy} = \tau'_{zy}$$

And in a similar manner, force equilibrium in the $z$ direction yields $\tau_{yz} = \tau'_{yz}$. Finally, taking moments about the $x$ axis,

$$\Sigma M_x = 0; \qquad -\tau_{zy}(\Delta x \Delta y)\Delta z + \tau_{zy}(\Delta x \Delta z)\Delta y = 0$$

$$\tau_{zy} = \tau_{yz}$$

so that

$$\tau_{zy} = \tau'_{zy} = \tau_{yz} = \tau'_{yz} = \tau$$

In other words, force and moment equilibrium requires the shear stress acting on the top face of the element, to be accompanied by shear stress acting on three other faces, Fig. 8–16b. Here *all four shear stresses must have equal magnitude and be directed either toward or away from each other at opposite edges of the element.* This is referred to as the *complementary property of shear,* and under the conditions shown in Fig. 8–16, the material is subjected to *pure shear.*

Although we have considered here a case of simple shear as caused by the *direct* action of a load, in later chapters we will show that shear stress can also arise *indirectly* due to the action of other types of loading.

## IMPORTANT POINTS

- If two parts which are *thin or small* are joined together, the applied loads can cause shearing of the material with negligible bending. If this is the case, it is generally suitable for engineering analysis to assume that an *average shear stress* acts over the cross-sectional area.

- Oftentimes fasteners, such as nails and bolts, are subjected to shear loads. The magnitude of a shear force on the fastener is greatest along a plane which passes through the surfaces being joined. A carefully drawn free-body diagram of a segment of the fastener will enable one to obtain the magnitude and direction of this force.

## PROCEDURE FOR ANALYSIS

The equation $\tau_{avg} = V/A$ is used to compute only the *average shear stress* in the material. Application requires the following steps.

*Internal Shear.*

- Section the member at the point where the average shear stress is to be determined.

- Draw the necessary free-body diagram, and calculate the internal shear force **V** acting at the section that is necessary to hold the part in equilibrium.

*Average Shear Stress.*

- Determine the sectioned area $A$, and compute the average shear stress $\tau_{avg} = V/A$.

- It is suggested that $\tau_{avg}$ be shown on a small volume element of material located at a point on the section where it is determined. To do this, first draw $\tau_{avg}$ on the face of the element, coincident with the sectioned area $A$. This shear stress acts in the same direction as **V**. The shear stresses acting on the three adjacent planes can then be drawn in their appropriate directions following the scheme shown in Fig. 8–16.

E X A M P L E   8.5

The bar shown in Fig. 8–17*a* has a square cross section for which the depth and thickness are 40 mm. If an axial force of 800 N is applied along the centroidal axis of the bar's cross-sectional area, determine the average normal stress and average shear stress acting on the material along (a) section plane *a–a* and (b) section plane *b–b*.

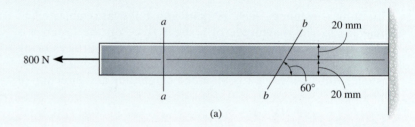

(a)

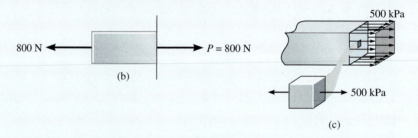

(b)

(c)

Fig. 8–17

**Solution**

*Part (a)*

*Internal Loading.* The bar is sectioned, Fig. 8–17b, and the internal resultant loading consists only of an axial force for which $P = 800$ N.

*Average Stress.* The average normal stress is determined from Eq.1–6.

$$\sigma = \frac{P}{A} = \frac{800 \text{ N}}{(0.04 \text{ m})(0.04 \text{ m})} = 500 \text{ kPa} \qquad Ans.$$

No shear stress exists on the section, since the shear force at the section is zero.

$$\tau_{avg} = 0 \qquad Ans.$$

The distribution of average normal stress over the cross section is shown in Fig. 8–17c.

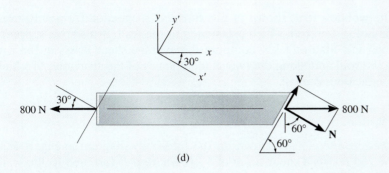

(d)

## Part (b)

*Internal Loading.*   If the bar is sectioned along *b–b*, the free-body diagram of the left segment is shown in Fig. 8–17*d*. Here both a normal force (**N**) and shear force (**V**) act on the sectioned area. Using *x, y* axes, we require

$$\xrightarrow{+}\Sigma F_x = 0; \qquad -800 \text{ N} + N \sin 60° + V \cos 60° = 0$$
$$+\uparrow\Sigma F_y = 0; \qquad V \sin 60° - N \cos 60° = 0$$

or, more directly, using *x′, y′* axes,

$$+\searrow\Sigma F_{x'} = 0; \qquad N - 800 \text{ N} \cos 30° = 0$$
$$+\nearrow\Sigma F_{y'} = 0; \qquad V - 800 \text{ N} \sin 30° = 0$$

Solving either set of equations,

$$N = 692.8 \text{ N}$$
$$V = 400 \text{ N}$$

*Average Stresses.*   In this case the sectioned area has a thickness and depth of 40 mm and 40 mm/sin 60° = 46.19 mm, respectively, Fig. 8–17*a*. Thus the average normal stress is

$$\sigma = \frac{N}{A} = \frac{692.8 \text{ N}}{(0.04 \text{ m})(0.04619 \text{ m})} = 375 \text{ kPa} \qquad Ans.$$

and the average shear stress is

$$\tau_{avg} = \frac{V}{A} = \frac{400 \text{ N}}{(0.04 \text{m})(0.04619 \text{ m})} = 217 \text{ kPa} \qquad Ans.$$

The stress distribution is shown in Fig. 8–17*e*.

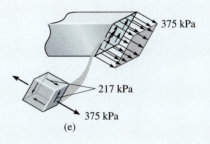

(e)

EXAMPLE 8.6

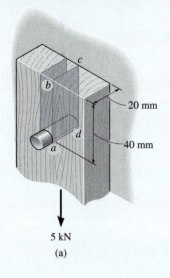

5 kN

(a)

The wooden strut shown in Fig. 8–18*a* is suspended from a 10-mm-diameter steel rod, which is fastened to the wall. If the strut supports a vertical load of 5 kN, compute the average shear stress in the rod at the wall and along the two shaded planes of the strut, one of which is indicated as *abcd*.

### Solution

*Internal Shear.* As shown on the free-body diagram in Fig. 8–18*b*, the rod resists a shear force of 5 kN where it is fastened to the wall. A free-body diagram of the sectioned segment of the strut that is in contact with the rod is shown in Fig. 8–18*c*. Here the shear force acting along each shaded plane is 2.5 kN.

*Average Shear Stress.* For the rod,

$$\tau_{avg} = \frac{V}{A} = \frac{5000 \text{ N}}{\pi(0.005 \text{ m})^2} = 63.7 \text{ MPa} \qquad Ans.$$

For the strut,

$$\tau_{avg} = \frac{V}{A} = \frac{2500 \text{ N}}{(0.04 \text{ m})(0.02 \text{ m})} = 3.12 \text{ MPa} \qquad Ans.$$

The average-shear-stress distribution on the sectioned rod and strut segment is shown in Fig. 8–18*d*, and 8–18*e*, respectively. Also shown with these figures is a typical volume element of the material taken at a point located on the surface of each section. Note carefully how the shear stress must act on each shaded face of these elements and then on the adjacent faces of the elements.

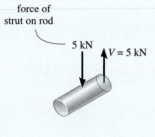

force of strut on rod

5 kN

V = 5 kN

(b)

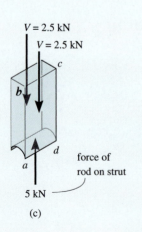

V = 2.5 kN
V = 2.5 kN

c

b

a    d

force of rod on strut

5 kN

(c)

5 kN

63.7 MPa

(d)

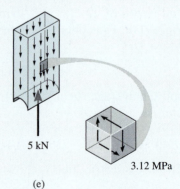

5 kN

3.12 MPa

(e)

**Fig. 8–18**

# EXAMPLE 8.7

The inclined member in Fig. 8–19a is subjected to a compressive force of 600 lb. Determine the average compressive stress along the smooth areas of contact defined by $AB$ and $BC$, and the average shear stress along the horizontal plane defined by $EDB$.

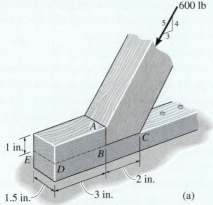

(a)

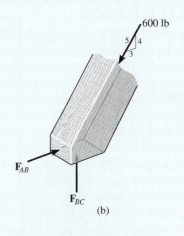

(b)

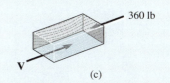

(c)

## Solution

***Internal Loadings.*** The free-body diagram of the inclined member is shown in Fig. 8–19b. The compressive forces acting on the areas of contact are

$$\xrightarrow{+} \Sigma F_x = 0; \qquad F_{AB} - 600 \text{ lb}(\tfrac{3}{5}) = 0 \qquad F_{AB} = 360 \text{ lb}$$

$$+\uparrow \Sigma F_y = 0; \qquad F_{BC} - 600 \text{ lb}(\tfrac{4}{5}) = 0 \qquad F_{BC} = 480 \text{ lb}$$

Also, from the free-body diagram of the top segment of the bottom member, Fig. 8–19c, the shear force acting on the sectioned horizontal plane $EDB$ is

$$\xrightarrow{+} \Sigma F_x = 0; \qquad V = 360 \text{ lb}$$

***Average Stress.*** The average compressive stresses along the horizontal and vertical planes of the inclined member are

$$\sigma_{AB} = \frac{360 \text{ lb}}{(1 \text{ in.})(1.5 \text{ in.})} = 240 \text{ psi} \qquad \textit{Ans.}$$

$$\sigma_{BC} = \frac{480 \text{ lb}}{(2 \text{ in.})(1.5 \text{ in.})} = 160 \text{ psi} \qquad \textit{Ans.}$$

These stress distributions are shown in Fig. 8–19d.

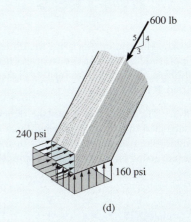

(d)

The average shear stress acting on the horizontal plane defined by $EDB$ is

$$\tau_{\text{avg}} = \frac{360 \text{ lb}}{(3 \text{ in.})(1.5 \text{ in.})} = 80.0 \text{ psi} \qquad \textit{Ans.}$$

This stress is shown distributed over the sectioned area in Fig. 8–19e.

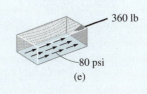

(e)

**Fig. 8–19**

# PROBLEMS

**8-1.** The column is subjected to an axial force of 8 kN at its top. If the cross-sectional area has the dimensions shown in the figure, determine the average normal stress acting at section *a–a*. Show this distribution of stress acting over the area's cross section.

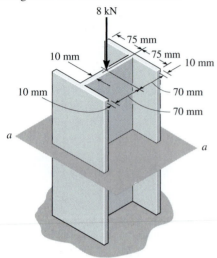

**Prob. 8–1**

**8-2.** The cinder block has the dimensions shown. If the material fails when the average normal stress reaches 120 psi, determine the largest centrally applied vertical load **P** it can support.

**8-3.** The cinder block has the dimensions shown. If it is subjected to a centrally applied force of **P** = 800 lb, determine the average normal stress in the material. Show the result acting on a differential volume element of the material.

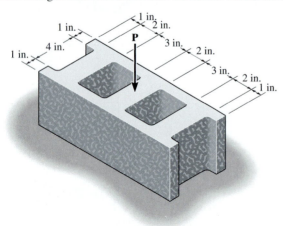

**Probs. 8–2/3**

*8-4.** The 50-lb lamp is supported by two steel rods connected by a ring at *A*. Determine which rod is subjected to the greater average normal stress and compute its value. Take $\theta = 60°$. The diameter of each rod is given in the figure.

**8-5.** Solve Prob. 8–4 for $\theta = 45°$.

**8-6.** The 50-lb lamp is supported by two steel rods connected by a ring at *A*. Determine the angle of orientation $\theta$ of *AC* such that the average normal stress in rod *AC* is twice the average normal stress in rod *AB*. What is the magnitude of this stress in each rod? The diameter of each rod is given in the figure.

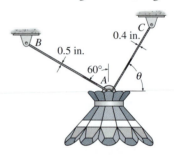

**Probs. 8–4/5/6**

**8-7.** The thrust bearing is subjected to the loads shown. Determine the average normal stress developed on cross sections through points *B*, *C*, and *D*. Sketch the results on a differential volume element located at each section.

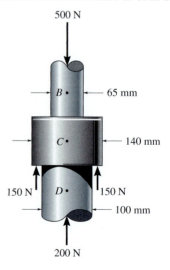

**Prob. 8–7**

**\*8–8.** The shaft is subjected to the axial force of 30 kN. If the shaft passes through the 53-mm diameter hole in the fixed support $A$, determine the bearing stress acting on the collar $C$. Also, what is the average shear stress acting along the inside surface of the collar where it is fixed connected to the 52-mm diameter shaft?

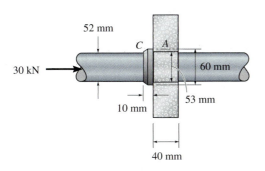

52 mm
$C$ $A$
30 kN
60 mm
53 mm
10 mm
40 mm

**Prob. 8–8**

**8–9.** The small block has a thickness of 5 mm. If the stress distribution at the support developed by the load varies as shown, determine the force $\mathbf{F}$ applied to the block, and the distance $d$ to where it is applied.

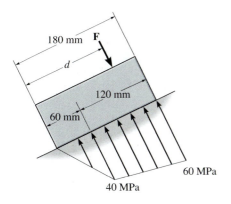

180 mm $\mathbf{F}$
$d$
120 mm
60 mm
60 MPa
40 MPa

**Prob. 8–9**

**8–10.** The supporting wheel on a scaffold is held in place on the leg using a 4-mm-diameter pin as shown. If the wheel is subjected to a normal force of 3 kN, determine the average shear stress developed in the pin. Neglect friction between the inner scaffold puller leg and the tube used on the wheel.

3 kN

**Prob. 8–10**

**8–11.** The pins on the frame at $B$ and $C$ each have a diameter of 0.25 in. If these pins are subjected to *double shear*, determine the average shear stress in each pin.

**\*8–12.** Solve Prob. 8–11 assuming that pins $B$ and $C$ are subjected to *single shear*.

**8–13.** The pins on the frame at $D$ and $E$ each have a diameter of 0.25 in. If these pins are subjected to *double shear*, determine the average shear stress in each pin.

**8–14.** Solve Prob. 8–13 assuming that pins $D$ and $E$ are subjected to *single shear*.

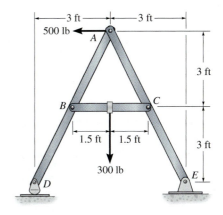

3 ft — 3 ft
500 lb
$A$
3 ft
$B$ $C$
1.5 ft 1.5 ft
3 ft
300 lb
$D$
$E$

**Probs. 8–11/12/13/14**

**8-15.** The pedestal has a triangular cross section as shown. If it is subjected to a compressive force of 500 lb, specify the $x$ and $y$ coordinates for the location of point $P(x, y)$, where the load must be applied on the cross section, so that the average normal stress is uniform. Compute the stress and sketch its distribution acting on the cross section at a location removed from the point of load application.

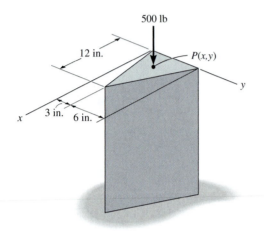

Prob. 8–15

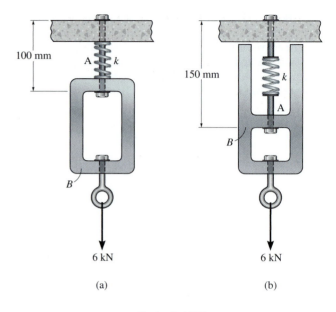

(a)

(b)

Probs. 8–16/17

**8-18.** The yoke is subjected to the force and couple moment. Determine the average shear stress in the bolt acting on the cross sections through $A$ and $B$. The bolt has a diameter of 0.25 in. *Hint:* The couple moment is resisted by a set of couple forces developed in the shank of the bolt.

***8-16.** Two designs for a shock support are shown in the figure. The spring has a stiffness of $k = 15$ kN/m and in (a) it is uncompressed, whereas in (b) it is originally stretched 0.2 m. Determine the average normal stress in the 5-mm-diameter bolt shank at $A$ when the 6-kN load is applied. In (b) the bracket $B$ is not connected to the support.

**8-17.** Two designs for a shock support are shown in the figure. The spring has a stiffness of $k = 15$ kN/m and in (a) it is uncompressed. Determine the maximum amount which the spring in (b) must be originally stretched so that the average normal stress in the 5-mm bolt shank at $A$ is equivalent for both designs when the load of 6 kN is applied. In (b) the bracket $B$ is not connected to the support. What is the bolt stress in both cases?

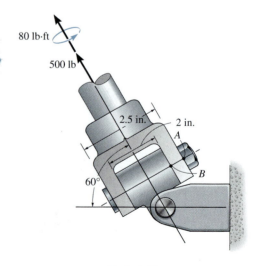

Prob. 8–18

**8-19.** The anchor bolt was pulled out of the concrete wall and the failure surface formed part of a frustum and cylinder. This indicates a shear failure occurred along the cylinder *BC* and tension failure along the frustum *AB*. If the shear and normal stresses along these surfaces have the magnitudes shown, determine the force **P** that must have been applied to the bolt.

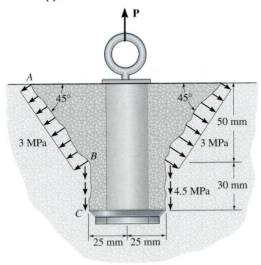

Prob. 8-19

*8-20.** The crimping tool is used to crimp the end of the wire *E*. If a force of 20 lb is applied to the handles, determine the average shear stress in the pin at *A*. The pin is subjected to double shear and has a diameter of 0.2 in. Only a vertical force is exerted on the wire.

**8-21.** Solve Prob. 8–20 for pin *B*. The pin is subjected to double shear and has a diameter of 0.2 in.

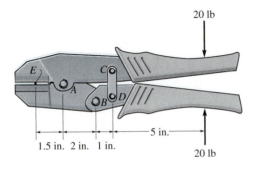

Probs. 8-20/21

**8-22.** The joint is subjected to the axial member force of 6 kip. Determine the average normal stress acting on sections *AB* and *BC*. Assume the member is smooth and is 1.5 in. thick.

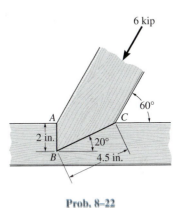

Prob. 8-22

**8-23.** The driver of the sports car applies the rear brakes and causes the tires to slip. If the normal force on each rear tire is 400 lb and the coefficient of kinetic friction between the tires and the pavement is $\mu_k = 0.5$, determine the average shear stress developed by the friction force on the tires. Assume the rubber of the tires is flexible and each tire is filled with an air pressure of 32 psi.

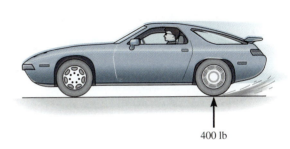

Prob. 8-23

**\*8-24.** The bar has a cross-sectional area $A$ and is subjected to the axial load $P$. Determine the average normal and average shear stresses acting over the shaded section, which is oriented at $\theta$ from the horizontal. Plot the variation of these stresses as a function of $\theta$ ($0 \le \theta \le 90°$).

**8-26.** The bars of the truss each have a cross-sectional area of 1.25 in². Determine the average normal stress in each member due to the loading $P = 8$ kip. State whether the stress is tensile or compressive.

**8-27.** The bars of the truss each have a cross-sectional area of 1.25 in². If the maximum average normal stress in any bar is not to exceed 20 ksi, determine the maximum magnitude $P$ of the loads that can be applied to the truss.

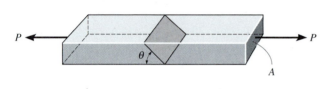

Prob. 8–24

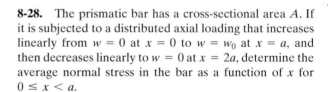

Probs. 8–26/27

**8-25.** The jaw clutch is used to transmit a torque of 450 lb · ft in only one direction. If each shaft has two teeth placed around its circumference as shown, determine the average shear stress along the root $AB$ of each of the teeth.

**8-28.** The prismatic bar has a cross-sectional area $A$. If it is subjected to a distributed axial loading that increases linearly from $w = 0$ at $x = 0$ to $w = w_0$ at $x = a$, and then decreases linearly to $w = 0$ at $x = 2a$, determine the average normal stress in the bar as a function of $x$ for $0 \le x < a$.

**8-29.** The prismatic bar has a cross-sectional area $A$. If it is subjected to a distributed axial loading that increases linearly from $w = 0$ at $x = 0$ to $w = w_0$ at $x = a$, and then decreases linearly to $w = 0$ at $x = 2a$, determine the average normal stress in the bar as a function of $x$ for $a < x \le 2a$.

Prob. 8–25

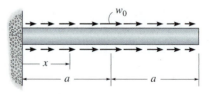

Probs. 8–28/29

**8-30.** The bar is subjected to a uniform distributed axial loading of 10 kN/m and a concentrated force of 1.5 kN at its midpoint as shown. Determine the maximum average normal stress in the bar and its location $x$.

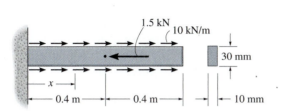

**Prob. 8–30**

**8-31.** The bar has a cross-sectional area of $400(10^{-6})$ m$^2$. If it is subjected to a uniform axial distributed loading along its length and to two concentrated loads as shown, determine the average normal stress in the bar as a function of $x$ for $0 < x \le 0.5$ m.

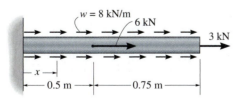

**Prob. 8–31**

*8-32.** The tapered rod has a radius of $r = (2 - x/6)$ in. and is subjected to the distributed loading of $w = (60 + 40x)$ lb/in. Determine the average normal stress at the center of the rod.

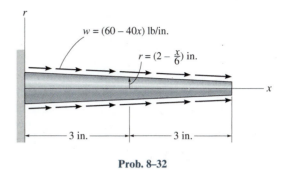

**Prob. 8–32**

**8-33.** Rods $AB$ and $BC$ have diameters of 25 mm and 18 mm, respectively. If a load of 6 kN is applied to the ring at $B$, determine the average normal stress in each rod if $\theta = 60°$.

**8-34.** Rods $AB$ and $BC$ each have a diameter of 4 mm. If a load of 6 kN is applied to the ring at $B$, determine the smallest angle $\theta$ of rod $BC$ so that the average normal stress in each rod is equivalent.

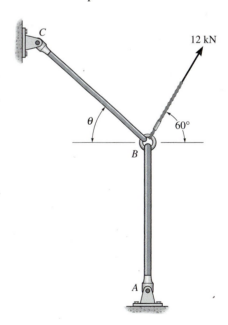

**Probs. 8–33/34**

## 8.5 Allowable Stress

Appropriate factors of safety must be considered when designing cranes and cables used to transfer heavy loads.

An engineer in charge of the *design* of a structural member or mechanical element must restrict the stress in the material to a level that will be safe. Furthermore, a structure or machine that is currently in use may, on occasion, have to be *analyzed* to see what additional loadings its members or parts can support. So again it becomes necessary to perform the calculations using a safe or allowable stress.

To ensure safety, it is necessary to choose an allowable stress that restricts the applied load to one that is *less* than the load the member can fully support. There are several reasons for this. For example, the load for which the member is designed may be different from actual loadings placed on it. The intended measurements of a structure or machine may not be exact due to errors in fabrication or in the assembly of its component parts. Unknown vibrations, impact, or accidental loadings can occur that may not be accounted for in the design. Atmospheric corrosion, decay, or weathering tend to cause materials to deteriorate during service. And lastly, some materials, such as wood, concrete, or fiber-reinforced composites, can show high variability in mechanical properties.

One method of specifying the allowable load for the design or analysis of a member is to use a number called the factor of safety. The *factor of safety* (F.S.) is a ratio of the failure load $F_{fail}$ divided by the allowable load, $F_{allow}$. Here $F_{fail}$ is found from experimental testing of the material, and the factor of safety is selected based on experience so that the above mentioned uncertainties are accounted for when the member is used under similar conditions of loading and geometry. Stated mathematically,

$$\text{F.S.} = \frac{F_{fail}}{F_{allow}} \qquad (8\text{--}5)$$

If the load applied to the member is *linearly related* to the stress developed within the member, as in the case of using $\sigma = P/A$ and $\tau_{avg} = V/A$, then we can express the factor of safety as a ratio of the failure stress $\sigma_{fail}$ (or $\tau_{fail}$) to the allowable stress $\sigma_{allow}$ (or $\tau_{allow}$);* that is,

$$\text{F.S.} = \frac{\sigma_{fail}}{\sigma_{allow}} \qquad (8\text{--}6)$$

or

$$\text{F.S.} = \frac{\tau_{fail}}{\tau_{allow}} \qquad (8\text{--}7)$$

*In some cases, such as columns, the applied load is *not* linearly related to stress and therefore only Eq. 8–5 can be used to determine the factor of safety. See Chapter 17.

In any of these equations, the factor of safety is chosen to be *greater* than 1 in order to avoid the potential for failure. Specific values depend on the types of materials to be used and the intended purpose of the structure or machine. For example, the F.S. used in the design of aircraft or space-vehicle components may be close to 1 in order to reduce the weight of the vehicle. On the other hand, in the case of a nuclear power plant, the factor of safety for some of its components may be as high as 3 since there may be uncertainties in loading or material behavior. In general, however, factors of safety and therefore the allowable loads or stresses for both structural and mechanical elements have become well standardized, since their design uncertainties have been reasonably evaluated. Their values, which can be found in design codes and engineering handbooks, are intended to form a balance of insuring public and environmental safety and providing a reasonable economic solution to design.

## 8.6   Design of Simple Connections

By making simplifying assumptions regarding the behavior of the material, the equations $\sigma = P/A$ and $\tau_{avg} = V/A$ can often be used to analyze or design a simple connection or a mechanical element. In particular, if a member is subjected to a *normal force* at a section, its required area at the section is determined from

$$A = \frac{P}{\sigma_{allow}} \qquad (8\text{--}8)$$

On the other hand, if the section is subjected to a *shear force*, then the required area at the section is

$$A = \frac{V}{\tau_{allow}} \qquad (8\text{--}9)$$

As discussed in Sec. 8.5, the allowable stress used in each of these equations is determined either by applying a factor of safety to a specified normal or shear stress or by finding these stresses directly from an appropriate design code.

We will now discuss four common types of problems for which the above equations can be used for design.

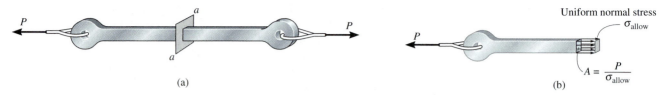

Fig. 8–20

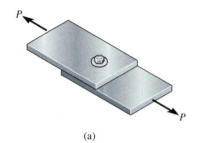

(a)

**Cross-Sectional Area of a Tension Member.** The cross-sectional area of a prismatic member subjected to a tension force can be determined *provided* the force has a line of action that passes through the centroid of the cross section. For example, consider the "eye bar" shown in Fig. 8–20a. At the intermediate section *a–a* the stress distribution is uniform over the cross section and the required area $A$ is determined, as shown in Fig. 8–20b.

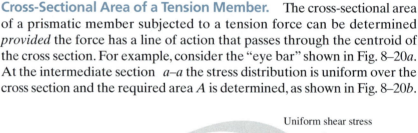

Fig. 8–21

**Cross-Sectional Area of a Connector Subjected to Shear.** Often bolts or pins are used to connect plates, boards, or several members together. For example, consider the lap joint shown in Fig. 8–21a. If the bolt is loose or the clamping force of the bolt is unknown, it is safe to assume that any frictional force *between* the plates is negligible. As a result, the free-body diagram for a section passing *between* the plates and through the bolt is shown in Fig. 8–21b. The bolt is subjected to a resultant internal shear force of $V = P$ at this cross section. Assuming that the shear stress causing this force is *uniformly distributed* over the cross section, the required $A$ is determined as shown in Fig. 8–21c.

**Required Area to Resist Bearing.** A normal stress that is produced by the compression of one surface against another is called a *bearing stress*. If this stress becomes large enough, it may crush or locally deform one or both of the surfaces. Hence, in order to prevent failure it is necessary to determine the proper bearing area for the material using an allowable bearing stress. For example, the area $A$ of the column base plate $B$ shown in Fig. 8–22 is determined from the allowable bearing stress of the concrete using $A = P/(\sigma_b)_{allow}$. This assumes, of course, that the allowable bearing stress for the concrete is smaller than that of the base plate material, and furthermore the bearing stress is uniformly distributed between the plate and the concrete as shown in the figure.

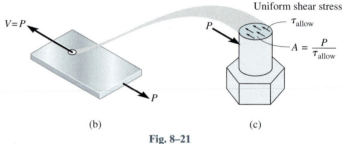

Fig. 8–22

**Required Area to Resist Shear Caused by Axial Load.** Occasionally rods or other members will be supported in such a way that shear stress can be developed in the member even though the member may be subjected to an axial load. An example of this situation would be a steel rod whose end is encased in concrete and loaded as shown in Fig. 8–23a. A free-body diagram of the rod, Fig. 8–23b, shows that *shear stress* acts over the area of contact of the rod with the concrete. This area is $(\pi d)l$, where $d$ is the rod's diameter and $l$ is the length of embedment. Although the actual shear-stress distribution along the rod would be difficult to determine, if we assume it is *uniform*, we can use $A = V/\tau_{allow}$ to calculate $l$, provided we know $d$ and $\tau_{allow}$, Fig. 8–23b.

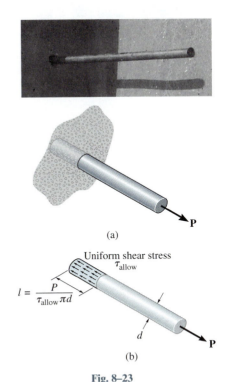

(a)

Uniform shear stress $\tau_{allow}$

$l = \dfrac{P}{\tau_{allow}\,\pi d}$

$d$

(b)

**Fig. 8–23**

## IMPORTANT POINTS

- Design of a member for strength is based on selecting an allowable stress that will enable it to safely support its intended load. There are many unknown factors that can influence the actual stress in a member and so, depending upon the intended uses of the member, a *factor of safety* is applied to obtain the allowable load the member can support.

- The four cases illustrated in this section represent just a few of the many applications of the average normal and shear stress formulas used for engineering design and analysis. Whenever these equations are applied, however, it is important to be aware that the stress distribution is assumed to be *uniformly distributed* or "averaged" over the section.

## PROCEDURE FOR ANALYSIS

When solving problems using the average normal and shear stress equations, a careful consideration should first be made as to the section over which the critical stress is acting. Once this section is made, the member must then be designed to have a sufficient area at the section to resist the stress that acts on it. To determine this area, application requires the following steps.

*Internal Loading.*

- Section the member through the area and draw a free-body diagram of a segment of the member. The internal resultant force at the section is then determined using the equations of equilibrium.

*Required Area.*

- Provided the allowable stress is known or can be determined, the required area needed to sustain the load at the section is then computed from $A = P/\sigma_{allow}$ or $A = V/\tau_{allow}$.

**E X A M P L E 8.8**

The two members are pinned together at $B$ as shown in Fig. 8–24a. Top views of the pin connections at $A$ and $B$ are also given in the figure. If the pins have an allowable shear stress of $\tau_{\text{allow}} = 12.5$ ksi and the allowable tensile stress of rod $CB$ is $(\sigma_t)_{\text{allow}} = 16.2$ ksi, determine to the nearest $\frac{1}{16}$ in. the smallest diameter of pins $A$ and $B$ and the diameter of rod $CB$ necessary to support the load.

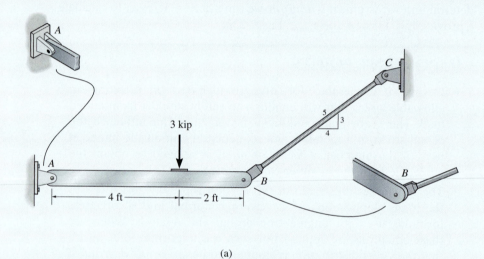

(a)

**Fig. 8–24**

**Solution**

Recognizing $CB$ to be a two-force member, the free-body diagram of member $AB$ along with the computed reactions at $A$ and $B$ is shown in Fig. 8–24b. As an exercise, verify the computations and notice that the *resultant force* at $A$ must be used for the design of pin $A$, since this is the shear force the pin resists.

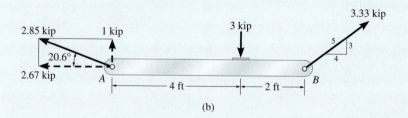

(b)

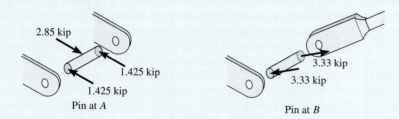

2.85 kip

1.425 kip

1.425 kip

Pin at A

3.33 kip

3.33 kip

Pin at B

**Diameter of the Pins.** From Fig. 8–24a and the free-body diagrams of the sectioned portion of each pin in contact with member AB, Fig. 8–24c, it is seen that pin A is subjected to double shear, whereas pin B is subjected to single shear. Thus,

$$A_A = \frac{V_A}{\tau_{allow}} = \frac{1.425 \text{ kip}}{12.5 \text{ kip/in}^2} = 0.1139 \text{ in}^2 = \pi\left(\frac{d_A^2}{4}\right) \quad d_A = 0.381 \text{ in.}$$

$$A_B = \frac{V_B}{\tau_{allow}} = \frac{3.333 \text{ kip}}{12.5 \text{ kip/in}^2} = 0.2667 \text{ in}^2 = \pi\left(\frac{d_B^2}{4}\right) \quad d_B = 0.583 \text{ in.}$$

Although these values represent the *smallest* allowable pin diameters, a *fabricated* or available pin size will have to be chosen. We will choose a size *larger* to the nearest $\frac{1}{16}$ in. as required.

$$d_A = \tfrac{7}{16} \text{ in.} = 0.4375 \text{ in.} \qquad\qquad \textit{Ans.}$$

$$d_B = \tfrac{5}{8} \text{ in.} = 0.625 \text{ in.} \qquad\qquad \textit{Ans.}$$

**Diameter of Rod.** The required diameter of the rod throughout its midsection is thus,

$$A_{BC} = \frac{P}{(\sigma_t)_{allow}} = \frac{3.333 \text{ kip}}{16.2 \text{ kip/in}^2} = 0.2058 \text{ in}^2 = \pi\left(\frac{d_{BC}^2}{4}\right)$$

$$d_{BC} = 0.512 \text{ in.}$$

We will choose

$$d_{BC} = \tfrac{9}{16} \text{ in.} = 0.5625 \text{ in.} \qquad\qquad \textit{Ans.}$$

EXAMPLE 8.9

The control arm is subjected to the loading shown in Fig. 8–25a. Determine to the nearest $\frac{1}{4}$ in. the required diameter of the steel pin at $C$ if the allowable shear stress for the steel is $\tau_{\text{allow}} = 8$ ksi. Note in the figure that the pin is subjected to double shear.

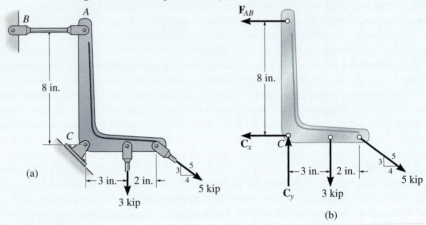

(a)

(b)

### Solution

**Internal Shear Force.**  A free-body diagram of the arm is shown in Fig. 8–25b. For equilibrium we have

$$\curvearrowleft + \Sigma M_C = 0; \qquad F_{AB}(8 \text{ in.}) - 3 \text{ kip}(3 \text{ in.}) - 5 \text{ kip}(\tfrac{3}{5})(5 \text{ in.}) = 0$$
$$F_{AB} = 3 \text{ kip}$$

$$\xrightarrow{+} \Sigma F_x = 0; \qquad -3 \text{ kip} - C_x + 5 \text{ kip}(\tfrac{4}{5}) = 0 \qquad C_x = 1 \text{ kip}$$
$$+\uparrow \Sigma F_y = 0; \qquad C_y - 3 \text{ kip} - 5 \text{ kip}(\tfrac{3}{5}) = 0 \qquad C_y = 6 \text{ kip}$$

The pin at $C$ resists the resultant force at $C$. Therefore,

$$F_C = \sqrt{(1 \text{ kip})^2 + (6 \text{ kip})^2} = 6.082 \text{ kip}$$

Since the pin is subjected to double shear, a shear force of 3.041 kip acts over its cross-sectional area *between* the arm and each supporting leaf for the pin, Fig. 8–25c.

**Required Area.** We have

$$A = \frac{V}{\tau_{\text{allow}}} = \frac{3.041 \text{ kip}}{8 \text{ kip/in}^2} = 0.3802 \text{ in}^2$$

$$\pi \left(\frac{d}{2}\right)^2 = 0.3802 \text{ in}^2$$

$$d = 0.696 \text{ in.}$$

Use a pin having a diameter of

$$d = \tfrac{3}{4} \text{ in.} = 0.750 \text{ in.} \qquad \qquad \textit{Ans.}$$

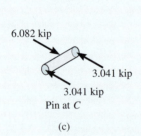

6.082 kip

3.041 kip

3.041 kip

Pin at $C$

(c)

**Fig. 8–25**

# E X A M P L E 8.10

The suspender rod is supported at its end by a fixed-connected circular disk as shown in Fig. 8–26a. If the rod passes through a 40-mm-diameter hole, determine the minimum required diameter of the rod and the minimum thickness of the disk needed to support the 20-kN load. The allowable normal stress for the rod is $\sigma_{allow} = 60$ MPa, and the allowable shear stress for the disk is $\tau_{allow} = 35$ MPa.

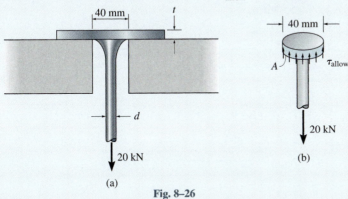

Fig. 8–26

## Solution

*Diameter of Rod.* By inspection, the axial force in the rod is 20 kN. Thus the required cross-sectional area of the rod is

$$A = \frac{P}{\sigma_{allow}} = \frac{20(10^3)\ \text{N}}{60(10^6)\ \text{N/m}^2} = 0.3333(10^{-3})\ \text{m}^2$$

So that

$$A = \pi\left(\frac{d^2}{4}\right) = 0.3333(10^{-3})\ \text{m}^2$$

$$d = 0.0206\ \text{m} = 20.6\ \text{mm} \qquad \text{Ans.}$$

*Thickness of Disk.* As shown on the free-body diagram of the core section of the disk, Fig. 8–26b, the material at the sectioned area must resist shear stress to prevent movement of the disk through the hole. If this shear stress is assumed to be distributed uniformly over the sectioned area, then, since $V = 20$ kN, we have

$$A = \frac{V}{\tau_{allow}} = \frac{20(10^3)\ \text{N}}{35(10^6)\ \text{N/m}^2} = 0.5714(10^{-3})\ \text{m}^2$$

Since the sectioned area $A = 2\pi(0.02\ \text{m})(t)$, the required thickness of the disk is

$$t = \frac{0.5714(10^{-3})\ \text{m}^2}{2\pi(0.02\ \text{m})} = 4.55(10^{-3})\ \text{m} = 4.55\ \text{mm} \qquad \text{Ans.}$$

E X A M P L E 8.11

An axial load on the shaft shown in Fig. 8–27a is resisted by the collar at $C$, which is attached to the shaft and located on the right side of the bearing at $B$. Determine the largest value of $P$ for the two axial forces at $E$ and $F$ so that the stress in the collar does not exceed an allowable bearing stress at $C$ of $(\sigma_b)_{\text{allow}} = 75$ MPa and the average normal stress in the shaft does not exceed an allowable tensile stress of $(\sigma_t)_{\text{allow}} = 55$ MPa.

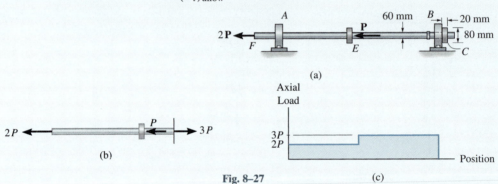

(a)

**Fig. 8–27**

(b)

(c)

(d)

**Solution**

To solve the problem we will determine $P$ for each possible failure condition. Then we will choose the *smallest* value. Why?

*Normal Stress.* Using the method of sections, the axial load within region $FE$ of the shaft is $2P$, whereas the *largest* axial load, $3P$, occurs within region $EC$, Fig. 8–27b. The variation of the internal loading is clearly shown on the normal-force diagram, Fig. 8–27c. Since the cross-sectional area of the entire shaft is constant, region $EC$ will be subjected to the maximum average normal stress. Applying Eq. 8–8, we have

$$\sigma_{\text{allow}} = \frac{P}{A} \qquad 55(10^6)\ \text{N/m}^2 = \frac{3P}{\pi(0.03\ \text{m})^2}$$

$$P = 51.8\ \text{kN}$$

*Bearing Stress.* As shown on the free-body diagram in Fig. 8–27d, the collar at $C$ must resist the load of $3P$, which acts over a bearing area of $A_b = [\pi(0.04\ \text{m})^2 - \pi(0.03\ \text{m})^2] = 2.199(10^{-3})\ \text{m}^2$. Thus,

$$A = \frac{P}{\sigma_{\text{allow}}}; \qquad 75(10^6)\ \text{N/m}^2 = \frac{3P}{2.199(10^{-3})\ \text{m}^2}$$

$$P = 55.0\ \text{kN}$$

By comparison, the largest load that can be applied to the shaft is $P = 51.8$ kN, since any load larger than this will cause the allowable normal stress in the shaft to be exceeded.

# EXAMPLE 8.12

The rigid bar $AB$ shown in Fig. 8–28a is supported by a steel rod $AC$ having a diameter of 20 mm and an aluminum block having a cross-sectional area of 1800 mm². The 18-mm-diameter pins at $A$ and $C$ are subjected to *single shear*. If the failure stress for the steel and aluminum is $(\sigma_{st})_{fail} = 680$ MPa and $(\sigma_{al})_{fail} = 70$ MPa, respectively, and the failure shear stress for each pin is $\tau_{fail} = 900$ MPa, determine the largest load $P$ that can be applied to the bar. Apply a factor of safety of F.S. = 2.

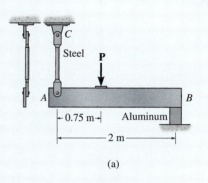

(a)

### Solution

Using Eqs. 8–6 and 8–7, the allowable stresses are

$$(\sigma_{st})_{allow} = \frac{(\sigma_{st})_{fail}}{F.S.} = \frac{680 \text{ MPa}}{2} = 340 \text{ MPa}$$

$$(\sigma_{al})_{allow} = \frac{(\sigma_{al})_{fail}}{F.S.} = \frac{70 \text{ MPa}}{2} = 35 \text{ MPa}$$

$$\tau_{allow} = \frac{\tau_{fail}}{F.S.} = \frac{900 \text{ MPa}}{2} = 450 \text{ MPa}$$

The free-body diagram for the bar is shown in Fig. 8–28b. There are three unknowns. Here we will apply the equations of equilibrium so as to express $F_{AC}$ and $F_B$ in terms of the applied load $P$. We have

$$\zeta + \Sigma M_B = 0; \qquad P(1.25 \text{ m}) - F_{AC}(2 \text{ m}) = 0 \qquad (1)$$

$$\zeta + \Sigma M_A = 0; \qquad F_B(2 \text{ m}) - P(0.75 \text{ m}) = 0 \qquad (2)$$

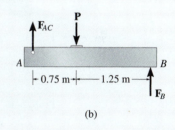

(b)

**Fig. 8–28**

We will now determine each value of $P$ that creates the allowable stress in the rod, block, and pins, respectively.

*Rod AC.* This requires

$$F_{AC} = (\sigma_{st})_{allow}(A_{AC}) = 340(10^6) \text{ N/m}^2 [\pi(0.01 \text{ m})^2] = 106.8 \text{ kN}$$

Using Eq. 1,

$$P = \frac{(106.8 \text{ kN})(2 \text{ m})}{1.25 \text{ m}} = 171 \text{ kN}$$

*Block B.* In this case,

$$F_B = (\sigma_{al})_{allow} A_B = 35(10^6) \text{ N/m}^2 [1800 \text{ mm}^2 (10^{-6}) \text{ m}^2/\text{mm}^2] = 63.0 \text{ kN}$$

Using Eq. 2,

$$P = \frac{(63.0 \text{ kN})(2 \text{ m})}{0.75 \text{ m}} = 168 \text{ kN}$$

*Pin A or C.* Here

$$V = F_{AC} = \tau_{allow} A = 450(10^6) \text{ N/m}^2 [\pi(0.09 \text{ m})^2] = 114.5 \text{ kN}$$

From Eq. 1,

$$P = \frac{114.5 \text{ kN}(2 \text{ m})}{1.25 \text{ m}} = 183 \text{ kN}$$

By comparison, when $P$ reaches its *smallest value* (168 kN), it develops the allowable normal stress in the aluminum block. Hence,

$$P = 168 \text{ kN} \qquad \textit{Ans.}$$

# PROBLEMS

**8-35.** Member $B$ is subjected to a compressive force of 800 lb. If $A$ and $B$ are both made of wood and are $\frac{3}{8}$ in. thick, determine to the nearest $\frac{1}{4}$ in. the smallest dimension $h$ of the support so that the average shear stress does not exceed $\tau_{\text{allow}} = 300$ psi.

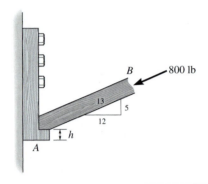

**Prob. 8–35**

**8-37.** The eye bolt is used to support the load of 5 kip. Determine its diameter $d$ to the nearest $\frac{1}{8}$ in. and the required thickness $h$ to the nearest $\frac{1}{8}$ in. of the support so that the washer will not penetrate or shear through it. The allowable normal stress for the bolt is $\sigma_{\text{allow}} = 21$ ksi and the allowable shear stress for the supporting material is $\tau_{\text{allow}} = 5$ ksi.

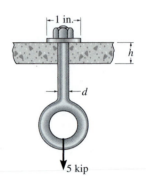

**Prob. 8–37**

*8-36.** The lever is attached to the shaft $A$ using a key that has a width $d$ and length of 25 mm. If the shaft is fixed and a vertical force of 200 N is applied perpendicular to the handle, determine the dimension $d$ if the allowable shear stress for the key is $\tau_{\text{allow}} = 35$ MPa.

**8-38.** Member $A$ of the timber step joint for a truss is subjected to a compressive force of 5 kN. Determine the required diameter $d$ of the steel rod at $C$ and the height $h$ of member $B$ if the allowable normal stress for the steel is $(\sigma_{\text{allow}})_{st} = 157$ MPa and the allowable normal stress for the wood is $(\sigma_{\text{allow}})_w = 2$ MPa. Member $B$ is 50 mm thick.

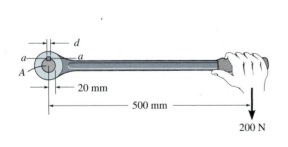

**Prob. 8–36**

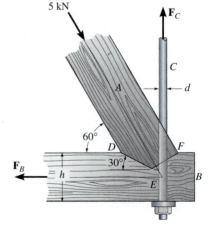

**Prob. 8–38**

**8-39.** The fillet weld size $a$ is determined by computing the average shear stress along the shaded plane, which has the smallest cross section. Determine the smallest size $a$ of the two welds if the force applied to the plate is $P = 20$ kip. The allowable shear stress for the weld material is $\tau_{allow} = 14$ ksi.

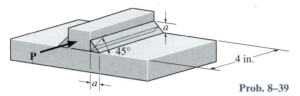

Prob. 8–39

**\*8-40.** The fillet weld size $a = 0.25$ in. If the joint is assumed to fail by shear on both sides of the block along the shaded plane, which is the smallest cross section, determine the largest force $P$ that can be applied to the plate. The allowable shear stress for the weld material is $\tau_{allow} = 14$ ksi.

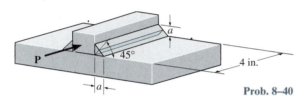

Prob. 8–40

**8-41.** The hanger is supported using the rectangular pin. Determine the magnitude of the allowable suspended load $P$ if the allowable bearing stress is $(\sigma_b)_{allow} = 220$ MPa, the allowable tensile stress is $(\sigma_t)_{allow} = 150$ MPa, and the allowable shear stress is $\tau_{allow} = 130$ MPa. Take $t = 6$ mm, $a = 5$ mm, and $b = 25$ mm.

**8-42.** The hanger is supported using the rectangular pin. Determine the required thickness $t$ of the hanger, and dimensions $a$ and $b$ if the suspended load is $P = 60$ kN. The allowable tensile stress is $(\sigma_t)_{allow} = 150$ MPa, the allowable bearing stress is $(\sigma_b)_{allow} = 290$ MPa, and the allowable shear stress is $\tau_{allow} = 125$ MPa.

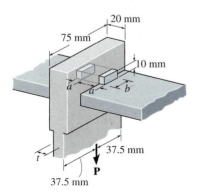

Probs. 8–41/42

**8-43.** The frame is subjected to the load of 1.5 kip. Determine the required diameter of the pins at $A$ and $B$ if the allowable shear stress for the material is $\tau_{allow} = 6$ ksi. Pin $A$ is subjected to double shear, whereas pin $B$ is subjected to single shear.

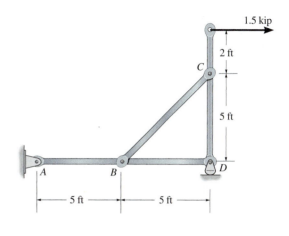

Prob. 8–43

**8-44.** Determine the required cross-sectional area of member $BC$ and the diameter of the pins at $A$ and $B$ if the allowable normal stress is $\sigma_{allow} = 3$ ksi and the allowable shear stress is $\tau_{allow} = 4$ ksi.

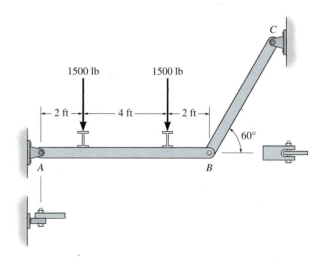

Prob. 8–44

**8-45.** The connection is made using a bolt and nut and two washers. If the allowable bearing stress for the boards is $(\sigma_b)_{allow} = 2$ ksi, and the allowable tensile stress for the bolt shank $S$ is $(\sigma_t)_{allow} = 18$ ksi, determine the maximum allowable tension in the bolt shank. The bolt shank has a diameter of 0.31 in., and the washers have an outer diameter of 0.75 in. and inner diameter (hole) of 0.50 in.

Prob. 8-45

**8-46.** The hangers support the joist uniformly, so that it is assumed the four nails on each hanger carry an equal portion of the load. If the joist is subjected to the loading shown, determine the average shear stress in each nail of the hanger at ends $A$ and $B$. Each nail has a diameter of 0.25 in. The hangers only support vertical loads.

**8-47.** The hangers support the joist uniformly, so that it is assumed the four nails on each hanger carry an equal portion of the load. Determine the smallest diameter of the nails at $A$ and at $B$ if the allowable shear stress for the nails is $\tau_{allow} = 4$ ksi. The hangers only support vertical loads.

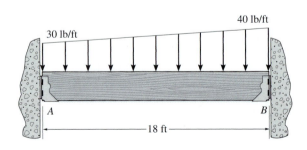

Probs. 8-46/47

*8-48.** The aluminum bracket $A$ is used to support the centrally applied load of 8 kip. If it has a constant thickness of 0.5 in., determine the smallest height $h$ in order to prevent a shear failure. The failure shear stress is $\tau_{fail} = 23$ ksi. Use a factor of safety for shear of F.S. $= 2.5$.

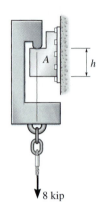

8 kip

Prob. 8-48

**8-49.** Determine the smallest dimensions of the circular shaft and circular end cap if the load it is required to support is $P = 150$ kN. The allowable tensile stress, bearing stress, and shear stress is $(\sigma_t)_{allow} = 175$ MPa, $(\sigma_b)_{allow} = 275$ MPa, and $\tau_{allow} = 115$ MPa.

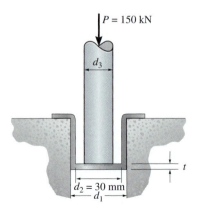

$P = 150$ kN

$d_3$

$d_2 = 30$ mm
$d_1$
$t$

Prob. 8-49

**8-50.** The hanger assembly is used to support a distributed loading of $w = 0.8$ kip/ft. Determine the average shear stress in the 0.40-in.-diameter bolt at $A$ and the average tensile stress in rod $AB$, which has a diameter of 0.5 in. If the yield shear stress for the bolt is $\tau_y = 25$ ksi, and the yield tensile stress for the rod is $\sigma_y = 38$ ksi, determine the factor of safety with respect to yielding in each case.

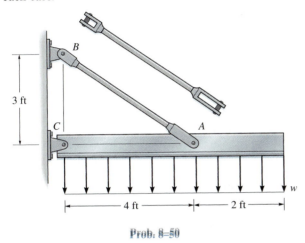

Prob. 8-50

**8-51.** Determine the intensity $w$ of the maximum distributed load that can be supported by the hanger assembly so that an allowable shear stress of $\tau_{allow} = 13.5$ ksi is not exceeded in the 0.40-in.-diameter bolts at $A$ and $B$, and an allowable tensile stress of $\sigma_{allow} = 22$ ksi is not exceeded in the 0.5-in.-diameter rod $AB$.

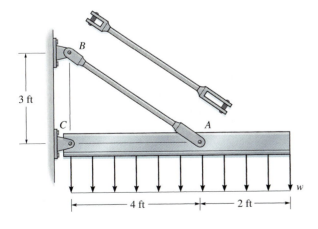

Prob. 8-51

**\*8-52.** The wood specimen is subjected to the pull of 10 kN in a tension testing machine. If the allowable normal stress for the wood is $(\sigma_t)_{allow} = 12$ MPa and the allowable shear stress is $\tau_{allow} = 1.2$ MPa, determine the required dimensions $b$ and $t$ so that the specimen reaches these stresses simultaneously. The specimen has a width of 25 mm.

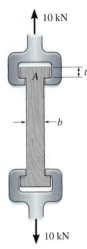

Prob. 8-52

**8-53.** The compound wooden beam is connected together by a bolt at $B$. Assuming that the connections at $A$, $B$, $C$, and $D$ exert only vertical forces on the beam, determine the required diameter of the bolt at $B$ and the required outer diameter of its washers if the allowable tensile stress for the bolt is $(\sigma_t)_{allow} = 150$ MPa and the allowable bearing stress for the wood is $(\sigma_b)_{allow} = 28$ MPa. Assume that the hole in the washers has the same diameter as the bolt.

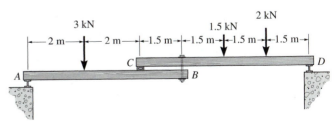

Prob. 8-53

# 8.7 Deformation

Note the before and after positions of three different line segments on this rubber membrane which is subjected to tension. The vertical line is lengthened, the horizontal line is shortened, and the inclined line changes its length and rotates.

Whenever a force is applied to a body, it will tend to change the body's shape and size. These changes are referred to as *deformation*, and they may be either highly visible or practically unnoticeable, without the use of equipment to make precise measurements. For example, a rubber band will undergo a very large deformation when stretched. On the other hand, only slight deformations of structural members occur when a building is occupied with people walking about. Deformation of a body can also occur when the temperature of the body is changed. A typical example is the thermal expansion or contraction of a roof caused by the weather.

In the general sense, the deformation of a body will not be uniform throughout its volume, and so the change in geometry of any line segment within the body may vary along its length. For example, one portion of the line may elongate, whereas another portion may contract. As shorter and shorter line segments are considered, however, they remain straighter after the deformation, and so to study deformational changes in a more uniform manner, we will consider the lines to be very short and located in the neighborhood of a point. In doing so, realize that any line segment located at one point in the body will change by a different amount from one located at some other point. Furthermore, these changes will also depend on the orientation of the line segment at the point. For example, a line segment may elongate if it is oriented in one direction, whereas it may contract if it is oriented in another direction.

# 8.8 Strain

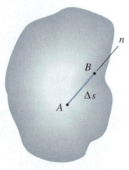

Undeformed body

(a)

**Fig. 8–29**

In order to describe the deformation by changes in length of line segments and the changes in the angles between them, we will develop the concept of strain. Measurements of strain are actually made by experiments, and once the strains are obtained, it will be shown in the next section how they can be related to the applied loads, or stresses, acting within the body.

**Normal Strain.** The elongation or contraction of a line segment per unit of length is referred to as *normal strain*. To develop a formalized definition of normal strain, consider the line $AB$, which is contained within the undeformed body shown in Fig. 8–29a. This line lies along the $n$ axis and has an original length of $\Delta s$. After deformation, points $A$ and $B$ are displaced to $A'$ and $B'$, and the line becomes a curve having a length of

$\Delta s'$, Fig. 8–29b. The change in length of the line is therefore $\Delta s' - \Delta s$. If we define the *average normal strain* using the symbol $\epsilon_{avg}$ (epsilon), then

$$\epsilon_{avg} = \frac{\Delta s' - \Delta s}{\Delta s} \qquad (8\text{–}10)$$

As point $B$ is chosen closer and closer to point $A$, the length of the line becomes shorter and shorter, such that $\Delta s \to 0$. Also, this causes $B'$ to approach $A'$, such that $\Delta s' \to 0$. Consequently, in the limit the normal strain at *point A* and in the direction of $n$ is

$$\epsilon = \lim_{B \to A \text{ along } n} \frac{\Delta s' - \Delta s}{\Delta s} \qquad (8\text{–}11)$$

If the normal strain is known, we can use this equation to obtain the approximate final length of a *short* line segment in the direction of $n$ after it is deformed. We have

$$\Delta s' \approx (1 + \epsilon) \, \Delta s \qquad (8\text{–}12)$$

Hence, when $\epsilon$ is positive the initial line will elongate, whereas if $\epsilon$ is negative the line contracts.

**Units.** Note that normal strain is a *dimensionless quantity*, since it is a ratio of two lengths. Although this is the case, it is common practice to state it in terms of a ratio of length units. If the SI system is used, then the basic units will be meters/meter (m/m). Ordinarily, for most engineering applications $\epsilon$ will be very small, so measurements of strain are in micrometers per meter ($\mu$m/m), where $1 \, \mu m = 10^{-6}$ m. In the Foot-Pound-Second system, strain can be stated in units of inches per inch (in./in.). Sometimes for experimental work, strain is expressed as a percent, e.g., 0.001 m/m = 0.1%. As an example, a normal strain of $480(10^{-6})$ can be reported as $480(10^{-6})$ in./in., 480 $\mu$m/m, or 0.0480%. Also, one can state this answer as simply 480 $\mu$ (480 "micros").

**Shear Strain.** The change in angle that occurs between two line segments that were originally *perpendicular* to one another is referred to as **shear strain**. This angle is denoted by $\gamma$ (gamma) and is measured in radians (rad). To show how it is developed, consider the line segments $AB$ and $AC$ originating from the same point $A$ in a body, and directed along the perpendicular $n$ and $t$ axes, Fig. 8–30a. After deformation, the ends of the lines are displaced, and the lines themselves become curves, such that the angle between them at $A$ is $\theta'$, Fig. 8–30b. Hence we define the shear strain at point $A$ that is associated with the $n$ and $t$ axes as

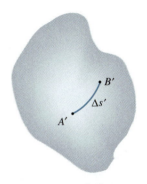

Deformed body

(b)

**Fig. 8–29**

$$\gamma_{nt} = \frac{\pi}{2} - \lim_{\substack{B \to A \text{ along } n \\ C \to A \text{ along } t}} \theta' \qquad (8\text{–}13)$$

Notice that if $\theta'$ is smaller than $\pi/2$ the shear strain is positive, whereas if $\theta'$ is larger than $\pi/2$ the shear strain is negative.

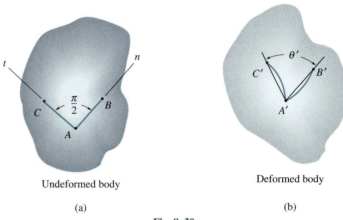

Undeformed body

(a)

Deformed body

(b)

**Fig. 8–30**

## Cartesian Strain Components.

Using the above definitions of normal and shear strain, we will now show how they can be used to describe the deformation of the body, Fig. 8–31a. To do so, imagine the body to be subdivided into small elements such as the one shown in Fig. 8–31b. This element is rectangular, has undeformed dimensions $\Delta x$, $\Delta y$, and $\Delta z$, and is located in the neighborhood of a point in the body, Fig. 8–31a. Assuming that the element's dimensions are very small, the deformed shape of the element will be a parallelepiped, Fig. 8–31c, since very small line segments will remain approximately straight after the body is deformed. In order to achieve this deformed shape, we must first consider how the normal strain changes the lengths of the sides of the rectangular element, and then how the shear strain changes the angles of each side. Hence, using Eq. 8–12, $\Delta s' \approx (1 + \epsilon) \Delta s$, in reference to the lines $\Delta x$, $\Delta y$, and $\Delta z$, the approximate lengths of the sides of the parallelepiped are

$$(1 + \epsilon_x) \Delta x \qquad (1 + \epsilon_y) \Delta y \qquad (1 + \epsilon_z) \Delta z$$

And the approximate angles between the sides, again originally defined by the sides $\Delta x$, $\Delta y$, and $\Delta z$, are

$$\frac{\pi}{2} - \gamma_{xy} \qquad \frac{\pi}{2} - \gamma_{yz} \qquad \frac{\pi}{2} - \gamma_{xz}$$

In particular, notice that the **normal strains cause a change in volume** of the rectangular element, whereas the **shear strains cause a change in its shape**. Of course, both of these effects occur simultaneously during the deformation.

(a)

$\frac{\pi}{2}$

$\Delta z$

$\frac{\pi}{2}$

$\Delta y$

$\frac{\pi}{2}$

$\Delta x$

Undeformed element

(b)

**Fig. 8–31**

In summary, then, the *state of strain* at a point in a body requires specifying three normal strains, $\epsilon_x$, $\epsilon_y$, $\epsilon_z$, and three shear strains, $\gamma_{xy}$, $\gamma_{yz}$, $\gamma_{xz}$. These strains completely describe the deformation of a rectangular volume element of material located at the point and oriented so that its sides are originally parallel to the *x, y, z* axes. Once these strains are defined at all points in the body, the deformed shape of the body can then be described. It should also be added that by knowing the state of strain at a point, defined by its six components, it is possible to determine the strain components on an element oriented at the point in any other direction. This is discussed in Chapter 15.

**Small Strain Analysis.** Most engineering design involves applications for which only *small deformations* are allowed. For example, almost all structures and machines appear to be rigid, and the deformations that occur during use are hardly noticeable. Furthermore, even if the deflection of a member such as a thin plate or slender rod may appear to be large, the material from which it is made may only be subjected to very small deformations. In this text, therefore, we will assume that the deformations that take place within a body are almost infinitesimal, so that the *normal strains* occurring within the material are *very small* compared to 1, that is, $\epsilon \ll 1$. This assumption, which is based on the magnitude of the strain, has wide practical application in engineering, and it is often referred to as a *small strain analysis*. For example, it allows us to approximate $\sin \theta \approx \theta$, $\cos \theta \approx 1$ and $\tan \theta \approx \theta$ provided $\theta$ is very small.

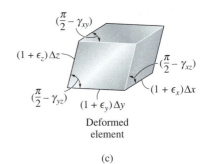

Deformed element

(c)

**Fig. 8–31**

The rubber bearing support under this concrete bridge girder is subjected to both normal and shear strain. The normal strain is caused by the weight and bridge loads on the girder, and the shear strain is caused by the horizontal movement of the girder due to temperature changes.

## IMPORTANT POINTS

- Loads will cause all material bodies to deform and, as a result, points in the body will undergo *displacements or changes in position*.

- *Normal strain* is a measure of the elongation or contraction of a small line segment in the body, whereas *shear strain* is a measure of the change in angle that occurs between two small line segments that are originally perpendicular to one another.

- The state of strain at a point is characterized by six strain components: three normal strains $\epsilon_x$, $\epsilon_y$, $\epsilon_z$ and three shear strains $\gamma_{xy}$, $\gamma_{yz}$, $\gamma_{xz}$. These components depend upon the orientation of the line segments and their location in the body.

- Strain is the geometrical quantity that is measured using experimental techniques. Once obtained, the stress in the body can then be determined from material property relations.

- Most engineering materials undergo small deformations, and so the normal strain $\epsilon \ll 1$. This assumption of "small strain analysis" allows the calculations for normal strain to be simplified since first-order approximations can be made about their size.

E X A M P L E   **8.13**

The slender rod shown in Fig. 8–32 is subjected to an increase of temperature along its axis, which creates a normal strain in the rod of $\epsilon_z = 40(10^{-3})z^{1/2}$, where $z$ is given in meters. Determine (a) the displacement of the end $B$ of the rod due to the temperature increase, and (b) the average normal strain in the rod.

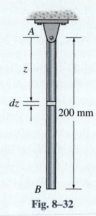

**Fig. 8–32**

**Solution**

*Part (a).*   Since the normal strain is reported at each point along the rod, a differential segment $dz$, located at position $z$, Fig. 8–32, has a deformed length that can be determined from Eq. 8–12; that is,

$$dz' = [1 + 40(10^{-3})z^{1/2}]\, dz$$

The sum total of these segments along the axis yields the *deformed length* of the rod, i.e.,

$$z' = \int_0^{0.2\text{ m}} [1 + 40(10^{-3})z^{1/2}]\, dz$$

$$= z + 40(10^{-3})(\tfrac{2}{3}z^{3/2}) \Big|_0^{0.2\text{ m}}$$

$$= 0.20239 \text{ m}$$

The displacement of the end of the rod is therefore

$$\Delta_B = 0.20239 \text{ m} - 0.2 \text{ m} = 0.00239 \text{ m} = 2.39 \text{ mm} \downarrow \qquad Ans.$$

*Part (b).*   The average normal strain in the rod is determined from Eq. 8–10, which assumes that the rod or "line segment" has an original length of 200 mm and a change in length of 2.39 mm. Hence,

$$\epsilon_{\text{avg}} = \frac{\Delta s' - \Delta s}{\Delta s} = \frac{2.39 \text{ mm}}{200 \text{ mm}} = 0.0119 \text{ mm/mm} \qquad Ans.$$

# E X A M P L E 8.14

A force acting on the grip of the lever arm shown in Fig. 8–33a causes the arm to rotate clockwise through an angle of $\theta = 0.002$ rad. Determine the average normal strain developed in the wire $BC$.

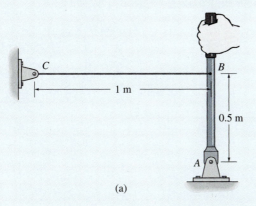

(a)

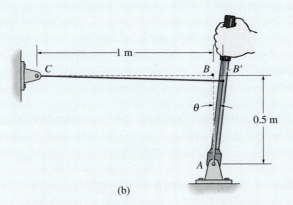

(b)

Fig. 8–33

## Solution

Since $\theta = 0.002$ rad is small, the stretch in the wire $CB$, Fig. 8–33b, is $BB' = \theta\,(0.5 \text{ m}) = (0.002 \text{ rad})(0.5 \text{ m}) = 0.001$ m. The average normal strain in the wire is therefore,

$$\epsilon_{\text{avg}} = \frac{BB'}{CB} = \frac{0.001}{1 \text{ m}} = 0.001 \text{ m/m} \qquad \textit{Ans.}$$

EXAMPLE 8.15

The plate is deformed into the dashed shape shown in Fig. 8–34a. If in this deformed shape horizontal lines on the plate remain horizontal and do not change their length, determine (a) the average normal strain along the side $AB$, and (b) the average shear strain in the plate relative to the $x$ and $y$ axes.

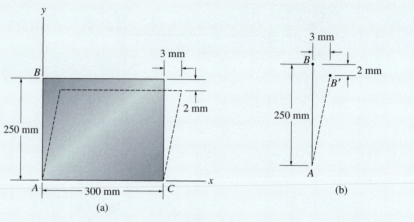

Fig. 8–34

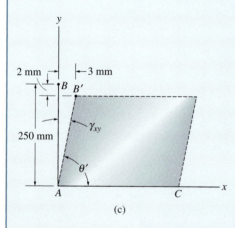

## Solution

*Part (a).*   Line $AB$, coincident with the $y$ axis, becomes line $AB'$ after deformation, as shown in Fig. 8–34b. The length of this line is

$$AB' = \sqrt{(250 - 2)^2 + (3)^2} = 248.018 \text{ mm}$$

The average normal strain for $AB$ is therefore

$$(\epsilon_{AB})_{avg} = \frac{AB' - AB}{AB} = \frac{248.018 \text{ mm} - 250 \text{ mm}}{250 \text{ mm}}$$

$$= -7.93(10^{-3}) \text{ mm/mm} \qquad \textit{Ans.}$$

The negative sign indicates the strain causes a contraction of $AB$.

*Part (b).*   As noted in Fig. 8–34c, the once 90° angle $BAC$ between the sides of the plate, referenced from the $x, y$ axes, changes to $\theta'$ due to the displacement of $B$ to $B'$. Since $\gamma_{xy} = \pi/2 - \theta'$, then $\gamma_{xy}$ is the angle shown in the figure. Thus,

$$\gamma_{xy} = \tan^{-1}\left(\frac{3 \text{ mm}}{250 \text{ mm} - 2 \text{ mm}}\right) = 0.0121 \text{ rad} \qquad \textit{Ans.}$$

# E X A M P L E 8.16

The plate shown in Fig. 8–35a is held in the rigid horizontal guides at its top and bottom, AD and BC. If its right side CD is given a uniform horizontal displacement of 2 mm, determine (a) the average normal strain along the diagonal AC, and (b) the shear strain at E relative to the x, y axes.

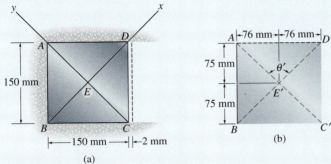

Fig. 8–35

## Solution

**Part (a).** When the plate is deformed, the diagonal AC becomes AC′, Fig. 8–35b. The length of diagonals AC and AC′ can be found from the Pythagorean theorem. We have

$$AC = \sqrt{(0.150)^2 + (0.150)^2} = 0.21213 \text{ m}$$
$$AC' = \sqrt{(0.150)^2 + (0.152)^2} = 0.21355 \text{ m}$$

Therefore the average normal strain along the diagonal is

$$(\epsilon_{AC})_{\text{avg}} = \frac{AC' - AC}{AC} = \frac{0.21355 \text{ m} - 0.21213 \text{ m}}{0.21213 \text{ m}}$$
$$= 0.00669 \text{ mm/mm} \qquad \textit{Ans.}$$

**Part (b).** To find the shear strain at E relative to the x and y axes, it is first necessary to find the angle θ′, which specifies the angle between these axes after deformation, Fig. 8–35b. We have

$$\tan\left(\frac{\theta'}{2}\right) = \frac{76 \text{ mm}}{75 \text{ mm}}$$

$$\theta' = 90.759° = \frac{\pi}{180°}(90.759°) = 1.58404 \text{ rad}$$

Applying Eq. 8–13, the shear strain at E is therefore

$$\gamma_{xy} = \frac{\pi}{2} - 1.58404 \text{ rad} = -0.0132 \text{ rad} \qquad \textit{Ans.}$$

According to the sign convention, the *negative sign* indicates that the angle θ′ is *greater than* 90°. Note that if the x and y axes were horizontal and vertical, then due to the deformation $\gamma_{xy} = 0$ at point E.

# PROBLEMS

**8-54.** The center portion of the rubber balloon has a diameter of $d = 4$ in. If the air pressure within it causes the balloon's diameter to become $d = 5$ in., determine the average normal strain in the rubber.

Prob. 8–54

**8–55.** A rubber band has an unstretched length of 10 in. If it is stretched around a pole having an outer diameter of 6 in., determine the average normal strain in the band.

**\*8-56.** The rigid beam is supported by a pin at $A$ and wires $BD$ and $CE$. If the load **P** on the beam is displaced 10 mm downward, determine the normal strain developed in wires $CE$ and $BD$.

**8-57.** The rigid beam is supported by a pin at $A$ and wires $BD$ and $CE$. If the maximum allowable normal strain in each wire is $\epsilon_{max} = 0.002$ mm/mm, determine the maximum vertical displacement of the load **P**.

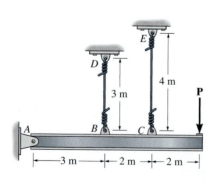

Probs. 8–56/57

**8-58.** Determine the approximate average normal strain in the wire $AB$ as a function of the rotation $\theta$ of the rigid bar $CA$ by assuming $\theta$ is small. What is this value if $\theta = 2°$?

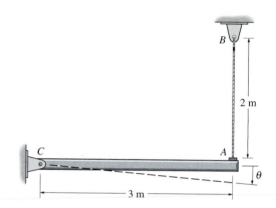

Prob. 8–58

**8-59.** The two wires are connected together at $A$. If the force **P** causes point $A$ to be displaced vertically 3 mm, determine the normal strain developed in each wire.

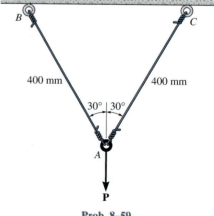

Prob. 8–59

*8-60. Two bars are used to support a load. When unloaded, $AB$ is 5 in. long, $AC$ is 8 in. long, and the ring at $A$ has coordinates $(0,0)$. If a load $\mathbf{P}$ acts on the ring at $A$, the normal strain in $AB$ becomes $\epsilon_{AB} = 0.02$ in./in., and the normal strain in $AC$ becomes $\epsilon_{AC} = 0.035$ in./in. Determine the coordinate position of the ring due to the load.

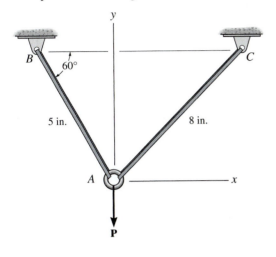

Prob. 8–60

8-61. Two bars are used to support a load $\mathbf{P}$. When unloaded, $AB$ is 5 in. long, $AC$ is 8 in. long, and the ring at $A$ has coordinates $(0, 0)$. If a load is applied to the ring at $A$, so that it moves it to the coordinate position $(0.25$ in., $-0.73$ in.), determine the normal strain in each bar.

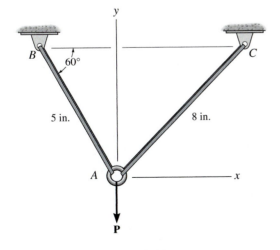

Prob. 8–61

8-62. The guy wire $AB$ of a building frame is originally unstretched. Due to an earthquake, the two columns of the frame tilt $\theta = 2°$. Determine the approximate normal strain in the wire when the frame is in this position. Assume the columns are rigid and rotate about their lower supports.

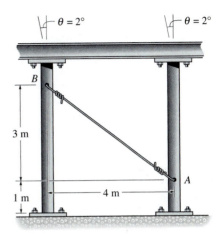

Prob. 8–62

8-63. Due to its weight, the rod is subjected to a normal strain that varies along its length such that $\epsilon = kz$, where $k$ is a constant. Determine the displacement $\Delta L$ of its end $B$ when it is suspended as shown.

Prob. 8–63

**\*8-64.** The rectangular membrane has an unstretched length $L_1$ and width $L_2$. If the sides are increased by small amounts $\Delta L_1$ and $\Delta L_2$, determine the normal strain along the diagonal $AB$.

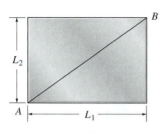

**Prob. 8-64**

**8-66.** The rectangular plate is subjected to the deformation shown by the dashed line. Determine the shear strain $\gamma_{x'y'}$ in the plate. The $x'$ axis is directed from $A$ through point $B$.

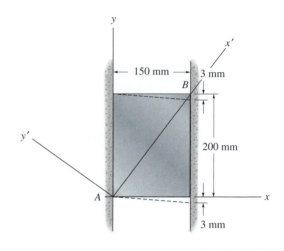

**Prob. 8-66**

**8-65.** The rectangular plate is subjected to the deformation shown by the dashed line. Determine the average shear strain $\gamma_{xy}$ of the plate.

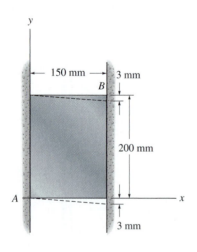

**Prob. 8-65**

**8-67.** The rectangular plate is subjected to the deformation shown by the dashed line. Determine the normal strains $\epsilon_x$, $\epsilon_y$, $\epsilon_{x'}$ $\epsilon_{y'}$.

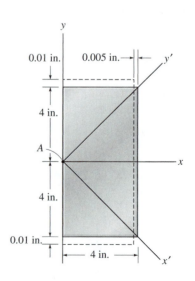

**Prob. 8-67**

*8-68. The rectangular plate is subjected to the deformation shown by the dashed line. Determine the shear strains $\gamma_{xy}$ and $\gamma_{x'y'}$ developed at point $A$.

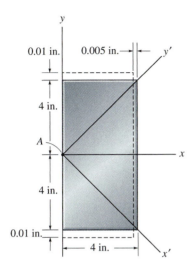

**Prob. 8–68**

8-69. The piece of plastic is originally rectangular. Determine the shear strain $\gamma_{xy}$ at corners $A$ and $B$ if the plastic distorts as shown by the dashed lines.

8-70. The piece of plastic is originally rectangular. Determine the shear strain $\gamma_{xy}$ at corners $D$ and $C$ if the plastic distorts as shown by the dashed lines.

8-71. The piece of plastic is originally rectangular. Determine the average normal strain that occurs along the diagonals $AC$ and $DB$.

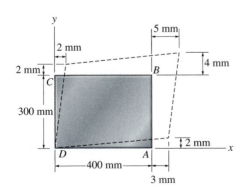

**Probs. 8–69/70/71**

*8-72. A square piece of material is deformed into the dashed position. Determine the shear strain $\gamma_{xy}$ at $A$.

8-73. A square piece of material is deformed into the dashed parallelogram. Determine the average normal strain that occurs along the diagonals $AC$ and $BD$.

8-74. A square piece of material is deformed into the dashed position. Determine the shear strain $\gamma_{xy}$ at $C$.

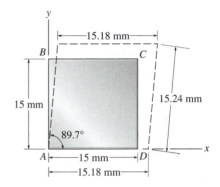

**Probs. 8–72/73/74**

8-75. The square deforms into the position shown by the dashed lines. Determine the average normal strain along each diagonal, $AB$ and $CD$. Side $DB$ remains horizontal.

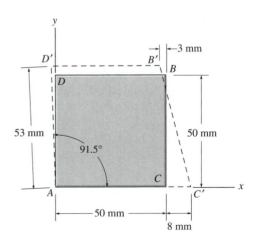

**Prob. 8–75**

**\*8-76.** The fiber $AB$ has a length $L$ and orientation $\theta$. If its ends $A$ and $B$ undergo very small displacements $u_A$ and $v_B$, respectively, determine the normal strain in the fiber when it is in position $A'B'$.

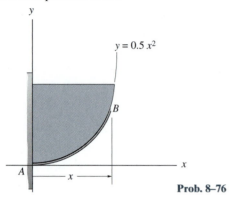

$$y = 0.5\,x^2$$

**Prob. 8–76**

**8-77.** If the normal strain is defined in reference to the final length, that is,

$$\epsilon'_n = \lim_{P \to P'} \left( \frac{\Delta s' - \Delta s}{\Delta s'} \right)$$

instead of in reference to the original length, Eq. 8–11, show that the difference in these strains is represented as a second-order term, namely, $\epsilon_n - \epsilon'_n = \epsilon_n\,\epsilon'_n$.

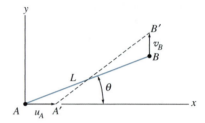

**Prob. 8–77**

# CHAPTER REVIEW

- **Internal Loadings.** The internal loadings in a body consist of a normal force, shear force, bending moment and torsional moment. They represent the resultants of a normal or shear stress distribution that acts over the cross section. To obtain these resultants, use the method of sections and the equations of equilibrium.

- **Average Normal Stress.** If a bar is made from homogeneous isotropic material and it is subjected to a series of external axial loads that pass through the centroid of the cross section, then a uniform normal stress distribution will act over the cross section. This average normal stress can be determined from $\sigma = P/A$, where $P$ is the internal axial load at the section.

- **Average Shear Stress.** The average shear stress can be determined using $\tau = V/A$, where $V$ is the resultant shear force on the cross sectional area $A$. This formula is often used to find the average shear stress in fasteners or in parts used for connections.

- *Connection Design.* The design of any simple connection requires that the average stress along any cross section not exceed a factor of safety or an allowable value of $\sigma_{allow}$ or $\tau_{allow}$. These values are reported in codes or standards and have been deemed safe on the basis of experiments or through experience.

- *Normal and Shear Strain.* The distortions within a body are measured by strain. Normal strain measures the elongation or contraction per unit length of a line segment, and shear strain measures the angular change between two line segments that are originally perpendicular to one another.

# REVIEW PROBLEMS

**8-78.** The truss is made from three pin-connected members having the cross-sectional areas shown in the figure. Determine the average normal stress developed in each member when the truss is subjected to the load shown. State whether the stress is tensile or compressive.

**8-79.** A force of 8 kN is applied at the *center* of the wooden post. If the post is mounted at the corner of its base plate $B$, can the bearing stress that the base plate exerts on the slab $S$ be assumed uniformly distributed? Why or why not? What is the average compressive stress in the wooden post?

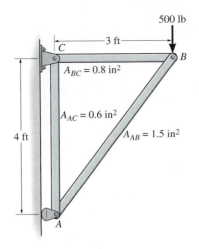

Prob. 8–78

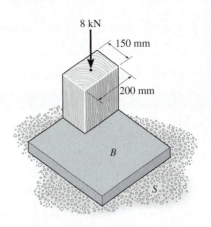

Prob. 8–79

**\*8-80.** The built-up shaft consists of a pipe $AB$ and solid rod $BC$. The pipe has an inner diameter of 20 mm and outer diameter of 28 mm. The rod has a diameter of 12 mm. Determine the normal stress at points $D$ and $E$ and represent the stress on a volume element located at each of these points.

**8-83.** The beam supports a distributed load and a couple moment. Determine the resultant internal loadings on the cross sections located just to the left and just to the right of the roller support at $B$.

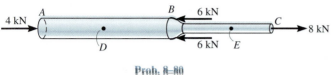

Prob. 8-80

**8-81.** The member $ABC$ is supported by a pin at $A$ and a rocker at $C$. Determine the resultant internal loadings acting on the cross sections located through points $D$ and $E$.

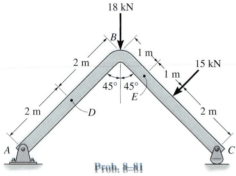

Prob. 8-81

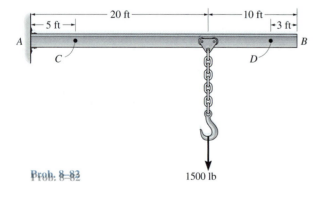

Prob. 8-83

**8-82.** The beam $AB$ is fixed to the wall and has a uniform weight of 80 lb/ft. If the trolley supports a load of 1500 lb, determine the resultant internal loadings on the cross sections through points $C$ and $D$.

**\*8-84.** The plastic block is subjected to an axial compressive force of 600 N. Assuming that the caps at the top and bottom distribute the load uniformly throughout the block, determine the average normal and average shear stress acting along section $a$–$a$.

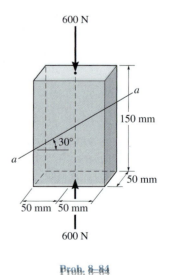

Prob. 8-84

Prob. 8-82

**8-85.** The lever is attached to the shaft $A$ using a key that has a width of 8 mm and length of 25 mm. If the shaft is fixed and a vertical force of 50 N is applied perpendicular to the handle, determine the average shear stress developed along section $a–a$ of the key.

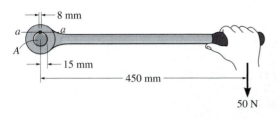

Prob. 8–85

**8-86.** Bar $ABC$ is originally in a horizontal position. If loads cause the end $A$ to be displaced downwards $\Delta_A = 0.002$ in. and the bar rotates $\theta = 0.2°$, determine the average normal strain in the rods $AD$, $BE$, and $CF$.

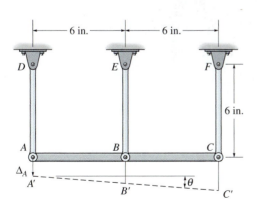

Prob. 8–86

**8-87.** The rectangular plate is distorted into a parallelogram as shown by the dashed lines. If the diagonal $AB'$ is 50.03 mm, determine the shear strain $\gamma_{xy}$.

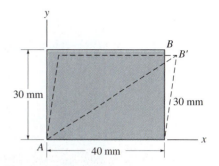

Prob. 8–87

**\*8-88.** The piece of rubber is originally rectangular. Determine the average shear strain $\gamma_{xy}$ if the corners $B$ and $D$ are subjected to the displacements that cause the rubber to distort as shown by the dashed lines.

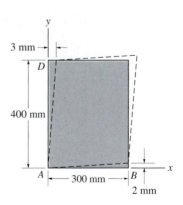

Prob. 8–88

The mechanical properties of a material must be known so that engineers can relate the measured strain in a material to its associated stress. Here the mechanical properties of bone are determined from a compression test.

# Mechanical Properties
of Materials

- To show how stress can be related to strain by using experimental methods to determine the stress–strain diagram for a particular material.
- To discuss the properties of the stress–strain diagram for materials commonly used in engineering.
- To discuss other mechanical properties and tests related to the development of the mechanics of materials.

## 9.1  The Tension and Compression Test

The strength of a material depends on its ability to sustain a load without undue deformation or failure. This property is inherent in the material itself and must be determined by *experiment*. One of the most important tests to perform in this regard is the ***tension or compression test***. Although many important mechanical properties of a material can be determined from this test, it is used primarily to determine the relationship between the average normal stress and average normal strain in many engineering materials such as metals, ceramics, polymers, and composites.

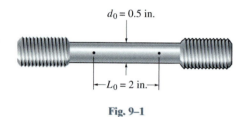

$d_0 = 0.5$ in.

$L_0 = 2$ in.

**Fig. 9–1**

Typical steel specimen with attached strain gauge.

To perform the tension or compression test a specimen of the material is made into a "standard" shape and size. Before testing, two small punch marks are identified along the specimen's length. These marks are located away from both ends of the specimen because the stress distribution at the ends is somewhat complex due to gripping at the connections where the load is applied. Measurements are taken of both the specimen's initial cross-sectional area, $A_0$, and the **gauge-length** distance $L_0$ between the punch marks. For example, when a metal specimen is used in a tension test it generally has an initial diameter of $d_0 = 0.5$ in. (13 mm) and a gauge length of $L_0 = 2$ in. (50 mm), Fig. 9–1. In order to apply an axial load with no bending of the specimen, the ends are usually seated into ball-and-socket joints. A testing machine like the one shown in Fig. 9–2 is then used to stretch the specimen at a very slow, constant rate until it reaches the breaking point. The machine is designed to read the load required to maintain this uniform stretching.

At frequent intervals during the test, data is recorded of the applied load $P$, as read on the dial of the machine or taken from a digital readout. Also, the elongation $\delta = L - L_0$ between the punch marks on the specimen may be measured using either a caliper or a mechanical or optical device called an **extensometer**. This value of $\delta$ (delta) is then used to calculate the average normal strain in the specimen. Sometimes, however, this measurement is not taken, since it is also possible to read the strain *directly* by using an **electrical-resistance strain gauge**, which looks like the one shown in Fig. 9–3. The operation of this gauge is based on the change in electrical resistance of a very thin wire or piece of metal foil under strain. Essentially the gauge is cemented to the specimen in a specified direction. If the cement is very strong in comparison to the gauge, then the gauge is in effect an integral part of the specimen, so that when the specimen is strained in the direction of the gauge, the wire and specimen will experience the same strain. By measuring the electrical resistance of the wire, the gauge may be calibrated to read values of normal strain directly.

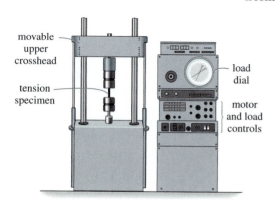

movable
upper
crosshead

tension
specimen

load
dial

motor
and load
controls

**Fig. 9–2**

Electrical–resistance
strain gauge

**Fig. 9–3**

## 9.2  The Stress–Strain Diagram

From the data of a tension or compression test, it is possible to compute various values of the stress and corresponding strain in the specimen and then plot the results. The resulting curve is called the **stress–strain diagram**, and there are two ways in which it is normally described.

**Conventional Stress–Strain Diagram.**  Using the recorded data, we can determine the **nominal** or **engineering stress** by dividing the applied load $P$ by the specimen's *original* cross-sectional area $A_0$. This calculation assumes that the stress is *constant* over the cross section and throughout the region between the gauge points. We have

$$\sigma = \frac{P}{A_0}$$

(9–1)

Likewise, the **nominal** or **engineering strain** is found directly from the strain gauge reading, or by dividing the change in the specimen's gauge length, $\delta$, by the specimen's original gauge length $L_0$. Here the strain is assumed to be constant throughout the region between the gauge points. Thus,

$$\epsilon = \frac{\delta}{L_0}$$

(9–2)

If the corresponding values of $\sigma$ and $\epsilon$ are plotted as a graph, for which the ordinate is the stress and the abscissa is the strain, the resulting curve is called a **conventional stress–strain diagram**. This diagram is very important in engineering since it provides the means for obtaining data about a material's tensile (or compressive) strength *without* regard for the material's physical size or shape, i.e., its geometry. Realize, however, that no two stress–strain diagrams for a particular material will be *exactly* the same, since the results depend on such variables as the material's composition, microscopic imperfections, the way it is manufactured, the rate of loading, and the temperature during the time of the test.

We will now discuss the characteristics of the conventional stress–strain curve as it pertains to *steel*, a commonly used material for fabricating both structural members and mechanical elements. Using the method described above, the characteristic stress–strain diagram for a steel specimen is shown in Fig. 9–4. From this curve we can identify four different ways in which the material behaves, depending on the amount of strain induced in the material.

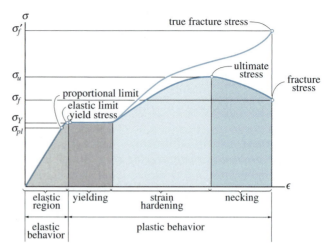

Conventional and true stress-strain diagrams
for ductile material (steel) (not to scale)

**Fig. 9–4**

**Elastic Behavior.**  Elastic behavior of the material occurs when the strains in the specimen are within the lightly shaded region shown in Fig. 9–4. It can be seen that the curve is actually a *straight line* throughout most of this region, so that stress is *proportional* to the strain. In other words, the material is said to be *linearly elastic*. The upper stress limit to this linear relationship is called the ***proportional limit***, $\sigma_{pl}$. If the stress slightly exceeds the proportional limit, the material may still respond elastically; however, the curve tends to bend and flatten out as shown. This continues until the stress reaches the ***elastic limit***. Upon reaching this point, if the load is removed the specimen will still return back to its original shape. Normally for steel, however, the elastic limit is seldom determined, since it is very close to the proportional limit and therefore rather difficult to detect.

**Yielding.**  A slight increase in stress above the elastic limit will result in a breakdown of the material and cause it to *deform permanently*. This behavior is called ***yielding***, and it is indicated by the dark-shaded region of the curve. The stress that causes yielding is called the ***yield stress or yield point***, $\sigma_Y$, and the deformation that occurs is called ***plastic deformation***. Although not shown in Fig. 9–4, for low-carbon steels or those that are hot rolled, the yield point is often distinguished by two values. The ***upper yield point*** occurs first, followed by a sudden decrease in load-carrying capacity to a ***lower yield point***. Once the yield point is reached, however, then as shown in Fig. 9–4, the specimen will continue to elongate (strain) *without any increase in load*. Realize that this figure is not drawn to scale. If it was, the induced strains due to yielding would be about 10 to 40 times greater than those produced up to the elastic limit. When the material is in this state, it is often referred to as being ***perfectly plastic***.

**Strain Hardening.** When yielding has ended, a further load can be applied to the specimen, resulting in a curve that rises continuously but becomes flatter until it reaches a maximum stress referred to as the ***ultimate stress***, $\sigma_u$. The rise in the curve in this manner is called ***strain hardening***, and it is identified in Fig. 9–4 as the region in light-shaded color. Throughout the test, while the specimen is elongating, its cross-sectional area will decrease. This decrease in area is fairly *uniform* over the specimen's entire gauge length, even up to the strain corresponding to the ultimate stress.

**Necking.** At the ultimate stress, the cross-sectional area begins to decrease in a *localized* region of the specimen, instead of over its entire length. This phenomenon is caused by slip planes formed within the material, and the actual strains produced are caused by shear stress. As a result, a constriction or "neck" gradually tends to form in this region as the specimen elongates further, Fig. 9–5a. Since the cross-sectional area within this region is continually decreasing, the smaller area can only carry an ever-decreasing load. Hence the stress–strain diagram tends to curve downward until the specimen breaks at the ***fracture stress***, $\sigma_f$, Fig. 9–5b. This region of the curve due to necking is indicated in dark color in Fig. 9–4.

Typical necking pattern which has occurred on this steel specimen just before fracture.

**True Stress–Strain Diagram.** Instead of always using the *original* cross-sectional area and specimen length to calculate the (engineering) stress and strain, we could have used the *actual* cross-sectional area and specimen length at the *instant* the load is measured. The values of stress and strain computed from these measurements are called *true stress* and *true strain*, and a plot of their values is called the ***true stress–strain diagram***. When this diagram is plotted it has a form shown by the light-colored curve in Fig. 9–4. Note that both the conventional and true $\sigma - \epsilon$ diagrams are practically coincident when the strain is small. The differences between the diagrams begin to appear in the strain-hardening range, where the magnitude of strain becomes more significant. In particular, there is a large divergence within the necking region. Here it can be seen from the conventional $\sigma - \epsilon$ diagram that the specimen *actually* supports a *decreasing load*, since $A_0$ is constant when calculating engineering stress, $\sigma = P/A_0$. However, from the true $\sigma - \epsilon$ diagram, the actual area $A$ within the necking region is always decreasing until fracture, $\sigma_{f'}$, and so the material actually sustains *increasing stress*, since $\sigma = P/A$.

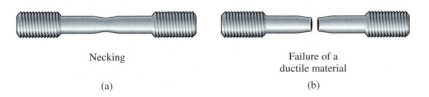

|  |  |
|:---:|:---:|
| Necking | Failure of a ductile material |
| (a) | (b) |

**Fig. 9–5**

Although the true and conventional stress–strain diagrams are different, most engineering design is done within the elastic range, since the distortion of the material is generally not severe within this range. Provided the material is "stiff," like most metals, the strain up to the elastic limit will remain small and the error in using the engineering values of $\sigma$ and $\epsilon$ is very small (about 0.1%) compared with their true values. This is one of the primary reasons for using conventional stress–strain diagrams.

The above concepts can be summarized with reference to Fig. 9–6, which shows an actual conventional stress–strain diagram for a mild steel specimen. In order to enhance the details, the elastic region of the curve has been shown in light color to an exaggerated strain scale, also shown in light color. Tracing the behavior, the proportional limit is reached at $\sigma_{pl} = 35$ ksi (241 MPa), where $\epsilon_{pl} = 0.0012$ in./in. This is followed by an upper yield point of $(\sigma_Y)_u = 38$ ksi (262 MPa), then suddenly a lower yield point of $(\sigma_Y)_l = 36$ ksi (248 MPa). The end of yielding occurs at a strain of $\epsilon_Y = 0.030$ in./in., which is 25 times greater than the strain at the proportional limit! Continuing, the specimen is strain hardened until it reaches the ultimate stress of $\sigma_u = 63$ ksi (435 MPa), then it begins to neck down until failure occurs, $\sigma_f = 47$ ksi (324 MPa). By comparison, the strain at failure, $\epsilon_f = 0.380$ in./in., is 317 times greater than $\epsilon_{pl}$!

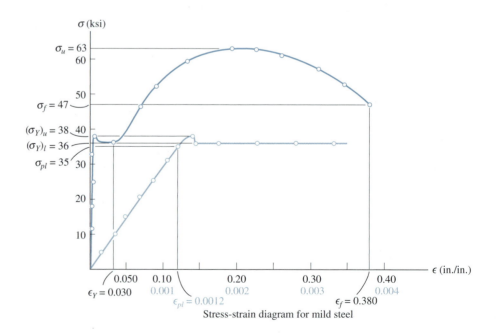

Stress-strain diagram for mild steel

**Fig. 9–6**

## 9.3   Stress–Strain Behavior of Ductile and Brittle Materials

Materials can be classified as either being ductile or brittle, depending on their stress–strain characteristics. Each will now be given separate treatment.

**Ductile Materials.**   Any material that can be subjected to large strains before it ruptures is called a *ductile material*. Mild steel, as discussed previously, is a typical example. Engineers often choose ductile materials for design because these materials are capable of absorbing shock or energy, and if they become overloaded, they will usually exhibit large deformation before failing.

One way to specify the ductility of a material is to report its percent elongation or percent reduction in area at the time of fracture. The *percent elongation* is the specimen's fracture strain expressed as a percent. Thus, if the specimen's original gauge-mark length is $L_0$ and its length at fracture is $L_f$, then

$$\text{Percent elongation} = \frac{L_f - L_0}{L_0}(100\%) \qquad (9\text{–}3)$$

As seen in Fig. 9–6, since $\epsilon_f = 0.380$, this value would be 38% for a mild steel specimen.

The *percent reduction in area* is another way to specify ductility. It is defined within the region of necking as follows:

$$\text{Percent reduction of area} = \frac{A_0 - A_f}{A_0}(100\%) \qquad (9\text{–}4)$$

Here $A_0$ is the specimen's original cross-sectional area and $A_f$ is the area at fracture. Mild steel has a typical value of 60%.

Besides steel, other metals such as brass, molybdenum, and zinc may also exhibit ductile stress–strain characteristics similar to steel, whereby they undergo elastic stress–strain behavior, yielding at constant stress, strain hardening, and finally necking until rupture. In most metals, however, constant yielding will *not occur* beyond the elastic range. One metal for which this is the case is aluminum. Actually, this metal often does not have a well-defined *yield point*, and consequently it is standard practice to define a *yield strength* for aluminum using a graphical procedure called the *offset method*. Normally a 0.2% strain (0.002 in./in.) is chosen, and from this point on the $\epsilon$ axis, a line parallel to the initial straight-line portion of the stress–strain diagram is drawn. The point where this line intersects the curve defines the yield strength. An example of the construction for determining the yield strength for an aluminum alloy is shown in Fig. 9–7. From the graph, the yield strength is $\sigma_{YS} = 51$ ksi (352 MPa).

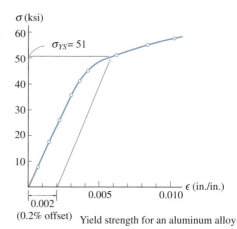

Yield strength for an aluminum alloy

**Fig. 9–7**

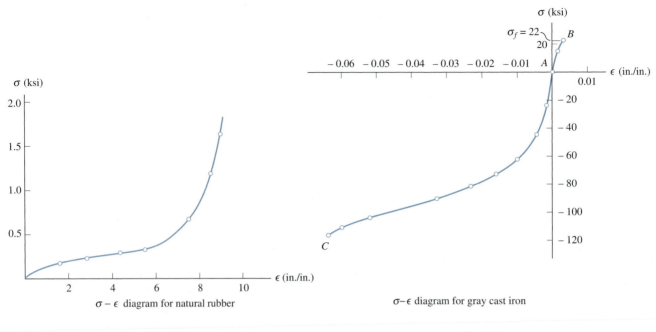

$\sigma - \epsilon$ diagram for natural rubber

Fig. 9–8

$\sigma-\epsilon$ diagram for gray cast iron

Fig. 9–9

Realize that the yield strength is not a physical property of the material, since it is a stress that caused a *specified* permanent strain in the material. In this text, however, we will assume that the yield strength, yield point, elastic limit, and proportional limit all *coincide* unless otherwise stated. An exception would be natural rubber, which in fact does not even have a proportional limit, since stress and strain are *not* linearly related, Fig. 9–8. Instead, this material, which is known as a polymer, exhibits *nonlinear elastic behavior*.

Wood is a material that is often moderately ductile, and as a result it is usually designed to respond only to elastic loadings. The strength characteristics of wood vary greatly from one species to another, and for each species they depend on the moisture content, age, and the size and arrangement of knots in the wood. Since wood is a fibrous material, its tensile or compressive characteristics will differ greatly when it is loaded either parallel or perpendicular to its grain. Specifically, wood splits easily when it is loaded in tension perpendicular to its grain, and consequently tensile loads are almost always intended to be applied parallel to the grain of wood members.

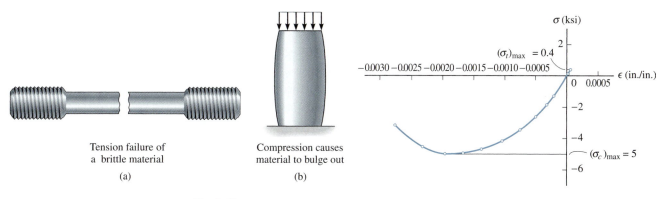

Tension failure of
a brittle material

(a)

Compression causes
material to bulge out

(b)

Fig. 9–10

$\sigma$-$\epsilon$ diagram for typical concrete mix

Fig. 9–11

## Brittle Materials.

Materials that exhibit little or no yielding before failure are referred to as ***brittle materials***. Gray cast iron is an example, having a stress–strain diagram in tension as shown by portion $AB$ of the curve in Fig. 9–9. Here fracture at $\sigma_f = 22$ ksi (152 MPa) took place initially at an imperfection or microscopic crack and then spread rapidly across the specimen, causing complete fracture. As a result of this type of failure, brittle materials do not have a well-defined tensile fracture stress, since the appearance of initial cracks in a specimen is quite random. Instead the *average* fracture stress from a set of observed tests is generally reported. A typical failed specimen is shown in Fig. 9–10a.

Compared with their behavior in tension, brittle materials, such as gray cast iron, exhibit a much higher resistance to axial compression, as evidenced by portion $AC$ of the curve in Fig. 9–9. For this case any cracks or imperfections in the specimen tend to close up, and as the load increases the material will generally bulge out or become barrel shaped as the strains become larger, Fig. 9–10b.

Like gray cast iron, concrete is classified as a brittle material, and it also has a low strength capacity in tension. The characteristics of its stress–strain diagram depend primarily on the mix of concrete (water, sand, gravel, and cement) and the time and temperature of curing. A typical example of a "complete" stress–strain diagram for concrete is given in Fig. 9–11. By inspection, its maximum compressive strength is almost 12.5 times greater than its tensile strength, $(\sigma_c)_{max} = 5$ ksi (34.5 MPa) versus $(\sigma_t)_{max} = 0.40$ ksi (2.76 MPa). For this reason, concrete is almost always reinforced with steel bars or rods whenever it is designed to support tensile loads.

It can generally be stated that most materials exhibit both ductile and brittle behavior. For example, steel has brittle behavior when it contains a high carbon content, and it is ductile when the carbon content is reduced. Also, at low temperatures materials become harder and more brittle, whereas when the temperature rises they become softer and more ductile. This effect is shown in Fig. 9–12 for a methacrylate plastic.

Steel rapidly loses its strength when heated. For this reason engineers often require main structural members to be insulated in case of fire.

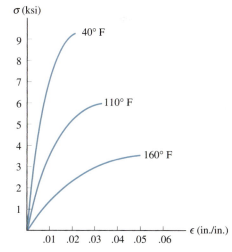

$\sigma - \epsilon$ diagrams for a methacrylate plastic

Fig. 9–12

## 9.4 Hooke's Law

As noted in the previous section, the stress–strain diagrams for most engineering materials exhibit a *linear relationship* between stress and strain within the elastic region. Consequently, an increase in stress causes a proportionate increase in strain. This fact was discovered by Robert Hooke in 1676 using springs and is known as *Hooke's law*. It may be expressed mathematically as

$$\sigma = E\epsilon \qquad\qquad (9\text{–}5)$$

Here $E$ represents the constant of proportionality, which is called the **modulus of elasticity** or **Young's modulus**, named after Thomas Young, who published an account of it in 1807.

Equation 9–5 actually represents the equation of the *initial straight-lined portion* of the stress–strain diagram up to the proportional limit. Furthermore, the modulus of elasticity represents the *slope* of this line. Since strain is dimensionless, from Eq. 9–5, $E$ will have units of stress, such as psi, ksi, or pascals. As an example of its calculation, consider the stress–strain diagram for steel shown in Fig. 9–6. Here $\sigma_{pl} = 35$ ksi and $\epsilon_{pl} = 0.0012$ in./in., so that

$$E = \frac{\sigma_{pl}}{\epsilon_{pl}} = \frac{35 \text{ ksi}}{0.0012 \text{ in./in.}} = 29(10^3) \text{ ksi}$$

As shown in Fig. 9–13, the proportional limit for a particular type of steel depends on its alloy content; however, most grades of steel, from the softest rolled steel to the hardest tool steel, have about the same modulus of elasticity, generally accepted to be $E_{st} = 29(10^3)$ ksi or 200 GPa. Common values of $E$ for other engineering materials are often tabulated in engineering code and reference books. Representative values are also listed in Appendix B. It should be noted that the modulus of elasticity is a mechanical property that indicates the *stiffness* of a material. Materials that are very stiff, such as steel, have large values of $E$ [$E_{st} = 29(10^3)$ ksi or 200 GPa], whereas spongy materials such as vulcanized rubber may have low values [$E_r = 0.10(10^3)$ ksi or 0.70 MPa].

The modulus of elasticity is one of the most important mechanical properties used in the development of equations presented in this text. It must always be remembered, though, that $E$ can be used only if a material has *linear-elastic* behavior. Also, if the stress in the material is *greater* than the proportional limit, the stress–strain diagram ceases to be a straight line and Eq. 9–5 is no longer valid.

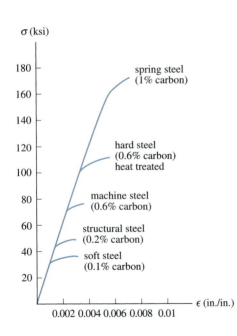

Fig. 9–13

## Strain Hardening.

If a specimen of ductile material, such as steel, is loaded into the *plastic region* and then unloaded, *elastic strain is recovered* as the material returns to its equilibrium state. The *plastic strain remains*, however, and as a result the material is subjected to a **permanent set**. For example, a wire when bent (plastically) will spring back a little (elastically) when the load is removed; however, it will not fully return to its original position. This behavior can be illustrated on the stress–strain diagram shown in Fig. 9–14a. Here the specimen is first loaded beyond its yield point A to the point A'. Since interatomic forces have to be overcome to elongate the specimen *elastically*, then these same forces pull the atoms back together when the load is removed, Fig. 9–14a. Consequently, the modulus of elasticity, E, is the same, and therefore the slope of line O'A' is the same as line OA.

If the load is reapplied, the atoms in the material will again be displaced until yielding occurs at or near the stress A', and the stress–strain diagram continues along the same path as before, Fig. 9–14b. It should be noted, however, that this new stress–strain diagram, defined by O'A'B, now has a *higher* yield point (A'), a consequence of strain-hardening. In other words, the material now has a *greater elastic region*; however, it has *less ductility*, a smaller plastic region, than when it was in its original state.

In the true sense some heat or *energy* may be *lost* as the specimen is unloaded from A' and then again loaded to this same stress. As a result, slight curves in the paths A' to O' and O' to A' will occur during a carefully measured cycle of loading. This is shown by the dashed curves in Fig. 9–14b. The colored area between these curves represents lost energy and is called **mechanical hysteresis**. It becomes an important consideration when selecting materials to serve as dampers for vibrating structural or mechanical equipment, although its effects will not be considered in this text.

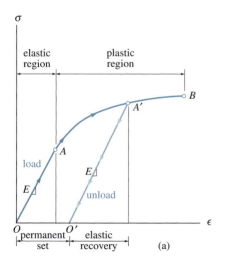

(a)

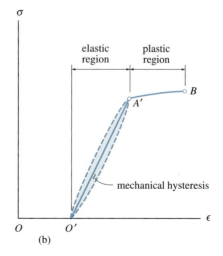

(b)

**Fig. 9–14**

## 9.5 Strain Energy

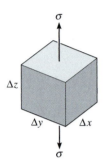

**Fig. 9–15**

As a material is deformed by an external loading, it tends to store energy *internally* throughout its volume. Since this energy is related to the strains in the material, it is referred to as **strain energy**. For example, when a tension-test specimen is subjected to an axial load, a volume element of the material is subjected to uniaxial stress as shown in Fig. 9–15. This stress develops a force $\Delta F = \sigma\,\Delta A = \sigma(\Delta x\,\Delta y)$ on the top and bottom faces of the element *after* the element undergoes a vertical displacement $\epsilon\,\Delta z$. By definition, *work* is determined by the product of the force and displacement in the direction of the force. Since the force is increased uniformly from zero to its final magnitude $\Delta F$ when the displacement $\epsilon\,\Delta z$ is attained, the work done on the element by the force is equal to the *average* force magnitude $(\Delta F/2)$ times the displacement $\epsilon\,\Delta z$. This "external work" is equivalent to the "internal work" or strain energy stored in the element—assuming that no energy is lost in the form of heat. Consequently, the strain energy $\Delta U$ is $\Delta U = (1/2\,\Delta F)\,\epsilon\,\Delta z = (1/2\,\sigma\,\Delta x\,\Delta y)\,\epsilon\,\Delta z$. Since the volume of the element is $\Delta V = \Delta x\,\Delta y\,\Delta z$, then $\Delta U = 1/2\,\sigma\epsilon\,\Delta V$.

It is sometimes convenient to formulate the strain energy per unit volume of material. This is called the **strain-energy density**, and it can be expressed as

$$u = \frac{\Delta U}{\Delta V} = \frac{1}{2}\sigma\epsilon \tag{9–6}$$

If the material behavior is *linear elastic*, then Hooke's law applies, $\sigma = E\epsilon$, and therefore we can express the strain-energy density in terms of the uniaxial stress as

$$u = \frac{1}{2}\frac{\sigma^2}{E} \tag{9–7}$$

**Modulus of Resilience.** In particular, when the stress $\sigma$ reaches the proportional limit, the strain-energy density, as calculated by Eq. 9–6 or 9–7, is referred to as the **modulus of resilience**, i.e.,

$$u_r = \frac{1}{2}\sigma_{pl}\epsilon_{pl} = \frac{1}{2}\frac{\sigma_{pl}^2}{E} \tag{9–8}$$

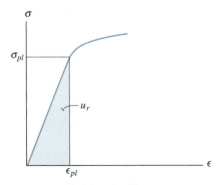

Modulus of resilience $u_r$

(a)

**Fig. 9–16**

From the elastic region of the stress–strain diagram, Fig. 9–16a, notice that $u_r$ is equivalent to the shaded *triangular area* under the diagram. Physically a material's resilience represents the ability of the material to absorb energy without any permanent damage to the material.

**Modulus of Toughness.** Another important property of a material is the **modulus of toughness**, $u_t$. This quantity represents the *entire area* under the stress–strain diagram, Fig. 9–16b, and therefore it indicates the strain-energy density of the material just before it fractures. This property becomes important when designing members that may be accidentally overloaded. Materials with a high modulus of toughness will distort greatly due to an overloading; however, they may be preferable to those with a low value, since materials having a low $u_t$ may suddenly fracture without warning of an approaching failure. Alloying metals can also change their resilience and toughness. For example, by changing the percentage of carbon in steel, the resulting stress–strain diagrams in Fig. 9–17 show how the degrees of resilience and toughness can be changed.

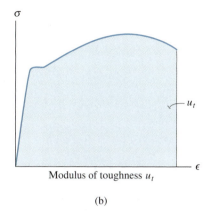

Modulus of toughness $u_t$

(b)

**Fig. 9–16**

## IMPORTANT POINTS

- A *conventional stress–strain diagram* is important in engineering since it provides a means for obtaining data about a material's tensile or compressive strength without regard for the material's physical size or shape.

- *Engineering stress and strain* are calculated using the *original* cross-sectional area and gauge length of the specimen.

- A *ductile material*, such as mild steel, has four distinct behaviors as it is loaded. They are *elastic behavior, yielding, strain hardening,* and *necking.*

- A material is *linear elastic* if the stress is proportional to the strain within the elastic region. This is referred to as *Hooke's law*, and the slope of the curve is called the *modulus of elasticity, E.*

- Important points on the stress–strain diagram are the *proportional limit, elastic limit, yield stress, ultimate stress,* and the *fracture stress.*

- The *ductility* of a material can be specified by the specimen's *percent elongation* or the *percent reduction in area.*

- If a material does not have a distinct yield point, a *yield strength* can be specified using a graphical procedure such as the *offset method.*

- *Brittle materials*, such as gray cast iron, have very little or no yielding and fracture suddenly.

- *Strain hardening* is used to establish a higher yield point for a material. This is done by straining the material beyond the elastic limit, then releasing the load. The modulus of elasticity remains the same; however, the material's ductility *decreases.*

- *Strain energy* is energy that is stored in a material due to its deformation. This energy per unit volume is called *strain-energy density*. If it is measured up to the proportional limit, it is referred to as the *modulus of resilience*, and if it is measured up to the point of fracture, it is called the *modulus of toughness.*

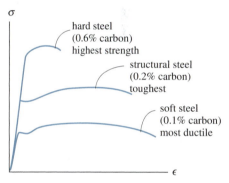

hard steel
(0.6% carbon)
highest strength

structural steel
(0.2% carbon)
toughest

soft steel
(0.1% carbon)
most ductile

**Fig. 9–17**

This nylon specimen exhibits a high degree of toughness, as noted by the large amount of necking that has occurred just before fracture.

**E X A M P L E** **9.1**

A tension test for a steel alloy results in the stress–strain diagram shown in Fig. 9–18. Calculate the modulus of elasticity and the yield strength based on a 0.2% offset. Identify on the graph the ultimate stress and the fracture stress.

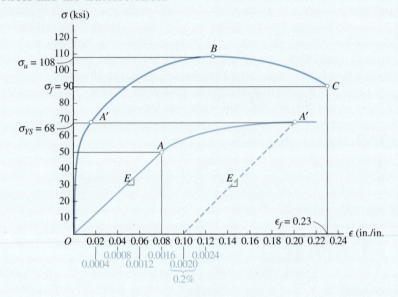

**Fig. 9–18**

**Solution**

*Modulus of Elasticity.* We must calculate the *slope* of the initial straight-line portion of the graph. Using the magnified curve and scale shown in color, this line extends from point $O$ to an estimated point $A$, which has coordinates of approximately (0.0016 in./in., 50 ksi). Therefore,

$$E = \frac{50 \text{ ksi}}{0.0016 \text{ in./in.}} = 31.2(10^3) \text{ ksi} \qquad Ans.$$

Note that the equation of the line $OA$ is thus $\sigma = 31.2(10^3)\epsilon$.

*Yield Strength.* For a 0.2% offset, we begin at a strain of 0.2% or 0.0020 in./in. and graphically extend a (dashed) line parallel to $OA$ until it intersects the $\sigma{-}\epsilon$ curve at $A'$. The yield strength is approximately

$$\sigma_{YS} = 68 \text{ ksi} \qquad Ans.$$

*Ultimate Stress.* This is defined by the peak of the $\sigma{-}\epsilon$ graph, point $B$ in Fig. 9–18.

$$\sigma_u = 108 \text{ ksi} \qquad Ans.$$

*Fracture Stress.* When the specimen is strained to its maximum of $\epsilon_f = 0.23$ in./in., it fractures at point $C$. Thus,

$$\sigma_f = 90 \text{ ksi} \qquad Ans.$$

# EXAMPLE 9.2

The stress–strain diagram for an aluminum alloy that is used for making aircraft parts is shown in Fig. 9–19. If a specimen of this material is stressed to 600 MPa, determine the permanent strain that remains in the specimen when the load is released. Also, compute the modulus of resilience both before and after the load application.

### Solution

*Permanent Strain.*   When the specimen is subjected to the load, it strain-hardens until point $B$ is reached on the $\sigma$–$\epsilon$ diagram, Fig. 9–19. The strain at this point is approximately 0.023 mm/mm. When the load is released, the material behaves by following the straight line $BC$, which is parallel to line $OA$. Since both lines have the same slope, the strain at point $C$ can be determined analytically. The slope of line $OA$ is the modulus of elasticity, i.e.,

$$E = \frac{450 \text{ MPa}}{0.006 \text{ mm/mm}} = 75.0 \text{ GPa}$$

From triangle $CBD$, we require

$$E = \frac{BD}{CD} = \frac{600(10^6) \text{ Pa}}{CD} = 75.0(10^9) \text{ Pa}$$

$$CD = 0.008 \text{ mm/mm}$$

This strain represents the amount of *recovered elastic strain*. The permanent strain, $\epsilon_{OC}$, is thus

$$\epsilon_{OC} = 0.023 \text{ mm/mm} - 0.008 \text{ mm/mm}$$

$$= 0.0150 \text{ mm/mm} \qquad\qquad Ans.$$

*Note:* If gauge marks on the specimen were originally 50 mm apart, then after the load is *released* these marks will be 50 mm + (0.0150)( 50 mm) = 50.75 mm apart.

*Modulus of Resilience.*   Applying Eq. 9–8, we have*

$$(u_r)_{\text{initial}} = \frac{1}{2}\sigma_{pl}\epsilon_{pl} = \frac{1}{2}(450 \text{ MPa})(0.006 \text{ mm/mm})$$

$$= 1.35 \text{ MJ/m}^3 \qquad\qquad Ans.$$

$$(u_r)_{\text{final}} = \frac{1}{2}\sigma_{pl}\epsilon_{pl} = \frac{1}{2}(600 \text{ MPa})(0.008 \text{ mm/mm})$$

$$= 2.40 \text{ MJ/m}^3 \qquad\qquad Ans.$$

By comparison, the effect of strain-hardening the material has caused an increase in the modulus of resilience; however, note that the modulus of toughness for the material has decreased since the area under the original curve, $OABF$, is larger than the area under curve $CBF$.

*Work in the SI system of units is measured in joules, where 1 J = 1 N · m.

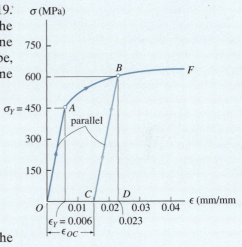

Fig. 9–19

# EXAMPLE 9.3

An aluminum rod shown in Fig. 9–20a has a circular cross section and is subjected to an axial load of 10 kN. If a portion of the stress–strain diagram for the material is shown in Fig. 9–20b, determine the approximate elongation of the rod when the load is applied. If the load is removed, what is the permanent elongation of the rod? Take $E_{al}$ = 70 GPa.

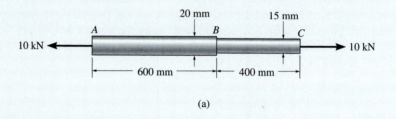

(a)

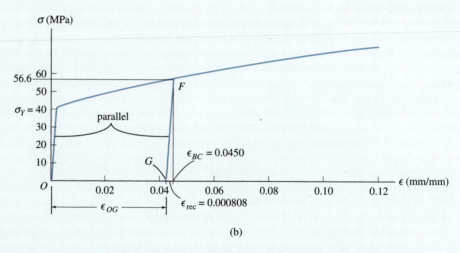

(b)

**Fig. 9–20**

## Solution

For the analysis we will neglect the *localized deformations* at the point of load application and where the rod's cross-sectional area suddenly changes. (These effects will be discussed in Secs. 10.1 and 10.6.) Throughout the midsection of each segment the normal stress and deformation are uniform.

In order to study the deformation of the rod, we must obtain the strain. This is done by first calculating the stress, then using the stress–strain diagram to obtain the strain. The normal stress within each segment is

$$\sigma_{AB} = \frac{P}{A} = \frac{10(10^3) \text{ N}}{\pi(0.01 \text{ m})^2} = 31.83 \text{ MPa}$$

$$\sigma_{BC} = \frac{P}{A} = \frac{10(10^3) \text{ N}}{\pi(0.0075 \text{ m})^2} = 56.59 \text{ MPa}$$

From the stress–strain diagram, the material in region $AB$ is strained *elastically* since $\sigma_Y = 40 \text{ MPa} > 31.83 \text{ MPa}$. Using Hooke's law,

$$\epsilon_{AB} = \frac{\sigma_{AB}}{E_{al}} = \frac{31.83(10^6) \text{ Pa}}{70(10^9) \text{ Pa}} = 0.0004547 \text{ mm/mm}$$

The material within region $BC$ is strained plastically, since $\sigma_Y = 40 \text{ MPa} < 56.59 \text{ MPa}$. From the graph, for $\sigma_{BC} = 56.59 \text{ MPa}$,

$$\epsilon_{BC} \approx 0.045 \text{ mm/mm}$$

The approximate elongation of the rod is therefore

$$\delta = \Sigma \epsilon L = 0.0004547(600 \text{ mm}) + 0.045(400 \text{ mm})$$
$$= 18.3 \text{ mm} \qquad\qquad\qquad \textit{Ans.}$$

When the 10-kN load is removed, segment $AB$ of the rod will be restored to its original length. Why? On the other hand, the material in segment $BC$ will recover elastically along line $FG$, Fig. 9–20b. Since the slope of $FG$ is $E_{al}$, the elastic strain recovery is

$$\epsilon_{\text{rec}} = \frac{\sigma_{BC}}{E_{al}} = \frac{56.59(10^6) \text{ Pa}}{70(10^9) \text{ Pa}} = 0.000808 \text{ mm/mm}$$

The remaining plastic strain in segment $BC$ is then

$$\epsilon_{OG} = 0.0450 - 0.000808 = 0.0442 \text{ mm/mm}$$

Therefore, when the load is removed the rod remains elongated by an amount

$$\delta' = \epsilon_{OG} L_{BC} = 0.0442(400 \text{ mm}) = 17.7 \text{ mm} \qquad \textit{Ans.}$$

# PROBLEMS

**9-1.** A concrete cylinder having a diameter of 6.00 in. and gauge length of 12 in. is tested in compression. The results of the test are reported in the table as load versus contraction. Draw the stress–strain diagram using scales of 1 in. = 0.5 ksi and 1 in. = $0.2(10^{-3})$ in./in. From the diagram, determine approximately the modulus of elasticity.

| Load (kip) | Contraction (in.) |
|---|---|
| 0 | 0 |
| 5.0 | 0.0006 |
| 9.5 | 0.0012 |
| 16.5 | 0.0020 |
| 20.5 | 0.0026 |
| 25.5 | 0.0034 |
| 30.0 | 0.0040 |
| 34.5 | 0.0045 |
| 38.5 | 0.0050 |
| 46.5 | 0.0062 |
| 50.0 | 0.0070 |
| 53.0 | 0.0075 |

**Prob. 9-1**

**9-2.** Data taken from a stress–strain test for a ceramic are given in the table. The curve is linear between the origin and the first point. Plot the diagram, and determine the modulus of elasticity and the modulus of resilience.

**9-3.** Data taken from a stress–strain test for a ceramic are given in the table. The curve is linear between the origin and the first point. Plot the diagram, and determine approximately the modulus of toughness. The rupture stress is $\sigma_r = 53.4$ ksi.

| $\sigma$ (ksi) | $\epsilon$ (in./in.) |
|---|---|
| 0 | 0 |
| 33.2 | 0.0006 |
| 45.5 | 0.0010 |
| 49.4 | 0.0014 |
| 51.5 | 0.0018 |
| 53.4 | 0.0022 |

**Probs. 9–2/3**

**\*9-4.** Data taken from a stress–strain test are given in the table. The curve is linear between the origin and the first point. Plot the diagram, and determine the modulus of elasticity and the modulus of resilience.

| $\sigma$ (ksi) | $\epsilon$ (in./in.) |
|---|---|
| 0 | 0 |
| 32.0 | 0.0016 |
| 33.5 | 0.0018 |
| 40.0 | 0.0030 |
| 41.2 | 0.0050 |

**Prob. 9–4**

**9-5.** A tension test was performed on a specimen having an original diameter of 12.5 mm and a gauge length of 50 mm. The data are listed in the table. Plot the stress–strain diagram, and determine approximately the modulus of elasticity, the ultimate stress, and the fracture stress. Use a scale of 20 mm = 50 MPa and 20 mm = 0.05 mm/mm. Redraw the linear-elastic region, using the same stress scale but a strain scale of 20 mm = 0.001 mm/mm.

**9-6.** A tension test was performed on a steel specimen having an original diameter of 12.5 mm and gauge length of 50 mm. Using the data listed in the table, plot the stress–strain diagram, and determine approximately the modulus of toughness. Use a scale of 20 mm = 50 MPa and 20 mm = 0.05 mm/mm.

| Load (kN) | Elongation (mm) |
|---|---|
| 0 | 0 |
| 11.1 | 0.0175 |
| 31.9 | 0.0600 |
| 37.8 | 0.1020 |
| 40.9 | 0.1650 |
| 43.6 | 0.2490 |
| 53.4 | 1.0160 |
| 62.3 | 3.0480 |
| 64.5 | 6.3500 |
| 62.3 | 8.8900 |
| 58.8 | 11.9380 |

**Probs. 9–5/6**

**9-7.** The stress–strain diagram for a steel alloy having an original diameter of 0.5 in. and a gauge length of 2 in. is given in the figure. If the specimen is loaded until it is stressed to 70 ksi, determine the approximate amount of elastic recovery and the permanent increase in the gauge length after it is unloaded.

**\*9-8.** The stress–strain diagram for a steel alloy having an original diameter of 0.5 in. and a gauge length of 2 in. is given in the figure. Determine approximately the modulus of elasticity for the material, the load on the specimen that causes yielding, and the ultimate load the specimen will support.

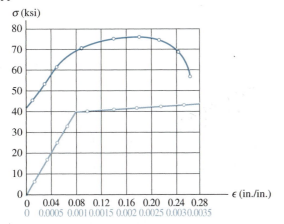

Probs. 9–7/8

**9-9.** The stress–strain diagram for a steel alloy having an original diameter of 0.5 in. and a gauge length of 2 in. is given in the figure. Determine approximately the modulus of resilience and the modulus of toughness for the material.

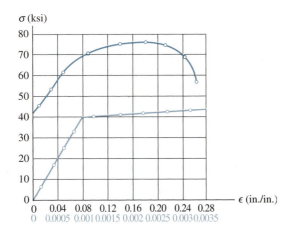

Prob. 9–9

**9-10.** The stress–strain diagram for a bar of steel alloy is shown in the figure. Determine approximately the modulus of elasticity, the proportional limit, the ultimate stress, and the modulus of resilience. If the bar is loaded until it is stressed to 360 MPa, determine the elastic strain recovery and the permanent set or strain in the bar when it is unloaded.

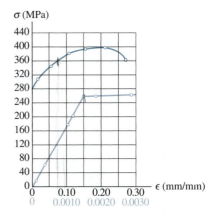

Prob. 9–10

**9-11.** An A-36 steel bar has a length of 50 in. and cross-sectional area of 0.7 in². Determine the length of the bar if it is subjected to an axial tension of 5000 lb. The material has linear-elastic behavior.

**\*9-12.** The stress–strain diagram for polyethylene, which is used to sheath coaxial cables, is determined from testing a specimen that has a gauge length of 10 in. If a load $P$ on the specimen develops a strain of $\epsilon = 0.024$ in./in., determine the approximate length of the specimen, measured between the gauge points, when the load is removed. Assume the specimen recovers elastically.

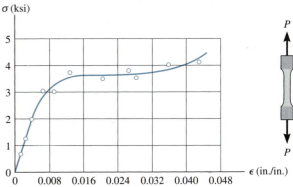

Probs. 9–11/12

**9-13.** The change in weight of an airplane is determined from reading the strain gauge $A$ mounted in the plane's aluminum wheel strut. *Before* the plane is loaded, the strain-gauge reading in a strut is $\epsilon_1 = 0.00100$ in./in., whereas after loading $\epsilon_2 = 0.00243$ in./in. Determine the change in the force on the strut if the cross-sectional area of the strut is 3.5 in$^2$. $E_{al} = 10(10^3)$ ksi.

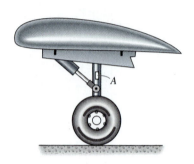

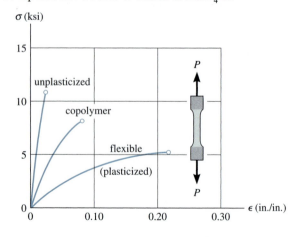

Prob. 9–13

**9-14.** By adding plasticizers to polyvinyl chloride, it is possible to reduce its stiffness. The stress–strain diagrams for three types of this material showing this effect are given below. Specify the type that should be used in the manufacture of a rod having a length of 5 in. and a diameter of 2 in., that is required to support at least an axial load of 20 kip and also be able to stretch at most $\frac{1}{4}$ in.

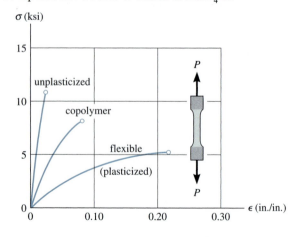

Prob. 9–14

**9-15.** A bar having a length of 5 in. and cross-sectional area of 0.7 in$^2$ is subjected to an axial force of 8000 lb. If the bar stretches 0.002 in., determine the modulus of elasticity of the material. The material has linear-elastic behavior.

8000 lb          5 in.          8000 lb

Prob. 9–15

*9-16.** A specimen is originally 1 ft long, has a diameter of 0.5 in., and is subjected to a force of 500 lb. When the force is increased from 500 lb to 1800 lb, the specimen elongates 0.009 in. Determine the modulus of elasticity for the material if it remains linear elastic.

**9-17.** A structural member in a nuclear reactor is made of a zirconium alloy. If an axial load of 4 kip is to be supported by the member, determine its required cross-sectional area. Use a factor of safety of 3 relative to yielding. What is the load on the member if it is 3 ft long and its elongation is 0.02 in.? $E_{zr} = 14(10^3)$ ksi, $\sigma_Y = 57.5$ ksi. The material has elastic behavior.

**9-18.** The steel wires $AB$ and $AC$ support the 200-kg mass. If the allowable axial stress for the wires is $\sigma_{allow} = 130$ MPa, determine the required diameter of each wire. Also, what is the new length of wire $AB$ after the load is applied? Take the unstretched length of $AB$ to be 750 mm. $E_{st} = 200$ GPa.

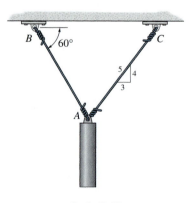

Prob. 9–18

## 9.6 Poisson's Ratio

When a deformable body is subjected to an axial tensile force, not only does it elongate but it also contracts laterally. For example, if a rubber band is stretched, it can be noted that both the thickness and width of the band are decreased. Likewise, a compressive force acting on a body causes it to contract in the direction of the force and yet its sides expand laterally. These two cases are illustrated in Fig. 9–21 for a bar having an original radius $r$ and length $L$.

When the load **P** is applied to the bar, it changes the bar's length by an amount $\delta$ and its radius by an amount $\delta'$. Strains in the longitudinal or axial direction and in the lateral or radial direction are, respectively,

$$\epsilon_{\text{long}} = \frac{\delta}{L} \quad \text{and} \quad \epsilon_{\text{lat}} = \frac{\delta'}{r}$$

In the early 1800s, the French scientist S. D. Poisson realized that within the *elastic range* the *ratio* of these strains is a *constant*, since the deformations $\delta$ and $\delta'$ are proportional. This constant is referred to as **Poisson's ratio**, $v$ (nu), and it has a numerical value that is unique for a particular material that is both *homogeneous and isotropic*. Stated mathematically it is

$$v = -\frac{\epsilon_{\text{lat}}}{\epsilon_{\text{long}}} \tag{9-9}$$

The negative sign is used here since *longitudinal elongation* (positive strain) causes *lateral contraction* (negative strain), and vice versa. Notice that this lateral strain is the *same* in all lateral (or radial) directions. Furthermore, this strain is caused only by the axial or longitudinal force; i.e., no force or stress acts in a lateral direction in order to strain the material in this direction.

Poisson's ratio is seen to be *dimensionless*, and for most nonporous solids it has a value that is generally between $\frac{1}{4}$ and $\frac{1}{3}$. Typical values of $v$ for common materials are listed in Appendix B. In particular, an ideal material having no lateral movement when it is stretched or compressed will have $v = 0$. Furthermore, it will be shown in Sec. 15.10 that the *maximum* possible value for Poisson's ratio is 0.5. Therefore $0 \leq v \leq 0.5$.

When the rubber block is compressed (negative strain) its sides will expand (positive strain). The ratio of these strains is constant.

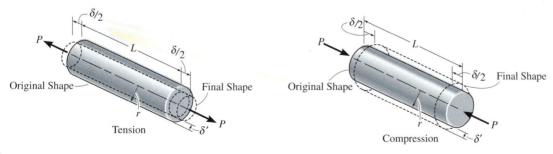

Fig. 9–21

E X A M P L E   9.4

A bar made of A-36 steel has the dimensions shown in Fig. 9–22. If an axial force of $P = 80$ kN is applied to the bar, determine the change in its length and the change in the dimensions of its cross section after applying the load. The material behaves elastically.

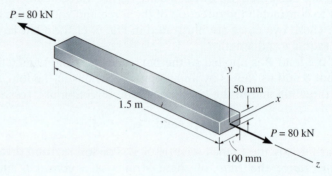

P = 80 kN

y

50 mm

1.5 m

x

P = 80 kN

100 mm

z

**Fig. 9–22**

**Solution**

The normal stress in the bar is

$$\sigma_z = \frac{P}{A} = \frac{80(10^3)\ \text{N}}{(0.1\ \text{m})(0.05\ \text{m})} = 16.0(10^6)\ \text{Pa}$$

From the tables in Appendix B, for A-36 steel, $E_{st} = 200$ GPa, and so the strain in the $z$ direction is

$$\epsilon_z = \frac{\sigma_z}{E_{st}} = \frac{16.0(10^6)\ \text{Pa}}{200(10^9)\ \text{Pa}} = 80(10^{-6})\ \text{mm/mm}$$

The axial elongation of the bar is therefore

$$\delta_z = \epsilon_z L_z = [80(10^{-6})](1.5\ \text{m}) = 120\ \mu\text{m} \qquad \qquad Ans.$$

Using Eq. 9–9, where $v_{st} = 0.32$ as found from the Appendix B, the contraction strains in *both* the $x$ and $y$ directions are

$$\epsilon_x = \epsilon_y = -v_{st}\epsilon_z = -0.32[80(10^{-6})] = -25.6\ \mu\text{m/m}$$

Thus the changes in the dimensions of the cross section are

$$\delta_x = \epsilon_x L_x = -[25.6(10^{-6})](0.1\ \text{m}) = -2.56\ \mu\text{m} \qquad Ans.$$

$$\delta_y = \epsilon_y L_y = -[25.6(10^{-6})](0.05\ \text{m}) = -1.28\ \mu\text{m} \qquad Ans.$$

$\ell$

## 9.7  The Shear Stress–Strain Diagram

In Sec. 8.4 it was shown that when an element of material is subjected to *pure shear*, equilibrium requires that equal shear stresses must be developed on four faces of the element. These stresses must be directed either toward or away from diagonally opposite corners of the element, Fig. 9–23a. Furthermore, if the material is *homogeneous* and *isotropic*, then this shear stress will distort the element uniformly, Fig. 9–23b. As mentioned in Sec. 8.8, the shear strain $\gamma_{xy}$ measures the angular distortion of the element relative to the sides originally along the $x$ and $y$ axes.

The behavior of a material subjected to pure shear can be studied in a laboratory by using specimens in the shape of thin circular tubes and subjecting them to a torsional loading. If measurements are made of the applied torque and the resulting angle of twist, then by the methods to be explained in Chapter 11, the data can be used to determine the shear stress and shear strain, and a shear stress–strain diagram plotted. An example of such a diagram for a ductile material is shown in Fig. 9–24. Like the tension test, this material when subjected to shear will exhibit linear-elastic behavior and it will have a defined *proportional limit* $\tau_{pl}$. Also, strain hardening will occur until an *ultimate shear stress* $\tau_u$ is reached. And finally, the material will begin to lose its shear strength until it reaches a point where it fractures, $\tau_f$.

For most engineering materials, like the one just described, the elastic behavior is *linear*, and so Hooke's law for shear can be written as

$$\tau = G\gamma \tag{9–10}$$

Here $G$ is called the ***shear modulus of elasticity*** or the ***modulus of rigidity***. Its value can be measured as the slope of the line on the $\tau$–$\gamma$ diagram, that is, $G = \tau_{pl}/\gamma_{pl}$. Typical values for common engineering materials are listed in Appendix B. Notice that the units of measurement for $G$ will be the *same* as those for $E$ (Pa or psi), since $\gamma$ is measured in radians, a dimensionless quantity.

It will be shown in Sec. 15.10 that the three material constants, $E$, $v$, and $G$ are actually *related* by the equation

$$G = \frac{E}{2(1 + v)} \tag{9–11}$$

Provided $E$ and $G$ are known, the value of $v$ can be determined from this equation rather than through experimental measurement. For example, in the case of A-36 steel, $E_{st} = 29(10^3)$ ksi and $G_{st} = 11.0(10^3)$ ksi, so that, from Eq. 9–11, $v_{st} = 0.32$.

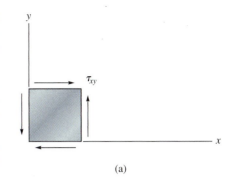

(a)

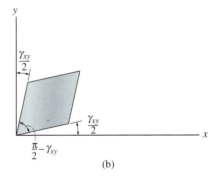

(b)

**Fig. 9–23**

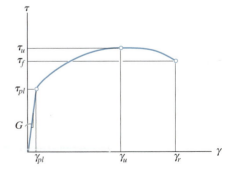

**Fig. 9–24**

# EXAMPLE 9.5

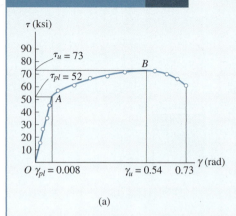

(a)

A specimen of titanium alloy is tested in torsion and the shear stress–strain diagram is shown in Fig. 9–25a. Determine the shear modulus $G$, the proportional limit, and the ultimate shear stress. Also, determine the maximum distance $d$ that the top of a block of this material, shown in Fig. 9–25b, could be displaced horizontally if the material behaves elastically when acted upon by a shear force $\mathbf{V}$. What is the magnitude of $\mathbf{V}$ necessary to cause this displacement?

## Solution

**Shear Modulus.** This value represents the slope of the straight-line portion $OA$ of the $\tau$–$\gamma$ diagram. The coordinates of point $A$ are (0.008 rad, 52 ksi). Thus,

$$G = \frac{52 \text{ ksi}}{0.008 \text{ rad}} = 6500 \text{ ksi} \qquad Ans.$$

The equation of line $OA$ is therefore $\tau = 6500\gamma$, which is Hooke's law for shear.

**Proportional Limit.** By inspection, the graph ceases to be linear at point $A$. Thus,

$$\tau_{pl} = 52 \text{ ksi} \qquad Ans.$$

**Ultimate Stress.** This value represents the maximum shear stress, point $B$. From the graph,

$$\tau_u = 73 \text{ ksi} \qquad Ans.$$

**Maximum Elastic Displacement and Shear Force.** Since the maximum elastic shear strain is 0.008 rad, a very small angle, the top of the block in Fig. 9–25b will be displaced horizontally:

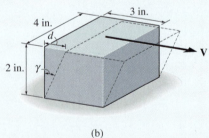

$$\tan(0.008 \text{ rad}) \approx 0.008 \text{ rad} = \frac{d}{2 \text{ in.}}$$

$$d = 0.016 \text{ in.} \qquad Ans.$$

The corresponding *average* shear stress in the block is $\tau_{pl} = 52$ ksi. Thus, the shear force $V$ needed to cause the displacement is

$$\tau_{avg} = \frac{V}{A}; \qquad 52 \text{ ksi} = \frac{V}{(3 \text{ in.})(4 \text{ in.})}$$

$$V = 624 \text{ kip} \qquad Ans.$$

(b)

**Fig. 9–25**

# E X A M P L E   9.6

An aluminum specimen shown in Fig. 9–26 has a diameter of $d_0 = 25$ mm and a gauge length of $L_0 = 250$ mm. If a force of 165 kN elongates the gauge length 1.20 mm, determine the modulus of elasticity. Also, determine by how much the force causes the diameter of the specimen to contract. Take $G_{al} = 26$ GPa and $\sigma_Y = 440$ MPa.

### Solution

*Modulus of Elasticity.* The average normal stress in the specimen is

$$\sigma = \frac{P}{A} = \frac{165(10^3)\ N}{(\pi/4)(0.025\ m)^2} = 336.1\ MPa$$

and the average normal strain is

$$\epsilon = \frac{\delta}{L} = \frac{1.20\ mm}{250\ mm} = 0.00480\ mm/mm$$

Since $\sigma < \sigma_Y = 440$ MPa, the material behaves elastically. The modulus of elasticity is

$$E_{al} = \frac{\sigma}{\epsilon} = \frac{336.1(10^6)\ Pa}{0.00480} = 70.0\ GPa \qquad \textit{Ans.}$$

*Contraction of Diameter.* First we will determine Poisson's ratio for the material using Eq. 9–11.

$$G = \frac{E}{2(1 + v)}$$

$$26\ GPa = \frac{70.0\ GPa}{2(1 + v)}$$

$$v = 0.346$$

Since $\epsilon_{long} = 0.00480$ mm/mm, then by Eq. 9–9,

$$v = -\frac{\epsilon_{lat}}{\epsilon_{long}}$$

$$0.346 = -\frac{\epsilon_{lat}}{0.00480\ mm/mm}$$

$$\epsilon_{lat} = -0.00166\ mm/mm$$

The contraction of the diameter is therefore

$$\delta' = (0.00166)(25\ mm)$$

$$= 0.0415\ mm \qquad \textit{Ans.}$$

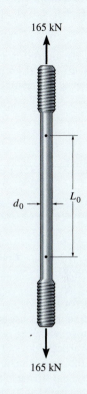

165 kN

$d_0$    $L_0$

165 kN

**Fig. 9–26**

# PROBLEMS

**9-19.** The plastic rod is made of Kevlar 49 and has a diameter of 10 mm. If an axial load of 80 kN is applied to it, determine the change in its length and the change in diameter.

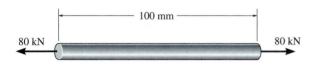

80 kN          100 mm          80 kN

Prob. 9–19

***9-20.** The support consists of three rigid plates, which are connected together using two symmetrically placed rubber pads. If a vertical force of 50 N is applied to plate $A$, determine the approximate vertical displacement of this plate due to shear strains in the rubber. Each pad has cross-sectional dimensions of 30 mm and 20 mm. $G_r = 0.20$ MPa.

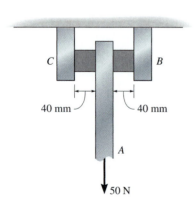

C          B

40 mm          40 mm

A

50 N

Prob. 9–20

**9-21.** A short cylindrical block of bronze C86100, having an original diameter of 1.5 in. and a length of 3 in., is placed in a compression machine and squeezed until its length becomes 2.98 in. Determine the new diameter of the block.

**9-22.** A short cylindrical block of 6061-T6 aluminum, having an original diameter of 20 mm and a length of 75 mm, is placed in a compression machine and squeezed until the axial load applied is 5 kN. Determine (a) the decrease in its length and (b) its new diameter.

**9-23.** The elastic portion of the stress–strain diagram for a steel alloy is shown in the figure. The specimen from which it was obtained had an original diameter of 13 mm and a gauge length of 50 mm. When the applied load on the specimen is 50 kN, the diameter is 12.99265 mm. Determine Poisson's ratio for the material.

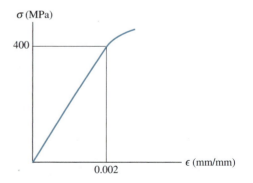

$\sigma$ (MPa)

400

0.002

$\epsilon$ (mm/mm)

Prob. 9–23

# CHAPTER REVIEW

- *Hooke's Law.* Many engineering materials exhibit initial linear elastic behavior, whereby stress is proportional to strain, defined by Hooke's law, $\sigma = E\epsilon$ where $E$, called the modulus of elasticity, is the slope of the line on the stress-strain diagram.

- *Stress-Strain Behavior.* When the material is stressed beyond the yield point, permanent deformation will occur. In particular, steel has a region of yielding, whereby the material will exhibit an increase in strain but no increase in stress. The region of strain hardening causes further yielding of the material with a corresponding increase in stress. Finally, at the ultimate stress, a localized region on the specimen will begin to constrict forming a neck. It is here that fracture occurs. Ductile materials, such as most metals, exhibit both elastic and plastic behavior. Ductility is usually specified by the permanent elongation to failure or by the permanent reduction in the cross-sectional area. Brittle materials exhibit little or no yielding before failure.

- *Strain Hardening.* The yield point of a material can be increased by strain hardening, which is accomplished by applying a load great enough to cause yielding, then releasing the load. The larger stress produced becomes the new yield point for the material.

- *Strain Energy.* When a load is applied, the deformations cause strain energy to be stored in the material. The strain energy per unit volume or strain energy density is equivalent to the area under the stress-strain curve. This area up to the yield point is called the modulus of resilience. And the entire area under the stress-strain diagram is called the modulus of toughness.

- *Possion's Ratio.* Poisson's ratio $\mu$ is a dimensionless material property that measures the lateral strain to the longitudinal strain. Its value is between $0 < \mu \le 0.5$.

- *Shearing Modulus.* Shear stress versus shear strain diagrams can also be established for a material. Within the elastic region, $\tau = G\gamma$, where $G$ is the shearing modulus, found from the slope of the line within the elastic region. The value of $G$ can also be obtained from the relationship that exists between $G$, $E$ and $\mu$, namely $G = E/[2(1+\mu)]$.

# REVIEW PROBLEMS

*9-28. The aluminum block has a rectangular cross section and is subjected to an axial compressive force of 8 kip. If the 1.5-in. side changed its length to 1.500132 in., determine Poisson's ratio and the new length of the 2-in. side. $E_{al} = 10(10^3)$ ksi.

9-29. While undergoing a tension test, a copper-alloy specimen having a gauge length of 2 in. is subjected to a strain of 0.40 in./in. when the stress is 70 ksi. If $\sigma_Y = 45$ ksi when $\epsilon_y = 0.0025$ in./in., determine the distance between the gauge points when the load is released.

9-30. An 8-mm-diameter brass rod has a modulus of elasticity of $E_{br} = 100$ GPa. If it is 3 m long and subjected to an axial load of 2 kN, determine its elongation. What is its elongation under the same load if its diameter is 6 mm?

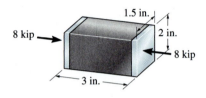

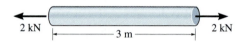

Prob. 9–28

Prob. 9–30

**9-31.** The rigid beam rests in the horizontal position on two aluminum cylinders having the *unloaded* lengths shown. If each cylinder has a diameter of 30 mm, determine the placement $x$ of the applied 80-kN load so that the beam remains horizontal. What is the new diameter of cylinder $A$ after the load is applied? $E_{al} = 70\,\text{GPa}$, $v_{al} = 0.33$.

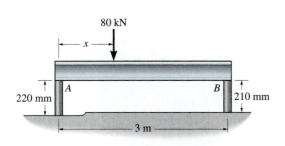

Prob. 9-31

*9-32.** The shear stress–strain diagram for a steel alloy is shown in the figure. If a bolt having a diameter of 0.25 in. is made of this material and used in the lap joint, determine the modulus of elasticity $E$ and the force $P$ required to cause the material to yield. Take $v = 0.3$.

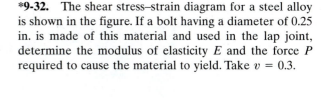

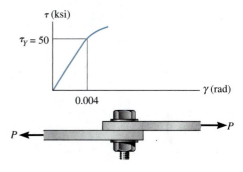

Prob. 9-32

The string of drill pipe suspended from this traveling block on an oil rig is subjected to extremely large loadings and axial deformations.

# CHAPTER
# 10

# Axial Load

## CHAPTER OBJECTIVES

In Chapter 8 we developed the method for finding the normal stress in axially loaded members. In this chapter, our objectives are:

- To determine the deformation of axially loaded members.
- To determine the support reactions when these reactions cannot be determined solely from the equations of equilibrium.
- To analyze the effects of thermal stresses.

## 10.1 Saint-Venant's Principle

In the previous chapters we have developed the concept of *stress* as a means of measuring the force distribution within a body and *strain* as a means of measuring a body's deformation. We have also shown that the mathematical relationship between stress and strain depends on the type of material from which the body is made. In particular, if the material behaves in a linear-elastic manner, then Hooke's law applies, and there is a proportional relationship between stress and strain.

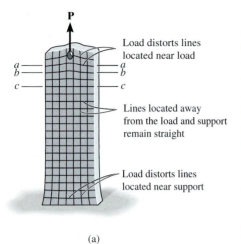

Load distorts lines
located near load

Lines located away
from the load and support
remain straight

Load distorts lines
located near support

(a)

**Fig. 10–1**

Using this idea, consider the manner in which a rectangular bar will deform elastically when the bar is subjected to a force **P** applied along its centroidal axis, Fig. 10–1*a*. Here the bar is fixed-connected at one end, with the force applied through a hole at its other end. Due to the loading, the bar deforms as indicated by the distortions of the once horizontal and vertical grid lines drawn on the bar. Notice the *localized deformation* that occurs at each end. This effect tends to *decrease* as measurements are taken farther and farther away from the ends. Furthermore, the deformations even out and become uniform throughout the midsection of the bar.

Since the deformation or strain is related to stress within the bar, we can state that stress will be distributed more uniformly throughout the cross-sectional area if the section is taken farther and farther from the point where the external load is applied. For example, consider a profile of the variation of the stress distribution acting at sections *a–a*, *b–b*, and *c–c*, each of which is shown in Fig. 10–1*b*. By comparison, the stress *almost* reaches a uniform value at section *c–c*, which is sufficiently removed from the end. In other words, section *c–c* is far enough away from the application of **P** so that the localized deformation caused by **P** *vanishes*. The minimum distance from the bar's end where this occurs can be determined using a mathematical analysis based on the theory of elasticity.

However, as a *general rule*, which applies as well to many other cases of loading and member geometry, we can consider this distance to be at least equal to the *largest dimension* of the loaded cross section. Hence, for the bar in Fig. 10–1*b*, section *c–c* should be located at a distance at least equal to the width (not the thickness) of the bar.[*] This rule is based on *experimental observation of material behavior*, and only in special cases, like the one discussed here, has it been validated mathematically. It should be noted, however, that this rule does not apply to every type of member and loading case. For example, members made from thin-walled elements, and subjected to loadings that cause large deflections, may create localized stresses and deformations that have an influence a considerable distance away from the point of application of loading.

---

[*]When section *c–c* is so located, the theory of elasticity predicts the maximum stress to be $\sigma_{\max} = 1.02\sigma_{\text{avg}}$.

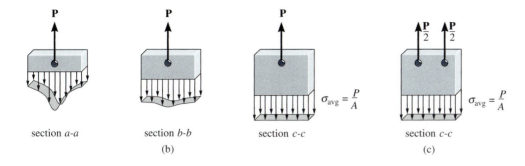

section *a-a*           section *b-b*           section *c-c*           section *c-c*

                (b)

$\sigma_{avg} = \dfrac{P}{A}$        $\sigma_{avg} = \dfrac{P}{A}$

                (c)

**Fig. 10–1**

At the support, in Fig. 10–1a notice how the bar is prevented from decreasing its width, which should occur due to the bar's lateral elongation—a consequence of the "Poisson effect," discussed in Sec. 9–6. By the same arguments given above, however, we could demonstrate that the stress distribution at the support will also even out and become uniform over the cross section at a short distance from the support; and furthermore, the magnitude of the resultant force created by this stress distribution must also equal $P$.

The fact that stress and deformation behave in this manner is referred to as *Saint-Venant's principle*, since it was first noticed by the French scientist Barré de Saint-Venant in 1855. Essentially it states that the stress and strain produced at points in a body sufficiently removed from the region of load application will be the *same* as the stress and strain produced by *any applied loadings* that have the same statically equivalent resultant and are applied to the body within the same region. For example, if two symmetrically applied forces $P/2$ act on the bar, Fig. 10–1c, the stress distribution at section *c–c*, which is sufficiently removed from the localized effects of these loads, will be uniform and therefore equivalent to $\sigma_{avg} = P/A$ as before.

To summarize, then, we do not have to consider the somewhat complex stress distributions that may actually develop at points of load application, or at supports, when studying the stress distribution in a body at sections *sufficiently removed* from the points of load application. Saint-Venant's principle claims that the localized effects caused by any load acting on the body will dissipate or smooth out within regions that are sufficiently removed from the location of the load. Furthermore, the resulting stress distribution at these regions will be the *same* as that caused by any other statically equivalent load applied to the body within the same localized area.

Notice how the lines on this rubber membrane distort after it is stretched. The localized distortions at the grips smooth out as expected. This is due to Saint-Venant's principle.

## 10.2 Elastic Deformation of an Axially Loaded Member

Using Hooke's law and the definitions of stress and strain, we will now develop an equation that can be used to determine the elastic deformation of a member subjected to axial loads. To generalize the development, consider the bar shown in Fig. 10–2*a*, which has a cross-sectional area that *gradually* varies along its length *L*. The bar is subjected to concentrated loads at its ends and a variable external load distributed along its length. This distributed load could, for example, represent the weight of a vertical bar, or friction forces acting on the bar's surface. Here we wish to find the **relative displacement** δ (delta) of one end of the bar with respect to the other end as caused by this loading. In the following analysis we will neglect the localized deformations that occur at points of concentrated loading and where the cross section suddenly changes. As noted in Sec. 10.1, these effects occur within small regions of the bar's length and will therefore have only a slight effect on the final result. For the most part, the bar will deform uniformly, so the normal stress will be uniformly distributed over the cross section.

Using the method of sections, a differential element (or wafer) of length *dx* and cross-sectional area $A(x)$ is isolated from the bar at the arbitrary position *x*. The free-body diagram of this element is shown in Fig. 10–2*b*. The resultant internal axial force is represented as $P(x)$, since the external loading will cause it to vary along the length of the bar. This load, $P(x)$, will deform the element into the shape indicated by the dashed outline, and therefore the displacement of one end of the element with respect to the other end is *d*δ. The stress and strain in the element are

$$\sigma = \frac{P(x)}{A(x)} \quad \text{and} \quad \epsilon = \frac{d\delta}{dx}$$

Provided these quantities do not exceed the proportional limit, we can relate them using Hooke's law; i.e.,

$$\sigma = E\epsilon$$

$$\frac{P(x)}{A(x)} = E\left(\frac{d\delta}{dx}\right)$$

$$d\delta = \frac{P(x)dx}{A(x)E}$$

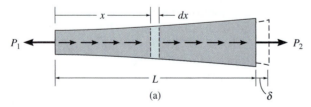

(a)

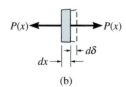

(b)

**Fig. 10–2**

For the entire length $L$ of the bar, we must integrate this expression to find the required end displacement. This yields

$$\delta = \int_0^L \frac{P(x)dx}{A(x)E} \qquad (10\text{–}1)$$

where

$\delta$ = displacement of one point on the bar relative to another point

$L$ = distance between the points

$P(x)$ = internal axial force at the section, located a distance $x$ from one end

$A(x)$ = cross-sectional area of the bar, expressed as a function of $x$

$E$ = modulus of elasticity for the material

**Constant Load and Cross-Sectional Area.**   In many cases the bar will have a constant cross-sectional area $A$; and the material will be homogeneous, so $E$ is constant. Furthermore, if a constant external force is applied at each end, Fig. 10–3, then the internal force $P$ throughout the length of the bar is also constant. As a result, Eq. 10–1 can be integrated to yield

$$\delta = \frac{PL}{AE} \qquad (10\text{–}2)$$

If the bar is subjected to several different axial forces, or the cross-sectional area or modulus of elasticity changes abruptly from one region of the bar to the next, the above equation can be applied to each *segment* of the bar where these quantities are all *constant*. The displacement of one end of the bar with respect to the other is then found from the *algebraic addition* of the end displacements of each segment. For this general case,

$$\delta = \sum \frac{PL}{AE} \qquad (10\text{–}3)$$

The vertical displacement at the top of these building columns depends upon the loading applied on the roof and from the floor attached to their midpoint.

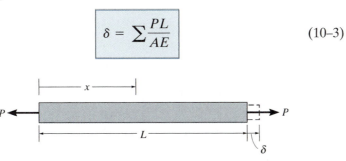

**Fig. 10–3**

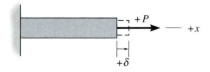

Positive sign convention for *P* and δ

**Fig. 10–4**

**Sign Convention.** In order to apply Eq. 10–3, we must develop a sign convention for the internal axial force and the displacement of one end of the bar with respect to the other end. To do so, we will consider both the force and displacement to be positive if they cause tension and elongation, respectively, Fig. 10–4; whereas a negative force and displacement will cause compression and contraction, respectively.

For example, consider the bar shown in Fig. 10–5a. The *internal axial forces* "*P*," are determined by the method of sections for each segment, Fig. 10–5b. They are $P_{AB} = +5$ kN, $P_{BC} = -3$ kN, $P_{CD} = -7$ kN. This variation in axial load is shown on the axial (or normal) force diagram for the bar, Fig. 10–5c. Applying Eq. 10–3 to obtain the displacement of end *A* relative to end *D*, we have

$$\delta_{A/D} = \sum \frac{PL}{AE} = \frac{(5 \text{ kN})L_{AB}}{AE} + \frac{(-3 \text{ kN})L_{BC}}{AE} + \frac{(-7 \text{ kN})L_{CD}}{AE}$$

If the other data are substituted and a positive answer is computed, it means that end *A* will move away from end *D* (the bar elongates), whereas a negative result would indicate that end *A* moves toward end *D* (the bar shortens). The double subscript notation is used to indicate this relative displacement ($\delta_{A/D}$); however, if the displacement is to be determined relative to a *fixed point*, then only a single subscript will be used. For example, if *D* is located at a *fixed* support, then the computed displacement will be denoted as simply $\delta_A$.

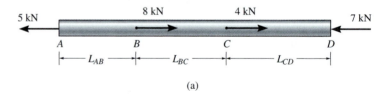

(a)

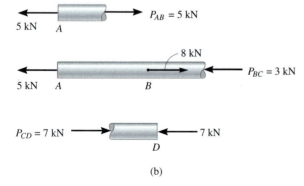

(b)

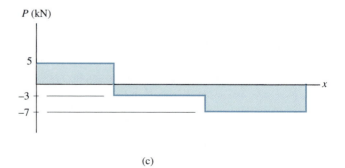

(c)

**Fig. 10–5**

## IMPORTANT POINTS

- *Saint-Venant's principle* states that both the localized deformation and stress which occur within the regions of load application or at the supports tend to "even out" at a distance sufficiently removed from these regions.
- The displacement of an axially loaded member is determined by relating the applied load to the stress using $\sigma = P/A$ and relating the displacement to the strain using $\epsilon = d\delta/dx$. Finally these two equations are combined using Hooke's law, $\sigma = E\epsilon$, which yields Eq. 10–1.
- Since Hooke's law has been used in the development of the displacement equation, it is important that the loads do not cause yielding of the material and that the material is homogeneous and behaves in a linear–elastic manner.

## PROCEDURE FOR ANALYSIS

The relative displacement between two points $A$ and $B$ on an axially loaded member can be determined by applying Eq. 10–1 (or Eq. 10–2). Application requires the following steps.

*Internal Force.*

- Use the method of sections to determine the internal axial force $P$ in the member.
- If this force varies along the member's length, a section should be made at the arbitrary location $x$ from one end of the member and the force represented as a function of $x$, i.e., $P(x)$.
- If several *constant external forces* act on the member, the internal force in each *segment* of the member, between any two external forces, must then be determined.
- For any segment, an internal *tensile force* is *positive* and an internal *compressive force* is *negative*. For convenience, the results of the internal loading can be shown graphically by constructing the normal-force diagram.

*Displacement.*

- When the member's cross-sectional area *varies* along its axis, the area must be expressed as a function of its position $x$, i.e., $A(x)$.
- If the cross-sectional area, the modulus of elasticity, or the internal loading *suddenly changes*, then Eq. 10–2 should be applied to each segment for which these quantities are constant.
- When substituting the data into Eqs. 10–1 through 10–3, be sure to account for the proper sign for $P$, tensile loadings are positive and compressive loadings are negative. Also, use a consistent set of units. For any segment, if the computed result is a *positive* numerical quantity, it indicates *elongation*; if it is *negative*, it indicates a *contraction*.

EXAMPLE 10.1

The composite A-36 steel bar shown in Fig. 10–6a is made from two segments, AB and BD, having cross-sectional areas of $A_{AB} = 1$ in² and $A_{BD} = 2$ in². Determine the vertical displacement of end A and the displacement of B relative to C.

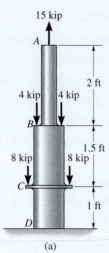

(a)

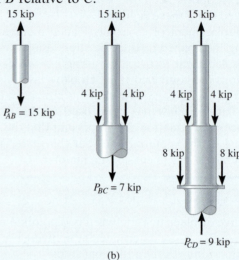

(b)

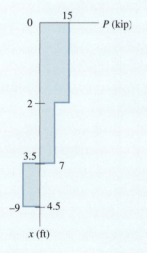

(c)

Fig. 10–6

## Solution

*Internal Force.* Due to the application of the external loadings, the *internal axial forces* in regions AB, BC, and CD will all be *different*. These forces are obtained by applying the method of sections and the equation of vertical force equilibrium as shown in Fig. 10–6b. This variation is plotted in Fig. 10–6c.

*Displacement.* From Appendix B, $E_{st} = 29(10^3)$ ksi. Using the sign convention, i.e., the internal tensile forces are positive and the compressive forces are negative, the vertical displacement of A relative to the *fixed* support D is

$$\delta_A = \sum \frac{PL}{AE} = \frac{[+15 \text{ kip}](2 \text{ ft})(12 \text{ in./ft})}{(1 \text{ in}^2)[29(10^3) \text{ kip/in}^2]} + \frac{[+7 \text{ kip}](1.5 \text{ ft})(12 \text{ in./ft})}{(2 \text{ in}^2)[29(10^3) \text{ kip/in}^2]}$$

$$+ \frac{[-9 \text{ kip}](1 \text{ ft})(12 \text{ in./ft})}{(2 \text{ in}^2)[29(10^3) \text{ kip/in}^2]}$$

$$= +0.0127 \text{ in.} \qquad Ans.$$

Since the result is *positive*, the bar *elongates* and so the displacement at A is upward.

Applying Eq. 10–2 between points B and C, we obtain,

$$\delta_{B/C} = \frac{P_{BC}L_{BC}}{A_{BC}E} = \frac{[+7 \text{ kip}](1.5 \text{ ft})(12 \text{ in./ft})}{(2 \text{ in}^2)[29(10^3) \text{ kip/in}^2]} = +0.00217 \text{ in.} \qquad Ans.$$

Here B moves away from C, since the segment elongates.

# E X A M P L E   10.2

The assembly shown in Fig. 10–7$a$ consists of an aluminum tube $AB$ having a cross-sectional area of 400 mm$^2$. A steel rod having a diameter of 10 mm is attached to a rigid collar and passes through the tube. If a tensile load of 80 kN is applied to the rod, determine the displacement of the end $C$ of the rod. Take $E_{st} = 200$ GPa, $E_{al} = 70$ GPa.

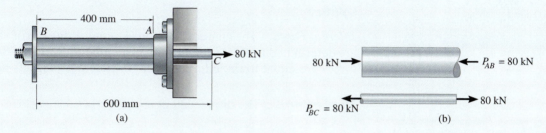

(a)

(b)

**Fig. 10–7**

## Solution

*Internal Force*   The free-body diagram of the tube and rod, Fig. 10–7$b$, shows that the rod is subjected to a tension of 80 kN and the tube is subjected to a compression of 80 kN.

*Displacement.*   We will first determine the displacement of end $C$ with respect to end $B$. Working in units of newtons and meters, we have

$$\delta_{C/B} = \frac{PL}{AE} = \frac{[+80(10^3)\text{N}](0.6\text{ m})}{\pi(0.005\text{ m})^2[200(10^9)\text{N/m}^2]} = +0.003056\text{ m} \rightarrow$$

The positive sign indicates that end $C$ moves *to the right* relative to end $B$, since the bar elongates.
   The displacement of end $B$ with respect to the *fixed* end $A$ is

$$\delta_B = \frac{PL}{AE} = \frac{[-80(10^3)\text{N}](0.4\text{ m})}{[400\text{ mm}^2(10^{-6})\text{m}^2/\text{mm}^2][70(10^9)\text{N/m}^2]}$$

$$= -0.001143\text{ m} = 0.001143\text{ m} \rightarrow$$

Here the negative sign indicates that the tube shortens, and so $B$ moves to the *right* relative to $A$.
   Since both displacements are to the right, the resultant displacement of $C$ relative to the fixed end $A$ is therefore

$$(\overset{+}{\rightarrow})\quad \delta_C = \delta_B + \delta_{C/B} = 0.001143\text{ m} + 0.003056\text{ m}$$

$$= 0.00420\text{ m} = 4.20\text{ mm} \rightarrow \qquad\qquad Ans.$$

**E X A M P L E 10.3**

A *rigid beam AB* rests on the two short posts shown in Fig. 10–8a. *AC* is made of steel and has a diameter of 20 mm, and *BD* is made of aluminum and has a diameter of 40 mm. Determine the displacement of point *F* on *AB* if a vertical load of 90 kN is applied over this point. Take $E_{st} = 200$ GPa, $E_{al} = 70$ GPa.

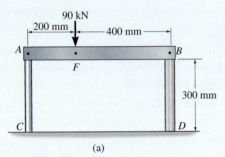

(a)

**Solution**

*Internal Force.* The compressive forces acting at the top of each post are determined from the equilibrium of member *AB*, Fig. 10–8b. These forces are equal to the internal forces in each post, Fig. 10–8c.

*Displacement.* The displacement of the top of each post is

*Post AC:*

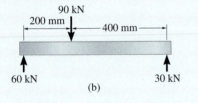

(b)

$$\delta_A = \frac{P_{AC}L_{AC}}{A_{AC}E_{st}} = \frac{[-60(10^3)\text{N}](0.300 \text{ m})}{\pi(0.010 \text{ m})^2[200(10^9) \text{ N/m}^2]} = -286(10^{-6}) \text{ m}$$

$$= 0.286 \text{ mm} \downarrow$$

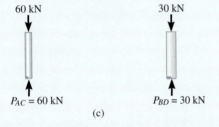

$P_{AC} = 60$ kN    $P_{BD} = 30$ kN

(c)

*Post BD:*

$$\delta_B = \frac{P_{BD}L_{BD}}{A_{BD}E_{at}} = \frac{[-30(10^3) \text{ N}](0.300 \text{ m})}{\pi(0.020 \text{ m})^2[70(10^9) \text{ N/m}^2]} = -102(10^{-6}) \text{ m}$$

$$= 0.102 \text{ mm} \downarrow$$

A diagram showing the centerline displacements at points *A*, *B*, and *F* on the beam is shown in Fig. 10–8d. By proportion of the shaded triangle, the displacement of point *F* is therefore

$$\delta_F = 0.102 \text{ mm} + (0.184 \text{ mm})\left(\frac{400 \text{ mm}}{600 \text{ mm}}\right) = 0.225 \text{ mm} \downarrow \qquad Ans.$$

0.102 mm | A | F |─ 600 mm ─| B

0.184 mm   $\delta_F$

0.286 mm

(d)

**Fig. 10–8**

## EXAMPLE 10.4

A member is made from a material that has a specific weight $\gamma$ and modulus of elasticity $E$. If it is formed into a *cone* having the dimensions shown in Fig. 10–9a, determine how far its end is displaced due to gravity when it is suspended in the vertical position.

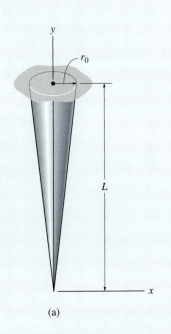

(a)

### Solution

*Internal Force.* The internal axial force varies along the member since it is dependent on the weight $W(y)$ of a segment of the member below any section, Fig. 10–9a. Hence, to calculate the displacement, we must use Eq. 10–1. At the section located at a distance $y$ from its bottom end, the radius $x$ of the cone as a function of $y$ is determined by proportion; i.e.,

$$\frac{x}{y} = \frac{r_0}{L}; \qquad x = \frac{r_0}{L}y$$

The volume of a cone having a base of radius $x$ and height $y$ is

$$V = \frac{\pi}{3}yx^2 = \frac{\pi r_0^2}{3L^2}y^3$$

Since $W = \gamma V$, the internal force at the section becomes

$$+\uparrow \Sigma F_y = 0; \qquad P(y) = \frac{\gamma \pi r_0^2}{3L^2}y^3$$

*Displacement.* The area of the cross section is also a function of position $y$, Fig. 10–9b. We have

$$A(y) = \pi x^2 = \frac{\pi r_0^2}{L^2}y^2$$

Applying Eq. 10–1 between the limits of $y = 0$ and $y = L$ yields

$$\delta = \int_0^L \frac{P(y)dy}{A(y)E} = \int_0^L \frac{[(\gamma \pi r_0^2/3L^2)y^3]dy}{[(\pi r_0^2/L^2)y^2]E}$$

$$= \frac{\gamma}{3E}\int_0^L ydy$$

$$= \frac{\gamma L^2}{6E} \qquad\qquad\qquad \textit{Ans.}$$

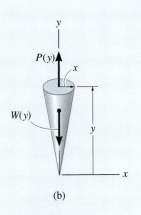

(b)

Fig. 10–9

As a partial check of this result, notice how the units of the terms, when canceled, give the displacement in units of length as expected.

# PROBLEMS

**10-1.** The ship is pushed through the water using an A-36 steel propeller shaft that is 8 m long, measured from the propeller to the thrust bearing $D$ at the engine. If it has an outer diameter of 400 mm and a wall thickness of 50 mm, determine the amount of axial contraction of the shaft when the propeller exerts a force on the shaft of 5 kN. The bearings at $B$ and $C$ are journal bearings.

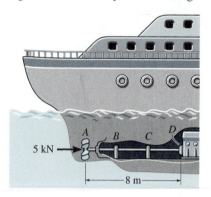

**Prob. 10–1**

**10-2.** The A-36 steel column is used to support the symmetric loads from the two floors of a building. Determine the vertical displacement of its top $A$ if $P_1 = 40$ kip, $P_2 = 62$ kip, and the column has a cross-sectional area of 23.4 in$^2$.

**10-3.** The A-36 steel column is used to support the symmetric loads from the two floors of a building. Determine the loads $P_1$ and $P_2$ if $A$ moves downward 0.12 in. and $B$ moves downward 0.09 in. when the loads are applied. The column has a cross-sectional area of 23.4 in$^2$.

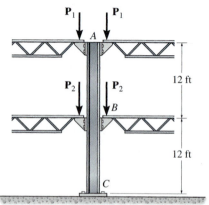

**Probs. 10–2/3**

*10-4.** The bronze C86100 shaft is subjected to the axial loads shown. Determine the displacement of end $A$ with respect to end $D$ if the diameters of each segment are $d_{AB} = 0.75$ in., $d_{BC} = 2$ in., and $d_{CD} = 0.5$ in.

**10-5.** Determine the displacement of end $A$ with respect to end $C$ of the shaft in Prob. 10–4.

**Probs. 10–4/5**

**10-6.** The assembly consists of an A-36 steel rod $CB$ and a 6061-T6 aluminum rod $BA$, each having a diameter of 1 in. If the rod is subjected to the axial loading $P_1 = 12$ kip at $A$ and $P_2 = 18$ kip at the coupling $B$, determine the displacement of the coupling $B$ and the end $A$. The unstretched length of each segment is shown in the figure. Neglect the size of the connections at $B$ and $C$, and assume that they are rigid.

**10-7.** The assembly consists of an A-36 steel rod $CB$ and a 6061-T6 aluminum rod $BA$, each having a diameter of 1 in. Determine the applied loads $P_1$ and $P_2$ if $A$ is displaced 0.08 in. to the right and $B$ is displaced 0.02 in. to the left when the loads are applied. The unstretched length of each segment is shown in the figure. Neglect the size of the connections at $B$ and $C$, and assume that they are rigid.

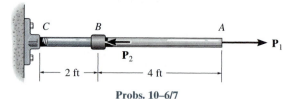

**Probs. 10–6/7**

*10-8.** The joint is made from three A-36 steel plates that are bonded together at their seams. Determine the displacement of end $A$ with respect to end $D$ when the joint is subjected to the axial loads shown. Each plate has a thickness of 6 mm.

**Prob. 10–8**

**10-9.** The assembly consists of two rigid bars that are originally horizontal. They are supported by pins and 0.25-in.-diameter A-36 steel rods *FC* and *EB*. If the vertical load of 5 kip is applied to the bottom bar *AB*, determine the displacement at *C*, *B*, and *E*.

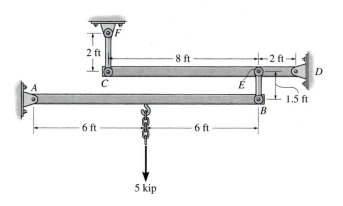

5 kip

**Prob. 10–9**

**10-10.** The truss is made from three A-36 steel members, each having a cross-sectional area of 400 mm². Determine the vertical displacement of the roller at *C* when the truss supports the load of *P* = 10 kN.

**10-11.** The truss is made from three A-36 steel members, each having a cross-sectional area of 400 mm². Determine the load *P* required to displace the roller downward 0.2 mm.

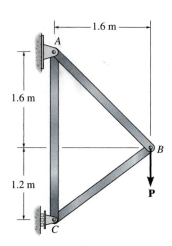

**Probs. 10–10/11**

***10-12.** The load is supported by the four 304 stainless steel wires that are connected to the rigid members *AB* and *DC*. Determine the vertical displacement of the 500-lb load if the members were horizontal when the load was originally applied. Each wire has a cross-sectional area of 0.025 in².

**10-13.** The load is supported by the four 304 stainless steel wires that are connected to the rigid members *AB* and *DC*. Determine the angle of tilt of each member after the 500-lb load is applied. The members were originally horizontal, and each wire has a cross-sectional area of 0.025 in².

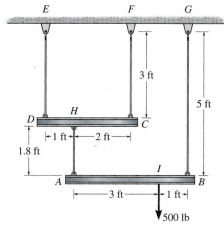

500 lb

**Probs. 10–12/13**

**10-14.** The linkage is made from three pin-connected 304 stainless steel members, each having a cross-sectional area of 0.75 in². If a horizontal force of *P* = 6 kip is applied to the end *B* of member *AB*, determine the horizontal displacement of point *B*.

**10-15.** The linkage is made from three pin-connected 304 stainless steel members, each having a cross-sectional area of 0.75 in². Determine the magnitude of the force **P** needed to displace point *B* 0.08 in. to the right.

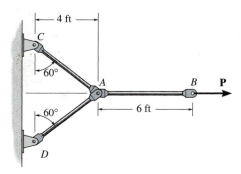

**Probs. 10–14/15**

**\*10-16.** A spring-supported pipe hanger consists of two springs which are originally unstretched and have a stiffness of $k = 60$ kN/m, three 304 stainless steel rods, $AB$ and $CD$, which have a diameter of 5 mm and $EF$, which has a diameter of 12 mm, and a rigid beam $GH$. If the pipe and the fluid it carries have a total weight of 4 kN, determine the displacement of the pipe when it is attached to the support.

**10-17.** A spring-supported pipe hanger consists of two springs, which are originally unstretched and have a stiffness of $k = 60$ kN/m, three 304 stainless steel rods, $AB$ and $CD$ which have a diameter of 5 mm and $EF$ which has a diameter of 12 mm, and a rigid beam $GH$. If the pipe is displaced 82 mm when it is filled with fluid, determine the weight of the fluid.

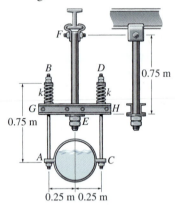

**Probs. 10–16/17**

**10-18.** The assembly consists of three titanium (Ti-6A1-4V) rods and a rigid bar $AC$. The cross-sectional area of each rod is given in the figure. If a force of 6 kip is applied to the ring, determine the horizontal displacement of point $F$.

**10-19.** The assembly consists of three titanium (Ti-6A1-4V) rods and a rigid bar $AC$. The cross-sectional area of each rod is given in the figure. If a force of 6 kip is applied to the ring $F$, determine the angle of tilt in radians of bar $AC$.

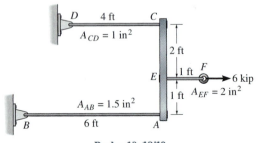

**Probs. 10–18/19**

**\*10-20.** The rigid beam is supported at its ends by two A-36 steel tie rods. If the allowable stress for the steel is $\sigma_{\text{allow}} = 16.2$ ksi, the load $w = 3$ kip/ft, and $x = 4$ ft, determine the diameter of each rod so that the beam remains in the horizontal position when it is loaded.

**10-21.** The rigid beam is supported at its ends by two A-36 steel tie rods. The rods have diameters $d_{AB} = 0.5$ in. and $d_{CD} = 0.3$ in. If the allowable stress for the steel is $\sigma_{\text{allow}} = 16.2$ ksi, determine the intensity of the distributed load $w$ and its length $x$ on the beam so that the beam remains in the horizontal position when it is loaded.

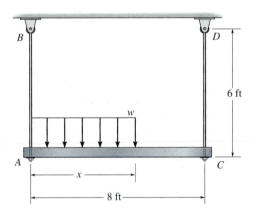

**Probs. 10–20/21**

**10-22.** The truss consists of three members, each made from A-36 steel and having a cross-sectional area of 0.75 in². Determine the greatest load $P$ that can be applied so that the roller support at $B$ is not displaced more than 0.03 in.

**10-23.** Solve Prob. 10–22 when the load $P$ acts vertically downward at $C$.

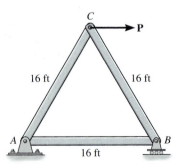

**Probs. 10–22/23**

## 10.3 Principle of Superposition

The principle of superposition is often used to determine the stress or displacement at a point in a member when the member is subjected to a complicated loading. By subdividing the loading into components, the ***principle of superposition*** states that the resultant stress or displacement at the point can be determined by first finding the stress or displacement caused by each component load acting *separately* on the member. The resultant stress or displacement is then determined by algebraically adding the contributions caused by each of the components.

The following two conditions must be valid if the principle of superposition is to be applied.

1.  ***The loading must be linearly related to the stress or displacement that is to be determined.*** For example, the equations $\sigma = P/A$ and $\delta = PL/AE$ involve a linear relationship between $P$ and $\sigma$ or $\delta$.

2.  ***The loading must not significantly change the original geometry or configuration of the member.*** If significant changes do occur, the direction and location of the applied forces and their moment arms will change, and consequently, application of the equilibrium equations will yield different results. For example, consider the slender rod shown in Fig. 10–10a, which is subjected to the load **P**. In Fig. 10–10b, **P** is replaced by two of its components, $\mathbf{P} = \mathbf{P}_1 + \mathbf{P}_2$. If **P** causes the rod to deflect a large amount, as shown, the moment of the load about its support, $Pd$, will *not* equal the sum of the moments of its component loads, $Pd \neq P_1d_1 + P_2d_2$, because $d_1 \neq d_2 \neq d$.

Most of the equations involving load, stress, and displacement that are developed in this text consist of linear relationships between these quantities. Also, members or bodies that are to be considered will be such that the loading will produce deformations that are so small that the change in position and direction of the loading will be insignificant and can be neglected. One exception to this rule, however, will be discussed in Chapter 17. It consists of a column that carries an axial load that is equivalent to the critical or buckling load. It will be shown that when this load increases only slightly, it will cause the column to have a large lateral deflection, even if the material remains linear-elastic. These deflections, associated with the components of any axial load, *cannot* be superimposed.

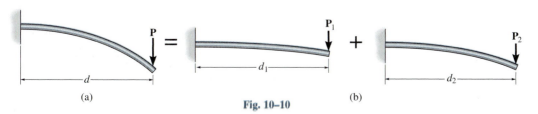

(a)      (b)

**Fig. 10–10**

## 10.4 Statically Indeterminate Axially Loaded Member

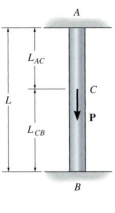

(a)

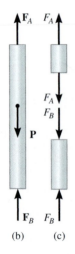

(b)      (c)

**Fig. 10–11**

When a bar is fixed-supported at only one end and is subjected to an axial force, the force equilibrium equation applied along the axis of the bar is *sufficient* to find the reaction at the fixed support. A problem such as this, where the reactions can be determined strictly from the equations of equilibrium, is called *statically determinate*. If the bar is fixed at *both ends*, however, as in Fig. 10–11a, then two unknown axial reactions occur, Fig. 10–11b, and the force equilibrium equation becomes

$$+ \uparrow \Sigma F = 0; \qquad\qquad F_B + F_A - P = 0$$

In this case the bar is called **statically indeterminate**, since the equilibrium equation(s) are not sufficient to determine the reactions.

In order to establish an additional equation needed for solution, it is necessary to consider the geometry of the deformation. Specifically, an equation that specifies the conditions for displacement is referred to as a **compatibility** or **kinematic condition**. A suitable compatibility condition would require the relative displacement of one end of the bar with respect to the other end to be equal to zero, since the end supports are fixed. Hence, we can write

$$\delta_{A/B} = 0$$

This equation can be expressed in terms of the applied loads by using a *load–displacement relationship*, which depends on the material behavior. For example, if linear-elastic behavior occurs, $\delta = PL/AE$ can be used. Realizing that the internal force in segment $AC$ is $+ F_A$, and in segment $CB$ the internal force is $-F_B$, Fig. 10–11c the compatibility equation can be written as

$$\frac{F_A L_{AC}}{AE} - \frac{F_B L_{CB}}{AE} = 0$$

Assuming that $AE$ is constant, we can solve the above two equations for the reactions, which gives

$$F_A = P\left(\frac{L_{CB}}{L}\right) \quad \text{and} \quad F_B = P\left(\frac{L_{AC}}{L}\right)$$

Both of these results are positive, so the reactions are shown correctly on the free-body diagram.

## IMPORTANT POINTS

- The *principle of superposition* is sometimes used to simplify stress and displacement problems having complicated loadings. This is done by subdividing the loading into components, then algebraically adding the results.
- Superposition requires that the loading be linearly related to the stress or displacement, and the loading does not significantly change the original geometry of the member.
- A member is *statically indeterminate* if the equations of equilibrium are not sufficient to determine the reactions on a member.
- *Compatibility conditions* specify the displacement constraints that occur at the supports or other points on a member.

## PROCEDURE FOR ANALYSIS

The unknown forces in statically indeterminate problems are determined by satisfying equilibrium, compatibility, and force-displacement requirements for the member.

*Equilibrium.*

- Draw a free–body diagram of the member in order to identify all the forces that act on it.
- The problem can be classified as statically indeterminate if the number of unknown reactions on the free-body diagram is greater than the number of available equations of equilibrium.
- Write the equations of equilibrium for the member.

*Compatibility.*

- To write the compatibility equations, consider drawing a displacement diagram in order to investigate the way the member will elongate or contract when subjected to the external loads.
- Express the compatibility conditions in terms of the displacements caused by the forces.
- Use a load–displacement relation, such as $\delta = PL/AE$, to relate the unknown displacements to the unknown reactions.
- Solve the equilibrium and compatibility equations for the unknown reactive forces. If any of the magnitudes has a negative numerical value, it indicates that this force acts in the opposite sense of direction to that indicated on the free-body diagram.

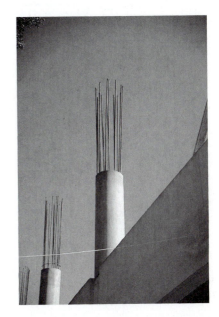

Most concrete columns are reinforced with steel rods; and since these two materials work together in supporting the applied load, the column becomes statically indeterminate.

# EXAMPLE 10.5

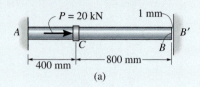

(a)

(b)

(c)

**Fig. 10–12**

The steel rod shown in Fig. 10–12a has a diameter of 5 mm. It is attached to the fixed wall at $A$, and before it is loaded, there is a gap between the wall at $B'$ and the rod of 1 mm. Determine the reactions at $A$ and $B'$ if the rod is subjected to an axial force of $P = 20$ kN as shown. Neglect the size of the collar at $C$. Take $E_{st} = 200$ GPa.

**Solution**

*Equilibrium.* As shown on the free-body diagram, Fig. 10–12b, we will assume that the force $P$ is large enough to cause the rod's end $B$ to contact the wall at $B'$. The problem is statically indeterminate since there are two unknowns and only one equation of equilibrium.

Equilibrium of the rod requires

$$\xrightarrow{+} \Sigma F_x = 0; \qquad -F_A - F_B + 20(10^3)\text{ N} = 0 \qquad (1)$$

*Compatibility.* The loading causes point $B$ to move to $B'$, with no further displacement. Therefore the compatibility condition for the rod is

$$\delta_{B/A} = 0.001 \text{ m}$$

This displacement can be expressed in terms of the unknown reactions by using the load–displacement relationship, Eq. 10–2, applied to segments $AC$ and $CB$, Fig. 10–12c. Working in units of newtons and meters, we have

$$\delta_{B/A} = 0.001 \text{ m} = \frac{F_A L_{AC}}{AE} - \frac{F_B L_{CB}}{AE}$$

$$0.001 \text{ m} = \frac{F_A(0.4 \text{ m})}{\pi(0.0025 \text{ m})^2[200(10^9)\text{ N/m}^2]}$$

$$- \frac{F_B(0.8 \text{ m})}{\pi(0.0025 \text{ m})^2[200(10^9)\text{ N/m}^2]}$$

or

$$F_A(0.4 \text{ m}) - F_B(0.8 \text{ m}) = 3927.0 \text{ N} \cdot \text{m} \qquad (2)$$

Solving Eqs. 1 and 2 yields

$$F_A = 16.6 \text{ kN} \qquad F_B = 3.39 \text{ kN} \qquad \textit{Ans.}$$

Since the answer for $F_B$ is *positive*, indeed the end $B$ contacts the wall at $B'$ as originally assumed. On the other hand, if $F_B$ were a negative quantity, the problem would be statically determinate, so that $F_B = 0$ and $F_A = 20$ kN.

EXAMPLE 10.6

The aluminum post shown in Fig. 10–13a is reinforced with a brass core. If this assembly supports a resultant axial compressive load of $P = 9$ kip, applied to the rigid cap, determine the average normal stress in the aluminum and the brass. Take $E_{al} = 10(10^3)$ksi and $E_{br} = 15(10^3)$ksi.

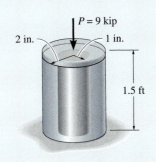

(a)

**Solution**

*Equilibrium.* The free-body diagram of the post is shown in Fig. 10–13b. Here the resultant axial force at the base is represented by the unknown components carried by the aluminum, $\mathbf{F}_{al}$, and brass, $\mathbf{F}_{br}$. The problem is statically indeterminate. Why?

Vertical force equilibrium requires

$$+\uparrow \Sigma F_y = 0; \qquad -9 \text{ kip} + F_{al} + F_{br} = 0 \qquad (1)$$

*Compatibility.* The rigid cap at the top of the post causes both the aluminum and brass to displace the same amount. Therefore,

$$\delta_{al} = \delta_{br}$$

Using the load–displacement relationships,

$$\frac{F_{al}L}{A_{al}E_{al}} = \frac{F_{br}L}{A_{br}E_{br}}$$

$$F_{al} = F_{br} \left( \frac{A_{al}}{A_{br}} \right) \left( \frac{E_{al}}{E_{br}} \right)$$

$$F_{al} = F_{br} \left[ \frac{\pi[(2 \text{ in.})^2 - (1 \text{ in.})^2]}{\pi(1 \text{ in.})^2} \right] \left[ \frac{10(10^3)\text{ksi}}{15(10^3)\text{ksi}} \right]$$

$$F_{al} = 2F_{br} \qquad (2)$$

Solving Eqs. 1 and 2 simultaneously yields

$$F_{al} = 6 \text{ kip} \qquad F_{br} = 3 \text{ kip}$$

Since the results are positive, indeed the stress will be compressive. The average normal stress in the aluminum and brass is therefore

$$\sigma_{al} = \frac{6 \text{ kip}}{\pi[(2 \text{ in.})^2 - (1 \text{ in.})^2]} = 0.637 \text{ ksi} \qquad \textit{Ans.}$$

$$\sigma_{br} = \frac{3 \text{ kip}}{\pi(1 \text{ in.})^2} = 0.955 \text{ ksi} \qquad \textit{Ans.}$$

The stress distributions are shown in Fig. 10–13c.

(b)

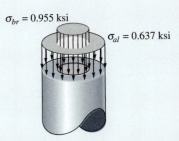

(c)

**Fig. 10–13**

## E X A M P L E  10.7

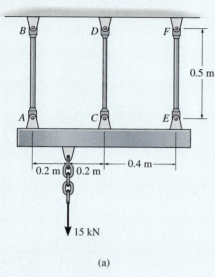

(a)

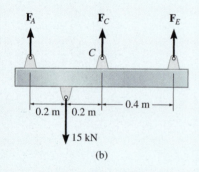

(b)

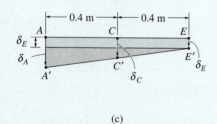

(c)

**Fig. 10–14**

The three A-36 steel bars shown in Fig. 10–14a are pin connected to a *rigid* member. If the applied load on the member is 15 kN, determine the force developed in each bar. Bars $AB$ and $EF$ each have a cross-sectional area of 25 mm², and bar $CD$ has a cross-sectional area of 15 mm².

### Solution

***Equilibrium.*** The free-body diagram of the rigid member is shown in Fig. 10–14b. This problem is statically indeterminate since there are three unknowns and only two available equilibrium equations. These equations are

$$+\uparrow \Sigma F_y = 0; \qquad F_A + F_C + F_E - 15 \text{ kN} = 0 \qquad (1)$$

$$\downarrow + \Sigma M_C = 0; \quad -F_A(0.4 \text{ m}) + 15 \text{ kN}(0.2 \text{ m}) + F_E(0.4 \text{ m}) = 0 \qquad (2)$$

***Compatibility.*** The applied load will cause the horizontal line $ACE$ shown in Fig. 10–14c to move to the inclined line $A'C'E'$. The displacements of points $A$, $C$, and $E$ can be related by proportional triangles. Thus the compatibility equation for these displacements is

$$\frac{\delta_A - \delta_E}{0.8 \text{ m}} = \frac{\delta_C - \delta_E}{0.4 \text{ m}}$$

$$\delta_C = \frac{1}{2}\delta_A + \frac{1}{2}\delta_E$$

Using the load–displacement relationship, Eq. 10–2, we have

$$\frac{F_C L}{(15 \text{ mm}^2)E_{st}} = \frac{1}{2}\left[\frac{F_A L}{(25 \text{ mm}^2)E_{st}}\right] + \frac{1}{2}\left[\frac{F_E L}{(25 \text{ mm}^2)E_{st}}\right]$$

$$F_C = 0.3F_A + 0.3F_E \qquad (3)$$

Solving Eqs. 1–3 simultaneously yields

$$F_A = 9.52 \text{ kN} \qquad \textit{Ans.}$$
$$F_C = 3.46 \text{ kN} \qquad \textit{Ans.}$$
$$F_E = 2.02 \text{ kN} \qquad \textit{Ans.}$$

# E X A M P L E 10.8

The bolt shown in Fig. 10–15a is made of 2014-T6 aluminum alloy and is tightened so it compresses a cylindrical tube made of Am 1004-T61 magnesium alloy. The tube has an outer radius of $\frac{1}{2}$ in., and it is assumed that both the inner radius of the tube and the radius of the bolt are $\frac{1}{4}$ in. The washers at the top and bottom of the tube are considered to be rigid and have a negligible thickness. Initially the nut is hand-tightened slightly; then, using a wrench, the nut is further tightened one-half turn. If the bolt has 20 threads per inch, determine the stress in the bolt.

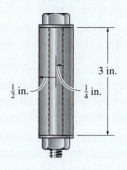

$\frac{1}{2}$ in.     $\frac{1}{4}$ in.

3 in.

(a)

## Solution

***Equilibrium.*** The free-body diagram of a section of the bolt and the tube, Fig. 10–15b, is considered in order to relate the force in the bolt $F_b$ to that in the tube, $F_t$. Equilibrium requires

$$+\uparrow \Sigma F_y = 0; \qquad\qquad F_b - F_t = 0 \qquad\qquad (1)$$

The problem is statically indeterminate since there are two unknowns in this equation.

***Compatibility.*** When the nut is tightened on the bolt, the tube will shorten $\delta_t$, and the bolt will *elongate* $\delta_b$, Fig. 10–15c. Since the nut undergoes one-half turn, it advances a distance of $\left(\frac{1}{2}\right)\left(\frac{1}{20}\,\text{in.}\right) = 0.025$ in. along the bolt. Thus, the compatibility of these displacements requires

$$(+\uparrow) \qquad\qquad \delta_t = 0.025 \text{ in.} - \delta_b$$

Taking the modulus of elasticity from the table in Appendix B, and applying Eq. 10–2, yields

$F_t$

$F_b$

(b)

$$\frac{F_t(3 \text{ in.})}{\pi[(0.5 \text{ in.})^2 - (0.25 \text{ in.})^2][6.48(10^3)\text{ksi}]} = 0.025 \text{ in.} - \frac{F_b(3 \text{ in.})}{\pi(0.25 \text{ in.})^2[10.6(10^3)\text{ksi}]}$$

$$0.78595F_t = 25 - 1.4414F_b \qquad\qquad (2)$$

Solving Eqs. 1 and 2 simultaneously, we get

$$F_b = F_t = 11.22 \text{ kip}$$

The stresses in the bolt and tube are therefore

$$\sigma_b = \frac{F_b}{A_b} = \frac{11.22 \text{ kip}}{\pi(0.25 \text{ in.})^2} = 57.2 \text{ ksi} \qquad\qquad Ans.$$

$$\sigma_s = \frac{F_t}{A_t} = \frac{11.22 \text{ kip}}{\pi[(0.5 \text{ in.})^2 - (0.25 \text{ in.})^2]} = 19.1 \text{ ksi} \qquad Ans.$$

These stresses are less than the reported yield stress for each material, $(\sigma_Y)_{al} = 60$ ksi and $(\sigma_Y)_{mg} = 22$ ksi (see Appendix B), and therefore this "elastic" analysis is valid.

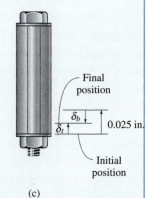

Final position

$\delta_b$

$\delta_t$     0.025 in.

Initial position

(c)

**Fig. 10–15**

# PROBLEMS

**\*10-24.** The A-36 steel column, having a cross-sectional area of 18 in², is encased in high-strength concrete as shown. If an axial force of 60 kip is applied to the column, determine the average compressive stress in the concrete and in the steel. How far does the column shorten? It has an original length of 8 ft.

**10-25.** The A-36 steel column is encased in high-strength concrete as shown. If an axial force of 60 kip is applied to the column, determine the required area of the steel so that the force is shared equally between the steel and concrete. How far does the column shorten? It has an original length of 8 ft.

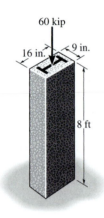

**Probs. 10–24/25**

**10-26.** The A-36 steel pipe has a 6061-T6 aluminum core. It is subjected to a tensile force of 200 kN. Determine the average normal stress in the aluminum and the steel due to this loading. The pipe has an outer diameter of 80 mm and an inner diameter of 70 mm.

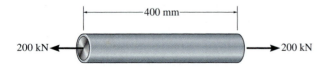

**Prob. 10–26**

**10-27.** The concrete column is reinforced using four steel reinforcing rods, each having a diameter of 18 mm. Determine the average normal stress in the concrete and the steel if the column is subjected to an axial load of 800 kN. $E_{st} = 200$ GPa, $E_c = 25$ GPa.

**\*10-28.** The column is constructed from high-strength concrete and four A-36 steel reinforcing rods. If it is subjected to an axial force of 800 kN, determine the required diameter of each rod so that one-fourth of the load is carried by the steel and three-fourths by the concrete.

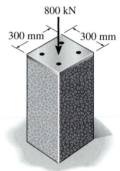

**Probs. 10–27/28**

**10-29.** The uniform bar is subjected to the load **P** at collar *B*. Determine the reactions at the pins *A* and *C*. Neglect the size of the collar.

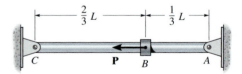

**Prob. 10–29**

**10-30.** The two pipes are made of the same material and are connected as shown. If the cross-sectional area of *BC* is *A* and that of *CD* is 2*A*, determine the reactions at *B* and *D* when a force **P** is applied at the junction *C*.

**Prob. 10–30**

**10-31.** The bolt $AB$ has a diameter of 20 mm and passes through a sleeve that has an inner diameter of 40 mm and an outer diameter of 50 mm. The bolt and sleeve are made of A-36 steel and are secured to the rigid brackets as shown. If the bolt length is 220 mm and the sleeve length is 200 mm, determine the tension in the bolt when a force of 50 kN is applied to the brackets.

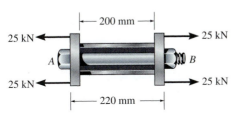

**Prob. 10–31**

*10-32.** The three A-36 steel wires each have a diameter of 2 mm and unloaded lengths of $L_{AC} = 1.60$ m and $L_{AB} = L_{AD} = 2.00$ m. Determine the force in each wire after the 150-kg mass is suspended from the ring at $A$.

**10-33.** The A-36 steel wires $AB$ and $AD$ each have a diameter of 2 mm and the unloaded lengths of each wire are $L_{AC} = 1.60$ m and $L_{AB} = L_{AD} = 2.00$ m. Determine the required diameter of wire $AC$ so that each wire is subjected to the same force caused by the 150-kg mass suspended from the ring at $A$.

**Probs. 10–32/33**

**10-34.** The three suspender bars are made of the same material and have equal cross-sectional areas $A$. Determine the average normal stress in each bar if the rigid beam $ACE$ is subjected to the force $P$.

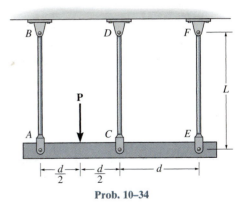

**Prob. 10–34**

**10-35.** The rigid bar is supported by the two short wooden posts and a spring. If each of the posts has an unloaded length of 500 mm and a cross-sectional area of 800 mm$^2$, and the spring has a stiffness of $k = 1.8$ MN/m and an unstretched length of 520 mm, determine the force in each post after the load is applied to the bar. $E_w = 11$ GPa.

*10-36.** The rigid bar is supported by the two short white spruce wooden posts and a spring. If each of the posts has an unloaded length of 500 mm and a cross-sectional area of 800 mm$^2$, and the spring has a stiffness of $k = 1.8$ MN/m and an unstretched length of 520 mm, determine the vertical displacement of $A$ and $B$ after the load is applied to the bar.

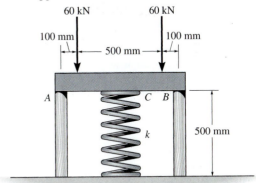

**Probs. 10–35/36**

**10-37.** The assembly consists of two posts $AB$ and $CD$ made from material 1 having a modulus of elasticity of $E_1$ and a cross-sectional area $A_1$ and a central post $EF$ made from material 2 having a modulus of elasticity $E_2$ and a cross-sectional area $A_2$. If posts $AB$ and $CD$ are to be replaced by those having a material 2, determine the required cross-sectional area of these new posts so that both assemblies deform the same amount when loaded. The support is also rigid.

## 10.5  Thermal Stress

A change in temperature can cause a material to change its dimensions. If the temperature increases, generally a material expands, whereas if the temperature decreases, the material will contract. Ordinarily this expansion or contraction is *linearly* related to the temperature increase or decrease that occurs. If this is the case, and the material is homogeneous and isotropic, it has been found from experiment that the deformation of a member having a length $L$ can be calculated using the formula

$$\delta_T = \alpha \Delta T L \qquad (10\text{–}4)$$

where

$\alpha$ = a property of the material, referred to as the ***linear coefficient of thermal expansion***. The units measure strain per degree of temperature. They are $1/°F$ (Fahrenheit) in the Foot-Pound-Second or FPS system, and $1/°C$ (Celsius) or $1/°F$ (Kelvin) in the SI system. Typical values are given in Appendix B

$\Delta T$ = the algebraic change in temperature of the member

$L$ = the original length of the member

$\delta_T$ = the algebraic change in length of the member

Most traffic bridges are designed with expansion joints to accommodate the thermal movement of the deck and thus avoid thermal stress.

If the change in temperature varies throughout the length of the member, i.e., $\Delta T = \Delta T(x)$, or if $\alpha$ varies along the length, then Eq. 10–4 applies for each segment having a length $dx$. In this case the change in the member's length is

$$\delta_T = \int_0^L \alpha \Delta T \, dx \qquad (10\text{–}5)$$

The change in length of a *statically determinate* member can readily be computed using Eq. 10–4 or 10–5, since the member is free to expand or contract when it undergoes a temperature change. However, in a *statically indeterminate* member, these thermal displacements can be constrained by the supports, producing ***thermal stresses*** that must be considered in design.

Computations of these thermal stresses can be made using the methods outlined in the previous sections. The following examples illustrate some applications.

# EXAMPLE 10.9

The A-36 steel bar shown in Fig. 10–16 is constrained to just fit between two fixed supports when $T_1 = 60°F$. If the temperature is raised to $T_2 = 120°F$, determine the average normal thermal stress developed in the bar.

0.5 in.

0.5 in.

A

2 ft

B

(a)

## Solution

*Equilibrium.* The free-body diagram of the bar is shown in Fig. 10–16b. Since there is no external load, the force at $A$ is equal but opposite to the force acting at $B$; that is,

$$+\uparrow \Sigma F_y = 0; \qquad\qquad F_A = F_B = F$$

The problem is statically indeterminate since this force cannot be determined from equilibrium.

*Compatibility.* Since $\delta_{A/B} = 0$, the thermal displacement $\delta_T$ at $A$ that would occur, Fig. 10–16c, is counteracted by the force $\mathbf{F}$ that would be required to push the bar $\delta_F$ back to its original position. The compatibility condition at $A$ becomes

$$(+\uparrow) \qquad\qquad \delta_{A/B} = 0 = \delta_T - \delta_F$$

F

F

(b)

Applying the thermal and load–displacement relationships, we have

$$0 = \alpha \Delta T L - \frac{FL}{AE}$$

Thus, from the data in Appendix B,

$$\begin{aligned} F &= \alpha \Delta T A E \\ &= [6.60(10^{-6})/°F](120°F - 60°F)(0.5 \text{ in.})^2[29(10^3)\text{kip/in}^2] \\ &= 2.87 \text{ kip} \end{aligned}$$

From the magnitude of $\mathbf{F}$, it should be apparent that changes in temperature can cause large reaction forces in statically indeterminate members.

Since $\mathbf{F}$ also represents the internal axial force within the bar, the average normal compressive stress is thus

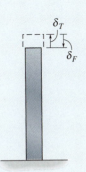

$\delta_T$

$\delta_F$

(c)

**Fig. 10–16**

$$\sigma = \frac{F}{A} = \frac{2.87 \text{ kip}}{(0.5 \text{ in.})^2} = 11.5 \text{ ksi} \qquad\qquad Ans.$$

A 2014-T6 aluminum tube having a cross-sectional area of 600 mm² is used as a sleeve for an A-36 steel bolt having a cross-sectional area of 400 mm², Fig. 10–17a. When the temperature is $T_1 = 15°C$, the nut holds the assembly in a snug position such that the axial force in the bolt is negligible. If the temperature increases to $T_2 = 80°C$, determine the average normal stress in the bolt and sleeve.

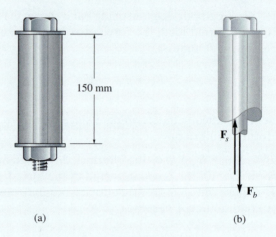

(a)                                    (b)

**Fig. 10–17**

### Solution

*Equilibrium.* A free-body diagram of a sectioned segment of the assembly is shown in Fig. 10–17b. The forces $F_b$ and $F_s$ are produced since the sleeve has a higher coefficient of thermal expansion than the bolt, and therefore the sleeve will expand more when the temperature is increased. The problem is statically indeterminate since these forces cannot be determined from equilibrium. However, it is required that

$$+\uparrow \Sigma F_y = 0; \qquad\qquad\qquad F_s = F_b \qquad\qquad\qquad (1)$$

*Compatibility.* The temperature increase causes the sleeve and bolt to expand $(\delta_s)_T$ and $(\delta_b)_T$, Fig. 10–17c. However, the redundant forces $F_b$ and $F_s$ elongate the bolt and shorten the sleeve. Consequently, the end of the assembly reaches a final position, which is not the same as the initial position. Hence, the compatibility condition becomes

$$(+\downarrow) \qquad\qquad \delta = (\delta_b)_T + (\delta_b)_F = (\delta_s)_T - (\delta_s)_F$$

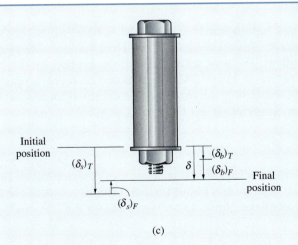

Initial
position
$(\delta_s)_T$
$(\delta_b)_T$
$\delta$
$(\delta_b)_F$
Final
position
$(\delta_s)_F$

(c)

Applying Eqs. 10–2 and 10–4, and using the mechanical properties from Appendix B, we have

$$[12(10^{-6})/°C](80°C - 15°C)(0.150 \text{ m})$$
$$+ \frac{F_b(0.150 \text{ m})}{(400 \text{ mm}^2)(10^{-6}\text{m}^2/\text{mm}^2)[200(10^9)\text{N/m}^2]}$$

$$=$$

$$[23(10^{-6})/°C](80°C - 15°C)(0.150 \text{ m})$$
$$- \frac{F_s(0.150 \text{ m})}{600 \text{ mm}^2(10^{-6}\text{m}^2/\text{mm}^2)[73.1(10^9)\text{N/m}^2]}$$

Using Eq. 1 and solving gives

$$F_s = F_b = 20.26 \text{ kN}$$

The average normal stress in the bolt and sleeve is therefore

$$\sigma_b = \frac{20.26 \text{ kN}}{400 \text{ mm}^2(10^{-6}\text{m}^2/\text{mm}^2)} = 50.6 \text{ MPa} \qquad Ans.$$

$$\sigma_s = \frac{20.26 \text{ kN}}{600 \text{ mm}^2(10^{-6}\text{m}^2/\text{mm}^2)} = 33.8 \text{ MPa} \qquad Ans.$$

Since linear–elastic material behavior was assumed in this analysis, the calculated stresses should be checked to make sure that they do not exceed the proportional limits for the material.

EXAMPLE 10.11

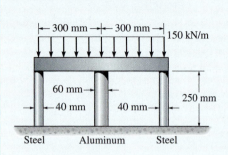

← 300 mm → ← 300 mm →  150 kN/m

60 mm →

40 mm    40 mm

250 mm

Steel    Aluminum    Steel

(a)

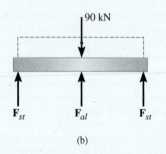

90 kN

$F_{st}$    $F_{al}$    $F_{st}$

(b)

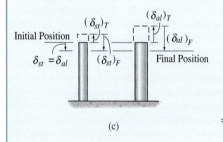

$(\delta_{st})_T$    $(\delta_{al})_T$

Initial Position

$(\delta_{al})_F$

$\delta_{st} = \delta_{al}$    $(\delta_{st})_F$    Final Position

(c)

Fig. 10–18

The rigid bar shown in Fig. 10–18a is fixed to the top of the three posts made of A-36 steel and 2014-T6 aluminum. The posts each have a length of 250 mm when no load is applied to the bar, and the temperature is $T_1 = 20°C$. Determine the force supported by each post if the bar is subjected to a uniform distributed load of 150 kN/m and the temperature is raised to $T_2 = 80°C$.

## Solution

*Equilibrium.* The free-body diagram of the bar is shown in Fig. 10–18b. Moment equilibrium about the bar's center requires the forces in the steel posts to be equal. Summing forces on the free-body diagram, we have

$$+\uparrow \Sigma F_y = 0; \qquad 2F_{st} + F_{al} - 90(10^3)\text{N} = 0 \qquad (1)$$

*Compatibility.* Due to load, geometry, and material symmetry, the top of each post is displaced by an equal amount. Hence,

$$(+\downarrow) \qquad \delta_{st} = \delta_{al} \qquad (2)$$

The final position of the top of each post is equal to its displacement caused by the temperature increase, plus its displacement caused by the internal axial compressive force, Fig. 10–18c. Thus, for a steel and aluminum post, we have

$$(+\downarrow) \qquad \qquad \delta_{st} = -(\delta_{st})_T + (\delta_{st})_F$$
$$(+\downarrow) \qquad \qquad \delta_{al} = -(\delta_{al})_T + (\delta_{al})_F$$

Applying Eq. 2 gives

$$-(\delta_{st})_T + (\delta_{st})_F = -(\delta_{al})_T + (\delta_{al})_F$$

Using Eqs. 10–2 and 10–4 and the material properties in Appendix B, we get

$$-[12(10^{-6})/°C](80°C - 20°C)(0.250\text{ m}) + \frac{F_{st}(0.250\text{ m})}{\pi(0.020\text{ m})^2[200(10^9)\text{N/m}^2]}$$

$$= -[23(10^{-6})/°C](80°C - 20°C)(0.250\text{ m}) + \frac{F_{al}(0.250\text{ m})}{\pi(0.03\text{ m})^2[73.1(10^9)\text{N/m}^2]}$$

$$F_{st} = 1.216F_{al} - 165.9(10^3) \qquad (3)$$

To be *consistent*, all numerical data has been expressed in terms of newtons, meters, and degrees Celsius. Solving Eqs. 1 and 3 simultaneously yields

$$F_{st} = -16.4\text{ kN} \qquad F_{al} = 123\text{ kN} \qquad \textit{Ans.}$$

The negative value for $F_{st}$ indicates that this force acts opposite to that shown in Fig. 10–18b. In other words, the steel posts are in tension and the aluminum post is in compression.

# PROBLEMS

**10-38.** Three bars each made of different materials are connected together and placed between two walls when the temperature is $T_1 = 12°C$. Determine the force exerted on the (rigid) supports when the temperature becomes $T_2 = 18°C$. The material properties and cross-sectional area of each bar are given in the figure.

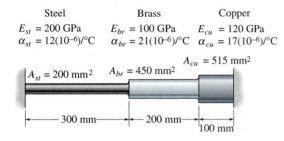

| Steel | Brass | Copper |
|---|---|---|
| $E_{st} = 200$ GPa | $E_{br} = 100$ GPa | $E_{cu} = 120$ GPa |
| $\alpha_{st} = 12(10^{-6})/°C$ | $\alpha_{br} = 21(10^{-6})/°C$ | $\alpha_{cu} = 17(10^{-6})/°C$ |

$A_{cu} = 515$ mm²
$A_{br} = 450$ mm²
$A_{st} = 200$ mm²

300 mm — 200 mm — 100 mm

**Prob. 10–38**

**10-39.** A 6-ft-long steam pipe is made of A-36 steel and is connected directly to two turbines $A$ and $B$ as shown. The pipe has an outer diameter of 4 in. and a wall thickness of 0.25 in. The connection was made at $T_1 = 70°F$. If the turbines' points of attachment are assumed rigid, determine the force the pipe exerts on the turbines when the steam and thus the pipe reach a temperature of $T_2 = 275°F$.

**\*10-40.** A 6-ft-long steam pipe is made of A-36 steel and is connected directly to two turbines $A$ and $B$ as shown. The pipe has an outer diameter of 4 in. and a wall thickness of 0.25 in. The connection was made at $T_1 = 70°F$. If the turbines' points of attachment are assumed to have a stiffness of $k = 80(10^3)$ determine the force the pipe exerts on the turbines when the steam and thus the pipe reach a temperature of $T_2 = 275°F$.

6 ft

**Probs. 10–39/40**

**10-41.** The 40-ft-long A-36 steel rails on a train track are laid with a small gap between them to allow for thermal expansion. Determine the required gap $\delta$ so that the rails just touch one another when the temperature is increased from $T_1 = -20°F$ to $T_2 = 90°F$. Using this gap, what would be the axial force in the rails if the temperature were to rise to $T_3 = 110°F$ The cross-sectional area of each rail is 5.10 in².

40 ft

**Prob. 10–41**

**10-42.** The 0.4-in.-diameter A-36 steel bolt is used to hold the (rigid) assembly together. Determine the clamping force that must be provided by the bolt when $T_1 = 90°F$ so that the clamping force it exerts when $T_2 = 175°F$ is 500 lb.

**10-43.** The 0.4-in.-diameter A-36 steel bolt is used to hold the (rigid) assembly together. If the nut is snug (no axial force in the bolt) when $T = 90°F$, determine the clamping force it exerts on the assembly when $T = 20°F$.

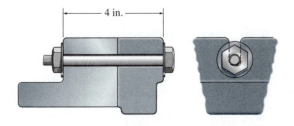

4 in.

**Probs. 10–42/43**

**\*10-44.** A thermo gate consists of a 6061-T6-aluminum plate $AB$ and an Am-1004-T61-magnesium plate $CD$, each having a width of 15 mm and fixed supported at their ends. If the gap between them is 1.5 mm when the temperature is $T_1 = 25°C$, determine the temperature required to just close the gap. Also, what is the axial force in each plate if the temperature becomes $T_2 = 100°C$? Assume bending or buckling will not occur.

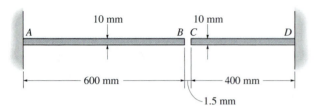

Prob. 10-44

**10-45.** The C83400-red-brass rod $AB$ and 2014-T6-aluminum rod $BC$ are joined at the collar $B$ and fixed connected at their ends. If there is no load in the members when $T_1 = 50°F$, determine the average normal stress in each member when $T_2 = 120°F$. Also, how far will the collar be displaced? The cross-sectional area of each member is 1.75 in². 

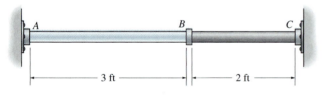

Prob. 10-45

**10-46.** The two circular rod segments, one of aluminum and the other of copper, are fixed to the rigid walls such that there is a gap of 0.008 in. between them when $T_1 = 60°F$. What larger temperature $T_2$ is required in order to just close the gap? Each rod has a diameter of 1.25 in., $\alpha_{al} = 13(10^{-6})/°F$, $E_{al} = 10(10^3)$ ksi, $\alpha_{cu} = 9.4(10^{-6})/°F$, $E_{cu} = 18(10^3)$ ksi. Determine the average normal stress in each rod if $T_2 = 200°F$.

**10-47.** The two circular rod segments, one of aluminum and the other of copper, are fixed to the rigid walls such that there is a gap of 0.008 in. between them when $T_1 = 60°F$. Each rod has a diameter of 1.25 in., $\alpha_{al} = 13(10^{-6})/°F$, $E_{al} = 10(10^3)$ ksi, $\alpha_{cu} = 9.4(10^{-6})/°F$, $E_{cu} = 18(10^3)$ ksi. Determine the average normal stress in each rod if $T_2 = 300°F$, and also calculate the new length of the aluminum segment.

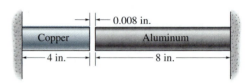

Probs. 10-46/47

**\*10-48.** The pipe is made of A-36 steel and is connected to the collars at $A$ and $B$. When the temperature is 60°F, there is no axial load in the pipe. If hot gas traveling through the pipe causes its temperature to rise by $\Delta T = (40 + 15x)°F$, where $x$ is in feet, determine the average normal stress in the pipe. The inner diameter is 2 in., the wall thickness is 0.15 in.

**10-49.** The bronze 86100 pipe has an inner radius of 0.5 in. and a wall thickness of 0.2 in. If the gas flowing through it changes the temperature of the pipe uniformly from $T_A = 200°F$ at $A$ to $T_B = 60°F$ at $B$, determine the axial force it exerts on the walls. The pipe was fitted between the walls when $T = 60°F$.

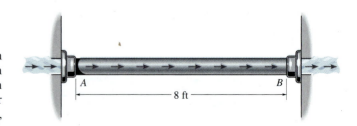

Probs. 10-48/49

**10-50.** The wires $AB$ and $AC$ are made of steel, and wire $AD$ is made of copper. Before the 150-lb force is applied, $AB$ and $AC$ are each 60 in. long and $AD$ is 40 in. long. If the temperature is increased by 80°F determine the force in each wire needed to support the load. Take $E_{st} = 29(10^3)$ ksi, $E_{cu} = 17(10^3)$ ksi, $\alpha_{st} = 8(10^{-6})/°F$, $\alpha_{cu} = 9.60(10^{-6})/°F$. Each wire has a cross-sectional area of 0.0123 in².

**10-51.** The steel bolt has a diameter of 7 mm and fits through an aluminum sleeve as shown. The sleeve has an inner diameter of 8 mm and an outer diameter of 10 mm. The nut at $A$ is adjusted so that it just presses up against the sleeve. If the assembly is originally at a temperature of $T_1 = 20°C$ and then is heated to a temperature of $T_2 = 100°C$, determine the average normal stress in the bolt and the sleeve. $E_{st} = 200$ GPa, $E_{al} = 70$ GPa, $\alpha_{st} = 14(10^{-6})/°C$, $\alpha_{al} = 23(10^{-6})/°C$.

**Prob. 10–51**

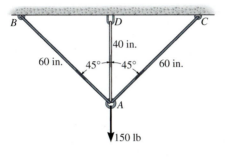

**Prob. 10–50**

# CHAPTER REVIEW

- *Saint-Venant's Principle.* When a loading is applied at a point on a body it tends to create a stress distribution within the body that becomes more uniformly distributed at regions removed from the point of application. This is called Saint-Venant's principle.

- The relative displacement at the end of an axially loaded member relative to the other end is determined from $\delta = \displaystyle\int_0^L \dfrac{P(x)\,dx}{AE}$. If a series of constant axial forces are applied to the member then $\delta = \Sigma \dfrac{PL}{AE}$. For application, it is necessary to use a sign convention for the internal load and to be sure the material does not yield, but remains linear elastic.

- *Principle of Superposition.* Superposition of load and displacement is possible provided the material remains linear elastic and no significant changes in geometry occur.

- *Statically Indeterminate Bar.* The reactions on a statically indeterminate bar can be determined by satisfying equilibrium requirements and the compatibility conditions at the supports. These two conditions are related using the load-displacement relationships that involves the material property $E$.

• *Thermal Stress.* Temperature can cause a member made from homogenous isotropic material to change its length by $\delta = \alpha \Delta T L$. If the member is confined, this expansion will produce thermal stress in the member.

# REVIEW PROBLEMS

**\*10-52.** The rigid beam is supported at its ends by steel tie rods. The rods have diameters $d_{AB} = 0.5$ in. and $d_{CD} = 0.4$ in. If the allowable stress for the steel is $\sigma_{allow} = 16.2$ ksi, determine the maximum value of $P$ and the position $x$ of this force on the beam so the beam remains in the horizontal position when it is loaded. $E_{st} = 29(10^3)$ ksi.

**10-53.** The 0.4-in.-diameter steel bolt is used to hold the (rigid) assembly together. Determine the clamping force that must be provided by the bolt when $T_1 = 80°F$, so that the clamping force it exerts when $T_2 = 180°F$ is 200 lb. $E_{st} = 29(10^3)$ ksi, $\alpha_{st} = 6(10^{-6})/°F$.

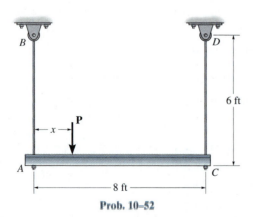

Prob. 10-52

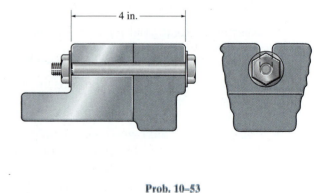

Prob. 10-53

**10-54.** The joint is made from three steel plates that are bonded together at their seams. Determine the displacement of end $A$ with respect to end $B$ when the joint is subjected to the axial loads shown. Each plate has a thickness of 5 mm. $E_{st} = 200$ GPa.

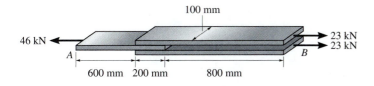

Prob. 10–54

**10-55.** The steel bar has the original dimensions shown in the figure. If it is subjected to an axial loading of 50 kN, determine the change in its length and its new cross-sectional dimensions at section $a$–$a$. $E_{st} = 200$ GPa, $v_{st} = 0.29$.

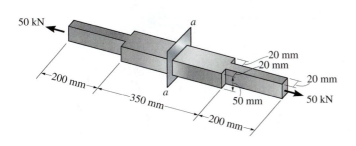

Prob. 10–55

**\*10-56.** The steel bolt has a diameter of 7 mm and fits through an aluminum sleeve as shown. The sleeve has an inner diameter of 8 mm and an outer diameter of 10 mm. The nut at $A$ is adjusted so that it just presses up against the sleeve. If it is then tightened one-half turn, determine the force in the bolt and the sleeve. The single-threaded screw on the bolt has a lead of 1.5 mm. $E_{st} = 200$ GPa, $E_{al} = 70$ GPa. *Note:* The lead represents the distance the nut advances along the bolt for one complete turn of the nut.

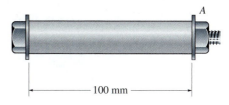

Prob. 10–56

**10-57.** The observation cage $C$ has a weight of 250 kip and, through a system of gears, travels upward at constant velocity along the steel column, which has a height of 200 ft. The column has an outer diameter of 3 ft and is made from steel plate having a thickness of 0.25 in. Neglect the weight of the column, and determine the normal stress in the column at its base, $B$, as a function of the cage's position $y$. Also, determine the relative displacement of end $A$ with respect to end $B$ as a function of $y$. $E_{st} = 29(10^3)$ ksi.

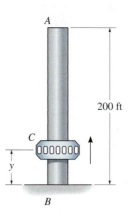

Prob. 10–57

The torsional stress developed within the drive shaft of this condensation fan depends upon the output of the motor.

# Torsion

- To determine the torsional deformation of a perfectly elastic circular shaft.
- To determine the support reactions when these reactions cannot be determined solely from the moment equilibrium equation.
- To determine the maximum power that can be transmitted by a shaft.

## 11.1   Torsional Deformation of a Circular Shaft

*Torque* is a moment that tends to twist a member about its longitudinal axis. Its effect is of primary concern in the design of axles or drive shafts used in vehicles and machinery. We can illustrate physically what happens when a torque is applied to a circular shaft by considering the shaft to be made of a highly deformable material such as rubber, Fig. 11–1a. When the torque is applied, the circles and longitudinal grid lines originally marked on the shaft tend to distort into the pattern shown in Fig. 11–1b. By inspection, twisting causes the circles to *remain circles*, and each longitudinal grid line deforms into a helix that intersects the circles at equal angles. Also, the cross sections at the *ends* of the shaft remain *flat*— that is, they do not warp or bulge in or out—and radial lines on these ends *remain straight* during the deformation, Fig. 11–1b. From these observations we can assume that if the angle of rotation is *small*, the *length of the shaft* and *its radius* will *remain unchanged*.

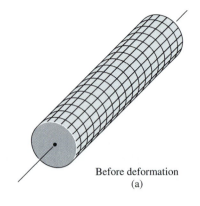

Before deformation
(a)

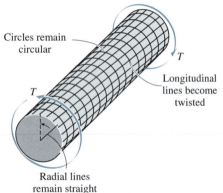

Circles remain circular

Longitudinal lines become twisted

$T$

$T$

Radial lines remain straight

After deformation
(b)

**Fig. 11–1**

Notice the deformation of the rectangular element when this rubber bar is subjected to a torque.

If the shaft is fixed at one end and a torque is applied to its other end, the shaded plane in Fig. 11–2 will distort into a skewed form as shown. Here a radial line located on the cross section at a distance $x$ from the fixed end of the shaft will rotate through an angle $\phi(x)$. The angle $\phi(x)$, so defined, is called the *angle of twist*. It depends on the position $x$ and will vary along the shaft as shown.

In order to understand how this distortion strains the material, we will now isolate a small element located at a radial distance $\rho$ (rho) from the axis of the shaft, Fig. 11–3. Due to the deformation as noted in Fig. 11–2, the front and rear faces of the element will undergo a rotation. The back face will rotate by $\phi(x)$, and the front face by $\phi(x) + \Delta\phi$. As a result, the *difference* in these rotations, $\Delta\phi$, causes the element to be subjected to a *shear strain*. To calculate this strain, note that before deformation the angle between the edges $AB$ and $AC$ is 90°; after deformation, however, the edges of the element are $AD$ and $AC$ and the angle between them is $\theta'$. From the definition of shear strain, Eq. 8–13, we have

$$\gamma = \frac{\pi}{2} - \lim_{\substack{C \to A \text{ along } CA \\ B \to A \text{ along } BA}} \theta'$$

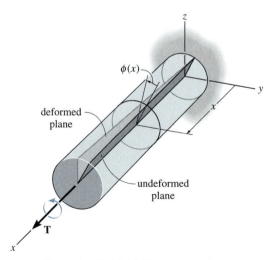

deformed plane

undeformed plane

$\phi(x)$

$z$

$y$

$x$

$x$

**T**

The angle of twist $\phi(x)$ increases as $x$ increases.

**Fig. 11–2**

This angle, $\gamma$, is indicated on the element. It can be related to the length $\Delta x$ of the element and the difference in the angle of rotation, $\Delta\phi$, between the shaded faces. If we let $\Delta x \to dx$ and $\Delta\phi \to d\phi$, we have

$$BD = \rho\, d\phi = dx\, \gamma$$

Therefore,

$$\gamma = \rho \frac{d\phi}{dx} \qquad (11\text{--}1)$$

Since $dx$ and $d\phi$ are the *same* for *all elements* located at points on the cross section at $x$, then $d\phi/dx$ is constant, and Eq. 11–1 states that the magnitude of the shear strain for any of these elements varies only with its radial distance $\rho$ from the axis of the shaft. In other words, the shear strain within the shaft varies linearly along any radial line, from zero at the axis of the shaft to a maximum $\gamma_{max}$ at its outer boundary, Fig. 11–4. Since $d\phi/dx = \gamma/\rho = \gamma_{max}/c$, then

$$\gamma = \left(\frac{\rho}{c}\right)\gamma_{max} \qquad (11\text{--}2)$$

The results obtained here are also valid for circular tubes. They depend only on the assumptions regarding the deformations mentioned above.

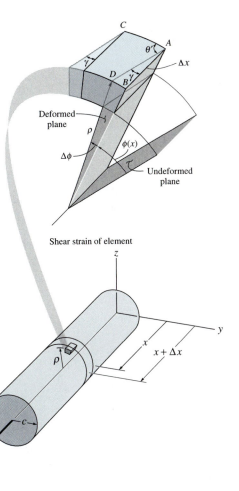

Shear strain of element

Fig. 11–3

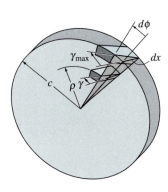

The shear strain for the material increases linearly with $\rho$, i.e., $\gamma = (\rho/c)\gamma_{max}$

Fig. 11–4

## 11.2 The Torsion Formula

When an external torque is applied to a shaft it creates a corresponding internal torque within the shaft. In this section, we will develop an equation that relates this internal torque to the shear stress distribution on the cross section of a circular shaft or tube.

If the material is linear-elastic, then Hooke's law applies, $\tau = G\gamma$, and consequently a ***linear variation in shear strain***, as noted in the previous section, leads to a corresponding ***linear variation in shear stress*** along any radial line on the cross section. Hence, like the shear-strain variation, for a solid shaft, $\tau$ will vary from zero at the shaft's longitudinal axis to a maximum value, $\tau_{max}$, at its outer surface. This variation is shown in Fig. 11–5 on the front faces of a selected number of elements, located at an intermediate radial position $\rho$ and at the outer radius $c$. Due to the proportionality of triangles, or by using Hooke's law ($\tau = G\gamma$) and Eq. 11–2 [$\gamma = (\rho/c)\gamma_{max}$], we can write

$$\tau = \left(\frac{\rho}{c}\right)\tau_{max} \tag{11–3}$$

This equation expresses the shear-stress distribution as a *function* of the radial position $\rho$ of the element; in other words, it defines the stress distribution over the cross section in terms of the geometry of the shaft. Using it, we will now apply the condition that requires the torque produced by the stress distribution over the entire cross section to be equivalent to the resultant internal torque $T$ at the section, which holds

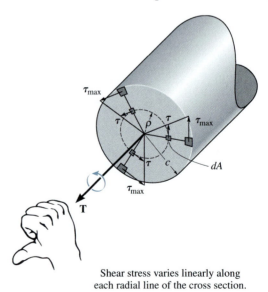

Shear stress varies linearly along
each radial line of the cross section.

**Fig. 11–5**

the shaft in equilibrium, Fig. 11–5. Specifically, each element of area $dA$, located at $\rho$, is subjected to a force of $dF = \tau\, dA$. The torque produced by this force is $dT = \rho(\tau\, dA)$. We therefore have for the entire cross section

$$T = \int_A \rho(\tau\, dA) = \int_A \rho\left(\frac{\rho}{c}\right)\tau_{max}\, dA \qquad (11\text{–}4)$$

Since $\tau_{max}/c$ is constant,

$$T = \frac{\tau_{max}}{c} \int_A \rho^2\, dA \qquad (11\text{–}5)$$

The integral in this equation depends only on the geometry of the shaft. It represents the *polar moment of inertia* of the shaft's cross-sectional area computed about the shaft's longitudinal axis. We will symbolize its value as $J$, and therefore the above equation can be written in a more compact form, namely,

$$\boxed{\tau_{max} = \frac{Tc}{J}} \qquad (11\text{–}6)$$

where

$\tau_{max}$ = the maximum shear stress in the shaft, which occurs at the outer surface

$T$ = the resultant internal torque acting at the cross section. Its value is determined from the method of sections and the equation of moment equilibrium applied about the shaft's longitudinal axis

$J$ = the polar moment of inertia of the cross-sectional area

$c$ = the outer radius of the shaft

Using Eqs. 11–3 and 11–6, the shear stress at the intermediate distance $\rho$ can be determined from a similar equation:

$$\boxed{\tau = \frac{T\rho}{J}} \qquad (11\text{–}7)$$

Either of the above two equations is often referred to as the *torsion formula*. Recall that it is used only if the shaft is circular and the material is homogeneous and behaves in a linear-elastic manner, since the derivation is based on the fact that the shear stress is proportional to the shear strain.

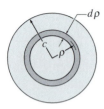

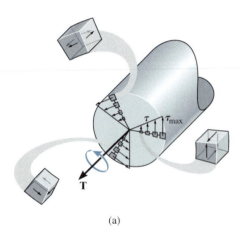

**Solid Shaft.** If the shaft has a solid circular cross section, the polar moment of inertia $J$ can be determined using an area element in the form of a *differential ring* or annulus having a thickness $d\rho$ and circumference $2\pi\rho$, Fig. 11–6. For this ring, $dA = 2\pi\rho\,d\rho$, so

$$J = \int_A \rho^2\,dA = \int_0^c \rho^2(2\pi\rho\,d\rho) = 2\pi\int_0^c \rho^3\,d\rho = 2\pi\left(\frac{1}{4}\right)\rho^4\bigg|_0^c$$

$$\boxed{J = \frac{\pi}{2}c^4} \tag{11–8}$$

**Fig. 11–6**

Note that $J$ is a *geometric property* of the circular area and is always positive. Common units used for its measurement are $mm^4$ or $in^4$.

The shear stress has been shown to vary linearly along each radial line of the cross section of the shaft. However, if a volume element of material on the cross section is isolated, then due to the complementary property of shear, equal shear stresses must also act on four of its adjacent faces as shown in Fig. 11–7a. Hence, **not only does the internal torque T develop a linear distribution of shear stress along each radial line in the plane of the cross-sectional area, but also an associated shear-stress distribution is developed along an axial plane**, Fig. 11–7b. It is interesting to note that because of this axial distribution of shear stress, shafts made from wood tend to *split* along the axial plane when subjected to excessive torque, Fig. 11–8. This is because wood is an anisotropic material. Its shear resistance parallel to its grains or fibers, directed along the axis of the shaft, is much less than its resistance perpendicular to the fibers, directed in the plane of the cross section.

(a)

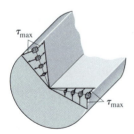

Shear stress varies linearly along
each radial line of the cross section.
(b)

**Fig. 11–7**

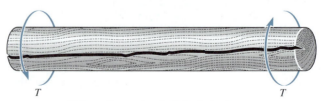

Failure of a wooden shaft due to torsion.

**Fig. 11–8**

## Tubular Shaft.

If a shaft has a tubular cross section, with inner radius $c_i$ and outer radius $c_o$, then from Eq. 11–8 we can determine its polar moment of inertia by subtracting $J$ for a shaft of radius $c_i$ from that determined for a shaft of radius $c_o$. The result is

$$\boxed{J = \frac{\pi}{2}(c_o^4 - c_i^4)} \qquad (11\text{–}9)$$

Like the solid shaft, the shear stress distributed over the tube's cross-sectional area varies linearly along any radial line, Fig. 11–9a. Furthermore, the shear stress varies along an axial plane in this same manner, Fig. 11–9b. Examples of the shear stress acting on typical volume elements are shown in Fig. 11–9a.

This tubular drive shaft for a truck was subjected to an overload resulting in failure caused by yielding of the material.

## Absolute Maximum Torsional Stress.

At any given cross section of the shaft the maximum shear stress occurs at the outer surface. However, if the shaft is subjected to a series of external torques, or the radius (polar moment of inertia) changes, then the maximum torsional stress within the shaft could be different from one section to the next. If the absolute maximum torsional stress is to be determined, then it becomes important to find the location where the ratio $Tc/J$ is a maximum. In this regard, it may be helpful to show the variation of the internal torque $T$ at each section along the axis of the shaft by drawing a *torque diagram*. Specifically, this diagram is a plot of the internal torque $T$ versus its position $x$ along the shaft's length. As a sign convention, $T$ will be positive if by the right-hand rule the thumb is directed outward from the shaft when the fingers curl in the direction of twist as caused by the torque, Fig. 11–5. Once the internal torque throughout the shaft is determined, the maximum ratio of $Tc/J$ can then be identified.

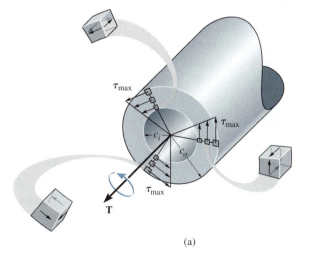

(a)

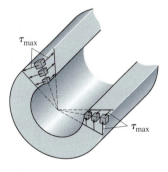

Shear stress varies linearly along each radial line of the cross section.

(b)

**Fig. 11–9**

## IMPORTANT POINTS

- When a shaft having a *circular cross section* is subjected to a torque, the cross section *remains plane* while radial lines rotate. This causes a *shear strain* within the material that *varies linearly* along any radial line, from zero at the axis of the shaft to a maximum at its outer boundary.
- For linearly elastic homogeneous material, due to Hooke's law, the *shear stress* along any radial line of the shaft also *varies linearly*, from zero at its axis to a maximum at its outer boundary. This maximum shear stress *must not* exceed the proportional limit.
- Due to the complementary property of shear, the linear shear stress distribution within the plane of the cross section is also distributed along an adjacent axial plane of the shaft.
- The torsion formula is based on the requirement that the resultant torque on the cross section is equal to the torque produced by the linear shear stress distribution about the longitudinal axis of the shaft. It is necessary that the shaft or tube have a *circular* cross section and that it is made of *homogeneous* material which has *linear-elastic* behavior.

## PROCEDURE FOR ANALYSIS

The torsion formula can be applied using the following procedure.

*Internal Loading.*

- Section the shaft perpendicular to its axis at the point where the shear stress is to be determined, and use the necessary free-body diagram and equations of equilibrium to obtain the internal torque at the section.

*Section Property.*

- Compute the polar moment of inertia of the cross-sectional area. For a solid section of radius $c$, $J = \pi c^4/2$, and for a tube of outer radius $c_o$ and inner radius $c_i$, $J = \pi(c_o^4 - c_i^4)/2$.

*Shear Stress.*

- Specify the radial distance $\rho$, measured from the center of the cross section to the point where the shear stress is to be found. Then apply the torsion formula $\tau = T\rho/J$, or if the maximum shear stress is to be determined use $\tau_{max} = Tc/J$. When substituting the data, make sure to use a consistent set of units.
- The shear stress acts on the cross section in a direction that is always perpendicular to $\rho$. The force it creates must contribute a torque about the axis of the shaft that is in the *same direction* as the internal resultant torque $\mathbf{T}$ acting on the section. Once this direction is established, a volume element located at the point where $\tau$ is determined can be isolated, and the direction of $\tau$ acting on the remaining three adjacent faces of the element can be shown.

# EXAMPLE 11.1

The stress distribution in a solid shaft has been plotted along three arbitrary radial lines as shown in Fig. 11–10a. Determine the resultant internal torque at the section.

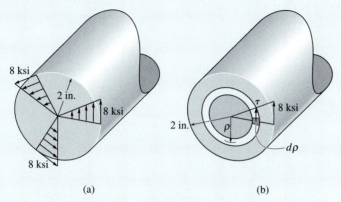

(a)                                      (b)

**Fig. 11–10**

## Solution I

The polar moment of inertia for the cross-sectional area is

$$J = \frac{\pi}{2}(2 \text{ in.})^4 = 25.13 \text{ in}^4$$

Applying the torsion formula, with $\tau_{max} = 8$ ksi, Fig. 11–10a, we have

$$\tau_{max} = \frac{Tc}{J}; \qquad 8 \text{ kip/in}^2 = \frac{T(2 \text{ in.})}{(25.13 \text{ in}^4)}$$

$$T = 101 \text{ kip} \cdot \text{in.} \qquad\qquad \textit{Ans.}$$

## Solution II

The same result can be obtained by finding the torque produced by the stress distribution about the centroidal axis of the shaft. First we must express $\tau = f(\rho)$. Using proportional triangles, Fig. 11–10b, we have

$$\frac{\tau}{\rho} = \frac{8 \text{ ksi}}{2 \text{ in.}}$$

$$\tau = 4\rho$$

This stress acts on all portions of the differential ring element that has an area $dA = 2\pi\rho \, d\rho$. Since the force created by $\tau$ is $dF = \tau \, dA$, the torque is

$$dT = \rho \, dF = \rho(\tau \, dA) = \rho(4\rho)2\pi\rho \, d\rho = 8\pi\rho^3 \, d\rho$$

For the entire area over which $\tau$ acts, we require

$$T = \int_0^2 8\pi\rho^3 \, d\rho = 8\pi\left(\frac{1}{4}\rho^4\right)\Big|_0^2 = 101 \text{ kip} \cdot \text{in.} \qquad \textit{Ans.}$$

## EXAMPLE 11.2

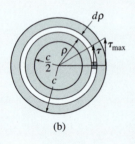

(a)

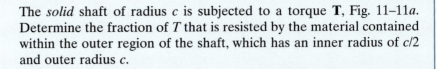

(b)

Fig. 11–11

The *solid* shaft of radius $c$ is subjected to a torque **T**, Fig. 11–11a. Determine the fraction of $T$ that is resisted by the material contained within the outer region of the shaft, which has an inner radius of $c/2$ and outer radius $c$.

### Solution

The stress in the shaft varies linearly, such that $\tau = (\rho/c)\tau_{max}$, Eq. 11–3. Therefore, the torque $dT'$ on the ring (area) located within the lighter-shaded region, Fig. 11–11b, is

$$dT' = \rho(\tau\, dA) = \rho(\rho/c)\tau_{max}(2\pi\rho\, d\rho)$$

For the entire lighter-shaded area the torque is

$$T' = \frac{2\pi\tau_{max}}{c}\int_{c/2}^{c}\rho^3\, d\rho$$

$$= \frac{2\pi\tau_{max}}{c}\cdot\frac{1}{4}\rho^4\Big|_{c/2}^{c}$$

So that

$$T' = \frac{15\pi}{32}\tau_{max}c^3 \qquad (1)$$

This torque $T'$ can be expressed in terms of the applied torque $T$ by first using the torsion formula to determine the maximum stress in the shaft. We have

$$\tau_{max} = \frac{Tc}{J} = \frac{Tc}{(\pi/2)c^4}$$

or

$$\tau_{max} = \frac{2T}{\pi c^3}$$

Substituting this into Eq. 1 yields

$$T' = \frac{15}{16}T \qquad \textit{Ans.}$$

Here, approximately 94% of the torque is resisted by the lighter-shaded region, and the remaining 6% of $T$ (or $\frac{1}{16}$) is resisted by the inner "core" of the shaft, $\rho = 0$ to $\rho = c/2$. As a result, the material located at the *outer region* of the shaft is highly effective in resisting torque, which justifies the use of tubular shafts as an efficient means for transmitting torque, and thereby saves material.

## EXAMPLE 11.3

The shaft shown in Fig. 11–12a is supported by two bearings and is subjected to three torques. Determine the shear stress developed at points A and B, located at section a–a of the shaft, Fig. 11–12b.

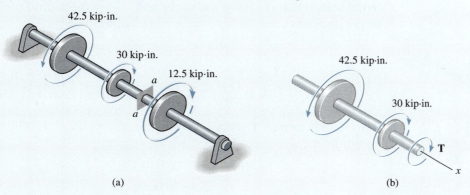

(a)                          (b)

**Fig. 11–12**

### Solution

*Internal Torque.* The bearing reactions on the shaft are zero, provided the shaft's weight is neglected. Furthermore, the applied torques satisfy moment equilibrium about the shaft's axis.

The internal torque at section a–a will be determined from the free-body diagram of the left segment, Fig. 11–12b. We have

$$\Sigma M_x = 0; \quad 42.5 \text{ kip} \cdot \text{in.} - 30 \text{ kip} \cdot \text{in.} - T = 0 \quad T = 12.5 \text{ kip} \cdot \text{in.}$$

*Section Property.* The polar moment of inertia for the shaft is

$$J = \frac{\pi}{2}(0.75 \text{ in.})^4 = 0.497 \text{ in}^4$$

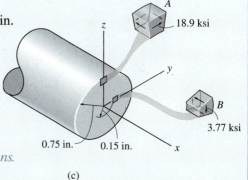

(c)

*Shear Stress.* Since point A is at $\rho = c = 0.75$ in.,

$$\tau_A = \frac{Tc}{J} = \frac{(12.5 \text{ kip} \cdot \text{in.})(0.75 \text{ in.})}{0.497 \text{ in}^4} = 18.9 \text{ ksi} \quad \textit{Ans.}$$

Likewise for point B, at $\rho = 0.15$ in., we have

$$\tau_B = \frac{T\rho}{J} = \frac{(12.5 \text{ kip} \cdot \text{in.})(0.15 \text{ in.})}{0.497 \text{ in}^4} = 3.77 \text{ ksi} \quad \textit{Ans.}$$

The directions of these stresses on each element at A and B, Fig. 11–12c, are established from the direction of the resultant internal torque **T**, shown in Fig. 11–12b. Note carefully how the shear stress acts on the planes of each of these elements.

**E X A M P L E  11.4**

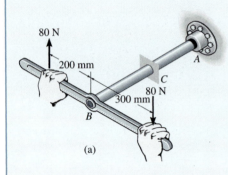

(a)

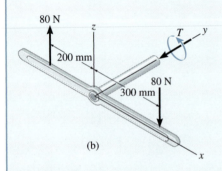

(b)

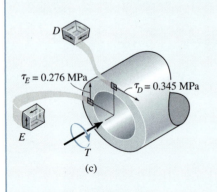

(c)

**Fig. 11–13**

The pipe shown in Fig. 11–13a has an inner diameter of 80 mm and an outer diameter of 100 mm. If its end is tightened against the support at $A$ using a torque wrench at $B$, determine the shear stress developed in the material at the inner and outer walls along the central portion of the pipe when the 80-N forces are applied to the wrench.

**Solution**

*Internal Torque.*   A section is taken at an intermediate location $C$ along the pipe's axis, Fig. 11–13b. The only unknown at the section is the internal torque **T**. Force equilibrium and moment equilibrium about the $x$ and $z$ axes are satisfied. We require

$$\Sigma M_y = 0; \qquad 80 \text{ N}(0.3 \text{ m}) + 80 \text{ N}(0.2 \text{ m}) - T = 0$$
$$T = 40 \text{ N} \cdot \text{m}$$

*Section Property.*   The polar moment of inertia for the pipe's cross-sectional area is

$$J = \frac{\pi}{2}[(0.05 \text{ m})^4 - (0.04 \text{ m})^4] = 5.80(10^{-6}) \text{ m}^4$$

*Shear Stress.*   For any point lying on the outside surface of the pipe, $\rho = c_o = 0.05$ m, we have

$$\tau_o = \frac{Tc_o}{J} = \frac{40 \text{ N} \cdot \text{m}(0.05 \text{ m})}{5.80(10^{-6}) \text{ m}^4} = 0.345 \text{ MPa} \qquad \textit{Ans.}$$

And for any point located on the inside surface, $\rho = c_i = 0.04$ m, so that

$$\tau_i = \frac{Tc_i}{J} = \frac{40 \text{ N} \cdot \text{m}(0.04 \text{ m})}{5.80(10^{-6}) \text{ m}^4} = 0.276 \text{ MPa} \qquad \textit{Ans.}$$

To show how these stresses act at representative points $D$ and $E$ on the cross-sectional area, we will first view the cross section from the front of segment $CA$ of the pipe, Fig. 11–13a. On this section, Fig. 11–13c, the resultant internal torque is equal but opposite to that shown in Fig. 11–13b. The shear stresses at $D$ and $E$ contribute to this torque and therefore act on the shaded faces of the elements in the directions shown. As a consequence, notice how the shear-stress components act on the other three faces. Furthermore, since the top face of $D$ and the inner face of $E$ are in stress-free regions taken from the pipe's outer and inner walls, no shear stress can exist on these faces or on the other corresponding faces of the elements.

## 11.3 Power Transmission

Shafts and tubes having circular cross sections are often used to transmit power developed by a machine. When used for this purpose, they are subjected to torques that depend on the power generated by the machine and the angular speed of the shaft. **Power** is defined as the work performed per unit of time. The work transmitted by a rotating shaft equals the torque applied times the angle of rotation. Therefore, if during an instant of time $dt$ an applied torque **T** causes the shaft to rotate $d\theta$, then the instantaneous power is

$$P = \frac{T \, d\theta}{dt}$$

Since the shaft's angular velocity $\omega = d\theta/dt$, we can also express the power as

$$P = T\omega \tag{11-10}$$

In the SI system, power is expressed in *watts* when torque is measured in newton-meters (N·m) and $\omega$ is in radians per second (rad/s) (1 W = 1 N·m/s). In the Foot-Pound-Second or FPS system, the basic units of power are foot-pounds per second (ft·lb/s); however, horsepower (hp) is often used in engineering practice, where

$$1 \text{ hp} = 550 \text{ ft} \cdot \text{lb/s}$$

For machinery, the *frequency* of a shaft's rotation, $f$, is often reported. This is a measure of the number of revolutions or cycles the shaft makes per second and is expressed in hertz (1 Hz = 1 cycle/s). Since 1 cycle = $2\pi$ rad, then $\omega = 2\pi f$, and the above equation for power becomes

$$P = 2\pi f T \tag{11-11}$$

**Shaft Design.** When the power transmitted by a shaft and its frequency of rotation are known, the torque developed in the shaft can be determined from Eq. 11–11, that is, $T = P/2\pi f$. Knowing $T$ and the allowable shear stress for the material, $\tau_{\text{allow}}$, we can determine the size of the shaft's cross section using the torsion formula, provided the material behavior is linear-elastic. Specifically, the design or geometric parameter $J/c$ becomes

$$\frac{J}{c} = \frac{T}{\tau_{\text{allow}}} \tag{11-12}$$

For a *solid shaft*, $J = (\pi/2)c^4$, and thus, upon substitution, a *unique value* for the shaft's radius $c$ can be determined. If the shaft is *tubular*, so that $J = (\pi/2)(c_o^4 - c_i^4)$, design permits a wide range of possibilities for the solution. This is because an *arbitrary choice* can be made for either $c_o$ or $c_i$ and the other radius can then be determined from Eq. 11–12.

The drive shaft of this cutting machine must be designed to meet the power requirements of its motor.

**E X A M P L E** **11.5**

A solid steel shaft $AB$ shown in Fig. 11–14 is to be used to transmit 5 hp from the motor $M$ to which it is attached. If the shaft rotates at $\omega = 175$ rpm and the steel has an allowable shear stress of $\tau_{\text{allow}} = 14.5$ ksi, determine the required diameter of the shaft to the nearest $\frac{1}{8}$ in.

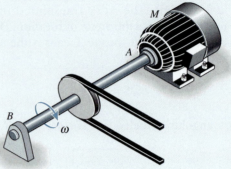

**Fig. 11–14**

**Solution**

The torque on the shaft is determined from Eq. 11–10, that is, $P = T\omega$. Expressing $P$ in foot-pounds per second and $\omega$ in radians/second, we have

$$P = 5 \text{ hp}\left(\frac{550 \text{ ft} \cdot \text{lb/s}}{1 \text{ hp}}\right) = 2750 \text{ ft} \cdot \text{lb/s}$$

$$\omega = \frac{175 \text{ rev}}{\text{min}}\left(\frac{2\pi \text{ rad}}{1 \text{ rev}}\right)\left(\frac{1 \text{ min}}{60 \text{ s}}\right) = 18.33 \text{ rad/s}$$

Thus,

$$P = T\omega; \qquad 2750 \text{ ft} \cdot \text{lb/s} = T(18.33 \text{ rad/s})$$
$$T = 150.1 \text{ ft} \cdot \text{lb}$$

Applying Eq. 11–12 yields

$$\frac{J}{c} = \frac{\pi}{2}\frac{c^4}{c} = \frac{T}{\tau_{\text{allow}}}$$

$$c = \left(\frac{2T}{\pi\tau_{\text{allow}}}\right)^{1/3} = \left(\frac{2(150.1 \text{ ft} \cdot \text{lb})(12 \text{ in./ft})}{\pi(14.500 \text{ lb/in}^2)}\right)^{1/3}$$

$$c = 0.429 \text{ in.}$$

Since $2c = 0.858$ in., select a shaft having a diameter of

$$d = \frac{7}{8}\text{in.} = 0.875 \text{ in.} \qquad \textit{Ans.}$$

# EXAMPLE 11.6

A tubular shaft, having an inner diameter of 30 mm and an outer diameter of 42 mm, is to be used to transmit 90 kW of power. Determine the frequency of rotation of the shaft so that the shear stress will not exceed 50 MPa.

### Solution

The maximum torque that can be applied to the shaft is determined from the torsion formula.

$$\tau_{max} = \frac{Tc}{J}$$

$$50(10^6) \text{ N/m}^2 = \frac{T(0.021 \text{ m})}{(\pi/2)[(0.021 \text{ m})^4 - (0.015 \text{ m})^4]}$$

$$T = 538 \text{ N} \cdot \text{m}$$

Applying Eq. 11–11, the frequency of rotation is

$$P = 2\pi f T$$

$$90(10^3) \text{ N} \cdot \text{m/s} = 2\pi f(538 \text{ N} \cdot \text{m})$$

$$f = 26.6 \text{ Hz} \qquad \qquad \textit{Ans.}$$

---

# PROBLEMS

**11-1.** A shaft is made of a steel alloy having an allowable shear stress of $\tau_{allow} = 12$ ksi. If the diameter of the shaft is 1.5 in., determine the maximum torque **T** that can be transmitted. What would be the maximum torque **T'** if a 1-in.-diameter hole is bored through the shaft? Sketch the shear-stress distribution along a radial line in each case.

**11-2.** The solid shaft of radius $r$ is subjected to a torque **T**. Determine the radius $r'$ of the inner core of the shaft that resists one-quarter of the applied torque $(T/4)$. Solve the problem two ways: (a) by using the torsion formula and (b) by finding the resultant of the shear-stress distribution.

**Prob. 11–1**

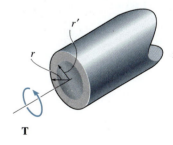

**Prob. 11–2**

**11-3.** The shaft has an outer diameter of 1.25 in. and an inner diameter of 1 in. If it is subjected to the applied torques as shown, determine the absolute maximum shear stress developed in the shaft. The smooth bearings at $A$ and $B$ do not resist torque.

**\*11-4.** The shaft has an outer diameter of 1.25 in. and an inner diameter of 1 in. If it is subjected to the applied torques as shown, plot the shear-stress distribution acting along a radial line lying within region $EA$ of the shaft. The smooth bearings at $A$ and $B$ do not resist torque.

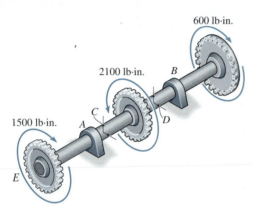

**Probs. 11–3/4**

**11-5.** The solid 30-mm-diameter shaft is used to transmit the torques applied to the gears. Determine the shear stress developed in the shaft at points $C$ and $D$. Indicate the shear stress on volume elements located at these points.

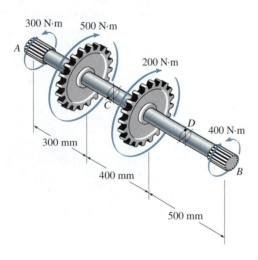

**Prob. 11–5**

**11-6.** The assembly consists of two sections of galvanized steel pipe connected together using a reducing coupling at $B$. The smaller pipe has an outer diameter of 0.75 in. and an inner diameter of 0.68 in., whereas the larger pipe has an outer diameter of 1 in. and an inner diameter of 0.86 in. If the pipe is tightly secured to the wall at $C$, determine the maximum shear stress developed in each section of the pipe when the couple shown is applied to the handles of the wrench.

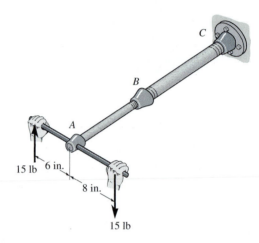

**Prob. 11–6**

**11-7.** The solid shaft has a diameter of 0.75 in. If it is subjected to the torques shown, determine the maximum shear stress developed in regions $BC$ and $DE$ of the shaft. The bearings at $A$ and $F$ allow free rotation of the shaft.

**\*11-8.** The solid shaft has a diameter of 0.75 in. If it is subjected to the torques shown, determine the maximum shear stress developed in regions $CD$ and $EF$ of the shaft. The bearings at $A$ and $F$ allow free rotation of the shaft.

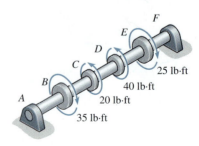

**Probs. 11–7/8**

**11-9.** A steel tube having an outer diameter of 2.5 in. is used to transmit 35 hp when turning at 2700 rev/min. Determine the inner diameter $d$ of the tube to the nearest $\frac{1}{8}$ in. if the allowable shear stress is $\tau_{\text{allow}} = 10$ ksi.

Prob. 11–9

**11-10.** The copper pipe has an outer diameter of 2.50 in. and an inner diameter of 2.30 in. If it is tightly secured to the wall at $C$ and a uniformly distributed torque is applied to it as shown, determine the shear stress developed at points $A$ and $B$. These points lie on the pipe's outer surface. Sketch the shear stress on volume elements located at $A$ and $B$.

**11-11.** The copper pipe has an outer diameter of 2.50 in. and an inner diameter of 2.30 in. If it is tightly secured to the wall at $C$ and it is subjected to the uniformly distributed torque along its entire length, determine the absolute maximum shear stress in the pipe. Discuss the validity of this result.

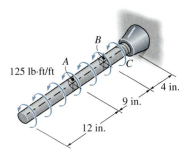

Probs. 11–10/11

**\*11-12.** The steel shaft is subjected to the torsional loading shown. Determine the shear stress developed at points $A$ and $B$ and sketch the shear stress on volume elements located at these points. The shaft where $A$ and $B$ are located has an outer radius of 60 mm.

**11-13.** The steel shaft is subjected to the torsional loading shown. Determine the absolute maximum shear stress in the shaft and sketch the shear-stress distribution along a radial line where it is maximum.

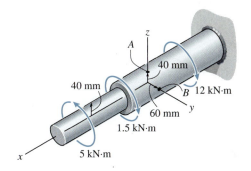

Probs. 11–12/13

**11-14.** Determine to the nearest $\frac{1}{8}$ in. the diameter of a solid shaft that is required to transmit 150 hp at 4000 rev/min. The material has an allowable shear stress of $\tau_{\text{allow}} = 8$ ksi.

**11-15.** The drilling pipe on an oil rig is made from steel pipe having an outer diameter of 4.5 in. and a thickness of 0.25 in. If the pipe is turning at 650 rev/min while being powered by a 15-hp motor, determine the maximum shear stress in the pipe.

**\*11-16.** The gear motor can develop 1/10 hp when it turns at 300 rev/min. If the shaft has a diameter of $\frac{1}{2}$ in., determine the maximum shear stress that will be developed in the shaft.

**11-17.** The gear motor can develop 1/10 hp when it turns at 80 rev/min. If the allowable shear stress for the shaft is $\tau_{\text{allow}} = 4$ ksi, determine the smallest diameter of the shaft to the nearest $\frac{1}{8}$ in. that can be used.

Probs. 11–16/17

**11-18.** The coupling is used to connect the two shafts together. Assuming that the shear stress in the bolts is *uniform*, determine the number of bolts necessary to make the maximum shear stress in the shaft equal to the shear stress in the bolts. Each bolt has a diameter $d$.

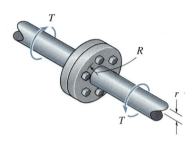

Prob. 11–18

**11-19.** The steel shafts are connected together using a fillet weld as shown. Determine the average shear stress in the weld along section $a$–$a$ if the torque applied to the shafts is $T = 60$ N·m. *Note:* The critical section where the weld fails is along section $a$–$a$.

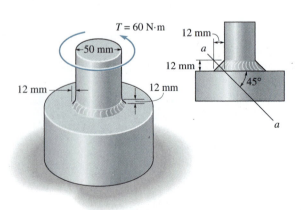

Prob. 11–19

**\*11-20.** A steel tube having an outer diameter of $d_1 = 2.5$ in. is used to transmit 35 hp when turning at 2700 rev/min. Determine the inner diameter $d_2$ of the tube to the nearest $\frac{1}{8}$ in. if the allowable shear stress is $\tau_{allow} = 10$ ksi.

Prob. 11–20

**11-21.** A steel tube having an outer diameter of $d_1 = 2.5$ in. and an inner diameter of $d_2 = 2$ in. is used to transmit 45 hp. Determine its maximum rate of rotation if the allowable shear stress is $\tau_{allow} = 12$ ksi.

Prob. 11–21

**11-22.** A ship has a propeller drive shaft that is turning at 1500 rev/min while developing 1800 hp. If it is 8 ft long and has a diameter of 4 in., determine the maximum shear stress in the shaft caused by torsion.

■**11-23.** The shaft has a diameter of 80 mm and due to friction at its surface within the hole, it is subjected to a variable torque described by the function $t = (25xe^{x^2})$ N·m/m, where $x$ is in meters. Determine the minimum torque $T_0$ needed to overcome friction and cause it to twist. Also, determine the absolute maximum stress in the shaft.

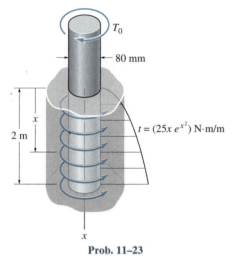

$T_0$

80 mm

2 m

$x$

$t = (25x\,e^{x^2})$ N·m/m

$x$

**Prob. 11–23**

*11-24. Determine the diameter $d$ of a solid shaft required to transmit a power $P$ if the allowable shear stress is $\tau_{allow}$ and the angular velocity of the shaft is $\omega$.

**11-25.** The drive shaft of an automobile is to be designed as a thin-walled tube. The engine delivers 150 hp when the shaft is turning at 1500 rev/min. Determine the minimum thickness of the tube's wall if the outer diameter is 2.5 in. The material has an allowable shear stress of $\tau_{allow} = 7$ ksi.

**11-26.** The motor delivers 50 hp while turning at a constant rate of 1350 rpm at $A$. Using the belt and pulley system this loading is delivered to the steel blower shaft $BC$. Determine to the nearest $\frac{1}{8}$ in. the smallest diameter of this shaft if the allowable shear stress for the steel is $\tau_{allow} = 12$ ksi.

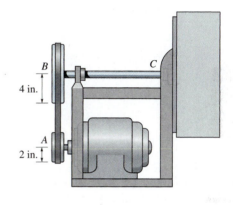

$B$

$C$

4 in.

$A$

2 in.

**Prob. 11–26**

## 11.4 Angle of Twist

Occasionally the design of a shaft depends on restricting the amount of rotation or twist that may occur when the shaft is subjected to a torque. Furthermore, being able to compute the angle of twist for a shaft is important when analyzing the reactions on statically indeterminate shafts.

In this section we will develop a formula for determining the **angle of twist** $\phi$ (phi) of one end of a shaft with respect to its other end. The shaft is assumed to have a circular cross section that can gradually vary along its length, Fig. 11–15a, and the material is assumed to be homogeneous and to behave in a linear-elastic manner when the torque is applied. As in the case of an axially loaded bar, we will neglect the localized deformations that occur at points of application of the torques and where the cross section changes abruptly. By Saint-Venant's principle, these effects occur within small regions of the shaft's length and generally have only a slight effect on the final result.

Oil wells are commonly drilled to depths exceeding a thousand meters. As a result, the total angle of twist of a string of drill pipe can be substantial and must be computed.

Using the method of sections, a differential disk of thickness $dx$, located at position $x$, is isolated from the shaft, Fig. 11–15b. The internal resultant torque is represented as $T(x)$, since the external loading may cause it to vary along the axis of the shaft. Due to $T(x)$, the disk will twist, such that the *relative rotation* of one of its faces with respect to the other face is $d\phi$, Fig. 11–15b. As a result an element of material located at an arbitrary radius $\rho$ within the disk will undergo a shear strain $\gamma$. The values of $\gamma$ and $d\phi$ are related by Eq. 11–1, namely,

$$d\phi = \gamma \frac{dx}{\rho} \tag{11–13}$$

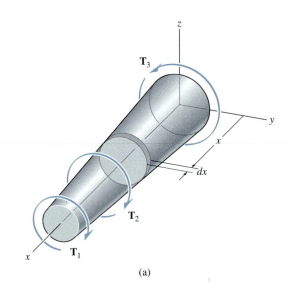

(a)

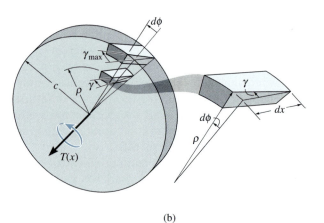

(b)

**Fig. 11–15**

Since Hooke's law, $\gamma = \tau/G$, applies and the shear stress can be expressed in terms of the applied torque using the torsion formula $\tau = T(x)\rho/J(x)$, then $\gamma = T(x)\rho/J(x)G$. Substituting this into Eq. 11–13, the angle of twist for the disk is

$$d\phi = \frac{T(x)}{J(x)G} dx$$

Integrating over the entire length $L$ of the shaft, we obtain the angle of twist for the entire shaft, namely,

$$\phi = \int_{o}^{L} \frac{T(x)\, dx}{J(x)G} \qquad (11\text{–}14)$$

Here

$\phi$ = the angle of twist of one end of the shaft with respect to the other end, measured in radians

$T(x)$ = the internal torque at the arbitrary position $x$, found from the method of sections and the equation of moment equilibrium applied about the shaft's axis

$J(x)$ = the shaft's polar moment of inertia expressed as a function of position $x$

$G$ = the shear modulus of elasticity for the material

When computing both the stress and the angle of twist of this soil auger, it is necessary to consider the variable loading which acts along its length.

**Constant Torque and Cross-Sectional Area.**   Usually in engineering practice the material is homogeneous so that $G$ is constant. Also, the shaft's cross-sectional area and the applied torque are constant along the length of the shaft, Fig. 11–16. If this is the case, the internal torque $T(x) = T$, the polar moment of inertia $J(x) = J$, and Eq. 11–14 can be integrated, which gives

$$\phi = \frac{TL}{JG} \qquad (11\text{–}15)$$

The similarities between the above two equations and those for an axially loaded bar ($\delta = \int P(x)\, dx/A(x)E$ and $\delta = PL/AE$) should be noted.

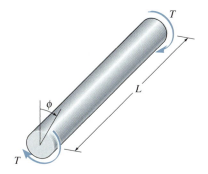

Fig. 11–16

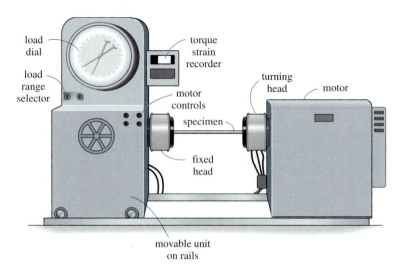

load dial

torque strain recorder

load range selector

turning head

motor

motor controls

specimen

fixed head

movable unit on rails

Fig. 11–17

We can use Eq. 11–15 to determine the shear modulus of elasticity $G$ of the material. To do so, a specimen of known length and diameter is placed in a torsion testing machine like the one shown in Fig. 11–17. The applied torque $T$ and angle of twist $\phi$ are then measured between a gauge length $L$. Using Eq. 11–15, $G = TL/J\phi$. Usually, to obtain a more reliable value of $G$, several of these tests are performed and the average value is used.

If the shaft is subjected to several different torques, or the cross-sectional area or shear modulus changes abruptly from one region of the shaft to the next, Eq. 11–15 can be applied to each segment of the shaft where these quantities are all constant. The angle of twist of one end of the shaft with respect to the other is then found from the vector addition of the angles of twist of each segment. For this case,

$$\phi = \sum \frac{TL}{JG} \qquad (11\text{–}16)$$

**Sign Convention.** In order to apply the above equation, we must develop a sign convention for the internal torque and the angle of twist of one end of the shaft with respect to the other end. To do this, we will use the right-hand rule, whereby both the torque and angle will be *positive*, provided the *thumb* is directed *outward* from the shaft when the fingers curl to give the tendency for rotation, Fig. 11–18.

To illustrate the use of this sign convention, consider the shaft shown in Fig. 11–19a, which is subjected to four torques. The angle of twist of end $A$ with respect to end $D$ is to be determined. For this problem, three segments of the shaft must be considered, since the internal torque

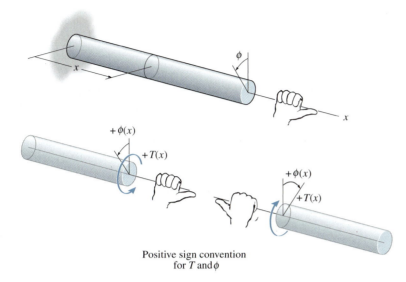

Positive sign convention
for $T$ and $\phi$

**Fig. 11–18**

changes at $B$ and $C$. Using the method of sections, the internal torques are found for each segment, Fig. 11–19$b$. By the right-hand rule, with positive torques directed away from the *sectioned end* of the shaft, we have $T_{AB} = +80$ N $\cdot$ m, $T_{BC} = -70$ N $\cdot$ m, and $T_{CD} = -10$ N $\cdot$ m. These results are also shown on the *torque diagram* for the shaft, Fig. 11–19$c$. Applying Eq. 11–16, we have

$$\phi_{A/D} = \frac{(+80 \text{ N} \cdot \text{m}) \, L_{AB}}{JG} + \frac{(-70 \text{ N} \cdot \text{m}) \, L_{BC}}{JG} + \frac{(-10 \text{ N} \cdot \text{m}) \, L_{CD}}{JG}$$

If the other data is substituted and the answer is found as a *positive* quantity, it means that end $A$ will rotate as indicated by the curl of the right-hand fingers when the thumb is directed *away* from the shaft, Fig. 11–19$a$. The double subscript notation is used to indicate this relative angle of twist ($\phi_{A/D}$); however, if the angle of twist is to be determined relative to a *fixed point*, then only a single subscript will be used. For example, if $D$ is located at a fixed support, then the computed angle of twist will be denoted as $\phi_A$.

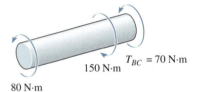

$T_{AB} = 80$ N·m

$T_{BC} = 70$ N·m

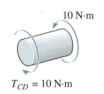

$T_{CD} = 10$ N·m

(b)

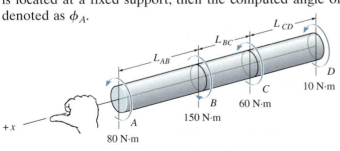

(a)

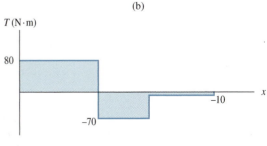

(c)

**Fig. 11–19**

## IMPORTANT POINTS

- The angle of twist is determined by relating the applied torque to the shear stress using the torsion formula, $\tau = T\rho/J$, and relating the relative rotation to the shear strain using $d\phi = \gamma \frac{dx}{\rho}$. Finally these equations are combined using Hooke's law, $\tau = G\gamma$ which yields Eq. 11–14.

- Since Hooke's law is used in the development of the formula for the angle of twist, it is important that the applied torques do not cause yielding of the material and that the material is homogeneous and behaves in a linear-elastic manner.

## PROCEDURE FOR ANALYSIS

The angle of twist of one end of a shaft or tube with respect to the other end can be determined by applying Eqs. 11–14 through 11–16.

*Internal Torque.*

- The internal torque is found at a point on the axis of the shaft by using the method of sections and the equation of moment equilibrium, applied along the shaft's axis.

- If the torque varies along the shaft's length, a section should be made at the arbitrary position $x$ along the shaft and the torque represented as a function of $x$, i.e., $T(x)$.

- If several constant external torques act on the shaft between its ends, the internal torque in each *segment* of the shaft, between any two external torques, must be determined. The results can be represented as a torque diagram.

*Angle of Twist.*

- When the circular cross-sectional area varies along the shaft's axis, the polar moment of inertia must be expressed as a function of its position $x$ along the axis, $J(x)$.

- If the polar moment of inertia or the internal torque *suddenly changes* between the ends of the shaft, then $\phi = \int (T(x)/J(x)G)\, dx$ or $\phi = TL/JG$ must be applied to *each segment* for which $J$, $G$, and $T$ are continuous or constant.

- When the internal torque in each segment is determined, be sure to use a consistent sign convention for the shaft, such as the one discussed above. Also make sure that a consistent set of units is used when substituting numerical data into the equations.

# EXAMPLE 11.7

The gears attached to the fixed-end steel shaft are subjected to the torques shown in Fig. 11–20a. If the shear modulus of elasticity is $G = 80$ GPa and the shaft has a diameter of 14 mm, determine the displacement of the tooth $P$ on gear $A$. The shaft turns freely within the bearing at $B$.

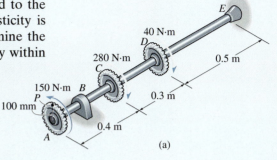

$T_{CD} = 130$ N·m

$T_{AC} = 150$ N·m

150 N·m   150 N·m

280 N·m

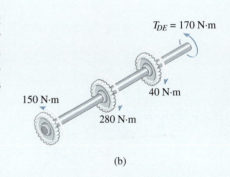

## Solution

**Internal Torque.** By inspection, the torques in segments $AC, CD,$ and $DE$ are different yet *constant* throughout each segment. Free-body diagrams of appropriate segments of the shaft along with the calculated internal torques are shown in Fig. 11–20b. Using the right-hand rule and the established sign convention that positive torque is directed away from the sectioned end of the shaft, we have

$$T_{AC} = +150 \text{ N} \cdot \text{m} \qquad T_{CD} = -130 \text{ N} \cdot \text{m} \qquad T_{DE} = -170 \text{ N} \cdot \text{m}$$

These results are also shown on the torque diagram, Fig. 11–20c.

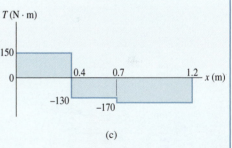

**Angle of Twist.** The polar moment of inertia for the shaft is

$$J = \frac{\pi}{2}(0.007 \text{ m})^4 = 3.77(10^{-9}) \text{ m}^4$$

Applying Eq. 11–16 to each segment and adding the results algebraically, we have

$$\phi_A = \sum \frac{TL}{JG} = \frac{(+150 \text{ N} \cdot \text{m})(0.4 \text{ m})}{3.77(10^{-9}) \text{ m}^4[80(10^9) \text{ N/m}^2]}$$
$$+ \frac{(-130 \text{ N} \cdot \text{m})(0.3 \text{ m})}{3.77(10^{-9}) \text{ m}^4[80(10^9) \text{ N/m}^2)]}$$
$$+ \frac{(-170 \text{ N} \cdot \text{m})(0.5 \text{ m})}{3.77(10^{-9}) \text{ m}^4[80(10^9) \text{ N/m}^2)]} = -0.212 \text{ rad}$$

Since the answer is negative, by the right-hand rule the thumb is directed *toward* the end $E$ of the shaft, and therefore gear $A$ will rotate as shown in Fig. 11–20d.

The displacement of tooth $P$ on gear $A$ is

$$s_P = \phi_A r = (0.212 \text{ rad})(100 \text{ mm}) = 21.2 \text{ mm} \qquad Ans.$$

Remember that this analysis is valid only if the shear stress does not exceed the proportional limit of the material.

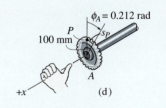

Fig. 11–20

The two solid steel shafts shown in Fig. 11–21a are coupled together using the meshed gears. Determine the angle of twist of end $A$ of shaft $AB$ when the torque $T = 45$ N·m is applied. Take $G = 80$ GPa. Shaft $AB$ is free to rotate within bearings $E$ and $F$, whereas shaft $DC$ is fixed at $D$. Each shaft has a diameter of 20 mm.

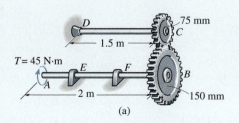

(a)

### Solution

***Internal Torque.*** Free-body diagrams for each shaft are shown in Fig. 11–21b and 11–21c. Summing moments along the $x$ axis of shaft $AB$ yields the tangential reaction between the gears of $F = 45$ N·m/0.15 m $= 300$ N. Summing moments about the $x$ axis of shaft $DC$, this force then creates a torque of $(T_D)_x = 300$ N(0.075 m) $=$ 22.5 N·m on shaft $DC$.

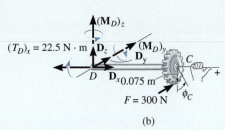

(b)

***Angle of Twist.*** To solve the problem, we will first calculate the rotation of gear $C$ due to the torque of 22.5 N·m in shaft $DC$, Fig. 11–21b. This angle of twist is

$$\phi_C = \frac{TL_{DC}}{JG} = \frac{(+22.5 \text{ N·m})(1.5 \text{ m})}{(\pi/2)(0.010 \text{ m})^4[80(10^9) \text{ N/m}^2]} = +0.0269 \text{ rad}$$

Since the gears at the end of the shaft are in mesh, the rotation $\phi_C$ of gear $C$ causes gear $B$ to rotate $\phi_B$, Fig. 11–21c, where

$$\phi_B(0.15 \text{ m}) = (0.0269 \text{ rad})(0.075 \text{ m})$$

$$\phi_B = 0.0134 \text{ rad}$$

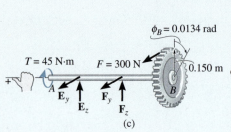

Fig. 11–21

We will now determine the angle of twist of end $A$ with respect to end $B$ of shaft $AB$ caused by the 45 N·m torque, Fig. 11–21c. We have

$$\phi_{A/B} = \frac{T_{AB}L_{AB}}{JG} = \frac{(+45 \text{ N·m})(2 \text{ m})}{(\pi/2)(0.010 \text{ m})^4[80(10^9) \text{ N/m}^2]} = +0.0716 \text{ rad}$$

The rotation of end $A$ is therefore determined by adding $\phi_B$ and $\phi_{A/B}$, since both angles are in the *same direction*, Fig. 11–21c. We have

$$\phi_A = \phi_B + \phi_{A/B} = 0.0134 \text{ rad} + 0.0716 \text{ rad} = +0.0850 \text{ rad} \qquad \textit{Ans.}$$

# E X A M P L E  11.9

The 2-in.-diameter solid cast-iron post shown in Fig. 11–22a is buried 24 in. in soil. If a torque is applied to its top using a rigid wrench, determine the maximum shear stress in the post and the angle of twist at its top. Assume that the torque is about to turn the post, and the soil exerts a uniform torsional resistance of $t$ lb · in./in. along its 24-in. buried length. $G = 5.5(10^3)$ ksi.

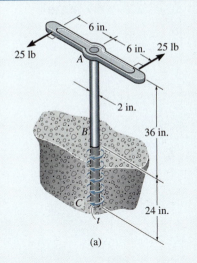

6 in.

6 in.   25 lb

25 lb

$A$

2 in.

$B$   36 in.

$C$

$t$

24 in.

(a)

### Solution

***Internal Torque.***   The internal torque in segment $AB$ of the post is constant. From the free-body diagram, Fig. 11–22b, we have

$$\Sigma M_z = 0; \qquad T_{AB} = 25 \text{ lb}(12 \text{ in.}) = 300 \text{ lb} \cdot \text{in.}$$

The magnitude of the uniform distribution of torque along the buried segment $BC$ can be determined from equilibrium of the entire post, Fig. 11–22c. Here

$$\Sigma M_z = 0; \qquad 25 \text{ lb}(12 \text{ in.}) - t(24 \text{ in.}) = 0$$
$$t = 12.5 \text{ lb} \cdot \text{in./in.}$$

Hence, from a free-body diagram of a section of the post located at the position $x$ within region $BC$, Fig. 11–22d, we have

$$\Sigma M_z = 0; \qquad T_{BC} - 12.5x = 0$$
$$T_{BC} = 12.5x$$

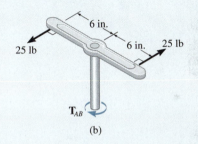

6 in.

25 lb

6 in.   25 lb

$T_{AB}$

(b)

***Maximum Shear Stress.***   The largest shear stress occurs in region $AB$, since the torque is largest there and $J$ is constant for the post. Applying the torsion formula, we have

$$\tau_{max} = \frac{T_{AB}c}{J} = \frac{(300 \text{ lb} \cdot \text{in.})(1 \text{ in.})}{(\pi/2)(1 \text{ in.})^4} = 191 \text{ psi} \qquad \textit{Ans.}$$

***Angle of Twist.***   The angle of twist at the top can be determined relative to the bottom of the post, since it is fixed and yet is about to turn. Both segments $AB$ and $BC$ twist, and so in this case we have

$$\phi_A = \frac{T_{AB}L_{AB}}{JG} + \int_0^{L_{BC}} \frac{T_{BC} \, dx}{JG}$$

$$= \frac{(300 \text{ lb} \cdot \text{in.})(36 \text{ in.})}{JG} + \int_0^{24 \text{ in.}} \frac{12.5x \, dx}{JG}$$

$$= \frac{10\,800 \text{ lb} \cdot \text{in}^2}{JG} + \frac{12.5[(24)^2/2] \text{ lb} \cdot \text{in}^2}{JG}$$

$$= \frac{14\,400 \text{ lb} \cdot \text{in}^2}{(\pi/2)(1 \text{ in.})^4 5500(10^3) \text{ lb/in}^2} = 0.00167 \text{ rad} \qquad \textit{Ans.}$$

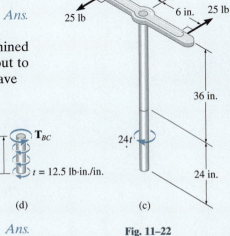

6 in.

6 in.   25 lb

25 lb

36 in.

$24t$

24 in.

(c)

$T_{BC}$

$x$

$t = 12.5 \text{ lb·in./in.}$

(d)

**Fig. 11–22**

EXAMPLE 11.10

The tapered shaft shown in Fig. 11–23a is made of a material having a shear modulus $G$. Determine the angle of twist of its end $B$ when subjected to the torque.

### Solution

*Internal Torque.* By inspection or from the free-body diagram of a section located at the arbitrary position $x$, Fig. 11–23b, the internal torque is $T$.

*Angle of Twist.* Here the polar moment of inertia varies along the shaft's axis and therefore we must express it in terms of the coordinate $x$. The radius $c$ of the shaft at $x$ can be determined in terms of $x$ by proportion of the slope of line $AB$ in Fig. 11–23c. We have

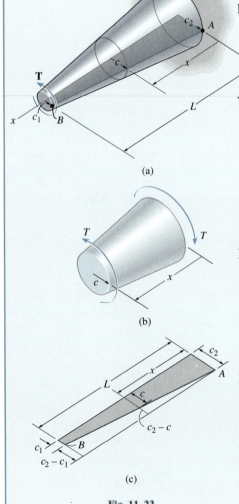

$$\frac{c_2 - c_1}{L} = \frac{c_2 - c}{x}$$

$$c = c_2 - x\left(\frac{c_2 - c_1}{L}\right)$$

Thus, at $x$,

$$J(x) = \frac{\pi}{2}\left[c_2 - x\left(\frac{c_2 - c_1}{L}\right)\right]^4$$

Applying Eq. 11–14, we have

$$\phi = \int_0^L \frac{T\,dx}{\left(\frac{\pi}{2}\right)\left[c_2 - x\left(\frac{c_2 - c_1}{L}\right)\right]^4 G} = \frac{2T}{\pi G}\int_0^L \frac{dx}{\left[c_2 - x\left(\frac{c_2 - c_1}{L}\right)\right]^4}$$

Performing the integration using an integral table, the result becomes

$$\phi = \left(\frac{2T}{\pi G}\right)\frac{1}{3\left(\frac{c_2 - c_1}{L}\right)\left[c_2 - x\left(\frac{c_2 - c_1}{L}\right)\right]^3}\Bigg|_0^L$$

$$= \frac{2T}{\pi G}\left(\frac{L}{3(c_2 - c_1)}\right)\left(\frac{1}{c_1^3} - \frac{1}{c_2^3}\right)$$

Rearranging terms yields

$$\phi = \frac{2TL}{3\pi G}\left(\frac{c_2^2 + c_1 c_2 + c_1^2}{c_1^3 c_2^3}\right) \qquad Ans.$$

To partially check this result, note that when $c_1 = c_2 = c$, then

$$\phi = \frac{TL}{[(\pi/2)c^4]G} = \frac{TL}{JG}$$

which is Eq. 11–15.

(a)

(b)

(c)

**Fig. 11–23**

# PROBLEMS

**11-27.** The propellers of a ship are connected to a solid A-36 steel shaft that is 60 m long and has an outer diameter of 340 mm and inner diameter of 260 mm. If the power output is 4.5 MW when the shaft rotates at 20 rad/s, determine the maximum torsional stress in the shaft and its angle of twist.

***11-28.** A shaft is subjected to a torque **T**. Compare the effectiveness of using the tube shown in the figure with that of a solid section of radius $c$. To do this, compute the percent increase in torsional stress and angle of twist per unit length for the tube versus the solid section.

**Prob. 11–28**

**11-29.** The hydrofoil boat has an A-36 steel propeller shaft that is 100 ft long. It is connected to an in-line diesel engine that delivers a maximum power of 2500 hp and causes the shaft to rotate at 1700 rpm. If the outer diameter of the shaft is 8 in. and the wall thickness is $\frac{3}{8}$ in., determine the maximum shear stress developed in the shaft. Also, what is the "wind up," or angle of twist in the shaft at full power?

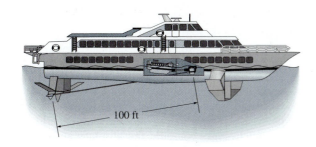

**Prob. 11–29**

**11-30.** The splined ends and gears attached to the A-36 steel shaft are subjected to the torques shown. Determine the angle of twist of end $B$ with respect to end $A$. The shaft has a diameter of 40 mm.

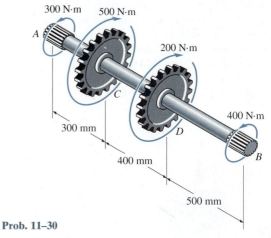

**Prob. 11–30**

**11-31.** The A-36 steel axle is made from tubes $AB$ and $CD$ and a solid section $BC$. It is supported on smooth bearings that allow it to rotate freely. If the ends are subjected to 85 N · m torques, determine the angle of twist of end $A$ relative to end $D$. The tubes have an outer diameter of 30 mm and an inner diameter of 20 mm. The solid section has a diameter of 40 mm.

***11-32.** The A-36 steel axle is made from tubes $AB$ and $CD$ and a solid section $BC$. It is supported on smooth bearings that allow it to rotate freely. If ends $A$ and $D$ are subjected to 85 N · m torques, determine the angle of twist of end $B$ of the solid section relative to end $C$. The tubes have an outer diameter of 30 mm and an inner diameter of 20 mm. The solid section has a diameter of 40 mm.

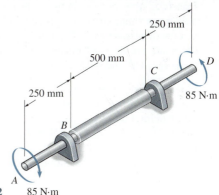

**Probs. 11–31/32**

**11-33.** The A-36 steel shaft is 2 m long and has an outer diameter of 40 mm. When it is rotating at 80 rad/s, it transmits 32 kW of power from the engine $E$ to the generator $G$. Determine the smallest thickness of the shaft if the allowable shear stress is $\tau_{allow} = 140$ MPa and the shaft is restricted not to twist more than 0.05 rad.

**11-34.** The A-36 solid steel shaft is 3 m long and has a diameter of 50 mm. It is required to transmit 35 kW of power from the engine $E$ to the generator $G$. Determine the smallest angular velocity the shaft can have if it is restricted not to twist more than 1°.

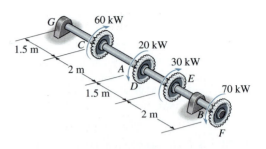

**Probs. 11–35/36**

**11-37.** The A-36 steel assembly consists of a tube having an outer radius of 1 in. and a wall thickness of 0.125 in. Using a rigid plate at $B$, it is connected to the solid 1-in.-diameter shaft $AB$. Determine the rotation of the tube's end $C$ if a torque of 200 lb·in. is applied to the tube at this end. The end $A$ of the shaft is fixed-supported.

**Probs. 11–33/34**

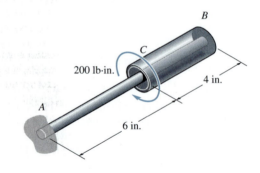

**Prob. 11–37**

**11-35.** The A-36 steel shaft rotates at $\omega = 125$ rad/s and transmits the power shown. Determine the absolute maximum shear stress in the shaft and the angle of twist of $C$ with respect to $F$. The inner and outer diameters of the shaft are $d_i = 30$ mm and $d_o = 40$ mm. The journal bearings at $B$ and $G$ are smooth.

***11-36.** The A-36 steel shaft rotates at $\omega = 125$ rad/s and transmits the power shown. Determine the inner and outer diameters of the shaft if $d_i/d_o = 0.6$, the allowable shear stress is $\tau_{allow} = 150$ MPa, and the allowable relative angle of twist is $\phi_{allow} = 2°/m$. The journal bearings at $B$ and $G$ are smooth.

**11-38.** The 6-in.-diameter L-2 steel shaft on the turbine is supported on journal bearings at $A$ and $B$. If $C$ is held fixed and the turbine blades create a torque on the shaft that increases linearly from zero at $C$ to 2000 lb·ft at $D$, determine the angle of twist of the shaft of end $D$ relative to end $C$. Also, compute the absolute maximum shear stress in the shaft. Neglect the size of the blades.

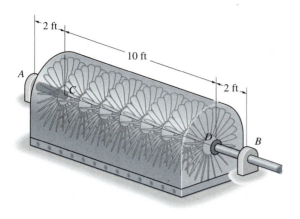

**Prob. 11–38**

**11-39.** The assembly is made of A-36 steel and consists of a solid rod 15 mm in diameter connected to the inside of a tube using a rigid disk at *B*. Determine the angle of twist at *A*. The tube has an outer diameter of 30 mm and wall thickness of 3 mm.

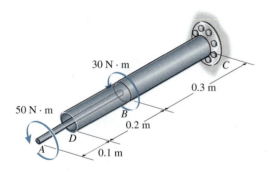

**Prob. 11–39**

*11-40.** The device serves as a compact torsional spring. It is made of A-36 steel and consists of a solid inner shaft *CB* which is surrounded by and attached to a tube *AB* using a rigid ring at *B*. The ring at *A* can also be assumed rigid and is fixed from rotating. If a torque of *T* = 2 kip·in. is applied to the shaft, determine the angle of twist at the end *C* and the maximum shear stress in the tube and shaft.

**11-41.** The device serves as a compact torsion spring. It is made of A-36 steel and consists of a solid inner shaft *CB* which is surrounded by and attached to a tube *AB* using a rigid ring at *B*. The ring at *A* can also be assumed rigid and is fixed from rotating. If the allowable shear stress for the material is $\tau_{allow} = 12$ ksi and the angle of twist at *C* is limited to $\phi_{allow} = 3°$, determine the maximum torque *T* that can be applied at the end *C*.

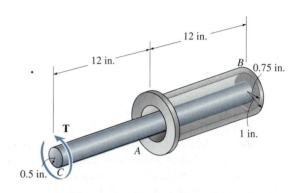

**Probs. 11–40/41**

**11-42.** The 60-mm-diameter solid shaft is made of A-36 steel and is subjected to the distributed and concentrated torsional loadings shown. Determine the angle of twist at the free end *A* of the shaft due to these loadings.

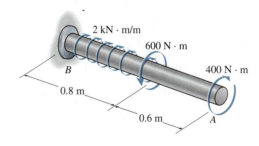

**Prob. 11–42**

**11-43.** The tapered shaft has a length $L$ and a radius $r$ at end $A$ and $2r$ at end $B$. If it is fixed at end $B$ and is subjected to a torque $T$, determine the angle of twist of end $A$. The shear modulus is $G$.

**11-45.** Show that when the bar is subjected to the torque $T$, the length $L$ shortens by an amount $L(1 - \sqrt{1 - (2T/\pi c^3 G)^2})$. The shaft is made of material having a shear modulus $G$.

Prob. 11–45

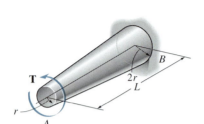

Prob. 11–43

**11-46.** A cylindrical spring consists of a rubber annulus bonded to a rigid ring and shaft. If the ring is held fixed and a torque $T$ is applied to the rigid shaft, determine the angle of twist of the shaft. The shear modulus of the rubber is $G$. *Hint:* As shown in the figure, the deformation of the element at radius $r$ can be determined from $r\,d\theta = dr\,\gamma$. Use this expression along with $\tau = T/(2\pi r^2 h)$, to obtain the result.

**\*11-44.** The solid steel shaft consists of two tapered ends, $AB$ and $CD$, and a central portion $BC$ having a constant diameter. It is supported by two smooth bearings, which allow it to rotate freely. If the gears fixed to its ends are subjected to counterbalancing torques of 1700 lb · in., determine the angle of twist of one end of the shaft relative to the other end. *Hint:* Use the results of Prob. 11–43. $G_{st} = 12(10^3)$ ksi.

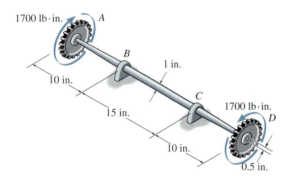

Prob. 11–44

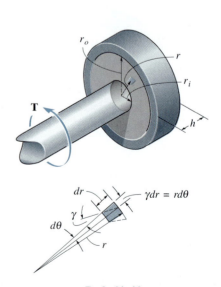

Prob. 11–46

## 11.5 Statically Indeterminate Torque-Loaded Members

A torsionally loaded shaft may be classified as statically indeterminate if the moment equation of equilibrium, applied about the axis of the shaft, is not adequate to determine the unknown torques acting on the shaft. An example of this situation is shown in Fig. 11–24a. As shown on the free-body diagram, Fig. 11–24b, the reactive torques at the supports A and B are unknown. We require that

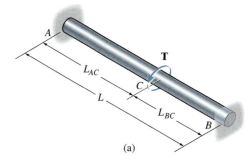

(a)

$$\Sigma M_x = 0; \qquad\qquad T - T_A - T_B = 0$$

Since only one equilibrium equation is relevant and there are two unknowns, this problem is statically indeterminate. In order to obtain a solution, we will use the method of analysis discussed in Sec. 10.4.

The necessary condition of compatibility, or the kinematic condition, requires the angle of twist of one end of the shaft with respect to the other end to be equal to zero, since the end supports are fixed. Therefore,

$$\phi_{A/B} = 0$$

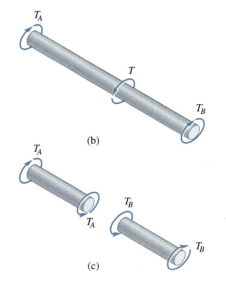

(b)

(c)

In order to write this equation in terms of the unknown torques, we will assume that the material behaves in a linear-elastic manner, so that the load–displacement relationship is expressed by $\phi = TL/JG$. Realizing that the internal torque in segment $AC$ is $+T_A$ and that in segment $CB$ the internal torque is $-T_B$, Fig. 11–24c, the above compatibility equation can be written as

$$\frac{T_A L_{AC}}{JG} - \frac{T_B L_{BC}}{JG} = 0$$

**Fig. 11–24**

Here $JG$ is assumed to be constant.

Solving the above two equations for the reactions, realizing that $L = L_{AC} + L_{BC}$, we get

$$TA = T\left(\frac{L_{BC}}{L}\right)$$

and

$$TB = T\left(\frac{L_{AC}}{L}\right)$$

Note that each of these reactive torques increases or decreases linearly with the placement $L_{AC}$ or $L_{BC}$ of the applied torque.

## PROCEDURE FOR ANALYSIS

The unknown torques in statically indeterminate shafts are determined by satisfying equilibrium, compatibility, and torque-displacement requirements for the shaft.

*Equilibrium.*

• Draw a free-body diagram of the shaft in order to identify all the torques that act on it. Then write the equations of moment equilibrium about the axis of the shaft.

*Compatibility.*

• To write the compatibility equation, investigate the way the shaft will twist when subjected to the external loads, and give consideration as to how the supports constrain the shaft when it is twisted.

• Express the compatibility condition in terms of the rotational displacements caused by the reactive torques, and then use a torque-displacement relation, such as $\phi = TL/JG$, to relate the unknown torques to the unknown displacements.

• Solve the equilibrium and compatibility equations for the unknown reactive torques. If any of the magnitudes have a negative numerical value, it indicates that this torque acts in the opposite sense of direction to that indicated on the free-body diagram.

# EXAMPLE 11.11

The solid steel shaft shown in Fig. 11–25a has a diameter of 20 mm. If it is subjected to the two torques, determine the reactions at the fixed supports A and B.

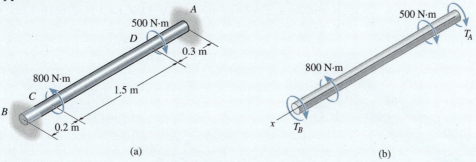

(a)                                                                        (b)

## Solution

*Equilibrium.*  By inspection of the free-body diagram, Fig. 11–25b, it is seen that the problem is statically indeterminate since there is only *one* available equation of equilibrium, whereas $\mathbf{T}_A$ and $\mathbf{T}_B$ are unknown. We require

$$\Sigma M_x = 0; \qquad -T_B + 800 \text{ N·m} - 500 \text{ N·m} - T_A = 0 \qquad (1)$$

*Compatibility.*  Since the ends of the shaft are fixed, the angle of twist of one end of the shaft with respect to the other must be zero. Hence, the compatibility equation can be written as

$$\phi_{A/B} = 0$$

This condition can be expressed in terms of the unknown torques by using the load–displacement relationship, $\phi = TL/JG$. Here there are three regions of the shaft where the internal torque is constant, $BC$, $CD$, and $DA$. On the free-body diagrams in Fig. 11–25c we have shown the internal torques acting on segments of the shaft which are sectioned in each of these regions. Using the sign convention established in Sec. 11.4, we have

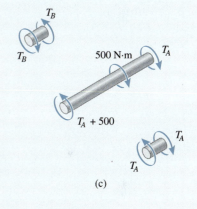

(c)

**Fig. 11–25**

$$\frac{-T_B(0.2 \text{ m})}{JG} + \frac{(T_A + 500 \text{ N·m})(1.5 \text{ m})}{JG} + \frac{T_A(0.3 \text{ m})}{JG} = 0$$

or

$$1.8 T_A - 0.2 T_B = -750 \qquad (2)$$

Solving Eqs. 1 and 2 yields

$$T_A = -345 \text{ N·m} \qquad T_B = 645 \text{ N·m} \qquad \textit{Ans.}$$

The negative sign indicates that $\mathbf{T}_A$ acts in the opposite direction of that shown in Fig. 11–25b.

E X A M P L E **11.12**

The shaft shown in Fig. 11–26a is made from a steel tube, which is bonded to a brass core. If a torque of $T = 250$ lb·ft is applied at its end, plot the shear-stress distribution along a radial line of its cross-sectional area. Take $G_{st} = 11.4(10^3)$ ksi, $G_{br} = 5.20(10^3)$ ksi.

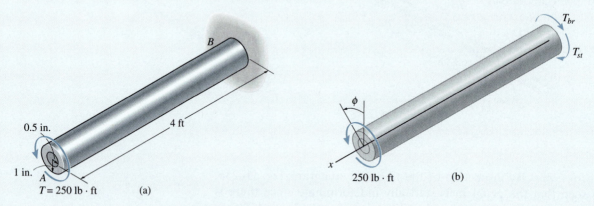

0.5 in.

4 ft

B

1 in.

A

T = 250 lb · ft

(a)

$T_{br}$

$T_{st}$

$\phi$

x

250 lb · ft

(b)

**Fig. 11–26**

**Solution**

*Equilibrium.* A free-body diagram of the shaft is shown in Fig. 11–26b. The reaction at the wall has been represented by the unknown amount of torque resisted by the steel, $T_{st}$, and by the brass, $T_{br}$. Working in units of pounds and inches, equilibrium requires

$$-T_{st} - T_{br} + 250 \text{ lb} \cdot \text{ft}(12 \text{ in./ft}) = 0 \qquad (1)$$

*Compatibility.* We require the angle of twist of end $A$ to be the same for both the steel and brass since they are bonded together. Thus,

$$\phi = \phi_{st} = \phi_{br}$$

Applying the load–displacement relationship, $\phi = TL/JG$, we have

$$\frac{T_{st}L}{(\pi/2)[(1 \text{ in.})^4 - (0.5 \text{ in.})^4]11.4(10^3) \text{ kip/in}^2} = \frac{T_{br}L}{(\pi/2)(0.5 \text{ in.})^4 5.20(10^3) \text{ kip/in}^2}$$

$$T_{st} = 32.88T_{br} \qquad (2)$$

Solving Eqs. 1 and 2, we get

$$T_{st} = 2911.0 \text{ lb} \cdot \text{in.} = 242.6 \text{ lb} \cdot \text{ft}$$
$$T_{br} = 88.5 \text{ lb} \cdot \text{in.} = 7.38 \text{ lb} \cdot \text{ft}$$

These torques act throughout the entire length of the shaft, since no external torques act at intermediate points along the shaft's axis. The shear stress in the brass core varies from zero at its center to a maximum at the interface where it contacts the steel tube. Using the torsion formula,

$$(\tau_{br})_{max} = \frac{(88.5 \text{ lb} \cdot \text{in.})(0.5 \text{ in.})}{(\pi/2)(0.5 \text{ in.})^4} = 451 \text{ psi}$$

For the steel, the minimum shear stress is also at this interface,

$$(\tau_{st})_{min} = \frac{(2911.0 \text{ lb} \cdot \text{in.})(0.5 \text{ in.})}{(\pi/2)[(1 \text{ in.})^4 - (0.5 \text{ in.})^4]} = 988 \text{ psi}$$

and the maximum shear stress is at the outer surface,

$$(\tau_{st})_{max} = \frac{(2911.0 \text{ lb} \cdot \text{in.})(1 \text{ in.})}{(\pi/2)[(1 \text{ in.})^4 - (0.5 \text{ in.})^4]} = 1977 \text{ psi}$$

The results are plotted in Fig. 11–26c. Note the discontinuity of *shear stress* at the brass and steel interface. This is to be expected, since the materials have different moduli of rigidity; i.e., steel is stiffer than brass ($G_{st} > G_{br}$) and thus it carries more shear stress at the interface. Although the shear stress is discontinuous here, the *shear strain* is not. Rather, the shear strain is the *same* for both the brass and the steel. This can be shown by using Hooke's law, $\gamma = \tau/G$. At the interface, Fig. 11–26d, the shear strain is

$$\gamma = \frac{\tau}{G} = \frac{451 \text{ psi}}{5.21(10^6)} = \frac{988 \text{ psi}}{11.4(10^6)} = 0.0867(10^{-6})$$

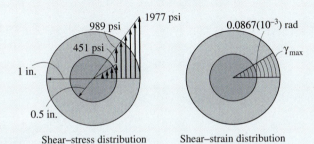

Shear–stress distribution

(c)

Shear–strain distribution

(d)

# PROBLEMS

**11-47.** The steel shaft has a diameter of 40 mm and is fixed at its ends $A$ and $B$. If it is subjected to the couple, determine the maximum shear stress in regions $AC$ and $CB$ of the shaft. $G_{st} = 10.8(10^3)$ ksi.

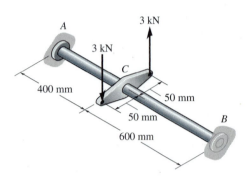

**Prob. 11–47**

**\*11-48.** The steel shaft is made from two segments: $AC$ has a diameter of 0.5 in., and $CB$ has a diameter of 1 in. If it is fixed at its ends $A$ and $B$ and subjected to a torque of 500 lb · ft, determine the maximum shear stress in the shaft. $G_{st} = 10.8(10^3)$ ksi.

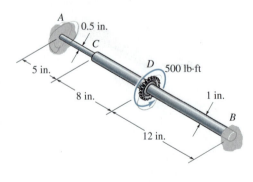

**Prob. 11–48**

**11-49.** The bronze C86100 pipe has an outer diameter of 1.5 in. and a thickness of 0.125 in. The coupling on it at $C$ is being tightened using a wrench. If the torque developed at $A$ is 125 lb · in., determine the magnitude $F$ of the couple forces. The pipe is fixed supported at end $B$.

**11-50.** The bronze C86100 pipe has an outer diameter of 1.5 in. and a thickness of 0.125 in. The coupling on it at $C$ is being tightened using a wrench. If the applied force is $F = 20$ lb, determine the maximum shear stress in the pipe.

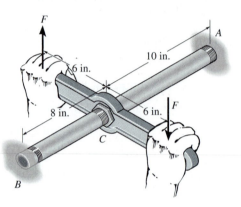

**Probs. 11–49/50**

**11-51.** A rod is made from two segments: $AB$ is A-36 steel and has a diameter of 30 mm and $BD$ is C83400 red brass and has a diameter of 50 mm. It is fixed at its ends and subjected to a torque of $T = 500$ N · m. Determine the torsional reactions at the walls $A$ and $D$.

**\*11-52.** Determine the absolute maximum shear stress in the shaft of Prob. 11–51.

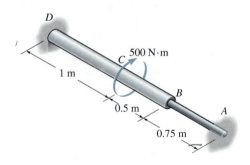

**Probs. 11–51/52**

**11-53.** The A-36 steel shaft is made from two segments: $AC$ has a diameter of 1 in. and $CB$ has a diameter of 2 in. If it is fixed at its ends $A$ and $B$ and subjected to a torque of $T = 500$ lb·ft, determine the absolute maximum shear stress in the shaft.

**11-54.** Determine the torsional reactions at the ends $A$ and $B$ of the shaft in Prob. 11–53.

**11-55.** The A-36 steel shaft is made from two segments: $AC$ has a diameter of 1 in. and $CB$ has a diameter of 2 in. If it is fixed at its ends $A$ and $B$, determine the magnitude of the applied torque $T$ if the reaction at $A$ is 50 lb·ft.

**11-57.** The two shafts are made of A-36 steel. Each has a diameter of 25 mm and they are connected using the gears fixed to their ends. Their other ends are attached to fixed supports at $A$ and $B$. They are also supported by journal bearings at $C$ and $D$, which allow free rotation of the shafts along their axes. If a torque of 500 N·m is applied to the gear at $E$ as shown, determine the reactions at $A$ and $B$.

**11-58.** Determine the rotation of the gear at $E$ in Prob. 11–57.

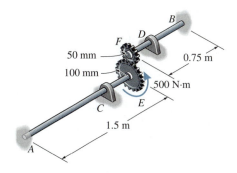

Probs. 11–57/58

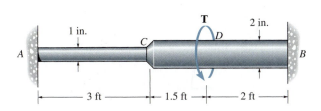

Probs. 11–53/54/55

*11-56.** The composite shaft consists of a mid-section that includes the 1-in.-diameter solid shaft and a tube that is welded to the rigid flanges at $A$ and $B$. Neglect the thickness of the flanges and determine the angle of twist of end $C$ of the shaft relative to end $D$. The shaft is subjected to a torque of 800 lb·ft. The material is A-36 steel.

**11-59.** The shaft is made from a solid steel section $AB$ and a tubular portion made of steel and having a brass core. If it is fixed to a rigid support at $A$, and a torque of $T = 50$ lb·ft is applied to it at $C$, determine the angle of twist that occurs at $C$ and compute the maximum shear and maximum shear strain in the brass and steel. Take $G_{st} = 11.5(10^3)$ ksi, $G_{br} = 5.6(10^3)$ ksi.

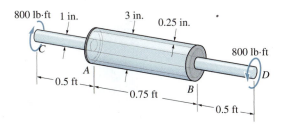

Prob. 11–56

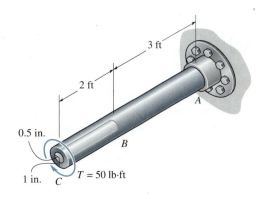

Prob. 11–59

## CHAPTER REVIEW

- *Torsion Formula.* Torque causes a shaft having a circular cross section to twist, such that the shear strain in the shaft is proportional to its radial distance from the center. Provided the material is homogeneous and Hooke's law applies, then the shear stress is determined from the torsion formula, $\tau = Tc/J$.

- *Design.* The design of a shaft requires finding the geometric parameter, $J/c = T/\tau_{allow}$. Often the power generated by a shaft is reported, in which case the torque is determined from $P = T\omega$.

- *Angle of Twist.* The angle of twist of a circular shaft is determined from $\phi = \int_0^L \dfrac{T(x)\,dx}{GJ}$, or if the torque is constant $\phi = \Sigma\dfrac{TL}{GJ}$.
  For application, it is necessary to use a sign convention for the internal torque and to be sure the material does not yield, but remains linear elastic.

- *Statically Indeterminate Shaft.* If the shaft is statically indeterminate, then the reactive torques are determined from equilibrium, compatibility of twist, and the torque-twist relationship.

## REVIEW PROBLEMS

*11-60. The glass tube is confined within a rubber stopper, such that when the tube is twisted at constant angular velocity the stopper creates a constant distribution of frictional torque along the contacting length $AB$ of the tube. If the tube has an inner diameter of 2 mm and an outer diameter of 4 mm, determine the shear stress developed at a point located at its inner and outer walls at a section through point $C$. Show the shear-stress distribution acting along a radial line segment at this section.

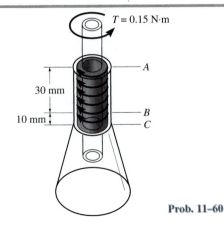

$T = 0.15$ N·m

30 mm

10 mm

Prob. 11–60

**11-61.** The tube is subjected to a torque of 600 N · m. Determine the amount of this torque that is resisted by the shaded section. Solve the problem two ways: (a) by using the torsion formula; (b) by finding the resultant of the shear-stress distribution.

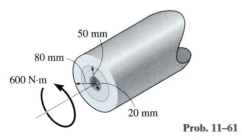

50 mm

80 mm

600 N·m

20 mm

**Prob. 11–61**

**11-62.** The motor of a fan delivers 150 W of power to the blade when it is turning at 18 rev/s. Determine the smallest diameter of shaft $A$ that can be used to connect the fan blade to the motor if the allowable shear stress is $\tau_{allow} = 80$ MPa.

$A$

**Prob. 11–62**

**11-63.** A shaft is subjected to a torque **T**. Compare the effectiveness of using the tube shown in the figure with that of a solid section of radius $c$. To do this, compute the percent increase in torsional stress and angle of twist per unit length for the tube versus the solid section.

$\frac{c}{3}$

$c$

**T**

**T**

$c$

**Prob. 11–63**

*__*11-64.** The solid shaft of radius $r$ is subjected to a torque **T**. Determine the radius $r'$ of the inner core of the shaft that resists one-half of the applied torque $(T/2)$. Solve the problem two ways: (a) by using the torsion formula, (b) by finding the resultant of the shear-stress distribution.

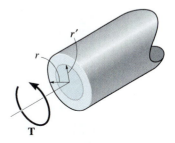

$r'$

$r$

**T**

**Prob. 11–64**

**11-65.** The contour of the surface of the shaft is defined by the equation $y = e^{ax}$, where $a$ is a constant. If the shaft is subjected to a torque $T$ at its ends, determine the angle of twist of end $A$ with respect to end $B$. The shear modulus is $G$.

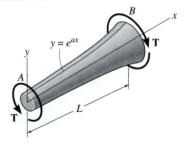

$B$

$x$

$y = e^{ax}$

$y$

$A$

**T**

**T**

$L$

**Prob. 11–65**

**11-66.** A cylindrical spring consists of a rubber annulus bonded to a rigid ring and shaft. If the ring is held fixed and a torque **T** is applied to the shaft, determine the maximum shear stress in the rubber.

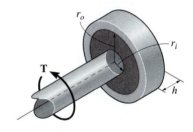

$r_o$

$r_i$

**T**

$h$

**Prob. 11–66**

Beams are important structural members used in building construction. Their design is often based upon their ability to resist bending stress, which forms the subject matter of this chapter.

# Bending

- To determine the stress in elastic symmetric members subject to bending.
- To develop methods to determine the stress in unsymmetric beams subject to bending.

## 12.1  Bending Deformation of a Straight Member

Members that are slender and support loadings applied perpendicular to their longitudinal axis are called *beams*. In this section we will discuss the deformations that occur when a straight prismatic beam, made of a homogeneous material, is subjected to bending. The discussion will be limited to beams having a cross-sectional area that is symmetrical with respect to an axis, and the bending moment is applied about an axis perpendicular to this axis of symmetry as shown in Fig. 12–1. The behavior of members that have unsymmetrical cross sections, or are made from several different materials, is based on similar observations and will be discussed separately in later sections of this chapter.

512 • CHAPTER 12 Bending

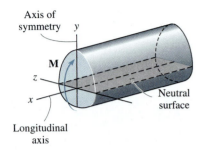

Axis of symmetry $y$

$M$

$z$

$x$

Neutral surface

Longitudinal axis

**Fig. 12–1**

By using a highly deformable material such as rubber, we can physically illustrate what happens when a straight prismatic member is subjected to a bending moment. Consider, for example, the undeformed bar in Fig. 12–2a, which has a square cross section and is marked with longitudinal and transverse grid lines. When a bending moment is applied, it tends to distort these lines into the pattern shown in Fig. 12–2b. Here it can be seen that the longitudinal lines become *curved* and the vertical transverse lines *remain straight* and yet undergo a *rotation*.

The behavior of any deformable bar subjected to a bending moment causes the material within the bottom portion of the bar to stretch and the material within the top portion to compress. Consequently, between these two regions there must be a surface, called the *neutral surface*, in which longitudinal fibers of the material will not undergo a change in length, Fig. 12–1.

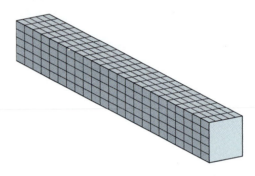

Before deformation

(a)

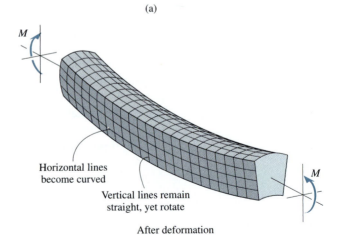

$M$

Horizontal lines become curved

Vertical lines remain straight, yet rotate

$M$

After deformation

(b)

**Fig. 12–2**

From these observations we will make the following three assumptions regarding the way the stress deforms the material. First, the *longitudinal axis x*, which lies within the neutral surface, Fig. 12–3a, does *not* experience any *change in length*. Rather the moment will tend to deform the beam so that this line *becomes a curve* that lies in the *x–y* plane of symmetry, Fig. 12–3b. Second, all **cross sections** of the beam **remain plane** and perpendicular to the longitudinal axis during the deformation. And third, any *deformation* of the *cross section* within its own plane, as noticed in Fig. 12–2b, will be *neglected*. In particular, the *z* axis, lying in the plane of the cross section and about which the cross section rotates, is called the *neutral axis*, Fig. 12–3b. Its location will be determined in the next section.

In order to show how this distortion will strain the material, we will isolate a segment of the beam that is located a distance *x* along the beam's length and has an undeformed thickness $\Delta x$, Fig. 12–3a. This element, taken from the beam, is shown in profile view in the undeformed and

Note the distortion of the lines due to bending of this rubber bar. The top line stretches, the bottom line compresses, and the center line remains the same length. Furthermore the vertical lines rotate and yet remain straight.

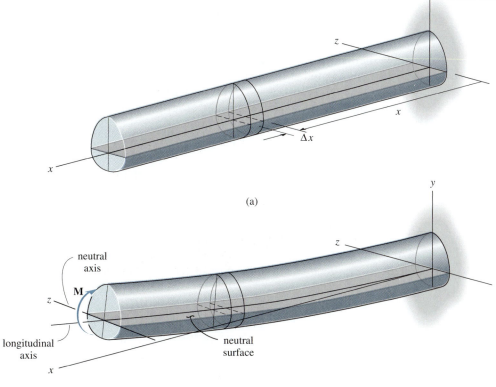

(a)

(b)

Fig. 12–3

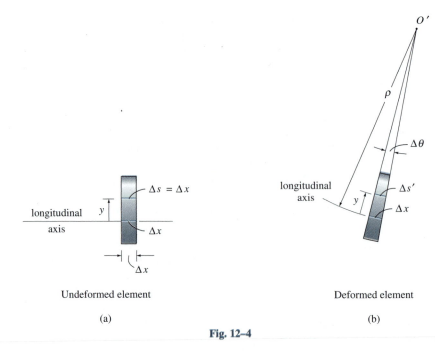

Undeformed element

(a)

Deformed element

(b)

**Fig. 12–4**

deformed positions in Fig. 12–4. Notice that any line segment $\Delta x$, located on the neutral surface, does not change its length, whereas any line segment $\Delta s$, located at the arbitrary distance $y$ above the neutral surface, will contract and become $\Delta s'$ after deformation. By definition, the normal strain along $\Delta s$ is determined from Eq. 8–11, namely,

$$\epsilon = \lim_{\Delta s \to 0} \frac{\Delta s' - \Delta s}{\Delta s}$$

We will now represent this strain in terms of the location $y$ of the segment and the radius of curvature $\rho$ of the longitudinal axis of the element. Before deformation, $\Delta s = \Delta x$, Fig. 12–4a. After deformation $\Delta x$ has a radius of curvature $\rho$, with center of curvature at point $O'$, Fig. 12–4b. Since $\Delta \theta$ defines the angle between the cross-sectional sides of the element, $\Delta x = \Delta s = \rho \, \Delta \theta$. In the same manner, the deformed length of $\Delta s$ becomes $\Delta s' = (\rho - y) \, \Delta \theta$. Substituting into the above equation, we get

$$\epsilon = \lim_{\Delta \theta \to 0} \frac{(\rho - y) \, \Delta \theta - \rho \, \Delta \theta}{\rho \, \Delta \theta}$$

or

$$\epsilon = -\frac{y}{\rho} \tag{12–1}$$

This important result indicates that the longitudinal normal strain of any element within the beam depends on its location $y$ on the cross

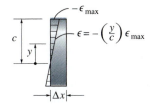

Normal strain distribution

**Fig. 12–5**

section and the radius of curvature of the beam's longitudinal axis at the point. In other words, for any specific cross section, the ***longitudinal normal strain will vary linearly*** with $y$ from the neutral axis. A contraction $(-\epsilon)$ will occur in fibers located above the neutral axis $(+y)$, whereas elongation $(+\epsilon)$ will occur in fibers located below the axis $(-y)$. This variation in strain over the cross section is shown in Fig. 12–5. Here the maximum strain occurs at the outermost fiber, located a distance $c$ from the neutral axis. Using Eq. 12–1, since $\epsilon_{max} = c/\rho$, then by division,

$$\frac{\epsilon}{\epsilon_{max}} = \frac{-y/\rho}{c/\rho}$$

So that

$$\epsilon = -\left(\frac{y}{c}\right)\epsilon_{max} \qquad (12\text{–}2)$$

This normal strain depends only on the assumptions made with regards to the *deformation*. Provided only a moment is applied to the beam, then it is reasonable to further assume that this moment causes a *normal stress only* in the longitudinal or $x$ direction. All the other components of normal and shear stress are zero, since the beam's surface is free of any other load. It is this uniaxial state of stress that causes the material to have the longitudinal normal strain component $\epsilon_x$, $(\sigma_x = E\epsilon_x)$, defined by Eq. 12–2. Furthermore, by Poisson's ratio, there must *also* be associated strain components $\epsilon_y = -\nu\epsilon_x$ and $\epsilon_z = -\nu\epsilon_x$, which deform the plane of the cross-sectional area, although here we have neglected these deformations. Such deformations will, however, cause the *cross-sectional dimensions* to become smaller below the neutral axis and larger above the neutral axis. For example, if the beam has a square cross section, it will actually deform as shown in Fig. 12–6.

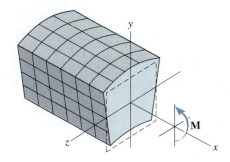

**Fig. 12–6**

## 12.2 The Flexure Formula

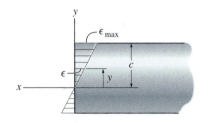

Normal strain variation
(profile view)

(a)

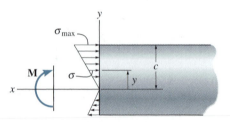

Bending stress variation
(profile view)

(b)

**Fig. 12–7**

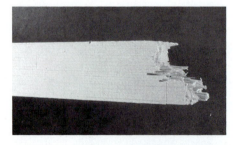

This wood specimen failed in bending due to its fibers being crushed at its top and torn apart at its bottom.

In this section we will develop an equation that relates the longitudinal stress distribution in a beam to the internal resultant bending moment acting on the beam's cross section. To do this we will assume that the material behaves in a linear-elastic manner so that Hooke's law applies, that is, $\sigma = E\epsilon$. A *linear variation of normal strain*, Fig. 12–7a, must then be the consequence of a *linear variation in normal stress*, Fig. 12–7b. Hence, like the normal strain variation, $\sigma$ will vary from zero at the member's neutral axis to a maximum value, $\sigma_{max}$, a distance $c$ farthest from the neutral axis. Because of the proportionality of triangles, Fig. 12–7b, or by using Hooke's law, $\sigma = E\epsilon$, and Eq. 12–2, we can write

$$\sigma = -\left(\frac{y}{c}\right)\sigma_{max} \tag{12–3}$$

This equation represents the stress distribution over the cross-sectional area. The sign convention established here is significant. For positive **M**, which acts in the $+z$ direction, positive values of $y$ give negative values for $\sigma$, that is, a compressive stress since it acts in the negative $x$ direction. Similarly, negative $y$ values will give positive or tensile values for $\sigma$. If a volume element of material is selected at a specific point on the cross section, only these tensile or compressive normal stresses will act on it. For example, the element located at $+y$ is shown in Fig. 12–7c.

We can locate the position of the neutral axis on the cross section by satisfying the condition that the *resultant force* produced by the stress distribution over the cross-sectional area must be equal to *zero*. Noting that the force $dF = \sigma\, dA$ acts on the arbitrary element $dA$ in Fig. 12–7c, we require

$$F_R = \Sigma F_x; \qquad 0 = \int_A dF = \int_A \sigma\, dA$$

$$= \int_A -\left(\frac{y}{c}\right)\sigma_{max}\, dA$$

$$= \frac{-\sigma_{max}}{c}\int_A y\, dA$$

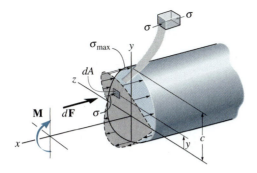

Bending stress variation

(c)

Since $\sigma_{max}/c$ is not equal to zero, then

$$\int_A y \, dA = 0 \qquad (12\text{–}4)$$

In other words, the first moment of the member's cross-sectional area about the neutral axis must be zero. This condition can only be satisfied if the *neutral axis* is also the horizontal *centroidal axis* for the cross section.[*] Consequently, once the centroid for the member's cross-sectional area is determined, the location of the neutral axis is known.

We can determine the stress in the beam from the requirement that the resultant internal moment $M$ must be equal to the moment produced by the stress distribution about the neutral axis. The moment of $d\mathbf{F}$ in Fig. 12–7c about the neutral axis is $dM = y \, dF$. This moment is *positive* since, by the right-hand rule, the thumb is directed along the positive $z$ axis when the fingers are curled with the sense of rotation caused by $d\mathbf{M}$. Since $dF = \sigma \, dA$, using Eq. 12–3, we have for the entire cross section,

$$(M_R)_z = \Sigma M_z; \quad M = \int_A y \, dF = \int_A y \, (\sigma \, dA) = \int_A y \left(\frac{y}{c}\sigma_{max}\right) dA$$

or

$$M = \frac{\sigma_{max}}{c} \int_A y^2 \, dA \qquad (12\text{–}5)$$

---

[*]Recall that the location $\bar{y}$ for the centroid of the cross-sectional area is defined from the equation $\bar{y} = \int y \, dA / \int dA$. If $\int y \, dA = 0$, then $\bar{y} = 0$, and so the centroid lies on the reference (neutral) axis. See Sec. 6.2.

Here the integral represents the *moment of inertia* of the cross-sectional area, computed about the neutral axis. We symbolize its value as $I$. Hence, Eq. 12–5 can be solved for $\sigma_{max}$ and written in general form as

$$\sigma_{max} = \frac{Mc}{I} \qquad (12\text{–}6)$$

Here

$\sigma_{max}$ = the maximum normal stress in the member, which occurs at a point on the cross-sectional area *farthest away* from the neutral axis

$M$ = the resultant internal moment, determined from the method of sections and the equations of equilibrium, and computed about the neutral axis of the cross section

$I$ = the moment of inertia of the cross-sectional area computed about the neutral axis

$c$ = the perpendicular distance from the neutral axis to a point farthest away from the neutral axis, where $\sigma_{max}$ acts

Since $\sigma_{max}/c = -\sigma/y$, Eq. 12–3, the normal stress at the intermediate distance $y$ can be determined from an equation similar to Eq. 12–6. We have

$$\sigma = -\frac{My}{I} \qquad (12\text{–}7)$$

Note that the negative sign is necessary since it agrees with the established $x$, $y$, $z$ axes. By the right-hand rule, $M$ is positive along the $+z$ axis, $y$ is positive upward, and $\sigma$ therefore must be negative (compressive) since it acts in the negative $x$ direction, Fig. 12–7c. Either of the above two equations is often referred to as the **flexure formula**. It is used to determine the normal stress in a straight member, having a cross section that is symmetrical with respect to an axis, and the moment is applied perpendicular to this axis. Although we have assumed that the member is prismatic, we can in most cases of engineering design also use the flexure formula to determine the normal stress in members that have a *slight taper*. For example, using a mathematical analysis based on the theory of elasticity, a member having a rectangular cross section and a length that is tapered 15° will have an actual maximum normal stress that is about 5.4% *less* than that calculated using the flexure formula.

## IMPORTANT POINTS

- The cross section of a straight beam *remains plane* when the beam deforms due to bending. This causes tensile stress on one side of the beam and compressive stress on the other side. The *neutral axis* is subjected to *zero stress.*

- Due to the deformation, the *longitudinal strain* varies *linearly* from zero at the neutral axis to a maximum at the outer fibers of the beam. Provided the material is homogeneous and Hooke's law applies, the *stress* also varies in a *linear* fashion over the cross section.

- For linear-elastic material the neutral axis passes through the *centroid* of the cross-sectional area. This conclusion is based on the fact that the resultant normal force acting on the cross section must be zero.

- The flexure formula is based on the requirement that the resultant moment on the cross section is equal to the moment produced by the linear normal stress distribution about the neutral axis.

## PROCEDURE FOR ANALYSIS

In order to apply the flexure formula, the following procedure is suggested.

*Internal Moment.*
- Section the member at the point where the bending or normal stress is to be determined, and obtain the internal moment $M$ at the section. The centroidal or neutral axis for the cross section must be known, since $M$ *must* be computed about this axis.
- If the absolute maximum bending stress is to be determined, then draw the moment diagram in order to determine the maximum moment in the beam.

*Section Property.*
- Determine the moment of inertia of the cross-sectional area about the neutral axis. A table listing values of $I$ for several common shapes is given in Appendix C.

*Normal Stress.*
- Specify the distance $y$, measured perpendicular to the neutral axis to the point where the normal stress is to be determined. Then apply the equation $\sigma = -My/I$, or if the maximum bending stress is to be calculated, use $\sigma_{max} = Mc/I$. When substituting the data, make sure the units are consistent.
- The stress acts in a direction such that the force it creates at the point contributes a moment about the neutral axis that is in the same direction as the internal moment **M**, Fig. 12–7c. In this manner the stress distribution acting over the entire cross section can be sketched, or a volume element of the material can be isolated and used to represent graphically the normal stress acting at the point.

EXAMPLE **12.1**

A beam has a rectangular cross section and is subjected to the stress distribution shown in Fig. 12–8a. Determine the internal moment **M** at the section caused by the stress distribution (a) using the flexure formula, (b) by finding the resultant of the stress distribution using basic principles.

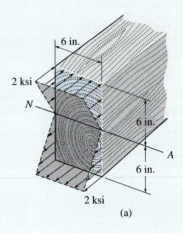

**Fig. 12–8**

**Solution**

*Part (a).* The flexure formula is $\sigma_{max} = Mc/I$. From Fig. 12–8a, $c = 6$ in. and $\sigma_{max} = 2$ ksi. The neutral axis is defined as line *NA*, because the stress is zero along this line. Since the cross section has a rectangular shape, the moment of inertia for the area about *NA* is determined from the formula for a rectangle given in Appendix C, i.e.,

$$I = \frac{1}{12}bh^3 = \frac{1}{12}(6\text{ in.})(12\text{ in.})^3 = 864\text{ in}^4$$

Therefore,

$$\sigma_{max} = \frac{Mc}{I}; \qquad 2\text{ kip/in}^2 = \frac{M(6\text{ in.})}{864\text{ in}^4}$$

$$M = 288\text{ kip}\cdot\text{in.} = 24\text{ kip}\cdot\text{ft} \qquad \textit{Ans.}$$

*Part (b).*   First we will show that the resultant force of the stress distribution is zero. As shown in Fig. 12–8b, the stress acting on the arbitrary element strip $dA = (6 \text{ in.}) \, dy$, located $y$ from the neutral axis, is

$$\sigma = \left(\frac{-y}{6 \text{ in.}}\right)(2 \text{ kip/in}^2)$$

The force created by this stress is $dF = \sigma \, dA$, and thus, for the entire cross section,

$$F_R = \int_A \sigma \, dA = \int_{-6 \text{ in.}}^{6 \text{ in.}} \left[\left(\frac{-y}{6 \text{ in.}}\right)(2 \text{ kip/in}^2)\right](6 \text{ in.}) \, dy$$

$$= (-1 \text{ kip/in}^2) y^2 \Big|_{-6 \text{ in.}}^{+6 \text{ in.}} = 0$$

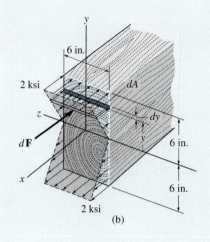

(b)

The resultant moment of the stress distribution about the neutral axis ($z$ axis) must equal $M$. Since the magnitude of the moment of $d\mathbf{F}$ about this axis is $dM = y \, dF$, and $d\mathbf{M}$ is *always positive*, Fig. 12–8b, then for the entire area,

$$M = \int_A y \, dF = \int_{-6 \text{ in.}}^{6 \text{ in.}} y\left[\left(\frac{y}{6 \text{ in.}}\right)(2 \text{ kip/in}^2)\right](6 \text{ in.}) \, dy$$

$$= \left(\frac{2}{3} \text{kip/in}^2\right) y^3 \Big|_{-6 \text{ in.}}^{+6 \text{ in.}}$$

$$= 288 \text{ kip} \cdot \text{in.} = 24 \text{ kip} \cdot \text{ft} \qquad \textit{Ans.}$$

The above result can *also* be determined without the need for integration. The resultant force for each of the two *triangular* stress distributions in Fig. 12–8c is graphically equivalent to the *volume* contained within each stress distribution. Thus, each volume is

$$F = \frac{1}{2}(6 \text{ in.})(2 \text{ kip/in}^2)(6 \text{ in.}) = 36 \text{ kip}$$

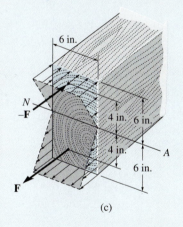

(c)

These forces, which form a couple, act in the same direction as the stresses within each distribution, Fig. 12–8c. Furthermore, they act through the *centroid* of each volume, i.e., $\frac{1}{3}(6 \text{ in.}) = 2 \text{ in.}$ from the top and bottom of the beam. Hence the distance between them is 8 in. as shown. The moment of the couple is therefore

$$M = 36 \text{ kip} (8 \text{ in.}) = 288 \text{ kip} \cdot \text{in.} = 24 \text{ kip} \cdot \text{ft} \qquad \textit{Ans.}$$

The simply supported beam in Fig. 12–9a has the cross-sectional area shown in Fig. 12–9b. Determine the bending stress that acts at points B and D located at section a–a.

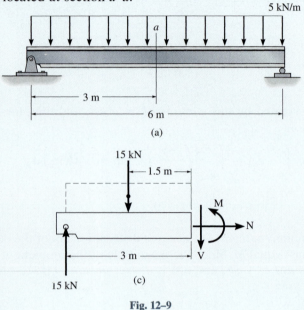

(a)

(c)

**Fig. 12–9**

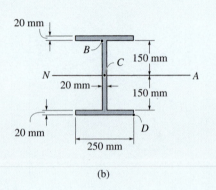

(b)

### Solution

***Internal Moment.*** Due to symmetry of both geometry and loading, the supports each exert a vertical reaction of 15 kN on the beam. Using the method of sections, the free-body diagram of the left segment of section a–a is shown in Fig. 12–9c. The moment is

$$M = 15 \text{ kN}(1.5\text{m}) = 22.5 \text{ kN} \cdot \text{m}$$

***Section Property.*** By reasons of symmetry, the centroid C and thus the neutral axis pass through the midheight of the beam, Fig. 12–9b. The area is subdivided into the three parts shown, and the moment of inertia of each part is computed about the neutral axis using the parallel-axis theorem. (See Sec. 6.7.) Choosing to work in meters, we have

$$I = \Sigma(\bar{I} + Ad^2)$$
$$= 2\left[\frac{1}{12}(0.25 \text{ m})(0.020 \text{ m})^3 + (0.25 \text{ m})(0.020 \text{ m})(0.160 \text{ m})^2\right]$$
$$+ \left[\frac{1}{12}(0.020 \text{ m})(0.300 \text{ m})^3\right]$$
$$= 301.3(10^{-6}) \text{ m}^4$$

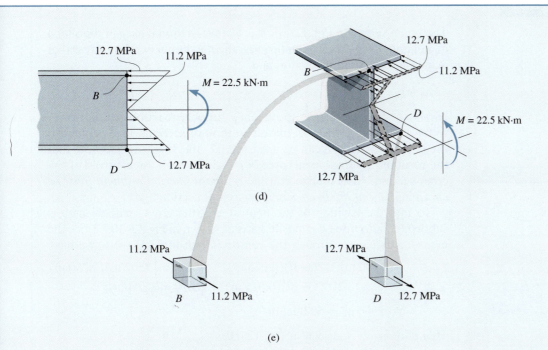

(d)

(e)

*Bending Stress.* Applying the flexure formula, with $c = 170$ mm, the absolute maximum bending stress is

$$\sigma_{max} = \frac{Mc}{I}; \qquad \sigma_{max} = \frac{22.5 \text{ kN} \cdot \text{m}(0.170 \text{ m})}{301.3(10^{-6}) \text{ m}^4} = 12.7 \text{ MPa} \qquad Ans.$$

Two-and-three-dimensional views of the stress distribution are shown in Fig. 12–9d. Notice how the stress at each point on the cross section develops a force that contributes a moment $d\mathbf{M}$ about the neutral axis such that it has the same direction as $\mathbf{M}$. Specifically, at point $B$, $y_B = 150$ mm, and so

$$\sigma_B = \frac{M y_B}{I}; \qquad \sigma_B = \frac{22.5 \text{ kN} \cdot \text{m}(0.150 \text{ m})}{301.3(10^{-6}) \text{ m}^4} = 11.2 \text{ MPa}$$

The normal stress acting on elements of material located at points $B$ and $D$ is shown in Fig. 12–9e.

EXAMPLE 12.3

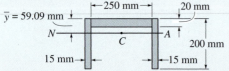

2.6 kN

$\frac{13}{5}\Big/12$

a

| 2 m | 1 m |

a

(a)

The beam shown in Fig. 12–10a has a cross-sectional area in the shape of a channel, Fig. 12–10b. Determine the maximum bending stress that occurs in the beam at section a–a.

**Solution**

*Internal Moment.* Here the beam's support reactions do not have to be determined. Instead, by the method of sections, the segment to the left of section a–a can be used, Fig. 12–10c. In particular, note that the resultant internal axial force **N** passes through the centroid of the cross section. Also, realize that *the resultant internal moment must be computed about the beam's neutral axis* at section a–a.

To find the location of the neutral axis, the cross-sectional area is subdivided into three composite parts as shown in Fig. 12–10d. Since the neutral axis passes through the centroid, then using Eq. 6–6, we have

$$\bar{y} = \frac{\Sigma \tilde{y} A}{\Sigma A} = \frac{2[0.100 \text{ m}](0.200 \text{ m})(0.015 \text{ m}) + [0.010 \text{ m}](0.02 \text{ m})(0.250 \text{ m})}{2(0.200 \text{ m})(0.015 \text{ m}) + 0.020 \text{ m}(0.250 \text{ m})}$$

$$= 0.05909 \text{ m} = 59.09 \text{ mm}$$

$\bar{y} = 59.09$ mm

|←250 mm→| 20 mm

N——

C    A

15 mm→    →←15 mm    200 mm

(b)

This dimension is shown in Fig. 12–10c.

Applying the moment equation of equilibrium about the neutral axis, we have

$$\zeta + \Sigma M_{NA} = 0; \quad 2.4 \text{ kN}(2 \text{ m}) + 1.0 \text{ kN}(0.05909 \text{ m}) - M = 0$$

$$M = 4.859 \text{ kN} \cdot \text{m}$$

*Section Property.* The moment of inertia about the neutral axis is determined using the parallel-axis theorem applied to each of the three composite parts of the cross-sectional area. Working in meters, we have

$$I = \left[\frac{1}{12}(0.250 \text{ m})(0.020 \text{ m})^3 + (0.250 \text{ m})(0.020 \text{ m})(0.05909 \text{ m} - 0.010 \text{ m})^2\right]$$

$$+ 2\left[\frac{1}{12}(0.015 \text{ m})(0.200 \text{ m})^3 + (0.015 \text{ m})(0.200 \text{ m})(0.100 \text{ m} - 0.05909 \text{ m})^2\right]$$

$$= 42.26(10^{-6}) \text{ m}^4$$

2.4 kN

1.0 kN    0.05909 m    V    M

N

C

| 2 m |

(c)

**Fig. 12–10**

*Maximum Bending Stress.* The maximum bending stress occurs at points farthest away from the neutral axis. This is at the bottom of the beam, $c = 0.200 \text{ m} - 0.05909 \text{ m} = 0.1409 \text{ m}$. Thus,

$$\sigma_{max} = \frac{Mc}{I} = \frac{4.859 \text{ kN} \cdot \text{m}(0.1409 \text{ m})}{42.26(10^{-6}) \text{ m}^4} = 16.2 \text{ MPa} \quad Ans.$$

Show that at the top of the beam the bending stress is $\sigma' = 6.79$ MPa. Note that in addition to this effect of bending, the normal force of $N = 1$ kN and shear force $V = 2.4$ kN will also contribute additional stress on the cross section. The superposition of all these effects will be discussed in a later chapter.

## E X A M P L E 12.4

The member having a rectangular cross section, Fig. 12–11a, is designed to resist a moment of 40 N · m. In order to increase its strength and rigidity, it is proposed that two small ribs be added at its bottom, Fig. 12–11b. Determine the maximum normal stress in the member for both cases.

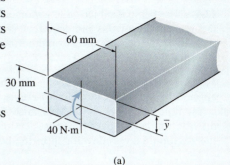

(a)

### Solution

*Without Ribs.* Clearly the neutral axis is at the center of the cross section, Fig. 12–11a, so $\bar{y} = c = 15$ mm $= 0.015$ m. Thus,

$$I = \frac{1}{12}bh^3 = \frac{1}{12}(0.06 \text{ m})(0.03 \text{ m})^3 = 0.135(10^{-6}) \text{ m}^4$$

Therefore the maximum normal stress is

$$\sigma_{max} = \frac{Mc}{I} = \frac{(40 \text{ N} \cdot \text{m})(0.015 \text{ m})}{0.135(10^{-6}) \text{ m}^4} = 4.44 \text{ MPa} \qquad Ans.$$

*With Ribs.* From Fig. 12–11b, segmenting the area into the large main rectangle and the bottom two rectangles (ribs), the location $\bar{y}$ of the centroid and the neutral axis is determined as follows:

$$\bar{y} = \frac{\Sigma \tilde{y} A}{\Sigma A}$$

$$= \frac{[0.015 \text{ m}](0.030 \text{ m})(0.060 \text{ m}) + 2[0.0325 \text{ m}](0.005 \text{ m})(0.010 \text{ m})}{(0.03 \text{ m})(0.060 \text{ m}) + 2(0.005 \text{ m})(0.010 \text{ m})}$$

$$= 0.01592 \text{ m}$$

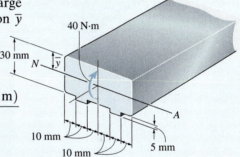

(b)

**Fig. 12–11**

This value does not represent $c$. Instead

$$c = 0.035 \text{ m} - 0.01592 \text{ m} = 0.01908 \text{ m}$$

Using the parallel-axis theorem, the moment of inertia about the neutral axis is

$$I = \left[ \frac{1}{12}(0.060 \text{ m})(0.030 \text{ m})^3 + (0.060 \text{ m})(0.030 \text{ m})(0.01592 \text{ m} - 0.015 \text{ m})^2 \right]$$

$$+ 2\left[ \frac{1}{12}(0.010 \text{ m})(0.005 \text{ m})^3 + (0.010 \text{ m})(0.005 \text{ m})(0.0325 \text{ m} - 0.01592 \text{ m})^2 \right]$$

$$= 0.1642(10^{-6}) \text{ m}^4$$

Therefore, the maximum normal stress is

$$\sigma_{max} = \frac{Mc}{I} = \frac{40 \text{ N} \cdot \text{m}(0.01908 \text{ m})}{0.1642(10^{-6}) \text{ m}^4} = 4.65 \text{ MPa} \qquad Ans.$$

This surprising result indicates that the addition of the ribs to the cross section will *increase* the normal stress rather than decrease it, and for this reason they should be omitted.

# PROBLEMS

**12-1.** The beam is subjected to a moment $M$. Determine the percentage of this moment that is resisted by the stresses acting on both the top and bottom boards, $A$ and $B$, of the beam.

**12-2.** Determine the moment $M$ that should be applied to the beam in order to create a compressive stress at point $D$ of $\sigma_D = 30$ MPa. Also sketch the stress distribution acting over the cross section and compute the maximum stress developed in the beam.

**\*12-4.** Determine the moment $M$ that will produce a maximum stress of 10 ksi on the cross section.

**12-5.** Determine the maximum tensile and compressive bending stress in the beam if it is subjected to a moment of $M = 4$ kip·ft.

**12-6.** Determine the resultant force the bending stresses produce on the horizontal top flange plate $AB$ of the beam if $M = 4$ kip·ft.

**12-7.** Determine the resultant force the bending stresses produce on the web $CD$ of the beam if $M = 4$ kip·ft.

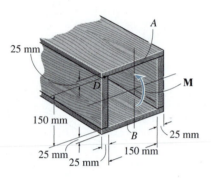

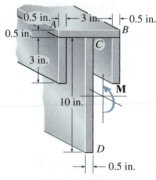

Probs. 12–1/2

Probs. 12–4/5/6/7

**12-3.** Two considerations have been proposed for the design of a beam. Determine which one will support a moment of $M = 150$ kN·m with the least amount of bending stress. What is that stress? By what percentage is it more effective?

**\*12-8.** The aluminum machine part is subjected to a moment of $M = 75$ N·m. Determine the bending stress created at points $B$ and $C$ on the cross section. Sketch the results on a volume element located at each of these points.

**12-9.** The aluminum machine part is subjected to a moment of $M = 75$ N·m. Determine the maximum tensile and compressive bending stresses in the part.

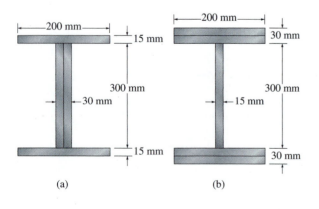

Prob. 12–3

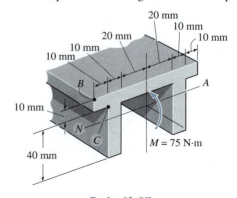

Probs. 12–8/9

**12-10.** The beam is made from three boards nailed together as shown. If the moment acting on the cross section is $M = 1\,\text{kip}\cdot\text{ft}$, determine the maximum bending stress in the beam. Sketch a three-dimensional view of the stress distribution acting over the cross section.

**12-11.** Determine the resultant force the bending stresses produce on the top board $A$ of the beam if $M = 1\,\text{kip}\cdot\text{ft}$.

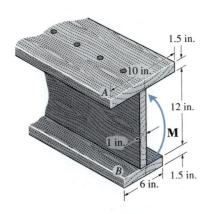

1.5 in.
10 in.
12 in.
A
M
1 in.
B
1.5 in.
6 in.

Probs. 12–10/11

**\*12-12.** The control lever is used on a riding lawn mower. Determine the maximum bending stress in the lever at section $a$–$a$ if a force of 20 lb is applied to the handle. The lever is supported by a pin at $A$ and a wire at $B$. Section $a$–$a$ is square, 0.25 in. by 0.25 in.

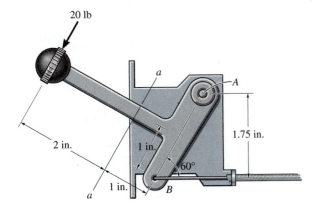

20 lb
a
A
2 in.
1 in.
1.75 in.
60°
1 in.
B
a

Prob. 12–12

**12-13.** Two considerations have been proposed for the design of a beam. Determine which one will support a moment of $M = 150\,\text{kN}\cdot\text{m}$ with the least amount of bending stress. What is that stress? By what percentage is it more effective?

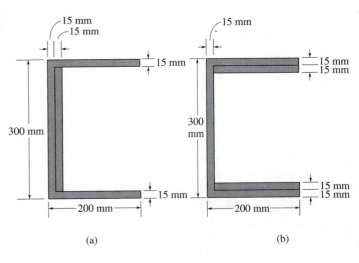

15 mm
15 mm
15 mm
300 mm
15 mm
300 mm
15 mm
15 mm
15 mm
15 mm
200 mm
15 mm
200 mm
15 mm
15 mm

(a)                                    (b)

Prob. 12–13

**12-14.** If the shaft has a diameter of 50 mm, determine the absolute maximum bending stress in the shaft.

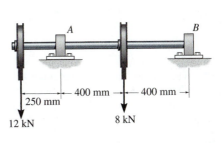

A
B
400 mm
400 mm
250 mm
12 kN
8 kN

Prob. 12–14

**12-15.** If the beam has a square cross section of 9 in. on each side, determine the absolute maximum bending stress in the beam.

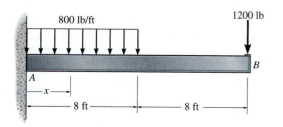

**Prob. 12-15**

*12-16.** Determine the absolute maximum bending stress in the 30-mm-diameter shaft which is subjected to the concentrated forces. The sleeve bearings at $A$ and $B$ support only vertical forces.

**12-17.** Determine the smallest allowable diameter of the shaft which is subjected to the concentrated forces. The sleeve bearings at $A$ and $B$ support only vertical forces, and the allowable bending stress is $\sigma_{allow} = 160$ MPa.

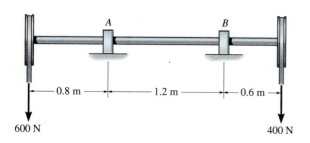

**Probs. 12-16/17**

**12-18.** The beam has a rectangular cross section as shown. Determine the largest load $P$ that can be supported on its overhanging ends so that the bending stress in the beam does not exceed $\sigma_{max} = 10$ MPa.

**12-19.** The beam has the rectangular cross section shown. If $P = 1.5$ kN, determine the maximum bending stress in the beam. Sketch the stress distribution acting over the cross section.

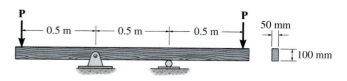

**Probs. 12-18/19**

*12-20.** Determine the absolute maximum bending stress in the 1.5-in.-diameter shaft which is subjected to the concentrated forces. The sleeve bearings at $A$ and $B$ support only vertical forces.

**12-21.** Determine the smallest allowable diameter of the shaft which is subjected to the concentrated forces. The sleeve bearings at $A$ and $B$ support only vertical forces, and the allowable bending stress is $\sigma_{allow} = 22$ ksi.

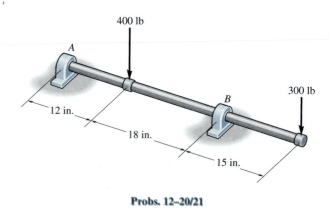

**Probs. 12-20/21**

**12-22.** The beam is subjected to the loading shown. Determine its required cross-sectional dimension $a$, if the allowable bending stress for the material is $\sigma_{allow} = 150$ MPa.

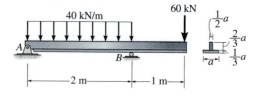

**Prob. 12-22**

**12-23.** Determine the magnitude of the maximum load **P** that can be applied to the beam if the beam is made of a material having an allowable bending stress of $(\sigma_{allow})_c = 16$ ksi in compression and $(\sigma_{allow})_t = 18$ ksi in tension.

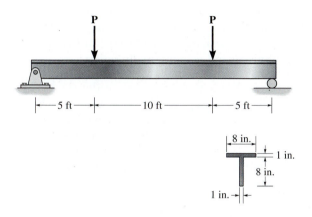

Prob. 12–23

**\*12-24.** The strut $CD$ on the utility pole supports the cable having a weight of 600 lb. Determine the absolute maximum bending stress in the strut if $A$, $B$, and $C$ are assumed to be pinned.

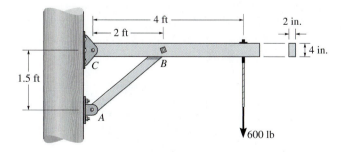

Prob. 12–24

**12-25.** The man has a mass of 78 kg and stands motionless at the end of the diving board. If the board has the cross section shown, determine the maximum normal strain developed in the board. The modulus of elasticity for the material is $E = 125$ GPa. Assume $A$ is a pin and $B$ is a roller.

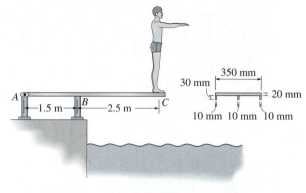

Prob. 12–25

**12-26.** The rod is supported by smooth journal bearings at $A$ and $B$ that only exert vertical reactions on the shaft. If $d = 90$ mm, determine the absolute maximum bending stress in the beam, and sketch the stress distribution acting over the cross section.

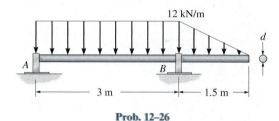

Prob. 12–26

**12-27.** The rod is supported by smooth journal bearings at $A$ and $B$ that only exert vertical reactions on the shaft. Determine its smallest diameter $d$ if the allowable bending stress is $\sigma_{allow} = 180$ MPa.

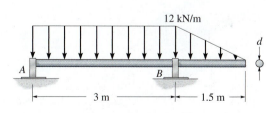

Prob. 12–27

## 12.3 Unsymmetric Bending

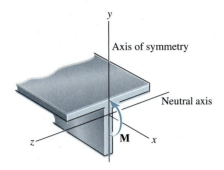

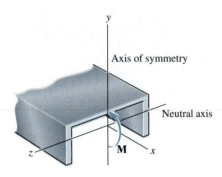

Axis of symmetry

Neutral axis

**M**

**Fig. 12–12**

When developing the flexure formula we imposed a condition that the cross-sectional area be *symmetric* about an axis perpendicular to the neutral axis; furthermore, the resultant internal moment **M** acts along the neutral axis. Such is the case for the "T" or channel sections shown in Fig. 12–12. These conditions, however, are unnecessary, and in this section we will show that the flexure formula can also be applied either to a beam having a cross-sectional area of any shape or to a beam having a resultant internal moment that acts in any direction.

**Moment Applied Along Principal Axis.** Consider the beam's cross section to have the unsymmetrical shape shown in Fig. 12–13a. As in Sec. 12.2, the right-handed $x, y, z$ coordinate system is established such that the origin is located at the centroid $C$ of the cross section, and the resultant internal moment **M** acts along the $+z$ axis. We require the stress distribution acting over the entire cross-sectional area to have a zero force resultant, the resultant internal moment about the $y$ axis to be zero, and the resultant internal moment about the $z$ axis to equal **M**.[*] These three conditions can be expressed mathematically by considering the force acting on the differential element $dA$ located at $(0, y, z)$, Fig. 12–13a. This force is $dF = \sigma\, dA$, and therefore we have

$$F_R = \Sigma F_x; \qquad\qquad 0 = \int_A \sigma\, dA \qquad (12\text{–}8)$$

$$(M_R)_y = \Sigma M_y; \qquad\qquad 0 = \int_A z\sigma\, dA \qquad (12\text{–}9)$$

$$(M_R)_z = \Sigma M_z; \qquad\qquad M = \int_A -y\sigma\, dA \qquad (12\text{–}10)$$

---

[*]The condition that moments about the $y$ axis be equal to zero was not considered in Sec. 12.2, since the bending-stress distribution was *symmetric* with respect to the $y$ axis and such a distribution of stress automatically produces zero moment about the $y$ axis. See Fig. 12–7c.

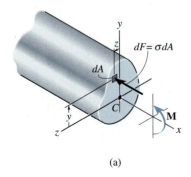

(a)

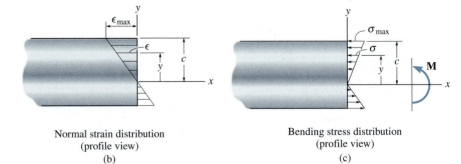

Normal strain distribution
(profile view)

(b)

Bending stress distribution
(profile view)

(c)

**Fig. 12–13**

As shown in Sec. 12.2, Eq. 12–8 is satisfied since the $z$ axis passes through the *centroid* of the cross-sectional area. Also, since the $z$ axis represents the *neutral axis* for the cross section, the normal strain will vary linearly from zero at the neutral axis, to a maximum at a point located the largest $y$ coordinate distance, $y = c$, from the neutral axis, Fig. 12–13$b$. Provided the material behaves in a linear-elastic manner, the normal-stress distribution over the cross-sectional area is *also* linear, so that $\sigma = -(y/c)\sigma_{max}$, Fig. 12–13$c$. When this equation is substituted into Eq. 12–10 and integrated, it leads to the flexure formula $\sigma_{max} = Mc/I$. When it is substituted into Eq. 12–9, we get

$$0 = \frac{-\sigma_{max}}{c} \int_A yz \, dA$$

which requires

$$\int_A yz \, dA = 0$$

This integral is called the *product of inertia* for the area. It will be zero provided the $y$ and $z$ axes are chosen as *principal axes of inertia* for the area. For an arbitrarily shaped area, like the one shown in Fig. 12–13$a$, the orientation of the principal axes can always be determined, using methods which are beyond the scope of this text. If the area has an axis of symmetry, however, the *principal axes* can easily be established since they will always be oriented *along the axis of symmetry* and *perpendicular* to it.

In summary, then, Eqs. 12–8 through 12–10 will *always* be satisfied regardless of the direction of the applied moment **M**. For example, consider the members shown in Fig. 12–14. In each of these cases, $y$ and $z$ define the principal axes of inertia for the cross section having the origin located at the area's centroid. Since **M** is applied about one of the principal axes ($z$ axis), the stress distribution is determined from the flexure formula, $\sigma = My/I_z$, and is shown for each case.

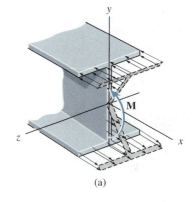

(a)

**Fig. 12–14**

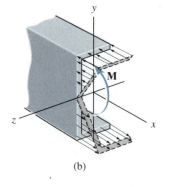

(b)

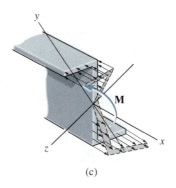

(c)

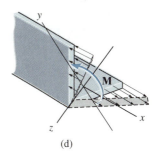

(d)

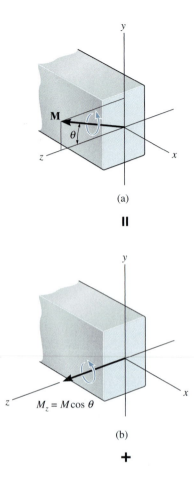

(a)

**II**

$M_z = M \cos \theta$

(b)

**+**

$M_y = M \sin \theta$

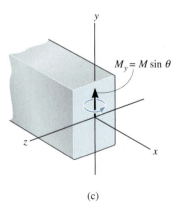

(c)

Fig. 12–15

**Moment Arbitrarily Applied.** Sometimes a member may be loaded such that the resultant internal moment does not act about one of the principal axes of the cross section. When this occurs, the moment should first be resolved into components directed along the principal axes. The flexure formula can then be used to determine the normal stress caused by each moment component. Finally, using the principle of superposition, the resultant normal stress at the point can be determined.

To show this, consider the beam to have a rectangular cross section and to be subjected to the moment **M**, Fig. 12–15a. Here **M** makes an angle $\theta$ with the *principal z* axis. We will assume $\theta$ is positive when it is directed from the $+z$ axis toward the $+y$ axis, as shown. Resolving **M** into components along the $z$ and $y$ axes, we have $M_z = M \cos \theta$ and $M_y = M \sin \theta$, respectively. Each of these components is shown separately on the cross section in Fig. 12–15b and 12–15c. The normal-stress distributions that produce **M** and its components $\mathbf{M}_z$ and $\mathbf{M}_y$ are shown in Fig. 12–15d, 12–15e, and 12–15f, respectively. Here it is assumed that $(\sigma_x)_{max} > (\sigma'_x)_{max}$. By inspection, the maximum tensile and compressive stresses $[(\sigma_x)_{max} + (\sigma'_x)_{max}]$ occur at two opposite corners of the cross section, Fig. 12–15d.

Applying the flexure formula to each moment component in Fig. 12–15b and 12–15c, we can express the resultant normal stress at any point on the cross section, Fig. 12–15d, in general terms as

$$\sigma = -\frac{M_z y}{I_z} + \frac{M_y z}{I_y}$$ 
(12–11)

where

$\sigma$ = the normal stress at the point

$y, z$ = the coordinates of the point measured from $x, y, z$ axes having their origin at the centroid of the cross-sectional area and forming a right-handed coordinate system. The $x$ axis is directed outward from the cross-section and the $y$ and $z$ axes represent respectively the principal axes of minimum and maximum moment of inertia for the area.

$M_y, M_z$ = the resultant internal moment components directed along the principal $y$ and $z$ axes. They are positive if directed along the $+y$ and $+z$ axes, otherwise they are negative. Or, stated another way, $M_y = M \sin \theta$ and $M_z = M \cos \theta$, where $\theta$ is measured positive from the $+z$ axis toward the $+y$ axis.

$I_y, I_z$ = the *principal moments of inertia* computed about the $y$ and $z$ axes, respectively. See Sec. 6.6. The principal moments of inertia are the moments of intertia about the principal axes of inertia.

As noted previously, it is *very important* that the x, y, z axes form a right-handed system and that the proper algebraic signs be assigned to the moment components and the coordinates when applying this equation. The resulting stress will be *tensile* if it is *positive* and *compressive* if it is *negative*.

## Orientation of the Neutral Axis.

The angle $\alpha$ of the neutral axis in Fig. 12–15d can be determined by applying Eq. 12–11 with $\sigma = 0$, since by definition no normal stress acts on the neutral axis. We have

$$y = \frac{M_y I_z}{M_z I_y} z$$

Since $M_z = M \cos\theta$ and $M_y = M \sin\theta$, then

$$y = \left(\frac{I_z}{I_y} \tan\theta\right) z \tag{12–12}$$

This is the equation of the line that defines neutral axis for the cross section. Since the slope of this line is $\tan\alpha = y/z$, then

$$\boxed{\tan\alpha = \frac{I_z}{I_y} \tan\theta} \tag{12–13}$$

Here it can be seen that for *unsymmetrical bending* the angle $\theta$, defining the direction of the moment **M**, Fig. 12–15a, is *not equal* to $\alpha$, the angle defining the inclination of the neutral axis, Fig. 12–15d, unless $I_z = I_y$. Instead, if as in Fig. 12–15a the y axis is chosen as the principal axis for the *minimum* moment of inertia, and the z axis is chosen as the principal axis for the *maximum* moment of inertia, so that $I_y < I_z$, then from Eq. 12–13 we can conclude that the angle $\alpha$, which is measured positive from the +z axis toward the +y axis, will lie *between* the line of action of **M** and the y axis, i.e., $\theta \le \alpha \le 90°$.

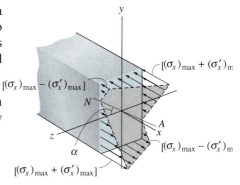

(d)

**=**

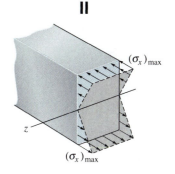

(e)

**+**

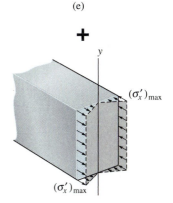

(f)

## IMPORTANT POINTS

- The flexure formula can be applied only when bending occurs about axes that represent the *principal axes of inertia* for the cross section. These axes have their origin at the centroid and are orientated along an axis of symmetry, if there is one, and perpendicular to it.

- If the moment is applied about some arbitrary axis, then the moment must be resolved into components along each of the principal axes, and the stress at a point is determined by superposition of the stress caused by each of the moment components.

EXAMPLE 12.5

The rectangular cross section shown in Fig. 12–16a is subjected to a bending moment of $M = 12$ kN·m. Determine the normal stress developed at each corner of the section, and specify the orientation of the neutral axis.

**Solution**

*Internal Moment Components.* By inspection it is seen that the $y$ and $z$ axes represent the principal axes of inertia since they are axes of symmetry for the cross section. As required we have established the $z$ axis as the principal axis for *maximum* moment of inertia. The moment is resolved into its $y$ and $z$ components, where

$$M_y = -\frac{4}{5}(12 \text{ kN} \cdot \text{m}) = -9.60 \text{ kN} \cdot \text{m}$$

$$M_z = \frac{3}{5}(12 \text{ kN} \cdot \text{m}) = 7.20 \text{ kN} \cdot \text{m}$$

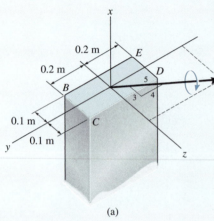

(a)

Fig. 12–16

*Section Properties.* The moments of inertia about the $y$ and $z$ axes are

$$I_y = \frac{1}{2}(0.4 \text{ m})(0.2 \text{ m})^3 = 0.2667(10^{-3}) \text{ m}^4$$

$$I_z = \frac{1}{12}(0.2 \text{ m})(0.4 \text{ m})^3 = 1.067(10^{-3}) \text{ m}^4$$

*Bending Stress.* Thus,

$$\sigma = -\frac{M_z y}{I_z} + \frac{M_y z}{I_y}$$

$$\sigma_B = -\frac{7.20(10^3) \text{ N} \cdot \text{m}(0.2 \text{ m})}{1.067(10^{-3}) \text{ m}^4} + \frac{-9.60(10^3) \text{ N} \cdot \text{m}(-0.1 \text{ m})}{0.2667(10^{-3}) \text{ m}^4} = 2.25 \text{ MPa} \qquad Ans.$$

$$\sigma_C = -\frac{7.20(10^3) \text{ N} \cdot \text{m}(0.2 \text{ m})}{1.067(10^{-3}) \text{ m}^4} + \frac{-9.60(10^3) \text{ N} \cdot \text{m}(0.1 \text{ m})}{0.2667(10^{-3}) \text{ m}^4} = -4.95 \text{ MPa} \qquad Ans.$$

$$\sigma_D = -\frac{7.20(10^3) \text{ N} \cdot \text{m}(-0.2 \text{ m})}{1.067(10^{-3}) \text{ m}^4} + \frac{-9.60(10^3) \text{ N} \cdot \text{m}(0.1 \text{ m})}{0.2667(10^{-3}) \text{ m}^4} = -2.25 \text{ MPa} \qquad Ans.$$

$$\sigma_E = -\frac{7.20(10^3) \text{ N} \cdot \text{m}(-0.2 \text{ m})}{1.067(10^{-3}) \text{ m}^4} + \frac{-9.60(10^3) \text{ N} \cdot \text{m}(-0.1 \text{ m})}{0.2667(10^{-3}) \text{ m}^4} = 4.95 \text{ MPa} \qquad Ans.$$

The resultant normal-stress distribution has been sketched using these values, Fig. 12–16b. Since superposition applies, the distribution is linear as shown.

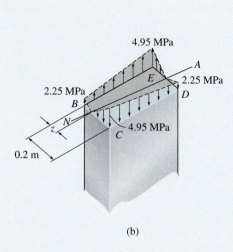

4.95 MPa

2.25 MPa

2.25 MPa

4.95 MPa

0.2 m

(b)

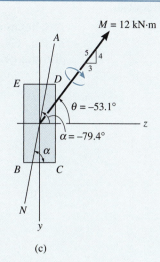

$M = 12$ kN·m

$\theta = -53.1°$

$\alpha = -79.4°$

(c)

*Orientation of Neutral Axis.* The location $z$ of the neutral axis ($NA$), Fig. 12–16b, can be established by proportion. Along the edge $BC$, we require

$$\frac{2.25 \text{ MPa}}{z} = \frac{4.95 \text{ MPa}}{(0.2 \text{ m} - z)}$$

$$0.450 - 2.25z = 4.95z$$

$$z = 0.0625 \text{ m}$$

In the same manner this is also the distance from $D$ to the neutral axis in Fig. 12–16b.

We can also establish the orientation of the $NA$ using Eq. 12–13, which is used to specify the angle $\alpha$ that the axis makes with the $z$ or *maximum* principal axis. According to our sign convention, $\theta$ must be measured from the $+z$ axis toward the $+y$ axis. By comparison, in Fig. 12–16c, $\theta = -\tan^{-1}\frac{4}{3} = -53.1°$ (or $\theta = +306.9°$). Thus,

$$\tan \alpha = \frac{I_z}{I_y}\tan \theta$$

$$\tan \alpha = \frac{1.067(10^{-3}) \text{ m}^4}{0.2667(10^{-3}) \text{ m}^4}\tan(-53.1°)$$

$$\alpha = -79.4° \hspace{2cm} Ans.$$

This result is shown in Fig. 12–16c. Using the value of $z$ calculated above, verify, using the geometry of the cross section, that one obtains the same answer.

**E X A M P L E 12.6**

A T-beam is subjected to the bending moment of 15 kN · m as shown in Fig. 12–17a. Determine the maximum normal stress in the beam and the orientation of the neutral axis.

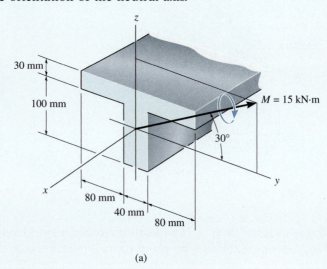

(a)

### Solution

***Internal Moment Components.*** The $y$ and $z$ axes are principal axes of inertia. Why? From Fig. 12–17a, both moment components are positive. We have

$$M_y = (15 \text{ kN} \cdot \text{m}) \cos 30° = 12.99 \text{ kN} \cdot \text{m}$$
$$M_z = (15 \text{ kN} \cdot \text{m}) \sin 30° = 7.50 \text{ kN} \cdot \text{m}$$

***Section Properties.*** With reference to Fig. 12–17b, working in units of meters, we have

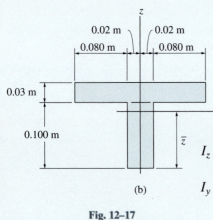

(b)

**Fig. 12–17**

$$\overline{z} = \frac{\Sigma \overline{z} A}{\Sigma A} = \frac{[0.05 \text{ m}](0.100 \text{ m})(0.04 \text{ m}) + [0.115 \text{ m}](0.03 \text{ m})(0.200 \text{ m})}{(0.100 \text{ m})(0.04 \text{ m}) + (0.03 \text{ m})(0.200 \text{ m})}$$

$$= 0.0890 \text{ m}$$

Using the parallel-axis theorem (Sec. 6.7), $I = \overline{I} + Ad^2$, the principal moments of inertia are thus

$$I_z = \frac{1}{12}(0.100 \text{ m})(0.04 \text{ m})^3 + \frac{1}{12}(0.03 \text{ m})(0.200 \text{ m})^3 = 20.53(10^{-6}) \text{ m}^4$$

$$I_y = \left[\frac{1}{12}(0.04 \text{ m})(0.100 \text{ m})^3 + (0.100 \text{ m})(0.04 \text{ m})(0.0890 \text{ m} - 0.05 \text{ m})^2\right]$$

$$+ \left[\frac{1}{12}(0.200 \text{ m})(0.03 \text{ m})^3 + (0.200 \text{ m})(0.03 \text{ m})(0.115 \text{ m} - 0.0890 \text{ m})^2\right]$$

$$= 13.92(10^{-6}) \text{ m}^4$$

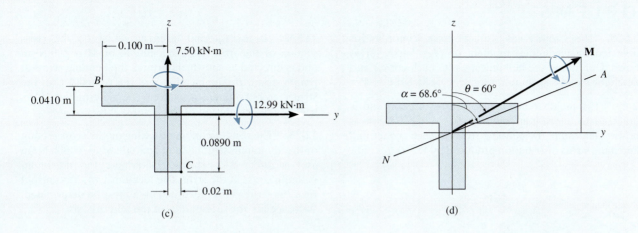

(c)                                                           (d)

*Maximum Bending Stress.*   The moment components are shown in Fig. 12–17c. By inspection, the largest *tensile* stress occurs at point $B$, since by superposition both moment components create a tensile stress there. Likewise, the greatest *compressive* stress occurs at point $C$. Thus,

$$\sigma = -\frac{M_z y}{I_z} + \frac{M_y z}{I_y}$$

$$\sigma_B = -\frac{7.50 \text{ kN} \cdot \text{m}(-0.100 \text{ m})}{20.53(10^{-6}) \text{ m}^4} + \frac{12.99 \text{ kN} \cdot \text{m}(0.0410 \text{ m})}{13.92(10^{-6}) \text{ m}^4}$$

$$= 74.8 \text{ MPa}$$

$$\sigma_C = -\frac{7.50 \text{ kN} \cdot \text{m}(0.020 \text{ m})}{20.53(10^{-6}) \text{ m}^4} + \frac{12.99 \text{ kN} \cdot \text{m}(-0.0890 \text{ m})}{13.92(10^{-6}) \text{ m}^4}$$

$$= -90.4 \text{ MPa} \qquad\qquad\qquad\qquad\qquad\qquad Ans.$$

By comparison, the largest normal stress is therefore compressive and occurs at point $C$.

*Orientation of Neutral Axis.*   When applying Eq. 12–6 it is important to be sure the angles $\alpha$ and $\theta$ are defined correctly. As previously stated, $y$ must represent the axis for *minimum* principal moment of inertia, and $z$ must represent the axis for *maximum* principal moment of inertia. These axes are properly positioned here since $I_y < I_z$. Using this setup, $\theta$ and $\alpha$ are measured positive from the $+z$ axis toward the $+y$ axis. Hence, from Fig. 12–17a, $\theta = +60°$. Thus,

$$\tan \alpha = \left(\frac{20.53(10^{-6}) \text{ m}^4}{13.92(10^{-6}) \text{ m}^4}\right) \tan 60°$$

$$\alpha = 68.6° \qquad\qquad\qquad\qquad Ans.$$

The neutral axis is shown in Fig. 12–17d. As expected, it lies between the $y$ axis and the line of action of **M**.

# PROBLEMS

*12-28. The member has a square cross section and is subjected to a resultant moment of $M = 850\,\text{N}\cdot\text{m}$ as shown. Determine the bending stress at each corner and sketch the stress distribution produced by **M**. Set $\theta = 45°$.

12-29. The member has a square cross section and is subjected to a resultant moment of $M = 850\,\text{N}\cdot\text{m}$ as shown. Determine the bending stress at each corner and sketch the stress distribution produced by **M**. Set $\theta = 30°$.

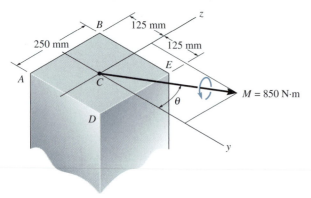

**Probs. 12–28/29**

12-30. Determine the maximum magnitude of the bending moment **M** so that the bending stress in the member does not exceed 24 ksi. The location $\bar{y}$ of the centroid $C$ must be determined.

12-31. The moment acting on the cross section of the T-beam has a magnitude of $M = 15\,\text{kip}\cdot\text{ft}$ and is directed as shown. Determine the bending stress at points $A$ and $B$. The location $\bar{y}$ of the centroid $C$ must be determined.

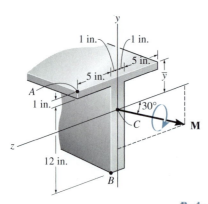

**Probs. 12–30/31**

*12-32. If the internal moment acting on the cross section of the strut has a magnitude of $M = 800\,\text{N}\cdot\text{m}$ and is directed as shown, determine the bending stress at points $A$ and $B$. The location $\bar{z}$ of the centroid $C$ of the strut's cross-sectional area must be determined. Also, specify the orientation of the neutral axis.

12-33. The resultant moment acting on the cross section of the aluminum strut has a magnitude of $M = 800\,\text{N}\cdot\text{m}$ and is directed as shown. Determine the maximum bending stress in the strut. The location $\bar{y}$ of the centroid $C$ of the strut's cross-sectional area must be determined. Also, specify the orientation of the neutral axis.

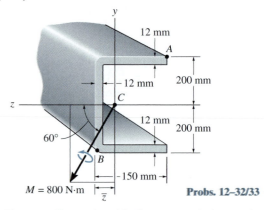

**Probs. 12–32/33**

12-34. The cantilevered wide-flange steel beam is subjected to the concentrated force **P** at its end. Determine the largest magnitude of this force so that the bending stress developed at section $A$ does not exceed $\sigma_{\text{allow}} = 180\,\text{MPa}$.

12-35. The cantilevered wide-flange steel beam is subjected to the concentrated force of $P = 600\,\text{N}$ at its end. Determine the maximum bending stress developed in the beam at section $A$.

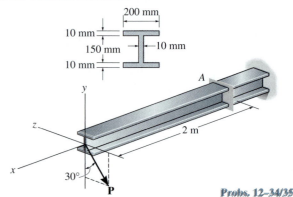

**Probs. 12–34/35**

*12-36. The steel beam has the cross-sectional area shown. Determine the largest intensity of distributed load $w$ that it can support so that the bending stress does not exceed $\sigma_{max} = 22$ ksi.

12-37. The steel beam has the cross-sectional area shown. If $w = 5$ kip/ft, determine the absolute maximum bending stress in the beam.

12-39. The shaft is subjected to the vertical and horizontal loadings of two pulleys $D$ and $E$ as shown. It is supported on two journal bearings at $A$ and $B$ which offer no resistance to axial loading. Furthermore, the coupling to the motor at $C$ can be assumed not to offer any support to the shaft. Determine the required diameter $d$ of the shaft if the allowable bending stress for the material is $\sigma_{allow} = 180$ MPa.

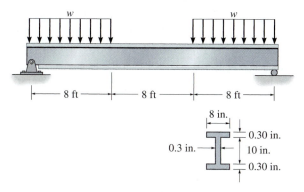

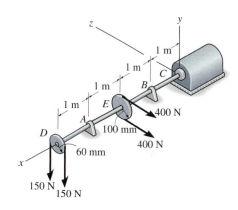

Probs. 12–36/37

Prob. 12–39

12-38. The 30-mm-diameter shaft is subjected to the vertical and horizontal loadings of two pulleys as shown. It is supported on two journal bearings at $A$ and $B$ which offer no resistance to axial loading. Furthermore, the coupling to the motor at $C$ can be assumed not to offer any support to the shaft. Determine the maximum bending stress developed in the shaft.

*12-40. The 65-mm-diameter steel shaft is subjected to the two loads that act in the directions shown. If the journal bearings at $A$ and $B$ do not exert an axial force on the shaft, determine the absolute maximum bending stress developed in the shaft.

12-41. The steel shaft is subjected to the two loads that act in the directions shown. If the journal bearings at $A$ and $B$ do not exert an axial force on the shaft, determine the required diameter of the shaft if the allowable bending stress is $\sigma_{allow} = 180$ MPa.

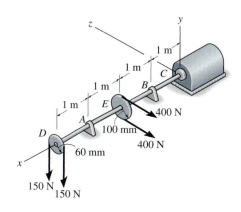

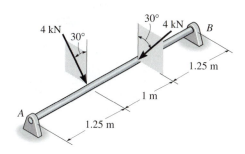

Prob. 12–38

Probs. 12–40/41

## CHAPTER REVIEW

• *Flexure Formula.* A bending moment tends to produce a linear variation of the normal strain within a beam. Provided the moment does not cause yielding of the material, then Hooke's law and equilibrium can be used to relate the internal moment in the beam to the stress distribution. The result is the flexure formula, $\sigma = Mc/I$, where $I$ and $c$ are determined from the neutral axis, that passes through the centroid of the cross section.

• *Unsymmetric Bending.* If the cross section of the beam is not symmetric about an axis, that is perpendicular to the neutral axis, then unsymmetric bending will occur. Formulas are available for this case, or the problem can be solved by considering the superposition of bending about two separate axes.

# REVIEW PROBLEMS

**12-42.** The beam is subjected to a moment of 15 kip · ft. Determine the resultant force the stress produces on the top flange $A$ and bottom flange $B$. Also compute the maximum stress developed in the beam.

**12-43.** The beam is subjected to a moment of 15 kip · ft. Determine the percent of this moment that is resisted by the web $D$ of the beam.

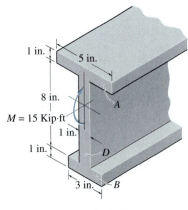

Probs. 12–42

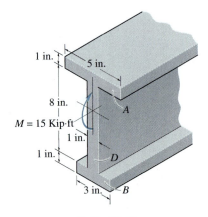

Probs. 12–43

**\*12-44.** The beam is made from three boards nailed together as shown. If the moment acting on the cross section is $M = 600 \, \text{N} \cdot \text{m}$, determine the maximum bending stress in the beam. Sketch a three-dimensional view of the stress distribution acting over the cross section.

**12-46.** Determine the maximum tensile and compressive bending stress in the beam if it is subjected to a moment of $M = 2 \, \text{kip} \cdot \text{ft}$.

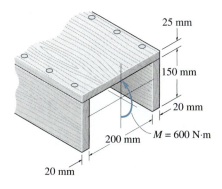

Prob. 12–44

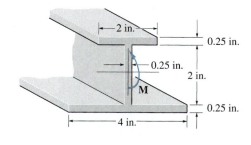

Probs. 12–46

**12-45.** Determine the moment $M$ that will produce a maximum stress of 12 ksi on the cross section.

**12-47.** The steel rod having a diameter of 1 in. is subjected to an internal moment of $M = 300 \, \text{lb} \cdot \text{ft}$. Determine the stress created at points $A$ and $B$. Also, sketch a three-dimensional view of the stress distribution acting over the cross-section.

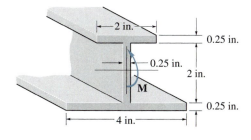

Probs. 12–45

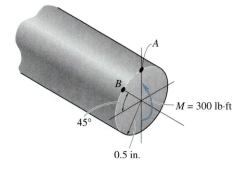

Prob. 12–47

**12-48.** The beam is constructed from four pieces of wood, glued together as shown. If the internal bending moment is $M = 80$ kip•ft, determine the maximum bending stress in the beam. Sketch a three-dimensional view of the stress distribution acting over the cross section.

**12–49.** The beam is constructed from four pieces of wood, glued together as shown. If the internal bending moment is $M = 80$ kip•ft, deteremine the resultant force the bending moment exerts on the top and bottom boards of the beam.

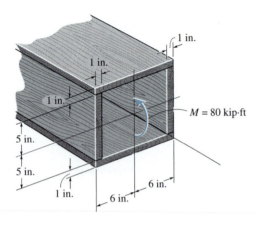

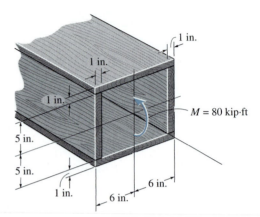

Prob. 12–48

Prob. 12–49

**12-50.** The axle of the freight car is subjected to wheel loadings of 20 kip. If it is supported by two journal bearings at $C$ and $D$, determine the maximum bending stress developed at the center of the axle, where the diameter is 5.5 in.

**12-51.** The pin is used to connect the three links together. Due to wear, the load is distributed over the top and bottom of the pin as shown on the free-body diagram. If the diameter of the pin is 0.40 in., determine the maximum bending stress on the cross-sectional area. For the solution it is first necessary to determine the load intensities $w_1$ and $w_2$ and then draw the moment diagram.

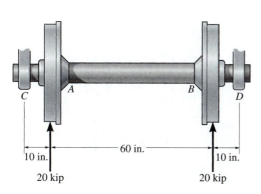

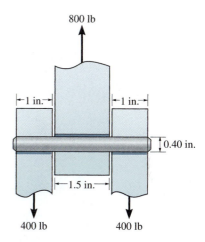

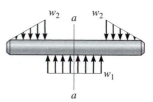

Prob. 12-50

Prob. 12-51

Railroad ties act as beams that support very large transverse shear loadings. As a result, wooden ties tend to split at their ends, where the shear loads are the largest.

# Transverse Shear

- To determine the shear stress in straight beams subject to transverse loading.
- To calculate the shear flow in beams composed of several members.

## 13.1 Shear in Straight Members

Beams generally support both shear and moment loadings. The shear $\mathbf{V}$ is the result of a transverse shear-stress distribution that acts over the beam's cross section, Fig. 13–1. Due to the complementary property of shear, notice that associated longitudinal shear stresses will also act along longitudinal planes of the beam. For example, a typical element removed from the interior point on the cross section is subjected to both transverse and longitudinal shear stress as shown in Fig. 13–1.

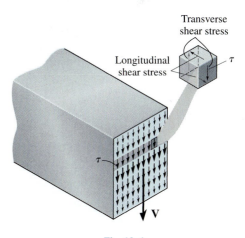

Fig. 13–1

Shear connectors are "tack welded" to this corrugated metal floor liner so that when the concrete floor is poured, the connectors will prevent the concrete slab from slipping on the liner surface. The two materials will thus act as a composite slab.

It is possible to physically illustrate why shear stress develops on the longitudinal planes of a beam by considering the beam to be made from three boards, Fig. 13–2a. If the top and bottom surfaces of each board are smooth, and the boards are not bonded together, then application of the load **P** will cause the boards to *slide* relative to one another, and so the beam will deflect as shown. On the other hand, if the boards are bonded together, then the longitudinal shear stresses between the boards will prevent the relative sliding of the boards, and consequently the beam will act as a single unit, Fig. 13–2b.

As a result of the shear stress, shear strains will be developed and these will tend to distort the cross section in a rather complex manner. To show this consider a bar made of a highly deformable material and marked with horizontal and vertical grid lines, Fig. 13–3a. When a shear **V** is applied, it tends to deform these lines into the pattern shown in Fig. 13–3b. This nonuniform shear-strain distribution over the cross section will cause the cross section to *warp*, that is, *not* to remain plane.

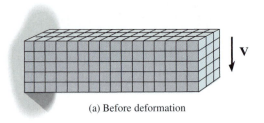

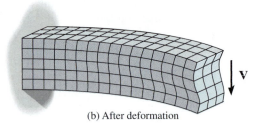

Fig. 13–3

Recall that in the development of the flexure formula, we assumed that cross sections must *remain plane* and perpendicular to the longitudinal axis of the beam after deformation. Although this is *violated* when the beam is subjected to *both* bending and shear, we can generally assume the cross-sectional warping described above is small enough so that it can be neglected. This assumption is particularly true for the most common case of a *slender beam*; that is, one that has a small depth compared with its length.

In the previous chapters we developed the axial load, torsion, and flexure formulas by first determining the strain distribution, based on assumptions regarding the deformation of the cross section. In the case of transverse shear, however, the shear-strain distribution throughout the depth of a beam *cannot* be easily expressed mathematically. For example, it is not uniform or linear for rectangular cross sections as we have shown. Therefore, the foregoing analysis of shear stress will be developed in a manner different from that used to study the previous loadings. Specifically, we will develop a formula for shear stress *indirectly*; that is, using the the flexure formula and the relationship between moment and shear $(V = dM/dx)$.

## 13.2 The Shear Formula

Development of a relationship between the shear-stress distribution, acting over the cross section of a beam, and the resultant shear force at the section is based on a study of the *longitudinal shear stress* and the results of Eq. 7–2, $V = dM/dx$. To show how this relationship is established, we will consider the *horizontal force equilibrium* of a portion of the element taken from the beam in Fig. 13–4a and shown in Fig. 13–4b. A free-body diagram of the *element* that shows *only* the normal-stress distribution acting on it is shown in Fig. 13–4c. This distribution is caused by the bending moments $M$ and $M + dM$. We have excluded the effects of $V, V + dV$, and $w(x)$ on the free-body diagram since these loadings are vertical and will therefore not be involved in a horizontal force summation. The element in Fig. 13–4c will indeed satisfy $\Sigma F_x = 0$ since the stress distribution on each side of the element forms only a couple moment and therefore a zero force resultant.

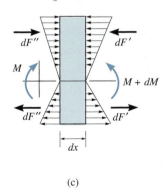
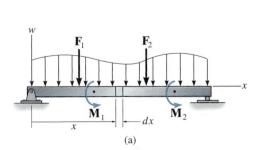
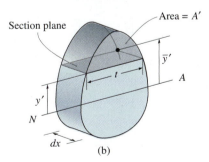

(a)　(b)　(c)

Fig. 13–4

Now consider the shaded top *segment* of the element that has been sectioned at $y'$ from the neutral axis, Fig. 13–4b. This segment has a width $t$ at the section, and the cross-sectional sides each have an area $A'$. Because the resultant moments on each side of the element differ by $dM$, it can be seen in Fig. 13–4d that $\Sigma F_x = 0$ will not be satisfied *unless* a longitudinal shear stress $\tau$ acts over the bottom face of the segment. In the following analysis, we will assume this shear stress is *constant* across the width $t$ of the bottom face. It acts on the area $t\,dx$. Applying the equation of horizontal force equilibrium, and using the flexure formula, Eq. 12–7, we have

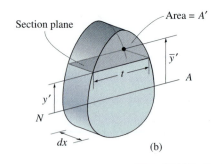

Section plane
Area = $A'$

$$\xrightarrow{+}\ \Sigma F_x = 0; \qquad \int_{A'} \sigma'\,dA - \int_{A'} \sigma\,dA - \tau(t\,dx) = 0$$

$$\int_{A'}\left(\frac{M + dM}{I}\right)y\,dA - \int_{A'}\left(\frac{M}{I}\right)y\,dA - \tau(t\,dx) = 0$$

$$\left(\frac{dM}{I}\right)\int_{A'} y\,dA = \tau(t\,dx) \qquad (13\text{–}1)$$

(b)

Solving for $\tau$, we get

$$\tau = \frac{1}{It}\left(\frac{dM}{dx}\right)\int_{A'} y\,dA$$

This equation can be simplified by noting that $V = dM/dx$ (Eq. 7–2). Also, the integral represents the first moment of the area $A'$ about the neutral axis. We will denote it by the symbol $Q$. Since the location of the

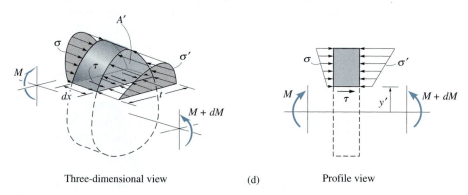

Three-dimensional view      (d)      Profile view

**Fig. 13–4**

centroid of the area $A'$ is determined from $\overline{y}' = \int_{A'} y\, dA / A'$, we can also write

$$Q = \int_{A'} y\, dA = \overline{y}' A' \qquad (13\text{--}2)$$

The final result is therefore

$$\boxed{\tau = \frac{VQ}{It}} \qquad (13\text{--}3)$$

Here

$\tau$ = the shear stress in the member at the point located a distance $y'$ from the neutral axis, Fig. 13–4b. This stress is assumed to be constant and therefore *averaged* across the width $t$ of the member, Fig. 13–4d

$V$ = internal resultant shear force, determined from the method of sections and the equations of equilibrium

$I$ = moment of inertia of the *entire* cross-sectional area computed about the neutral axis

$t$ = width of the member's cross-sectional area, measured at the point where $\tau$ is to be determined

$Q = \int_{A'} y\, dA' = \overline{y}' A'$, where $A'$ is the top (or bottom) portion of the member's cross-sectional area, defined from the section where $t$ is measured, and $\overline{y}'$ is the distance to the centroid of $A'$, measured from the neutral axis

The above equation is referred to as the *shear formula*. Although in the derivation we considered only the shear stresses acting on the beam's longitudinal plane, the formula applies as well for finding the transverse shear stress on the beam's cross-sectional area. This, of course, is because the transverse and longitudinal shear stresses are complementary and numerically equal.

Since Eq. 13–3 was derived indirectly from the flexure formula, it is necessary that the material behave in a linear-elastic manner and have a modulus of elasticity that is the *same* in tension as it is in compression.

## 13.3   Shear Stresses in Beams

In order to develop some insight as to the method of applying the shear formula and also discuss some of its limitations, we will now study the shear-stress distributions in a few common types of beam cross sections. Numerical applications of the shear formula will then be given in the examples that follow.

Typical shear failure of this wooden beam occurred at the support and through the approximate center of its cross section.

**Rectangular Cross Section.**   Consider the beam to have a rectangular cross section of width $b$ and height $h$ as shown in Fig. 13–5a. The distribution of the shear stress throughout the cross section can be determined by computing the shear stress at an *arbitrary height y* from the neutral axis, Fig. 13–5b, and then plotting this function. Here the dark color shaded area $A'$ will be used for computing $\tau$.* Hence

$$Q = \bar{y}'A' = \left[ y + \frac{1}{2}\left( \frac{h}{2} - y \right) \right]\left( \frac{h}{2} - y \right)b$$

$$= \frac{1}{2}\left( \frac{h^2}{4} - y^2 \right)b$$

Applying the shear formula, we have

$$\tau = \frac{VQ}{It} = \frac{V(\frac{1}{2})[(h^2/4) - y^2]b}{(\frac{1}{12}bh^3)b}$$

or

$$\tau = \frac{6V}{bh^3}\left( \frac{h^2}{4} - y^2 \right) \tag{13–4}$$

This result indicates that the shear-stress distribution over the cross section is ***parabolic***. As shown in Fig. 13–5c, the intensity varies from zero at the top and bottom, $y = \pm h/2$, to a maximum value at the neutral axis, $y = 0$. Specifically, since the area of the cross section is $A = bh$, then at $y = 0$ we have, from Eq. 13–4,

$$\tau_{max} = 1.5\frac{V}{A} \tag{13–5}$$

*The area below $y$ can also be used [$A' = b(h/2 + y)$], but doing so involves a bit more algebraic manipulation.

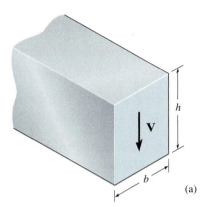

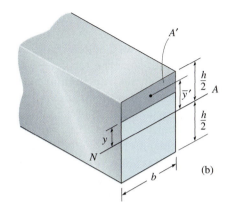

(a)

(b)

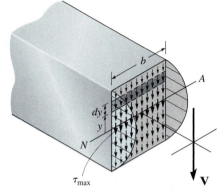

Shear–stress distribution

(c)

This same value for $\tau_{max}$ can be obtained directly from the shear formula, $\tau = VQ/It$, by realizing that $\tau_{max}$ occurs where $Q$ is *largest*, since $V$, $I$, and $t$ are *constant*. By inspection, $Q$ will be a maximum when the area above (or below) the neutral axis is considered; that is, $A' = bh/2$ and $\bar{y}' = h/4$. Thus,

$$\tau_{max} = \frac{VQ}{It} = \frac{V(h/4)(bh/2)}{[\frac{1}{12}bh^3]b} = 1.5\frac{V}{A}$$

By comparison, $\tau_{max}$ is 50% greater than the *average* shear stress determined from Eq. 8–4; that is, $\tau_{avg} = V/A$.

It is important to remember that for every $\tau$ acting on the cross-sectional area in Fig. 13–5c, there is a corresponding $\tau$ acting in the longitudinal direction along the beam. For example, if the beam is sectioned by a longitudinal plane through its neutral axis, then as noted above, the *maximum shear stress* acts on this plane, Fig. 13–5d. It is this stress that can cause a timber beam to fail as shown in Fig. 13–6. Here horizontal splitting of the wood starts to occur through the neutral axis at the beam's ends, since there the vertical reactions subject the beam to large shear stress and wood has a low resistance to shear along its grains, which are oriented in the longitudinal direction.

It is instructive to show that when the shear-stress distribution, Eq. 13–4, is integrated over the cross section it yields the resultant shear $V$. To do this, a differential strip of area $dA = b\,dy$ is chosen, Fig. 13–5c, and since $\tau$ acts uniformly over this strip, we have

$$\int_A \tau\, dA = \int_{-h/2}^{h/2} \frac{6V}{bh^3}\left(\frac{h^2}{4} - y^2\right)b\, dy$$

$$= \frac{6V}{h^3}\left[\frac{h^2}{4}y - \frac{1}{3}y^3\right]_{-h/2}^{h/2}$$

$$= \frac{6V}{h^3}\left[\frac{h^2}{4}\left(\frac{h}{2} + \frac{h}{2}\right) - \frac{1}{3}\left(\frac{h^3}{8} + \frac{h^3}{8}\right)\right] = V$$

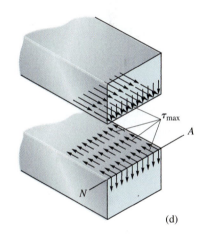

(d)

Fig. 13–5

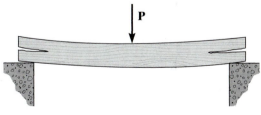

Fig. 13–6

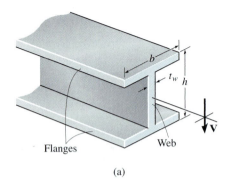

Flanges  Web

(a)

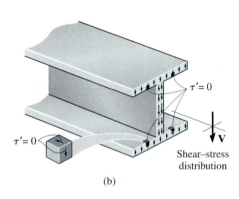

$\tau' = 0$

$\tau' = 0$

Shear–stress
distribution

(b)

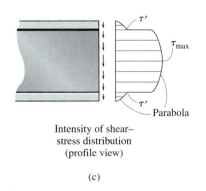

$\tau'$

$\tau_{max}$

$\tau'$
Parabola

Intensity of shear–
stress distribution
(profile view)

(c)

**Fig. 13–7**

**Wide-Flange Beam.** A *wide-flange beam* consists of two (wide) "flanges" and a "web" as shown in Fig. 13–7a. Using an analysis similar to that just given we can determine the shear-stress distribution acting over the cross section. The results are depicted graphically in Fig. 13–7b and 13–7c. Like the rectangular cross section, the shear stress varies *parabolically* over the beam's depth, since the cross section can be treated like the rectangular section, first having the width of the top flange, $b$, then the thickness of the web, $t_w$, and again the width of the bottom flange, $b$. In particular, notice that the shear stress will vary *only slightly* throughout the web, and also, a *jump* in shear stress occurs at the flange–web junction since the cross-sectional thickness changes at this point, or in other words, $t$ in the shear formula changes. By comparison, the web will carry significantly more of the shear force than the flanges. This will be illustrated numerically in Example 13–2.

**Limitations on the Use of the Shear Formula.** One of the major assumptions used in the development of the shear formula is that the shear stress is *uniformly* distributed over the *width t* at the section where the shear stress is determined. In other words, the *average* shear stress is computed across the width. We can test the accuracy of this assumption by comparing it with a more exact mathematical analysis based on the theory of elasticity. In this regard, if the beam's cross section is rectangular, the *actual* shear-stress distribution across the neutral axis as calculated from the theory of elasticity varies as shown in Fig. 13–8. The maximum value, $\tau'_{max}$, occurs at the *edges* of the cross section, and its magnitude depends on the ratio $b/h$ (width/depth). For sections having a $b/h = 0.5, \tau'_{max}$ is only about 3% greater than the shear stress calculated from the shear formula, Fig. 13–8a. However, for *flat sections*, say $b/h = 2, \tau'_{max}$ is about 40% greater than $\tau_{max}$, Fig. 13–8b. The error becomes even greater as the section becomes flatter, or as the $b/h$ ratio increases. Errors of this magnitude are certainly intolerable if one uses the shear formula to determine the shear stress in the *flange* of a wide-flange beam, as discussed above.

It should also be pointed out that the shear formula will not give accurate results when used to determine the shear stress at the flange–web junction of a wide-flange beam, since this is a point of sudden cross-sectional change and therefore a *stress concentration* occurs here. Furthermore, the inner regions of the flanges are free boundaries, Fig. 13–7b, and as a result the shear stress on these boundaries must be zero. If the shear formula is applied to determine the shear stress at these boundaries, however, one obtains a value of $\tau'$ that is *not* equal to zero, Fig. 13–7c. Fortunately, these limitations for applying the shear formula to the flanges of a wide-flange beam are not important in engineering practice. Most often engineers must only calculate the *average maximum shear stress*, which occurs at the neutral axis, where the $b/h$ (width/depth) ratio is *very small*, and therefore the calculated result is very close to the *actual* maximum shear stress as explained above.

Another important limitation on the use of the shear formula can be illustrated with reference to Fig. 13–9a, which shows a beam having a cross section with an irregular or nonrectangular boundary. If we apply the shear formula to determine the (average) shear stress $\tau$ along the line $AB$, it will be directed as shown in Fig. 13–9b. Consider now an element of material taken from the boundary point $B$, such that one of its faces is located on the outer surface of the beam, Fig. 13–9c. Here the calculated shear stress $\tau$ on the front face of the element is resolved into components, $\tau'$ and $\tau''$. By inspection, the component $\tau'$ must be equal to zero since its corresponding longitudinal component $\tau'$, acting on the stress-free boundary surface, must be zero. To satisfy this boundary condition therefore, the shear stress acting on the element at the boundary must be directed tangent to the boundary. The shear-stress distribution across line $AB$ would then be directed as shown in Fig. 13–9d. Due to the greatest inclination of the shear stresses at $A$ and $B$, the maximum shear stress will occur at these points. Specific values for this shear stress must be obtained using the principles of the theory of elasticity. Note, however, that we can apply the shear formula to obtain the shear stress acting across each of the colored lines in Fig. 13–9a. Here these lines intersect the tangents to the boundary of the cross section at *right angles*, and as shown in Fig. 13–9e, the transverse shear stress is vertical and constant along each line.

To summarize the above points, the shear formula does not give accurate results when applied to members having cross sections that are *short or flat*, or at points where the cross section suddenly changes. Nor should it be applied across a section that intersects the boundary of the member at an angle other than 90°. Instead, for these cases the shear stress should be determined using more advanced methods based on the theory of elasticity.

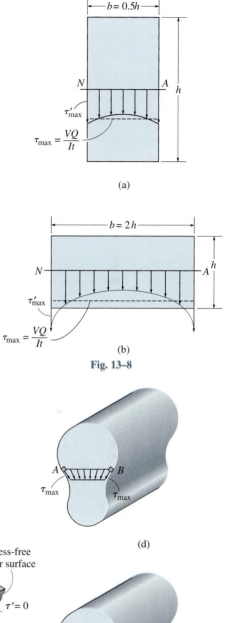

(a)

(b)

**Fig. 13–8**

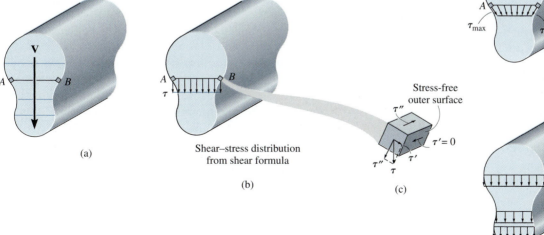

(a)

Shear–stress distribution
from shear formula

(b)

(c)

(d)

(e)

**Fig. 13–9**

## IMPORTANT POINTS

- Shear forces in beams cause *nonlinear shear-strain* distributions over the cross section, causing it to *warp*.
- Due to the complementary property of shear stress, the shear stress developed in a beam acts on both the cross section and on longitudinal planes.
- The *shear formula* was derived by considering horizontal force equilibrium of the longitudinal shear stress and bending-stress distributions acting on a portion of a differential segment of the beam.
- The shear formula is to be used on straight prismatic members made of homogeneous material that has linear-elastic behavior. Also, the internal resultant shear force must be directed along an axis of symmetry for the cross-sectional area.
- For a beam having a rectangular cross section, the *shear stress varies parabolically* with depth. The maximum shear stress is along the neutral axis.
- The shear formula should not be used to determine the shear stress on cross sections that are short or flat, or at points of sudden cross-sectional changes, or at a point on an inclined boundary.

## PROCEDURE FOR ANALYSIS

In order to apply the shear formula, the following procedure is suggested.

*Internal Shear.*

- Section the member perpendicular to its axis at the point where the shear stress is to be determined, and obtain the internal shear **V** at the section.

*Section Properties.*

- Determine the location of the neutral axis, and determine the moment of inertia $I$ of the *entire cross-sectional area* about the neutral axis.
- Pass an imaginary horizontal section through the point where the shear stress is to be determined. Measure the width $t$ of the area at this section.
- The portion of the area lying either above or below this section is $A'$. Determine $Q$ either by integration, $Q = \int_{A'} y \, dA'$, or by using $Q = \overline{y}' A'$. Here $\overline{y}'$ is the distance to the centroid of $A'$, measured from the neutral axis. It may be helpful to realize that $A'$ is the portion of the member's cross-sectional area that is being "held onto the member" by the longitudinal shear stresses, Fig. 13–4*d*.

*Shear Stress.*

- Using a consistent set of units, substitute the data into the shear formula and compute the shear stress $\tau$.
- It is suggested that the proper direction of the transverse shear stress $\tau$ be established on a volume element of material located at the point where it is computed. This can be done by realizing that $\tau$ acts on the cross section in the same direction as **V**. From this, corresponding shear stresses acting on the other three planes of the element can then be established.

# EXAMPLE 13.1

The beam shown in Fig. 13–10a is made of wood and is subjected to a resultant internal vertical shear force of $V = 3$ kip. (a) Determine the shear stress in the beam at point $P$, and (b) compute the maximum shear stress in the beam.

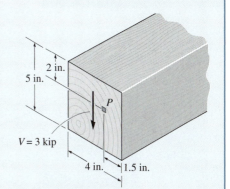

(a)

## Solution

### Part (a)

*Section Properties.* The moment of inertia of the cross-sectional area computed about the neutral axis is

$$I = \frac{1}{12}bh^3 = \frac{1}{12}(4 \text{ in.})(5 \text{ in.})^3 = 41.7 \text{ in}^4$$

A horizontal section line is drawn through point $P$ and the partial area $A'$ is shown shaded in Fig. 13–10b. Hence

$$Q = \bar{y}'A' = \left[0.5 \text{ in.} + \frac{1}{2}(2 \text{ in.})\right](2 \text{ in.})(4 \text{ in.}) = 12 \text{ in}^3$$

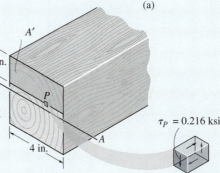

*Shear Stress.* The shear force at the section is $V = 3$ kip. Applying the shear formula, we have

$$\tau_P = \frac{VQ}{It} = \frac{(3 \text{ kip})(12 \text{ in}^3)}{(41.7 \text{ in}^4)(4 \text{ in.})} = 0.216 \text{ ksi} \qquad Ans.$$

(b)                    (c)

Since $\tau_P$ contributes to $V$, it acts downward at $P$ on the cross section. Consequently, a volume element of the material at this point would have shear stresses acting on it as shown in Fig. 13–10c.

### Part (b)

*Section Properties.* Maximum shear stress occurs at the neutral axis, since $t$ is constant throughout the cross section and $Q$ is largest for this case. For the dark shaded area $A'$ in Fig. 13–10d, we have

$$Q = \bar{y}'A' = \left[\frac{2.5 \text{ in.}}{2}\right](4 \text{ in.})(2.5 \text{ in.}) = 12.5 \text{ in}^3$$

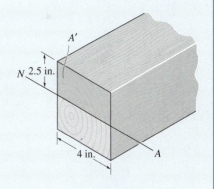

(d)

**Fig. 13–10**

*Shear Stress.* Applying the shear formula yields

$$\tau_{max} = \frac{VQ}{It} = \frac{(3 \text{ kip})(12.5 \text{ in}^3)}{(41.7 \text{ in}^4)(4 \text{ in.})} = 0.225 \text{ ksi} \qquad Ans.$$

Note that this is equivalent to

$$\tau_{max} = 1.5\frac{V}{A} = 1.5\frac{3 \text{ kip}}{(4 \text{ in.})(5 \text{ in.})} = 0.225 \text{ ksi} \qquad Ans.$$

**E X, A M P L E 13.2**

A steel wide-flange beam has the dimensions shown in Fig. 13–11a. If it is subjected to a shear of $V = 80$ kN, (a) plot the shear-stress distribution acting over the beam's cross-sectional area, and (b) determine the shear force resisted by the web.

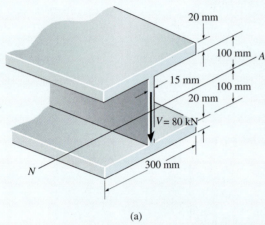

20 mm

100 mm — A

15 mm

100 mm

20 mm

$V = 80$ kN

300 mm

N

(a)

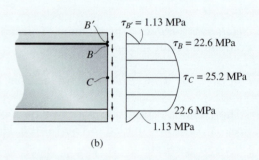

$B'$ $\tau_{B'} = 1.13$ MPa

$\tau_B = 22.6$ MPa

$B$

$\tau_C = 25.2$ MPa

$C$

22.6 MPa

1.13 MPa

(b)

**Solution**

*Part (a).* The shear-stress distribution will be parabolic and varies in the manner shown in Fig. 13–11b. Due to symmetry, only the shear stresses at points $B'$, $B$, and $C$ have to be computed. To show how these values are obtained, we must first determine the moment of inertia of the cross-sectional area about the neutral axis. Working in meters, we have

$$I = \left[\frac{1}{2}(0.015 \text{ m})(0.200 \text{ m})^3\right]$$

$$+ 2\left[\frac{1}{12}(0.300 \text{ m})(0.02 \text{ m})^3 + (0.300 \text{ m})(0.02 \text{ m})(0.110 \text{ m})^2\right]$$

$$= 155.6(10^{-6}) \text{ m}^4$$

For point $B'$, $t_{B'} = 0.300$ m, and $A'$ is the dark shaded area shown in Fig. 13–11c. Thus,

$$Q_{B'} = \overline{y}' A' = [0.110 \text{ m}](0.300 \text{ m})(0.02 \text{ m}) = 0.660(10^{-3}) \text{ m}^3$$

so that

$$\tau_{B'} = \frac{V Q_{B'}}{I t_{B'}} = \frac{80 \text{ kN}(0.660(10^{-3}) \text{ m}^3)}{155.6(10^{-6}) \text{ m}^4(0.300 \text{ m})} = 1.13 \text{ MPa}$$

For point $B$, $t_B = 0.015$ m and $Q_B = Q_{B'}$, Fig. 13–11c. Hence

$$\tau_B = \frac{V Q_B}{I t_B} = \frac{80 \text{ kN}(0.660(10^{-3}) \text{ m}^3)}{155.6(10^{-6}) \text{ m}^4(0.015 \text{ m})} = 22.6 \text{ MPa}$$

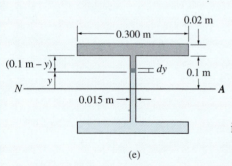

0.02 m

0.300 m

$(0.1 \text{ m} - y)$

$= dy$

0.1 m

$y$

N

A

0.015 m

(e)

**Fig. 13–11**

Note from the discussion of "Limitations on the Use of the Shear Formula" that the calculated value for both $\tau_{B'}$ and $\tau_B$ will actually be very misleading. Why?

For point $C$, $t_C = 0.015$ m and $A'$ is the dark shaded area shown in Fig. 13–11d. Considering this area to be composed of two rectangles, we have

$$Q_C = \Sigma \bar{y}'A' = [0.110 \text{ m}](0.300 \text{ m})(0.02 \text{ m})$$
$$+[0.05 \text{ m}](0.015 \text{ m})(0.100 \text{ m})$$
$$= 0.735(10^{-3}) \text{ m}^3$$

Thus,

$$\tau_C = \tau_{max} = \frac{VQ_C}{It_C} = \frac{80 \text{ kN}[0.735(10^{-3}) \text{ m}^3]}{155.6(10^{-6}) \text{ m}^4(0.015 \text{ m})} = 25.2 \text{ MPa}$$

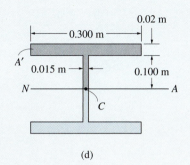

(d)

**Part (b).** The shear force in the web will be determined by first formulating the shear stress at the *arbitrary* location $y$ within the web, Fig. 13–11e. Using units of meters, we have

$$I = 155.6(10^{-6}) \text{ m}^4$$
$$t = 0.015 \text{ m}$$
$$A' = (0.300 \text{ m})(0.02 \text{ m}) + (0.015 \text{ m})(0.1 \text{ m} - y)$$
$$Q = \Sigma \bar{y}'A' = (0.11 \text{ m})(0.300 \text{ m})(0.02 \text{ m})$$
$$+ [y + \tfrac{1}{2}(0.1 \text{ m} - y)](0.015 \text{ m})(0.1 \text{ m} - y)$$
$$= (0.735 - 7.50 \, y^2)(10^{-3}) \text{ m}^3$$

so that

$$\tau = \frac{VQ}{It} = \frac{80 \text{ kN}(0.735 - 7.50 \, y^2)(10^{-3}) \text{ m}^3}{(155.6(10^{-6}) \text{ m}^4)(0.015 \text{ m})}$$
$$= (25.192 - 257.07 \, y^2) \text{ MPa}$$

This stress acts on the area strip $dA = 0.015 \, dy$ shown in Fig. 13–11e, and therefore the shear force resisted by the web is

$$V_w = \int_{A_w} \tau \, dA = \int_{-0.1 \text{ m}}^{0.1 \text{ m}} (25.192 - 257.07 \, y^2)(10^6)(0.015 \text{ m}) \, dy$$

$$V_w = 73.0 \text{ kN} \qquad Ans.$$

Note that by comparison, the web supports 91% of the total shear (80 kN), whereas the flanges support the remaining 9%. Try solving this problem by finding the force in one of the flanges (3.496 kN) using the same method. Then $V_w = V - 2V = 80 \text{ kN} - 2(3.496 \text{kN}) = 73.0 \text{ kN}$.

**E X A M P L E** 13.3

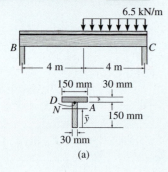

(a)

The beam shown in Fig. 13–12a is made from two boards. Determine the maximum shear stress in the glue necessary to hold the boards together along the seam where they are joined. The supports at $B$ and $C$ exert only vertical reactions on the beam.

### Solution

***Internal Shear.*** The support reactions and the shear diagram for the beam are shown in Fig. 13–12b. It is seen that the maximum shear in the beam is 19.5 kN.

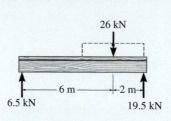

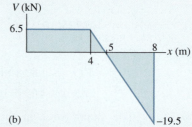

(b)

***Section Properties.*** The centroid and therefore the neutral axis will be determined from the reference axis placed at the bottom of the cross-sectional area, Fig. 13–12a. Working in units of meters, we have

$$\bar{y} = \frac{\Sigma \tilde{y} A}{\Sigma A} = \frac{[0.075 \text{ m}](0.150 \text{ m})(0.030 \text{ m}) + [0.165 \text{ m}](0.030 \text{ m})(0.150 \text{ m})}{(0.150 \text{ m})(0.030 \text{ m}) + (0.030 \text{ m})(0.150 \text{ m})} = 0.120 \text{ m}$$

The moment of inertia, computed about the neutral axis, Fig. 13–12a, is therefore

$$I = \left[\frac{1}{12}(0.030 \text{ m})(0.150 \text{ m})^3 + (0.150 \text{ m})(0.030 \text{ m})(0.120 \text{ m} - 0.075 \text{ m})^2\right]$$
$$+ \left[\frac{1}{12}(0.150 \text{ m})(0.030 \text{ m})^3 + (0.030 \text{ m})(0.150 \text{ m})(0.165 \text{ m} - 0.120 \text{ m})^2\right] = 27.0(10^{-6}) \text{ m}^4$$

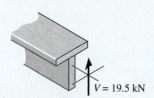

Plane containing glue

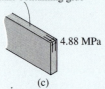

4.88 MPa

(c)

**Fig. 13–12**

The top board (flange) is being held onto the bottom board (web) by the glue, which is applied over the thickness $t = 0.03$ m. Consequently $A'$ is defined as the area of the top board, Fig. 13–12a. We have

$$Q = \bar{y}'A' = [0.180 \text{ m} - 0.015 \text{ m} - 0.120 \text{ m}](0.03 \text{ m})(0.150 \text{ m})$$
$$= 0.2025(10^{-3}) \text{ m}^3$$

***Shear Stress.*** Using the above data and applying the shear formula yields

$$\tau_{max} = \frac{VQ}{It} = \frac{19.5 \text{ kN}(0.2025(10^{-3}) \text{ m}^3)}{27.0(10^{-6}) \text{ m}^4(0.030 \text{ m})} = 4.88 \text{ MPa} \qquad \textit{Ans.}$$

The shear stress acting at the top of the bottom board is shown in Fig. 13–12c. Note that it is the glue's resistance to this lateral or *horizontal shear* stress that is necessary to hold the boards from slipping at the support $C$.

# PROBLEMS

**13-1.** The beam is fabricated from three steel plates, and it is subjected to a shear force of $V = 150$ kN. Determine the shear stress at points $A$ and $C$ where the plates are joined. Show $\bar{y} = 0.080196$ m from the bottom and $I_{NA} = 4.8646(10^{-6})$ m$^4$.

**13-2.** The beam is fabricated from three steel plates, and it is subjected to a shear force of $V = 150$ kN. Determine the shear stress at point $B$ where the plates are joined. Show $\bar{y} = 0.080196$ m from the bottom and $I_{NA} = 4.8646(10^{-6})$ m$^4$.

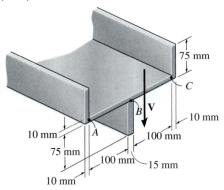

**Probs. 13-1/2**

**13-3.** If the T-beam is subjected to a vertical shear of $V = 12$ kip, determine the maximum shear stress in the beam. Also, compute the shear-stress jump at the flange–web junction $AB$. Sketch the variation of the shear-stress intensity over the entire cross section.

***13-4.** If the T-beam is subjected to a vertical shear of $V = 12$ kip, determine the vertical shear force resisted by the flange.

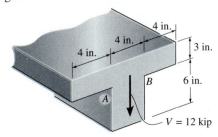

**Probs. 13-3/4**

**13-5.** If the T-beam is subjected to a vertical shear of $V = 10$ kip, determine the maximum shear stress in the beam. Also, compute the shear-stress jump at the flange–web junction $AB$. Sketch the variation of the shear-stress intensity over the entire cross section. Show that $I_{NA} = 532.04$ in$^4$.

**13-6.** If the T-beam is subjected to a vertical shear of $V = 10$ kip, determine the vertical shear force resisted by the flange. Show that $I_{NA} = 532.04$ in$^4$.

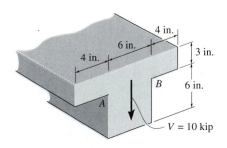

**Probs. 13-5/6**

**13-7.** Determine the shear stress at point $B$ on the web of the cantilevered strut at section $a$–$a$.

***13-8.** Determine the maximum shear stress acting at section $a$–$a$ of the cantilevered strut.

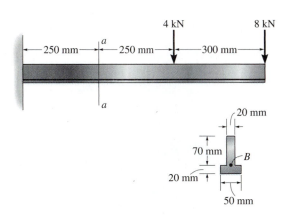

**Probs. 13-7/8**

**13-9.** Sketch the intensity of the shear-stress distribution acting over the beam's cross-sectional area, and determine the resultant shear force acting on the segment $AB$. The shear acting at the section is $V = 35$ kip. Show that $I_{NA} = 872.49$ in⁴.

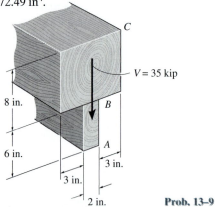

Prob. 13–9

**13-10.** Determine the largest shear force $V$ that the member can sustain if the allowable shear stress is $\tau_{allow} = 8$ ksi.

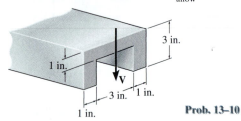

Prob. 13–10

**13-11.** Determine the maximum shear stress in the strut if it is subjected to a shear force of $V = 20$ kN.

**\*13-12.** Determine the maximum shear force $V$ that the strut can support if the allowable shear stress for the material is $\tau_{allow} = 40$ MPa.

**13-13.** Plot the intensity of the shear stress distributed over the cross section of the strut if it is subjected to a shear force of $V = 15$ kN.

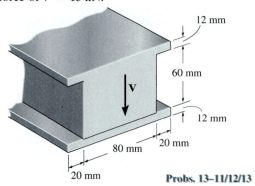

Probs. 13–11/12/13

**13-14.** Determine the largest end forces $P$ that the member can support if the allowable shear stress is $\tau_{allow} = 10$ ksi. The supports at $A$ and $B$ only exert vertical reactions on the beam.

**13-15.** If the force $P = 800$ lb, determine the maximum shear stress in the beam at the critical section. The supports at $A$ and $B$ only exert vertical reactions on the beam.

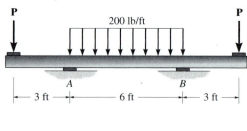

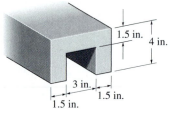

Probs. 13–14/15

**\*13-16.** The T-beam is subjected to the loading shown. Determine the maximum transverse shear stress in the beam at the critical section.

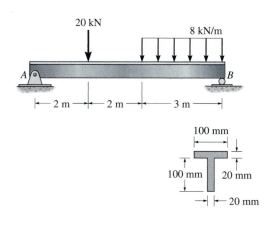

Prob. 13–16

**13-17.** The supports at *A* and *B* exert vertical reactions on the wood beam. If the distributed load *w* = 4 kip/ft, determine the maximum shear stress in the beam at section *a–a*.

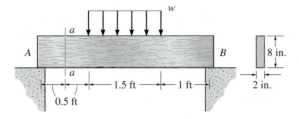

Prob. 13–17

**13-18.** The supports at *A* and *B* exert vertical reactions on the wood beam. If the allowable shear stress is $\tau_{allow}$ = 400 psi, determine the intensity *w* of the largest distributed load that can be applied to the beam.

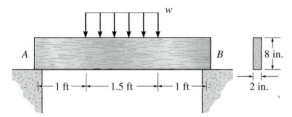

Prob. 13–18

**13-19.** Railroad ties must be designed to resist large shear loadings. If the tie is subjected to the 30-kip rail loadings and the gravel bed exerts a distributed reaction as shown, determine the intensity *w* for equilibrium, and find the maximum shear stress in the tie.

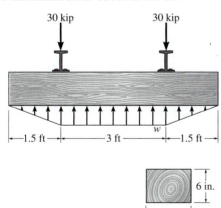

Prob. 13–19

**\*13-20.** The beam is made from three plastic pieces glued together at the seams *A* and *B*. If it is subjected to the loading shown, determine the shear stress developed in the glued joints at the critical section. The supports at *C* and *D* exert only vertical reactions on the beam.

**13-21.** The beam is made from three plastic pieces glued together at the seams *A* and *B*. If it is subjected to the loading shown, determine the vertical shear force resisted by the top flange of the beam at the critical section. The supports at *C* and *D* exert only vertical reactions on the beam.

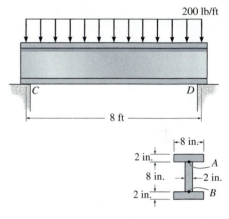

Probs. 13–20/21

**13-22.** The beam is made from three boards glued together at the seams *A* and *B*. If it is subjected to the loading shown, determine the maximum vertical shear force resisted by the top flange of the beam. The supports at *C* and *D* exert only vertical reactions on the beam.

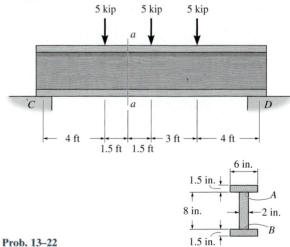

Prob. 13–22

## 13.4   Shear Flow in Built-Up Members

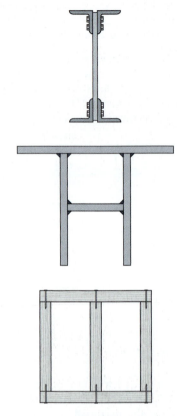

Occasionally in engineering practice members are "built up" from several composite parts in order to achieve a greater resistance to loads. Some examples are shown in Fig. 13–13. If the loads cause the members to bend, fasteners such as nails, bolts, welding material, or glue may be needed to keep the component parts from sliding relative to one another, Fig. 13–2. In order to design these fasteners it is necessary to know the shear force that must be resisted by the fastener along the member's *length*. This loading, when measured as a force per unit length, is referred to as the **shear flow q.**

The magnitude of the shear flow along any longitudinal section of a beam can be obtained using a development similar to that for finding the shear stress in the beam. To show this, we will consider finding the shear flow along the juncture where the composite part in Fig. 13–14a is connected to the flange of the beam. As shown in Fig. 13–14b, three horizontal forces must act on this part. Two of these forces, $F$ and $F + dF$, are developed by normal stresses caused by the moments $M$ and $M + dM$, respectively. The third force, which for equilibrium equals $dF$, acts at the juncture and is to be supported by the fastener. Realizing that $dF$ is the result of $dM$, then, as in the case of the shear formula, Eq. 13–1, we have

$$dF = \frac{dM}{I} \int_{A'} y \, dA'$$

Fig. 13–13

The integral represents $Q$, that is, the moment of the light colored area $A'$ in Fig. 13–14b about the neutral axis for the cross section. Since the segment has a length $dx$, the shear flow, or force per unit length along the beam, is $q = dF/dx$. Hence dividing both sides by $dx$ and noting that $V = dM/dx$, Eq. 7–2, we can write

$$q = \frac{VQ}{I} \qquad (13\text{–}6)$$

Here

$q$ = the shear flow, measured as a force per unit length along the beam

$V$ = the internal resultant shear force, determined from the method of sections and the equations of equilibrium

$I$ = the moment of inertia of the *entire* cross-sectional area computed about the neutral axis

$Q = \int_{A'} y \, dA' = \bar{y}' A'$, where $A'$ is the cross-sectional area of the segment that is connected to the beam at the juncture where the shear flow is to be calculated, and $\bar{y}'$ is the distance from the neutral axis to the centroid of $A'$

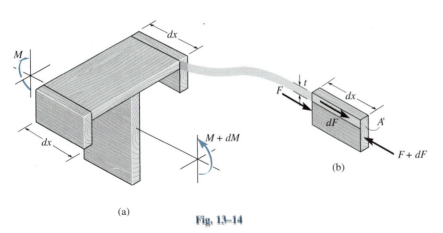

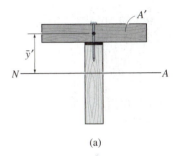

Fig. 13–14

(a)

(b)

(a)

Application of this equation follows the same "procedure for analysis" as outlined in Sec. 13.3 for the shear formula. In this regard it is very important to correctly identify the proper value for $Q$ when determining the shear flow at a particular junction on the cross section. A few examples should serve to illustrate how this is done. Consider the beam cross sections shown in Fig. 13–15. The shaded composite parts are connected to the beam by fasteners. At the planes of connection, the necessary shear flow $q$ is determined by using a value of $Q$ computed from $A'$ and $\bar{y}'$ indicated in each figure. Notice that this value of $q$ will be resisted by a *single* fastener in Fig. 13–15a and 13–15b, by *two* fasteners in Fig. 13–15c, and by *three* fasteners in Fig. 13–15d. In other words, the fastener in Figs. 13–15a and 13–15b supports the calculated value of $q$ and in Figs. 13–15c and 13–15d, each fastener supports $q/2$ and $q/3$, respectively.

## IMPORTANT POINTS

- Shear flow is a measure of the force per unit length along a longitudinal axis of a beam. This value is found from the shear formula and is used to determine the shear force developed in fasteners and glue that holds the various segments of a beam together.

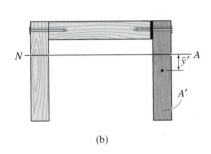

(b)

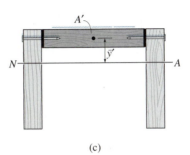

(c)

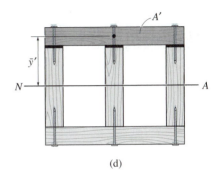

(d)

Fig. 13–15

**E X A M P L E  13.4**

The beam is constructed from four boards glued together as shown in Fig. 13–16a. If it is subjected to a shear of $V = 850$ kN, determine the shear flow at $B$ and $C$ that must be resisted by the glue.

**Solution**

*Section Properties.*   The neutral axis (centroid) will be located from the bottom of the beam, Fig. 13–16a. Working in units of meters, we have

$$\bar{y} = \frac{\Sigma \bar{y} A}{\Sigma A} = \frac{2[0.15 \text{ m}](0.3 \text{ m})(0.01 \text{ m}) + [0.205 \text{ m}](0.125 \text{ m})(0.01 \text{ m}) + [0.305 \text{ m}](0.250 \text{ m})(0.01 \text{ m})}{2(0.3 \text{ m})(0.01 \text{ m}) + 0.125 \text{ m}(0.01 \text{ m}) + 0.250 \text{ m}(0.01 \text{ m})}$$

$$= 0.1968 \text{ m}$$

The moment of inertia computed about the neutral axis is thus

$$I = 2\left[\frac{1}{12}(0.01 \text{ m})(0.3 \text{ m})^3 + (0.01 \text{ m})(0.3 \text{ m})(0.1968 \text{ m} - 0.150 \text{ m})^2\right]$$

$$+ \left[\frac{1}{12}(0.125 \text{ m})(0.01 \text{ m})^3 + (0.125 \text{ m})(0.01 \text{ m})(0.205 \text{ m} - 0.1968 \text{ m})^2\right]$$

$$+ \left[\frac{1}{12}(0.250 \text{ m})(0.01 \text{ m})^3 + (0.250 \text{ m})(0.01 \text{ m})(0.305 \text{ m} - 0.1968 \text{ m})^2\right]$$

$$= 87.52(10^{-6}) \text{ m}^4$$

Since the glue at $B$ and $B'$ holds the top board to the beam, Fig. 13–16b, we have

$$Q_B = \bar{y}'_B A'_B = [0.305 \text{ m} - 0.1968 \text{ m}](0.250 \text{ m})(0.01 \text{ m})$$
$$= 0.270(10^{-3}) \text{ m}^3$$

Likewise, the glue at $C$ and $C'$ holds the inner board to the beam, Fig. 13–16b, and so

$$Q_C = \bar{y}'_C A'_C = [0.205 \text{ m} - 0.1968 \text{ m}](0.125 \text{ m})(0.01 \text{ m})$$
$$= 0.01025(10^{-3}) \text{ m}^3$$

*Shear Flow.*   For $B$ and $B'$ we have

$$q'_B = \frac{VQ_B}{I} = \frac{850 \text{ kN}(0.270(10^{-3}) \text{ m}^3)}{87.52(10^{-6}) \text{ m}^4} = 2.62 \text{ MN/m}$$

And for $C$ and $C'$,

$$q'_C = \frac{VQ_C}{I} = \frac{850 \text{ kN}(0.01025(10^{-3}) \text{ m}^3)}{87.52(10^{-6}) \text{ m}^4} = 0.0995 \text{ MN/m}$$

Since *two seams* are used to secure each board, the glue per meter length of beam at each seam must be strong enough to resist *one-half* of each calculated value of $q'$. Thus,

$$q_B = 1.31 \text{ MN/m} \quad \text{and} \quad q_C = 0.0498 \text{ MN/m} \quad Ans.$$

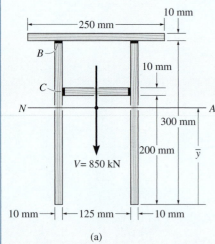

(a)

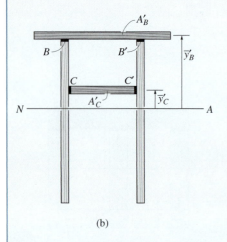

(b)

**Fig. 13–16**

# E X A M P L E 13.5

A box beam is to be constructed from four boards nailed together as shown in Fig. 13–17a. If each nail can support a shear force of 30 lb, determine the maximum spacing s of nails at B and at C so that the beam will support the vertical force of 80 lb.

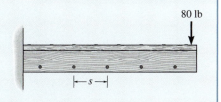

## Solution

**Internal Shear.** If the beam is sectioned at an *arbitrary point* along its length, the internal shear required for equilibrium is always V = 80 lb, and so the shear diagram is shown in Fig. 13–17b.

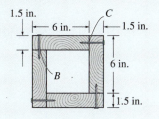

(a)

**Section Properties.** The moment of inertia of the cross-sectional area about the neutral axis can be determined by considering a 7.5-in. × 7.5-in. square minus a 4.5-in. × 4.5-in. square.

$$I = \frac{1}{12}(7.5 \text{ in.})(7.5 \text{ in.})^3 - \frac{1}{12}(4.5 \text{ in.})(4.5 \text{ in.})^3 = 229.5 \text{ in}^4$$

The shear flow at B is determined using $Q_B$ found from the darker shaded area shown in Fig. 13–17c. It is this "symmetric" portion of the beam that is to be "held" onto the rest of the beam by nails on the left side and by the fibers of the board on the right side. Thus,

$$Q_B = \bar{y}'A' = [3 \text{ in.}](7.5 \text{ in.})(1.5 \text{ in.}) = 33.75 \text{ in}^3$$

Likewise, the shear flow at C can be determined using the "symmetric" shaded area shown in Fig. 13–17d. We have

$$Q_C = \bar{y}'A' = [3 \text{ in.}](4.5 \text{ in.})(1.5 \text{ in.}) = 20.25 \text{ in}^3$$

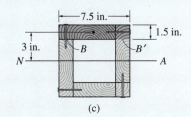

(c)

**Shear Flow.**

$$q_B = \frac{VQ_B}{I} = \frac{80 \text{ lb}(33.75 \text{ in}^3)}{229.5 \text{ in}^4} = 11.76 \text{ lb/in.}$$

$$q_C = \frac{VQ_C}{I} = \frac{80 \text{ lb}(20.25 \text{ in}^3)}{229.5 \text{ in}^4} = 7.059 \text{ lb/in.}$$

These values represent the shear force per unit length of the beam that must be resisted by the nails at B and the fibers at B', Fig. 13–17c, and the nails at C and the fibers at C', Fig. 13–17d, respectively. Since in each case the shear flow is resisted at *two* surfaces and each nail can resist 30 lb, for B the spacing is

$$s_B = \frac{30 \text{ lb}}{(11.76/2) \text{ lb/in.}} = 5.10 \text{ in.} \quad \text{Use } s_B = 5 \text{ in.} \quad \textit{Ans.}$$

And for C,

$$s_C = \frac{30 \text{ lb}}{(7.059/2) \text{ lb/in.}} = 8.50 \text{ in.} \quad \text{Use } s_C = 8.5 \text{ in.} \quad \textit{Ans.}$$

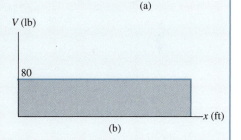

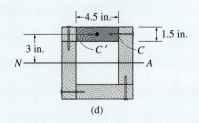

(d)

**Fig. 13–17**

EXAMPLE 13.6

Nails having a total shear strength of 40 lb are used in a beam that can be constructed either as in Case I or as in Case II, Fig. 13–18. If the nails are spaced at 9 in., determine the largest vertical shear that can be supported in each case so that the fasteners will not fail.

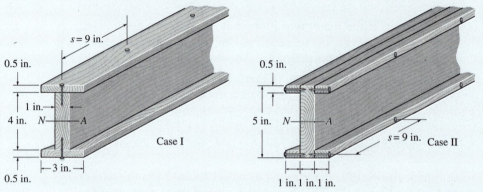

Fig. 13–18

**Solution**

Since the geometry is the same in both cases, the moment of inertia about the neutral axis is

$$I = \frac{1}{12}(3 \text{ in.})(5 \text{ in.})^3 - 2\left[\frac{1}{12}(1 \text{ in.})(4 \text{ in.})^3\right] = 20.58 \text{ in}^4$$

*Case I.*   For this design a single row of nails holds the top or bottom flange onto the web. For one of these flanges,

$$Q = \overline{y}'A' = [2.25 \text{ in.}](3 \text{ in.}(0.5 \text{ in.})) = 3.375 \text{ in}^3$$

so that

$$q = \frac{VQ}{I}$$

$$\frac{40 \text{ lb}}{9 \text{ in.}} = \frac{V(3.375 \text{ in}^3)}{20.58 \text{ in}^4}$$

$$V = 27.1 \text{ lb} \qquad\qquad Ans.$$

*Case II.*   Here a single row of nails holds one of the side boards onto the web. Thus,

$$Q = \overline{y}'A' = [2.25 \text{ in.}](1 \text{ in.}(0.5 \text{ in.})) = 1.125 \text{ in}^3$$

$$q = \frac{VQ}{I}$$

$$\frac{40 \text{ lb}}{9 \text{ in.}} = \frac{V(1.125 \text{ in}^3)}{20.58 \text{ in}^4}$$

$$V = 81.3 \text{ lb} \qquad\qquad Ans.$$

# PROBLEMS

**13-23.** The beam is constructed from three boards. Determine the maximum shear $V$ that it can sustain if the allowable shear stress for the wood is $\tau_{allow} = 400$ psi. What is the required spacing $s$ of the nails if each nail can resist a shear force of 400 lb?

**13-25.** A beam is constructed from five boards bolted together as shown. Determine the maximum shear force developed in each bolt if the bolts are spaced $s = 250$ mm apart and the applied shear is $V = 35$ kN.

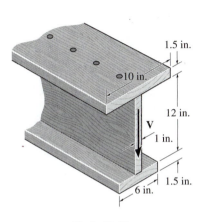

Prob. 13-23

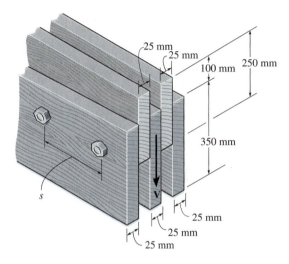

Prob. 13-25

**\*13-24.** The beam is constructed from two boards fastened together at the top and bottom with two rows of nails spaced every 6 in. If an internal shear force of $V = 600$ lb is applied to the boards, determine the shear force resisted by each nail.

**13-26.** The box beam is made from four pieces of plastic that are glued together as shown. If the shear $V = 2$ kip, determine the shear stress resisted by the seam at each of the glued joints.

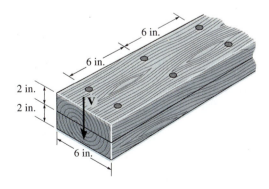

Prob. 13-24

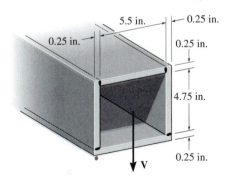

Prob. 13-26

**13-27.** A beam is constructed from three boards bolted together as shown. Determine the shear force developed in each bolt if the bolts are spaced $s = 250$ mm apart and the applied shear is $V = 35$ kN.

**13-29.** The beam is fabricated from two equivalent structural tees and two plates. Each plate has a height of 6 in. and a thickness of 0.5 in. If the bolts are spaced at $s = 8$ in., determine the maximum shear force **V** that can be applied to the cross section. Each bolt can resist a shear force of 15 kip.

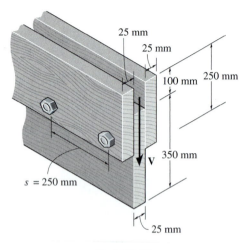

Prob. 13–27

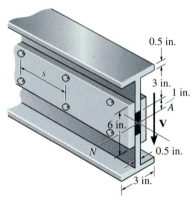

Prob. 13–29

**\*13-28.** The beam is fabricated from two equivalent structural tees and two plates. Each plate has a height of 6 in. and a thickness of 0.5 in. If a shear of $V = 50$ kip is applied to the cross section, determine the maximum spacing of the bolts. Each bolt can resist a shear force of 15 kip.

**13-30.** The beam is made from three polystyrene strips that are glued together as shown. If the glue has a shear strength of 80 kPa, determine the maximum load $P$ that can be applied without causing the glue to lose its bond.

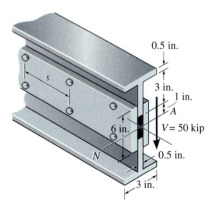

Prob. 13–28

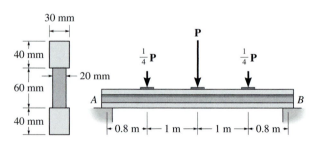

Prob. 13–30

## CHAPTER REVIEW

- *Shear Formula.* Transverse shear stress in beams is determined indirectly by using the flexure formula and the relationship between moment and shear ($V = dM/dx$). The result is a shear formula $\tau = VQ/It$. In particular, the value for $Q$ is the moment of the area $A'$ about the neutral axis. This area is the portion of the cross-sectional area that is "held on" to the beam above the thickness $t$ where $\tau$ is to be determined. If the beam has a rectangular cross section then the shear stress will vary parabolically, obtaining a maximum value at the neutral axis.

- *Shear Flow.* Fasteners, glue, or weld are used to connect the composite parts of a "built-up" section. The strength of these fasteners is determined from the shear flow, or force per unit length, that must be carried by the beam. It is $q = VQ/I$.

# REVIEW PROBLEMS

**13-31.** The beam is subjected to the loading shown, where $P = 7$ kN. Determine the average shear stress developed in the nails within region $AB$ of the beam. The nails are located on each side of the beam and are spaced 100 mm apart. Each nail has a diameter of 5 mm.

**\*13-32.** The beam is constructed from four boards which are nailed together. If the nails are on both sides of the beam and each can resist a shear of 3 kN, determine the maximum load $P$ that can be applied to the end of the beam.

**13-33.** Show that the maximum shear stress in the shaft, which has a cross section of radius $r$ and area $A = \pi r^2$, is $\tau_{max} = 4V/3A$.

Prob. 13–33

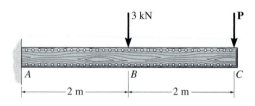

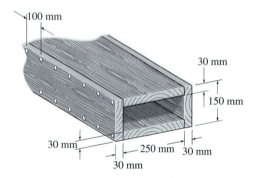

**13-34.** Develop an expression for the average vertical component of shear stress acting on the horizontal plane through the shaft, located a distance $y$ from the neutral axis.

Prob. 13–34

Probs. 13–31/32

**13-35.** If the applied shear force $V = 18$ kip, determine the maximum shear stress in the member.

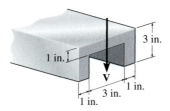

Prob. 13–35

*13-36. The box beam is constructed from four boards that are fastened together using nails spaced along the beam every 2 in. If each nail can resist a shear force of 50 lb, determine the greatest shear force $V$ that can be applied to the beam without causing failure of the nails.

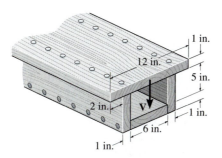

Prob. 13–36

**13-37.** The beam supports a vertical shear of $V = 7$ kip. Determine the resultant force this develops in segment $AB$ of the beam.

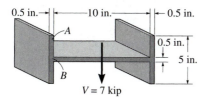

Prob. 13–37

**13-38.** The member consists of two triangular plastic strips bonded together along $AB$. If the glue can support an allowable shear stress of $\tau_{allow} = 600$ psi, determine the maximum vertical shear $V$ that can be applied to the member based on the strength of the glue.

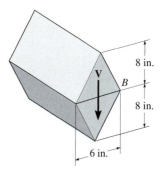

Prob. 13–38

**13-39.** If the pipe is subjected to a shear of $V = 15$ kip, determine the maximum shear stress in the pipe.

Prob. 13–39

The offset column supporting this sign is subjected to the combined loadings of normal force, shear force, bending moment, and torsion.

# Combined Loadings

- To analyze the stresses in thin-walled pressure vessels.
- To develop methods to analyze the stresses in members subject to combined loadings (e.g. tension or compression, shear, torsion, bending moments).

## 14.1  Thin-Walled Pressure Vessels

Cylindrical or spherical vessels are commonly used in industry to serve as boilers or tanks. When under pressure, the material of which they are made is subjected to a loading from all directions. Although this is the case, the vessel can be analyzed in a simpler manner provided it has a thin wall. In general, "***thin wall***" refers to a vessel having an inner-radius-to-wall-thickness ratio of 10 or more ($r/t \geq 10$) Specifically, when $r/t = 10$ the results of a thin-wall analysis will predict a stress that is approximately 4% less than the actual maximum stress in the vessel. For larger $r/t$ ratios this error will be even smaller.

When the vessel wall is "thin," the stress distribution throughout its thickness will not vary significantly, and so we will assume that it is *uniform or constant*. Using this assumption, we will now analyze the state of stress in thin-walled cylindrical and spherical pressure vessels. In both cases, the pressure in the vessel is understood to be the *gauge pressure*, since it measures the pressure *above* atmospheric pressure, which is assumed to exist both inside and outside the vessel's wall.

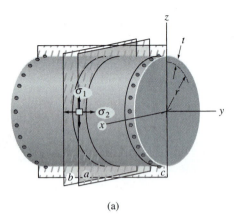

(a)

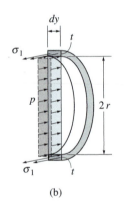

(b)

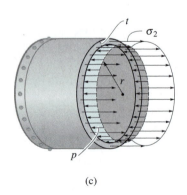

(c)

**Fig. 14–1**

**Cylindrical Vessels.** Consider the cylindrical vessel having a wall thickness $t$ and inner radius $r$ as shown in Fig. 14–1a. A gauge pressure $p$ is developed within the vessel by a contained gas or fluid, which is assumed to have negligible weight. Due to the uniformity of this loading, an element of the vessel that is sufficiently removed from the ends and oriented, as shown, is subjected to normal stresses $\sigma_1$ in the **circumferential or hoop direction** and $\sigma_2$ in the **longitudinal or axial direction**. Both of these stress components exert tension on the material. We wish to determine the magnitude of each of these components in terms of the vessel's geometry and the internal pressure. To do this requires using the method of sections and applying the equations of force equilibrium.

For the hoop stress, consider the vessel to be sectioned by planes $a$, $b$, and $c$. A free-body diagram of the back segment along with the contained gas or fluid is shown in Fig. 14–1b. Here only the loadings in the $x$ direction are shown. These loadings are developed by the uniform hoop stress $\sigma_1$, acting throughout the vessel's wall, and the pressure acting on the vertical face of the sectioned gas or fluid. For equilibrium in the $x$ direction, we require

$$\Sigma F_x = 0; \qquad 2[\sigma_1(t\,dy)] - p(2r\,dy) = 0$$

$$\boxed{\sigma_1 = \frac{pr}{t}} \qquad (14\text{–}1)$$

In order to obtain the longitudinal stress $\sigma_2$, we will consider the left portion of section $b$ of the cylinder, Fig. 14–1a. As shown in Fig. 14–1c, $\sigma_2$ acts uniformly throughout the wall, and $p$ acts on the section of gas or fluid. Since the mean radius is approximately equal to the vessel's inner radius, equilibrium in the $y$ direction requires

$$\Sigma F_y = 0; \qquad \sigma_2(2\pi r t) - p(\pi r^2) = 0$$

$$\boxed{\sigma_2 = \frac{pr}{2t}} \qquad (14\text{–}2)$$

In the above equations,

$\sigma_1$, $\sigma_2$ = the normal stress in the hoop and longitudinal directions, respectively. Each is assumed to be *constant* throughout the wall of the cylinder, and each subjects the material to tension

$p$ = the internal gauge pressure developed by the contained gas or fluid

$r$ = the inner radius of the cylinder

$t$ = the thickness of the wall ($r/t \geq 10$)

Comparing Eqs. 14–1 and 14–2, it should be noted that the hoop or circumferential stress is twice as large as the longitudinal or axial stress. Consequently, when fabricating cylindrical pressure vessels from rolled-formed plates, the longitudinal joints must be designed to carry twice as much stress as the circumferential joints.

**Spherical Vessels.** We can analyze a spherical pressure vessel in a similar manner. For example, consider the vessel to have a wall thickness $t$ and inner radius $r$ and to be subjected to an internal gauge pressure $p$, Fig. 14–2a. If the vessel is sectioned in half using section $a$, the resulting free-body diagram is shown in Fig. 14–2b. Like the cylinder, equilibrium in the $y$ direction requires

Shown is the barrel of a shotgun which was clogged with debris just before firing. Gas pressure from the charge increased the circumferential stress within the barrel enough to cause the rupture.

$$\Sigma F_y = 0; \qquad \sigma_2(2\pi rt) - p(\pi r^2) = 0$$

$$\sigma_2 = \frac{pr}{2t} \qquad (14\text{–}3)$$

By comparison, this is the *same result* as that obtained for the longitudinal stress in the cylindrical pressure vessel. Furthermore, from the analysis, this stress will be the same *regardless* of the orientation of the hemispheric free-body diagram. Consequently, an element of the material is subjected to the state of stress shown in Fig. 14–2a.

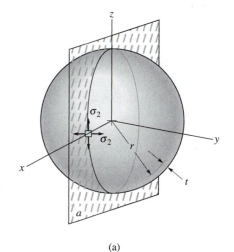

(a)

The above analysis indicates that an element of material taken from either a cylindrical or a spherical pressure vessel is subjected to **biaxial stress**, i.e., normal stress existing in only two directions. Actually, material of the vessel is also subjected to a **radial stress**, $\sigma_3$, which acts along a radial line. This stress has a maximum value equal to the pressure $p$ at the interior wall and decreases through the wall to zero at the exterior surface of the vessel, since the gauge pressure there is zero. For thin-walled vessels, however, we will *ignore* the radial-stress component, since our limiting assumption of $r/t = 10$ results in $\sigma_2$ and $\sigma_1$ being, respectively, 5 and 10 times *higher* than the maximum radial stress, $(\sigma_3)_{max} = p$. Lastly, realize that the above formulas should be used only for vessels subjected to an internal gauge pressure. If the vessel is subjected to an external pressure, the compressive stress developed within the thin wall may cause the vessel to become unstable, and collapse may occur by buckling.

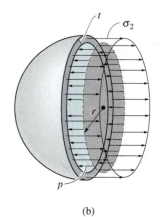

(b)

Fig. 14–2

# EXAMPLE 14.1

A cylindrical pressure vessel has an inner diameter of 4 ft and a thickness of $\frac{1}{2}$ in. Determine the maximum internal pressure it can sustain so that neither its circumferential nor its longitudinal stress component exceeds 20 ksi. Under the same conditions, what is the maximum internal pressure that a similar-size spherical vessel can sustain?

## Solution

*Cylindrical Pressure Vessel.* The maximum stress occurs in the circumferential direction. From Eq. 14–1 we have

$$\sigma_1 = \frac{pr}{t}; \qquad\qquad 20 \text{ kip/in}^2 = \frac{p(24 \text{ in.})}{\frac{1}{2}\text{ in.}}$$

$$p = 417 \text{ psi} \qquad\qquad Ans.$$

Note that when this pressure is reached, from Eq. 14–2, the stress in the longitudinal direction will be $\sigma_2 = \frac{1}{2}(20 \text{ ksi}) = 10$ ksi. Furthermore, the *maximum stress* in the *radial direction* occurs on the material at the inner wall of the vessel and is $(\sigma_3)_{max} = p = 417$ psi. This value is 48 times smaller than the circumferential stress (20 ksi), and as stated earlier, its effects will be neglected.

*Spherical Vessel.* Here the maximum stress occurs in any two perpendicular directions on an element of the vessel, Fig. 14–2a. From Eq. 14–3, we have

$$\sigma_2 = \frac{pr}{2t}; \qquad\qquad 20 \text{ kip/in}^2 = \frac{p(24 \text{ in.})}{2(\frac{1}{2}\text{ in.})}$$

$$p = 833 \text{ psi} \qquad\qquad Ans.$$

Although it is more difficult to fabricate, the spherical pressure vessel will carry twice as much internal pressure as a cylindrical vessel.

# PROBLEMS

**14-1.** A spherical gas tank has an inner radius of $r = 1.5$ m. If it is subjected to an internal pressure of $p = 300$ kPa, determine its required thickness if the maximum normal stress is not to exceed 12 MPa.

**14-2.** The open-ended polyvinyl chloride pipe has an inner diameter of 4 in. and thickness of 0.2 in. If it carries flowing water at 60 psi pressure, determine the state of stress in the walls of the pipe.

**14-3.** If the flow of water within the pipe in Prob. 14–2 is stopped due to the closing of a valve, determine the state of stress in the walls of the pipe. Neglect the weight of the water. Assume the supports only exert vertical forces on the pipe.

**Probs. 14–1/2/3**

***14-4.** The open-ended pipe has a wall thickness of 2 mm and an internal diameter of 40 mm. Calculate the pressure that ice exerted on the interior wall of the pipe to cause it to burst in the manner shown. The maximum stress that the material can support at freezing temperatures is $\sigma_{max} = 360$ MPa. Show the stress acting on a small element of material just before the pipe fails.

**Prob. 14–4**

**14-5.** Two hemispheres having an inner radius of 2 ft and wall thickness of 0.25 in. are fitted together, and the inside gauge pressure is reduced to $-10$ psi. If the coefficient of static friction is $\mu_s = 0.5$ between the hemispheres, determine (a) the torque $T$ needed to initiate the rotation of the top hemisphere relative to the bottom one, (b) the vertical force needed to pull the top hemisphere off the bottom one, and (c) the horizontal force needed to slide the top hemisphere off the bottom one.

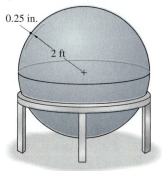

0.25 in.

2 ft

**Prob. 14–5**

**14-6.** The A-36-steel band is 2 in. wide and is secured around the smooth rigid cylinder. If the bolts are tightened so that the tension in them is 400 lb, determine the normal stress in the band, the pressure exerted on the cylinder, and the distance half the band stretches.

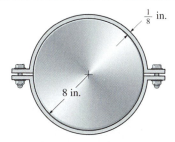

$\frac{1}{8}$ in.

8 in.

**Prob. 14–6**

**14-7.** A pressure-vessel head is fabricated by gluing the circular plate to the end of the vessel as shown. If the vessel sustains an internal pressure of 450 kPa, determine the average shear stress in the glue and the state of stress in the wall of the vessel.

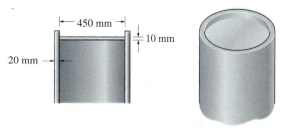

450 mm

10 mm

20 mm

**Prob. 14–7**

***14-8.** A wood pipe having an inner diameter of 3 ft is bound together using steel hoops having a cross-sectional area of 0.2 in². If the allowable stress for the hoops is $\sigma_{allow} = 12$ ksi, determine their maximum spacing $s$ along the section of pipe so that the pipe can resist an internal gauge pressure of 4 psi. Assume each hoop supports the pressure loading acting along the length $s$ of the pipe.

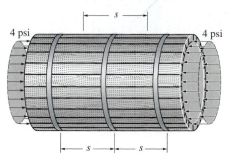

4 psi

4 psi

$s$

$s$    $s$

**Prob. 14–8**

**14-9.** The ring, having the dimensions shown, is placed over a flexible membrane which is pumped up with a pressure $p$. Determine the change in the internal radius of the ring after this pressure is applied. The modulus of elasticity for the ring is $E$.

**14-10.** A boiler is constructed of 8-mm steel plates that are fastened together at their ends using a butt joint consisting of two 8-mm cover plates and rivets having a diameter of 10 mm and spaced 50 mm apart as shown. If the steam pressure in the boiler is 1.35 MPa, determine (a) the circumferential stress in the boiler's plate apart from the seam, (b) the circumferential stress in the outer cover plate along the rivet line $a$–$a$, and (c) the shear stress in the rivets.

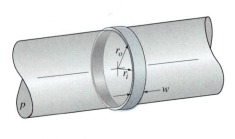

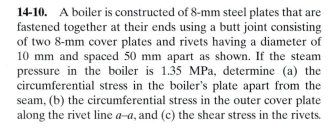

Prob. 14-9

Prob. 14-10

## 14.2   State of Stress Caused by Combined Loadings

This chimney is subjected to the combined loading of wind and weight. It is important to investigate the tensile stress in the chimney since masonry is weak in tension.

In previous chapters we developed methods for determining the stress distributions in a member subjected to either an internal axial force, a shear force, a bending moment, or a torsional moment. Most often, however, the cross section of a member is subjected to several of these types of loadings *simultaneously*, and as a result, the method of superposition, if it applies, can be used to determine the *resultant* stress distribution caused by the loads. For application, the stress distribution due to *each loading* is first determined, and then these distributions are superimposed to determine the resultant stress distribution. As stated in Sec. 10.3, the principle of superposition can be used for this purpose provided a *linear relationship* exists between the *stress* and the *loads*. Also, the geometry of the member should *not* undergo *significant change* when the loads are applied. This is necessary in order to ensure that the stress produced by one load is not related to the stress produced by any other load. The discussion will be confined to meet these two criteria.

## PROCEDURE FOR ANALYSIS

The following procedure provides a general means for establishing the normal and shear stress components at a point in a member when the member is subjected to several different types of loadings simultaneously. It is assumed that the material is homogeneous and behaves in a linear-elastic manner. Also, Saint-Venant's principle requires that the point where the stress is to be determined is far removed from any discontinuities in the cross section or points of applied load.

### Internal Loading.

- Section the member perpendicular to its axis at the point where the stress is to be determined and obtain the resultant internal normal and shear force components and the bending and torsional moment components.
- The force components should act through the *centroid* of the cross section, and the moment components should be computed about *centroidal axes*, which represent the principal axes of inertia for the cross section.

### Average Normal Stress.

- Compute the stress component associated with *each* internal loading. For each case, represent the effect either as a distribution of stress acting over the entire cross-sectional area, or show the stress on an element of the material located at a specified point on the cross section.

NORMAL FORCE.   The internal normal force is developed by a uniform normal-stress distribution determined from $\sigma = P/A$.

SHEAR FORCE.   The internal shear force in a member that is subjected to bending is developed by a shear-stress distribution determined from the shear formula, $\tau = VQ/It$. Special care, however, must be exercised when applying this equation, as noted in Sec. 13.3.

BENDING MOMENT.   For *straight members* the internal bending moment is developed by a normal-stress distribution that varies linearly from zero at the neutral axis to a maximum at the outer boundary of the member. The stress distribution is determined from the flexure formula, $\sigma = -My/I$.

TORSIONAL MOMENT.   For circular shafts and tubes the internal torsional moment is developed by a shear-stress distribution that varies linearly from the central axis of the shaft to a maximum at the shaft's outer boundary. The shear-stress distribution is determined from the torsional formula, $\tau = T\rho/J$.

THIN-WALLED PRESSURE VESSELS.   If the vessel is a thin-walled cylinder, the internal pressure $p$ will cause a biaxial state of stress in the material such that the hoop or circumferential stress component is $\sigma_1 = pr/t$ and the longitudinal stress component is $\sigma_2 = pr/2t$. If the vessel is a thin-walled sphere, then the biaxial state of stress is represented by two equivalent components, each having a magnitude of $\sigma_2 = pr/2t$.

### Superposition.

- Once the normal and shear stress components for each loading have been calculated, use the principle of superposition and determine the resultant normal and shear stress components.
- Represent the results on an element of material located at the point, or show the results as a distribution of stress acting over the member's cross-sectional area.

Problems in this section, which involve combined loadings, serve as a basic *review* of the application of many of the important stress equations mentioned above. A thorough understanding of how these equations are applied, as indicated in the previous chapters, is necessary if one is to successfully solve the problems at the end of this section. The following examples should be carefully studied before proceeding to solve the problems.

**E X A M P L E 14.2**

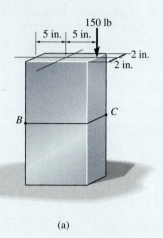

150 lb

5 in. | 5 in.

2 in.
2 in.

B

C

(a)

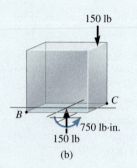

150 lb

C

B

750 lb·in.

150 lb

(b)

**Fig. 14–3**

A force of 150 lb is applied to the edge of the member shown in Fig. 14–3a. Neglect the weight of the member and determine the state of stress at points $B$ and $C$.

**Solution**

*Internal Loadings.* The member is sectioned through $B$ and $C$. For equilibrium at the section there must be an axial force of 150 lb acting through the centroid and a bending moment of 750 lb · in. about the centroidal or principal axis, Fig. 14–3b.

*Stress Components.*

**NORMAL FORCE.** The uniform normal-stress distribution due to the normal force is shown in Fig. 14–3c. Here

$$\sigma = \frac{P}{A} = \frac{150 \text{ lb}}{(10 \text{ in.})(4 \text{ in.})} = 3.75 \text{ psi}$$

**BENDING MOMENT.** The normal-stress distribution due to the bending moment is shown in Fig. 14–3d. The maximum stress is

$$\sigma_{\max} = \frac{M_C}{I} = \frac{750 \text{ lb} \cdot \text{in.}(5 \text{ in.})}{\frac{1}{12}(4 \text{ in.})(10 \text{ in.})^3} = 11.25 \text{ psi}$$

*Superposition.* If the above normal-stress distributions are added algebraically, the resultant stress distribution is shown in Fig. 14–3e. Although it is not needed here, the location of the line of zero stress can be determined by proportional triangles; i.e.,

$$\frac{7.5 \text{ psi}}{x} = \frac{15 \text{ psi}}{10 \text{ in.} - x}$$

$$x = 3.33 \text{ in.}$$

Elements of material at $B$ and $C$ are subjected only to normal or *uniaxial stress* as shown in Fig. 14–3f and 14–3g. Hence,

$$\sigma_B = 7.5 \text{ psi} \quad \text{(tension)} \qquad \qquad \textit{Ans.}$$

$$\sigma_C = 15 \text{ psi} \quad \text{(compression)} \qquad \textit{Ans.}$$

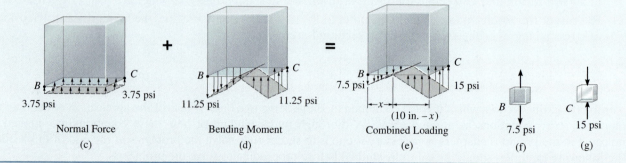

| Normal Force | Bending Moment | Combined Loading | | |
| --- | --- | --- | --- | --- |
| (c) | (d) | (e) | (f) | (g) |

## EXAMPLE 14.3

The tank in Fig. 14–4a has an inner radius of 24 in. and a thickness of 0.5 in. It is filled to the top with water having a specific weight of $\gamma_w = 62.4 \text{ lb/ft}^3$. If it is made of steel having a specific weight of $\gamma_{st} = 490 \text{ lb/ft}^3$, determine the state of stress at point $A$. The tank is open at the top.

(a)

### Solution

*Internal Loadings.* The free-body diagram of the section of both the tank and the water above point A is shown in Fig. 14–4b. Notice that the weight of the water is supported by the water surface just *below* the section, *not* by the walls of the tank. In the vertical direction, the walls simply hold up the weight of the tank. This weight is

$$W_{st} = \gamma_{st} V_{st} = (490 \text{ lb/ft}^3)\left[ \pi\left(\frac{24.5}{12}\text{ft}\right)^2 - \pi\left(\frac{24}{12}\text{ft}\right)^2 \right](3 \text{ ft}) = 777.7 \text{ lb}$$

The stress in the circumferential direction is developed by the water pressure at level $A$. To obtain this pressure we must use *Pascal's law*, which states that the pressure at a point located a depth $z$ in the water is $p = \gamma_w z$. Consequently, the pressure on the tank at level $A$ is

$$p = \gamma_w z = (62.4 \text{ lb/ft}^3)(3 \text{ ft}) = 187.2 \text{ lb/ft}^2 = 1.30 \text{ psi}$$

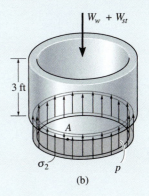

(b)

### Stress Components.

**CIRCUMFERENTIAL STRESS.** Applying Eq. 14–1, using the inner radius $r = 24$ in., we have

$$\sigma_1 = \frac{pr}{t} = \frac{1.30 \text{ lb/in}^2 (24 \text{ in.})}{0.5 \text{ in.}} = 62.4 \text{ psi} \qquad Ans.$$

**LONGITUDINAL STRESS.** Since the weight of the tank is supported uniformly by the walls, we have

$$\sigma_2 = \frac{W_{st}}{A_{st}} = \frac{777.7 \text{ lb}}{\pi[(24.5 \text{ in.})^2 - (24 \text{ in.})^2]} = 10.2 \text{ psi} \qquad Ans.$$

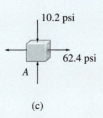

(c)

Note that Eq. 14–2, $\sigma_2 = pr/2t$, does *not apply* here, since the tank is open at the top and therefore, as stated previously, the water cannot develop a loading on the walls in the longitudinal direction.

Point $A$ is therefore subjected to the biaxial stress shown in Fig. 14–4c.

**Fig. 14–4**

EXAMPLE 14.4

The member shown in Fig. 14–5a has a rectangular cross section. Determine the state of stress that the loading produces at point C.

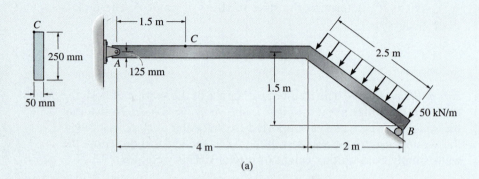

(a)

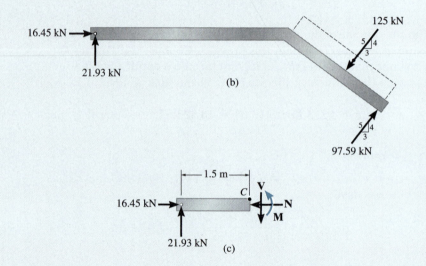

(b)

(c)

Fig. 14–5

## Solution

**Internal Loadings.** The support reactions on the member have been determined and are shown in Fig. 14–5b. If the left segment AC of the member is considered, Fig. 14–5c, the resultant internal loadings at the section consist of a normal force, a shear force, and a bending moment. Solving,

$$N = 16.45 \text{ kN} \qquad V = 21.93 \text{ kN} \qquad M = 32.89 \text{ kN} \cdot \text{m}$$

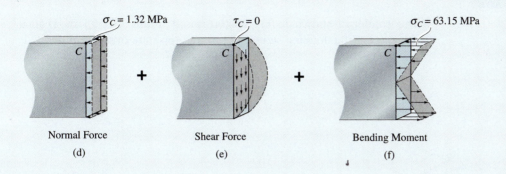

Normal Force
(d)

Shear Force
(e)

Bending Moment
(f)

## Stress Components.

**NORMAL FORCE.** The uniform normal-stress distribution acting over the cross section is produced by the normal force, Fig. 14–5d. At point $C$,

$$\sigma_C = \frac{P}{A} = \frac{16.45 \text{ kN}}{(0.050 \text{ m}) (0.250 \text{ m})} = 1.32 \text{ MPa}$$

**SHEAR FORCE.** Here the area $A' = 0$, since point $C$ is located at the top of the member. Thus $Q = \bar{y}' A' = 0$ and for $C$, Fig. 14–5e, the shear stress

$$\tau_C = 0$$

**BENDING MOMENT.** Point $C$ is located at $y = c = 125$ mm from the neutral axis, so the normal stress at $C$, Fig. 14–5f, is

$$\sigma_C = \frac{Mc}{I} = \frac{(32.89 \text{ kN} \cdot \text{m}) (0.125 \text{ m})}{\frac{1}{12}(0.050 \text{ m}) (0.250)^3} = 63.15 \text{ MPa}$$

**Superposition.** The shear stress is zero. Adding the normal stresses determined above gives a compressive stress at $C$ having a value of

$$\sigma_C = 1.32 \text{ MPa} + 63.15 \text{ MPa} = 64.5 \text{ MPa} \qquad \textit{Ans.}$$

This result, acting on an element at $C$, is shown in Fig. 14–5g.

64.5 MPa

(g)

The solid rod shown in Fig. 14–6*a* has a radius of 0.75 in. If it is subjected to the loading shown, determine the state of stress at point *A*.

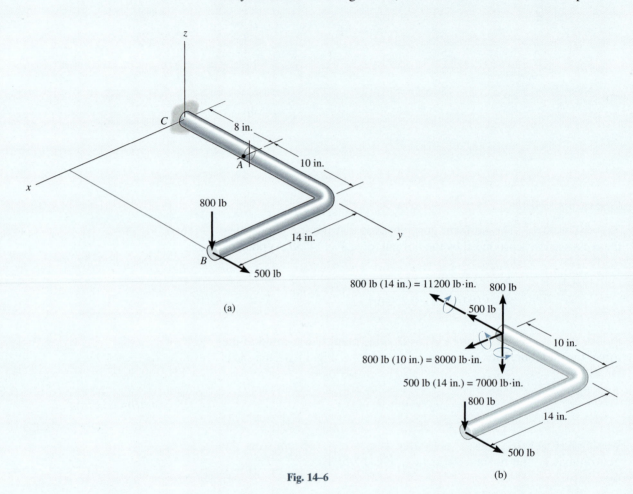

Fig. 14–6

### Solution

*Internal Loadings.* The rod is sectioned through point *A*. Using the free-body diagram of segment *AB*, Fig. 14–6*b*, the resultant internal loadings can be determined from the six equations of equilibrium. Verify these results. The normal force (500 lb) and shear force (800 lb) must act through the centroid of the cross section and the bending-moment components (8000 lb · in. and 7000 lb · in.) are applied about centroidal (principal) axes. In order to better "visualize" the stress distributions due to each of these loadings, we will consider the *equal but opposite resultants* acting on *AC*, Fig. 14–6*c*.

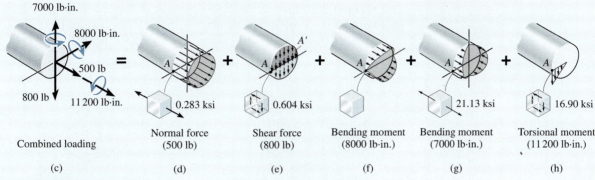

| Combined loading | Normal force (500 lb) | Shear force (800 lb) | Bending moment (8000 lb·in.) | Bending moment (7000 lb·in.) | Torsional moment (11 200 lb·in.) |

(c)      (d)      (e)      (f)      (g)      (h)

## Stress Components.

**NORMAL FORCE.** The normal-stress distribution is shown in Fig. 14–6d. For point $A$, we have

$$\sigma_A = \frac{P}{A} = \frac{500 \text{ lb}}{\pi(0.75 \text{ in.})^2} = 283 \text{ psi} = 0.283 \text{ ksi}$$

**SHEAR FORCE.** The shear-stress distribution is shown in Fig. 14–6e. For point $A$, $Q$ is determined from the shaded *semicircular* area. Using the table in Appendix C, we have

$$Q = \bar{y}'A' = \frac{4(0.75 \text{ in.})}{3\pi}\left[\frac{1}{2}\pi\left(0.75 \text{ in.}\right)^2\right] = 0.2813 \text{ in}^3$$

so that

$$\tau_A = \frac{VQ}{It} = \frac{800 \text{ lb}(0.2813 \text{ in}^3)}{[\frac{1}{2}\pi(0.75 \text{ in.})^4]2(0.75 \text{ in.})} = 604 \text{ psi} = 0.604 \text{ ksi}$$

**BENDING MOMENTS.** For the 8000-lb · in. component, point $A$ lies on the neutral axis, Fig. 14–6f, so the normal stress is

$$\sigma_A = 0$$

For the 7000-lb · in. moment, $c = 0.75$ in., so the normal stress at point $A$, Fig. 14–6g, is

$$\sigma_A = \frac{Mc}{I} = \frac{7000 \text{ lb} \cdot \text{in.}(0.75 \text{ in.})}{\frac{1}{4}\pi(0.75 \text{ in.})^4} = 21\,126 \text{ psi} = 21.13 \text{ ksi}$$

**TORSIONAL MOMENT.** At point $A$, $\rho_A = c = 0.75$ in., Fig. 14–6h. Thus the shear stress is

$$\tau_A = \frac{Tc}{J} = \frac{11\,200 \text{ lb} \cdot \text{in.}(0.75 \text{ in.})}{\frac{1}{2}\pi(0.75 \text{ in.})^4} = 16\,901 \text{ psi} = 16.90 \text{ ksi}$$

*Superposition.* When the above results are superimposed, it is seen that an element of material at $A$ is subjected to both normal and shear stress components, Fig. 14–6i.

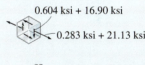

0.604 ksi + 16.90 ksi

0.283 ksi + 21.13 ksi

or

17.5 ksi

21.4 ksi

(i)

**E X A M P L E   14.6**

The rectangular block of negligible weight in Fig. 14–7a is subjected to a vertical force of 40 kN, which is applied to its corner. Determine the normal-stress distribution acting on a section through $ABCD$.

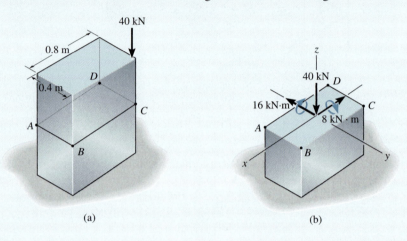

(a)                                    (b)

**Fig. 14–7**

### Solution

*Internal Loadings.* If we consider the equilibrium of the bottom segment of the block, Fig. 14–7b, it is seen that the 40-kN force must act through the centroid of the cross section and *two* bending-moment components must also act about the centroidal or principal axes of inertia for the section. Verify these results.

*Stress Components.*

**NORMAL FORCE.** The uniform normal-stress distribution is shown in Fig. 14–7c. We have

$$\sigma = \frac{P}{A} = \frac{40 \text{ kN}}{(0.8 \text{ m})(0.4 \text{ m})} = 125 \text{ kPa}$$

**BENDING MOMENTS.** The normal-stress distribution for the 8-kN·m moment is shown in Fig. 14–7d. The maximum stress is

$$\sigma_{max} = \frac{M_x c_y}{I_x} = \frac{8 \text{ kN} \cdot \text{m}(0.2 \text{ m})}{\frac{1}{12}(0.8 \text{ m})(0.4 \text{ m})^3} = 375 \text{ kPa}$$

Likewise, for the 16-kN·m moment, Fig. 14–7e, the maximum normal stress is

$$\sigma_{max} = \frac{M_y c_x}{I_y} = \frac{16 \text{ kN} \cdot \text{m}(0.4 \text{ m})}{\frac{1}{12}(0.4 \text{ m})(0.8 \text{ m})^3} = 375 \text{ kPa}$$

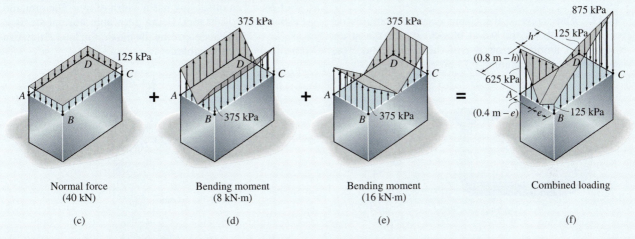

Normal force
(40 kN)

(c)

Bending moment
(8 kN·m)

(d)

Bending moment
(16 kN·m)

(e)

Combined loading

(f)

**Superposition.** The normal stress at each corner point can be determined by algebraic addition. Assuming that tensile stress is positive, we have

$$\sigma_A = -125 \text{ kPa} + 375 \text{ kPa} + 375 \text{ kPa} = 625 \text{ kPa}$$
$$\sigma_B = -125 \text{ kPa} - 375 \text{ kPa} + 375 \text{ kPa} = -125 \text{ kPa}$$
$$\sigma_C = -125 \text{ kPa} - 375 \text{ kPa} - 375 \text{ kPa} = -875 \text{ kPa}$$
$$\sigma_D = -125 \text{ kPa} + 375 \text{ kPa} - 375 \text{ kPa} = -125 \text{ kPa}$$

Since the stress distributions due to bending moment are linear, the resultant stress distribution is also linear and therefore looks like that shown in Fig. 14–7f. The line of zero stress can be located along each side by proportional triangles. From the figure we require

$$\frac{0.4 \text{ m} - e}{625 \text{ kPa}} = \frac{e}{125 \text{ kPa}}$$
$$e = 0.0667 \text{ m}$$

and

$$\frac{0.8 \text{ m} - h}{625 \text{ kPa}} = \frac{h}{125 \text{ kPa}}$$
$$h = 0.133 \text{ m}$$

# PROBLEMS

**14-11.** The screw of the clamp exerts a compressive force of 500 lb on the wood blocks. Determine the maximum normal stress developed along section $a$–$a$. The cross section there is rectangular, 0.75 in. by 0.50 in.

*****14-12.** The screw of the clamp exerts a compressive force of 500 lb on the wood blocks. Sketch the stress distribution along section $a$–$a$ of the clamp. The cross section there is rectangular, 0.75 in. by 0.50 in.

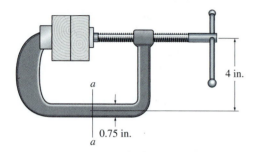

Probs. 14–11/12

**14-13.** The joint is subjected to a force of 250 lb as shown. Sketch the normal-stress distribution acting over section $a$–$a$ if the member has a rectangular cross section of width 0.5 in. and thickness 0.75 in.

**14-14.** The joint is subjected to a force of 250 lb as shown. Determine the state of stress at points $A$ and $B$, and sketch the results on differential elements located at these points. The member has a rectangular cross-sectional area of width 0.5 in. and thickness 0.75 in.

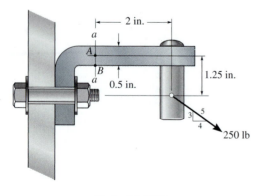

Probs. 14–13/14

**14-15.** The offset link supports the loading of $P = 30$ kN. Determine its required width $w$ if the allowable normal stress is $\sigma_{allow} = 73$ MPa. The link has a thickness of 40 mm.

*****14-16.** The offset link has a width of $w = 200$ mm and a thickness of 40 mm. If the allowable normal stress is $\sigma_{allow} = 75$ MPa, determine the maximum load $P$ that can be applied to the cables.

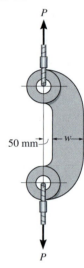

Probs. 14–15/16

**14-17.** Determine the maximum and minimum normal stress in the bracket at section $a$ when the load is applied at $x = 0$.

**14-18.** Determine the maximum and minimum normal stress in the bracket at section $a$ when the load is applied at $x = 50$ mm.

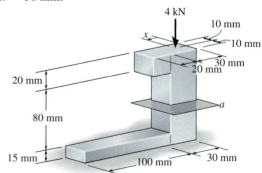

Probs. 14–17/18

**14-19.** The bent link is subjected to the cable load of $P = 500$ N. Determine its required diameter $d$ if the allowable normal stress for the material is $\sigma_{allow} = 175$ MPa. Consider the critical section to be at $A$.

***14-20.** The bent link has a diameter of $d = 15$ mm and is made of a material having an allowable normal stress of $\sigma_{allow} = 175$ MPa. Determine the maximum load $P$ it will safely support. Consider the critical section to be at $A$.

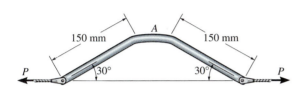

**Probs. 14–19/20**

**14-23.** The stepped support is subjected to the bearing load of 50 kN. Determine the maximum and minimum compressive stress in the material.

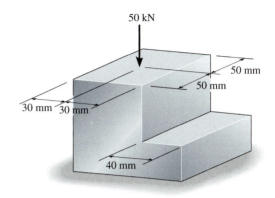

**Prob. 14–23**

**14-21.** The bar has a diameter of 40 mm. If it is subjected to a force of 800 N as shown, determine the stress components that act at point $A$ and show the results on a volume element located at this point.

**14-22.** Solve Prob. 14–21 for point $B$.

***14-24.** The pin support is made from a steel rod and has a diameter of 20 mm. Determine the stress components at points $A$ and $B$ and represent the results on a volume element located at each of these points.

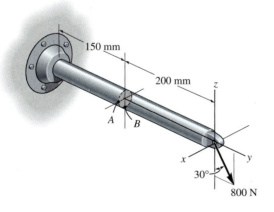

**Probs. 14–21/22**

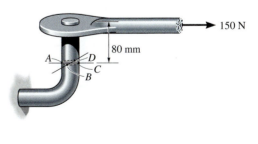

**Prob. 14–24**

**14-25.** Since concrete can support little or no tension, this problem can be avoided by using wires or rods to *prestress* the concrete once it is formed. Consider the simply supported beam shown, which has a rectangular cross section of 18 in. by 12 in. If concrete has a specific weight of 150 lb/ft³, determine the required tension in rod $AB$, which runs through the beam so that no tensile stress is developed in the concrete at its center section $a$–$a$. Neglect the size of the rod and any deflection of the beam.

**14-26.** Solve Prob. 14–25 if the rod has a diameter of 0.5 in. Use the transformed area method $E_{st} = 29(10^3)$ ksi, $E_c = 3.60(10^3)$ ksi.

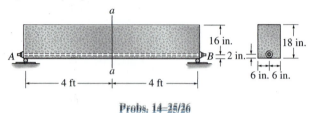

Probs. 14–25/26

**14-27.** The chimney is subjected to the uniform wind loading of $w = 150$ lb/ft and has a weight of 2200 lb/ft. If the mortar between the bricks cannot support a tensile stress, determine if the chimney is safe. Take $d = 6$ ft. The thickness of the brick wall is 1 ft.

**\*14-28.** The chimney is subjected to the uniform wind pressure of $p = 25$ lb/ft². It is to be constructed with 1-ft-thick brick walls. If the bricks and mortar have a specific weight of 145 lb/ft³, determine the smallest outer diameter $d$ of the chimney so that no tensile stress is developed in the material. The wind loading can be approximated by $w = pd$.

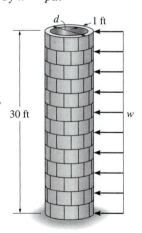

Probs. 14–27/28

**14-29.** The block is subjected to the two axial loads shown. Determine the normal stress developed at points $A$ and $B$. Neglect the weight of the block.

**14-30.** The block is subjected to the two axial loads shown. Sketch the normal stress distribution acting over the cross section at section $a$–$a$. Neglect the weight of the block.

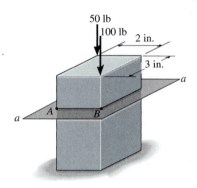

Probs. 14–29/30

**14-31.** The frame supports the distributed load shown. Determine the state of stress acting at point $D$. Show the results on a differential element located at this point.

**\*14-32.** The frame supports the distributed load shown. Determine the state of stress acting at point $E$. Show the results on a differential element located at this point.

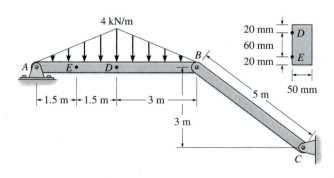

Probs. 14–31/32

**14-33.** Determine the state of stress at point $A$ when the beam is subjected to the cable force of 4 kN. Indicate the result as a differential volume element.

**14-34.** Determine the state of stress at point $B$ when the beam is subjected to the cable force of 4 kN. Indicate the result as a differential volume element.

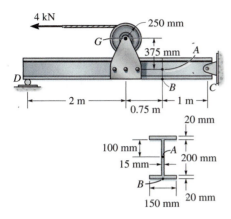

Probs. 14-33/34

**14-35.** The sign is subjected to the uniform wind loading. Determine the stress components at points $A$ and $B$ on the 100-mm-diameter supporting post. Show the results on a volume element located at each of these points.

**\*14-36.** The sign is subjected to the uniform wind loading. Determine the stress components at points $C$ and $D$ on the 100-mm-diameter supporting post. Show the results on a volume element located at each of these points..

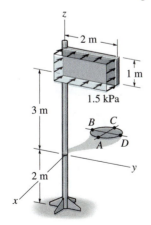

Probs. 14-35/36

**14-37.** The solid rod is subjected to the loading shown. Determine the state of stress developed in the material at point $A$, and show the results on a differential volume element at this point.

**14-38.** The solid rod is subjected to the loading shown. Determine the state of stress at point $B$, and show the results on a differential volume element located at this point.

**14-39.** The solid rod is subjected to the loading shown. Determine the state of stress at point $C$, and show the results on a differential volume element located at this point.

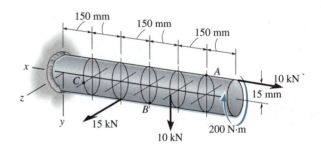

Probs. 14-37/38/39

**\*14-40.** The bent shaft is fixed in the wall at $A$. If a force $\mathbf{F}$ is applied at $B$, determine the stress components at points $D$ and $E$. Show the results on a differential element located at each of these points. Take $F = 12$ lb and $\theta = 90°$.

**14-41.** The bent shaft is fixed in the wall at $A$. If a force $\mathbf{F}$ is applied at $B$, determine the stress components at points $D$ and $E$. Show the results on a volume element located at each of these points. Take $F = 12$ lb and $\theta = 45°$.

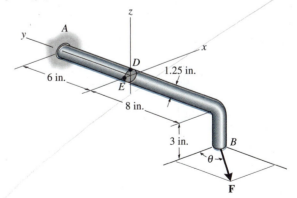

Probs. 14-40/41

**14-42.** The caster wheel supports a reactive load of 180 N. Determine the state of stress at points $A$ and $B$ on one of the two supporting leaves. Show the results on a differential volume element located at each point.

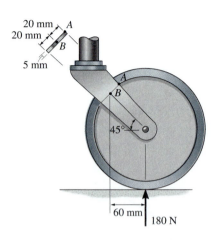

**Prob. 14-42**

**14-43.** The crane boom is subjected to the load of 500 lb. Determine the state of stress at points $A$ and $B$. Show the results on a differential volume element located at each of these points.

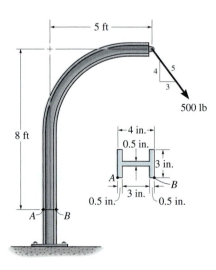

**Prob. 14-43**

*14-44.** The beam supports the loading shown. Determine the state of stress at points $E$ and $F$ at section $a$–$a$, and represent the results on a differential volume element located at each of these points.

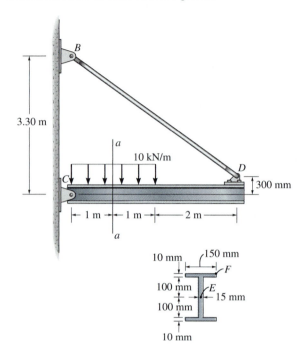

**Prob. 14-44**

**14-45.** The symmetrically loaded spreader bar is used to lift the 2000-lb tank. Determine the state of stress at points $A$ and $B$, and indicate the results on a differential volume elements.

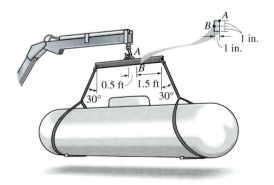

**Prob. 14-45**

# CHAPTER REVIEW

- *Pressure Vessels.*   A pressure vessel is considered to have a thin wall provided $r/t \leq 10$. For a thin-walled cylindrical vessel the circumferential or hoop stress is $\sigma_1 = pr/t$. This stress is twice as great as the longitundial stess, $\sigma_2 = pr/2t$. Thin-walled spherical vessels have the same stress within their walls in all directions so that $\sigma_1 = \sigma_2 = pr/2t$.

- *Superposition of Stress Components.*   Superposition of stress components can be used to determine the normal and shear stress at a point in a member subjected to a combined loading. To solve, it is first necessary to determine the resultant axial and shear force and the resultant torsional and bending moment at the section where the point is located. Then the resultant stress components are determined due to each of these loadings.  The results are added algebraically to find the resultant normal and shear stress components.

# REVIEW PROBLEMS

**14-46.** The beveled gear is subjected to the loads shown. Determine the stress components acting on the shaft at point $A$, and show the results on a volume element located at this point. The shaft has a diameter of 1 in. and is fixed to the wall at $C$.

**14-47.** The beveled gear is subjected to the loads shown. Determine the stress components acting on the shaft at point $B$, and show the results on a volume element located at this point. The shaft has a diameter of 1 in. and is fixed to the wall at $C$.

**\*14-48.** The frame supports a centrally applied distributed load of 1.8 kip/ft. Determine the stress components developed at points $A$ and $B$ on member $CD$ and indicate the results on a volume element located at each of these points. The pins at $C$ and $D$ are at the same location as the neutral axis for the cross section.

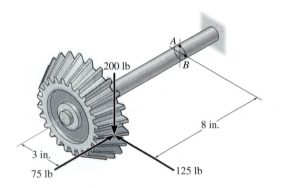

Probs. 14–46/47

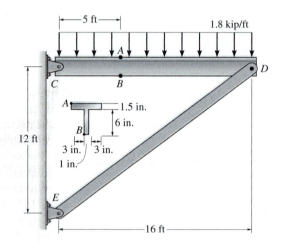

Prob. 14–48

**14-49.** The C-frame is used in a riveting machine. If the force at the ram on the dolly $D$ is 8 kN, sketch the stress distribution acting over the section $a$–$a$.

**14-50.** Plot the distribution of stress acting over the cross section $a$–$a$ of the offset link.

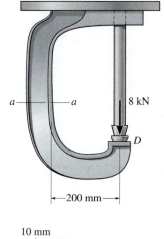

10 mm

40 mm

60 mm

10 mm

**Prob. 14–49**

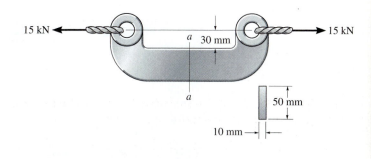

15 kN

$a$ 30 mm

$a$

15 kN

50 mm

10 mm

**Prob. 14–50**

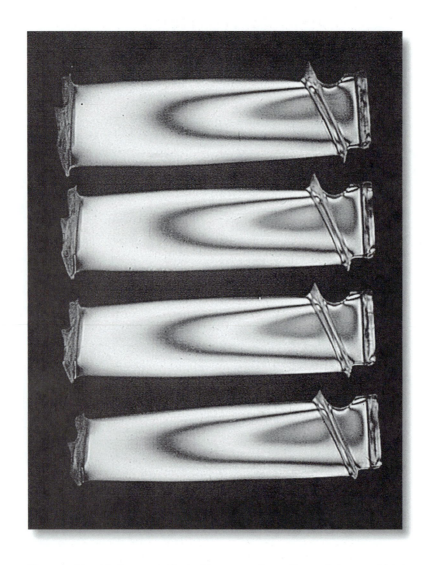

These turbine blades are subjected to a complex pattern of stress, which is illustrated by the shaded bands that appear on the blades when they are made of transparent material and viewed through polarized light. For proper design, engineers must be able to determine where and in what direction the maximum stress occurs. *(Courtesy of Measurements Group, Inc., Raleigh, North Carolina 27611, USA.)*

# Stress and Strain Transformation

- To develop the transformation of stress components from one orientation of the coordinate system to another orientation.
- To determine the principal stress and maximum in-plane shear stress at a point.
- To develop the transformation of plane strain components from one orientation of the coordinate system to another orientation.
- To determine the principal strain and maximum in-plane shear strain at a point.
- To develop Mohr's circle for analyzing stress and strain components.
- To discuss strain rosettes for strain components.
- To present the relationships between material properties, such as the elastic modulus, shear modulus, and Poisson's ratio.

## 15.1  Plane-Stress Transformation

It was shown in Sec. 8.2 that the general state of stress at a point is characterized by *six* independent normal and shear stress components, which act on the faces of an element of material located at the point, Fig. 15–1a. This state of stress, however, is not often encountered in engineering practice. Instead, engineers frequently make approximations or simplifications of the loadings on a body in order that the stress produced in a structural member or mechanical element can be analyzed in a *single plane*. When this is the case, the material is said to be subjected to *plane stress*, Fig. 15–1b. For example, if there is no load on the surface of a body, then the normal and shear stress components will be zero on the face of an element that lies on the surface. Consequently, the corresponding stress components on the opposite face will also be zero, and so the material at the point will be subjected to plane stress.

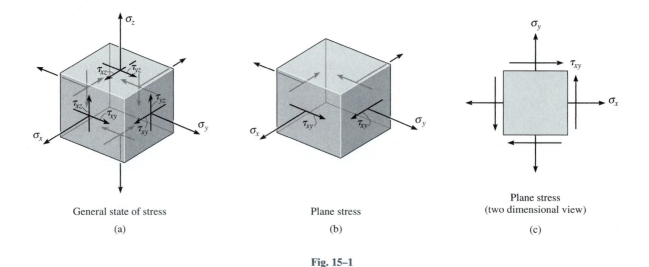

General state of stress

(a)

Plane stress

(b)

Plane stress
(two dimensional view)

(c)

**Fig. 15–1**

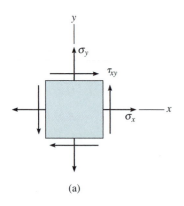

(a)

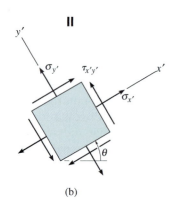

(b)

**Fig. 15–2**

The general state of ***plane stress*** at a point is therefore represented by a combination of two normal-stress components, $\sigma_x$, $\sigma_y$, and one shear-stress component, $\tau_{xy}$, which act on four faces of the element. For convenience, in this text we will view this state of stress in the $x$–$y$ plane, Fig. 15–1$c$. Realize that if the state of stress at a point is defined by the three stress components shown on the element in Fig. 15–2$a$, then an element having a different orientation, such as in Fig. 15–2$b$, will be subjected to three different stress components. In other words, ***the state of plane stress at the point is uniquely represented by three components acting on an element that has a specific orientation at the point***.

We will now show how to *transform* the stress components from one orientation of an element to an element having a different orientation. That is, if the state of stress is defined by the components $\sigma_x$, $\sigma_y$, $\tau_{xy}$, oriented along the $x$, $y$ axes, Fig. 15–2$a$, we will show how to obtain the components $\sigma_{x'}$, $\sigma_{y'}$, $\tau_{x'y'}$, oriented along the $x'$, $y'$ axes, Fig. 15–2$b$, so that they represent the *same* state of stress at the point. This is like knowing two force components, say, $\mathbf{F}_x$ and $\mathbf{F}_y$, directed along the $x$, $y$ axes, that produce a resultant force $\mathbf{F}_R$, and then trying to find the force components $\mathbf{F}_{x'}$ and $\mathbf{F}_{y'}$, directed along the $x'$, $y'$ axes, so they produce the *same* resultant. The transformation of stress components, however, is more difficult than that of force components, since for *stress*, the transformation must account for the magnitude and direction of each stress component *and* the orientation of the area upon which each component acts. For force, the transformation must account only for the force component's magnitude and direction.

## PROCEDURE FOR ANALYSIS

If the state of stress at a point is known for a given orientation of an element of material, Fig. 15–3a, then the state of stress for some other orientation, Fig. 15–3b, can be determined using the following procedure.

- To determine the normal and shear stress components $\sigma_{x'}$, $\tau_{x'y'}$ acting on the $x'$ face of the element, Fig. 15–3b, section the element in Fig. 15–3a as shown in Fig. 15–3c. If it is assumed the sectioned area is $\Delta A$, then the adjacent areas of the segment will be $\Delta A \sin \theta$ and $\Delta A \cos \theta$.

- Draw the free-body diagram of the segment, which requires showing the *forces* that act on the element. This is done by multiplying the stress components on each face by the area upon which they act.

- Apply the equations of force equilibrium in the $x'$ and $y'$ directions to obtain the two unknown stress components $\sigma_{x'}$ and $\tau_{x'y'}$.

- If $\sigma_{y'}$, acting on the $+y'$ face of the element in Fig. 15–3b, is to be determined, then it is necessary to consider a segment of the element as shown in Fig. 15–3d and follow the same procedure just described. Here, however, the shear stress $\tau_{x'y'}$ does not have to be determined if it was previously calculated, since it is complementary, that is, it has the same magnitude on each of the four faces of the element, Fig. 15–3b.

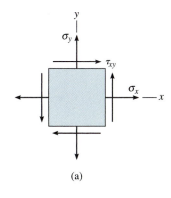

(a)

‖

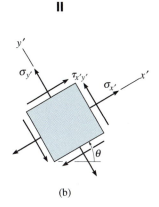

(b)

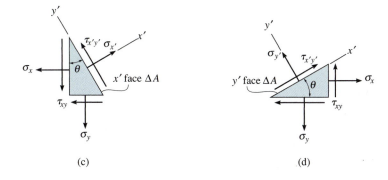

(c)　　　　(d)

**Fig. 15–3**

**E X A M P L E  15.1**

The state of plane stress at a point on the surface of the airplane fuselage is represented on the element oriented as shown in Fig. 15–4a. Represent the state of stress at the point on an element that is oriented 30° clockwise from the position shown.

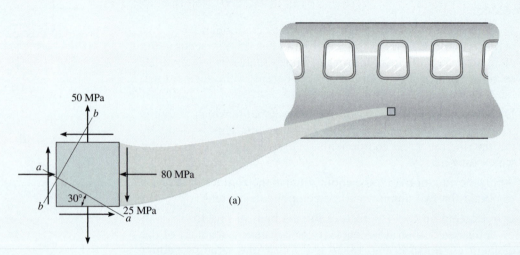

(a)

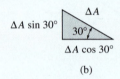

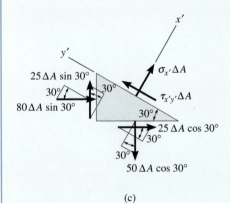

(b)

(c)

**Fig. 15–4**

### Solution

The element is sectioned by the line a–a in Fig. 15–4a, the bottom segment is removed, and assuming the sectioned (inclined) plane has an area $\Delta A$, the horizontal and vertical planes have the areas shown in Fig. 15–4b. The free-body diagram of the segment is shown in Fig. 15–4c. Applying the equations of force equilibrium in the $x'$ and $y'$ directions to avoid a simultaneous solution for the two unknowns $\sigma_{x'}$ and $\tau_{x'y'}$, we have

$$+\nearrow \Sigma F_{x'} = 0; \quad \sigma_{x'} \Delta A - (50 \, \Delta A \cos 30°) \cos 30°$$
$$+ (25 \, \Delta A \cos 30°) \sin 30° + (80 \, \Delta A \sin 30°) \sin 30°$$
$$+ (25 \, \Delta A \sin 30°) \cos 30° = 0$$
$$\sigma_{x'} = -4.15 \text{ MPa} \qquad \textit{Ans.}$$

$$+\nwarrow \Sigma F_{y'} = 0; \quad \tau_{x'y'} \Delta A - (50 \, \Delta A \cos 30°) \sin 30°$$
$$- (25 \, \Delta A \cos 30°) \cos 30° - (80 \, \Delta A \sin 30°) \cos 30°$$
$$+ (25 \, \Delta A \sin 30°) \sin 30° = 0$$
$$\tau_{x'y'} = 68.8 \text{ MPa} \qquad \textit{Ans.}$$

Since $\sigma_{x'}$ is negative, it acts in the opposite direction of that shown in Fig. 15–4c. The results are shown on the *top* of the element in Fig. 15–4d, since this surface is the one considered in Fig. 15–4c.

We must now repeat the procedure to obtain the stress on the *perpendicular* plane b–b. Sectioning the element in Fig. 15–4a along b–b results in a segment having sides with areas shown in Fig. 15–4e. Orientating the $+x'$ axis outward, perpendicular to the sectioned face, the associated free-body diagram is shown in Fig. 15–4f. Thus,

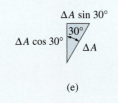

(e)

$$+\searrow\Sigma F_{x'} = 0; \quad \sigma_{x'}\,\Delta A - (25\,\Delta A\cos 30°)\sin 30°$$

$$+(80\,\Delta A\cos 30°)\cos 30° - (25\,\Delta A\sin 30°)\cos 30°$$

$$-(50\,\Delta A\sin 30°)\sin 30° = 0$$

$$\sigma_{x'} = -25.8 \text{ MPa} \qquad\qquad Ans.$$

$$+\nearrow\Sigma F_{y'} = 0; \quad -\tau_{x'y'}\,\Delta A + (25\,\Delta A\cos 30°)\cos 30°$$

$$+(80\,\Delta A\cos 30°)\sin 30° - (25\,\Delta A\sin 30°)\sin 30°$$

$$+(50\,\Delta A\sin 30°)\cos 30° = 0$$

$$\tau_{x'y'} = 68.8 \text{ MPa} \qquad\qquad Ans.$$

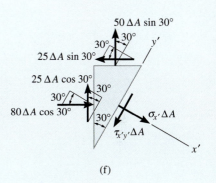

(f)

Since $\sigma_{x'}$ is a negative quantity, it acts opposite to its direction shown in Fig. 15–4f. The stress components are shown acting on the *right side* of the element in Fig. 15–4d.

From this analysis we may therefore conclude that the state of stress at the point can be represented by choosing an element oriented as shown in Fig. 15–4a, or by choosing one oriented as shown in Fig. 15–4d. In other words, the states of stress are equivalent.

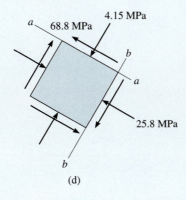

(d)

## 15.2 General Equations of Plane-Stress Transformation

The method of transforming the normal and shear stress components from the $x$, $y$ to the $x'$, $y'$ coordinate axes, as discussed in the previous section, will now be developed in a general manner and expressed as a set of stress-transformation equations.

**Sign Convention.** Before the transformation equations are derived, we must first establish a sign convention for the stress components. Here we will adopt the same one used in Sec. 8.2. Briefly stated, once the $x$, $y$ or $x'$, $y'$ axes have been established, a normal or shear stress component is *positive* provided it acts in the *positive* coordinate direction on the *positive* face of the element, or it acts in the *negative* coordinate direction on the *negative* face of the element, Fig. 15–5a. For example, $\sigma_x$ is positive since it acts to the right on the right-hand vertical face, and it acts to the left ($-x$ direction) on the left-hand vertical face. The shear stress in Fig. 15–5a is shown acting in the positive direction on all four faces of the element. On the right-hand face, $\tau_{xy}$ acts upward ($+y$ direction); on the bottom face, $\tau_{xy}$ acts to the left ($-x$ direction), and so on.

All the stress components shown in Fig. 15–5a maintain equilibrium of the element, and because of this, knowing the direction of $\tau_{xy}$ on one face of the element defines its direction on the other three faces. Hence, the above sign convention can also be remembered by simply noting that *positive normal stress acts outward from all faces and positive shear stress acts upward on the right-hand face of the element.*

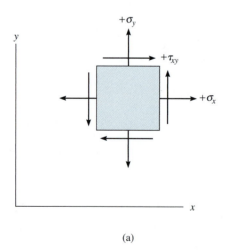

(a)

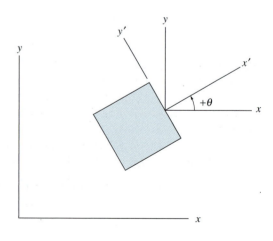

(b)

Positive Sign Convention

**Fig. 15–5**

Given the state of plane stress shown in Fig. 15–5a, the orientation of the inclined plane on which the normal and shear stress components are to be determined will be defined using the angle $\theta$. To show this angle properly, it is first necessary to establish a positive $x'$ axis, *directed outward, perpendicular* or normal to the plane, and an associated $y'$ axis, directed along the plane, Fig. 15–5b. Notice that the unprimed and primed sets of axes both form right-handed coordinate systems; that is, the positive $z$ (or $z'$) axis is established by the right-hand rule. Curling the fingers from $x$ (or $x'$) toward $y$ (or $y'$) gives the direction for the positive $z$ (or $z'$) axis that points outward. The *angle* $\theta$ is measured from the positive $x$ to the positive $x'$ axis. It is *positive* provided it follows the curl of the right-hand fingers, i.e., counterclockwise as shown in Fig. 15–5b.

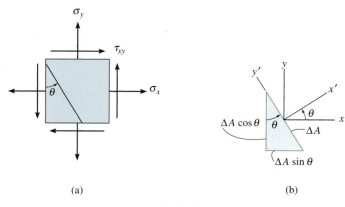

(a)                          (b)

**Fig. 15–6**

## Normal and Shear Stress Components.

Using the established sign convention, the element in Fig. 15–6a is sectioned along the inclined plane and the segment shown in Fig. 15–6b is isolated. Assuming the sectioned area is $\Delta A$, then the horizontal and vertical faces of the segment have an area of $\Delta A \sin \theta$ and $\Delta A \cos \theta$, respectively.

The resulting *free-body diagram* of the segment is shown in Fig. 15–6c. Applying the equations of force equilibrium to determine the unknown normal and shear stress components $\sigma_{x'}$ and $\tau_{x'y'}$, we obtain

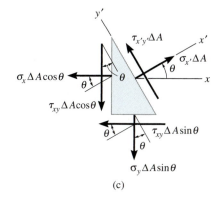

(c)

$$+\nearrow \Sigma F_{x'} = 0; \quad \sigma_{x'} \Delta A - (\tau_{xy} \Delta A \sin \theta) \cos \theta - (\sigma_y \Delta A \sin \theta) \sin \theta$$
$$-(\tau_{xy} \Delta A \cos \theta) \sin \theta - (\sigma_x \Delta A \cos \theta) \cos \theta = 0$$
$$\sigma_{x'} = \sigma_x \cos^2 \theta + \sigma_y \sin^2 \theta + \tau_{xy}(2 \sin \theta \cos \theta)$$

$$+\nwarrow \Sigma F_{y'} = 0; \quad \tau_{x'y'} \Delta A + (\tau_{xy} \Delta A \sin \theta) \sin \theta - (\sigma_y \Delta A \sin \theta) \cos \theta$$
$$-(\tau_{xy} \Delta A \cos \theta) \cos \theta + (\sigma_x \Delta A \cos \theta) \sin \theta = 0$$
$$\tau_{x'y'} = (\sigma_y - \sigma_x) \sin \theta \cos \theta + \tau_{xy}(\cos^2 \theta - \sin^2 \theta)$$

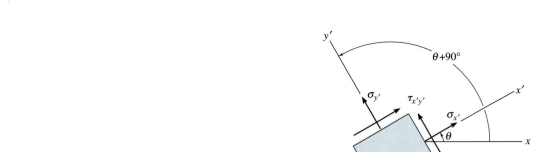

(d)

**Fig. 15–6**

These two equations may be simplified by using the trigonometric identities $\sin 2\theta = 2 \sin \theta \cos \theta$, $\sin^2 \theta = (1 - \cos 2\theta)/2$, and $\cos^2 \theta = (1 + \cos 2\theta)/2$, in which case,

$$\sigma_{x'} = \frac{\sigma_x + \sigma_y}{2} + \frac{\sigma_x - \sigma_y}{2} \cos 2\theta + \tau_{xy} \sin 2\theta \qquad (15\text{–}1)$$

$$\tau_{x'y'} = -\frac{\sigma_x - \sigma_y}{2} \sin 2\theta + \tau_{xy} \cos 2\theta \qquad (15\text{–}2)$$

If the normal stress acting in the $y'$ direction is needed, it can be obtained by simply substituting $(\theta = \theta + 90°)$ for $\theta$ into Eq. 15–1, Fig. 15–6d. This yields

$$\sigma_{y'} = \frac{\sigma_x + \sigma_y}{2} - \frac{\sigma_x - \sigma_y}{2} \cos 2\theta - \tau_{xy} \sin 2\theta \qquad (15\text{–}3)$$

If $\sigma_{y'}$ is calculated as a positive quantity, this indicates that it acts in the positive $y'$ direction as shown in Fig. 15–6d.

## PROCEDURE FOR ANALYSIS

To apply the stress transformation Eqs. 15–1 and 15–2, it is simply necessary to substitute in the known data for $\sigma_x$, $\sigma_y$, $\tau_{xy}$, and $\theta$ in accordance with the established sign convention, Fig. 15–5. If $\sigma_{x'}$ and $\tau_{x'y'}$ are calculated as positive quantities, then these stresses act in the positive direction of the $x'$ and $y'$ axes.

For convenience these equations can easily be programmed on a pocket calculator.

# E X A M P L E  15.2

The state of plane stress at a point is represented by the element shown in Fig. 15–7a. Determine the state of stress at the point on another element oriented 30° clockwise from the position shown.

### Solution

This problem was solved in Example 15–1 using basic principles. Here we will apply Eqs. 15–1 and 15–2. From the established sign convention, Fig. 15–5, it is seen that

$$\sigma_x = -80 \text{ MPa} \qquad \sigma_y = 50 \text{ MPa} \qquad \tau_{xy} = -25 \text{ MPa}$$

*Plane CD.*  To obtain the stress components on plane $CD$, Fig. 15–7b, the positive $x'$ axis is directed outward, perpendicular to $CD$, and the associated $y'$ axis is directed along $CD$. The angle measured from the $x$ to the $x'$ axis is $\theta = -30°$ (clockwise). Applying Eqs. 15–1 and 15–2 yields

$$\sigma_{x'} = \frac{\sigma_x + \sigma_y}{2} + \frac{\sigma_x - \sigma_y}{2}\cos 2\theta + \tau_{xy}\sin 2\theta$$

$$= \frac{-80 + 50}{2} + \frac{-80 - 50}{2}\cos 2(-30°) + (-25)\sin 2(-30°)$$

$$= -25.8 \text{ MPa} \qquad\qquad\qquad Ans.$$

$$\tau_{x'y'} = -\frac{\sigma_x - \sigma_y}{2}\sin 2\theta + \tau_{xy}\cos 2\theta$$

$$= -\frac{-80 - 50}{2}\sin 2(-30°) + (-25)\cos 2(-30°)$$

$$= -68.8 \text{ MPa} \qquad\qquad\qquad Ans.$$

The negative signs indicate that $\sigma_{x'}$ and $\tau_{x'y'}$ act in the negative $x'$ and $y'$ directions, respectively. The results are shown acting on the element in Fig. 15–7d.

*Plane BC.*  In a similar manner, the stress components acting on face $BC$, Fig. 15–7c, are obtained using $\theta = 60°$. Applying Eqs. 15–1 and 15–2[*], we get

$$\sigma_{x'} = \frac{-80 + 50}{2} + \frac{-80 - 50}{2}\cos 2(60°) + (-25)\sin 2(60°)$$

$$= -4.15 \text{ MPa} \qquad\qquad\qquad Ans.$$

$$\tau_{x'y'} = -\frac{-80 - 50}{2}\sin 2(60°) + (-25)\cos 2(60°)$$

$$= 68.8 \text{ MPa} \qquad\qquad\qquad Ans.$$

Here $\tau_{x'y'}$ has been computed twice in order to provide a check. The negative sign for $\sigma_{x'}$ indicates that this stress acts in the negative $x'$ direction, Fig. 15–7c. The results are shown on the element in Fig. 15–7d.

[*]Alternatively, we could apply Eq. 15–3 with $\theta = -30°$ rather than Eq. 15–1.

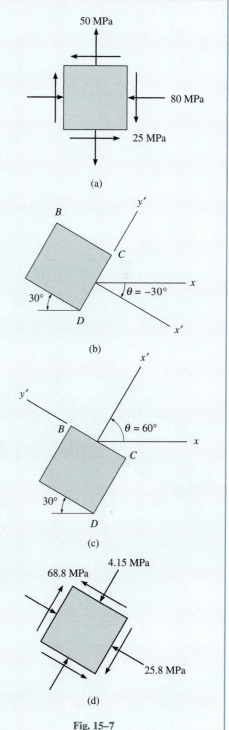

(a)

(b)

(c)

(d)

**Fig. 15–7**

## 15.3 Principal Stresses and Maximum In-Plane Shear Stress

From Eqs. 15–1 and 15–2, it can be seen that $\sigma_{x'}$ and $\tau_{x'y'}$ depend on the angle of inclination $\theta$ of the planes on which these stresses act. In engineering practice it is often important to determine the orientation of the planes that causes the normal stress to be a maximum and a minimum and the orientation of the planes that causes the shear stress to be a maximum. In this section each of these problems will be considered.

**In-Plane Principal Stresses.** To determine the maximum and minimum *normal stress* we must differentiate Eq. 15–1 with respect to $\theta$ and set the result equal to zero. This gives

$$\frac{d\sigma_{x'}}{d\theta} = -\frac{\sigma_x - \sigma_y}{2}(2 \sin 2\theta) + 2\tau_{xy} \cos 2\theta = 0$$

Solving this equation we obtain the orientation $\theta = \theta_p$ of the planes of maximum and minimum normal stress.

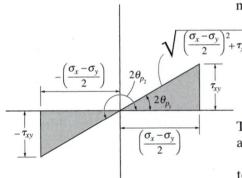

**Fig. 15–8**

$$\tan 2\theta_p = \frac{\tau_{xy}}{(\sigma_x - \sigma_y)/2} \tag{15–4}$$

The solution has two roots, $\theta_{p_1}$, and $\theta_{p_2}$. Specifically, the values of $2\theta_{p_1}$ and $2\theta_{p_2}$ are 180° apart, so $\theta_{p_1}$ and $\theta_{p_2}$ will be 90° apart.

The values of $\theta_{p_1}$ and $\theta_{p_2}$ must be substituted into Eq. 15–1 if we are to obtain the required normal stresses. We can obtain the necessary sine and cosine of $2\theta_{p_1}$ and $2\theta_{p_2}$ from the shaded triangles shown in Fig. 15–8. The construction of these triangles is based on Eq. 15–4, assuming that $\tau_{xy}$ and $(\sigma_x - \sigma_y)$ are both positive or both negative quantities. We have for $\theta_{p_1}$,

$$\sin 2\theta_{p_1} = \tau_{xy} \Big/ \sqrt{\left(\frac{\sigma_x - \sigma_y}{2}\right)^2 + \tau_{xy}^2}$$

$$\cos 2\theta_{p_1} = \left(\frac{\sigma_x - \sigma_y}{2}\right) \Big/ \sqrt{\left(\frac{\sigma_x - \sigma_y}{2}\right)^2 + \tau_{xy}^2}$$

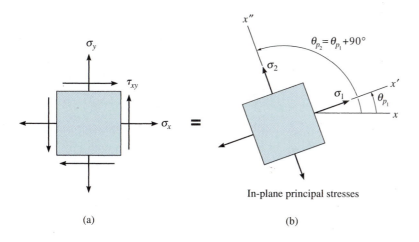

In-plane principal stresses

(a)          (b)

**Fig. 15–9**

for $\theta_{p_2}$,

$$\sin 2\theta_{p_2} = -\tau_{xy} \bigg/ \sqrt{\left(\frac{\sigma_x - \sigma_y}{2}\right)^2 + \tau_{xy}^2}$$

$$\cos 2\theta_{p_2} = -\left(\frac{\sigma_x - \sigma_y}{2}\right) \bigg/ \sqrt{\left(\frac{\sigma_x - \sigma_y}{2}\right)^2 + \tau_{xy}^2}$$

The cracks in this concrete beam were caused by tension stress, even though the beam was subjected to both an internal moment and shear. The stress transformation equations can be used to predict the direction of the cracks, and the principal normal stresses that caused it.

If either of these two sets of trigonometric relations are substituted into Eq. 15–1 and simplified, we obtain

$$\sigma_{1,2} = \frac{\sigma_x + \sigma_y}{2} \pm \sqrt{\left(\frac{\sigma_x - \sigma_y}{2}\right)^2 + \tau_{xy}^2} \qquad (15\text{–}5)$$

Depending upon the sign chosen, this result gives the maximum or minimum in-plane normal stress acting at a point, where $\sigma_1 \geq \sigma_2$. This particular set of values are called the in-plane ***principal stresses***, and the corresponding planes on which they act are called the ***principal planes*** of stress, Fig. 15–9b. Furthermore, if the trigonometric relations for $\theta_{p_1}$ and $\theta_{p_2}$ are substituted into Eq. 15–2, it can be seen that $\tau_{x'y'} = 0$; that is, ***no shear stress acts on the principal planes***.

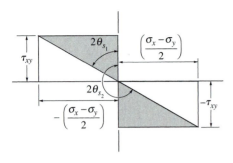

**Fig. 15–10**

**Maximum In-Plane Shear Stress.** The orientation of an element that is subjected to maximum shear stress on its faces can be determined by taking the derivative of Eq. 15–2 with respect to $\theta$ and setting the result equal to zero. This gives

$$\tan 2\theta_s = \frac{-(\sigma_x - \sigma_y)/2}{\tau_{xy}} \qquad (15\text{–}6)$$

The two roots of this equation, $\theta_{s_1}$ and $\theta_{s_2}$, can be determined from the shaded triangles shown in Fig. 15–10. By comparison with Fig. 15–8, each root of $2\theta_s$ is $90°$ from $2\theta_p$. Thus, the roots $\theta_s$ and $\theta_p$ are $45°$ apart, and as a result *the planes for maximum shear stress can be determined by orienting an element $45°$ from the position of an element that defines the planes of principal stress.*

Using either one of the roots $\theta_{s_1}$ or $\theta_{s_2}$, the maximum shear stress can be found by taking the trigonometric values of $\sin 2\theta_s$ and $\cos 2\theta_s$ from Fig. 15–10 and substituting them into Eq. 15–2. The result is

$$\tau_{\substack{\max \\ \text{in-plane}}} = \sqrt{\left(\frac{\sigma_x - \sigma_y}{2}\right)^2 + \tau_{xy}^2} \qquad (15\text{–}7)$$

The value of $\tau_{\substack{\max \\ \text{in-plane}}}$ as calculated by Eq. 15–7 is referred to as the *maximum in-plane shear stress* because it acts on the element in the $x$–$y$ plane.

Substituting the values for $\sin 2\theta_s$ and $\cos 2\theta_s$ into Eq. 15–1, we see that there is also a normal stress on the planes of maximum in-plane shear stress. We get

$$\sigma_{\text{avg}} = \frac{\sigma_x + \sigma_y}{2} \qquad (15\text{–}8)$$

Like the stress-transformation equations, it may be convenient to program the above equations so they can be used on a pocket calculator.

## IMPORTANT POINTS

- The *principal stresses* represent the maximum and minimum normal stress at the point.

- When the state of stress is represented by the principal stresses, *no shear stress* will act on the element.

- The state of stress at the point can also be represented in terms of the *maximum in-plane shear stress*. In this case an *average normal stress* will also act on the element.

- The element representing the maximum in-plane shear stress with the associated average normal stresses is *oriented $45°$ from the* element representing the principal stresses.

# EXAMPLE 15.3

When the torsional loading $T$ is applied to the bar in Fig. 15–11a, it produces a state of pure shear stress in the material. Determine (a) the maximum in-plane shear stress and the associated average normal stress, and (b) the principal stress.

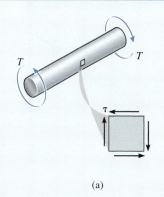

## Solution

From the established sign convention,

$$\sigma_x = 0 \qquad \sigma_y = 0 \qquad \tau_{xy} = -\tau$$

(a)

**Maximum In-Plane Shear Stress.** Applying Eqs. 15–7 and 15–8, we have

$$\tau_{\substack{max \\ \text{in-plane}}} = \sqrt{\left(\frac{\sigma_x - \sigma_y}{2}\right)^2 + \tau_{xy}^2} = \sqrt{(0)^2 + (-\tau)^2} = \pm\tau \qquad Ans.$$

$$\sigma_{avg} = \frac{\sigma_x + \sigma_y}{2} = \frac{0 + 0}{2} = 0 \qquad Ans.$$

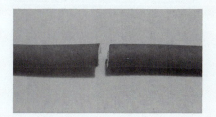

Thus, as expected, the maximum in-plane shear stress is represented by the element in Fig. 15–11a.

Through experiment it has been found that materials that are *ductile* will *fail* due to *shear stress*. As a result, if a torque is applied to a bar made from mild steel, the maximum in-plane shear stress will cause it to fail as shown in the adjacent photo.

**Principal Stress.** Applying Eqs. 15–4 and 15–5 yields

$$\tan 2\theta_p = \frac{\tau_{xy}}{(\sigma_x - \sigma_y)/2} = \frac{-\tau}{(0-0)/2}, \qquad \sigma_{p_2} = 45°, \qquad \sigma_{p_1} = 135°$$

$$\sigma_{1,2} = \frac{\sigma_x + \sigma_y}{2} \pm \sqrt{\left(\frac{\sigma_x - \sigma_y}{2}\right)^2 + \tau_{xy}^2} = 0 \pm \sqrt{(0)^2 + \tau^2} = \pm\tau \\ Ans.$$

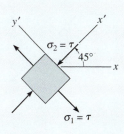

(b)

**Fig. 15–11**

If we now apply Eq. 15–1 with $\theta_{p_2} = 45°$, then

$$\sigma_{x'} = \frac{\sigma_x + \sigma_y}{2} + \frac{\sigma_x - \sigma_y}{2} \cos 2\theta + \tau_{xy} \sin 2\theta = 0+0+(-\tau)\sin 90° = -\tau$$

Thus, $\sigma_2 = -\tau$ acts at $\theta_{p_2} = 45°$ as shown in Fig. 15–11b, and $\sigma_1 = \tau$ acts on the other face, $\theta_{p_1} = 135°$.

Materials that are *brittle* fail due to *normal stress*. That is why when a brittle material, such as cast iron, is subjected to torsion it will fail in tension at a 45° inclination as seen in the adjacent photo.

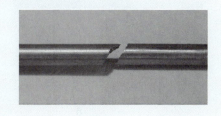

**E X A M P L E   15.4**

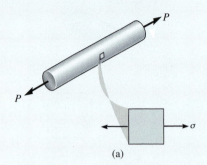

(a)

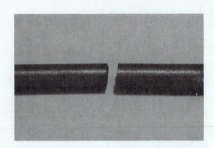

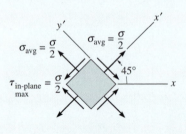

$\sigma_{avg} = \dfrac{\sigma}{2}$

$\sigma_{avg} = \dfrac{\sigma}{2}$

$\tau_{\substack{in\text{-}plane \\ max}} = \dfrac{\sigma}{2}$

45°

(b)

**Fig. 15–12**

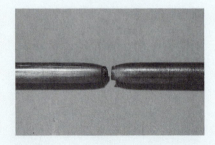

When the axial loading $P$ is applied to the bar in Fig. 15–12a, it produces a tensile stress in the material. Determine (a) the principal stress and (b) the maximum in-plane shear stress and associated average normal stress.

**Solution**

From the established sign convention,

$$\sigma_x = \sigma \qquad \sigma_y = 0 \qquad \tau_{xy} = 0$$

*Principal Stress.* By observation, the element orientated as shown in Fig. 15–12a illustrates a condition of principal stress since no shear stress acts on this element. This can also be shown by direct substitution of the above values into Eqs. 15–4 and 15–5. Thus,

$$\sigma_1 = \sigma \qquad \sigma_2 = 0 \qquad \textit{Ans.}$$

Since experiments have shown that normal stress causes brittle materials to fail, then if the bar is made from *brittle material*, such as cast iron, it will cause failure as shown in the adjacent photo.

*Maximum In-Plane Shear Stress.* Applying Eqs. 15–6, 15–7, and 15–8, we have

$$\tan 2\theta_s = \frac{-(\sigma_x - \sigma_y)/2}{\tau_{xy}} = \frac{-(\sigma - 0)/2}{0}; \qquad \theta_{s_1} = 45°, \qquad \theta_{s_2} = 135°$$

$$\tau_{\substack{max \\ in\text{-}plane}} = \sqrt{\left(\frac{\sigma_x - \sigma_y}{2}\right)^2 + \tau_{xy}{}^2} = \sqrt{\left(\frac{\sigma - 0}{2}\right)^2 + (0)^2} = \pm\frac{\sigma}{2} \quad \textit{Ans.}$$

$$\sigma_{avg} = \frac{\sigma_x + \sigma_y}{2} = \frac{\sigma + 0}{2} = \frac{\sigma}{2} \qquad \textit{Ans.}$$

To determine the proper orientation of the element, apply Eq. 15–2.

$$\tau_{x'y'} = -\frac{\sigma_x - \sigma_y}{2}\sin 2\theta + \tau_{xy}\cos 2\theta = -\frac{\sigma - 0}{2}\sin 90° + 0 = -\frac{\sigma}{2}$$

This negative shear stress acts on the $x'$ face, in the negative $y'$ direction as shown in Fig. 15–12b.

If the bar is made from a *ductile material* such as mild steel then shear stress will cause it to fail when it is subjected to *tension*. This can be noted in the adjacent photo, where within the region of necking, shear stress has caused "slipping" along the steel's crystalline boundaries, resulting in a plane of failure that has formed a *cone* around the bar oriented at approximately 45° as calculated above.

# EXAMPLE 15.5

The state of plane stress at a point on a body is shown on the element in Fig. 15–13a. Represent this stress state in terms of the principal stresses.

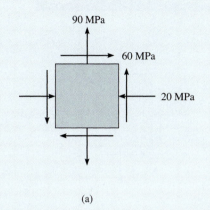

(a)

### Solution

From the established sign convention, we have

$$\sigma_x = -20 \text{ MPa} \qquad \sigma_y = 90 \text{ MPa} \qquad \tau_{xy} = 60 \text{ MPa}$$

*Orientation of Element.*  Applying Eq. 15–4, we have

$$\tan 2\theta_p = \frac{\tau_{xy}}{(\sigma_x - \sigma_y)/2} = \frac{60}{(-20 - 90)/2}$$

Solving, and referring to this root as $\theta_{p_2}$, as will be shown below, yields

$$2\theta_{p_2} = -47.49° \qquad \theta_{p_2} = -23.7°$$

Since the difference between $2\theta_{p_1}$ and $2\theta_{p_2}$ is 180°, we have

$$2\theta_{p_1} = 180° + 2\theta_{p_2} = 132.51° \qquad \theta_{p_1} = 66.3°$$

Recall that $\theta$ is measured positive *counterclockwise* from the $x$ axis to the outward normal ($x'$ axis) on the face of the element, and so the results are shown in Fig. 15–13b.

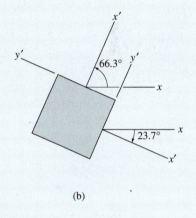

(b)

*Principal Stresses.*  We have

$$\sigma_{1,2} = \frac{\sigma_x + \sigma_y}{2} \pm \sqrt{\left(\frac{\sigma_x - \sigma_y}{2}\right)^2 + \tau_{xy}^2}$$

$$= \frac{-20 + 90}{2} \pm \sqrt{\left(\frac{-20 - 90}{2}\right)^2 + (60)^2}$$

$$= 35.0 \pm 81.4$$

$$\sigma_1 = 116 \text{ MPa} \hspace{3cm} \textit{Ans.}$$

$$\sigma_2 = -46.4 \text{ MPa} \hspace{2.5cm} \textit{Ans.}$$

The principal plane on which each normal stress acts can be determined by applying Eq. 15–1 with, say, $\theta = \theta_{p_2} = -23.7°$. We have

$$\sigma_{x'} = \frac{\sigma_x + \sigma_y}{2} + \frac{\sigma_x - \sigma_y}{2}\cos 2\theta + \tau_{xy}\sin 2\theta$$

$$= \frac{-20 + 90}{2} + \frac{-20 - 90}{2}\cos 2(-23.7°) + 60 \sin 2(-23.7°)$$

$$= -46.4 \text{ MPa}$$

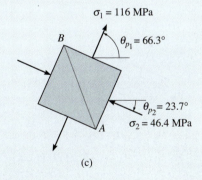

(c)

**Fig. 15–13**

Hence, $\sigma_2 = -46.4$ MPa acts on the plane defined by $\theta_{p_2} = -23.7°$, whereas $\sigma_1 = 116$ MPa acts on the plane defined by $\theta_{p_1} = 66.3°$. The results are shown on the element in Fig. 15–13c. Recall that no shear stress acts on this element.

**E X A M P L E  15.6**

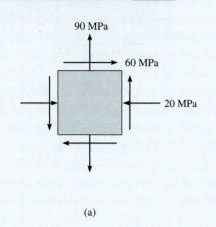

90 MPa

60 MPa

20 MPa

(a)

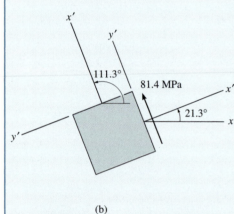

$x'$

$y'$

111.3°

81.4 MPa

$x'$

21.3°

$x$

$y'$

(b)

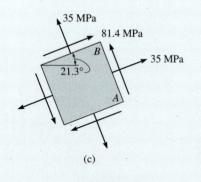

35 MPa

81.4 MPa

$B$

35 MPa

21.3°

$A$

(c)

**Fig. 15–14**

The state of plane stress at a point on a body is represented on the element shown in Fig. 15–14a. Represent this stress state in terms of the maximum in-plane shear stress and associated average normal stress.

**Solution**

*Orientation of Element.* Since $\sigma_x = -20$ MPa, $\sigma_y = 90$ MPa, and $\tau_{xy} = 60$ MPa, applying Eq 15–6, we have

$$\tan 2\theta_s = \frac{-(\sigma_x - \sigma_y)/2}{\tau_{xy}} = \frac{-(-20 - 90)/2}{60}$$

$$2\theta_{s_2} = 42.5° \qquad \theta_{s_2} = 21.3°$$

$$2\theta_{s_1} = 180° + 2\theta_{s_2} \qquad \theta_{s_1} = 111.3°$$

Note that these angles shown in Fig. 15–14b are 45° away from the principal planes of stress, which was determined in Example 15.5.

*Maximum In-Plane Shear Stress.* Applying Eq. 15–7,

$$\tau_{\substack{\max \\ \text{in-plane}}} = \sqrt{\left(\frac{\sigma_x - \sigma_y}{2}\right)^2 + \tau_{xy}^2} = \sqrt{\left(\frac{-20 - 90}{2}\right)^2 + (60)^2}$$

$$= 81.4 \text{ MPa} \qquad\qquad\qquad Ans.$$

The proper direction of $\tau_{\substack{\max \\ \text{in-plane}}}$ on the element can be determined by considering $\theta = \theta_{s_2} = 21.3°$, and applying Eq. 15–2. We have

$$\tau_{x'y'} = -\left(\frac{\sigma_x - \sigma_y}{2}\right) \sin 2\theta + \tau_{xy} \cos 2\theta$$

$$= -\left(\frac{-20 - 90}{2}\right) \sin 2(21.3°) + 60 \cos 2(21.3°)$$

$$= 81.4 \text{ MPa}$$

Thus, $\tau_{\substack{\max \\ \text{in-plane}}} = \tau_{x'y'}$ acts in the *positive* $y'$ direction on this face ($\theta = 21.3°$), Fig. 15–14b. The shear stresses on the other three faces are directed as shown in Fig. 15–14c.

*Average Normal Stress.* Besides the maximum shear stress, as calculated above, the element is also subjected to an average normal stress determined from Eq. 15–8; that is,

$$\sigma_{\text{avg}} = \frac{\sigma_x + \sigma_y}{2} = \frac{-20 + 90}{2} = 35 \text{ MPa} \qquad Ans.$$

This is a tensile stress. The results are shown in Fig. 15–14c.

# PROBLEMS

**15-1.** Prove that the sum of the normal stresses $\sigma_x + \sigma_y = \sigma_{x'} + \sigma_{y'}$ is constant.

**15-2.** The state of stress at a point in a member is shown on the element. Determine the stress components acting on the inclined plane $AB$. Solve the problem using the method of equilibrium described in Sec. 15.1.

**15-3.** Solve Prob. 15–2 using the stress-transformation equations developed in Sec. 15.2.

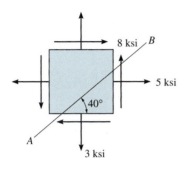

Probs. 15–2/3

**\*15-4.** The state of stress at a point in a member is shown on the element. Determine the stress components acting on the inclined plane $AB$. Solve the problem using the method of equilibrium described in Sec. 15.1.

**15-5.** Solve Prob. 15–4 using the stress-transformation equations developed in Sec. 15.2. Show the result on a sketch.

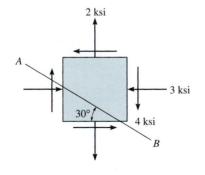

Probs. 15–4/5

**15-6.** The state of stress at a point in a member is shown on the element. Determine the stress components acting on the inclined plane $AB$. Solve the problem using the method of equilibrium described in Sec. 15.1.

**15-7.** Solve Prob. 15–6 using the stress-transformation equations developed in Sec. 15.2.

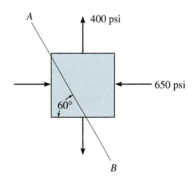

Probs. 15–6/7

**\*15-8.** Determine the equivalent state of stress on an element if the element is oriented 60° clockwise from the element shown.

**15-9.** Determine the equivalent state of stress on an element if the element is oriented 30° counterclockwise from the element shown.

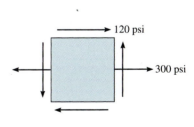

Probs. 15–8/9

**15-10.** The state of stress at a point is shown on the element. Determine (a) the principal stresses and (b) the maximum in-plane shear stress and average normal stress at the point. Specify the orientation of the element in each case.

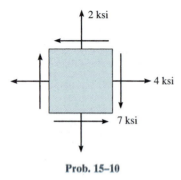

Prob. 15–10

**15-11.** Determine the equivalent state of stress on an element if it is oriented 50° counterclockwise from the element shown. Use the stress-transformation equations.

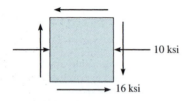

Prob. 15–11

**\*15-12.** The state of stress at a point is shown on the element. Determine (a) the principal stresses and (b) the maximum in-plane shear stress and average normal stress at the point. Specify the orientation of the element in each case.

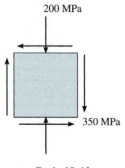

Prob. 15–12

**15-13.** The state of stress at a point on the upper surface of the airplane wing is shown on the element. Determine (a) the principal stresses and (b) the maximum in-plane shear stress and average normal stress at the point. Specify the orientation of the element in each case.

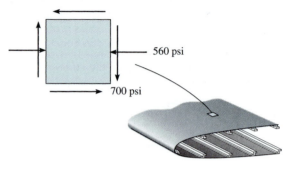

Prob. 15–13

**15-14.** The steel bar has a thickness of 0.5 in. and is subjected to the edge loading shown. Determine the principal stresses developed in the bar.

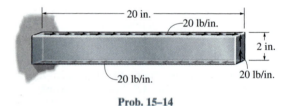

Prob. 15–14

**15-15.** The steel plate has a thickness of 10 mm and is subjected to the edge loading shown. Determine the maximum in-plane shear stress and the average normal stress developed in the steel.

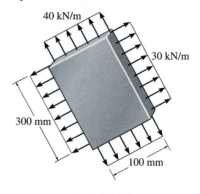

Prob. 15–15

*■**15-16.** Consider the general case of plane stress as shown. Write a computer program that can be used to determine the normal and shear stresses, $\sigma_{x'}$ and $\tau_{x'y'}$, on the plane of an element oriented at an angle $\theta$ from the horizontal. Also, compute the principal stresses, the maximum in-plane shear stress, the average normal stress, and the element's orientation.

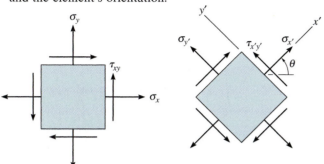

**Prob. 15–16**

**15-17.** A point on a thin plate is subjected to the two successive states of stress shown. Determine the resultant state of stress represented on the element oriented, as shown on the right.

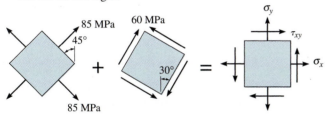

**Prob. 15–17**

**15-18.** The stress acting on two planes at a point is indicated. Determine the shear stress on plane $a$–$a$ and the principal stresses at the point.

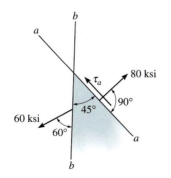

**Prob. 15–18**

**15-19.** The stress acting on two planes at a point is indicated. Determine the normal stress $\sigma_b$ and the principal stresses at the point.

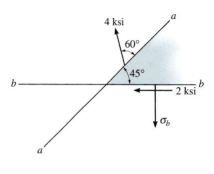

**Prob. 15–19**

The following problems involve material covered in Chapter 14.

*■**15-20.** The wooden beam is subjected to a load of 12 kN. If grains of wood in the beam at point $A$ make an angle of 25° with the horizontal as shown, determine the normal and shear stresses that act perpendicular and parallel to the grains due to the loading.

**15-21.** The wooden beam is subjected to a load of 12 kN. Determine the principal stresses at point $A$ and specify the orientation of the element.

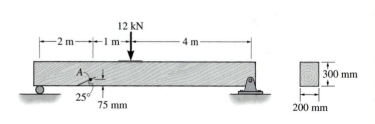

**Probs. 15–20/21**

**15-22.** The clamp bears down on the smooth surface at *E* by tightening the bolt. If the tensile force in the bolt is 40 kN, determine the principal stresses at points *A* and *B* and show the results on elements located at each of these points. The cross-sectional area at *A* and *B* is shown in the adjacent figure.

**15-23.** Solve Prob. 15–22 for points *C* and *D*.

**15-25.** The T-beam is subjected to the distributed loading that is applied along its centerline. Determine the principal stresses at points *A* and *B* and show the results on elements located at each of these points.

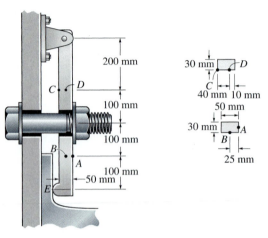

Probs. 15–22/23

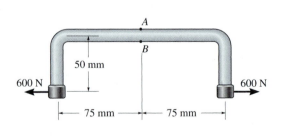

Prob. 15–25

*15-24.** The square steel plate has a thickness of 10 mm and is subjected to the edge loading shown. Determine the maximum in-plane shear stress and the average normal stress developed in the steel.

**15-26.** The bent rod has a diameter of 15 mm and is subjected to the force of 600 N. Determine the principal stresses and the maximum in-plane shear stress that are developed at point *A* and point *B*. Show the results on properly oriented elements located at these points.

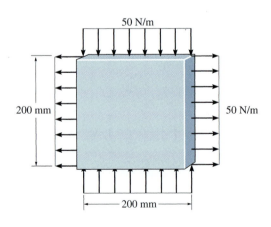

Prob. 15–24

Prob. 15–26

**15-27.** The beam has a rectangular cross section and is subjected to the loading shown. Determine the principal stresses and the maximum in-plane shear stress that are developed at point $A$ and point $B$. Show the results on properly oriented elements located at these points.

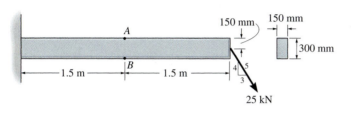

Prob. 15–27

**15-29.** The wide-flange beam is subjected to the loading shown. Determine the principal stress in the beam at point $A$ and at point $B$. These points are located at the top and bottom of the web, respectively. Although it is not very accurate, use the shear formula to compute the shear stress.

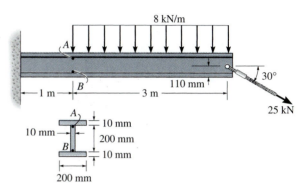

Prob. 15–29

**\*15-28.** The clamp bears down on the smooth surfaces at $C$ and $D$ when the bolt is tightened. If the tensile force in the bolt is 40 kN, determine the principal stresses at points $A$ and $B$ and show the results on elements located at each of these points. The cross-sectional area at $A$ and $B$ is shown in the adjacent figure.

**15-30.** The shaft has a diameter $d$ and is subjected to the loadings shown. Determine the principal stresses and the maximum in-plane shear stress that is developed anywhere on the surface of the shaft.

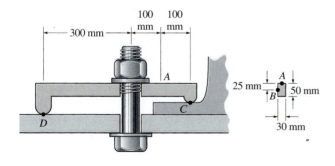

Prob. 15–28

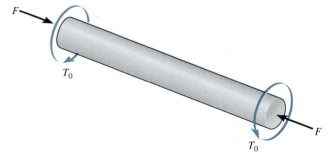

Prob. 15–30

## 15.4  Mohr's Circle—Plane Stress

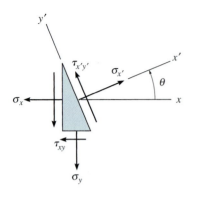

(a)

In this section we will show that the equations for plane stress transformation have a graphical solution that is often convenient to use and easy to remember. Furthermore, this approach will allow us to *visualize* how the normal and shear stress components $\sigma_{x'}$ and $\tau_{x'y'}$ vary as the plane on which they act is oriented in different directions, Fig. 15–15a.

Equations 15–1 and 15–2 can be rewritten in the form

$$\sigma_{x'} - \left(\frac{\sigma_x + \sigma_y}{2}\right) = \left(\frac{\sigma_x - \sigma_y}{2}\right)\cos 2\theta + \tau_{xy}\sin 2\theta \quad (15\text{–}9)$$

$$\tau_{x'y'} = -\left(\frac{\sigma_x - \sigma_y}{2}\right)\sin 2\theta + \tau_{xy}\cos 2\theta \quad (15\text{–}10)$$

The parameter $\theta$ can be *eliminated* by squaring each equation and adding the equations together. The result is

$$\left[\sigma_{x'} - \left(\frac{\sigma_x + \sigma_y}{2}\right)\right]^2 + \tau_{x'y'}^2 = \left(\frac{\sigma_x - \sigma_y}{2}\right)^2 + \tau_{xy}^2$$

For a specific problem, $\sigma_x$, $\sigma_y$, $\tau_{xy}$ are *known constants*. Thus the above equation can be written in a more compact form as

$$(\sigma_{x'} - \sigma_{avg})^2 + \tau_{x'y'}^2 = R^2 \quad (15\text{–}11)$$

where

$$\sigma_{avg} = \frac{\sigma_x + \sigma_y}{2}$$

$$R = \sqrt{\left(\frac{\sigma_x - \sigma_y}{2}\right)^2 + \tau_{xy}^2} \quad (15\text{–}12)$$

If we establish coordinate axes, $\sigma$ *positive to the right* and $\tau$ *positive downward*, and then plot Eq. 15–11, it will be seen that this equation represents a *circle* having a radius $R$ and center on the $\sigma$ axis at point $C(\sigma_{avg}, 0)$, Fig. 15–15b. This circle is called *Mohr's circle*, because it was developed by the German engineer Otto Mohr.

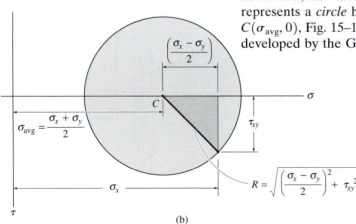

(b)

**Fig. 15–15**

To draw Mohr's circle it is necessary to first establish the $\sigma$ and $\tau$ axis, Fig. 15–16c. Since the stress components $\sigma_x$, $\sigma_y$, $\tau_{xy}$ are known, the center of the circle can then be plotted, $C(\sigma_{avg}, 0)$. To obtain the radius, we need to know at least one point on the circle. Consider the case when the $x'$ axis is coincident with the $x$ axis as shown in Fig. 15–16a. Then $\theta = 0°$ and $\sigma_{x'} = \sigma_x$, $\tau_{x'y'} = \tau_{xy}$. We will refer to this as the *reference point A* and plot its coordinates $A(\sigma_x, \tau_{xy})$, Fig. 15–16c. Applying the Pythagorean theorem to the shaded triangle, the radius $R$ can now be determined, which checks with Eq. 15–12. With points $C$ and $A$ known, the circle can be drawn as shown.

Now consider rotating the $x'$ axis 90° counterclockwise, Fig. 15–16b. Then $\sigma_{x'} = \sigma_y$, $\tau_{x'y'} = -\tau_{xy}$. These values are the coordinates of point $G(\sigma_y, -\tau_{xy})$ on the circle, Fig. 15–16c. Hence, the radial line $CG$ is 180° counterclockwise from the *reference line CA*. In other words, a rotation $\theta$ of the $x'$ axis on the element will correspond to a rotation $2\theta$ on the circle in the *same direction*.[*]

Once established, Mohr's circle can be used to determine the principal stresses, the maximum in-plane shear stress and associated average normal stress, or the stress on any arbitrary plane. The method for doing this is explained in the following procedure for analysis.

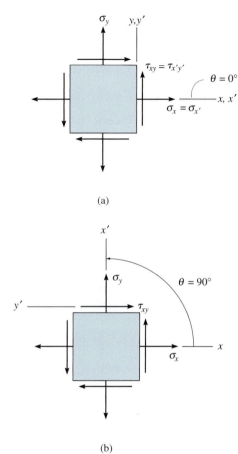

(a)

(b)

---

[*]If instead the $\tau$ axis is constructed *positive upwards,* then the angle $2\theta$ on the circle would be measured in the *opposite direction* to the orientation $\theta$ of the plane.

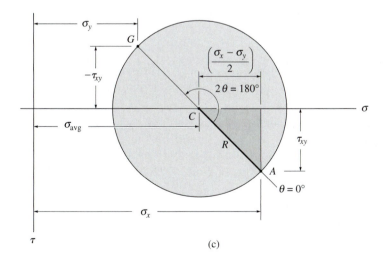

(c)

**Fig. 15–16**

## PROCEDURE FOR ANALYSIS

The following steps are required to draw and use Mohr's circle.

*Construction of the Circle.*

• Establish a coordinate system such that the abscissa represents the normal stress $\sigma$, with *positive to the right*, and the ordinate represents the shear stress $\tau$, with *positive downward*, Fig. 15–17a.

• Using the positive sign convention for $\sigma_x$, $\sigma_y$, $\tau_{xy}$, as shown in Fig. 15–17b, plot the center of the circle $C$, which is located on the $\sigma$ axis at a distance $\sigma_{avg} = (\sigma_x + \sigma_y)/2$ from the origin, Fig. 15–17a.*

• Plot the *reference point A* having coordinates $A(\sigma_x, \tau_{xy})$. This point represents the normal and shear stress components on the element's right-hand vertical face, and since the $x'$ axis coincides with the $x$ axis, this represents $\theta = 0°$, Fig. 15–17b.

• Connect point $A$ with the center $C$ of the circle and determine $CA$ by trigonometry. This distance represents the radius $R$ of the circle, Fig. 15–17a.

• Once $R$ has been determined, sketch the circle.

*Principal Stress.*

• The principal stresses $\sigma_1$ and $\sigma_2$ ($\sigma_1 \geq \sigma_2$) are represented by the two points $B$ and $D$ where the circle intersects the $\sigma$ axis, i.e., where $\tau = 0$, Fig. 15–17a.

• These stresses act on planes defined by angles $\theta_{p_1}$ and $\theta_{p_2}$, Fig. 15–17c. They are represented on the circle by angles $2\theta_{p_1}$ (shown) and $2\theta_{p_2}$ (not shown) and are measured *from* the radial reference line $CA$ to lines $CB$ and $CD$, respectively.

• Using trigonometry, only one of these angles needs to be calculated from the circle, since $\theta_{p_1}$ and $\theta_{p_2}$ are 90° apart. Remember that the direction of rotation $2\theta_p$ on the circle (here counterclockwise) represents the *same* direction of rotation $\theta_p$ from the reference axis $(+x)$ to the principal plane $(+x')$, Fig. 15–17c.*

### Maximum In-Plane Shear Stress.

- The average normal stress and maximum in-plane shear stress components are determined from the circle as the coordinates of either point $E$ or $F$, Fig. 15–17a.

- In this case the angles $\theta_{s_1}$ and $\theta_{s_2}$ give the orientation of the planes that contain these components, Fig. 15–17d. The angle $2\theta_{s_1}$ is shown in Fig. 15–17a and can be determined using trigonometry. Here the rotation is clockwise, and so $\theta_{s_1}$ must be clockwise on the element, Fig. 15–17d.*

### Stresses on Arbitrary Plane.

- The normal and shear stress components $\sigma_{x'}$ and $\tau_{x'y'}$ acting on a specified plane defined by the angle $\theta$, Fig. 15–17e, can be obtained from the circle using trigonometry to determine the coordinates of point $P$, Fig. 15–17a.

- To locate $P$, the known angle $\theta$ for the plane (in this case counterclockwise), Fig. 15–17e, must be measured on the circle in the *same direction* $2\theta$ (counterclockwise), *from* the radial reference line $CA$ *to* the radial line $CP$, Fig. 15–17a.*

*If instead the $\tau$ axis is constructed *positive upwards*, then the angle $2\theta$ on the circle would be measured in the *opposite direction* to the orientation $\theta$ of the plane.

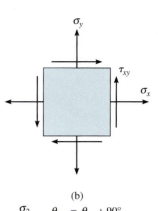

(b)

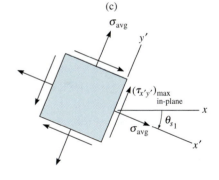

(c)

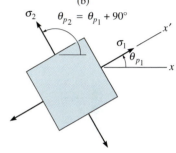

(d)

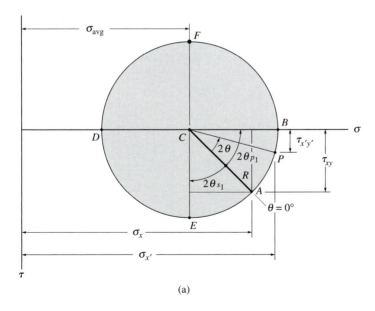

(a)

**Fig. 15–17**

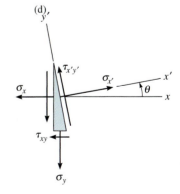

(e)

**E X A M P L E　15.7**

The axial loading $P$ produces the state of stress in the material as shown in Fig. 15–18a. Draw Mohr's circle for this case.

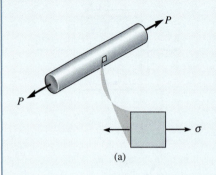

(a)

**Solution**

*Construction of the Circle.* From Fig. 15–18a,

$$\sigma_x = \sigma \qquad \sigma_y = 0 \qquad \tau_{xy} = 0$$

The $\sigma$ and $\tau$ axes are established in Fig. 15–18b. The center of the circle $C$ is on the $\sigma$ axis at

$$\sigma_{avg} = \frac{\sigma_x + \sigma_y}{2} = \frac{\sigma + 0}{2} = \frac{\sigma}{2}$$

From the right-hand face of the element, Fig. 15–18a, the reference point for $\theta = 0°$ has coordinates $A(\sigma, 0)$. Hence the radius of the circle $CA$ is $R = \sigma/2$, Fig. 15–18b.

*Stresses.* Note that the principal stresses are at points $A$ and $D$.

$$\sigma_1 = \sigma \qquad \sigma_2 = 0$$

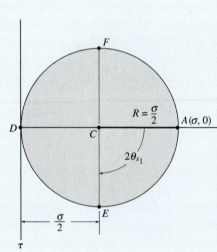

(b)

The element in Fig. 15–18a represents this principal state of stress.

The maximum in-plane shear stress and associated average normal stress is identified on the circle as point $E$ or $F$, Fig. 15–18b. At $E$ we have

$$\tau_{\substack{max \\ in\text{-}plane}} = \frac{\sigma}{2}$$

$$\sigma_{avg} = \frac{\sigma}{2}$$

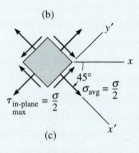

(c)

**Fig. 15–18**

By observation, the clockwise angle $2\theta_{s_1} = 90°$. Therefore $\theta_{s_1} = 45°$, so that the $x'$ axis is oriented $45°$ clockwise from the $x$ axis; Fig. 15–18c. Since $E$ has positive coordinates, then $\sigma_{avg}$ and $\tau_{\substack{max \\ in\text{-}plane}}$ act in the positive $x'$ and $y'$ directions, respectively.

# E X A M P L E 15.10

The state of plane stress at a point is shown on the element in Fig. 15–21a. Determine the maximum in-plane shear stresses and the orientation of the element upon which they act.

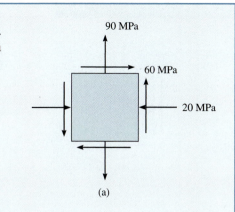

(a)

## Solution

*Construction of the Circle.* From the problem data,

$$\sigma_x = -20 \text{ MPa} \qquad \sigma_y = 90 \text{ MPa} \qquad \tau_{xy} = 60 \text{ MPa}$$

The $\sigma$, $\tau$ axes are established in Fig. 15–21b. The center of the circle $C$ is located on the $\sigma$ axis, at the point

$$\sigma_{avg} = \frac{-20 + 90}{2} = 35 \text{ MPa}$$

Point $C$ and the reference point $A(-20, 60)$ are plotted. Applying the Pythagorean theorem to the shaded triangle to determine the circle's radius $CA$, we have

$$R = \sqrt{(60)^2 + (55)^2} = 81.4 \text{ MPa}$$

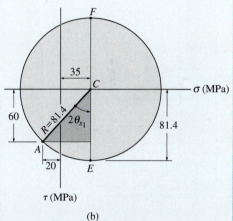

(b)

*Maximum In-Plane Shear Stress.* The maximum in-plane shear stress and the average normal stress are identified by point $E$ or $F$ on the circle. In particular, the coordinates of point $E(35, 81.4)$ give

$$\tau_{\substack{max \\ \text{in-plane}}} = 81.4 \text{ MPa} \qquad\qquad\qquad \textit{Ans.}$$

$$\sigma_{avg} = 35 \text{ MPa} \qquad\qquad\qquad \textit{Ans.}$$

The *counterclockwise* angle $\theta_{s_1}$ can be found from the circle, identified as $2\theta_{s_1}$. We have

$$2\theta_{s_1} = \tan^{-1}\left(\frac{20 + 35}{60}\right) = 42.5°$$

$$\theta_{s_1} = 21.3° \qquad\qquad\qquad \textit{Ans.}$$

This *counterclockwise* angle defines the direction of the $x'$ axis, Fig. 15–21c. Since point $E$ has *positive* coordinates, then the average normal stress and the maximum in-plane shear stress both act in the *positive* $x'$ and $y'$ directions as shown.

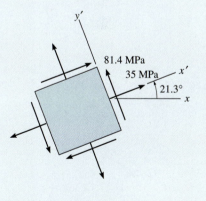

(c)

**Fig. 15–21**

E X A M P L E **15.11**

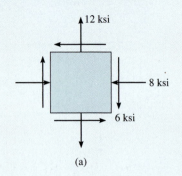

(a)

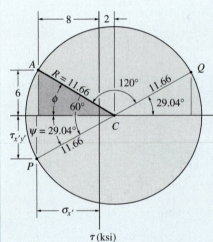

(b)

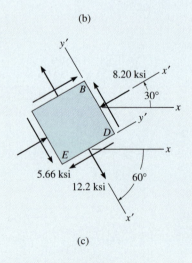

(c)

**Fig. 15–22**

The state of plane stress at a point is shown on the element in Fig. 15–22a. Represent this state of stress on an element oriented 30° counterclockwise from the position shown.

**Solution**

*Construction of the Circle.* From the problem data,

$$\sigma_x = -8 \text{ ksi} \qquad \sigma_y = 12 \text{ ksi} \qquad \tau_{xy} = -6 \text{ ksi}$$

The $\sigma$ and $\tau$ axes are established in Fig. 15–22b. The center of the circle $C$ is on the $\sigma$ axis at

$$\sigma_{\text{avg}} = \frac{-8 + 12}{2} = 2 \text{ ksi}$$

The initial point for $\theta = 0°$ has coordinates $A(-8, -6)$. Hence from the shaded triangle the radius $CA$ is

$$R = \sqrt{(10)^2 + (6)^2} = 11.66$$

*Stresses on 30° Element.* Since the element is to be rotated 30° *counterclockwise*, we must construct a radial line $CP$, $2(30°) = 60°$ *counterclockwise*, measured from $CA$ ($\theta = 0°$), Fig. 15–22b. The coordinates of point $P$ ($\sigma_{x'}, \tau_{x'y'}$) must now be obtained. From the geometry of the circle,

$$\phi = \tan^{-1}\frac{6}{10} = 30.96° \qquad \psi = 60° - 30.96° = 29.04°$$

$$\sigma_{x'} = 2 - 11.66 \cos 29.04° = -8.20 \text{ ksi} \qquad \qquad \textit{Ans.}$$

$$\tau_{x'y'} = 11.66 \sin 29.04° = 5.66 \text{ ksi} \qquad \qquad \textit{Ans.}$$

These two stress components act on face $BD$ of the element shown in Fig. 15–22c since the $x'$ axis for this face is oriented 30° counterclockwise from the $x$ axis.

The stress components acting on the adjacent face $DE$ of the element, which is 60° *clockwise* from the positive $x$ axis, Fig. 15–22c, are represented by the coordinates of point $Q$ on the circle. This point lies on the radial line $CQ$, which is 180° from $CP$. The coordinates of point $Q$ are

$$\sigma_{x'} = 2 + 11.66 \cos 29.04° = 12.2 \text{ ksi} \qquad \qquad \textit{Ans.}$$

$$\tau_{x'y'} = -(11.66 \sin 29.04) = -5.66 \text{ ksi} \qquad \text{(check)}$$

Note that here $\tau_{x'y'}$ acts in the $-y'$ direction.

## 15.5  Absolute Maximum Shear Stress

When a point in a body is subjected to a general three-dimensional state of stress, an element of material has a normal-stress and two shear-stress components acting on each of its faces, Fig. 15–23a. Like the case of plane stress, it is possible to develop stress-transformation equations that can be used to determine the normal and shear stress components $\sigma$ and $\tau$ acting on *any* skewed plane of the element, Fig. 15–23b. Furthermore, at the point it is also possible to determine the unique orientation of an element having only principal stresses acting on its faces. As shown in Fig. 15–23c, these principal stresses are assumed to have magnitudes of maximum, intermediate, and minimum intensity, i.e., $\sigma_{max} \geq \sigma_{int} \geq \sigma_{min}$.

A discussion of the transformation of stress in three dimensions is beyond the scope of this text; however, it is discussed in books related to the theory of elasticity. For our purposes, we will assume that the principal orientation of the element and the principal stresses are known, Fig. 15–23c. This is a condition known as ***triaxial stress***. If we view this element in two dimensions, that is, in the $y'-z'$, $x'-z'$, and $x'-y'$ planes, Fig. 15–24a, 15–24b, 15–24c, we can then use Mohr's circle to determine the *maximum in-plane shear stress* for each case. For example, the diameter of Mohr's circle extends between the principal stresses $\sigma_{int}$ and $\sigma_{min}$ for the case shown in Fig. 15–24a. From this circle, Fig. 15–24d, the maximum in-plane shear stress is $(\tau_{y'z'})_{max} = (\sigma_{int} - \sigma_{min})/2$, and the associated average normal stress is $(\sigma_{int} + \sigma_{min})/2$. As shown in Fig. 15–24e, the element having these stress components on it must be oriented 45° from the position of the element shown in Fig. 15–24a. Mohr's circles for the elements in Fig. 15–24b and 15–24c have also been constructed in Fig. 15–24d. The corresponding elements having a 45° orientation and subjected to maximum in-plane shear and average normal stress components are shown in Fig. 15–24f and 15–24g, respectively.

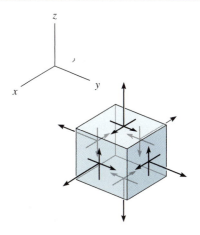

(a)

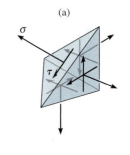

(b)

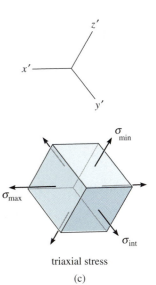

triaxial stress

(c)

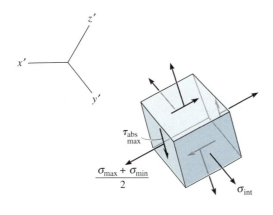

(d)

**Fig. 15–23**

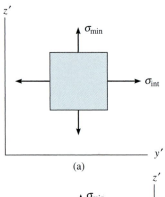

(a)

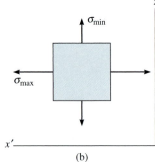

(b)

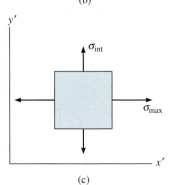

(c)

Comparing the three circles in Fig. 15–24d, it is seen that the **absolute maximum shear stress**, $\tau_{max}^{abs}$, is defined by the circle having the largest radius, which occurs for the element shown in Fig. 15–24b. In other words, the element in Fig. 15–24f is oriented by a rotation of 45° about the $y'$ axis from the element in Fig. 15–23b. Notice that this condition can also be *determined directly* by simply choosing the maximum and minimum principal stresses from Fig. 15–23c, in which case the absolute maximum shear stress will be

$$\tau_{max}^{abs} = \frac{\sigma_{max} - \sigma_{min}}{2} \qquad (15\text{–}13)$$

And the associated average normal stress will be

$$\sigma_{avg} = \frac{\sigma_{max} + \sigma_{min}}{2} \qquad (15\text{–}14)$$

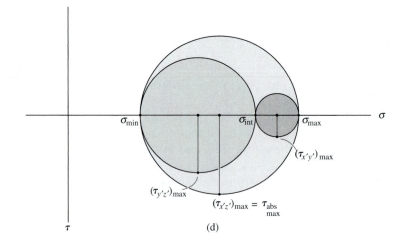

(d)

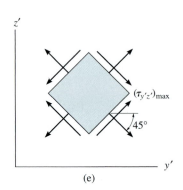

(e)

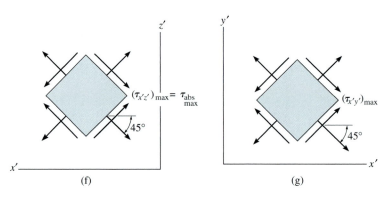

(f)                    (g)

**Fig. 15–24**

The analysis considered only the stress components acting on elements located in positions found from rotations about the $x'$, $y'$, or $z'$ axis. If we had used the three-dimensional stress-transformation equations of the theory of elasticity to obtain values of the normal and shear stress components acting on any arbitrary skewed plane at the point, as in Fig. 15–23b, it could be shown that regardless of the orientation of the plane, specific values of the shear stress $\tau$ on the plane will *always be less* than the absolute maximum shear stress found from Eq. 15–13. Furthermore, the normal stress $\sigma$ acting on any plane will have a value lying between the maximum and minimum principal stresses, that is, $\sigma_{max} \geq \sigma \geq \sigma_{min}$.

**Plane Stress.** The above results have an important implication for the case of plane stress, particularly when the in-plane principal stresses have the *same sign*, i.e., they are both tensile or both compressive. For example, consider the material to be subjected to plane stress such that the in-plane principal stresses are represented as $\sigma_{max}$ and $\sigma_{int}$, in the $x'$ and $y'$ directions, respectively; while the out-of-plane principal stress in the $z'$ direction is $\sigma_{min} = 0$, Fig. 15–25a. Mohr's circles that describe this state of stress for element orientations about each of the three coordinate axes are shown in Fig. 15–25b. Here it is seen that although the maximum in-plane shear stress is $(\tau_{x'y'})_{max} = (\sigma_{max} - \sigma_{int})/2$, this value is *not* the absolute maximum shear stress to which the material is subjected. Instead, from Eq. 15–13 or Fig. 15–25b,

$$\tau_{\underset{max}{abs}} = (\tau_{x'z'})_{max} = \frac{\sigma_{max} - 0}{2} = \frac{\sigma_{max}}{2} \qquad (15\text{–}15)$$

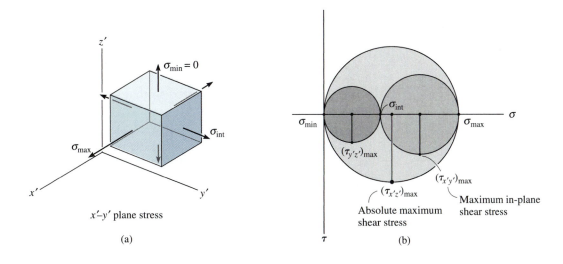

$x'$–$y'$ plane stress

(a)

$(\tau_{y'z'})_{max}$

$(\tau_{x'z'})_{max}$

$(\tau_{x'y'})_{max}$

Maximum in-plane shear stress

Absolute maximum shear stress

(b)

Fig. 15–25

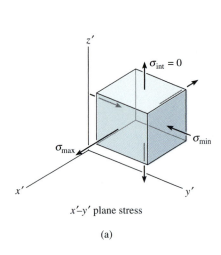

x′–y′ plane stress

(a)

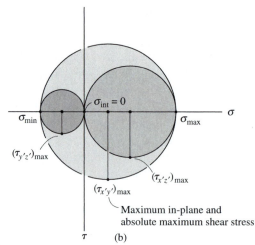

Maximum in-plane and
absolute maximum shear stress

(b)

**Fig. 15–26**

In the case where one of the in-plane principal stresses has the *opposite sign* of that of the other, then these stresses will be represented as $\sigma_{max}$ and $\sigma_{min}$, and the out-of-plane principal stress $\sigma_{int} = 0$, Fig. 15–26a. Mohr's circles that describe this state of stress for element orientations about each coordinate axis are shown in Fig. 15–26b. Clearly, in this case

$$\tau_{\substack{abs \\ max}} = (\tau_{x'y'})_{max} = \frac{\sigma_{max} - \sigma_{min}}{2} \qquad (15\text{--}16)$$

Calculation of the absolute maximum shear stress as indicated here is important when designing members made of a ductile material, since the strength of the material depends on its ability to resist shear stress.

## IMPORTANT POINTS

- The general three-dimensional state of stress at a point can be represented by an element oriented so that only three principal stresses act on it.

- From this orientation, the orientation of the element representing the absolute maximum shear stress can be obtained by rotating the element 45° about the axis defining the direction of $\sigma_{int}$.

- If the in-plane principal stresses both have the *same sign*, the *absolute maximum shear stress* will occur *out of the plane* and has a value of $\tau_{\substack{abs \\ max}} = \sigma_{max}/2$.

- If the in-plane principal stresses are of *opposite signs*, then the *absolute maximum shear stress equals the maximum in-plane shear stress*; that is, $\tau_{\substack{abs \\ max}} = (\sigma_{max} - \sigma_{min})/2$.

# EXAMPLE 15.12

Due to the applied loading, the element at the point on the frame in Fig. 15–27a is subjected to the state of plane stress shown. Determine the principal stresses and the absolute maximum shear stress at the point.

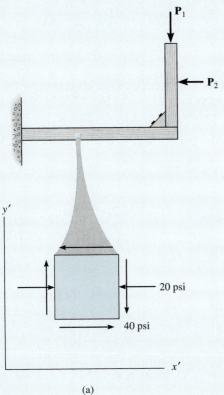

(a)

Fig. 15–27

## Solution

*Principal Stresses.* The in-plane principal stresses can be determined from Mohr's circle. The center of the circle is on the $\sigma$ axis at $\sigma_{avg} = (-20 + 0)/2 = -10$ psi. Plotting the controlling point $A(-20, -40)$, the circle can be drawn as shown in Fig. 15–27b. The radius is

$$R = \sqrt{(20 - 10)^2 + (40)^2} = 41.2 \text{ psi}$$

The principal stresses are at the points where the circle intersects the $\sigma$ axis; i.e.,

$$\sigma_{max} = -10 + 41.2 = 31.2 \text{ psi}$$
$$\sigma_{min} = -10 - 41.2 = -51.2 \text{ psi}$$

From the circle, the *counterclockwise* angle $2\theta$, measured from $CA$ to the $-\sigma$ axis, is

$$2\theta = \tan^{-1}\left(\frac{40}{(20 - 10)}\right) = 76.0°$$

Thus,

$$\theta = 38.0°$$

This *counterclockwise* rotation defines the direction of the $x'$ axis or $\sigma_{min}$ and its associated principal plane, Fig. 15–27c. Since there is no principal stress on the element in the $z$ direction, we have

$$\sigma_{max} = 31.2 \text{ psi} \qquad \sigma_{int} = 0 \qquad \sigma_{min} = -51.2 \text{ psi} \qquad \textit{Ans.}$$

**Absolute Maximum Shear Stress.** Applying Eqs. 15–13 and 15–14, we have

$$\tau_{\substack{abs \\ max}} = \frac{\sigma_{max} - \sigma_{min}}{2} = \frac{31.2 - (-51.2)}{2} = 41.2 \text{ psi} \qquad \textit{Ans.}$$

$$\sigma_{avg} = \frac{\sigma_{max} + \sigma_{min}}{2} = \frac{31.2 - 51.2}{2} = -10 \text{ psi}$$

These same results can also be obtained by drawing Mohr's circle for each orientation of an element about the $x'$, $y'$, and $z'$ axes, Fig. 15–27d. Since $\sigma_{max}$ and $\sigma_{min}$ are of *opposite signs*, then the absolute maximum shear stress equals the maximum in-plane shear stress. This results from a 45° rotation of the element in Fig. 15–27c about the $z'$ axis, so that the properly oriented element is shown in Fig. 15–27e.

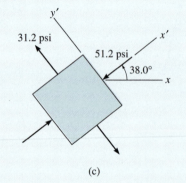

(c)

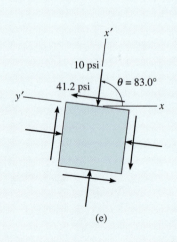

(e)

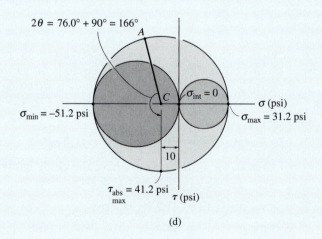

(d)

E X A M P L E **15.13**

The point on the surface of the cylindrical pressure vessel in Fig. 15–28a is subjected to the state of plane stress. Determine the absolute maximum shear stress at this point.

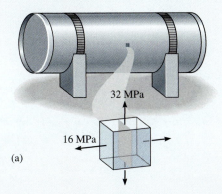

(a)

### Solution

The principal stresses are $\sigma_{max} = 32$ MPa, $\sigma_{int} = 16$ MPa, and $\sigma_{min} = 0$. If these stresses are plotted along the $\sigma$ axis, the three Mohr's circles can be constructed that describe the stress state viewed in each of the three perpendicular planes, Fig. 15–28b. The largest circle has a radius of 16 MPa and describes the state of stress in the plane containing $\sigma_{max} = 32$ MPa and $\sigma_{min} = 0$, shown shaded in Fig. 15–28a. An orientation of an element 45° within this plane yields the state of absolute maximum shear stress and the associated average normal stress, namely,

$$\tau_{\substack{abs \\ max}} = 16 \text{ MPa} \qquad \textit{Ans.}$$
$$\sigma_{avg} = 16 \text{ MPa}$$

These same results can be obtained from direct application of Eqs. 15–13 and 15–14; that is,

$$\tau_{\substack{abs \\ max}} = \frac{\sigma_{max} - \sigma_{min}}{2} = \frac{32 - 0}{2} = 16 \text{ MPa} \qquad \textit{Ans.}$$

$$\sigma_{avg} = \frac{\sigma_{max} + \sigma_{min}}{2} = \frac{32 + 0}{2} = 16 \text{ MPa}$$

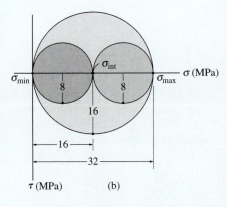

**Fig. 15–28**

By comparison, the maximum in-plane shear stress can be determined from the Mohr's circle drawn between $\sigma_{max} = 32$ MPa and $\sigma_{int} = 16$ MPa, Fig. 15–28a. This gives a value of

$$\tau_{\substack{max \\ in\text{-plane}}} = \frac{32 - 16}{2} = 8 \text{ MPa}$$

$$\sigma_{avg} = 16 + \frac{32 - 16}{2} = 24 \text{ MPa}$$

# PROBLEMS

**15-31.** Draw Mohr's circle that describes each of the following states of stress.

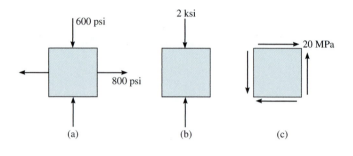

(a)     (b)     (c)

**Prob. 15–31**

**15-33.** Draw Mohr's circle that describes each of the following states of stress.

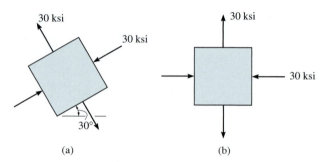

(a)     (b)

**Prob. 15–33**

**15-34.** Mohr's circle for the state of stress in Fig. 15–15*a* is shown in Fig. 15–15*b*. Show that finding the coordinates of point $P(\sigma_{x'}, \tau_{x'y'})$ on the circle gives the same value as the stress-transformation Eqs. 15–1 and 15–2.

**15-35.** The cantilevered rectangular bar is subjected to the force of 5 kip. Determine the principal stresses at point *A*.

***15-36.** Solve Prob. 15–35 for the principal stresses at point *B*.

***15-32.** A point on a thin plate is subjected to two successive states of stress as shown. Determine the resulting state of stress with reference to an element oriented as shown at the right.

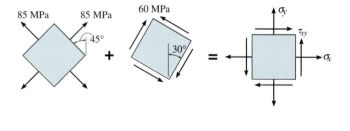

**Prob. 15–32**

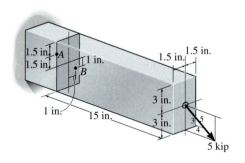

**Probs. 15–35/36**

**15-37.** The pipe has an inner radius of 25 mm and an outer radius of 27 mm. If it is subjected to an internal pressure of 8 MPa and a torsional moment of 500 N · m, determine the principal stresses and the maximum in-plane shear stress at point $A$, which lies on the pipe's outer surface.

**15-39.** The thin-walled pipe has an inner diameter of 0.5 in. and a thickness of 0.025 in. If it is subjected to an internal pressure of 500 psi and the axial tension and torsional loadings shown, determine the principal stresses at a point on the surface of the pipe.

**Prob. 15–39**

**Prob. 15–37**

**\*15-40.** The frame supports the distributed loading of 200 N/m. Determine the normal and shear stresses at point $E$ that act perpendicular and parallel, respectively, to the grains. The grains at this point make an angle of 60° with the horizontal as shown.

**15-38.** The cylindrical pressure vessel has an inner radius of 1.25 m and a wall thickness of 15 mm. It is made from steel plates that are welded along the 45° seam. Determine the normal and shear stress components along this seam if the vessel is subjected to an internal pressure of 8 MPa.

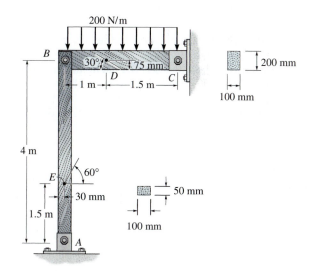

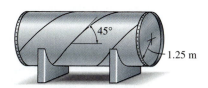

**Prob. 15–38**

**Prob. 15–40**

**15-41.** Draw the three Mohr's circles that describe each of the following states of stress.

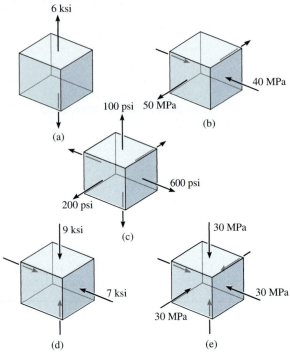

Prob. 15-41

**15-42.** The principal stresses acting at a point in a body are shown. Draw the three Mohr's circles that describe this state of stress, and find the maximum in-plane shear stresses and associated average normal stresses for the x–y, y–z, and x–z planes. For each case, show the results on the element oriented in the appropriate direction.

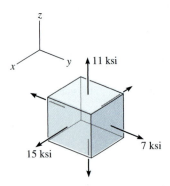

Prob. 15-42

**15-43.** The stress at a point is shown on the element. Determine the principal stresses and the absolute maximum shear stress.

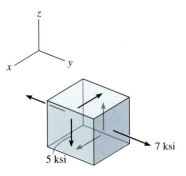

Prob. 15-43

**\*15-44.** The stress at a point is shown on the element. Determine the principal stresses and the absolute maximum shear stress.

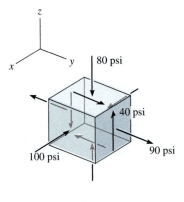

Prob. 15-44

**15-45.** The principal stresses acting at a point in a body are shown. Draw the three Mohr's circles that describe this state of stress and find the maximum in-plane shear stresses and associated average normal stresses for the x–y, y–z, and x–z planes. For each case, show the results on the element oriented in the appropriate direction.

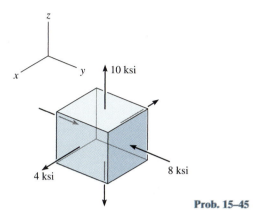

Prob. 15–45

The following problems involve material covered in Chapter 14.

**15-46.** The solid cylinder having a radius r is placed in a sealed container and subjected to a pressure p. Determine the stress components acting at point a located on the center line of the cylinder. Draw Mohr's circles for the element at this point.

Prob. 15–46

**15-47.** Determine the principal stresses and the absolute maximum shear stress at point A on the frame. The cross-sectional area at this point is shown.

**\*15-48.** Solve Prob. 15–47 for point B.

**15-49.** The bolt is fixed to its support at C. If a force of 18 lb is applied to the wrench to tighten it, determine the principal stresses and the absolute maximum shear stress developed in the bolt shank at point A. Represent the results on an element located at this point. The shank has a diameter of 0.25 in.

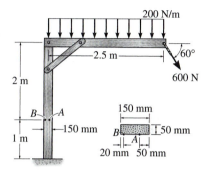

Probs. 15–48/49

**15-50.** Solve Prob. 15–49 for point B.

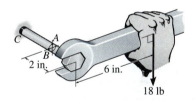

Probs. 15–49/50

## 15.6 Plane Strain

As outlined in Sec. 8.8, the general state of strain at a point in a body is represented by a combination of three components of normal strain, $\epsilon_x$, $\epsilon_y$, $\epsilon_z$, and three components of shear strain $\gamma_{xy}$, $\gamma_{xz}$, $\gamma_{yz}$. These six components tend to deform each face of an element of the material, and like stress, the normal and shear strain *components* at the point will vary according to the orientation of the element. The *strain* components at a point are often determined by using strain gauges, which measure these components in *specified directions*. For both analysis and design, however, engineers must sometimes transform this data in order to obtain the strain components in other directions.

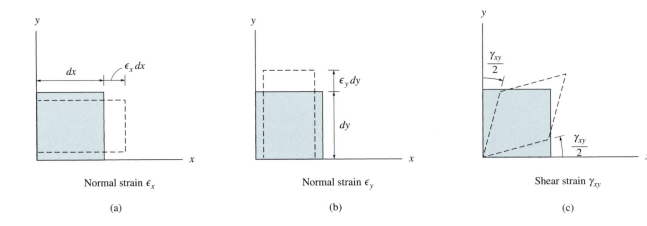

Normal strain $\epsilon_x$

(a)

Normal strain $\epsilon_y$

(b)

Shear strain $\gamma_{xy}$

(c)

**Fig. 15–29**

The rubber specimen is constrained between the two fixed supports, and so it will undergo plane strain when loads are applied to it in the horizontal plane.

To understand how this is done, we will first confine our attention to a study of *plane strain*. Specifically, we will not consider the effects of the components $\epsilon_z$, $\gamma_{xz}$, and $\gamma_{yz}$. In general, then, a plane-strained element is subjected to two components of normal strain, $\epsilon_x$, $\epsilon_y$, and one component of shear strain, $\gamma_{xy}$. The deformations of an element caused by each of these strains are shown graphically in Fig. 15–29. Note that the normal strains are produced by *changes in length* of the element in the $x$ and $y$ directions, and the shear strain is produced by the *relative rotation* of two adjacent sides of the element.

Although plane strain and plane stress each have three components lying in the same plane, realize that plane stress *does not* necessarily cause plane strain or vice versa. The reason for this has to do with the Poisson effect discussed in Sec. 9.6. For example, if the element in Fig. 15–30 is subjected to plane stress $\sigma_x$ and $\sigma_y$, not only are normal strains $\epsilon_x$ and $\epsilon_y$ produced, but there is *also* an associated normal strain, $\epsilon_z$. This is obviously *not* a case of plane strain. In general, then, unless $v = 0$, the Poisson effect will *prevent* the simultaneous occurrence of plane strain and plane stress. It should also be pointed out that since shear stress and shear strain are *not* affected by Poisson's ratio, a condition of $\tau_{xz} = \tau_{yz} = 0$ requires $\gamma_{xz} = \gamma_{yz} = 0$.

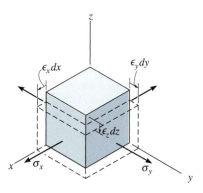

Plane stress, $\sigma_x$, $\sigma_y$, does not cause plane
strain in the $x$–$y$ plane since $\epsilon_z \neq 0$.

**Fig. 15–30**

## 15.7  General Equations of Plane-Strain Transformation

It is important in plane-strain analysis to establish transformation equations that can be used to determine the $x'$, $y'$ components of normal and shear strain at a point, provided the $x, y$ components of strain are known. Essentially this problem is one of geometry and requires relating the deformations and rotations of differential line segments, which represent the sides of differential elements that are parallel to each set of axes.

**Sign Convention.**  Before the strain-transformation equations can be developed, we must first establish a sign convention for the strains. This convention is the same as that established in Sec. 8.8 and will be restated here for the condition of plane strain. With reference to the differential element shown in Fig. 15–31a, *normal strains* $\epsilon_x$ and $\epsilon_y$ are *positive* if they cause *elongation* along the $x$ and $y$ axes, respectively, and the *shear strain* $\gamma_{xy}$ is *positive* if the interior angle $AOB$ *becomes smaller* than 90°. This sign convention also follows the corresponding one used for plane stress, Fig. 15–5a, that is, positive $\sigma_x$, $\sigma_y$, $\tau_{xy}$ will cause the element to *deform* in the positive $\epsilon_x$, $\epsilon_y$, $\gamma_{xy}$ directions, respectively.

The problem here will be to determine at a point the normal and shear strains $\epsilon_{x'}$, $\epsilon_{y'}$, $\gamma_{x'y'}$, measured relative to the $x'$, $y'$ axes, if we know $\epsilon_x$, $\epsilon_y$, $\gamma_{xy}$, measured relative to the $x, y$ axes. If the angle between the $x$ and $x'$ axes is $\theta$, then, like the case of plane stress, $\theta$ will be *positive* provided it follows the curl of the right-hand fingers, i.e., counterclockwise, as shown in Fig. 15–31b.

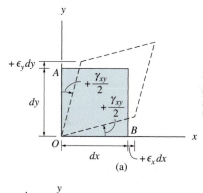

(a)

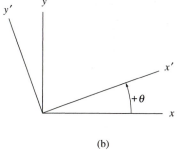

(b)

Positive sign convention

**Fig. 15–31**

**Normal and Shear Strains.** In order to develop the strain-transformation equation for determining $\epsilon_{x'}$, we must determine the elongation of a line segment $dx'$ that lies along the $x'$ axis and is subjected to strain components $\epsilon_x$, $\epsilon_y$, $\gamma_{xy}$. As shown in Fig. 15–32a, the components of the line $dx'$ along the $x$ and $y$ axes are

$$dx = dx' \cos \theta$$
$$dy = dx' \sin \theta \qquad (15\text{–}17)$$

When the positive normal strain $\epsilon_x$ occurs, Fig. 15–32b, the line $dx$ is elongated $\epsilon_x \, dx$, which causes line $dx'$ to elongate $\epsilon_x \, dx \cos \theta$. Likewise, when $\epsilon_y$ occurs, Fig. 15–32c, line $dy$ elongates $\epsilon_y \, dy$, $dy$, which causes line $dx'$ to elongate $\epsilon_y \, dy \sin \theta$. Lastly, assuming that $dx$ remains fixed in position, the shear strain $\gamma_{xy}$, which is the change in angle between $dx$ and $dy$, causes the top of line $dy$ to be displaced $\gamma_{xy} \, dy$ to the right, as shown in Fig. 15–32d. This causes $dx'$ to elongate $\gamma_{xy} \, dy \cos \theta$. If all three of these elongations are added together, the resultant elongation of $dx'$ is then

$$\delta_{x'} = \epsilon_x \, dx \cos \theta + \epsilon_y \, dy \sin \theta + \gamma_{xy} \, dy \cos \theta$$

From Eq. 8–11, the normal strain along the line $dx'$ is $\epsilon_{x'} = \delta x'/dx'$. Using Eq. 15–17, we therefore have

$$\epsilon_{x'} = \epsilon_x \cos^2 \theta + \epsilon_y \sin^2 \theta + \gamma_{xy} \sin \theta \cos \theta \qquad (15\text{–}18)$$

The strain-transformation equation for determining $\gamma_{x'y'}$ can be developed by considering the amount of rotation each of the line segments $dx'$ and $dy'$ undergo when subjected to the strain components $\epsilon_x$, $\epsilon_y$, $\gamma_{xy}$. First we will consider the rotation of $dx'$, which is defined by the counterclockwise angle $\alpha$ shown in Fig. 15–32e. It can be determined from the displacement $\delta y'$ using $\alpha = \delta y'/dx'$. To obtain $\delta y'$, consider the following three displacement components acting in the $y'$ direction: one from $\epsilon_x$, giving $-\epsilon_x \, dx \sin \theta$, Fig. 15–32b; another from $\epsilon_y$, giving $\epsilon_y \, dy \cos \theta$, Fig. 15–32c; and the last from $\gamma_{xy}$, giving $-\gamma_{xy} \, dy \sin \theta$, Fig. 15–32d. Thus, $\delta y'$, as caused by all three strain components, is

$$\delta y' = -\epsilon_x \, dx \sin \theta + \epsilon_y \, dy \cos \theta - \gamma_{xy} \, dy \sin \theta$$

Using Eq. 15–17, with $\alpha = \delta y'/dx'$, we have

$$\alpha = (-\epsilon_x + \epsilon_y) \sin \theta \cos \theta - \gamma_{xy} \sin^2 \theta \qquad (15\text{–}19)$$

As shown in Fig. 15–32e, the line $dy'$ rotates by an amount $\beta$. We can determine this angle by a similar analysis, or by simply substituting $\theta + 90°$ for $\theta$ into Eq. 15–19. Using the identities $\sin(\theta + 90°) = \cos \theta$, $\cos(\theta + 90°) = -\sin \theta$, we have

$$\beta = (-\epsilon_x + \epsilon_y) \sin(\theta + 90°) \cos(\theta + 90°) - \gamma_{xy} \sin^2(\theta + 90°)$$
$$= -(-\epsilon_x + \epsilon_y) \cos \theta \sin \theta - \gamma_{xy} \cos^2 \theta$$

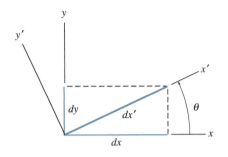

Before deformation

(a)

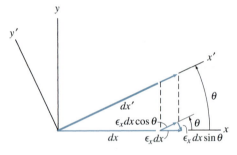

Normal strain $\epsilon_x$

(b)

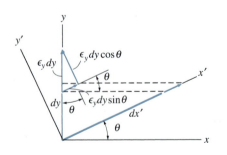

Normal strain $\epsilon_y$

(c)

**Fig. 15–32**

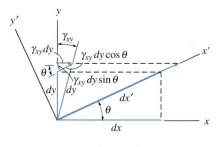

Shear strain $\gamma_{xy}$

(d)

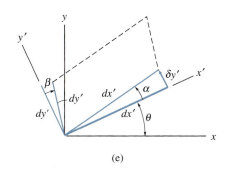

(e)

Fig. 15–32

Since $\alpha$ and $\beta$ represent the rotation of the sides $dx'$ and $dy'$ of a differential element whose sides were originally oriented along the $x'$ and $y'$ axes, and $\beta$ is in the opposite direction to $\alpha$, Fig. 15–32e, the element is then subjected to a shear strain of

$$\gamma_{x'y'} = \alpha - \beta = -2(\epsilon_x - \epsilon_y) \sin\theta \cos\theta + \gamma_{xy}(\cos^2\theta - \sin^2\theta) \quad (15\text{–}20)$$

Using the trigonometric identities $\sin 2\theta = 2 \sin\theta \cos\theta$, $\cos^2\theta = (1 + \cos 2\theta)/2$, and $\sin^2\theta + \cos^2\theta = 1$, we can rewrite Eqs. 15–18 and 15–20 in the final form

$$\epsilon_{x'} = \frac{\epsilon_x + \epsilon_y}{2} + \frac{\epsilon_x - \epsilon_y}{2}\cos 2\theta + \frac{\gamma_{xy}}{2}\sin 2\theta \quad (15\text{–}21)$$

$$\frac{\gamma_{x'y'}}{2} = -\left(\frac{\epsilon_x - \epsilon_y}{2}\right)\sin 2\theta + \frac{\gamma_{xy}}{2}\cos 2\theta \quad (15\text{–}22)$$

These strain-transformation equations give the normal strain $\epsilon_{x'}$ in the $x'$ direction and the shear strain $\gamma_{x'y'}$ of an element oriented at an angle $\theta$, as shown in Fig. 15–33. According to the established sign convention, if $\epsilon_{x'}$ is *positive*, the element *elongates* in the positive $x'$ direction, Fig. 15–33a, and if $\gamma_{x'y'}$ is positive, the element deforms as shown in Fig. 15–33b. Note that these deformations occur as if positive normal stress $\sigma_{x'}$ and positive shear stress $\tau_{x'y'}$ act on the element.

If the normal strain in the $y'$ direction is required, it can be obtained from Eq. 15–21 by simply substituting $(\theta + 90°)$ for $\theta$. The result is

$$\epsilon_{y'} = \frac{\epsilon_x + \epsilon_y}{2} - \frac{\epsilon_x - \epsilon_y}{2}\cos 2\theta - \frac{\gamma_{xy}}{2}\sin 2\theta \quad (15\text{–}23)$$

The similarity between the above three equations and those for plane-stress transformation, Eqs. 15–1, 15–2, and 15–3, should be noted. By comparison, $\sigma_x$, $\sigma_y$, $\sigma_{x'}$, $\sigma_{y'}$ correspond to $\epsilon_x$, $\epsilon_y$, $\epsilon_{x'}$, $\epsilon_{y'}$; and $\tau_{xy}$, $\tau_{x'y'}$ correspond to $\gamma_{xy}/2$, $\gamma_{x'y'}/2$.

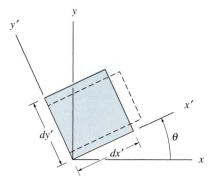

Positive normal strain, $\epsilon_{x'}$

(a)

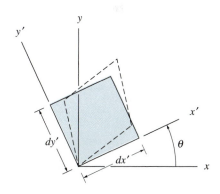

Positive shear strain, $\gamma_{x'y'}$

(b)

Fig. 15–33

**Principal Strains.** Like stress, the orientation of an element at a point can be determined such that the element's deformation is represented by normal strains, with *no* shear strain. When this occurs the normal strains are referred to as *principal strains*, and if the material is isotropic, the axes along which these strains occur coincide with the axes that define the planes of principal stress.

From Eqs. 15–4 and 15–5, and the correspondence between stress and strain mentioned above, the direction of the axes and the two values of the principal strains $\epsilon_1$ and $\epsilon_2$ are determined from

$$\tan 2\theta_p = \frac{\gamma_{xy}}{\epsilon_x - \epsilon_y} \qquad (15\text{–}24)$$

$$\epsilon_{1,2} = \frac{\epsilon_x + \epsilon_y}{2} \pm \sqrt{\left(\frac{\epsilon_x - \epsilon_y}{2}\right)^2 + \left(\frac{\gamma_{xy}}{2}\right)^2} \qquad (15\text{–}25)$$

**Maximum In-Plane Shear Strain.** Using Eqs. 15–6, 15–7, and 15–8, the direction of the axis, and the maximum in-plane shear strain and associated average normal strain are determined from the following equations:

$$\tan 2\theta_s = -\left(\frac{\epsilon_x - \epsilon_y}{\gamma_{xy}}\right) \qquad (15\text{–}26)$$

$$\frac{\gamma_{\substack{\text{max} \\ \text{in-plane}}}}{2} = \sqrt{\left(\frac{\epsilon_x - \epsilon_y}{2}\right)^2 + \left(\frac{\gamma_{xy}}{2}\right)^2} \qquad (15\text{–}27)$$

$$\epsilon_{\text{avg}} = \frac{\epsilon_x + \epsilon_y}{2} \qquad (15\text{–}28)$$

Complex stresses are often developed at the joints where vessels are connected together. The stresses are determined by making measurements of strain.

## IMPORTANT POINTS

- Due to the Poisson effect, the state of plane strain is not a state of plane stress, and vice versa.

- A point on a body is subjected to plane stress when the surface of the body is stress free. Plane-strain analysis may be used within the plane of the stresses to analyze the results from the gauges. Remember, though, there is a normal strain that is perpendicular to the gauges.

- When the state of strain is represented by the principal strains, no shear strain will act on the element.

- The state of strain at the point can also be represented in terms of the maximum in-plane shear strain. In this case an average normal strain will also act on the element.

- The element representing the maximum in-plane shear strain and its associated average normal strains is 45° from the element representing the principal strains.

## EXAMPLE 15.14

A differential element of material at a point is subjected to a state of plane strain $\epsilon_x = 500(10^{-6})$, $\epsilon_y = -300(10^{-6})$, $\gamma_{xy} = 200(10^{-6})$, which tends to distort the element as shown in Fig. 15–34a. Determine the equivalent strains acting on an element oriented at the point, *clockwise* 30° from the original position.

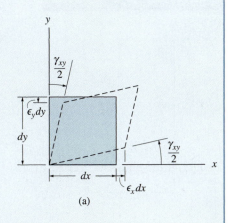

(a)

### Solution

The strain-transformation Eqs. 15–21 and 15–22 will be used to solve the problem. Since $\theta$ is *positive counterclockwise*, then for this problem $\theta = -30°$. Thus,

$$\epsilon_{x'} = \frac{\epsilon_x + \epsilon_y}{2} + \frac{\epsilon_x - \epsilon_y}{2}\cos 2\theta + \frac{\gamma_{xy}}{2}\sin 2\theta$$

$$= \left[\frac{500 + (-300)}{2}\right](10^{-6}) + \left[\frac{500 - (-300)}{2}\right](10^{-6})\cos(2(-30°))$$

$$+ \left[\frac{200(10^{-6})}{2}\right]\sin(2(-30°))$$

$$\epsilon_{x'} = 213(10^{-6}) \qquad\qquad Ans.$$

$$\frac{\gamma_{x'y'}}{2} = -\left(\frac{\epsilon_x - \epsilon_y}{2}\right)\sin 2\theta + \frac{\gamma_{xy}}{2}\cos 2\theta$$

$$= -\left[\frac{500 - (-300)}{2}\right](10^{-6})\sin(2(-30°)) + \frac{200(10^{-6})}{2}\cos(2(-30°))$$

$$\gamma_{x'y'} = 793(10^{-6}) \qquad\qquad Ans.$$

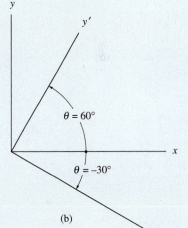

(b)

The strain in the $y'$ direction can be obtained from Eq. 15–23 with $\theta = -30°$. However, we can also obtain $\epsilon_{y'}$ using Eq. 15–21 with $\theta = 60°(\theta = -30° + 90°)$, Fig. 15–34b. We have with $\epsilon_{y'}$ replacing $\epsilon_{x'}$,

$$\epsilon_{y'} = \frac{\epsilon_x + \epsilon_y}{2} + \frac{\epsilon_x - \epsilon_y}{2}\cos 2\theta + \frac{\gamma_{xy}}{2}\sin 2\theta$$

$$= \left[\frac{500 + (-300)}{2}\right](10^{-6}) + \left[\frac{500 - (-300)}{2}\right](10^{-6})\cos(2(60°))$$

$$+ \frac{200(10^{-6})}{2}\sin(2(60°))$$

$$\epsilon_{y'} = -13.4(10^{-6}) \qquad\qquad Ans.$$

These results tend to distort the element as shown in Fig. 15–34c.

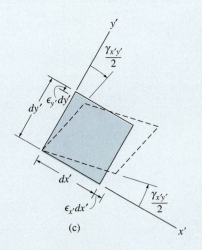

(c)

Fig. 15–34

E X A M P L E  **15.15**

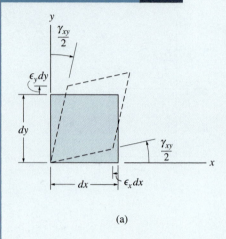

(a)

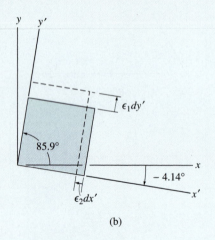

(b)

**Fig. 15–35**

A differential element of material at a point is subjected to a state of plane strain defined by $\epsilon_x = -350(10^{-6})$, $\epsilon_y = 200(10^{-6})$, $\gamma_{xy} = 80(10^{-6})$, which tends to distort the element as shown in Fig. 15–35a. Determine the principal strains at the point and the associated orientation of the element.

**Solution**

*Orientation of the Element.* From Eq. 15–24 we have

$$\tan 2\theta_p = \frac{\gamma_{xy}}{\epsilon_x - \epsilon_y}$$

$$= \frac{80(10^{-6})}{(-350 - 200)(10^{-6})}$$

Thus, $2\theta_p = -8.28°$ and $-8.28° + 180° = 172°$, so that

$$\theta_p = -4.14° \text{ and } 85.9° \qquad Ans.$$

Each of these angles is measured *positive counterclockwise*, from the $x$ axis to the outward normals on each face of the element, Fig. 15–35b.

*Principal Strains.* The principal strains are determined from Eq. 15–25. We have

$$\epsilon_{1,2} = \frac{\epsilon_x + \epsilon_y}{2} \pm \sqrt{\left(\frac{\epsilon_x - \epsilon_y}{2}\right)^2 + \left(\frac{\gamma_{xy}}{2}\right)^2}$$

$$= \frac{(-350 + 200)(10^{-6})}{2} \pm \left[\sqrt{\left(\frac{-350 - 200}{2}\right)^2 + \left(\frac{80}{2}\right)^2}\right](10^{-6})$$

$$= -75.0(10^{-6}) \pm 277.9(10^{-6})$$

$$\epsilon_1 = 203(10^{-6}) \qquad \epsilon_2 = -353(10^{-6}) \qquad Ans.$$

We can determine which of these two strains deforms the element in the $x'$ direction by applying Eq. 15–21 with $\theta = -4.14°$. Thus,

$$\epsilon_{x'} = \frac{\epsilon_x + \epsilon_y}{2} + \frac{\epsilon_x - \epsilon_y}{2}\cos 2\theta + \frac{\gamma_{xy}}{2}\sin 2\theta$$

$$= \left(\frac{-350 + 200}{2}\right)(10^{-6}) + \left(\frac{-350 - 200}{2}\right)(10^{-6})\cos 2(-4.14°)$$

$$+ \frac{80(10^{-6})}{2}\sin 2(-4.14°)$$

$$\epsilon_{x'} = -353(10^{-6})$$

Hence $\epsilon_{x'} = \epsilon_2$. When subjected to the principal strains, the element is distorted as shown in Fig. 15–35b.

# EXAMPLE 15.16

A differential element of material at a point is subjected to a state of plane strain defined by $\epsilon_x = -350(10^{-6})$, $\epsilon_y = 200(10^{-6})$, $\gamma_{xy} = 80(10^{-6})$, which tends to distort the element as shown in Fig. 15–36a. Determine the maximum in-plane shear strain at the point and the associated orientation of the element.

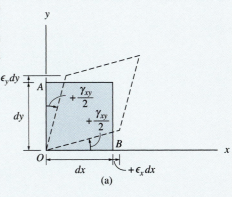

(a)

## Solution

*Orientation of the Element.* From Eq. 15–26 we have

$$\tan 2\theta_s = -\left(\frac{\epsilon_x - \epsilon_y}{\gamma_{xy}}\right) = \frac{(-350 - 200)(10^{-6})}{80(10^{-6})}$$

Thus, $2\theta_s = 81.72°$ and $81.72° + 180° = 261.72°$, so that

$$\theta_s = 40.9° \text{ and } 130.9°$$

Note that this orientation is 45° from that shown in Fig. 15–35b in Example 15.15 as expected.

*Maximum In-Plane Shear Strain.* Applying Eq. 15–27 gives

$$\frac{\gamma_{\substack{\max \\ \text{in-plane}}}}{2} = \sqrt{\left(\frac{\epsilon_x - \epsilon_y}{2}\right)^2 + \left(\frac{\gamma_{xy}}{2}\right)^2}$$

$$= \left[\sqrt{\left(\frac{-350 - 200}{2}\right)^2 + \left(\frac{80}{2}\right)^2}\right](10^{-6})$$

$$\gamma_{\substack{\max \\ \text{in-plane}}} = 556(10^{-6}) \qquad \textit{Ans.}$$

The proper sign of $\gamma_{\substack{\max \\ \text{in-plane}}}$ can be obtained by applying Eq. 15–22 with $\theta_s = 40.9°$. We have

$$\frac{\gamma_{x'y'}}{2} = -\frac{\epsilon_x - \epsilon_y}{2}\sin 2\theta + \frac{\gamma_{xy}}{2}\cos 2\theta$$

$$= -\left(\frac{-350 - 200}{2}\right)(10^{-6})\sin 2(40.9°) + \frac{80(10^{-6})}{2}\cos 2(40.9°)$$

$$\gamma_{x'y'} = 556(10^{-6})$$

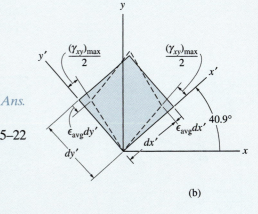

(b)

**Fig. 15–36**

Thus, $\gamma_{\substack{\max \\ \text{in-plane}}}$ tends to distort the element so that the right angle between $dx'$ and $dy'$ is decreased (positive sign convention), Fig. 15–36b.

Also, there are associated average normal strains imposed on the element that are determined from Eq. 15–28:

$$\epsilon_{avg} = \frac{\epsilon_x + \epsilon_y}{2} = \frac{-350 + 200}{2}(10^{-6}) = -75(10^{-6})$$

These strains tend to cause the element to contract, Fig. 15–36b.

## 15.8 Mohr's Circle—Plane Strain

Since the equations of plane-strain transformation are mathematically similar to the equations of plane-stress transformation, we can also solve problems involving the transformation of strain using Mohr's circle. This approach has the advantage of making it possible to see graphically how the normal and shear strain components at a point vary from one orientation of the element to the next.

Like the case for stress, the parameter $\theta$ in Eqs. 15–21 and 15–22 can be eliminated and the result rewritten in the form

$$(\epsilon_x - \epsilon_{avg})^2 + \left(\frac{\gamma_{xy}}{2}\right)^2 = R^2 \tag{15–29}$$

where

$$\epsilon_{avg} = \frac{\epsilon_x + \epsilon_y}{2}$$

$$R = \sqrt{\left(\frac{\epsilon_x - \epsilon_y}{2}\right)^2 + \left(\frac{\gamma_{xy}}{2}\right)^2}$$

Equation 15–29 represents the equation of Mohr's circle for strain. It has a center on the $\epsilon$ axis at point $C(\epsilon_{avg}, 0)$ and a radius $R$.

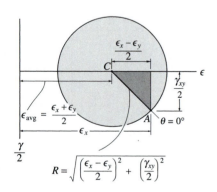

$$R = \sqrt{\left(\frac{\epsilon_x - \epsilon_y}{2}\right)^2 + \left(\frac{\gamma_{xy}}{2}\right)^2}$$

**Fig. 15–37**

### PROCEDURE FOR ANALYSIS

The procedure for drawing Mohr's circle for strain follows the same one established for stress.

*Construction of the Circle.*

• Establish a coordinate system such that the abscissa represents the normal strain $\epsilon$, with *positive to the right*, and the ordinate represents *half* the value of the shear strain, $\gamma/2$, with *positive downward*, Fig. 15–37.

• Using the positive sign convention for $\epsilon_x$, $\epsilon_y$, $\gamma_{xy}$, as shown in Fig. 15–31, determine the center of the circle $C$, which is located on the $\epsilon$ axis at a distance $\epsilon_{avg} = (\epsilon_x + \epsilon_y)/2$ from the origin, Fig. 15–37.

• Plot the reference point $A$ having coordinates $A(\epsilon_x, \gamma_{xy}/2)$. This point represents the case for which the $x'$ axis coincides with the $x$ axis. Hence $\theta = 0°$, Fig. 15–37.

• Connect point $A$ with the center $C$ of the circle and from the shaded triangle determine the radius $R$ of the circle, Fig. 15–37.

• Once $R$ has been determined, sketch the circle.

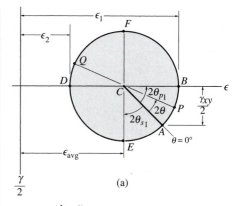

(a)

*Principal Strains.*

- The principal strains $\epsilon_1$ and $\epsilon_2$ are determined from the circle as the coordinates of points $B$ and $D$, that is where $\gamma/2 = 0$, Fig. 15–38a.

- The orientation of the plane on which $\epsilon_1$ acts can be determined from the circle by calculating $2\theta_{p_1}$ using trigonometry. Here this angle is measured counterclockwise *from* the radial reference line $CA$ to line $CB$, Fig. 15–38a. Remember that the *rotation* of $\theta_{p_1}$ must be in this *same direction*, from the element's reference axis $x$ to the $x'$ axis, Fig. 15–38b.*

- When $\epsilon_1$ and $\epsilon_2$ are indicated as being positive as in Fig. 15–38a, the element in Fig. 15–38b will elongate in the $x'$ and $y'$ directions as shown by the dashed outline.

*Maximum In-Plane Shear Strain.*

- The average normal strain and half the maximum in-plane shear strain are determined from the circle as the coordinates of points $E$ and $F$, Fig. 15–38a.

- The orientation of the plane on which $\gamma_{\text{max in-plane}}$ and $\epsilon_{\text{avg}}$ act can be determined from the circle by calculating $2\theta_{s_1}$ using trigonometry. Here this angle is measured clockwise from the radial reference line $CA$ to line $CE$, Fig. 15–38a. Remember that the rotation of $\theta_{s_1}$ must be in this same direction, from the element's reference axis $x$ to the $x'$ axis, Fig. 15–38c.*

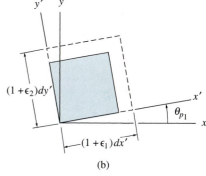

(b)

*Strains on Arbitrary Plane.*

- The normal and shear strain components $\epsilon_{x'}$ and $\gamma_{x'y'}$ for a plane specified at an angle $\theta$, Fig. 15–38d, can be obtained from the circle using trigonometry to determine the coordinates of point $P$, Fig. 15–38a.

- To locate $P$, the known angle $\theta$ of the $x'$ axis is measured on the circle as $2\theta$. This measurement is made *from* the radial reference line $CA$ *to* the radial line $CP$. Remember that measurements for $2\theta$ on the circle must be in the same direction as $\theta$ for the $x'$ axis.*

- If the value of $\epsilon_{y'}$ is required, it can be determined by calculating the $\epsilon$ coordinate of point $Q$ in Fig. 15–38a. The line $CQ$ lies $180°$ away from $CP$ and thus represents a rotation of $90°$ of the $x'$ axis.

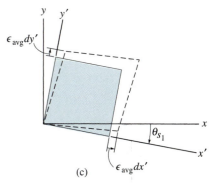

(c)

*If instead the $\gamma/2$ axis is constructed *positive upwards*, then the angle $2\theta$ on the circle would be measured in the *opposite direction* to the orientation $\theta$ of the plane.

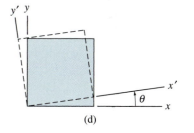

(d)

**Fig. 15–38**

## E X A M P L E 15.17

The state of plane strain at a point is represented by the components $\epsilon_x = 250(10^{-6})$, $\epsilon_y = -150(10^{-6})$, and $\gamma_{xy} = 120(10^{-6})$. Determine the principal strains and the orientation of the element.

### Solution

*Construction of the Circle.* The $\epsilon$ and $\gamma/2$ axes are established in Fig. 15–39a. Remember that the *positive* $\gamma/2$ axis must be directed *downward* so that *counterclockwise* rotations of the element correspond to *counterclockwise* rotation around the circle, and vice versa. The center of the circle $C$ is located on the $\epsilon$ axis at

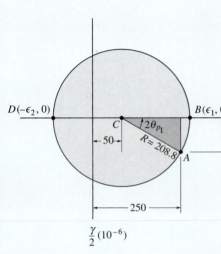

(a)

$$\epsilon_{\text{avg}} = \frac{250 + (-150)}{2}(10^{-6}) = 50(10^{-6})$$

Since $\gamma_{xy}/2 = 60(10^{-6})$, the reference point $A$ ($\theta = 0°$) has coordinates $A(250(10^{-6}), 60(10^{-6}))$. From the shaded triangle in Fig. 15–39a, the radius of the circle is $CA$; that is,

$$R = \left[\sqrt{(250 - 50)^2 + (60)^2}\right](10^{-6}) = 208.8(10^{-6})$$

*Principal Strains.* The $\epsilon$ coordinates of points $B$ and $D$ represent the principal strains. They are

$$\epsilon_1 = (50 + 208.8)(10^{-6}) = 259(10^{-6}) \qquad \textit{Ans.}$$

$$\epsilon_2 = (50 - 208.8)(10^{-6}) = -159(10^{-6}) \qquad \textit{Ans.}$$

The direction of the positive principal strain $\epsilon_1$ is defined by the *counterclockwise* angle $2\theta_{p_1}$, measured from the radial reference line $CA$ to the line $CB$. We have

$$\tan 2\theta_{p_1} = \frac{60}{(250 - 50)}$$

$$\theta_{p_1} = 8.35° \qquad \textit{Ans.}$$

Hence the side $dx'$ of the element is oriented *counterclockwise* 8.35° as shown in Fig. 15–39a. This also defines the direction of $\epsilon_1$. The deformation of the element is also shown in the figure.

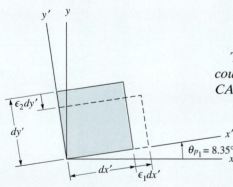

(b)

**Fig. 15–39**

# EXAMPLE 15.18

The state of plane strain at a point is represented by the components $\epsilon_x = 250(10^{-6})$, $\epsilon_y = -150(10^{-6})$, and $\gamma_{xy} = 120(10^{-6})$. Determine the maximum in-plane shear strains and the orientation of an element.

## Solution

The circle has been established in the previous example and is shown in Fig. 15–40a.

*Maximum In-Plane Shear Strain.* Half the maximum in-plane shear strain and average normal strain are represented by the coordinates of point $E$ or $F$ on the circle. From the coordinates of point $E$,

$$\frac{(\gamma_{x'y'})_{\substack{\text{max} \\ \text{in-plane}}}}{2} = 208.8(10^{-6})$$

$$(\gamma_{x'y'})_{\substack{\text{max} \\ \text{in-plane}}} = 418(10^{-6}) \qquad Ans.$$

$$\epsilon_{\text{avg}} = 50(10^{-6})$$

To orient the element, we can determine the clockwise angle $2\theta_{s_1}$ from the circle.

$$2\theta_{s_1} = 90° - 2(8.35°)$$

$$\theta_{s_1} = 36.6° \qquad Ans.$$

This angle is shown in Fig. 15–40b. Since the shear strain defined from point $E$ on the circle has a positive value and the average normal strain is also positive, corresponding positive shear stress and positive average normal stress deform the element into the dashed shape shown in the figure.

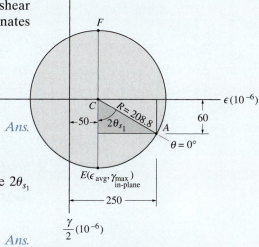

(a)

Fig. 15–40

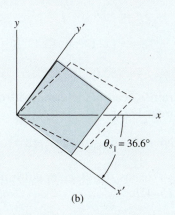

(b)

## E X A M P L E 15.19

The state of plane strain at a point is represented on an element having components $\epsilon_x = -300(10^{-6})$, $\epsilon_y = -100(10^{-6})$, and $\gamma_{xy} = 100(10^{-6})$. Determine the state of strain on an element oriented 20° clockwise from this reported position.

### Solution

***Construction of the Circle.*** The $\epsilon$ and $\gamma/2$ axes are established in Fig. 15–41a. The center of the circle is on the $\epsilon$ axis at

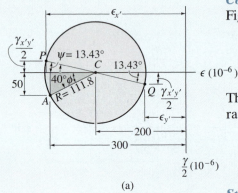

$$\epsilon_{avg} = \left(\frac{-300 - 100}{2}\right)(10^{-6}) = -200(10^{-6})$$

The reference point $A$ has coordinates $A(-300(10^{-6}), 50(10^{-6}))$. The radius $CA$ determined from the shaded triangle is therefore

$$R = \left[\sqrt{(300 - 200)^2 + (50)^2}\right](10^{-6}) = 111.8(10^{-6})$$

(a)

***Strains on Inclined Element.*** Since the element is to be oriented 20° *clockwise*, we must establish a radial line $CP$, $2(20°) = 40°$ *clockwise*, measured from $CA$ ($\theta = 0°$), Fig. 15–41a. The coordinates of point $P$ ($\epsilon_{x'}$, $-\gamma_{x'y'}/2$) are obtained from the geometry of the circle. Note that

$$\phi = \tan^{-1}\left(\frac{50}{(300 - 200)}\right) = 26.57°, \quad \psi = 40° - 26.57° = 13.43°$$

Thus,

$$\epsilon_{x'} = -(200 + 111.8 \cos 13.43°)(10^{-6})$$
$$= -309(10^{-6}) \qquad\qquad Ans.$$

$$\frac{\gamma_{x'y'}}{2} = -(111.8 \sin 13.43°)(10^{-6})$$

$$\gamma_{x'y'} = -52.0(10^{-6}) \qquad\qquad Ans.$$

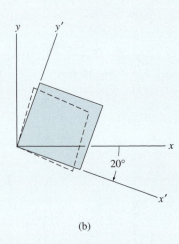

(b)

**Fig. 15–41**

The normal strain $\epsilon_{y'}$ can be determined from the $\epsilon$ coordinate of point $Q$ on the circle, Fig. 15–41a. Why?

$$\epsilon_{y'} = -(200 - 111.8 \cos 13.43°)(10^{-6}) = -91.3(10^{-6}) \qquad Ans.$$

As a result of these strains, the element deforms relative to $x'$, $y'$ axes as shown in Fig. 15–41b.

# PROBLEMS

**15-51.** Prove that the sum of the normal strains in perpendicular directions is constant.

**\*15-52.** The state of strain at the point on the bracket has components $\epsilon_x = -200(10^{-6})$, $\epsilon_y = -650(10^{-6})$, $\gamma_{xy} = -175(10^{-6})$. Use the strain-transformation equations to determine the equivalent in-plane strains on an element oriented at an angle of $\theta = 20°$ counterclockwise from the original position. Sketch the deformed element due to these strains within the x–y plane.

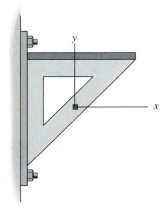

Prob. 15–52

**15-53.** A differential element on the bracket is subjected to plane strain that has the following components: $\epsilon_x = 150(10^{-6})$, $\epsilon_y = 200(10^{-6})$, $\gamma_{xy} = -700(10^{-6})$. Use the strain-transformation equations and determine the equivalent in-plane strains on an element oriented at an angle of $\theta = 60°$ counterclockwise from the original position. Sketch the deformed element within the x–y plane due to these strains.

**15-54.** Solve Prob. 15–57 for an element oriented $\theta = 30°$ clockwise.

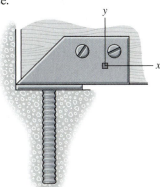

Probs. 15–53/54

**15-55.** Due to the load **P**, the state of strain at the point on the bracket has components of $\epsilon_x = 500(10^{-6})$, $\epsilon_y = 350(10^{-6})$, and $\gamma_{xy} = -430(10^{-6})$. Use the strain-transformation equations to determine the equivalent in-plane strains on an element oriented at an angle of $\theta = 30°$ clockwise from the original position. Sketch the deformed element due to these strains with in the x–y plane.

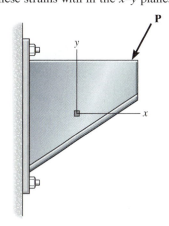

Prob. 15–55

**\*15-56.** The state of strain at the point on the boom of the hydraulic engine crane has components of $\epsilon_x = 250(10^{-6})$, $\epsilon_y = 300(10^{-6})$, and $\gamma_{xy} = -180(10^{-6})$. Use the strain-transformation equations to determine (a) the in-plane principal strains and (b) the maximum in-plane shear strain and average normal strain. In each case, specify the orientation of the element and show how the strains deform the element within the x–y plane.

Prob. 15–56

**15-57.** The state of strain at the point on the gear tooth has components $\epsilon_x = 850(10^{-6})$, $\epsilon_y = 480(10^{-6})$, $\gamma_{xy} = 650(10^{-6})$. Use the strain-transformation equations to determine (a) the in-plane principal strains and (b) the maximum in-plane shear strain and average normal strain. In each case, specify the orientation of the element and show how the strains deform the element within the $x$–$y$ plane.

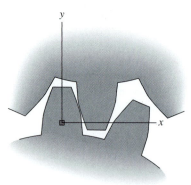

**Prob. 15–57**

**15-58.** The state of strain at the point on the fan blade has components of $\epsilon_x = 250(10^{-6})$, $\epsilon_y = -450(10^{-6})$, and $\gamma_{xy} = -825(10^{-6})$. Use the strain transformation equations to determine (a) the in-plane principal strains and (b) the maximum in-plane shear strain and average normal strain. In each case specify the orientation of the element and show how the strains deform the element within the $x$–$y$ plane.

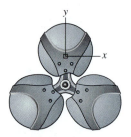

**Prob. 15–58**

**15-59.** The state of strain at the point on the spanner wrench has components of $\epsilon_x = 260(10^{-6})$, $\epsilon_y = 320(10^{-6})$, and $\gamma_{xy} = 180(10^{-6})$. Use the strain-transformation equations to determine (a) the in-plane principal strains and (b) the maximum in-plane shear strain and average normal strain. In each case, specify the orientation of the element and show how the strains deform the element within the $x$–$y$ plane.

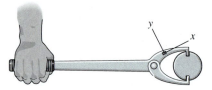

**Prob. 15–59**

**\*15-60.** The state of strain at the point on the wrench has components $\epsilon_x = 120(10^{-6})$, $\epsilon_y = -180(10^{-6})$, $\gamma_{xy} = 150(10^{-6})$. Use the strain-transformation equations to determine (a) the in-plane principal strains and (b) the maximum in-plane shear strain and average normal strain. In each case, specify the orientation of the element and show how the strains deform the element within the $x$–$y$ plane.

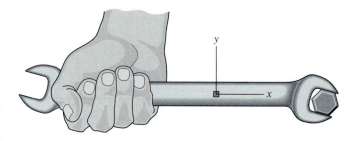

**Prob. 15–60**

■**15-61.** Consider the general case of plane strain where $\epsilon_x$, $\epsilon_y$, and $\gamma_{xy}$ are known. Write a computer program that can be used to determine the normal and shear strains, $\epsilon_{x'}$ and $\gamma_{x'y'}$, on the plane of an element oriented $\theta$ from the horizontal. Also, compute the principal strains and the element's orientation, and the maximum in-plane shear strain, the average normal strain, and the element's orientation.

**15-62.** Solve Prob. 15–52 using Mohr's circle.

**15-63.** Solve Prob. 15–53 using Mohr's circle.

**\*15-64.** Solve Prob. 15–54 using Mohr's circle.

**15-65.** Solve Prob. 15–55 using Mohr's circle.

**15-66.** Solve Prob. 15–56 using Mohr's circle.

**15-67.** Solve Prob. 15–57 using Mohr's circle.

**\*15-68.** Solve Prob. 15–58 using Mohr's circle.

**15-69.** Solve Prob. 15–59 using Mohr's circle.

## 15.9 Strain Rosettes

It was mentioned in Sec. 9.1 that the normal strain in a tension-test specimen can be measured using an *electrical-resistance strain gauge*, which consists of a wire grid or piece of metal foil bonded to the specimen. However, for a general loading on a body, the *normal strains* at a point on its free surface are often determined using a cluster of three electrical-resistance strain gauges, arranged in a specified pattern. This pattern is referred to as a *strain rosette*, and once the readings on the three gauges are made, the data can then be used to specify the state of strain at the point. It should be noted, however, that these strains are measured *only* in the plane of the gauges, and since the body is stress-free on its surface, the gauges may be subjected to *plane stress* but *not* plane strain. In this regard, the normal line to the free surface is a principal axis of strain, and so the principal normal strain along this axis is *not* measured by the strain rosette. What is important here is that the out-of-plane displacement caused by this principal strain will *not* affect the in-plane measurements of the gauges.

In the general case, the axes of the three gauges are arranged at the angles $\theta_a$, $\theta_b$, $\theta_c$ as shown in Fig. 15–42a. If the readings $\epsilon_a$, $\epsilon_b$, $\epsilon_c$ are taken, we can determine the strain components $\epsilon_x$, $\epsilon_y$, $\gamma_{xy}$ at the point by applying the strain-transformation equation, Eq. 15–18, for each gauge. We have

$$\epsilon_a = \epsilon_x \cos^2 \theta_a + \epsilon_y \sin^2 \theta_a + \gamma_{xy} \sin \theta_a \cos \theta_a$$
$$\epsilon_b = \epsilon_x \cos^2 \theta_b + \epsilon_y \sin^2 \theta_b + \gamma_{xy} \sin \theta_b \cos \theta_b$$
$$\epsilon_c = \epsilon_x \cos^2 \theta_c + \epsilon_y \sin^2 \theta_c + \gamma_{xy} \sin \theta_c \cos \theta_c \qquad (15–30)$$

The values of $\epsilon_x$, $\epsilon_y$, $\gamma_{xy}$ are determined by solving these three equations simultaneously.

Strain rosettes are often arranged in 45° or 60° patterns. In the case of the 45° or (rectangular) strain rosette shown in Fig. 15–42b, $\theta_a = 0°$, $\theta_b = 45°$, $\theta_c = 90°$, so that Eq. 15–30 gives

$$\epsilon_x = \epsilon_a$$
$$\epsilon_y = \epsilon_c$$
$$\gamma_{xy} = 2\epsilon_b - (\epsilon_a + \epsilon_c)$$

And for the 60° strain rosette in Fig. 15–42c, $\theta_a = 0°$, $\theta_b = 60°$, $\theta_c = 120°$. Here Eq. 15–30 gives

$$\epsilon_x = \epsilon_a$$
$$\epsilon_y = \frac{1}{3}(2\epsilon_b + 2\epsilon_c - \epsilon_a)$$
$$\gamma_{xy} = \frac{2}{\sqrt{3}}(\epsilon_b - \epsilon_c) \qquad (15–31)$$

Once $\epsilon_x$, $\epsilon_y$, $\gamma_{xy}$ are determined, the transformation equations of Sec. 15.7 or Mohr's circle can then be used to determine the principal in-plane strains and the maximum in-plane shear strain at the point.

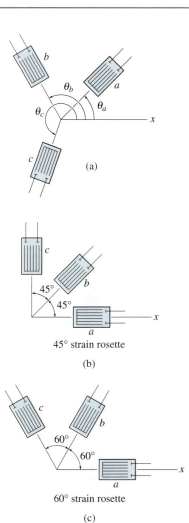

(a)

45° strain rosette

(b)

60° strain rosette

(c)

**Fig. 15–42**

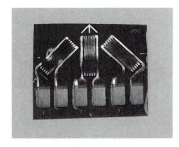

Typical electrical resistance 45° strain rosette.

**E X A M P L E 15.20**

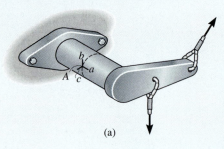

(a)

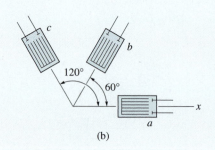

(b)

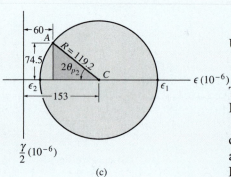

(c)

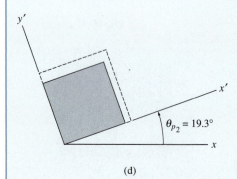

(d)

**Fig. 15–43**

The state of strain at point $A$ on the bracket in Fig. 15–43$a$ is measured using the strain rosette shown in Fig. 15–43$b$. Due to the loadings, the readings from the gauges give $\epsilon_a = 60(10^{-6})$, $\epsilon_b = 135(10^{-6})$, and $\epsilon_c = 264(10^{-6})$. Determine the in-plane principal strains at the point and the directions in which they act.

**Solution**

We will use Eq. 15–30 for the solution. Establishing an $x$ axis as shown in Fig. 15–43$b$ and measuring the angles counterclockwise from the $+x$ axis to the center-lines of each gauge, we have $\theta_a = 0°$, $\theta_b = 60°$, and $\theta_c = 120°$. Substituting these results, along with the problem data, into Eq. 15–30 gives

$$60(10^{-6}) = \epsilon_x \cos^2 0° + \epsilon_y \sin^2 0° + \gamma_{xy} \sin 0° \cos 0°$$
$$= \epsilon_x \qquad (1)$$

$$135(10^{-6}) = \epsilon_x \cos^2 60° + \epsilon_y \sin^2 60° + \gamma_{xy} \sin 60° \cos 60°$$
$$= 0.25\epsilon_x + 0.75\epsilon_y + 0.433\gamma_{xy} \qquad (2)$$

$$264(10^{-6}) = \epsilon_x \cos^2 120° + \epsilon_y \sin^2 120° + \gamma_{xy} \sin 120° \cos 120°$$
$$= 0.25\epsilon_x + 0.75\epsilon_y - 0.433\gamma_{xy} \qquad (3)$$

Using Eq. 1 and solving Eqs. 2 and 3 simultaneously, we get

$$\epsilon_x = 60(10^{-6}) \qquad \epsilon_y = 246(10^{-6}) \qquad \gamma_{xy} = -149(10^{-6})$$

These same results can also be obtained in a more direct manner from Eq. 15–31.

The in-plane principal strains can be determined using Mohr's circle. The reference point on the circle is at $A (60(10^{-6}), -74.5(10^{-6}))$ and the center of the circle, $C$, is on the $\epsilon$ axis at $\epsilon_{\text{avg}} = 153(10^{-6})$, Fig. 15–43$c$. From the shaded triangle, the radius is

$$R = \left[\sqrt{(153 - 60)^2 + (74.5)^2}\right](10^{-6}) = 119.2(10^{-6})$$

The in-plane principal strains are thus

$$\epsilon_1 = 153(10^{-6}) + 119.2(10^{-6}) = 272(10^{-6}) \qquad \textit{Ans.}$$
$$\epsilon_2 = 153(10^{-6}) - 199.2(10^{-6}) = 33.8(10^{-6}) \qquad \textit{Ans.}$$

$$2\theta_{p2} = \tan^{-1}\frac{74.5}{(153 - 60)} = 38.7°$$

$$\theta_{p2} = 19.3° \qquad \textit{Ans.}$$

The deformed element is shown in the dashed position in Fig. 15–43$d$. Realize that, due to the Poisson effect, the element is *also* subjected to an out-of-plane strain, i.e., in the $z$ direction, although this value does not influence the calculated results.

# PROBLEMS

**15-70.** The 45° strain rosette is mounted on a steel shaft. The following readings are obtained from each gauge: $\epsilon_a = 300(10^{-6})$, $\epsilon_b = 180(10^{-6})$, and $\epsilon_c = -250(10^{-6})$. Determine the in-plane principal strains and their orientation.

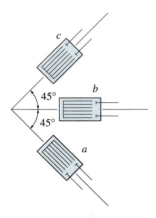

**Prob. 15–70**

**15-71.** The 45° strain rosette is mounted on the link of the backhoe. The following readings are obtained from each gauge: $\epsilon_a = 650(10^{-6})$, $\epsilon_b = -300(10^{-6})$, $\epsilon_c = 480(10^{-6})$. Determine (a) the in-plane principal strains and (b) the maximum in-plane shear strain and associated average normal strain.

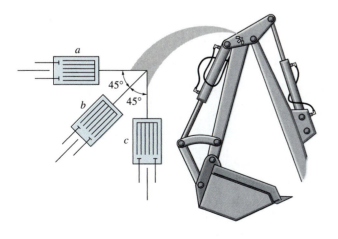

**Prob. 15–71**

**\*15-72.** The 60° strain rosette is mounted on the surface of an aluminum plate. The following readings are obtained from each gauge: $\epsilon_a = 950(10^{-6})$, $\epsilon_b = 380(10^{-6})$, $\epsilon_c = -220(10^{-6})$. Determine the in-plane principal strains and their orientation.

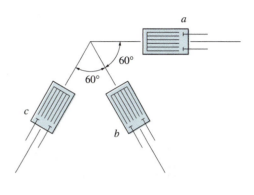

**Prob. 15–72**

**■15-73.** Consider the general orientation of three strain gauges at a point as shown. Write a computer program that can be used to determine the principal in-plane strains and the maximum in-plane shear strain at the point. Show an application of the program using the values and $\theta_a = 40°$, $\epsilon_a = 160(10^{-6})$, $\theta_b = 125°$, $\epsilon_b = 100(10^{-6})$, $\theta_c = 220°$, and $\epsilon_c = 80(10^{-6})$.

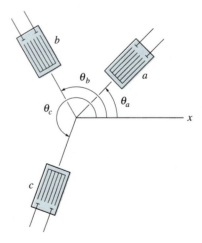

**Prob. 15–73**

## 15.10 Material-Property Relationships

Now that the general principles of multiaxial stress and strain have been presented, we will use these principles to develop some important relationships involving the material's properties. To do so we will assume that the material is homogeneous and isotropic and behaves in a linear-elastic manner.

**Generalized Hooke's Law.** If the material at a point is subjected to a state of triaxial stress, $\sigma_x, \sigma_y, \sigma_z$, Fig. 15–44a, associated normal strains $\epsilon_x$, $\epsilon_y, \epsilon_z$ are developed in the material. The stresses can be related to the strains by using the principle of superposition, Poisson's ratio, $\epsilon_{lat} = -v\epsilon_{long}$, and Hooke's law, as it applies in the uniaxial direction, $\epsilon = \sigma/E$. To show how this is done we will first consider the normal strain of the element in the $x$ direction, caused by separate application of each normal stress. When $\sigma_x$ is applied, Fig. 15–44b, the element elongates in the $x$ direction and the strain $\epsilon'_x$ in this direction is

$$\epsilon'_x = \frac{\sigma_x}{E}$$

Application of $\sigma_y$ causes the element to contract with a strain $\epsilon''_x$ in the $x$ direction, Fig. 15–44c. Here

$$\epsilon''_x = -v\frac{\sigma_y}{E}$$

Likewise, application of $\sigma_z$, Fig. 15–44d, causes a contraction in the $x$ direction such that

$$\epsilon'''_x = -v\frac{\sigma_z}{E}$$

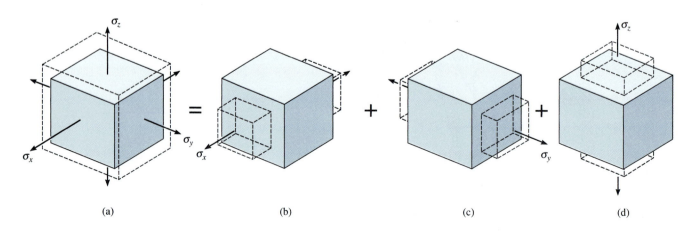

(a)    (b)    (c)    (d)

Fig. 15–44

When these three normal strains are superimposed, the normal strain $\epsilon_x$ is determined for the state of stress in Fig. 15–44a. Similar equations can be developed for the normal strains in the $y$ and $z$ directions. The final results can be written as

$$\epsilon_x = \frac{1}{E}[\sigma_x - v(\sigma_y + \sigma_z)]$$
$$\epsilon_y = \frac{1}{E}[\sigma_y - v(\sigma_x + \sigma_z)] \qquad (15\text{--}32)$$
$$\epsilon_z = \frac{1}{E}[\sigma_z - v(\sigma_x + \sigma_y)]$$

These three equations express Hooke's law in a general form for a triaxial state of stress. As noted in the derivation, they are valid only if the principle of superposition applies, which requires a *linear-elastic* response of the material and application of strains that do not severely alter the shape of the material—i.e., small deformations are required. When applying these equations, note that tensile stresses are considered positive quantities, and compressive stresses are negative. If a resulting normal strain is *positive*, it indicates that the material *elongates*, whereas a *negative* normal strain indicates the material *contracts*.

Since the material is isotropic, the element in Fig. 15–44a will *remain a rectangular block* when subjected to the normal stresses; i.e., *no shear strains* will be produced in the material. If we now apply a shear stress $\tau_{xy}$ to the element, Fig. 15–45a, experimental observations indicate that the material will deform *only* due to a shear strain $\gamma_{xy}$; that is, $\tau_{xy}$ will not cause other strains in the material. Likewise, $\tau_{yz}$ and $\tau_{xz}$ will only cause shear strains $\gamma_{yz}$ and $\gamma_{xz}$, respectively. Hooke's law for shear stress and shear strain can therefore be written as

$$\gamma_{xy} = \frac{1}{G}\tau_{xy} \qquad \gamma_{yz} = \frac{1}{G}\tau_{yz} \qquad \gamma_{xz} = \frac{1}{G}\tau_{xz} \qquad (15\text{--}33)$$

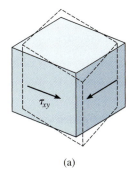

(a)

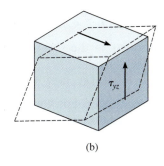

(b)

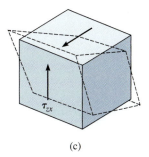

(c)

Fig. 15–45

## Relationship Involving E, v, and G.

In Sec. 9.7 we stated that the modulus of elasticity $E$ is related to the shear modulus $G$ by Eq. 9–11, namely,

$$G = \frac{E}{2(1 + v)} \tag{15–34}$$

One way to derive this relationship is to consider an element of the material to be subjected to pure shear ($\sigma_x = \sigma_y = \sigma_z = 0$), Fig. 15–46a. Applying Eq. 15–5 to obtain the principal stresses yields $\sigma_{max} = \tau_{xy}$ and $\sigma_{min} = -\tau_{xy}$. From Eq. 15–4, the element must be oriented $\theta_{p_1} = 45°$ counterclockwise from the $x$ axis in order to define the direction of the plane on which $\sigma_{max}$ acts, Fig. 15–46b. If the three principal stresses $\sigma_{max} = \tau_{xy}$, $\sigma_{int} = 0$, and $\sigma_{min} = -\tau_{xy}$ are substituted into the first of Eq. 15–32, the principal strain $\epsilon_{max}$ can be related to the shear stress $\tau_{xy}$. The result is

$$\epsilon_{max} = \frac{\tau_{xy}}{E}(1 + v) \tag{15–35}$$

This strain, which deforms the element along the $x'$ axis, can also be related to the shear strain $\gamma_{xy}$ using the strain transformation equations or Mohr's circle for strain. To do this, first note that since $\sigma_x = \sigma_y = \sigma_z = 0$, then from Eq. 15–32 $\epsilon_x = \epsilon_y = 0$. Substituting these results into the transformation Eq. 15–25, we get

$$\epsilon_1 = \epsilon_{max} = \frac{\gamma_{xy}}{2}$$

By Hooke's law, $\gamma_{xy} = \tau_{xy}/G$, so that $\epsilon_{max} = \tau_{xy}/2G$. Substituting into Eq. 15–35 and rearranging terms gives the final result, namely, Eq. 15–34.

## Dilatation and Bulk Modulus.

When an elastic material is subjected to normal stress, its volume will change. In order to compute this change, consider a volume element which is subjected to the principal stresses $\sigma_x$, $\sigma_y$, $\sigma_z$. The sides of the element are originally dx, dy, dz, Fig. 15–47a; however, after application of the stress they become $(1 + \epsilon_x)\,dx$, $(1 + \epsilon_y)\,dy$, $(1 + \epsilon_z)\,dz$, respectively, Fig. 15–47b. The change in volume of the element is therefore

$$\delta V = (1 + \epsilon_x)(1 + \epsilon_y)(1 + \epsilon_z)\,dx\,dy\,dz - dx\,dy\,dz$$

Neglecting the products of the strains since the strains are very small, we have

$$\delta V = (\epsilon_x + \epsilon_y + \epsilon_z)\,dx\,dy\,dz$$

The change in volume per unit volume is called the *volumetric strain* or the **dilatation** e. It can be written as

$$e = \frac{\delta V}{dV} = \epsilon_x + \epsilon_y + \epsilon_z \tag{15–36}$$

By comparison, the shear strains will *not* change the volume of the element, rather they will only change its rectangular shape.

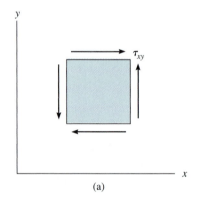

(a)

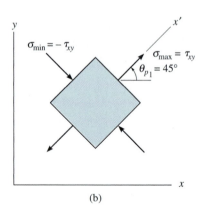

(b)

Fig. 15–46

If we use the generalized Hooke's law, as defined by Eq. 15–32, we can write the dilatation in terms of the applied stress. We have

$$e = \frac{1-2v}{E}(\sigma_x + \sigma_y + \sigma_z) \qquad (15\text{–}37)$$

When a volume element of material is subjected to the uniform pressure $p$ of a liquid, the pressure on the body is the same in all directions and is always normal to any surface on which it acts. Shear stresses are *not present*, since the shear resistance of a liquid is zero. This state of *hydrostatic* loading requires the normal stresses to be equal in any and all directions, and therefore an element of the body is subjected to principal stresses $\sigma_x = \sigma_y = \sigma_z = -p$, Fig. 15–48. Substituting into Eq. 15–37 and rearranging terms yields

$$\frac{p}{e} = -\frac{E}{3(1-2v)} \qquad (15\text{–}38)$$

The term on the right consists *only* of the material's properties $E$ and $v$. It is equal to the ratio of the uniform normal stress $p$ to the dilatation or volumetric strain. Since this ratio is *similar* to the ratio of linear-elastic stress to strain, which defines $E$, i.e., $\sigma/\epsilon = E$, the terms on the right are called the *volume modulus of elasticity* or the **bulk modulus**. It has the same units as stress and will be symbolized by the letter $k$; that is,

$$k = \frac{E}{3(1-2v)} \qquad (15\text{–}39)$$

Note that for most metals $v \approx \frac{1}{3}$ so $k \approx E$. If a material existed that did not change its volume then $\delta V = 0$, and so $k$ would have to be infinite. From Eq. 15–39 the theoretical *maximum* value for Poisson's ratio is therefore $v = 0.5$. Also, during yielding, no actual volume change of the material is observed, and so $v = 0.5$ is used when plastic yielding occurs.

## IMPORTANT POINTS

- When a homogeneous and isotropic material is subjected to a state of triaxial stress, the strain in one of the stress directions is influenced by the strains produced by *all* the stresses. This is the result of the Poisson effect, and results in the form of a generalized Hooke's law.

- A shear stress applied to homogeneous and isotropic material will only produce shear strain in the same plane.

- The material constants, $E$, $G$, and $v$, are related mathematically.

- *Dilatation*, or *volumetric strain*, is caused only by normal strain, not shear strain.

- The *bulk modulus* is a measure of the stiffness of a volume of material. This material property provides an upper limit to Poisson's ratio of $v = 0.5$, which remains at this value while plastic yielding occurs.

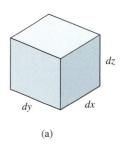

(a)

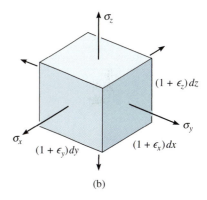

(b)

**Fig. 15–47**

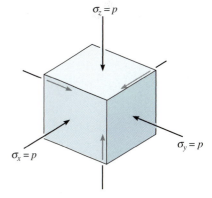

Hydrostatic stress

**Fig. 15–48**

E X A M P L E 15.21

The bracket in Example 15.20, Fig. 15–49a, is made of steel for which $E_{st} = 200$ GPa, $\nu_{st} = 0.3$. Determine the principal stresses at point $A$.

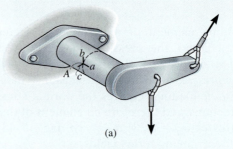

(a)

Fig. 15–49

**Solution I**

From Example 15.20 the principal strains have been determined as

$$\epsilon_1 = 272(10^{-6})$$
$$\epsilon_2 = 33.8(10^{-6})$$

Since point $A$ is on the *surface* of the bracket for which there is no loading, the stress on the surface is zero, and so point $A$ is subjected to plane stress. Applying Hooke's law with $\sigma_3 = 0$, we have

$$\epsilon_1 = \frac{\sigma_1}{E} - \frac{\nu}{E}\sigma_2; \quad 272(10^{-6}) = \frac{\sigma_1}{200(10^9)} - \frac{0.3}{200(10^9)}\sigma_2$$

$$54.4(10^6) = \sigma_1 - 0.3\sigma_2 \tag{1}$$

$$\epsilon_2 = \frac{\sigma_2}{E} - \frac{\nu}{E}\sigma_1; \quad 33.8(10^{-6}) = \frac{\sigma_2}{200(10^9)} - \frac{0.3}{200(10^9)}\sigma_1$$

$$6.76(10^6) = \sigma_2 - 0.3\sigma_1 \tag{2}$$

Solving Eqs. 1 and 2 simultaneously yields

$$\sigma_1 = 62.0 \text{ MPa} \qquad Ans.$$
$$\sigma_2 = 25.4 \text{ MPa} \qquad Ans.$$

**Solution II**

It is also possible to solve the problem using the given state of strain,

$$\epsilon_x = 60(10^{-6}) \qquad \epsilon_y = 246(10^{-6}) \qquad \gamma_{xy} = -149(10^{-6})$$

as specified in Example 15.20. Applying Hooke's law in the $x$–$y$ plane, we have

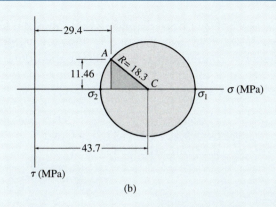

(b)

$$\epsilon_x = \frac{\sigma_x}{E} - \frac{v}{E}\sigma_y; \quad 60(10^{-6}) = \frac{\sigma_x}{200(10^9)\ \text{Pa}} - \frac{0.3\sigma_y}{200(10^9)\ \text{Pa}}$$

$$\epsilon_y = \frac{\sigma_y}{E} - \frac{v}{E}\sigma_x; \quad 246(10^{-6}) = \frac{\sigma_y}{200(10^9)\ \text{Pa}} - \frac{0.3\sigma_x}{200(10^9)\ \text{Pa}}$$

$$\sigma_x = 29.4\ \text{MPa} \qquad \sigma_y = 58.0\ \text{MPa}$$

The shear stress is determined using Hooke's law for shear. First, however, we must calculate $G$.

$$G = \frac{E}{2(1 + v)} = \frac{200\ \text{GPa}}{2(1 + 0.3)} = 76.9\ \text{GPa}$$

Thus,

$$\tau_{xy} = G\gamma_{xy}; \quad \tau_{xy} = 76.9(10^9)[-149(10^{-6})] = -11.46\ \text{MPa}$$

The Mohr's circle for this state of plane stress has a reference point $A(29.4\ \text{MPa}, -11.46\ \text{MPa})$ and center at $\sigma_{\text{avg}} = 43.7\ \text{MPa}$, Fig. 15–49b. The radius is determined from the shaded triangle.

$$R = \sqrt{(43.7 - 29.4)^2 + (11.46)^2} = 18.3\ \text{MPa}$$

Therefore,

$$\sigma_1 = 43.7\ \text{MPa} + 18.3\ \text{MPa} = 62.0\ \text{MPa} \qquad \textit{Ans.}$$

$$\sigma_2 = 43.7\ \text{MPa} - 18.3\ \text{MPa} = 25.4\ \text{MPa} \qquad \textit{Ans.}$$

Note that each of these solutions is valid provided the material is both linear elastic and isotropic, since then the principal planes of stress and strain coincide.

## E X A M P L E | 15.22

The copper bar in Fig. 15–50 is subjected to a uniform loading along its edges as shown. If it has a length $a = 300$ mm, width $b = 50$ mm, and thickness $t = 20$ mm before the load is applied, determine its new length, width, and thickness after application of the load. Take $E_{cu} = 120$ GPa, $v_{cu} = 0.34$.

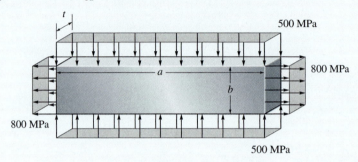

**Fig. 15–50**

### Solution

By inspection, the bar is subjected to a state of plane stress. From the loading we have

$$\sigma_x = 800 \text{ MPa} \qquad \sigma_y = -500 \text{ MPa} \qquad \tau_{xy} = 0 \qquad \sigma_z = 0$$

The associated normal strains are determined from the generalized Hooke's law, Eq. 15–32; that is,

$$\epsilon_x = \frac{\sigma_x}{E} - \frac{v}{E}(\sigma_y + \sigma_z)$$

$$= \frac{800 \text{ MPa}}{120(10^3) \text{ MPa}} - \frac{0.34}{120(10^3) \text{ MPa}}(-500 \text{ MPa}) = 0.00808$$

$$\epsilon_y = \frac{\sigma_y}{E} - \frac{v}{E}(\sigma_x + \sigma_z)$$

$$= \frac{-500 \text{ MPa}}{120(10^3) \text{ MPa}} - \frac{0.34}{120(10^3) \text{ MPa}}(800 \text{ MPa} + 0) = -0.00643$$

$$\epsilon_z = \frac{\sigma_z}{E} - \frac{v}{E}(\sigma_x + \sigma_y)$$

$$= 0 - \frac{0.34}{120(10^3) \text{ MPa}}(800 \text{ MPa} - 500 \text{ MPa}) = -0.000850$$

The new bar length, width, and thickness are therefore

$$a' = 300 \text{ mm} + 0.00808(300 \text{ mm}) = 302.4 \text{ mm} \qquad \textit{Ans.}$$

$$b' = 50 \text{ mm} + (-0.00643)(50 \text{ mm}) = 49.68 \text{ mm} \qquad \textit{Ans.}$$

$$t' = 20 \text{ mm} + (-0.000850)(20 \text{ mm}) = 19.98 \text{ mm} \qquad \textit{Ans.}$$

# EXAMPLE 15.23

If the rectangular block shown in Fig. 15–51 is subjected to a uniform pressure of $p = 20$ psi, determine the dilatation and the change in length of each side. Take $E = 600$ psi, $v = 0.45$.

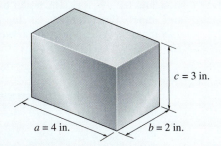

$c = 3$ in.

$a = 4$ in.  $b = 2$ in.

**Fig. 15–51**

## Solution

*Dilatation.* The dilatation can be determined using Eq. 15–37 with $\sigma_x = \sigma_y = \sigma_z = -20$ psi. We have

$$e = \frac{1 - 2v}{E}(\sigma_x + \sigma_y + \sigma_z)$$

$$= \frac{1 - 2(0.45)}{600 \text{ psi}}[3(-20 \text{ psi})]$$

$$= -0.01 \text{ in}^3/\text{in}^3 \qquad \qquad Ans.$$

*Change in Length.* The normal strain on each side can be determined from Hooke's law, Eq. 15–32; that is,

$$\epsilon = \frac{1}{E}[\sigma_x - v(\sigma_y + \sigma_z)]$$

$$= \frac{1}{600 \text{ psi}}[-20 \text{ psi} - (0.45)(-20 \text{ psi} - 20 \text{ psi})] = -0.00333 \text{ in./in.}$$

Thus the change in length of each side is

$$\delta a = -0.00333(4 \text{ in.}) = -0.0133 \text{ in.} \qquad Ans.$$
$$\delta b = -0.00333(2 \text{ in.}) = -0.00667 \text{ in.} \qquad Ans.$$
$$\delta c = -0.00333(3 \text{ in.}) = -0.0100 \text{ in.} \qquad Ans.$$

The negative signs indicate that each dimension is decreased.

# PROBLEMS

**15-74.** For the case of plane stress, show that Hooke's law can be written as

$$\sigma_x = \frac{E}{(1 - v^2)}(\epsilon_x + v\epsilon_y), \quad \sigma_y = \frac{E}{(1 - v^2)}(\epsilon_y + v\epsilon_x)$$

**15-75.** Use Hooke's law, Eq. 15–32, to develop the strain-transformation equations, Eqs. 15–21 and 15–22, from the stress-transformation equations, Eqs. 15–1 and 15–2.

**\*15-76.** Determine the bulk modulus for gray cast iron if $E_{fe} = 14(10^3)$ ksi and $v_{fe} = 0.20$.

**15-77.** Determine the bulk modulus for hard rubber if $E_r = 0.68(10^3)$ ksi and $v_r = 0.43$.

**15-78.** The polyvinyl chloride bar is subjected to an axial force of 900 lb. If it has the original dimensions shown, determine the value of Poisson's ratio if the angle $\theta$ decreases by $\Delta\theta = 0.01°$ after the load is applied. $E_{pvc} = 800(10^3)$ psi.

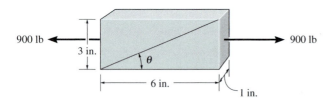

**Prob. 15–78**

**15-79.** The rod is made of aluminum 2014-T6. If it is subjected to the tensile load of 700 N and has a diameter of 20 mm, determine the principal strains at a point on the surface of the rod.

**Prob. 15–79**

**\*15-80.** The strain gauge is placed on the surface of a thin-walled steel boiler as shown. If it is 0.5 in. long, determine the pressure in the boiler when the gauge elongates $0.2(10^{-3})$ in. The boiler has a thickness of 0.5 in. and inner diameter of 60 in. Also, determine the maximum $x$, $y$ in-plane shear strain in the material. $E_{st} = 29(10^3)$ ksi, $v_{st} = 0.3$.

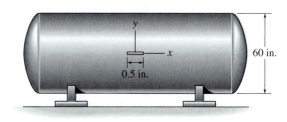

**Prob. 15–80**

**15-81.** The shaft has a radius of 15 mm and is made of L2 tool steel. Determine the strains in the $x'$ and $y'$ directions if a torque $T = 2$ kN·m is applied to the shaft.

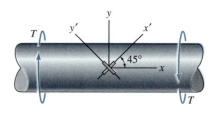

**Prob. 15–81**

**15-82.** Determine the principal strains that occur at a point on a steel member where the principal stresses are $\sigma_{max} = 18$ ksi, $\sigma_{int} = 15$ ksi, $\sigma_{min} = -28$ ksi, $E_{st} = 29(10^3)$ ksi and $v_{st} = 0.3$.

**15-83.** A bar of plastic having a diameter of 0.5 in. is loaded in a tension machine, and it is determined that $\epsilon_x = 530(10^{-6})$ when the load is 80 lb. Determine the modulus of elasticity, $E_p$, and the dilatation, $e_p$, of the plastic $v_p = 0.26$.

**\*15-84.** A rod has a radius of 10 mm. If it is subjected to an axial load of 15 N such that the axial strain in the rod is $\epsilon_x = 2.75(10^{-6})$, determine the modulus of elasticity $E$ and the change in its diameter. $v = 0.23$.

**15-85.** From experiment, the principal strains in a plane at a point on a steel shell are $\epsilon_1 = 350(10^{-6})$ and $\epsilon_2 = -250(10^{-6})$. If $E_{st} = 200$ GPa and $v_{st} = 0.3$, determine the principal plane stresses in this plane.

**15-86.** The principal plane stresses and associated strains in a plane at a point are $\sigma_1 = 40$ ksi, $\sigma_2 = 25$ ksi, $\epsilon_1 = 1.15(10^{-3})$, and $\epsilon_2 = 0.450(10^{-3})$. If this is a case of plane stress, determine the modulus of elasticity and Poisson's ratio.

**15-87.** The spherical pressure vessel has an inner diameter of 2 m and a thickness of 10 mm. A strain gauge having a length of 20 mm is attached to it, and it is observed to increase in length by 0.012 mm when the vessel is pressurized. Determine the pressure causing this deformation, and find the maximum in-plane shear stress, and the absolute maximum shear stress at a point on the outer surface of the vessel. The material is steel, for which $E_{st} = 200$ GPa and $v_{st} = 0.3$.

20 mm

Prob. 15-87

**\*15-88.** The principal strains in a plane, measured experimentally at a point on the aluminum fuselage of a jet aircraft, are $\epsilon_1 = 630(10^{-6})$ and $\epsilon_2 = 350(10^{-6})$. If this is a case of plane stress, determine the associated principal stresses at the point in the same plane. $E_{al} = 10(10^3)$ ksi and $v_{al} = 0.33$.

**15-89.** A single strain gauge, placed in the vertical plane on the outer surface and at an angle of 60° to the axis of the pipe, gives a reading at point $A$ of $\epsilon_A = -250(10^{-6})$. Determine the vertical force $P$ if the pipe has an outer diameter of 1 in. and an inner diameter of 0.6 in. The pipe is made of C86100 bronze.

**15-90.** A single strain gauge, placed in the vertical plane on the outer surface and at an angle of 60° to the axis of the pipe, gives a reading at point $A$ of $\epsilon_A = -250(10^{-6})$. Determine the principal strains in the pipe at point $A$. The pipe has an outer diameter of 1 in. and an inner diameter of 0.6 in. and is made of C86100 bronze.

**15-91.** Air is pumped into the steel thin-walled pressure vessel at $C$. If the ends of the vessel are closed using two pistons connected by a rod $AB$, determine the increase in the diameter of the pressure vessel when the internal gauge pressure is 5 MPa. Also, what is the tensile stress in rod $AB$ if it has a diameter of 100 mm? The inner radius of the vessel is 400 mm, and its thickness is 10 mm. $E_{st} = 200$ GPa and $v_{st} = 0.3$.

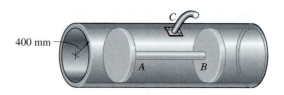

**Prob. 15–91**

**\*15-92.** The shaft has a radius of 15 mm and is made of L2 tool steel. Determine the strains in the $x'$ and $y'$ directions if a torque of $T = 2$ kN · m is applied to the shaft.

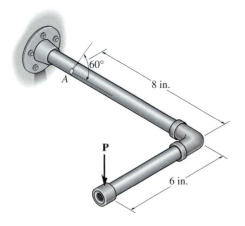

**Probs. 15–89/90**

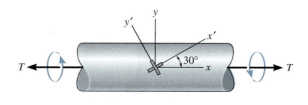

**Prob. 15–92**

**15-93.** The shaft has a radius of 15 mm and is made of L2 tool steel. Determine the torque $T$ in the shaft if the two strain gauges, attached to the surface of the shaft, report strains of $\epsilon_{x'} = -45(10^{-6})$ and $\epsilon_{y'} = 45(10^{-6})$. Also, compute the strains acting in the $x$ and $y$ directions.

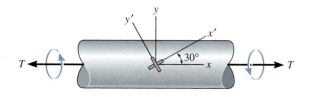

**Prob. 15–93**

**15-94.** Determine the change in volume of the tapered plate when it is subjected to the axial load **P**. The material has a thickness $t$, a modulus of elasticity $E$, and Poisson's ratio is $v$.

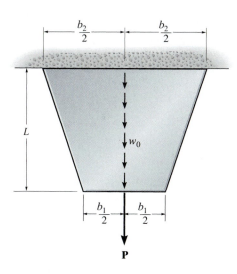

**Prob. 15–94**

**15-95.** A thin-walled cylindrical pressure vessel has an inner radius $r$, thickness $t$, and length $L$. If it is subjected to an internal pressure $p$, show that the increase in its inner radius is $\delta r = pr^2(2 - v)/2Et$ and the increase in its length is $\delta L = pLr(1 - 2v)/2Et$. Using these results, show that the change in internal volume becomes $\delta V = \pi r^2(1 + \epsilon_1)^2(1 + \epsilon_2)L - \pi r^2 L$. Since $\epsilon_1$ and $\epsilon_2$ are small quantities, show further that the change in volume per unit volume, called *volumetric strain*, can be written as $\delta V/V = (pr/2Et)(5 - 4v)$.

**\*15-96.** The A-36 steel pipe is subjected to the axial loading of 60 kN. Determine the change in volume of the material after the load is applied.

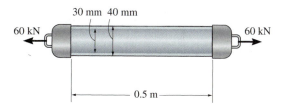

**Prob. 15–96**

**15-97.** A soft material is placed within the confines of a rigid cylinder which rests on a rigid support. Determine the factor by which the apparent modulus of elasticity will be increased from not being confined when a load is applied. Take $v = 0.3$ for the material.

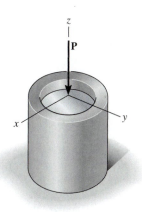

**Prob. 15–97**

**15-98.** A thin-walled spherical pressure vessel having an inner radius $r$ and thickness $t$ is subjected to an internal pressure $p$. Show that the increase in volume within the vessel is $\delta V = (2p\pi r^4/Et)(1 - v)$. Use a small-strain analysis.

**15-99.** The thin-walled cylindrical pressure vessel of inner radius $r$ and thickness $t$ is subjected to an internal pressure $p$. If the material constants are $E$ and $v$, determine the strains in the circumferential and longitudinal directions. Using these results, compute the increase in both the diameter and the length of a steel pressure vessel filled with air and having an internal gauge pressure of 20 MPa. The vessel is 2 m long and has an inner radius of 0.4 m and a thickness of 10 mm. $E_{st} = 200$ GPa, and $v_{st} = 0.3$.

**\*15-100.** Estimate the increase in volume of the tank in Prob. 15–99. *Suggestion:* Use the results of Prob. 15–95 as a check.

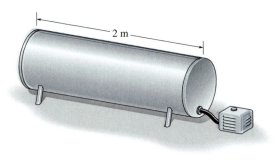

Probs. 15–99/100

**15-101.** The smooth rigid-body cavity is filled with liquid 6061-T6 aluminum. When cooled it is 0.012 in. from the top of the cavity. If the top of the cavity is covered and the temperature is increased by 200° F, determine the stress components $\sigma_x$, $\sigma_y$, and $\sigma_z$ in the aluminum. *Hint:* Use Eq. 15–32 with an additional strain term of $\alpha \Delta T$ (Eq. 10–4).

**15-102.** The smooth rigid-body cavity is filled with liquid 6061-T6 aluminum. When cooled it is 0.012 in. from the top of the cavity. If the top of the cavity is not covered and the temperature is increased by 200° F, determine the strain components $\epsilon_x$, $\epsilon_y$, and $\epsilon_z$ in the aluminum. *Hint:* Use Eq. 15–32 with an additional strain term of $\alpha \Delta T$ (Eq. 10–4).

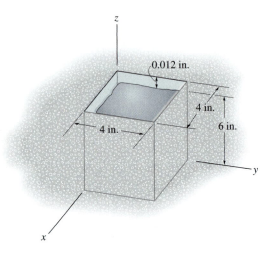

Probs. 15–101/102

**15-103.** The block is fitted between the fixed supports. If the glued joint can resist a maximum shear stress of $\tau_{\text{allow}} = 2$ ksi, determine the temperature rise that will cause the joint to fail. Take $E = 10(10^3)$ksi, $v = 0.2$, and $\alpha = 6.0(10^{-6})/°$F. *Hint:* Use Eq. 15–32 with an additional strain term of $\alpha \Delta T$ (Eq. 10–4).

Prob. 15–103

# CHAPTER REVIEW

- *Plane Stress.* Plane stress occurs when the material at a point is subjected to two normal stress components $\sigma_x$ and $\sigma_y$ and a shear stress $\tau_{xy}$. Provided these components are known, then the stress components acting on an element having a different orientation can be determined using the force equations of equilibrium or the equations of stress transformation. For design, it is important to determine the orientations of the element that produces the maximum principal normal stress and maximum in-plane shear stress. This can be done using the equations of stress transformation. It is found that no shear stress acts on the planes of principal stress. The planes of maximum in-plane shear stress are oriented 45° from this orientation, and on these shear planes there is an associated average normal stress $(\sigma_x + \sigma_y)/2$. Mohr's circle provides a graphical aid for finding the stress on any plane, the principal normal stresses, or the maximum in-plane shear stress. To draw the circle, the $\sigma$ and $\tau$ axes are established, the center of the circle $[(\sigma_x + \sigma_y)/2, 0]$ and the controlling point $(\sigma_x, \tau_{xy})$ are plotted. The radius of the circle extends between these two points and is determined from trigonometry.

- *Absolute Maximum Shear Stress.* The absolute maximum shear stress will be equal to the maximum in-plane shear stress provided the in-plane principal stresses have the same sign. If they are of opposite sign, then the absolute maximum shear stress will lie out of plane. Its value is $\tau_{\text{abs}_{\max}} = (\sigma_{\max} - \sigma_{\min})/2$.

- *Plane Strain.* When an element of material is subjected to deformations that only occurs in a single plane, then it undergoes plane strain. Like stress, the strain components can be transformed onto different planes.

- *Material Properties.* Hooke's law can be expressed in three dimensions, where each strain is related to the three normal stress components using the material properties $E$ and $\mu$. The dilatation is a measure of volumetric strain, and the bulk modulus is used to measure the stiffness of a volume of material.

# REVIEW PROBLEMS

**\*15-104.** A rod has a circular cross section with a diameter of 2 in. It is subjected to a torque of 12 kip · in. and a bending moment **M**. The greater principal stress at the point of maximum flexural stress is 15 ksi. Determine the magnitude of the bending moment.

**Prob. 15–104**

**15-105.** A steel pipe has an inner diameter of 2.75 in. and an outer diameter of 3 in. If it is fixed at $C$ and subjected to the horizontal force acting on the handle of the pipe wrench at its end, determine the principal stresses in the pipe at point $A$.

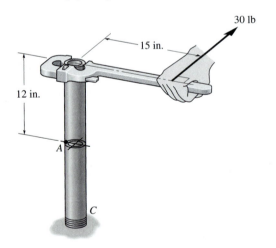

**Prob. 15–105**

**15-106.** The state of stress at a point on the upper surface of the wing is shown on the element. Determine (a) the principal stresses and (b) the maximum in-plane shear stress and average normal stress at the point. Specify the orientation of the element in each case.

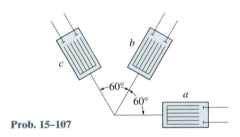

**Prob. 15–106**

**15-107.** The 60° strain rosette is mounted on the surface of a dome. The following readings are obtained for each gauge: $\epsilon_a = -780(10^{-6})$, $\epsilon_b = 400(10^{-6})$, $\epsilon_c = 500(10^{-6})$. Determine (a) the principal strains and (b) the maximum in-plane shear strain and associated average strain. In each case, specify the orientation of the element and show how the strain deforms the element.

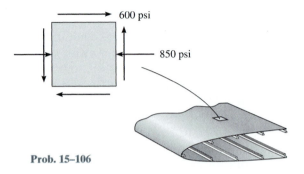

**Prob. 15–107**

**\*15-108.** The state of stress at a point is shown on the element. Determine (a) the principal stresses and (b) the maximum in-plane shear stress and average normal stress on the element. Specify the orientation of the element in each case. Use the stress transformation equations.

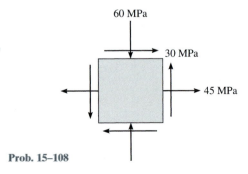

**Prob. 15–108**

**15-109.** The drill pipe has an outer diameter of 3 in. and a wall thickness of 0.25 in. If it is subjected to a torque and axial load as shown, determine (a) the principal stresses and (b) the maximum in-plane shear stress at a point on its surface.

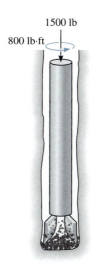

**Prob. 15–109**

**15-110.** The beam is subjected to the two forces shown. Determine the principal stresses at point *A*.

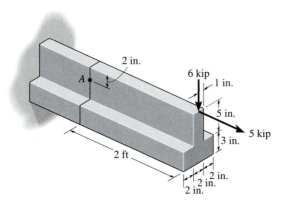

**Prob. 15–110**

**15-111.** Determine the equivalent state of stress if the element is oriented 30° clockwise from its position shown. Use Mohr's circle.

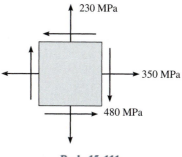

**Prob. 15–111**

***15-112.** The state of stress at a point in shown on the element. Determine (a) the principal stress and (b) the maximum in-plane shear stress and the average normal stress at the point. Specify the orientation of the element in each case.

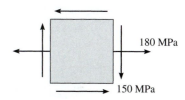

**Prob. 15–112**

**15-113.** The square steel plate has a thickness of 0.5 in. and is subjected to the edge loading shown. Determine the principal stresses developed in the steel.

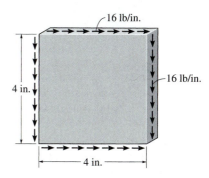

**Prob. 15–113**

Beams are important structural members that are used to support roof and floor loadings.

# Design of Beams

- To develop methods for designing beams to resist both bending and shear loads.
- To develop methods for determining the deflection of beams subject to bending and shear loads.
- To use beam deflection methods to solve statically indeterminate equilibrium problems.

## 16.1  Basis for Beam Design

*Beams* are structural members designed to support loadings applied perpendicular to their longitudinal axes. Because of these loadings, beams develop an internal shear force and bending moment that, in general, vary from point to point along the axis of the beam. Some beams may also be subjected to an internal axial force; however, the effects of this force are often neglected in design, since the axial stress is generally much smaller than the stresses developed by shear and bending. A beam that is chosen to resist both shear and bending stresses is said to be designed on the *basis of strength*. To design a beam in this way requires the use of the shear and flexure formulas developed in Chapters 12 and 13. Application of these formulas, however, is limited to beams made of a homogeneous material that has linear-elastic behavior.

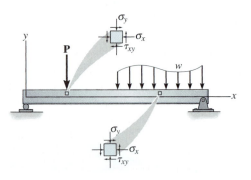

**Fig. 16–1**

The stress analysis of a beam generally neglects the effects caused by external distributed loadings and concentrated forces applied to the beam. As shown in Fig. 16–1, these loadings will create additional stresses in the beam *directly under the load*. Notably, a compressive stress $\sigma_y$ will be developed, in addition to the bending stress $\sigma_x$ and shear stress $\tau_{xy}$ discussed previously. Using advanced methods of analysis, as treated in the theory of elasticity, it can be shown, however, that the stress $\sigma_y$ diminishes rapidly throughout the beam's depth, and for *most* beam span-to-depth ratios used in engineering practice, the maximum value of $\sigma_y$ generally represents only a small percentage compared to the bending stress $\sigma_x$, that is, $\sigma_x \gg \sigma_y$. Furthermore, the direct application of concentrated loads is generally avoided in beam design. Instead, *bearing plates* are used to spread these loads more evenly onto the surface of the beam.

Although beams are designed mainly for strength, they must also be braced properly along their sides so that they do not buckle or suddenly become unstable. Furthermore, in some cases beams must be designed to resist a limited amount of *deflection*, as when they support ceilings made of brittle materials such as plaster. Methods for finding beam deflections will be discussed later in this chapter, and limitations placed on beam buckling are often discussed in codes on structural or mechanical design.

Beams normally support a large shear load at their supports. For this reason, a metal stiffener is often connected to the web of the beam, as shown here, to prevent localized deformation of the beam.

## 16.2   Stress Variations Throughout a Prismatic Beam

Since beams resist both internal shear and moment loadings, the stress analysis of a beam requires application of the shear and flexure formulas. Here we will discuss the general results obtained when these equations are applied to various points in a cantilevered beam that has a rectangular cross section and supports a load **P** at its end, Fig. 16–2a.

In general, at an arbitrary section *a–a* along the beam's axis, Fig. 16–2b, the internal shear **V** and moment **M** are developed from a *parabolic* shear-stress distribution, and a *linear* normal-stress distribution, Fig. 16–2c. As a result, the stresses acting on elements located at points 1 through 5 along the section will be as shown in Fig. 16–2d. Note that elements 1 and 5 are subjected only to the maximum normal stress, whereas element 3, which is on the neutral axis, is subjected only to the maximum shear stress. The intermediate elements 2 and 4 resist *both* normal and shear stress.

In each case the state of stress can be transformed into *principal stresses*, using either the stress-transformation equations or Mohr's circle. The results are shown in Fig. 16–2e. Here each successive element, 1 through 5, undergoes a counterclockwise orientation. Specifically, relative to element 1, considered to be at the 0° position, element 3 is oriented at 45° and element 5 is oriented at 90°. Also, the *maximum tensile stress* acting on the vertical faces of element 1 becomes smaller on the corresponding faces of each of the successive elements, until it is zero on the horizontal faces of element 5. In a similar manner, the *maximum compressive stress* on the vertical faces of element 5 reduces to zero on the horizontal faces of element 1.

If this analysis is extended to many vertical sections along the beam other than *a–a*, a profile of the results can be represented by curves called *stress trajectories*. Each of these curves indicates the *direction* of a principal stress having a constant magnitude. Some of these trajectories are shown for the cantilevered beam in Fig. 16–3. Here the solid lines represent the direction of the tensile principal stresses and the dashed lines represent the direction of the compressive principal stresses. As expected, the lines intersect the neutral axis at 45° angles, and the solid and dashed lines always intersect at 90°. Why? Knowing the direction of these lines can help engineers decide where to reinforce a beam so that it does not crack or become unstable.

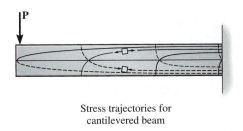

Stress trajectories for
cantilevered beam

**Fig. 16–3**

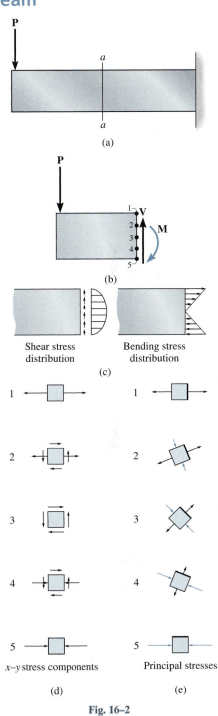

**Fig. 16–2**

# E X A M P L E 16.1

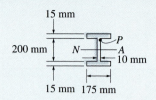

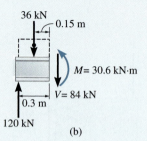

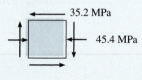

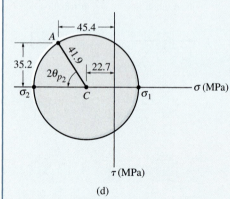

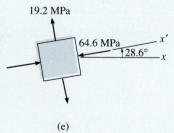

Fig. 16–4

The beam shown in Fig. 16–4a is subjected to the distributed loading of $w = 120$ kN/m. Determine the principal stresses in the beam at point $P$, which lies at the top of the web. Neglect the size of the fillets and stress concentrations at this point. $I = 67.4(10^{-6})\text{m}^4$.

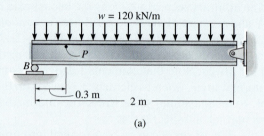

(a)

## Solution

*Internal Loadings.* The support reaction on the beam at $B$ is determined, and equilibrium of the sectioned beam shown in Fig. 16–4b yields

$$V = 84 \text{ kN} \qquad M = 30.6 \text{ kN} \cdot \text{m}$$

*Stress Components.* At point $P$,

$$\sigma = \frac{-My}{I} = \frac{30.6(10^3) \text{ N} \cdot \text{m} \,(0.100 \text{ m})}{67.4(10^{-6}) \text{ m}^4} = -45.4 \text{ MPa} \qquad Ans.$$

$$\tau = \frac{Q}{It} = \frac{84(10^3) \text{ N} [(0.1075 \text{ m})(0.175 \text{ m})(0.015 \text{ m})]}{67.4(10^{-6}) \text{ m}^4 (0.010 \text{ m})}$$

$$= 35.2 \text{ MPa} \qquad Ans.$$

These results are shown in Fig. 16–4c.

*Principal Stresses.* Using Mohr's circle the principal stresses at $P$ can be determined. As shown in Fig. 16–4d, the center of the circle is at $(-45.4 + 0)/2 = -22.7$, and point $A$ has coordinates of $A(-45.4, -35.2)$. Show that the radius is $R = 41.9$, and therefore

$$\sigma_1 = (41.9 - 22.7) = 19.2 \text{ MPa}$$
$$\sigma_2 = -(22.7 + 41.9) = -64.6 \text{ MPa}$$

The counterclockwise angle $2\theta_{p_2} = 57.2°$, so that

$$\theta_{p_2} = 28.6°$$

These results are shown in Fig. 16–4e.

## 16.3 Prismatic Beam Design

In order to design a beam on the basis of *strength*, it is required that the actual bending stress and shear stress in the beam do not exceed the allowable bending and shear stress for the material as defined by structural or mechanical codes. If the suspended span of the beam is relatively long, so that the internal moments become large, the engineer will first consider a design based upon bending and then check the shear strength. A bending design requires a determination of the beam's *section modulus*, which is the ratio of $I$ and $c$, that is, $S = I/c$. Using the flexure formula, $\sigma = Mc/I$, we have

$$S_{\text{req'd}} = \frac{M}{\sigma_{\text{allow}}} \qquad\qquad (16\text{--}1)$$

Here $M$ is determined from the beam's moment diagram, and the allowable bending stress, $\sigma_{\text{allow}}$, is specified in a design code. In many cases the beam's unknown weight will be small and can be neglected in comparison with the loads the beam must carry. However, if the additional moment caused by the weight is to be included in the design, a selection for $S$ is made so that it slightly *exceeds* $S_{\text{req'd}}$.

Once $S_{\text{req'd}}$ is known, if the beam has a simple cross-sectional shape, such as a square, a circle, or a rectangle of known width-to-height proportions, its *dimensions* can be determined directly from $S_{\text{req'd}}$, since by definition $S_{\text{req'd}} = I/c$. However, if the cross section is made from several elements, such as a wide-flange section, then an infinite number of web and flange dimensions can be determined that satisfy the value of $S_{\text{req'd}}$. In practice, however, engineers choose a particular beam meeting the requirement that $S > S_{\text{req'd}}$ from a handbook that lists the standard shapes available from manufacturers. Often several beams that have the same section modulus can be selected from these tables. If deflections are not restricted, usually the beam having the smallest cross-sectional area is chosen, since it is made of less material and is therefore both lighter and more economical than the others.

The above discussion assumes that the material's allowable bending stress is the *same* for both tension and compression. If this is the case, then a beam having a cross section that is *symmetric* with respect to the neutral axis should be chosen. However, if the allowable tensile and compressive bending stresses are *not* the same, then the choice of an unsymmetric cross section may be more efficient. Under these circumstances the beam must be designed to resist *both* the largest positive and the largest negative moment in the span.

The two floor beams are connected to the girder, which transmits the load to the columns of this building frame. For a force analysis, the connections can be considered to act as pins.

Once the beam has been selected, the shear formula $\tau_{\text{allow}} \geq VQ/It$ can then be used to check that the allowable shear stress is not exceeded. Often this requirement will not present a problem. However, if the beam is "short" and supports large concentrated loads, the shear-stress limitation may dictate the size of the beam. This limitation is particularly important in the design of wood beams, because wood tends to split along its grain due to shear (see Fig. 13–6).

**Fabricated Beams.** Since beams are often made of steel or wood, we will now discuss some of the tabulated properties of beams made from these materials.

**Steel Sections.** Most manufactured steel beams are produced by rolling a hot ingot of steel until the desired shape is formed. These so-called *rolled shapes* have properties that are tabulated in the American Institute of Steel Construction (AISC) manual. A representative listing for wide-flange beams taken from this manual is given in Appendix D. As noted in this appendix, the wide-flange shapes are designated by their depth and weight per unit length; for example, W18 × 46 indicates a wide-flange cross section (W) having a depth of 18 in. and a weight of 46 lb>ft, Fig. 16–5. For any given section, the weight per length, dimensions, cross-sectional area, moment of inertia, and section modulus are reported. Also included is the radius of gyration $r$, which is a geometric property related to the section's buckling strength. This will be discussed in Chapter 17.

Typical profile view of a steel wide-flange beam.

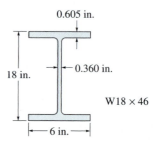

**Fig. 16–5**

**Wood Sections.** Most beams made of wood have rectangular cross sections because such beams are easy to manufacture and handle. Manuals, such as that of the National Forest Products Association, list the dimensions of lumber often used in the design of wood beams. Often, both the nominal and actual dimensions are reported. Lumber is identified by its *nominal* dimensions, such as 2 × 4 (2 in. by 4 in.); however, its actual or "dressed" dimensions are smaller, being 1.5 in. by 3.5 in. The reduction in the dimensions occurs due to the requirement of obtaining smooth surfaces from lumber that is rough sawn. Obviously, the *actual dimensions* must be used whenever stress calculations are performed on wood beams.

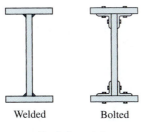

Welded          Bolted

Steel plate girders

**Fig. 16–6**

**Built-up Sections.**    A *built-up section* is constructed from two or more parts joined together to form a single unit. As indicated by Eq. 16–1, the capacity of the beam to resist a moment will vary directly with its section modulus $S$, and since $S = I/c$, then $S$ is *increased* if $I$ is *increased*. In order to increase $I$, *most of the material* should be placed as far *away* from the neutral axis as practical. This, of course, is what makes a deep wide-flange beam so efficient in resisting a moment. For very large loads, however, an available rolled-steel section may not have a section modulus great enough to support a given moment. Rather than using several available beams to support the load, engineers will usually "build up" a beam made from plates and angles. A deep I-shaped section having this form is called a *plate girder*. For example, the steel plate girder in Fig. 16–6 has two flange plates that are either welded or, using angles, bolted to the web plate.

Wood beams are also "built up," usually in the form of a box beam section, Fig. 16–7a. They may be made having plywood webs and larger boards for the flanges. For very large spans, *glulam beams* are used. These members are made from several boards glue-laminated together to form a single unit, Fig. 16–7b.

Just as in the case of rolled sections or beams made from a single piece, the design of built-up sections requires that the bending and shear stresses be checked. In addition, the shear stress in the fasteners, such as weld, glue, nails, etc., must be checked to be certain the beam acts as a single unit. The principles for doing this were outlined in Sec. 13.4.

Wooden box beam

(a)

Glulam beam

(b)

**Fig. 16–7**

## IMPORTANT POINTS

- Beams support loadings that are applied perpendicular to their axes. If they are designed on the basis of strength, they must resist allowable shear and bending stresses.
- The maximum bending stress in the beam is assumed to be much greater than the localized stresses caused by the application of loadings on the surface of the beam.

## PROCEDURE FOR ANALYSIS

Based on the previous discussion, the following procedure provides a rational method for the design of a beam on the basis of strength.

*Shear and Moment Diagrams.*

- Determine the maximum shear and moment in the beam. Often this is done by constructing the beam's shear and moment diagrams.
- For built-up beams, shear and moment diagrams are useful for identifying *regions* where the shear and moment are excessively large and may require additional structural reinforcement or fasteners.

*Average Normal Stress.*

- If the beam is relatively long, it is designed by finding its section modulus using the flexure formula, $S_{req'd} = M_{max}/\sigma_{allow}$.
- Once $S_{req'd}$ is determined, the cross-sectional dimensions for simple shapes can then be computed, since $S_{req'd} = I/c$.
- If rolled-steel sections are to be used, several possible values of $S$ may be selected from the tables in Appendix D. Of these, choose the one having the smallest cross-sectional area, since this beam has the least weight and is therefore the most economical.
- Make sure that the selected section modulus, $S$, is *slightly greater* than $S_{req'd}$, so that the additional moment created by the beam's weight is considered.

*Shear Stress.*

- Normally beams that are short and carry large loads, especially those made of wood, are first designed to resist shear and then later checked against the allowable-bending-stress requirements.
- Using the shear formula, check to see that the allowable shear stress is not exceeded; that is, use $\tau_{allow} \geq V_{max} Q/It$.
- If the beam has a solid *rectangular* cross section, the shear formula becomes $\tau_{allow} \geq 1.5(V_{max}/A)$, Eq. 13–5, and if the cross section is a *wide flange*, it is generally appropriate to assume that the shear stress is *constant* over the cross-sectional area of the beam's web so that $\tau_{allow} \geq V_{max}/A_{web}$, where $A_{web}$ is determined from the product of the beam's depth and the web's thickness. (See Sec. 13.3.)

*Adequacy of Fasteners.*

- The adequacy of fasteners used on built-up beams depends upon the shear stress the fasteners can resist. Specifically, the required spacing of nails or bolts of a particular size is determined from the allowable shear flow, $q_{allow} = VQ/I$, calculated at points on the cross section where the fasteners are located. (See Sec. 13.4.)

# E X A M P L E 16.2

A beam is to be made of steel that has an allowable bending stress of $\sigma_{allow} = 24$ ksi and an allowable shear stress of $\tau_{allow} = 14.5$ ksi. Select an appropriate W shape that will carry the loading shown in Fig. 16–8a.

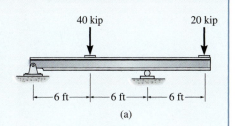

(a)

## Solution

*Shear and Moment Diagrams.* The support reactions have been calculated, and the shear and moment diagrams are shown in Fig. 16–8b. From these diagrams, $V_{max} = 30$ kip and $M_{max} = 120$ kip · ft.

*Bending Stress.* The required section modulus for the beam is determined from the flexure formula,

$$S_{req'd} = \frac{M_{max}}{\sigma_{allow}} = \frac{120 \text{ kip} \cdot \text{ft}(12 \text{ in./ft})}{24 \text{ kip/in}^2} = 60 \text{ in}^3$$

Using the table in Appendix D, the following beams are adequate:

| | |
|---|---|
| W18 × 40 | $S = 68.4 \text{ in}^3$ |
| W16 × 45 | $S = 72.7 \text{ in}^3$ |
| W14 × 43 | $S = 62.7 \text{ in}^3$ |
| W12 × 50 | $S = 64.7 \text{ in}^3$ |
| W10 × 54 | $S = 60.0 \text{ in}^3$ |
| W8 × 67 | $S = 60.4 \text{ in}^3$ |

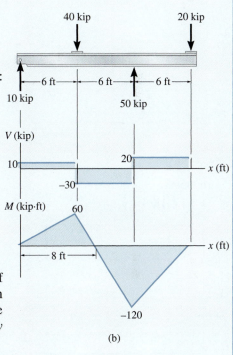

(b)

Fig. 16–8

The beam having the least weight per foot is chosen, i.e.,

$$\text{W18} \times 40$$

The *actual* maximum moment $M_{max}$, which includes the weight of the beam, can be computed and the adequacy of the selected beam can be checked. In comparison with the applied loads, however, the beam's weight, $(0.040 \text{ kip/ft})(18 \text{ ft}) = 0.720 \text{ kip}$, will only *slightly increase* $S_{req'd}$. In spite of this,

$$S_{req'd} = 60 \text{ in}^3 < 68.4 \text{ in}^3 \qquad \text{OK}$$

*Shear Stress.* Since the beam is a *wide-flange section*, the *average shear stress* within the web will be considered. Here the web is assumed to extend from the very top to the very bottom of the beam. From Appendix D, for a W18 × 40, $d = 17.90$ in, $t_w = 0.315$ in. Thus,

$$\tau_{avg} = \frac{V_{max}}{A_w} = \frac{30 \text{ kip}}{(17.90 \text{ in.})(0.315 \text{ in.})} = 5.32 \text{ ksi} < 14.5 \text{ ksi} \qquad \text{OK}$$

Use a W18 × 40. *Ans.*

## E X A M P L E  16.3

The wooden T-beam shown in Fig. 16–9a is made from two 200 mm × 30 mm boards. If the allowable bending stress is $\sigma_{allow} = 12$ MPa and the allowable shear stress is $\tau_{allow} = 0.8$ MPa, determine if the beam can safely support the loading shown. Also, specify the maximum spacing of nails needed to hold the two boards together if each nail can safely resist 1.50 kN in shear.

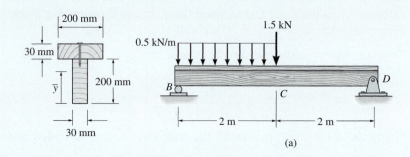

(a)

### Solution

***Shear and Moment Diagrams.***  The reactions on the beam are shown, and the shear and moment diagrams are drawn in Fig. 16–9b. Here $V_{max} = 1.5$ kN, $M_{max} = 2$ kN·m.

***Bending Stress.***  The neutral axis (centroid) will be located from the bottom of the beam. Working in units of meters, we have

$$\bar{y} = \frac{\Sigma \tilde{y} A}{\Sigma A}$$

$$= \frac{(0.1 \text{ m})(0.03 \text{ m})(0.2 \text{ m}) + 0.215 \text{ m}(0.03 \text{ m})(0.2 \text{ m})}{0.03 \text{ m}(0.2 \text{ m}) + 0.03 \text{ m}(0.2 \text{ m})} = 0.1575 \text{ m}$$

Thus,

$$I = \left[\frac{1}{12}(0.03 \text{ m})(0.2 \text{ m})^3 + (0.03 \text{ m})(0.2 \text{ m})(0.1575 \text{ m} - 0.1 \text{ m})^2\right]$$

$$+ \left[\frac{1}{12}(0.2 \text{ m})(0.03 \text{ m})^3 + (0.03 \text{ m})(0.2 \text{ m})(0.215 \text{ m} - 0.1575 \text{ m})^2\right]$$

$$= 60.125(10^{-6}) \text{ m}^4$$

Since $c = 0.1575$ m (not 0.230 m − 0.1575 m = 0.0725 m), we require

$$\sigma_{allow} \geq \frac{M_{max}c}{I}$$

$$12(10^3) \text{ kPa} \geq \frac{2 \text{ kN} \cdot \text{m}(0.1575 \text{ m})}{60.125(10^{-6}) \text{ m}^4} = 5.24(10^3) \text{ kPa} \qquad \text{OK}$$

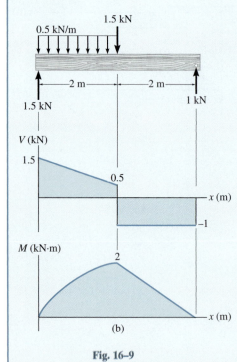

**Fig. 16–9**

***Shear Stress.*** Maximum shear stress in the beam depends upon the magnitude of $Q$ and $t$. It occurs at the neutral axis, since $Q$ is a maximum there and the neutral axis is in the web, where the thickness $t = 0.03$ m is smallest for the cross section. For simplicity, we will use the rectangular area below the neutral axis to calculate $Q$, rather than a two-part composite area above this axis, Fig. 16–9c. We have

$$Q = \bar{y}'A' = \left(\frac{0.1575 \text{ m}}{2}\right)[(0.1575 \text{ m})(0.03 \text{ m})] = 0.372(10^{-3}) \text{ m}^3$$

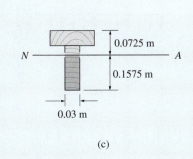

(c)

So that

$$\tau_{\text{allow}} \geq \frac{V_{\text{max}}Q}{It}$$

$$800 \text{ kPa} \geq \frac{1.5 \text{ kN}[0.372(10^{-3})] \text{ m}^3}{60.125(10^{-6}) \text{ m}^4(0.03 \text{ m})} = 309 \text{ kPa} \quad \text{OK}$$

***Nail Spacing.*** From the shear diagram it is seen that the shear varies over the entire span. Since the nail spacing depends on the magnitude of shear in the beam, for simplicity (and to be conservative), we will design the spacing on the basis of $V = 1.5$ kN for region $BC$ and $V = 1$ kN for region $CD$. Since the nails join the flange to the web, Fig. 16–9d, we have

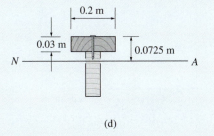

(d)

$$Q = \bar{y}'A' = (0.0725 \text{ m} - 0.015 \text{ m})[(0.2 \text{ m})(0.03 \text{ m})] = 0.345(10^{-3}) \text{ m}^3$$

The shear flow for each region is therefore

$$q_{BC} = \frac{V_{BC}Q}{I} = \frac{1.5 \text{ kN}[0.345(10^{-3})] \text{ m}^3}{60.125(10^{-6}) \text{ m}^4} = 8.61 \text{ kN/m}$$

$$q_{CD} = \frac{V_{CD}Q}{I} = \frac{1 \text{ kN}[0.345(10^{-3})] \text{ m}^3}{60.125(10^{-6}) \text{ m}^4} = 5.74 \text{ kN/m}$$

One nail can resist 1.50 kN in shear, so the spacing becomes

$$s_{BC} = \frac{1.50 \text{ kN}}{8.61 \text{ kN/m}} = 0.174 \text{ m}$$

$$s_{CD} = \frac{1.50 \text{ kN}}{5.74 \text{ kN/m}} = 0.261 \text{ m}$$

For ease of measuring, use

$$s_{BC} = 150 \text{ mm} \qquad \qquad \textit{Ans.}$$

$$s_{CD} = 250 \text{ mm} \qquad \qquad \textit{Ans.}$$

**E X A M P L E** 16.4

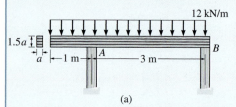

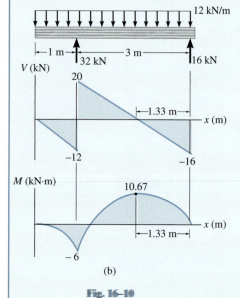

(a)

(b)

**Fig. 16–10**

The laminated wooden beam shown in Fig. 16–10$a$ supports a uniform distributed loading of 12 kN/m. If the beam is to have a height-to-width ratio of 1.5, determine its smallest width. The allowable bending stress is $\sigma_{allow}$ = 9 MPa and the allowable shear stress is $\tau_{allow}$ = 0.6 MPa. Neglect the weight of the beam.

**Solution**

***Shear and Moment Diagrams.***  The support reactions at $A$ and $B$ have been calculated and the shear and moment diagrams are shown in Fig. 16–10$b$. Here $V_{max}$ = 20 kN, $M_{max}$ = 10.67 kN·m.

***Bending Stress.***  Applying the flexure formula yields

$$S_{req'd} = \frac{M_{max}}{\sigma_{allow}} = \frac{10.67 \text{ kN} \cdot \text{m}}{9(10^3) \text{ kN/m}^2} = 0.00119 \text{ m}^3$$

Assuming that the width is $a$, then the height is $h = 1.5a$, Fig. 16–10$a$. Thus,

$$S_{req'd} = \frac{I}{c} = \frac{\frac{1}{12}(a)(1.5a)^3}{(0.75a)} = 0.00119 \text{ m}^3$$

$$a^3 = 0.003160 \text{ m}^3$$

$$a = 0.147 \text{ m}$$

***Shear Stress.***  Applying the shear formula for rectangular sections (which is a special case of $\tau_{max} = VQ/It$), we have

$$\tau_{max} = 1.5\frac{V_{max}}{A} = (1.5)\frac{20 \text{ kN}}{(0.147 \text{ m})(1.5)(0.147 \text{ m})}$$

$$= 0.929 \text{ MPa} > 0.6 \text{ MPa}$$

**Equation**

Since the shear criterion fails, the beam must be redesigned on the basis of shear.

$$\tau_{allow} = \frac{3}{2}\frac{V_{max}}{A}$$

$$600 \text{ kN/m}^2 = \frac{3}{2}\frac{20 \text{ kN}}{(a)(1.5a)}$$

$$a = 0.183 \text{ m} = 183 \text{ mm} \qquad Ans.$$

This larger section will also adequately resist the normal stress.

# PROBLEMS

**16-1.** The wooden beam has a rectangular cross section and is used to support a load of 1200 lb. If the allowable bending stress is $\sigma_{allow} = 2$ ksi and the allowable shear stress is $\tau_{allow} = 750$ psi, determine the height $h$ of the cross section to the nearest $\frac{1}{4}$ in. if it is to be rectangular and have a width of $b = 3$ in. Assume the supports at $A$ and $B$ only exert vertical reactions on the beam.

**16-2.** Solve Prob. 16-1 if the cross section has an unknown width but is to be square, i.e., $h = b$.

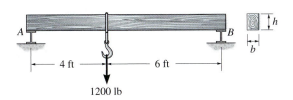

Probs. 16–1/2

**16-3.** The beam is made of Douglas fir having an allowable bending stress of $\sigma_{allow} = 1.1$ ksi and an allowable shear stress of $\tau_{allow} = 0.70$ ksi. Determine the width $b$ of the beam if the height $h = 2b$.

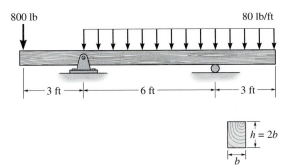

Prob. 16–3

**\*16-4.** Select the lightest-weight steel wide-flange beam from Appendix D that will safely support the machine loading shown. The allowable bending stress is $\sigma_{allow} = 24$ ksi and the allowable shear stress is $\tau_{allow} = 14$ ksi.

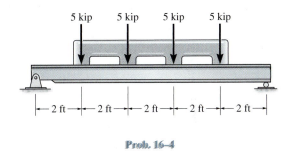

Prob. 16–4

**16-5.** Select the lightest-weight steel wide-flange beam from Appendix D that will safely support the loading shown, where $w = 6$ kip/ft and $P = 5$ kip. The allowable bending stress is $\sigma_{allow} = 24$ ksi, and the allowable shear stress is $\tau_{allow} = 14$ ksi.

**16-6.** Select the lightest-weight steel wide-flange beam having the shortest height from Appendix D that will safely support the loading shown, where $w = 0$ and $P = 10$ kip. The allowable bending stress is $\sigma_{allow} = 24$ ksi, and the allowable shear stress is $\tau_{allow} = 14$ ksi.

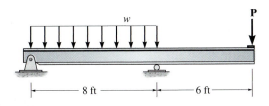

Probs. 16–5/6

**16-7.** The spreader beam $AB$ is used to lift slowly the 3000-lb pipe that is centrally located on the straps at $C$ and $D$. If the beam is a W12 × 45, determine if it can safely support the load. The allowable bending stress is $\sigma_{allow} = 22$ ksi and the allowable shear stress is $\tau_{allow} = 12$ ksi.

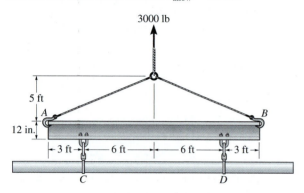

3000 lb

5 ft

A

12 in.

B

← 3 ft → ← 6 ft → ← 6 ft → ← 3 ft →

C

D

**Prob. 16–7**

**\*16-8.** Select the lightest-weight steel structural wide-flange beam with the shortest depth from Appendix D that will safely support the loading shown. The allowable bending stress is $\sigma_{allow} = 24$ ksi and the allowable shear stress is $\tau_{allow} = 14$ ksi.

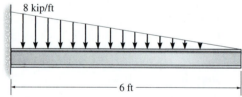

8 kip/ft

← 6 ft →

**Prob. 16–8**

**16-9.** The beam is made of a ceramic material having an allowable bending stress of $\sigma_{allow} = 735$ psi and an allowable shear stress of $\tau_{allow} = 400$ psi. Determine the width $b$ of the beam if the height $h = 2b$.

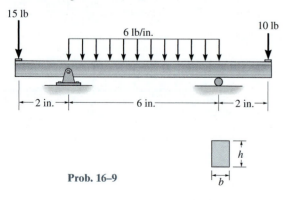

15 lb

6 lb/in.

10 lb

← 2 in. → ← 6 in. → ← 2 in. →

h

b

**Prob. 16–9**

**16-10.** The simply supported beam is composed of two W12 × 22 sections built up as shown. Determine the maximum uniform loading $w$ the beam will support if the allowable bending stress is $\sigma_{allow} = 22$ ksi and the allowable shear stress is $\tau_{allow} = 14$ ksi.

**16-11.** The simply supported beam is composed of two W12 × 22 sections built up as shown. Determine if the beam will safely support a loading of $w = 2$ kip/ft. The allowable bending stress is $\sigma_{allow} = 22$ ksi and the allowable shear stress is $\tau_{allow} = 14$ ksi.

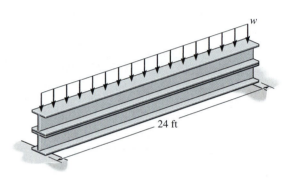

w

24 ft

**Probs. 16–10/11**

**\*16-12.** Select the lightest-weight steel wide-flange beam from Appendix D that will safely support the loading shown. The allowable bending stress is $\sigma_{allow} = 24$ ksi and the allowable shear stress is $\tau_{allow} = 14$ ksi.

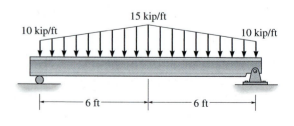

15 kip/ft

10 kip/ft

10 kip/ft

← 6 ft → ← 6 ft →

**Prob. 16–12**

**16-13.** Draw the shear and moment diagrams for the W12 × 14 beam and check if the beam will safely support the loading. The allowable bending stress is $\sigma_{allow} = 22$ ksi and the allowable shear stress is $\tau_{allow} = 12$ ksi.

**16-14.** Select the lightest-weight steel wide-flange beam from Appendix D that will safely support the loading shown. The allowable bending stress is $\sigma_{allow} = 22$ ksi and the allowable shear stress is $\tau_{allow} = 12$ ksi.

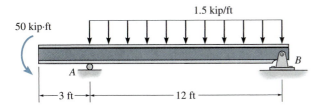

Probs. 16–13/14

**16-15.** Determine the smallest diameter rod that will safely support the loading shown. The allowable bending stress is $\sigma_{allow} = 167$ MPa and the allowable shear stress is $\tau_{allow} = 97$ MPa.

**\*16-16.** The pipe has an outer diameter of 15 mm. Determine the smallest inner diameter so that it will safely support the loading shown. The allowable bending stress is $\sigma_{allow} = 167$ MPa and the allowable shear stress is $\tau_{allow} = 97$ MPa.

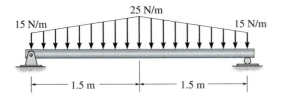

Probs. 16–15/16

**16-17.** Two acetyl plastic members are to be glued together and used to support the loading shown. If the allowable bending stress for the plastic is $\sigma_{allow} = 13$ ksi and the allowable shear stress is $\tau_{allow} = 4$ ksi, determine the greatest load $P$ that can be supported and specify the required shear stress capacity of the glue.

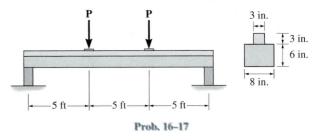

Prob. 16–17

**16-18.** The timber beam has a rectangular cross section. If the width of the beam is 6 in., determine its height $h$ so that it simultaneously reaches its allowable bending stress of $\sigma_{allow} = 1.50$ ksi and an allowable shear stress of $\tau_{allow} = 50$ psi. Also, what is the maximum load $P$ that the beam can then support?

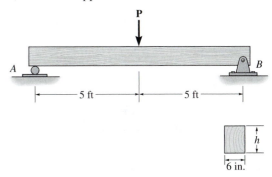

Prob. 16–18

**16-19.** Determine the minimum width $b$ of the beam to the nearest $\frac{1}{4}$ in. that will safely support the loading of $P = 8$ kip. The allowable bending stress is $\sigma_{allow} = 24$ ksi and the allowable shear stress is $\tau_{allow} = 15$ ksi.

**\*16-20.** Solve Prob. 16–19 if $P = 10$ kip.

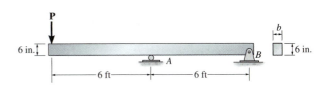

Probs. 16–19/20

**16-21.** The brick wall exerts a uniform distributed load of 1.20 kip/ft on the beam. If the allowable bending stress is $\sigma_{allow}$ = 22 ksi, determine the required width $b$ of the flange to the nearest $\frac{1}{4}$ in.

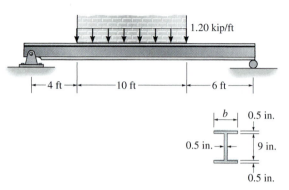

Prob. 16-21

**16-22.** The brick wall exerts a uniform distributed load of 1.20 kip/ft on the beam. If the allowable bending stress is $\sigma_{allow}$ = 22 ksi and the allowable shear stress is $\tau_{allow}$ = 12 ksi, select the lightest wide-flange section with the shortest depth from Appendix D that will safely support the load.

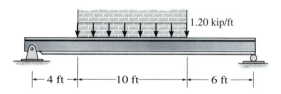

Prob. 16-22

**16-23.** The simply supported joist is used in the construction of a floor for a building. In order to keep the floor low with respect to the sill beams $C$ and $D$, the ends of the joists are notched as shown. If the allowable shear stress for the wood is $\tau_{allow}$ = 350 psi and the allowable bending stress is $\sigma_{allow}$ = 1500 psi, determine the height $h$ that will cause the beam to reach both allowable stresses at the same time. Also, what load $P$ causes this to happen? Neglect the stress concentration at the notch.

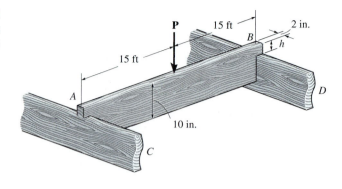

Prob. 16-23

**\*16-24.** The joist $AB$ used in housing construction is to be made from 8-in. by 1.5-in. Southern-pine boards. If the design loading on each board is placed as shown, determine the largest room width $L$ that the boards can span. The allowable bending stress for the wood is $\sigma_{allow}$ = 2 ksi and the allowable shear stress is $\tau_{allow}$ = 180 psi. Assume that the beam is simply supported from the walls at $A$ and $B$.

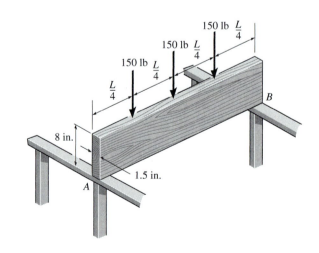

Prob. 16-24

## 16.4  The Elastic Curve

Before the slope or the displacement at a point on a beam is determined, it is often helpful to sketch the deflected shape of the beam when it is loaded, in order to visualize any computed results and thereby partially check these results. The deflection diagram of the longitudinal axis that passes through the centroid of each cross-sectional area of the beam is called the *elastic curve*. For most beams the elastic curve can be sketched without much difficulty. When doing so, however, it is necessary to know how the slope or displacement is restricted at various types of supports. In general, supports that resist a *force*, such as a pin, restrict *displacement*, and those that resist a *moment*, such as a fixed wall, restrict *rotation* or *slope* as well as displacement. With this in mind, two typical examples of the elastic curves for loaded beams (or shafts), sketched to a greatly exaggerated scale, are shown in Fig. 16–11.

If the elastic curve for a beam seems difficult to establish, it is suggested that the moment diagram for the beam be drawn first. Using the beam sign convention, a positive internal moment tends to bend the beam concave upward, Fig. 16–12a. Likewise, a negative moment tends to bend the beam concave downward, Fig. 16–12b. Therefore, if the moment diagram is *known*, it will be easy to construct the elastic curve.

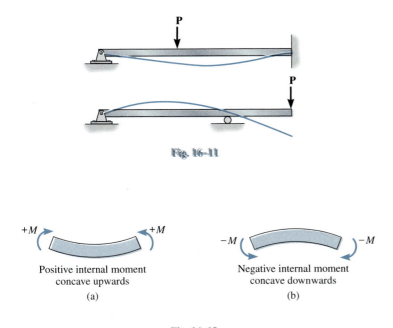

Fig. 16–11

| Positive internal moment | Negative internal moment |
| concave upwards | concave downwards |
| (a) | (b) |

Fig. 16–12

For example, consider the beam in Fig. 16–13*a* with its associated moment diagram shown in Fig. 16–13*b*. Due to the roller and pin supports, the displacement at *B* and *D* must be zero. Within the region of negative moment, *AC*, Fig. 16–13*b*, the elastic curve must be concave downward, and within the region of positive moment, *CD*, the elastic curve must be concave upward. Hence, there must be an *inflection point* at point *C*, where the curve changes from concave up to concave down, since this is a point of zero moment. Using these facts, the beam's elastic curve is sketched to a greatly exaggerated scale in Fig. 16–13*c*. It should also be noted that the displacements $\Delta_A$ and $\Delta_E$ are especially critical. At point *E* the *slope* of the elastic curve is *zero*, and there the beam's *deflection* may be a *maximum*. Whether $\Delta_E$ is actually greater than $\Delta_A$ depends on the relative magnitudes of $\mathbf{P}_1$ and $\mathbf{P}_2$ and the location of the roller at *B*.

Following these same principles, note how the elastic curve in Fig. 16–14 was constructed. Here the beam is cantilevered from a fixed support at *A* and therefore the elastic curve must have both zero displacement and zero slope at this point. Also, the largest displacement will occur either at *D*, where the slope is zero, or at *C*.

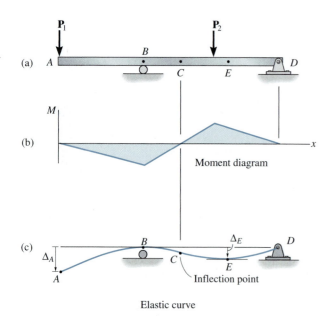

(a)

(b) Moment diagram

(c) Elastic curve

**Fig. 16–13**

(a)

(b) Moment diagram

(c) Inflection point    Elastic curve

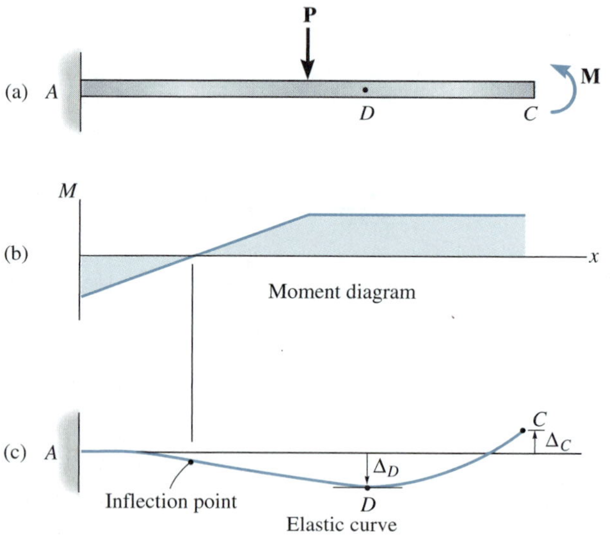

**Fig. 16–14**

## Moment–Curvature Relationship.

We will now develop an important relationship between the internal moment in the beam and the radius of curvature $\rho$ (rho) of the elastic curve at a point. The resulting equation will be used throughout the chapter as a basis for establishing each of the methods presented for finding the slope and displacement of the elastic curve for a beam.

The following analysis, here and in the next section, will require the use of three coordinates. As shown in Fig. 16–15a, the $x$ axis extends positive to the right, along the initially straight longitudinal axis of the beam. It is used to locate the differential element, having an undeformed width $dx$. The $v$ axis extends *positive upward* from the $x$ axis. It measures the *displacement* of the centroid on the cross-sectional area of the element. With these two coordinates, we will later define the equation of the elastic curve, $v$, as a function of $x$. Lastly, a "localized" $y$ coordinate is used to specify the position of a fiber in the beam element. It is measured *positive upward* from the neutral axis, as shown in Fig. 16–15b. Recall that this same sign convention for $x$ and $y$ was used in the derivation of the flexure formula.

To derive the relationship between the internal moment and $\rho$, we will limit the analysis to the most common case of an initially straight beam that is elastically deformed by loads applied perpendicular to the beam's $x$ axis and lying in the $x-v$ plane of symmetry for the beam's cross-sectional area. Due to the loading, the deformation of the beam is caused by both the internal shear force and bending moment. If the beam has a length that is much greater than its depth, the greatest deformation will be caused by bending, and therefore we will direct our attention to its effects. Deflections caused by shear will be discussed later in the chapter.

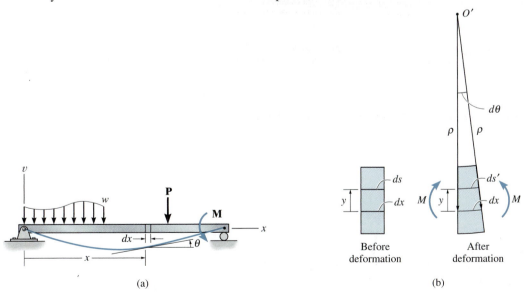

Before deformation    After deformation

(a)                (b)

**Fig. 16–15**

When the internal moment $M$ deforms the element of the beam, the angle between the cross sections becomes $d\theta$, Fig. 16–15$b$. The arc $dx$ represents a portion of the elastic curve that intersects the neutral axis for each cross section. The *radius of curvature* for this arc is defined as the distance $\rho$, which is measured from the *center of curvature O'* to $dx$. Any arc on the element other than $dx$ is subjected to a normal strain. For example, the strain in arc $ds$, located at a position $y$ from the neutral axis, is $\epsilon = (ds' - ds)/ds$. However, $ds = dx = \rho\, d\theta$ and $ds' = (\rho - y)\, d\theta$, and so $\epsilon = [(\rho - y)\, d\theta - \rho\, d\theta]/\rho\, d\theta$ or

$$\frac{1}{\rho} = -\frac{\epsilon}{y} \tag{16-2}$$

If the material is homogeneous and behaves in a linear-elastic manner, then Hooke's law applies, $\epsilon = \sigma/E$. Also, since the flexure formula applies, $\sigma = -My/I$. Combining these equations and substituting into the above equation, we have

$$\boxed{\frac{1}{\rho} = \frac{M}{EI}} \tag{16-3}$$

where

$\rho$ = the radius of curvature at a specific point on the elastic curve ($1/\rho$ is referred to as the *curvature*)

$M$ = the internal moment in the beam at the point where $\rho$ is to be determined

$E$ = the material's modulus of elasticity

$I$ = the beam's moment of inertia computed about the neutral axis

The product $EI$ in this equation is referred to as the *flexural rigidity*, and it is always a positive quantity. The sign for $\rho$ therefore depends on the direction of the moment. As shown in Fig. 16–16, when $M$ is *positive*, $\rho$ extends *above* the beam, i.e., in the positive $v$ direction; when $M$ is *negative*, $\rho$ extends *below* the beam, or in the negative $v$ direction.

Using the flexure formula, $\sigma = -My/I$, we can also express the curvature in terms of the stress in the beam, namely,

$$\frac{1}{\rho} = -\frac{\sigma}{Ey} \tag{16-4}$$

Both Eqs. 16–3 and 16–4 are valid for either small or large radii of curvature. However, the value of $\rho$ is almost always calculated as a *very large quantity*. For example, consider an A-36 steel beam made from a W14 × 53 (Appendix D), where $E_{st} = 29(10^3)$ ksi and $\sigma_Y = 36$ ksi. When the material at the outer fibers, $y = \pm 7$ in., is about to *yield*, then, from Eq. 16–4, $\rho = \pm 5639$ in. Values of $\rho$ calculated at other points along the beam's elastic curve may be even *larger*, since $\sigma$ cannot exceed $\sigma_Y$ at the outer fibers.

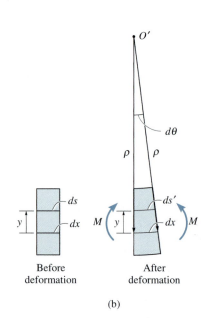

Before deformation

After deformation

(b)

**Fig. 16–15**

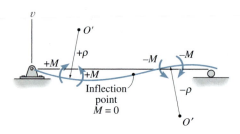

**Fig. 16–16**

## 16.5  Slope and Displacement by Integration

The elastic curve for a beam can be expressed mathematically as $v = f(x)$. To obtain this equation, we must first represent the curvature $(1/\rho)$ in terms of $v$ and $x$. In most calculus books it is shown that this relationship is

$$\frac{1}{\rho} = \frac{d^2v/dx^2}{[1 + (dv/dx)^2]^{3/2}}$$

Substituting into Eq. 16–3, we get

$$\frac{d^2v/dx^2}{[1 + (dv/dx)^2]^{3/2}} = \frac{M}{EI} \qquad (16\text{–}5)$$

This equation represents a nonlinear second-order differential equation. Its solution, which is called the *elastica*, gives the exact shape of the elastic curve, assuming, of course, that beam deflections occur only due to bending. Through the use of higher mathematics, elastica solutions have been obtained only for simple cases of beam geometry and loading.

In order to facilitate the solution of a greater number of deflection problems, Eq. 16–5 can be modified. Most engineering design codes specify *limitations* on deflections for tolerance or esthetic purposes, and as a result the elastic deflections for the majority of beams and shafts form a shallow curve. Consequently, the slope of the elastic curve which is determined from $dv/dx$ will be *very small*, and its square will be negligible compared with unity.* Therefore the curvature, as defined above, can be approximated by $1/\rho = d^2v/dx^2$. Using this simplification, Eq. 16–5 can now be written as

The moment of inertia of this bridge support varies along its length and this must be taken into account when computing its deflection.

$$\frac{d^2v}{dx^2} = \frac{M}{EI} \qquad (16\text{–}6)$$

It is also possible to write this equation in two alternative forms. If we differentiate each side with respect to $x$ and substitute $V = dM/dx$, we get

$$\frac{d}{dx}\left( EI\frac{d^2v}{dx^2} \right) = V(x) \qquad (16\text{–}7)$$

Differentiating again, using $-w = dV/dx$, yields

$$\frac{d^2}{dx^2}\left( EI\frac{d^2v}{dx^2} \right) = -w(x) \qquad (16\text{–}8)$$

*See Example 16.5.

For most problems the flexural rigidity will be constant along the length of the beam. Assuming this to be the case, the above results may be reordered into the following set of equations:

$$EI\frac{d^4v}{dx^4} = -w(x) \tag{16-9}$$

$$EI\frac{d^3v}{dx^3} = V(x) \tag{16-10}$$

$$EI\frac{d^2v}{dx^2} = M(x) \tag{16-11}$$

Solution of any of these equations requires successive integrations to obtain the deflection $v$ of the elastic curve. For each integration it is necessary to introduce a "constant of integration" and then solve for all the constants to obtain a unique solution for a particular problem. For example, if the distributed load is expressed as a function of $x$ and Eq. 16–9 is used, then four constants of integration must be evaluated; however, if the internal moment $M$ is determined and Eq. 16–11 is used, only two constants of integration must be found. The choice of which equation to start with depends on the problem. Generally, however, it is easier to determine the internal moment $M$ as a function of $x$, integrate twice, and evaluate only two integration constants.

If the loading on a beam is discontinuous, that is, consists of a series of several distributed and concentrated loads, then several functions must be written for the internal moment, each valid within the region between the discontinuities. Also, for convenience in writing each moment expression, *the origin* for each $x$ coordinate can be *selected arbitrarily*. For example, consider the beam shown in Fig. 16–17a. The internal moment in regions $AB$, $BC$, and $CD$ can be written in terms of the $x_1$, $x_2$, and $x_3$ coordinates selected, as shown in either Fig. 16–17b or 16–17c, or in fact in any manner that will yield $M = f(x)$ in as simple a form as possible. Once these functions are integrated through the use of Eq. 16–11 and the constants of integration determined, the functions will give the slope and deflection (elastic curve) for each region of the beam for which they are valid.

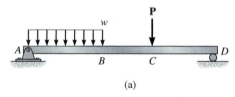

(a)

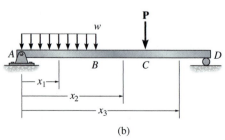

(b)

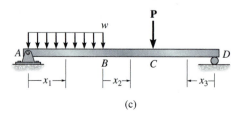

(c)

**Fig. 16–17**

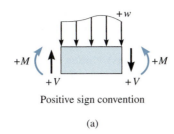

Positive sign convention

(a)

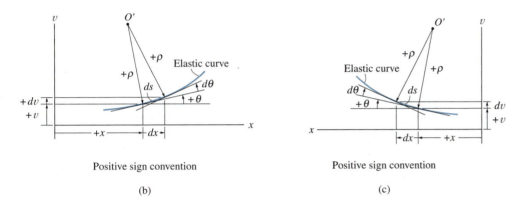

Positive sign convention

(b)

Positive sign convention

(c)

**Fig. 16–18**

**Sign Convention and Coordinates.** When applying Eqs. 16–9 through 16–11, it is important to use the proper signs for $M$, $V$, or $w$ as established by the sign convention that was used in the derivation of these equations. For review, these terms are shown in their *positive directions* in Fig. 16–18*a*. Furthermore, recall that *positive deflection, v*, is *upward*, and as a result, the *positive slope angle* $\theta$ will be measured *counterclockwise* from the $x$ axis when $x$ is *positive to the right*. The reason for this is shown in Fig. 16–18*b*. Here positive increases $dx$ and $dv$ in $x$ and $v$ create an increased $\theta$ that is counterclockwise. On the other hand, if *positive x* is directed to the *left*, then $\theta$ will be *positive clockwise*, Fig. 16–18*c*.

It should be pointed out that by assuming $dv/dx$ to be very small, the original horizontal length of the beam's axis and the arc of its elastic curve will be about the same. In other words, $ds$ in Fig. 16–18*b* and 16–18*c* is approximately equal to $dx$, since $ds = \sqrt{(dx)^2 + (dv)^2} = \sqrt{1 + (dv/dx)^2}\, dx \approx dx$. As a result, points on the elastic curve are assumed to be *displaced vertically*, and not horizontally. Also, since the *slope angle* $\theta$ will be *very small*, its value in radians can be determined directly from $\theta \approx \tan\theta = dv/dx$.

The design of a roof system requires a careful consideration of deflection. For example, rain can accumulate on areas of the roof, which then causes ponding, leading to further deflection and possible failure of the roof.

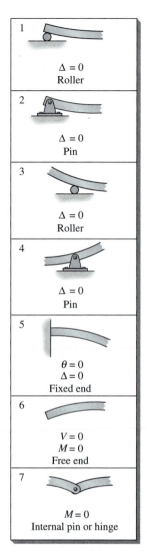

**Table 16–1**

**Boundary and Continuity Conditions.** The constants of integration are determined by evaluating the functions for shear, moment, slope, or displacement at a particular point on the beam where the value of the function is known. These values are called **boundary conditions**. Several possible boundary conditions that are often used to solve beam (or shaft) deflection problems are listed in Table 16–1. For example, if the beam is supported by a roller or pin (1, 2, 3, 4), then it is required that the displacement be *zero* at these points. Furthermore, if these supports are located at the *ends of the beam* (1, 2), the internal moment in the beam must also be zero. At the fixed support (5), the slope and displacement are both zero, whereas the free-ended beam (6) has both zero moment and zero shear. Lastly, if two segments of a beam are connected by an *internal* pin or hinge (7), the moment must be zero at this connection.

If a single *x* coordinate cannot be used to express the equation for the beam's slope or the elastic curve, then **continuity conditions** must be used to evaluate some of the integration constants. For example, consider the beam in Fig. 16–19a. Here the *x* coordinates are both chosen with origins at *A*. Each is valid only within the regions $0 \leq x_1 \leq a$ and $a \leq x_2 \leq (a + b)$. Once the functions for the slope and deflection are obtained, they must give the *same values* for the slope and deflection at point *B* so the elastic curve is physically *continuous*. Expressed mathematically, this requires that $\theta_1(a) = \theta_2(a)$ and $v_1(a) = v_2(a)$. These equations can then be used to evaluate two constants of integration. On the other hand, if the elastic curve is expressed in terms of the coordinates $0 \leq x_1 \leq a$ and $0 \leq x_2 \leq b$, shown in Fig. 16–19b, then the continuity of slope and deflection at *B* requires $\theta_1(a) = -\theta_2(b)$ and $v_1(a) = v_2(b)$. In this particular case, a *negative* sign is necessary to match the slopes at *B* since $x_1$ extends positive to the right, whereas $x_2$ extends positive to the left. Consequently, $\theta_1$ is positive counterclockwise, and $\theta_2$ is positive clockwise. See Fig. 16–18b and 16–18c.

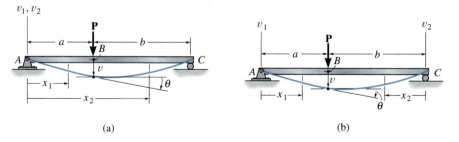

(a)                                              (b)

**Fig. 16–19**

## PROCEDURE FOR ANALYSIS

The following procedure provides a method for determining the slope and deflection of a beam (or shaft) using the method of integration.

*Elastic Curve.*

• Draw an exaggerated view of the beam's elastic curve. Recall that zero slope and zero displacement occur at all fixed supports, and zero displacement occurs at all pin and roller supports.

• Establish the $x$ and $v$ coordinate axes. The $x$ axis must be parallel to the undeflected beam and can have an origin at any point along the beam, with a positive direction either to the right or to the left.

• If several discontinuous loads are present, establish $x$ coordinates that are valid for each region of the beam between the discontinuities. Choose these coordinates so that they will simplify subsequent algebraic work.

• In all cases, the associated positive $v$ axis should be directed upward.

*Load or Moment Function.*

• For each region in which there is an $x$ coordinate, express the loading $w$ or the internal moment $M$ as a function of $x$. In particular, *always* assume that $M$ acts in the *positive direction* when applying the equation of moment equilibrium to determine $M = f(x)$.

*Slope and Elastic Curve.*

• Provided $EI$ is constant, apply either the load equation $EI\, d^4v/dx^4 = -w(x)$, which requires four integrations to get $v = v(x)$, or the moment equation $EI\, d^2v/dx^2 = M(x)$, which requires only two integrations. For each integration it is important to include a constant of integration.

• The constants are evaluated using the boundary conditions for the supports (Table 16–1) and the continuity conditions that apply to slope and displacement at points where two functions meet. Once the constants are evaluated and substituted back into the slope and deflection equations, the slope and displacement at *specific points* on the elastic curve can then be determined.

• The numerical values obtained can be checked graphically by comparing them with the sketch of the elastic curve. Realize that *positive* values for *slope* are *counterclockwise* if the $x$ axis extends *positive* to the *right*, and *clockwise* if the $x$ axis extends *positive* to the *left*. In either of these cases, *positive displacement* is *upward*.

E X A M P L E **16.5**

The cantilevered beam shown in Fig. 16–20a is subjected to a vertical load **P** at its end. Determine the equation of the elastic curve. $EI$ is constant.

**Solution I**

*Elastic Curve.* The load tends to deflect the beam as shown in Fig. 16–20a. By inspection, the internal moment can be represented throughout the beam using a single $x$ coordinate.

*Moment Function.* From the free-body diagram, with **M** acting in the *positive direction*, Fig. 16–20b, we have

$$M = -Px$$

*Slope and Elastic Curve.* Applying Eq. 16–11 and integrating twice yields

$$EI\frac{d^2v}{dx^2} = -Px \qquad (1)$$

$$EI\frac{dv}{dx} = -\frac{Px^2}{2} + C_1 \qquad (2)$$

$$EIv = -\frac{Px^3}{6} + C_1x + C_2 \qquad (3)$$

Using the boundary conditions $dv/dx = 0$ at $x = L$ and $v = 0$ at $x = L$, Eqs. 2 and 3 become

$$0 = -\frac{PL^2}{2} + C_1$$

$$0 = -\frac{PL^3}{6} + C_1L + C_2$$

Thus, $C_1 = PL^2/2$ and $C_2 = -PL^3/3$. Substituting these results into Eqs. 2 and 3 with $\theta = dv/dx$, we get

$$\theta = \frac{P}{2EI}(L^2 - x^2)$$

$$v = \frac{P}{6EI}(-x^3 + 3L^2x - 2L^3) \qquad \textit{Ans.}$$

Maximum slope and displacement occur at $A$ ($x = 0$), for which

$$\theta_A = \frac{PL^2}{2EI} \qquad (4)$$

$$v_A = -\frac{PL^3}{3EI} \qquad (5)$$

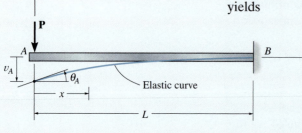

(a)

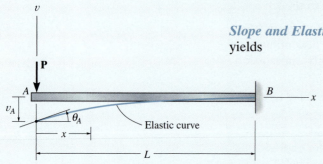

(b)

Fig. 16–20

The *positive* result for $\theta_A$ indicates *counterclockwise* rotation and the *negative* result for $v_A$ indicates that $v_A$ is *downward*. This agrees with the results sketched in Fig. 16–20a.

In order to obtain some idea as to the actual *magnitude* of the slope and displacement at the end $A$, consider the beam in Fig. 16–20a to have a length of 15 ft, support a load of $P = 6$ kip, and be made of A-36 steel having $E_{st} = 29(10^3)$ ksi. Using the methods of Sec. 16.3, if this beam was designed without a factor of safety by assuming the allowable normal stress is equal to the yield stress $\sigma_{allow} = 36$ ksi, then a W12 × 26 would be found to be adequate ($I = 204$ in⁴). From Eqs. 4 and 5 we get

$$\theta_A = \frac{6 \text{ kip}(15 \text{ ft})^2(12 \text{ in./ft})^2}{2[29(10^3) \text{ kip/in}^2](204 \text{ in}^4)} = 0.0164 \text{ rad}$$

$$v_A = -\frac{6 \text{ kip}(15 \text{ ft})^3(12 \text{ in./ft})^3}{3[29(10^3) \text{ kip/in}^2](204 \text{ in}^4)} = -1.97 \text{ in.}$$

Since $\theta_A^2 = (dv/dx)^2 = 0.000270 \text{ rad}^2 \ll 1$, this justifies the use of Eq. 16–11, rather than applying the more exact Eq. 16–5, for computing the deflection of beams. Also, since this numerical application is for a *cantilevered beam*, we have obtained *larger values* for $\theta$ and $v$ than would have been obtained if the beam was supported using pins, rollers, or other fixed supports.

## Solution II

This problem can also be solved using Eq. 16–9, $EI \, d^4v/dx^4 = -w(x)$. Here $w(x) = 0$ for $0 \leq x \leq L$, Fig. 16–20a, so that upon integrating once we get the form of Eq. 16–11, i.e.,

$$EI\frac{d^4v}{dx^4} = 0$$

$$EI\frac{d^3v}{dx^3} = C_1' = V$$

The shear constant $C_1'$ can be evaluated at $x = 0$, since $V_A = -P$ (negative according to the beam sign convention, Fig. 16–18a.) Thus, $C_1' = -P$. Integrating again yields the form of Eq. 16–11, i.e.,

$$EI\frac{d^3v}{dx^3} = -P$$

$$EI\frac{d^2v}{dx^2} = -Px + C_2' = M$$

Here $M = 0$ at $x = 0$, so $C_2' = 0$, and as a result one obtains Eq. 1 and the solution proceeds as before.

# E X A M P L E 16.6

The simply supported beam shown in Fig. 16–21*a* supports the triangular distributed loading. Determine its maximum deflection. *EI* is constant.

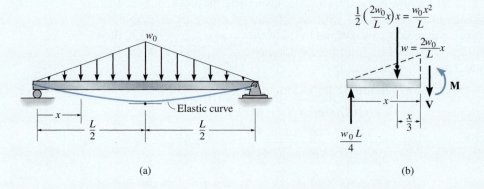

Fig. 16–21

## Solution I

*Elastic Curve.* Due to symmetry, only one *x* coordinate is needed for the solution, in this case $0 \leq x \leq L/2$. The beam deflects as shown in Fig. 16–21*a*. Notice that maximum deflection occurs at the center since the slope is zero at this point.

*Moment Function.* The distributed load acts downward, and therefore it is positive according to our sign convention. A free-body diagram of the segment on the left is shown in Fig. 16–21*b*. The equation for the distributed loading is

$$w = \frac{2w_0}{L}x \qquad (1)$$

Hence,

$$\zeta+ \ \Sigma M_{NA} = 0; \qquad M + \frac{w_0 x^2}{L}\left(\frac{x}{3}\right) - \frac{w_0 L}{4}(x) = 0$$

$$M = -\frac{w_0 x^3}{3L} + \frac{w_0 L}{4}x$$

*Slope and Elastic Curve.* Using Eq. 16–11 and integrating twice, we have

$$EI\frac{d^2v}{dx^2} = M = -\frac{w_0}{3L}x^3 + \frac{w_0L}{4}x \qquad (2)$$

$$EI\frac{dv}{dx} = -\frac{w_0}{12L}x^4 + \frac{w_0L}{8}x^2 + C_1$$

$$EIv = -\frac{w_0}{60L}x^5 + \frac{w_0L}{24}x^3 + C_1x + C_2$$

The constants of integration are obtained by applying the boundary condition $v = 0$ at $x = 0$ and the symmetry condition that $dv/dx = 0$ at $x = L/2$. This leads to

$$C_1 = -\frac{5w_0L^3}{192} \qquad C_2 = 0$$

Hence,

$$EI\frac{dv}{dx} = -\frac{w_0}{12L}x^4 + \frac{w_0L}{8}x^2 - \frac{5w_0L^3}{192}$$

$$EIv = -\frac{w_0}{60L}x^5 + \frac{w_0L}{24}x^3 - \frac{5w_0L^3}{192}x$$

Determining the maximum deflection at $x = L/2$, we have

$$v_{max} = -\frac{w_0L^4}{120EI} \qquad \textit{Ans.}$$

## Solution II

Starting with the distributed loading, Eq. 1, and applying Eq. 16–9, we have

$$EI\frac{d^4v}{dx^4} = -\frac{2w_0}{L}x$$

$$EI\frac{d^3v}{dx^3} = V = -\frac{w_0}{L}x^2 + C_1'$$

Since $V = +w_0L/4$ at $x = 0$, then $C_1' = w_0L/4$. Integrating again yields

$$EI\frac{d^3v}{dx^3} = V = -\frac{w_0}{L}x^2 + \frac{w_0L}{4}$$

$$EI\frac{d^2v}{dx^2} = M = -\frac{w_0}{3L}x^3 + \frac{w_0L}{4}x + C_2'$$

Here $M = 0$ at $x = 0$, so $C_2' = 0$. This yields Eq. 2. The solution now proceeds as before.

**EXAMPLE 16.7**

The simply supported beam shown in Fig. 16–22a is subjected to the concentrated force **P**. Determine the maximum deflection of the beam. $EI$ is constant.

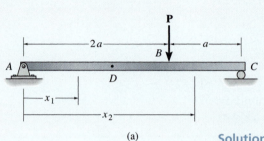

(a)

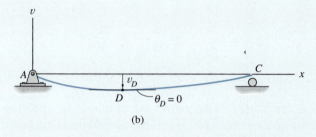

(b)

**Solution**

*Elastic Curve.* The beam deflects as shown in Fig. 16–22b. Two coordinates must be used, since the moment becomes discontinuous at $P$. Here we will take $x_1$ and $x_2$, having the *same origin* at $A$, so that $0 \le x_1 < 2a$ and $2a < x_2 \le 3a$.

*Moment Function.* From the free-body diagrams shown in Fig. 16–22c,

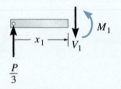

$$M_1 = \frac{P}{3}x_1$$

$$M_2 = \frac{P}{3}x_2 - P(x_2 - 2a) = \frac{2P}{3}(3a - x_2)$$

*Slope and Elastic Curve.* Applying Eq. 16–11 for $M_1$ and integrating twice yields

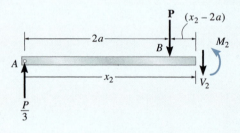

(c)

**Fig. 16–22**

$$EI\frac{d^2v_1}{dx_1^2} = \frac{P}{3}x_1$$

$$EI\frac{dv_1}{dx_1} = \frac{P}{6}x_1^2 + C_1 \qquad (1)$$

$$EIv_1 = \frac{P}{18}x_1^3 + C_1x_1 + C_2 \qquad (2)$$

Likewise for $M_2$,

$$EI\frac{d^2v_2}{dx_2^2} = \frac{2P}{3}(3a - x_2)$$

$$EI\frac{dv_2}{dx_2} = \frac{2P}{3}\left(3ax_2 - \frac{x_2^2}{2}\right) + C_3 \qquad (3)$$

$$EIv_2 = \frac{2P}{3}\left(\frac{3}{2}ax_2^2 - \frac{x_2^3}{6}\right) + C_3x_2 + C_4 \qquad (4)$$

The four constants are evaluated using *two* boundary conditions, namely, $x_1 = 0$, $v_1 = 0$ and $x_2 = 3a$, $v_2 = 0$. Also, *two* continuity conditions must be applied at $B$, that is, $dv_1/dx_1 = dv_2/dx_2$ at $x_1 = x_2 = 2a$ and $v_1 = v_2$ at $x_1 = x_2 = 2a$. Substitution as specified results in the following four equations:

$v_1 = 0$ at $x_1 = 0$;     $0 = 0 + 0 + C_2$

$v_2 = 0$ at $x_2 = 3a$;     $0 = \dfrac{2P}{3}\left(\dfrac{3}{2}a(3a)^2 - \dfrac{(3a)^3}{6}\right) + C_3(3a) + C_4$

$\dfrac{dv_1(2a)}{dx_1} = \dfrac{dv_2(2a)}{dx_2}$;     $\dfrac{P}{6}(2a)^2 + C_1 = \dfrac{2P}{3}\left(3a(2a) - \dfrac{(2a)^2}{2}\right) + C_3$

$v_1(2a) = v_2(2a)$;     $\dfrac{P}{18}(2a)^3 + C_1(2a) + C_2 = \dfrac{2P}{3}\left(\dfrac{3}{2}a(2a)^2 - \dfrac{(2a)^3}{6}\right) + C_3(2a) + C_4$

Solving these equations we get

$$C_1 = -\frac{4}{9}Pa^2 \qquad C_2 = 0$$

$$C_3 = -\frac{22}{9}Pa^2 \qquad C_4 = \frac{4}{3}Pa^3$$

Thus Eqs. 1–4 become

$$\frac{dv_1}{dx_1} = \frac{P}{6EI}x_1{}^2 - \frac{4}{9}\frac{Pa^2}{EI} \qquad (5)$$

$$v_1 = \frac{P}{18EI}x_1{}^3 - \frac{4}{9}\frac{Pa^2}{EI}x_1 \qquad (6)$$

$$\frac{dv_2}{dx_2} = \frac{2Pa}{EI}x_2 - \frac{P}{3EI}x_2{}^2 - \frac{22}{9}\frac{Pa^2}{EI} \qquad (7)$$

$$v_2 = \frac{Pa}{EI}x_2{}^2 - \frac{P}{9EI}x_2{}^3 - \frac{22}{9}\frac{Pa^2}{EI}x_2 + \frac{4}{3}\frac{Pa^3}{EI} \qquad (8)$$

By inspection of the elastic curve, Fig. 16–22b, the maximum deflection occurs at $D$, somewhere within region $AB$. Here the slope must be zero. From Eq. 5,

$$\frac{1}{6}x_1{}^2 - \frac{4}{9}a^2 = 0$$

$$x_1 = 1.633a$$

Substituting into Eq. 6,

$$v_{max} = -0.484\frac{Pa^3}{EI} \qquad \qquad Ans.$$

The negative sign indicates that the deflection is downward.

**E X A M P L E  16.8**

The beam in Fig. 16–23a is subjected to a load **P** at its end. Determine the displacement at *C*. *EI* is constant.

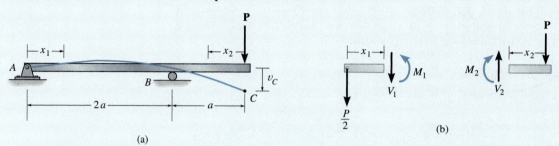

**Fig. 16–23**

**Solution**

*Elastic Curve.* The beam deflects into the shape shown in Fig. 16–23a. Due to the loading, two *x* coordinates will be considered, namely, $0 \le x_1 < 2a$ and $0 \le x_2 < a$, where $x_2$ is directed to the left from *C*, since the internal moment is easy to formulate.

*Moment Functions.* Using the free-body diagrams shown in Fig. 16–23b, we have

$$M_1 = -\frac{P}{2}x_1 \qquad M_2 = -Px_2$$

*Slope and Elastic Curve.* Applying Eq. 16–11,

for $0 \le x_1 < 2a$,

$$EI\frac{d^2v_1}{dx_1^2} = -\frac{P}{2}x_1$$

$$EI\frac{dv_1}{dx_1} = -\frac{P}{4}x_1^2 + C_1 \tag{1}$$

$$EIv_1 = -\frac{P}{12}x_1^3 + C_1x_1 + C_2 \tag{2}$$

For $0 \le x_2 < a$,

$$EI\frac{d^2v_2}{dx_2^2} = -Px_2$$

$$EI\frac{dv_2}{dx_2} = -\frac{P}{2}x_2^2 + C_3 \tag{3}$$

$$EIv_2 = -\frac{P}{6}x_2^3 + C_3x_2 + C_4 \tag{4}$$

The *four* constants of integration are determined using *three* boundary conditions, namely, $v_1 = 0$ at $x_1 = 0$, $v_1 = 0$ at $x_1 = 2a$, and $v_2 = 0$ at $x_2 = a$ and *one* continuity equation. Here the continuity of slope at the roller requires $dv_1/dx_1 = -dv_2/dx_2$ at $x_1 = 2a$ and $x_2 = a$. Why is there a negative sign in this equation? (Note that continuity of displacement at $B$ has been indirectly considered in the boundary conditions, since $v_1 = v_2 = 0$ at $x_1 = 2a$ and $x_2 = a$.) Applying these four conditions yields

$$v_1 = 0 \text{ at } x_1 = 0; \qquad 0 = 0 + 0 + C_2$$

$$v_1 = 0 \text{ at } x_1 = 2a; \qquad 0 = -\frac{P}{12}(2a)^3 + C_1(2a) + C_2$$

$$v_2 = 0 \text{ at } x_2 = a; \qquad 0 = -\frac{P}{6}a^3 + C_3 a + C_4$$

$$\frac{dv_1(2a)}{dx_1} = -\frac{dv_2(a)}{dx_2}; \qquad -\frac{P}{4}(2a)^2 + C_1 = -\left(-\frac{P}{2}(a)^2 + C_3\right)$$

Solving, we obtain

$$C_1 = \frac{Pa^2}{3} \qquad C_2 = 0 \qquad C_3 = \frac{7}{6}Pa^2 \qquad C_4 = -Pa^3$$

Substituting $C_3$ and $C_4$ into Eq. 4 gives

$$v_2 = -\frac{P}{6EI}x_2{}^3 + \frac{7Pa^2}{6EI}x_2 - \frac{Pa^3}{EI}$$

The displacement at $C$ is determined by setting $x_2 = 0$. We get

$$v_C = -\frac{Pa^3}{EI} \qquad\qquad\qquad \textit{Ans.}$$

# PROBLEMS

**16-25.** An A-36 steel strap having a thickness of 10 mm and a width of 20 mm is bent into a circular arc of radius $\rho = 10$ m. Determine the maximum bending stress in the strap.

**16-26.** A picture is taken of a man performing a pole vault, and the minimum radius of curvature of the pole is estimated by measurement to be 4.5 m. If the pole is 40 mm in diameter and it is made of a glass-reinforced plastic for which $E_g = 131$ GPa, determine the maximum bending stress in the pole.

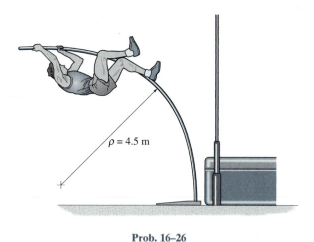

**Prob. 16–26**

**16-27.** Determine the equations of the elastic curve using the $x_1$ and $x_2$ coordinates. $EI$ is constant.

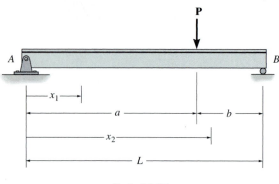

**Prob. 16–27**

*16-28.** The shaft supports the two pulley loads shown. Determine the equation of the elastic curve. The bearings at $A$ and $B$ exert only vertical reactions on the shaft. $EI$ is constant.

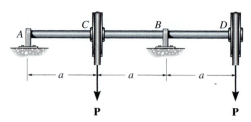

**Prob. 16–28**

**16-29.** Determine the equations of the elastic curve for the beam using the $x_1$ and $x_2$ coordinates. Specify the slope at $A$ and the maximum deflection. $EI$ is constant.

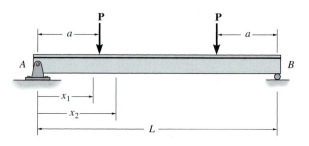

**Prob. 16–29**

**16-30.** Determine the equations of the elastic curve for the beam using the $x_1$ and $x_2$ coordinates. Specify the beam's maximum deflection. $EI$ is constant.

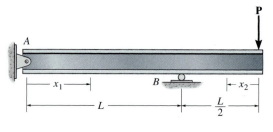

**Prob. 16–30**

**16-31.** The shaft is supported at $A$ by a journal bearing that exerts only vertical reactions on the shaft, and at $C$ by a thrust bearing that exerts horizontal and vertical reactions on the shaft. Determine the equations of the elastic curve using the coordinates $x_1$ and $x_2$. $EI$ is constant.

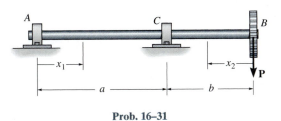

Prob. 16–31

***16-32.** The fence board weaves between the three smooth fixed posts. If the posts remain along the same line, determine the maximum bending stress in the board. The board has a width of 6 in. and a thickness of 0.5 in. $E_w = 1.60(10^3)$ ksi. Assume the displacement of each end of the board relative to its center is 3 in.

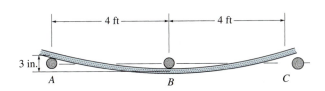

Prob. 16–32

**16-33.** The shaft supports the three pulley loads shown. Determine the equation of the elastic curve. The bearings at $A$ and $B$ exert only vertical reactions on the shaft. What is the maximum deflection? $EI$ is constant.

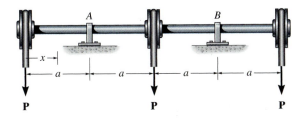

Prob. 16–33

**16-34.** Determine the maximum deflection of the beam and the slope at $A$. $EI$ is constant.

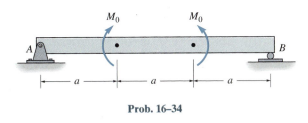

Prob. 16–34

**16-35.** The A-36 steel beam has a depth of 10 in. and is subjected to a constant moment $\mathbf{M}_0$, which causes the stress at the outer fibers to become $\sigma_Y = 36$ ksi. Determine the radius of curvature of the beam and the maximum slope and deflection.

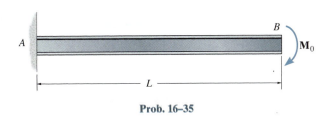

Prob. 16–35

***16-36.** The shaft is supported at $A$ by a journal bearing that exerts only vertical reactions on the shaft and at $B$ by a thrust bearing that exerts horizontal and vertical reactions on the shaft. Draw the bending-moment diagram for the shaft and then, from this diagram, sketch the deflection or elastic curve for the shaft's centerline. Determine the equations of the elastic curve using the coordinates $x_1$ and $x_2$. $EI$ is constant.

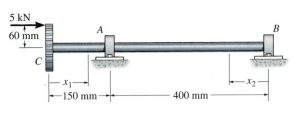

Prob. 16–36

**16-37.** Determine the elastic curve for the cantilevered beam, which is subjected to the couple moment $M_0$. Also compute the maximum slope and maximum deflection of the beam. $EI$ is constant.

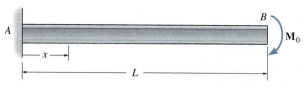

Prob. 16-37

**16-38.** Determine the maximum slope and maximum deflection of the simply-supported beam which is subjected to the couple moment $M_0$. $EI$ is constant.

Prob. 16-38

**16-39.** Determine the elastic curve for the simply supported beam, which is subjected to the couple moments $M_0$. Also, compute the maximum slope and the maximum deflection of the beam. $EI$ is constant.

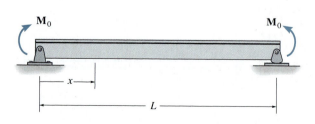

Prob. 16-39

**\*16-40.** Determine the equation of the elastic curve using the coordinate $x$, and specify the slope at point $A$ and the deflection at point $C$. $EI$ is constant.

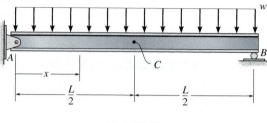

Prob. 16-40

**16-41.** The bar is supported by a roller constraint at $B$, which allows vertical displacement but resists axial load and moment. If the bar is subjected to the loading shown, determine the slope at $A$ and the deflection at $C$. $EI$ is constant.

**16-42.** Determine the deflection at $B$ of the bar in Prob. 16-41.

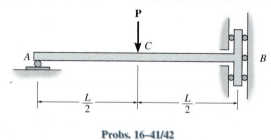

Probs. 16-41/42

**16-43.** Wooden posts used for a retaining wall have a diameter of 3 in. If the soil pressure along a post varies uniformly from zero at the top $A$ to a maximum of 300 lb/ft at the bottom $B$, determine the slope and displacement at the top of the post. $E_w = 1.6(10^3)$ ksi.

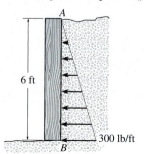

Prob. 16-43

## 16.6 Method of Superposition

The differential equation $EI \, d^4v/dx^4 = -w(x)$ satisfies the two necessary requirements for applying the principle of superposition; i.e., the load $w(x)$ is linearly related to the deflection $v(x)$, and the load is assumed not to change significantly the original geometry of the beam or shaft. As a result, the deflections for a series of separate loadings acting on a beam may be superimposed. For example, if $v_1$ is the deflection for one load and $v_2$ is the deflection for another load, the total deflection for both loads acting together is the algebraic sum $v_1 + v_2$. Using tabulated results for various beam loadings, such as the ones listed in Appendix B, or those found in various engineering handbooks, it is therefore possible to find the slope and displacement at a point on a beam subjected to several different loadings by algebraically adding the effects of its various component parts.

The following examples illustrate how to use the method of superposition to solve deflection problems, where the deflection is caused not only by beam deformations, but also by rigid-body displacements, which can occur when the beam is supported by springs or portions of a segmented beam are supported by hinges.

The resultant deflection at any point on this beam can be determined from the superposition of the deflections caused by each of the separate loadings acting on the beam.

E X A M P L E 16.9

Determine the displacement at point $C$ and the slope at the support $A$ of the beam shown in Fig. 16–24a. $EI$ is constant.

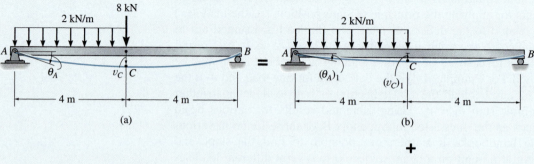

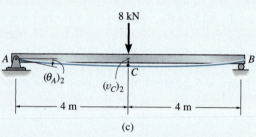

**Fig. 16–24**

**Solution**

The loading can be separated into two component parts as shown in Figs. 16–24b and 16–24c. The displacement at $C$ and slope at $A$ are found using the table in Appendix B for each part.

For the distributed loading,

$$(\theta_A)_1 = \frac{3wL^3}{128EI} = \frac{3(2\text{ kN/m})(8\text{ m})^3}{128EI} = \frac{24\text{ kN}\cdot\text{m}^2}{EI}\;\downarrow$$

$$(v_C)_1 = \frac{5wL^4}{768EI} = \frac{5(2\text{ kN/m})(8\text{ m})^4}{768EI} = \frac{53.33\text{ kN}\cdot\text{m}^3}{EI}\;\downarrow$$

For the 8-kN concentrated force,

$$(\theta_A)_2 = \frac{PL^2}{16EI} = \frac{8\text{ kN}(8\text{ m})^2}{16EI} = \frac{32\text{ kN}\cdot\text{m}^2}{EI}\;\downarrow$$

$$(v_C)_2 = \frac{PL^3}{48EI} = \frac{8\text{ kN}(8\text{ m})^3}{48EI} = \frac{85.33\text{ kN}\cdot\text{m}^3}{EI}\;\downarrow$$

The total displacement at $C$ and the slope at $A$ are the algebraic sums of these components. Hence

$$(+\downarrow)\qquad \theta_A = (\theta_A)_1 + (\theta_A)_2 = \frac{56\text{ kN}\cdot\text{m}^2}{EI}\;\downarrow \qquad\qquad Ans.$$

$$(+\downarrow)\qquad v_C = (v_C)_1 + (v_C)_2 = \frac{139\text{ kN}\cdot\text{m}^3}{EI}\;\downarrow \qquad\qquad Ans.$$

# E X A M P L E 16.10

Determine the displacement at the end $C$ of the overhanging beam shown in Fig. 16–25a. $EI$ is constant.

### Solution

Since the table in Appendix B *does not* include beams with overhangs, the beam will be separated into a simply supported and a cantilevered portion. First we will calculate the slope at $B$, as caused by the distributed load acting on the simply supported span, Fig. 16–25b.

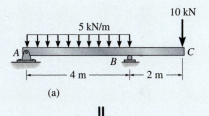

(a)

$$(\theta_B)_1 = \frac{wL^3}{24EI} = \frac{5 \text{ kN/m}(4 \text{ m})^3}{24EI} = \frac{13.33 \text{ kN} \cdot \text{m}^2}{EI} \quad\text{↱}$$

Since this angle is *small*, $(\theta_B)_1 \approx \tan(\theta_B)_1$, and the vertical displacement at point $C$ is

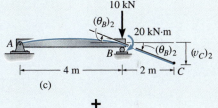

(b)

$$(v_C)_1 = (2 \text{ m})\left(\frac{13.33 \text{ kN} \cdot \text{m}^2}{EI}\right) = \frac{26.67 \text{ kN} \cdot \text{m}^3}{EI} \uparrow$$

Next, the 10-kN load on the overhang causes a statically equivalent force of 10 kN and couple moment of 20 kN · m at the support $B$ of the simply supported span, Fig. 16–25c. The 10-kN force does not cause a displacement or slope at $B$; however, the 20-kN · m couple moment does cause a slope. The slope at $B$ due to this moment is

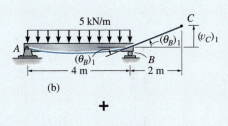

(c)

$$(\theta_B)_2 = \frac{M_0 L}{3EI} = \frac{20 \text{ kN} \cdot \text{m}(4 \text{ m})}{3EI} = \frac{26.67 \text{ kN} \cdot \text{m}^2}{EI} \quad\text{↴}$$

So that the extended point $C$ is displaced

$$(v_C)_2 = (2 \text{ m})\left(\frac{26.7 \text{ kN} \cdot \text{m}^2}{EI}\right) = \frac{53.33 \text{ kN} \cdot \text{m}^3}{EI} \downarrow$$

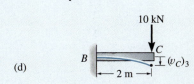

(d)

Finally, the cantilevered portion $BC$ is displaced by the 10-kN force, Fig. 16–25d. We have

$$(v_C)_3 = \frac{PL^3}{3EI} = \frac{10 \text{ kN}(2 \text{ m})^3}{3EI} = \frac{26.67 \text{ kN} \cdot \text{m}^3}{EI} \downarrow$$

**Fig. 16–25**

Summing these results algebraically, we obtain the final displacement of point $C$,

$$(+\downarrow) \qquad v_C = -\frac{26.7}{EI} + \frac{53.3}{EI} + \frac{26.7}{EI} = \frac{53.3 \text{ kN} \cdot \text{m}^3}{EI} \downarrow \qquad\qquad \textit{Ans.}$$

**E X A M P L E** 16.11

Determine the displacement at point $C$ and the slope at the support $A$ of the beam shown in Fig. 16–26a. $EI$ is constant.

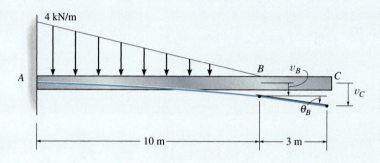

4 kN/m

$A$

$B$

$v_B$

$C$

$v_C$

$\theta_B$

10 m

3 m

**Fig. 16–26**

**Solution**

Using the table in Appendix B for the triangular loading, the slope and displacement at point $B$ are

$$\theta_B = \frac{w_0 L^3}{24EI} = \frac{4 \text{ kN/m } (10\text{m})^3}{24EI} = \frac{166.67 \text{ kN} \cdot \text{m}^2}{EI}$$

$$v_B = \frac{w_0 L^4}{30EI} = \frac{4 \text{ kN/m}(10\text{m})^4}{30EI} = \frac{1333.33 \text{ kN} \cdot \text{m}^3}{EI}$$

The unloaded region $BC$ of the beam remains straight, as shown in Fig. 16–26. Since $\theta_B$ is small, the displacement at $C$ becomes

$$(+\downarrow) \qquad\qquad v_C = v_B + \theta_B \,(3 \text{ m})$$

$$= \frac{1333.33 \text{ kN} \cdot \text{m}^3}{EI} + \frac{166.67 \text{ kN} \cdot \text{m}^2}{EI}(3 \text{ m})$$

$$= \frac{1833 \text{ kN} \cdot \text{m}^3}{EI} \downarrow \qquad\qquad\qquad\qquad \textit{Ans.}$$

# E X A M P L E  16.12

The steel bar shown in Fig. 16–27a is supported by two springs at its ends $A$ and $B$. Each spring has a stiffness of $k = 15$ kip/ft and is originally unstretched. If the bar is loaded with a force of 3 kip at point $C$, determine the vertical displacement of the force. Neglect the weight of the bar and take $E_{st} = 29(10^3)$ ksi, $I = 12$ in$^4$.

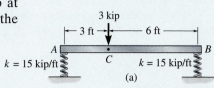

(a)

## Solution

The end reactions at $A$ and $B$ are computed and shown in Fig. 16–27b. Each spring deflects by an amount

$$(v_A)_1 = \frac{2 \text{ kip}}{15 \text{ kip/ft}} = 0.1333 \text{ ft}$$

$$(v_B)_1 = \frac{1 \text{ kip}}{15 \text{ kip/ft}} = 0.0667 \text{ ft}$$

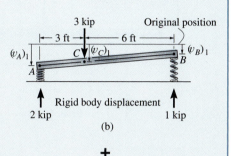

(b)

If the bar is considered to be *rigid*, these displacements cause it to move into the position shown in Fig. 16–27b. For this case, the vertical displacement at $C$ is

$$(v_C)_1 = (v_B)_1 + \frac{6 \text{ ft}}{9 \text{ ft}}[(v_A)_1 - (v_B)_1]$$

$$= 0.0667 \text{ ft} + \frac{2}{3}[0.1333 \text{ ft} - 0.0667 \text{ ft}] = 0.1111 \text{ ft} \downarrow$$

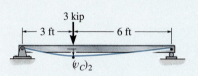

Deformable body displacement

(c)

**Fig. 16–27**

We can find the displacement at $C$ caused by the *deformation* of the bar, Fig. 16–27c, by using the table in Appendix B. We have

$$(v_C)_2 = \frac{Pab}{6EIL}(L^2 - b^2 - a^2)$$

$$= \frac{3 \text{ kip}(3 \text{ ft})(6 \text{ ft})[(9 \text{ ft})^2 - (6 \text{ ft})^2 - (3 \text{ ft})^2]}{6[29(10^3)] \text{ kip/in.}^2(144 \text{ in.}^2/1 \text{ ft}^2)\, 12 \text{ in.}^4(1 \text{ ft}^4/20\,736 \text{ in.}^4)(9 \text{ ft})}$$

$$= 0.0149 \text{ ft} \downarrow$$

Adding the two displacement components, we get

$$(+\downarrow) \quad v_C = 0.1111 \text{ ft} + 0.0149 \text{ ft} = 0.126 \text{ ft} = 1.51 \text{ in.} \downarrow \qquad \textit{Ans.}$$

# PROBLEMS

**\*16-44.** The W8 × 48 cantilevered beam is made of A-36 steel and is subjected to the loading shown. Determine the deflection at its end $A$.

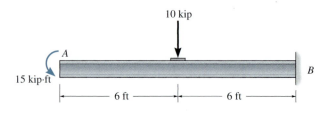

10 kip

$A$

15 kip·ft

$B$

6 ft — 6 ft

**Prob. 16–44**

**16-46.** Determine the moment $M_0$ in terms of the load $P$ and dimension $a$ so that the deflection at the center of the beam is zero. $EI$ is constant.

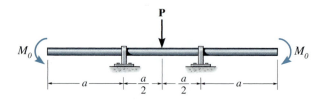

P

$M_0$ $M_0$

$a$ — $\dfrac{a}{2}$ — $\dfrac{a}{2}$ — $a$

**Prob. 16–46**

**16-45.** The W12 × 45 simply supported beam is made of A-36 steel and is subjected to the loading shown. Determine the deflection at its center $C$.

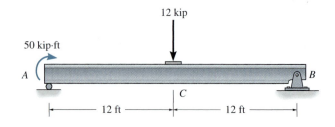

12 kip

50 kip·ft

$A$ $B$

$C$

12 ft — 12 ft

**Prob. 16–45**

**16-47.** The W24 × 104 beam is made of A-36 steel and is subjected to the loading shown. Determine the displacement at its end $A$.

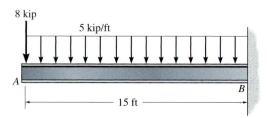

8 kip

5 kip/ft

$A$

$B$

15 ft

**Prob. 16–47**

**\*16-48.** The relay switch consists of a thin metal strip or armature $AB$ that is made of red brass C83400 and is attracted to the solenoid $S$ by a magnetic field. Determine the smallest force $F$ required to attract the armature at $C$ in order that contact is made at the free end $B$. Also, what should the distance $a$ be for this to occur? The armature is fixed at $A$ and has a moment of inertia of $I = 0.18(10^{-12})$ m$^4$.

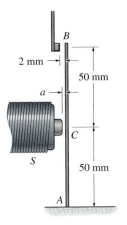

Prob. 16–48

**16-49.** The simply supported beam carries a uniform load of 2 kip/ft. Code restrictions, due to a plaster ceiling, require the maximum deflection not to exceed 1/360 of the span length. Select the lightest-weight A-36 steel wide-flange beam from Appendix D that will satisfy this requirement and safely support the load. The allowable bending stress is $\sigma_{allow} = 24$ ksi and the allowable shear stress is $\tau_{allow} = 14$ ksi. Assume $A$ is a pin and $B$ a roller support.

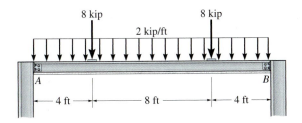

Prob. 16–49

**16-50.** The rod is pinned at its end $A$ and attached to a torsional spring having a stiffness $k$, which measures the torque per radian of rotation of the spring. If a force $\mathbf{P}$ is always applied perpendicular to the end of the rod, determine the displacement of the force. $EI$ is constant.

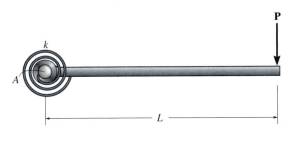

Prob. 16–50

**16-51.** The pipe assembly consists of three equal-sized pipes with flexibility stiffness $EI$ and torsional stiffness $GJ$. Determine the vertical deflection at point $A$.

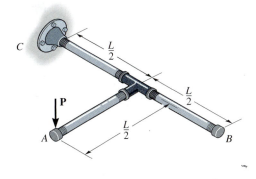

Prob. 16–51

**\*16-52.** Determine the vertical deflection and slope at the end $A$ of the bracket. Assume that the bracket is fixed supported at its base, and neglect the axial deformation of segment $AB$. $EI$ is constant.

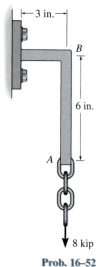

Prob. 16–52

**16-53.** The framework consists of two A-36 steel cantilevered beams $CD$ and $BA$ and a simply supported beam $CB$. If each beam is made of steel and has a moment of inertia about its principal axis of $I_x = 118$ in$^4$, determine the deflection at the center $G$ of beam $CB$.

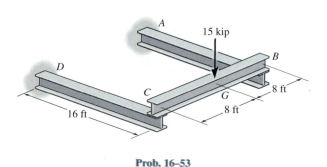

Prob. 16–53

# 16.7 Statically Indeterminate Beams

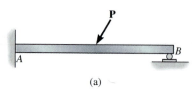

(a)

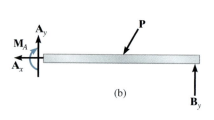

(b)

Fig. 16–28

The analysis of statically indeterminate axially loaded bars and torsionally loaded shafts has been discussed in Secs. 10.4 and 11.5, respectively. In this section we will illustrate a general method for determining the reactions on statically indeterminate beams. Specifically, a member of any type is classified as *statically indeterminate* if the number of unknown reactions *exceeds* the available number of equilibrium equations.

The additional support reactions on the beam or shaft that are *not needed* to keep it in stable equilibrium are called *redundants*. The number of these redundants is referred to as the *degree of indeterminacy*. For example, consider the beam shown in Fig. 16–28a. If the free-body diagram is drawn, Fig. 16–28b, there will be four unknown support reactions, and since three equilibrium equations are available for solution, the beam is classified as being indeterminate to the first degree. Either $\mathbf{A}_y$, $\mathbf{B}_y$, or $\mathbf{M}_A$ can be classified as the redundant, for if any one of these reactions is removed, the beam remains stable and in equilibrium ($\mathbf{A}_x$ cannot be classified as the redundant, for if it were removed, $\Sigma F_x = 0$ would not be satisfied.) In a similar manner, the *continuous beam* in Fig. 16–29a is indeterminate to the second degree, since there are five unknown reactions and only three available equilibrium equations, Fig. 16–29b. Here the two redundant support reactions can be chosen among $\mathbf{A}_y$, $\mathbf{B}_y$, $\mathbf{C}_y$, and $\mathbf{D}_y$.

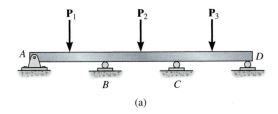

(a)

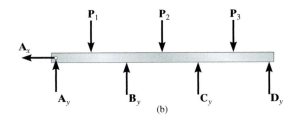

(b)

**Fig. 16–29**

To determine the reactions on a beam (or shaft) that is statically indeterminate, it is first necessary to specify the redundant reactions. We can determine these redundants from conditions of geometry known as *compatibility conditions*. Once determined, the redundants are then applied to the beam, and the remaining reactions are determined from the equations of equilibrium.

In the following section we will illustrate this procedure for the method of superposition.

An example of a statically indeterminate beam used to support a bridge deck.

# 16.8 Statically Indeterminate Beams and Shafts— Method of Superposition

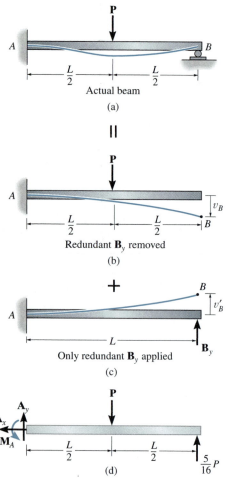

**Actual beam**
(a)

$\parallel$

**Redundant B$_y$ removed**
(b)

$+$

**Only redundant B$_y$ applied**
(c)

(d)

**Fig. 16–30**

The method of superposition has been used previously to solve for the redundant loadings on axially loaded bars and torsionally loaded shafts. In order to apply this method to the solution of statically indeterminate beams (or shafts), it is first necessary to identify the redundant support reactions as explained in Sec. 16.7. By *removing* them from the beam we obtain the so-called *primary beam*, which is statically determinate and stable, and is subjected *only* to the external load. If we add to this beam a succession of similarly supported beams, each loaded with a *separate* redundant, then by the principle of superposition, we obtain the actual loaded beam. Finally, in order to solve for the redundants, we must write the *conditions of compatibility* that exist at the supports where each of the redundants act. Since the redundant forces are determined directly in this manner, this method of analysis is sometimes called the *force method*. Once the redundants are obtained, the other reactions on the beam are then determined from the three equations of equilibrium.

To clarify these concepts, consider the beam shown in Fig. 16–30a. If we choose the reaction **B**$_y$ at the roller as the redundant, then the primary beam is shown in Fig. 16–30b, and the beam with the redundant **B**$_y$ acting on it is shown in Fig. 16–30c. The displacement at the roller is to be zero, and since the displacement of point $B$ on the primary beam is $v_B$, and **B**$_y$ causes point $B$ to be displaced upward $v'_B$, we can write the compatibility equation at $B$ as

$$(+\uparrow) \qquad\qquad 0 = -v_B + v'_B$$

The displacements $v_B$ and $v'_B$ can be obtained using any one of the methods discussed in Secs. 16.5 through 16.6. Here we will obtain them directly from the table Appendix B. We have

$$v_B = \frac{5PL^3}{48EI} \quad \text{and} \quad v'_B = \frac{B_y L^3}{3EI}$$

Substituting into the compatibility equation, we get

$$0 = -\frac{5PL^3}{48EI} + \frac{B_y L^3}{3EI}$$

$$B_y = \frac{5}{16}P$$

Now that **B**$_y$ is known, the reactions at the wall are determined from the three equations of equilibrium applied to the entire beam, Fig. 16–30d. The results are

$$A_x = 0 \qquad A_y = \frac{11}{16}P$$

$$M_A = \frac{3}{16}PL$$

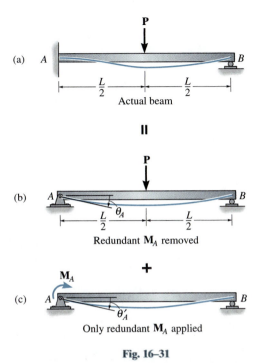

Fig. 16–31

As stated in Sec. 16.7, choice of the redundant is *arbitrary*, provided the primary beam remains stable. For example, the moment at $A$ for the beam in Fig. 16–31a can also be chosen as the redundant. In this case the capacity of the beam to resist $\mathbf{M}_A$ is removed, and so the primary beam is then pin supported at $A$, Fig. 16–31b. Also, the redundant at $A$ acts alone on this beam, Fig. 16–31c. Referring to the slope at $A$ caused by the load $\mathbf{P}$ as $\theta_A$, and the slope at $A$ caused by the redundant $\mathbf{M}_A$ as $\theta'_A$, the compatibility equation for the slope at $A$ requires

$$(\curvearrowright+) \qquad\qquad 0 = \theta_A + \theta'_A$$

Again using the table Appendix B, we have

$$\theta_A = \frac{PL^2}{16EI} \quad \text{and} \quad \theta'_A = \frac{M_A L}{3EI}$$

Thus

$$0 = \frac{PL^2}{16EI} + \frac{M_A L}{3EI}$$

$$M_A = -\frac{3}{16}PL$$

This is the same result computed previously. Here the negative sign for $M_A$ simply means that $\mathbf{M}_A$ acts in the opposite sense of direction of that shown in Fig. 16–31c.

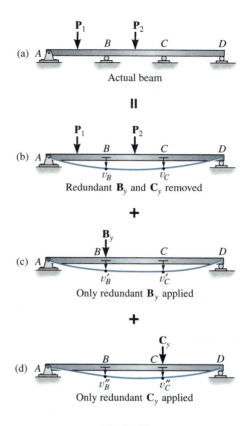

**Fig. 16–32**

Another example that illustrates this method is given in Fig. 16–32a. In this case the beam is indeterminate to the second degree and therefore *two* compatibility equations will be necessary for the solution. We will choose the forces at the roller supports $B$ and $C$ as redundants. The primary (statically determinate) beam deforms as shown in Fig. 16–32b when the redundants are removed. Each redundant force deforms this beam as shown in Figs. 16–32c and 16–32d, respectively. By superposition, the compatibility equations for the displacements at $B$ and $C$ are

$$(+\downarrow) \qquad\qquad 0 = v_B + v'_B + v''_B$$
$$(+\downarrow) \qquad\qquad 0 = v_C + v'_C + v''_C \qquad\qquad (16\text{--}12)$$

Here the displacement components $v'_B$ and $v'_C$ will be expressed in terms of the unknown $\mathbf{B}_y$, and the components $v''_B$ and $v''_C$ will be expressed in terms of the unknown $\mathbf{C}_y$. When these displacements have been determined and substituted into Eq. 16–12, these equations may then be solved simultaneously for the two unknowns $\mathbf{B}_y$ and $\mathbf{C}_y$.

## PROCEDURE FOR ANALYSIS

The following procedure provides a means for applying the method of superposition (or the force method) to determine the reactions on statically indeterminate beams or shafts.

### Elastic Curve.

- Specify the unknown redundant forces or moments that must be removed from the beam in order to make it statically determinate and stable.

- Using the principle of superposition, draw the statically indeterminate beam and show it equal to a sequence of corresponding *statically determinate beams*.

- The first of these beams, the primary beam, supports the same external loads as the statically indeterminate beam, and each of the other beams "added" to the primary beam shows the beam loaded with a separate redundant force or moment.

- Sketch the deflection curve for each beam and indicate symbolically the displacement or slope at the point of each redundant force or moment.

### Compatibility Equations.

- Write a compatibility equation for the displacement or slope at each point where there is a redundant force or moment.

- Determine all the displacements or slopes using an appropriate method as explained in Secs. 16.5 and 16.6.

- Substitute the results into the compatibility equations and solve for the unknown redundants.

- If a numerical value for a redundant is *positive*, it has the *same sense of direction* as originally assumed. Similarly, a *negative* numerical value indicates the redundant acts *opposite* to its assumed *sense of direction*.

### Equilibrium Equations.

- Once the redundant forces and/or moments have been determined, the remaining unknown reactions can be found from the equations of equilibrium applied to the loadings shown on the beam's free-body diagram.

The following examples illustrate application of this procedure. For brevity, all displacements and slopes have been found using the table Appendix B.

E X A M P L E  **16.13**

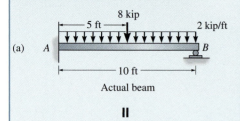

(a)  A  B

| 8 kip |
| 5 ft |
| 2 kip/ft |

10 ft

Actual beam

**||**

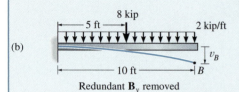

(b)

10 ft  B

$v_B$

Redundant $\mathbf{B}_y$ removed

**+**

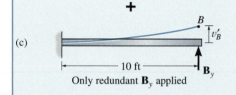

(c)

B  $v_B'$

10 ft  $\mathbf{B}_y$

Only redundant $\mathbf{B}_y$ applied

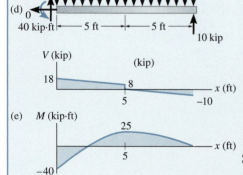

(d)

18 kip  8 kip  2 kip/ft

0

40 kip·ft  5 ft  5 ft

10 kip

V (kip)

(kip)

18  8

5  x (ft)

−10

(e)  M (kip·ft)

25

5  x (ft)

−40

**Fig. 16–33**

Determine the reactions at the roller support $B$ of the beam shown in Fig. 16–33a, then draw the shear and moment diagrams. $EI$ is constant.

**Solution**

*Principle of Superposition.* By inspection, the beam is statically indeterminate to the first degree. The roller support at $B$ will be chosen as the redundant so that $\mathbf{B}_y$ will be determined *directly*. Figures 16–33b and 16–33c show application of the principle of superposition. Here we have assumed that $\mathbf{B}_y$ acts upward on the beam.

*Compatibility Equation.* Taking positive displacement as downward, the compatibility equation at $B$ is

$$(+\downarrow) \qquad\qquad 0 = v_B - v_B' \qquad\qquad (1)$$

These displacements can be obtained directly from the table Appendix B.

$$v_B = \frac{wL^4}{8EI} + \frac{5PL^3}{48EI}$$

$$= \frac{2 \text{ kip/ft } (10 \text{ ft})^4}{8EI} + \frac{5(8 \text{ kip})(10 \text{ ft})^3}{48EI} = \frac{3333 \text{ kip} \cdot \text{ft}^3}{EI}\downarrow$$

$$v_B' = \frac{PL^3}{3EI} = \frac{B_y (10 \text{ ft})^3}{3EI} = \frac{333.3 \text{ ft}^3 B_y}{EI}\uparrow$$

Substituting into Eq. 1 and solving yields

$$0 = \frac{3333}{EI} - \frac{333.3B_y}{EI}$$

$$B_y = 10 \text{ kip} \qquad\qquad Ans.$$

*Equilibrium Equations.* Using this result and applying the three equations of equilibrium, we obtain the results shown on the beam's free-body diagram in Fig. 16–33d. The shear and moment diagrams are shown in Fig. 16–33e.

# EXAMPLE 16.14

Determine the reactions on the beam shown in Fig. 16–34a. Due to the loading and poor construction, the roller support at $B$ settles 12 mm. Take $E = 200$ GPa and $I = 80 (10^6)$ mm$^4$.

## Solution

*Principle of Superposition.* By inspection, the beam is indeterminate to the first degree. The roller support at $B$ will be chosen as the redundant. The principle of superposition is shown in Figs. 16–34b and 16–34c. Here $\mathbf{B}_y$ is assumed to act upward on the beam.

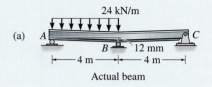

(a)

Actual beam

*Compatibility Equation.* With reference to point $B$, using units of meters, we require

$(+\downarrow)$ $\qquad\qquad 0.012\ m = v_B - v'_B \qquad\qquad$ (1)

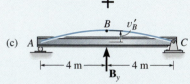

(b)

Redundant $\mathbf{B}_y$ removed

Using the table Appendix B. The displacements are

$$v_B = \frac{5wL^4}{768EI} = \frac{5(24\ \text{kN/m})(8\ \text{m})^4}{768EI} = \frac{640\ \text{kN}\cdot\text{m}^3}{EI}\downarrow$$

$$v'_B = \frac{PL^3}{48EI} = \frac{B_y(8\ \text{m})^3}{48EI} = \frac{10.67\ \text{m}^3 B_y}{EI}\uparrow$$

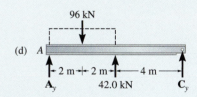

(c)

Only redundant $\mathbf{B}_y$ applied

Thus Eq. 1 becomes

$$0.012EI = 640 - 10.67B_y$$

Expressing $E$ and $I$ in units of kN/m$^2$ and m$^4$, respectively, we have

$$0.012(200)(10^6)[80(10^{-6})] = 640 - 10.67B_y$$
$$B_y = 42.0\ \text{kN}\uparrow \qquad\qquad\qquad Ans.$$

(d)

Fig. 16–34

*Equilibrium Equations.* Applying this result to the beam, Fig. 16–34d, we can calculate the reactions at $A$ and $C$ using the equations of equilibrium. We obtain

$\zeta + \Sigma M_A = 0;$ $\qquad -96\ \text{kN}(2\ \text{m}) + 42.0\ \text{kN}(4\ \text{m}) + C_y(8\ \text{m}) = 0$

$\qquad\qquad\qquad C_y = 3.00\ \text{kN}\uparrow \qquad\qquad Ans.$

$+\uparrow \Sigma F_y = 0;$ $\qquad A_y - 96\ \text{kN} + 42.0\ \text{kN} + 3.00\ \text{kN} = 0$

$\qquad\qquad\qquad A_y = 51\ \text{kN}\uparrow \qquad\qquad Ans.$

**E X A M P L E** **16.15**

The beam in Fig. 16–35a is fixed supported to the wall at *A* and pin connected to a $\frac{1}{2}$-in.-diameter rod *BC*. If $E = 29(10^3)$ ksi for both members, determine the force developed in the rod due to the loading. The moment of inertia of the beam about its neutral axis is $I = 475$ in⁴.

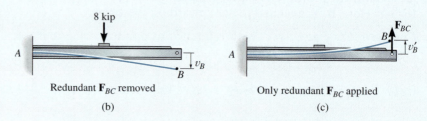

Actual beam and rod
(a)

Redundant $\mathbf{F}_{BC}$ removed
(b)

Only redundant $\mathbf{F}_{BC}$ applied
(c)

**Fig. 16–35**

### Solution I

*Principle of Superposition.* By inspection, this problem is indeterminate to the first degree. Here *B* will undergo an unknown displacement $v_B''$, since the rod will stretch. The rod will be treated as the redundant and hence the force of the rod is removed from the beam at *B*, Fig. 16–35b, and then reapplied, Fig. 16–35c.

*Compatibility Equation.* At point *B* we require

$$(+\downarrow) \qquad v_B'' = v_B - v_B' \tag{1}$$

The displacements $v_B$ and $v_B'$ are determined from the table Appendix B. $v_B''$ is calculated from Eq. 16–2. Working in kilopounds and inches, we have

$$v_B'' = \frac{PL}{AE} = \frac{F_{BC}(8\text{ ft})(12\text{ in./ft})}{(\pi/4)(\frac{1}{2}\text{in.})^2[29(10^3)\text{ kip/in}^2]} = 0.01686 F_{BC} \downarrow$$

$$v_B = \frac{5PL^3}{48EI} = \frac{5(8\text{ kip})(10\text{ ft})^3(12\text{ in./ft})^3}{48[29(10^3)\text{ kip/in}^2](475\text{ in}^4)} = 0.1045\text{ in.} \downarrow$$

$$v_B' = \frac{PL^3}{3EI} = \frac{F_{BC}(10\text{ ft})^3(12\text{ in./ft})^3}{3[29(10^3)\text{ kip/in}^2](475\text{ in}^4)} = 0.04181 F_{BC} \uparrow$$

Thus, Eq. 1 becomes

$$(+\downarrow) \quad 0.01686 F_{BC} = 0.1045 - 0.04181 F_{BC}$$
$$F_{BC} = 1.78\text{ kip} \qquad \textit{Ans.}$$

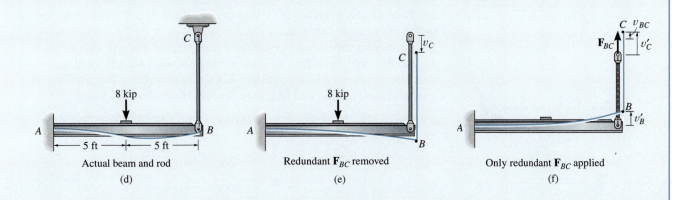

| Actual beam and rod | Redundant $\mathbf{F}_{BC}$ removed | Only redundant $\mathbf{F}_{BC}$ applied |
|:---:|:---:|:---:|
| (d) | (e) | (f) |

## Solution II

*Principle of Superposition.* We can also solve this problem by removing the pin support at $C$ and keeping the rod attached to the beam. In this case the 8-kip load will cause points $B$ and $C$ to be displaced downward the *same amount* $v_C$, Fig. 16–35e, since no force exists in rod $BC$. When the redundant force $\mathbf{F}_{BC}$ is applied at point $C$, it causes the end $C$ of the rod to be displaced upward $v'_C$ and the end $B$ of the beam to be displaced upward $v'_B$, Fig. 16–35f. The difference in these two displacements, $v_{BC}$, represents the stretch of the rod due to $\mathbf{F}_{BC}$, so that $v'_C = v_{BC} + v'_B$. Hence, from Figs. 16–35d, 16–35e, and 16–35f, the compatibility of displacement at point $C$ is

$$(+\downarrow) \qquad\qquad 0 = v_C - (v_{BC} + v'_B) \qquad\qquad (2)$$

From Solution I, we have

$$v_C = v_B = 0.1045 \text{ in. } \downarrow$$
$$v_{BC} = v''_B = 0.01686 F_{BC} \uparrow$$
$$v'_B = 0.04181 F_{BC} \uparrow$$

Therefore, Eq. 2 becomes

$$(+\downarrow) \qquad 0 = 0.1045 - (0.01686 F_{BC} + 0.04181 F_{BC})$$
$$F_{BC} = 1.78 \text{ kip} \qquad\qquad \textit{Ans.}$$

**E X A M P L E  16.16**

Determine the moment at $B$ for the beam shown in Fig. 16–36a. $EI$ is constant. Neglect the effects of axial load.

### Solution

*Principle of Superposition.* Since the axial load on the beam is neglected, there will be a vertical force and moment at $A$ and $B$. Here there are only two available equations of equilibrium ($\Sigma M = 0$, $\Sigma F_y = 0$), and so the problem is indeterminate to the second degree. We will assume that $\mathbf{B}_y$ and $\mathbf{M}_B$ are redundant, so that by the principle of superposition, the beam is represented as a cantilever, loaded *separately* by the distributed load and reactions $\mathbf{B}_y$ and $\mathbf{M}_B$, Figs. 16–36b, 16–36c and 16–36d.

*Compatibility Equations.* Referring to the displacement and slope at $B$, we require

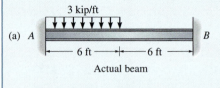

(a) Actual beam

$$(\curvearrowleft+) \qquad\qquad 0 = \theta_B + \theta'_B + \theta''_B \qquad\qquad (1)$$

$$(+\downarrow) \qquad\qquad 0 = v_B + v'_B + v''_B \qquad\qquad (2)$$

Using the table on the back cover to compute the slopes and displacements, we have

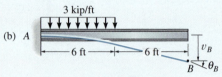

(b) Redundants $\mathbf{M}_B$ and $\mathbf{B}_y$ removed

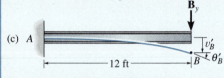

(c) Only redundant $\mathbf{B}_y$ applied

$$\theta_B = \frac{wL^3}{48EI} = \frac{3\text{ kip/ft}(12\text{ ft})^3}{48EI} = \frac{108}{EI} \downarrow$$

$$v_B = \frac{7wL^4}{384EI} = \frac{7(3\text{ kip/ft})(12\text{ ft})^4}{384EI} = \frac{1134}{EI} \downarrow$$

$$\theta'_B = \frac{PL^2}{2EI} = \frac{B_y(12\text{ ft})^2}{2EI} = \frac{72B_y}{EI} \downarrow$$

$$v'_B = \frac{PL^3}{3EI} = \frac{B_y(12\text{ ft})^3}{3EI} = \frac{576B_y}{EI} \downarrow$$

$$\theta''_B = \frac{ML}{EI} = \frac{M_B(12\text{ ft})}{EI} = \frac{12M_B}{EI} \downarrow$$

$$v''_B = \frac{ML^2}{2EI} = \frac{M_B(12\text{ ft})^2}{2EI} = \frac{72M_B}{EI} \downarrow$$

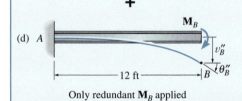

(d) Only redundant $\mathbf{M}_B$ applied

**Fig. 16–36**

Substituting these values into Eqs. 1 and 2 and canceling out the common factor $EI$, we get

$$(\curvearrowleft+) \qquad\qquad 0 = 108 + 72B_y + 12M_B$$

$$(+\downarrow) \qquad\qquad 0 = 1134 + 576B_y + 72M_B$$

Solving these equations simultaneously gives

$$B_y = -3.375\text{ kip}$$

$$M_B = 11.25\text{ kip}\cdot\text{ft} \qquad\qquad Ans.$$

# PROBLEMS

**16-54.** The assembly consists of a steel and an aluminum bar, each of which is 1 in. thick, fixed at its ends $A$ and $B$, and pin connected to the right short link $CD$. If a horizontal force of 80 lb is applied to the link as shown, determine the moments created at $A$ and $B$. $E_{st} = 29(10^3)$ ksi, $E_{al} = 10(10^3)$ ksi.

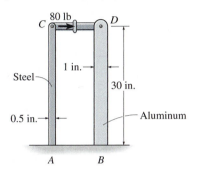

**Prob. 16–54**

**16-55.** The A-36 steel beam and rod are used to support the load of 8 kip. If it is required that the allowable normal stress for the steel is $\sigma_{allow} = 18$ ksi, and the maximum deflection not exceed 0.05 in., determine the smallest diameter rod that should be used. The beam is rectangular, having a height of 5 in. and a thickness of 3 in.

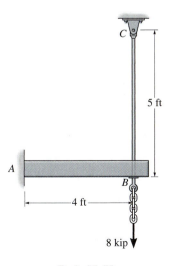

**Prob. 16–55**

**\*16-56.** Determine the reactions at the supports $A$, $B$, and $C$, then draw the shear and moment diagrams. $EI$ is constant.

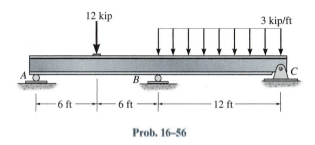

**Prob. 16–56**

**16-57.** The beam is used to support the 20-kip load. Determine the reactions at the supports. Assume $A$ is fixed and $B$ is a roller.

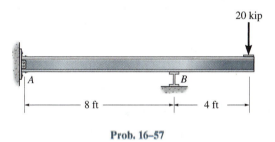

**Prob. 16–57**

**16-58.** The beam has a constant $E_1 I_1$ and is supported by the fixed wall at $B$ and the rod $AC$. If the rod has a cross-sectional area $A_2$ and the material has a modulus of elasticity $E_2$, determine the force in the rod.

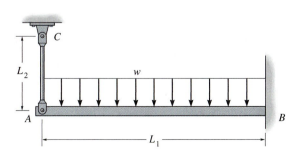

**Prob. 16–58**

**16-59.** Determine the reactions at support *C*. *EI* is constant for both beams.

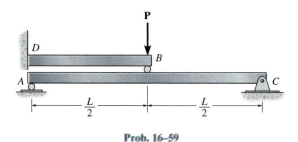

Prob. 16–59

**\*16-60.** Determine the reactions at the supports *A* and *B*. *EI* is constant.

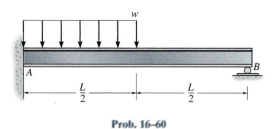

Prob. 16–60

**16-61.** The assembly consists of three simply supported beams for which the bottom of the top beam rests on the top of the bottom two. If a uniform load of 3 kN/m is applied to the top beam, determine the vertical reactions at each of the supports. *EI* is constant.

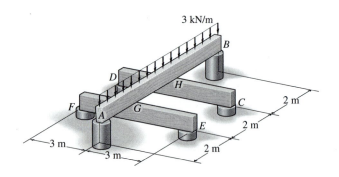

Prob. 16–61

**16-62.** Determine the reactions at the supports, then draw the shear and moment diagrams. *EI* is constant.

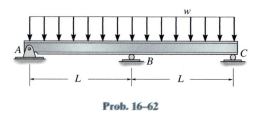

Prob. 16–62

**16-63.** The beam is supported by a pin at *A*, a spring having a stiffness *k* at *B*, and a roller at *C*. Determine the force the spring exerts on the beam. *EI* is constant.

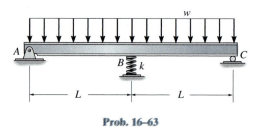

Prob. 16–63

**\*16-64.** Each of the two members is made from 6061-T6 aluminum and has a square cross section 1 in. × 1 in. They are pin connected at their ends and a jack is placed between them and opened until the force it exerts on each member is 50 lb. Determine the greatest force *P* that can be applied to the center of the top member without causing either of the two members to yield. For the analysis neglect the axial force in each member. Assume the jack is rigid.

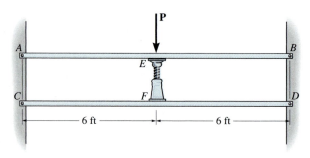

Prob. 16–64

**16-65.** The 1-in.-diameter A-36 steel shaft is supported by unyielding bearings at $A$ and $C$. The bearing at $B$ rests on a simply supported steel wide-flange beam having a moment of inertia of $I = 500$ in⁴. If the belt loads on the pulley are 400 lb each, determine the vertical reactions at $A$, $B$, and $C$.

**16-66.** Determine the force in the spring. $EI$ is constant.

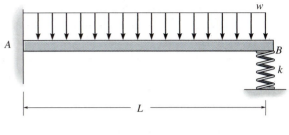

**Prob. 16-66**

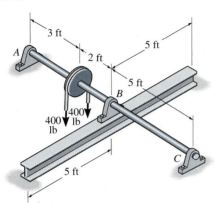

**Prob. 16-65**

## CHAPTER REVIEW

- **Beam Design.** Failure of a beam occurs when the internal shear or moment in the beam is a maximum. To resist these loadings it is therefore important that the associated maximum shear and bending stress not exceed allowable values as stated in codes. Normally the cross section of a beam is first designed to resist the allowable bending stress, then the allowable shear stress is checked. For rectangular sections, $\tau_{allow} \geq 1.5(V_{max}/A)$ and for wide-flange sections is appropriate to use $\tau_{allow} \geq V_{max}/A_{web}$. For built-up beams the spacing of fasteners or the strength of glue or weld is determined using an allowable shear flow $q_{allow} = VQ/I$.

- **Beam Deflection.** The equation of the elastic curve and its slope can be obtained by first finding the internal moment in the member as a function of $x$. If several loadings act on the member, then separate moment functions must be determined between each of the loadings. Integrating these functions once using $EI\dfrac{d^2v}{dx^2} = M(x)$ gives the equation of a slope of the elastic curve, and integrating again gives the equation for the deflection. The constants of integration are determined from the boundary conditions at the supports, or in cases where several moment functions are invloved, continuity of slope and deflection at points where these functions join must be satisfied. The deflection or slope at a point on a member can also be determined using the method of superposition of loads and displacements. The table in the back of the book is available for this purpose.

• *Statically Indeterminate Members.* Statically indeterminate beams and shafts have more unknown support reactions than available equations of equilibrium. To solve, one first identifies the redundant reactions, and the other unknown reactions are written in terms of these redundants. The method of integration can then be used to solve for the unknown redundants. It is also possible to determine the redundants by using the method of superposition, where one considers the continuity of displacement at the redundant. Here the displacement due to the external loading is determined with the redundant removed, and again with the redundant applied and the external loading removed. The tables in the back of the book can be used to determine these necessary displacements.

# REVIEW PROBLEMS

**16-67.** The compound beam is made from two sections, which are pinned together at $B$. Use Appendix D and select the lightest wide-flange beam that would be safe for each section if the allowable bending stress is $\sigma_{\text{allow}} = 24$ ksi and the allowable shear stress is $\tau_{\text{allow}} = 14$ ksi. The beam supports a pipe loading of 4 kip and 6 kip as shown.

*16-68.** Determine the equation of the elastic curve for the beam using the $x$ coordinate. Specify the slope at $A$ and the maximum deflection. $EI$ is constant.

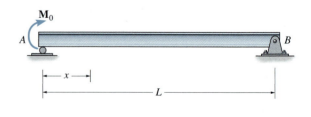

Prob. 16-68

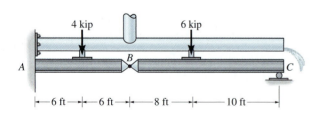

Prob. 16-67

**16-69.** The steel shaft is subjected to the loadings developed in the belts passing over the two pulleys. If the bearings at $A$ and $B$ exert vertical reactions on the shaft, determine the slope at $A$. The shaft has a diameter of 0.75 in. Use the moment-area method. $E_{st} = 29(10^3)$ ksi.

**16-71.** The beam is subjected to the linearly varying distributed load. Determine the maximum deflection of the beam. Use the method of integration. $EI$ is constant.

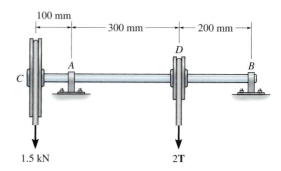

**Prob. 16–71**

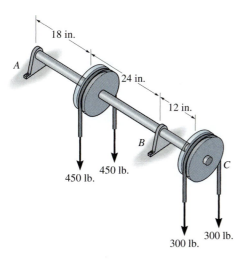

**Prob. 16–69**

*16-72.** The 25-mm-diameter steel shaft is supported at $A$ and $B$ by bearings. If the tension in the belt on the pulley at $C$ is 0.75 kN, determine the largest belt tension $T$ on the pulley at $D$ so that the slope of the shaft at $A$ or $B$ does not exceed 0.02 rad. The bearings exert only vertical reactions on the shaft. Use the moment-area method. $E_{st} = 200$ GPa.

**16-70.** Determine the deflection at $C$ and the slope of the beam at $A$, $B$, and $C$. Use the moment-area method. $EI$ is constant.

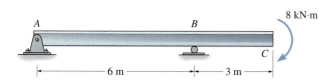

**Prob. 16–70**

**Prob. 16–72**

The columns for this building are used to support the floor loading. Engineers design these members to resist buckling. *(Chris Baker/Tony Stone Images.)*

# CHAPTER
# 17

# Buckling of Columns

## CHAPTER OBJECTIVES

- To develop methods to calculate the critical loads causing columns to buckle in the elastic region of the stress-strain curve.
- To consider the effect of different supports on the critical load for buckling.
- To introduce approximate methods for critical buckling loads in the inelastic region.

## 17.1   Critical Load

Whenever a member is designed, it is necessary that it satisfy specific strength, deflection, and stability requirements. In the preceding chapters we have discussed some of the methods used to determine a member's strength and deflection, while assuming that the member was always in stable equilibrium. Some members, however, may be subjected to compressive loadings, and if these members are long and slender the loading may be large enough to cause the member to deflect laterally or sidesway. To be specific, long slender members subjected to an axial compressive force are called *columns*, and the lateral deflection that occurs is called *buckling*. Quite often the buckling of a column can lead to a sudden and dramatic failure of a structure or mechanism, and as a result, special attention must be given to the design of columns so that they can safely support their intended loadings without buckling.

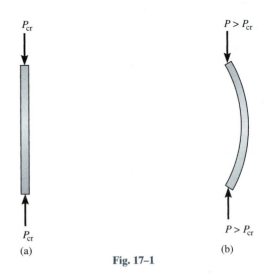

$P_{cr}$

$P > P_{cr}$

$P_{cr}$

$P > P_{cr}$

(a)

(b)

**Fig. 17–1**

The maximum axial load that a column can support when it is on the *verge* of buckling is called the **critical load**, $P_{cr}$, Fig. 17–1a. Any additional loading will cause the column to buckle and therefore deflect laterally as shown in Fig. 17–1b. In order to better understand the nature of this instability, consider a two-bar mechanism consisting of weightless bars that are rigid and pin connected at their ends, Fig. 17–2a. When the bars are in the vertical position, the spring, having a stiffness $k$, is unstretched, and a *small* vertical force **P** is applied at the top of one of the bars. We can upset this equilibrium position by displacing the pin at $A$ by a small amount $\Delta$, Fig. 17–2b. As shown on the free-body diagram of the pin when the bars are displaced, Fig. 17–2c, the spring will produce a restoring force $F = k\Delta$, while the applied load **P** develops two horizontal components, $P_x = P \tan \theta$, which tend to push the pin (and the bars) further out of equilibrium. Since $\theta$ is small, $\Delta = \theta(L/2)$ and $\tan \theta \approx \theta$. Thus the *restoring* spring force becomes $F = k\theta L/2$, and the *disturbing* force is $2P_x = 2P\theta$.

If the restoring force is greater than the disturbing force, that is, $k\theta L/2 > 2P\theta$, then, noticing that $\theta$ cancels out, we can solve for $P$, which gives

$$P < \frac{kL}{4} \qquad \text{stable equilibrium}$$

This is a condition for *stable equilibrium* since the force developed by the spring would be adequate to restore the bars back to their vertical position. On the other hand, if $kL\theta/2 < 2P\theta$, or

$$P > \frac{kL}{4} \qquad \text{unstable equilibrium}$$

then the mechanism would be in *unstable equilibrium*. In other words, if this load $P$ is applied, and a slight displacement occurs at $A$, the mechanism will tend to move out of equilibrium and not be restored to its original position.

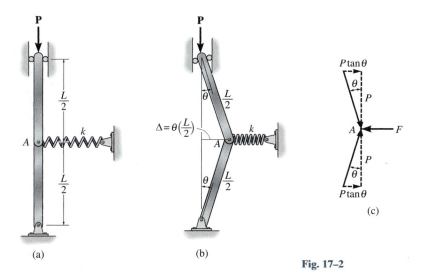

**Fig. 17–2**

The intermediate value of $P$, defined by requiring $kL\theta/2 = 2P\theta$, is the *critical load*. Here

$$P_{cr} = \frac{kL}{4} \qquad \text{neutral equilibrium}$$

This loading represents a case of the mechanism being in *neutral equilibrium*. Since $P_{cr}$ is *independent* of the (small) displacement $\theta$ of the bars, any slight disturbance given to the mechanism will not cause it to move further out of equilibrium, nor will it be restored to its original position. Instead, the bars will *remain* in the deflected position.

These three different states of equilibrium are represented graphically in Fig. 17–3. The transition point where the load is equal to the critical value $P = P_{cr}$ is called the *bifurcation point*. At this point the mechanism will be in equilibrium for any *small value* of $\theta$, measured either to the right or to the left of the vertical. Physically, $P_{cr}$ represents the load for which the mechanism is on the verge of buckling. It is quite valid to determine this value by assuming *small displacements* as done here; however, it should be understood that $P_{cr}$ may *not* be the largest value of $P$ that the mechanism can support. Indeed, if a larger load is placed on the bars, then the mechanism may have to undergo a further deflection before the spring is compressed or elongated enough to hold the mechanism in equilibrium.

Like the two-bar mechanism just discussed, the critical buckling loads on columns supported in various ways can be obtained, and the method used to do this will be explained in the next section. Although in engineering design the critical load may be considered to be the largest load the column can support, realize that, like the two-bar mechanism in the deflected or buckled position, a column may actually support an

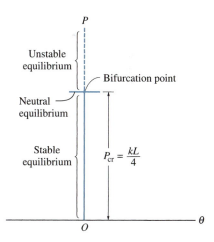

**Fig. 17–3**

even greater load than $P_{cr}$. Unfortunately, however, this loading may require the column to undergo a *large* deflection, which is generally not tolerated in engineering structures or machines. For example, it may take only a few newtons of force to buckle a meterstick, but the additional load it may support can be applied only after the stick undergoes a relatively large lateral deflection.

## 17.2 Ideal Column with Pin Supports

Some pin-connected members used in moving machinery, such as this short link, are subjected to compressive loads and thus act as columns.

In this section we will determine the critical buckling load for a column that is pin supported as shown in Fig. 17–4a. The column to be considered is an *ideal column*, meaning one that is perfectly straight before loading, is made of homogeneous material, and upon which the load is applied through the centroid of the cross section. It is further assumed that the material behaves in a linear-elastic manner and that the column buckles or bends in a single plane. In reality, the conditions of column straightness and load application are never accomplished; however, the analysis to be performed on an "ideal column" is similar to that used to analyze initially crooked columns or those having an eccentric load application. These more realistic cases will be discussed later in this chapter.

Since an ideal column is straight, theoretically the axial load $P$ could be increased until failure occurs by either fracture or yielding of the material. However, when the critical load $P_{cr}$ is reached, the column is on the verge of becoming unstable, so that a small lateral force $F$, Fig. 17–4b, will cause the column to remain in the deflected position when $F$ is removed, Fig. 17–4c. Any slight reduction in the axial load $P$ from $P_{cr}$ will allow the column to straighten out, and any slight increase in $P$, beyond $P_{cr}$, will cause further increases in lateral deflection.

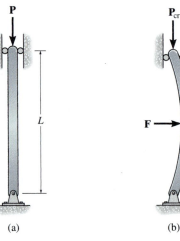

(a)        (b)        (c)

**Fig. 17–4**

Whether or not a column will remain stable or become unstable when subjected to an axial load will depend on its ability to restore itself, which is based on its resistance to bending. Hence, in order to determine the critical load and the buckled shape of the column, we will apply Eq. 16–11, which relates the internal moment in the column to its deflected shape, i.e.,

$$EI\frac{d^2v}{dx^2} = M \qquad (17\text{–}1)$$

Recall that this equation assumes that the slope of the elastic curve is small* and that deflections occur only by bending. When the column is in its deflected position, Fig. 17–5a, the internal bending moment can be determined by using the method of sections. The free-body diagram of a segment in the deflected position is shown in Fig. 17–5b. Here both the deflection $v$ and the internal moment $M$ are shown in the *positive direction* according to the sign convention used to establish Eq. 17–1. Summing moments, the internal moment is $M = -Pv$. Thus Eq. 17–1 becomes

$$EI\frac{d^2v}{dx^2} = -Pv$$

$$\frac{d^2v}{dx^2} + \left(\frac{P}{EI}\right)v = 0 \qquad (17\text{–}2)$$

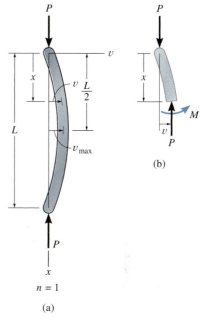

Fig. 17–5a,b

This is a homogeneous, second-order, linear differential equation with constant coefficients. It can be shown by using the methods of differential equations, or by direct substitution into Eq. 17–2, that the general solution is

$$v = C_1 \sin\left(\sqrt{\frac{P}{EI}}x\right) + C_2 \cos\left(\sqrt{\frac{P}{EI}}x\right) \qquad (17\text{–}3)$$

The two constants of integration are determined from the boundary conditions at the ends of the column. Since $v = 0$ at $x = 0$, then $C_2 = 0$. And since $v = 0$ at $x = L$, then

$$C_1 \sin\left(\sqrt{\frac{P}{EI}}L\right) = 0$$

This equation is satisfied if $C_1 = 0$; however, then $v = 0$, which is a *trivial solution* that requires the column to always remain straight, even though the load causes the column to become unstable. The other possibility is for

$$\sin\left(\sqrt{\frac{P}{EI}}L\right) = 0$$

which is satisfied if

$$\sqrt{\frac{P}{EI}}L = n\pi$$

---

*If large deflections are to be considered, the more accurate differential equation, Eq. 16–5, $EI(d^2v/dx^2)/[1 + (dv/dx)^2]^{3/2} = M$ must be used.

or

$$P = \frac{n^2\pi^2 EI}{L^2} \quad n = 1, 2, 3, \ldots \qquad (17\text{-}4)$$

The *smallest value* of $P$ is obtained when $n = 1$, so the *critical load* for the column is therefore

$$P_{cr} = \frac{\pi^2 EI}{L^2}$$

This load is sometimes referred to as the *Euler load*, after the Swiss mathematician Leonhard Euler, who originally solved this problem in 1757. The corresponding buckled shape is defined by the equation

$$v = C_1 \sin\frac{\pi x}{L}$$

Here the constant $C_1$ represents the maximum deflection, $v_{max}$, which occurs at the midpoint of the column, Fig. 17–5a. Specific values for $C_1$ cannot be obtained, since the exact deflected form for the column is unknown once it has buckled. It has been assumed, however, that this deflection is small.

Realize that $n$ in Eq. 17–4 represents the number of waves in the deflected shape of the column. For example, if $n = 2$, then from Eqs. 17–3 and 17–4, *two waves* will appear in the buckled shape, Fig. 17–5c, and the column will support a critical load that is $4P_{cr}$ just prior to buckling. Since this value is four times the critical load and the deflected shape is unstable, this form of buckling, practically speaking, will not exist.

Like the two-bar mechanism discussed in Sec. 17.1, we can represent the load-deflection characteristics of the ideal column by the graph shown in Fig. 17–6. The bifurcation point represents the state of *neutral equilibrium*, at which point the *critical load* acts on the column. Here the column is on the verge of impending buckling.

It should be noted that the critical load is independent of the strength of the material; rather it depends only on the column's dimensions ($I$ and $L$) and the material's stiffness or modulus of elasticity $E$. For this reason, as far as elastic buckling is concerned, columns made, for example, of high-strength steel offer no advantage over those made of lower-strength steel, since the modulus of elasticity for both is approximately the same. Also note that the load-carrying capacity of a column will increase as the moment of inertia of the cross section increases. Thus, efficient columns are designed so that most of the column's cross-sectional area is located as far away as possible from the principal centroidal axes for the section. This is why hollow sections such as tubes are more economical than solid sections. Furthermore, wide-flange sections, and columns that are "built up" from channels, angles, plates, etc., are better than sections that are solid and rectangular.

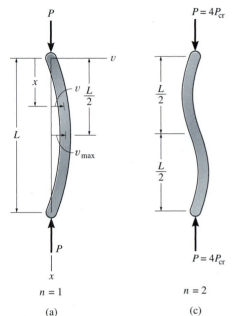

**Fig. 17–5a,c**

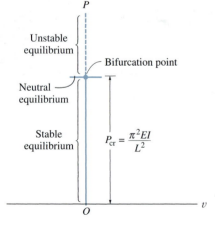

**Fig. 17–6**

It is also important to realize that a column will buckle about the principal axis of the cross section having the *least moment of inertia* (the weakest axis). For example, a column having a rectangular cross section, like a meter stick, as shown in Fig. 17–7, will buckle about the *a–a* axis, not the *b–b* axis. As a result, engineers usually try to achieve a balance, keeping the moments of inertia the same in all directions. Geometrically, then, circular tubes would make excellent columns. Also, square tubes or those shapes having $I_x \approx I_y$ are often selected for columns.

Summarizing the above discussion, the buckling equation for a pin-supported long slender column can be rewritten, and the terms defined as follows:

$$P_{cr} = \frac{\pi^2 EI}{L^2} \qquad (17\text{–}5)$$

where

$P_{cr}$ = critical or maximum axial load on the column just before it begins to buckle. This load must *not* cause the stress in the column to exceed the proportional limit

$E$ = modulus of elasticity for the material

$I$ = *least* moment of inertia for the column's cross-sectional area

$L$ = unsupported length of the column, whose ends are pinned

For purposes of design, Eq. 17–5 can also be written in a more useful form by expressing $I = Ar^2$, where $A$ is the cross-sectional area and $r$ is the *radius of gyration* of the cross-sectional area. Thus,

$$P_{cr} = \frac{\pi^2 E(Ar^2)}{L^2}$$

$$\left(\frac{P}{A}\right)_{cr} = \frac{\pi^2 E}{(L/r)^2}$$

or

$$\sigma_{cr} = \frac{\pi^2 E}{(L/r)^2} \qquad (17\text{–}6)$$

Here

$\sigma_{cr}$ = critical stress, which is an *average stress* in the column just before the column buckles. This stress is an *elastic stress* and therefore $\sigma_{cr} \le \sigma_Y$

$E$ = modulus of elasticity for the material

$L$ = unsupported length of the column, whose ends are pinned

$r$ = *smallest* radius of gyration of the column, determined from $r = \sqrt{I/A}$, where $I$ is the *least* moment of inertia of the column's cross-sectional area $A$

The geometric ratio $L/r$ in Eq. 17–6 is known as the **slenderness ratio**. It is a measure of the column's flexibility, and as will be discussed later, it serves to classify columns as long, intermediate, or short.

**Fig. 17–7**

Typical interior steel pipe columns used to support the roof of a single story building.

It is possible to graph Eq. 17–6 using axes that represent the critical stress versus the slenderness ratio. Examples of this graph for columns made of a typical structural steel and aluminum alloy are shown in Fig. 17–8. Note that the curves are hyperbolic and are valid only for critical stresses below the material's yield point (proportional limit), since the material must behave elastically. For the steel the yield stress is $(\sigma_Y)_{st} = 36$ ksi $[E_{st} = 29(10^3)$ ksi$]$, and for the aluminum it is $(\sigma_Y)_{al} = 27$ ksi $[E_{al} = 10(10^3)$ ksi$]$. Substituting $\sigma_{cr} = \sigma_Y$ into Eq. 17–6, the *smallest* acceptable slenderness ratios for the steel and aluminum columns are therefore $(L/r)_{st} = 89$ and $(L/r)_{al} = 60.5$, respectively. Thus, for a steel column, if $(L/r)_{st} \geq 89$, Euler's formula can be used to determine the buckling load since the stress in the column remains elastic. On the other hand, if $(L/r)_{st} < 89$, the column's stress will exceed the yield point before buckling can occur, and therefore the Euler formula is not valid in this case.

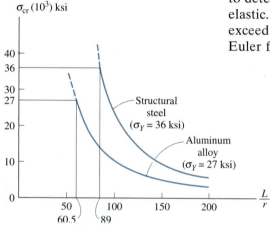

Fig. 17–8

## IMPORTANT POINTS

- *Columns* are long slender members that are subjected to axial loads.

- The *critical load* is the maximum axial load that a column can support when it is on the verge of buckling. This loading represents a case of *neutral equilibrium*.

- An *ideal column* is initially perfectly straight, made of homogeneous material, and the load is applied through the centroid of the cross section.

- A pin-connected column will buckle about the principal axis of the cross section having the *least* moment of inertia.

- The *slenderness ratio* is L/r, where r is the smallest radius of gyration of the cross section. Buckling will occur about the axis where this ratio gives the greatest value.

## EXAMPLE 17.1

A 24-ft-long A-36 steel tube having the cross section shown in Fig. 17–9 is to be used as a pin-ended column. Determine the maximum allowable axial load the column can support so that it does not buckle.

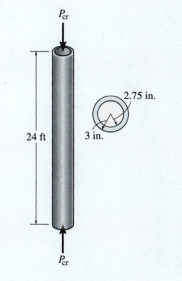

### Solution
Using Eq. 17–5 to obtain the critical load with $E_{st} = 29(10^3)$ ksi,

$$P_{cr} = \frac{\pi^2 EI}{L^2}$$

$$= \frac{\pi^2[29(10^3) \text{ kip/in}^2](\frac{1}{4}\pi(3)^4 - \frac{1}{4}\pi(2.75)^4) \text{ in}^4}{[24 \text{ ft}(12 \text{ in./ft})]^2}$$

$$= 64.5 \text{ kip}$$

This force creates an average compressive stress in the column of

$$\sigma_{cr} = \frac{P_{cr}}{A} = \frac{64.5 \text{ kip}}{[\pi(3)^2 - \pi(2.75)^2] \text{ in}^2} = 14.3 \text{ ksi}$$

Since $\sigma_{cr} < \sigma_Y = 36$ ksi, application of Euler's equation is appropriate.

Fig. 17–9

## EXAMPLE 17.2

The A-36 steel W8 × 31 member shown in Fig. 17–10 is to be used as a pin-connected column. Determine the largest axial load it can support before it either begins to buckle or the steel yields.

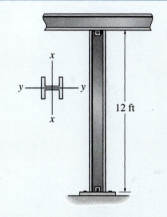

### Solution
From the table in Appendix D, the column's cross-sectional area and moments of inertia are $A = 9.13$ in$^2$, $I_x = 110$ in$^4$, and $I_y = 37.1$ in$^4$. By inspection, buckling will occur about the $y$–$y$ axis. Why? Applying Eq. 17–5, we have

$$P_{cr} = \frac{\pi^2 EI}{L^2} = \frac{\pi^2[29(10^3) \text{ kip/in}^2](37.1 \text{ in}^4)}{[12 \text{ ft}(12 \text{ in./ft})]^2} = 512 \text{ kip}$$

When fully loaded, the average compressive stress in the column is

$$\sigma_{cr} = \frac{P_{cr}}{A} = \frac{512 \text{ kip}}{9.13 \text{ in}^2} = 56.1 \text{ ksi}$$

Fig. 17–10

Since this stress exceeds the yield stress (36 ksi), the load $P$ is determined from simple compression:

$$36 \text{ ksi} = \frac{P}{9.13 \text{ in}^2}; \qquad\qquad P = 329 \text{ kip} \qquad\qquad Ans.$$

In actual practice, a factor of safety would be placed on this loading.

## 17.3 Columns Having Various Types of Supports

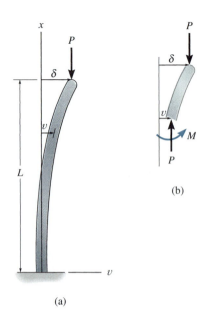

L

(a)

**Fig. 17–11**

The tubular columns used to support this water tank have been braced at three locations along their length to prevent them from buckling.

In Sec. 17.2 we derived the Euler load for a column that is pin connected or free to rotate at its ends. Oftentimes, however, columns may be supported in some other way. For example, consider the case of a column fixed at its base and free at the top, Fig. 17–11a. Determination of the buckling load on this column follows the same procedure as that used for the pinned column. From the free-body diagram in Fig. 17–11b, the internal moment at the arbitrary section is $M = P(\delta - v)$. Consequently, the differential equation for the deflection curve is

$$EI\frac{d^2v}{dx^2} = P(\delta - v)$$

$$\frac{d^2v}{dx^2} + \frac{P}{EI}v = \frac{P}{EI}\delta \qquad (17\text{–}7)$$

Unlike Eq. 17–2, this equation is nonhomogeneous because of the nonzero term on the right side. The solution consists of both a complementary and particular solution, namely,

$$v = C_1 \sin\left(\sqrt{\frac{P}{EI}}x\right) + C_2 \cos\left(\sqrt{\frac{P}{EI}}x\right) + \delta$$

The constants are determined from the boundary conditions. At $x = 0$, $v = 0$, so that $C_2 = -\delta$. Also,

$$\frac{dv}{dx} = C_1\sqrt{\frac{P}{EI}}\cos\left(\sqrt{\frac{P}{EI}}x\right) - C_2\sqrt{\frac{P}{EI}}\sin\left(\sqrt{\frac{P}{EI}}x\right)$$

At $x = 0$, $dv/dx = 0$, so that $C_1 = 0$. The deflection curve is therefore

$$v = \delta\left[1 - \cos\left(\sqrt{\frac{P}{EI}}x\right)\right] \qquad (17\text{–}8)$$

Since the deflection at the top of the column is $\delta$, that is, at $x = L$, $v = \delta$, we require

$$\delta \cos\left(\sqrt{\frac{P}{EI}}L\right) = 0$$

The trivial solution $\delta = 0$ indicates that no buckling occurs, regardless of the load $P$. Instead,

$$\cos\left(\sqrt{\frac{P}{EI}}L\right) = 0 \qquad \text{or} \qquad \sqrt{\frac{P}{EI}}L = \frac{n\pi}{2}$$

The smallest critical load occurs when $n = 1$, so that

$$P_{cr} = \frac{\pi^2 EI}{4L^2} \qquad (17\text{–}9)$$

By comparison with Eq. 17–5 it is seen that a column fixed-supported at its base will carry only one-fourth the critical load that can be applied to a pin-supported column.

Other types of supported columns are analyzed in much the same way and will not be covered in detail here.[*] Instead, we will tabulate the results for the most common types of column support and show how to apply these results by writing Euler's formula in a general form.

**Effective Length.** As stated previously, the Euler formula, Eq. 17–5, was developed for the case of a column having ends that are pinned or free to rotate. In other words, $L$ in the equation represents the unsupported distance between the points of zero moment. If the column is supported in other ways, then Euler's formula can be used to determine the critical load provided $L$ represents the distance between the zero-moment points. This distance is called the column's *effective length*, $L_e$. Obviously, for a pin-ended column $L_e = L$, Fig. 17–12a. For the fixed and free-ended column analyzed above, the deflection curve was found to be one-half that of a column that is pin-connected and has a length of $2L$, Fig. 17–12b. Thus the effective length between the points of zero moment is $L_e = 2L$. Examples for two other columns with different end supports are also shown in Fig. 17–12. The column fixed at its ends, Fig. 17–12c, has inflection points or points of zero moment $L/4$ from each support. The effective length is therefore represented by the middle half of its length, that is, $L_e = 0.5L$. Lastly, the pin- and fixed-ended column, Fig. 17–12d, has an inflection point at approximately $0.7L$ from its pinned end, so that $L_e = 0.7L$.

Rather than specifying the column's effective length, many design codes provide column formulas that employ a dimensionless coefficient $K$ called the *effective-length factor*. $K$ is defined from

$$L_e = KL \qquad (17\text{--}10)$$

Specific values of $K$ are also given in Fig. 17–12. Based on this generality, we can therefore write Euler's formula as

$$P_{cr} = \frac{\pi^2 EI}{(KL)^2} \qquad (17\text{--}11)$$

or

$$\sigma_{cr} = \frac{\pi^2 E}{(KL/r)^2} \qquad (17\text{--}12)$$

Here $(KL/r)$ is the column's *effective-slenderness ratio*. For example, note that for the column fixed at its base and free at its end, we have $K = 2$, and therefore Eq. 17–11 gives the same result as Eq. 17–9.

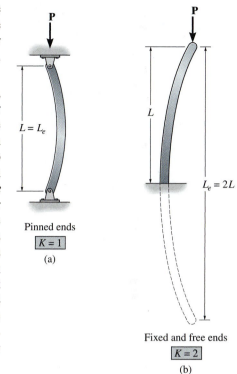

Pinned ends
$K = 1$
(a)

Fixed and free ends
$K = 2$
(b)

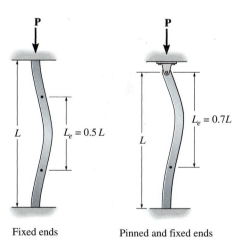

Fixed ends
$K = 0.5$
(c)

Pinned and fixed ends
$K = 0.7$
(d)

Fig. 17–12

[*]See Problems 17–42, 17–43, 17–44, and 17–45.

EXAMPLE 17.3

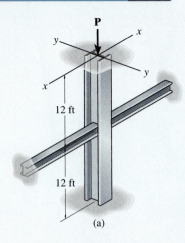

**P**

12 ft

12 ft

(a)

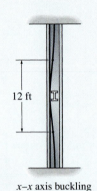

12 ft

*x–x* axis buckling

(b)

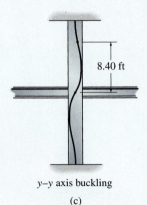

8.40 ft

*y–y* axis buckling

(c)

**Fig. 17–13**

A W6 × 15 steel column is 24 ft long and is fixed at its ends as shown in Fig. 17–13*a*. Its load-carrying capacity is increased by bracing it about the *y–y* (weak) axis using struts that are assumed to be pin-connected to its midheight. Determine the load it can support so that the column does not buckle nor the material exceed the yield stress. Take $E_{st} = 29(10^3)$ ksi and $\sigma_Y = 60$ ksi.

**Solution**

The buckling behavior of the column will be *different* about the *x* and *y* axes due to the bracing. The buckled shape for each of these cases is shown in Figs. 17–13*b* and 17–13*c*. From Fig. 17–13*b*, the effective length for buckling about the *x–x* axis is $(KL)_x = 0.5(24 \text{ ft}) = 12 \text{ ft} = 144 \text{ in.}$, and from Fig. 17–13*c*, for buckling about the *y–y* axis, $(KL)_y = 0.7(24 \text{ ft}/2) = 8.40 \text{ ft} = 100.8 \text{ in.}$ The moments of inertia for a W6 × 15 are determined from the table in Appendix B. We have $I_x = 29.1 \text{ in}^4$, $I_y = 9.32 \text{ in}^4$.

Applying Eq. 17–11, we have

$$(P_{cr})_x = \frac{\pi^2 E I_x}{(KL)_x^2} = \frac{\pi^2[29(10^3) \text{ ksi}]29.1 \text{ in}^4}{(144 \text{ in.})^2} = 401.7 \text{ kip} \qquad (1)$$

$$(P_{cr})_y = \frac{\pi^2 E I_y}{(KL)_y^2} = \frac{\pi^2[29(10^3) \text{ ksi}]9.32 \text{ in}^4}{(100.8 \text{ in.})^2} = 262.5 \text{ kip} \qquad (2)$$

By comparison, buckling will occur about the *y–y* axis.

The area of the cross section is 4.43 $\text{in}^2$, so the average compressive stress in the column will be

$$\sigma_{cr} = \frac{P_{cr}}{A} = \frac{262.5 \text{ kip}}{4.43 \text{ in}^2} = 59.3 \text{ ksi}$$

Since this stress is less than the yield stress, buckling will occur before the material yields. Thus,

$$P_{cr} = 263 \text{ kip} \qquad \qquad Ans.$$

*Note:* From Eq. 17–11 it can be seen that buckling will always occur about the column axis having the *largest* slenderness ratio, since a large slenderness ratio will give a small critical load. Thus, using the data for the radius of gyration from the table in Appendix D, we have

$$\left(\frac{KL}{r}\right)_x = \frac{144 \text{ in.}}{2.56 \text{ in.}} = 56.2$$

$$\left(\frac{KL}{r}\right)_y = \frac{100.8 \text{ in.}}{1.46 \text{ in.}} = 69.0$$

Hence, *y–y* axis buckling will occur, which is the same conclusion reached by comparing Eqs. 1 and 2.

E X A M P L E 17.4

The aluminum column is fixed at its bottom and is braced at its top by cables so as to prevent movement at the top along the $x$ axis, Fig. 17–14a. If it is assumed to be fixed at its base, determine the largest allowable load $P$ that can be applied. Use a factor of safety for buckling of F.S. = 3.0. Take $E_{al}$ = 70 GPa, $\sigma_Y$ = 215 MPa, $A = 7.5(10^{-3})$ m², $I_x = 61.3(10^{-6})$ m⁴, $I_y = 23.2(10^{-6})$ m⁴.

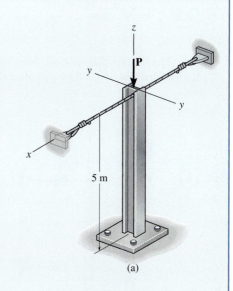

(a)

## Solution

Buckling about the $x$ and $y$ axes is shown in Fig. 17–14b and 17–14c, respectively. Using Fig. 17–12, for $x$–$x$ axis buckling, $K = 2$, so $(KL)_x = 2(5 \text{ m}) = 10$ m. Also, for $y$–$y$ axis buckling, $K = 0.7$, so $(KL)_y = 0.7(5 \text{ m}) = 3.5$ m.

Applying Eq. 17–11, the critical loads for each case are

$$(P_{cr})_x = \frac{\pi^2 E I_x}{(KL)_x^2} = \frac{\pi^2 [70(10^9) \text{ N/m}^2](61.3(10^{-6}) \text{ m}^4)}{(10 \text{ m})^2}$$
$$= 424 \text{ kN}$$

$$(P_{cr})_y = \frac{\pi^2 E I_y}{(KL)_y^2} = \frac{\pi^2 [70(10^9) \text{ N/m}^2](23.2(10^{-6}) \text{ m}^4)}{(3.5 \text{ m})^2}$$
$$= 1.31 \text{ MN}$$

By comparison, as $P$ is increased the column will buckle about the $x$–$x$ axis. The allowable load is therefore

$$P_{allow} = \frac{P_{cr}}{\text{F.S.}} = \frac{424 \text{ kN}}{3.0} = 141 \text{ kN} \qquad Ans.$$

Since

$$\sigma_{cr} = \frac{P_{cr}}{A} = \frac{424 \text{ kN}}{7.5(10^{-3}) \text{ m}^2} = 56.5 \text{ MPa} < 215 \text{ MPa}$$

Euler's equation can be applied.

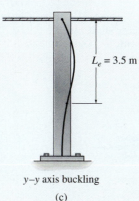

$L_e = 3.5$ m

y–y axis buckling

(c)

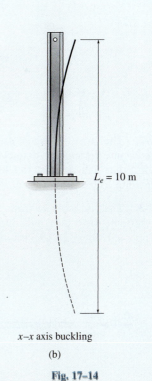

$L_e = 10$ m

x–x axis buckling

(b)

Fig. 17–14

## PROBLEMS

**17-1.** Determine the critical buckling load for the column. The material can be assumed rigid.

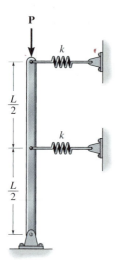

**Prob. 17-1**

**17-2.** The rod is made from an A-36 steel rod. Determine the smallest diameter of the rod, to the nearest $\frac{1}{16}$ in., that will support the load of $P = 5$ kip without buckling. The ends are roller supported.

**17-3.** The rod is made from a 1-in.-diameter steel rod. Determine the critical buckling load if the ends are roller supported. $E_{st} = 29(10^3)$ ksi, $\sigma_Y = 50$ ksi.

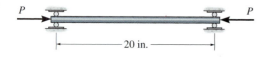

**Probs. 17-2/3**

*17-4.** The column consists of a rigid member that is pinned at its bottom and attached to a spring at its top. If the spring is unstretched when the column is in the vertical position, determine the critical load that can be placed on the column.

**Prob. 17-4**

**17-5.** The aircraft link is made from an A-36 steel rod. Determine the smallest diameter of the rod, to the nearest $\frac{1}{16}$ in., that will support the load of 4 kip without buckling. The ends are pin connected.

**Prob. 17-5**

**17-6.** An A-36 steel column has a length of 4 m and is pinned at both ends. If the cross sectional area has the dimensions shown, determine the critical load.

**17-7.** Solve Prob. 17-6 if the column is fixed at its bottom and pinned at its top.

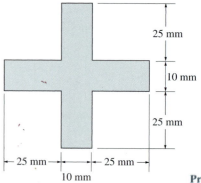

**Probs. 17-6/7**

**\*17-8.** The W8 × 67 is used as a structural A-36 steel column that can be assumed fixed at its base and pinned at its top. Determine the largest axial force $P$ that can be applied without causing it to buckle.

**17-9.** Solve Prob. 17–8 if the column is assumed fixed at its bottom and free at its top.

**Probs. 17–8/9**

**17-10.** A steel column has a length of 9 m and is fixed at both ends. If the cross-sectional area has the dimensions shown, determine the critical load. $E_{st} = 200$ GPa, $\sigma_Y = 250$ MPa.

**17-11.** Solve Prob. 17–10 if the column is pinned at its top and bottom.

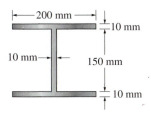

**Probs. 17–10/11**

**\*17-12.** A rod made from polyurethane has a stress–strain diagram in compression as shown. If the rod is pinned at its ends and is 37 in. long, determine its smallest diameter so it does not fail from elastic buckling.

**17-13.** A rod made from polyurethane has a stress–strain diagram in compression as shown. If the rod is pinned at its top and fixed at its base, and is 37 in. long, determine its smallest diameter so it does not fail from elastic buckling.

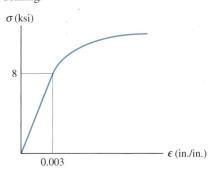

**Probs. 17–12/13**

**17-14.** The 10-ft wooden rectangular column has the dimensions shown. Determine the critical load if the ends are assumed to be pin connected. $E_w = 1.6(10^3)$ ksi, $\sigma_Y = 5$ ksi.

**17-15.** The 10-ft column has the dimensions shown. Determine the critical load if the bottom is fixed and the top is pinned. $E_w = 1.6(10^3)$ ksi, $\sigma_Y = 5$ ksi.

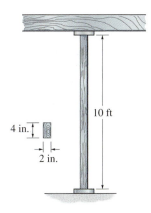

**Probs. 17–14/15**

**\*17-16.** An L-2 tool steel link in a forging machine is pin connected to the forks at its ends as shown. Determine the maximum load $P$ it can carry without buckling. Use a factor of safety with respect to buckling of F.S. = 1.75. Note from the figure on the left that the ends are pinned for buckling, whereas from the figure on the right the ends are fixed.

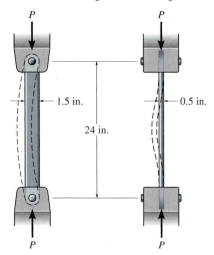

1.5 in.

0.5 in.

24 in.

**Prob. 17–16**

**17-17.** The W12 × 87 structural A-36 steel column has a length of 12 ft. If its bottom end is fixed supported while its top is free, and it is subjected to an axial load of $P$ = 380 kip, determine the factor of safety with respect to buckling.

**17-18.** The W12 × 87 structural A-36 steel column has a length of 12 ft. If its bottom end is fixed supported while its top is free, determine the largest axial load it can support. Use a factor of safety with respect to buckling of 1.75.

P

12 ft

**Probs. 17–17/18**

**17-19.** The handle is used to operate a simple press used to crush cans. Determine the maximum force $P$ that can be applied to the handle so that the rod $BC$ does not buckle. The rod is made of steel and has a diameter of `0.5 in. It is pin connected at its ends. $E_{st} = 29(10^3)$ ksi, $\sigma_Y$ = 36 ksi.

**\*17-20.** The handle is used to operate a simple press used to crush cans. Determine the smallest diameter steel rod $BC$, to the nearest $\frac{1}{8}$ in., that can be used if the maximum force $P$ applied to the handle is $P$ = 60 lb. The rod is pin connected at its ends. $E_{st} = 29(10^3)$ ksi, $\sigma_Y$ = 36 ksi.

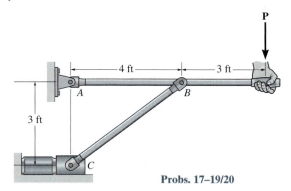

P

4 ft

3 ft

A

B

3 ft

C

**Probs. 17–19/20**

**17-21.** The 12-ft A-36 steel pipe column has an outer diameter of 3 in. and a thickness of 0.25 in. Determine the critical load if the ends are assumed to be pin connected.

**17-22.** The 12-ft A-36 steel column has an outer diameter of 3 in. and a thickness of 0.25 in. Determine the critical load if the bottom is fixed and the top is pinned.

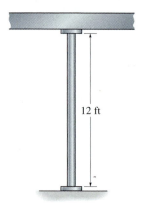

12 ft

**Probs. 17–21/22**

**17-23.** The A-36 steel pipe has an outer diameter of 2 in. and a thickness of 0.5 in. If it is held in place by a guywire, determine the largest horizontal force $P$ that can be applied without causing the pipe to buckle. Assume that the ends of the pipe are pin connected.

**\*17-24.** The A-36 steel pipe has an outer diameter of 2 in. If it is held in place by a guywire, determine the pipe's required inner diameter to the nearest $\frac{1}{8}$ in., so that it can support a maximum horizontal load of $P = 4$ kip without causing the pipe to buckle. Assume the ends of the pipe are pin connected.

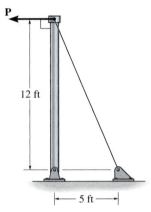

Probs. 17–23/24

**17-25.** The truss is made from A-36 steel bars, each of which has a circular cross section with a diameter of 1.5 in. Determine the maximum force $P$ that can be applied without causing any of the members to buckle. The members are pin connected at their ends.

**17-26.** The truss is made from A-36 steel bars, each of which has a circular cross section. If the applied load $P = 10$ kip, determine the diameter of member $AB$ to the nearest $\frac{1}{8}$ in. that will prevent this member from buckling. The members are pin supported at their ends.

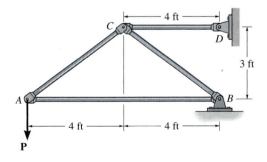

Probs. 17–25/26

**17-27.** The members of the truss are assumed to be pin connected. If member $GF$ is an A-36 steel rod having a diameter of 2 in., determine the greatest magnitude of load **P** that can be supported by the truss without causing this member to buckle.

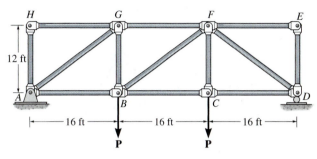

Prob. 17–27

**\*17-28.** The members of the truss are assumed to be pin connected. If member $AG$ is an A-36 steel rod having a diameter of 2 in., determine the greatest magnitude of load **P** that can be supported by the truss without causing this member to buckle.

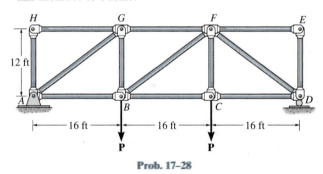

Prob. 17–28

**17-29.** Determine the maximum force $P$ that can be applied to the handle so that the A-36 steel control rod $BC$ does not buckle. The rod has a diameter of 25 mm.

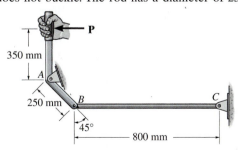

Prob. 17–29

**17-30.** The linkage is made using two A-36 steel rods, each having a circular cross section. If each rod has a diameter of $\frac{3}{4}$ in., determine the largest load it can support without causing any rod to buckle. Assume that the rods are pin connected at their ends.

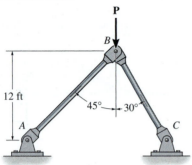

**Prob. 17–30**

**17-31.** The A-36 steel bar $AB$ has a square cross section. If it is pin-connected at its ends, determine the maximum allowable load $P$ that can be applied to the frame. Use a factor of safety with respect to buckling of 2.

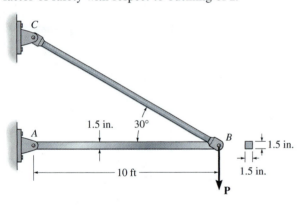

**Prob. 17–31**

**\*17-32.** The 50-mm diameter C86100 bronze rod is fixed supported at $A$ and has a gap of 2 mm from the wall at $B$. Determine the increase in temperature $\Delta T$ that will cause the rod to buckle. Assume that the contact at $B$ acts as a pin.

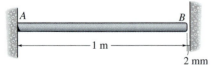

**Prob. 17–32**

**17-33.** Determine if the frame can support a load of $w = 6$ kN/m if the factor of safety with respect to buckling of member $AB$ is 3. Assume that $AB$ is made of steel and is pinned at its ends for $x$–$x$ axis buckling and fixed at its ends for $y$–$y$ axis buckling. $E_{st} = 200$ GPa, $\sigma_Y = 360$ MPa.

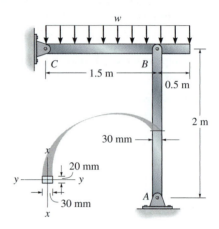

**Prob. 17–33**

**17-34.** The A-36 steel bar $AB$ of the frame is pin connected at its ends. If $P = 30$ kip, determine the factor of safety with respect to buckling about the $y$–$y$ axis due to the applied loading.

**17-35.** The A-36 steel bar $AB$ of the frame is pin connected at its ends. Determine the largest load $P$ that can be applied to the frame without causing it to buckle about the $y$–$y$ axis.

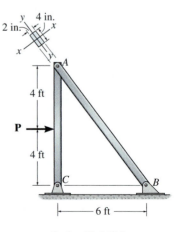

**Probs. 17–34/35**

# *17.4 Inelastic Buckling

In engineering practice, columns are generally classified according to the type of stresses developed within the column at the time of failure. *Long slender columns* will become unstable when the compressive stress remains elastic. The failure that occurs is referred to as *elastic instability*. *Intermediate columns* fail due to *inelastic instability*, meaning that the compressive stress at failure is greater than the material's proportional limit. And *short columns*, sometimes called *posts*, do not become unstable; rather the material simply yields or fractures.

Application of the Euler equation requires that the stress in the column remain *below* the material's yield point (actually the proportional limit) when the column buckles, and so this equation applies only to long columns. In practice, however, most columns are selected to have intermediate lengths. The behavior of these columns can be studied by modifying the Euler equation so that it applies for inelastic buckling. To show how this can be done, consider the material to have a stress–strain diagram as shown in Fig. 17–15. Here the proportional limit is $\sigma_{pl}$, and the modulus of elasticity, or slope of the line $AB$, is $E$. A plot of Euler's hyperbola, Fig. 17–8, is shown in Fig. 17–16. This equation is valid for a column having a slenderness ratio as small as $(KL/r)_{pl}$, since at this point the axial stress in the column becomes $\sigma_{cr} = \sigma_{pl}$.

If the column has a slenderness ratio that is *less* than $(KL/r)_{pl}$, then the critical stress in the column must be greater than $\sigma_Y$. For example, suppose a column has a slenderness ratio of $(KL/r)_1 < (KL/r)_{pl}$, with corresponding critical stress $\sigma_D > \sigma_{pl}$ needed to cause instability. When the column is *about to buckle*, the change in strain that occurs in the column is within a *small range* $\Delta\epsilon$, so that the modulus of elasticity or stiffness for the material can be taken as the ***tangent modulus*** $E_t$, defined as the slope of the $\sigma-\epsilon$ diagram at point $D$, Fig. 17–15. In other words, at the time of failure, the column behaves as if it were made from a material that has a *lower stiffness* than when it behaves elastically, $E_t < E$.

This crane boom failed by buckling caused by an overload. Note the region of localized collapse.

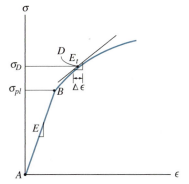

Fig. 17–15

In general, therefore, as the slenderness ratio decreases, the *critical stress* for a column continues to rise; and from the $\sigma-\epsilon$ diagram, the *tangent modulus* for the material *decreases*. Using this idea, we can modify Euler's equation to include these cases of inelastic buckling by substituting the material's tangent modulus $E_t$ for $E$, so that

$$\sigma_{\text{cr}} = \frac{\pi^2 E_t}{(KL/r)^2} \tag{17-13}$$

This is the so-called *tangent modulus* or *Engesser equation*, proposed by F. Engesser in 1889. A plot of this equation for intermediate and short-length columns of a material defined by the $\sigma-\epsilon$ diagram in Fig. 17–15 is shown in Fig. 17–16.

No *actual column* can be considered to be either perfectly straight or loaded along its centroidal axis, as assumed here, and therefore it is indeed very difficult to develop an expression that will provide a full analysis of this phenomenon. It should also be pointed out that other methods of describing the inelastic buckling of columns have been considered. One of these methods was developed by the aeronautical engineer F. R. Shanley and is called the *Shanley theory* of inelastic buckling. Although it provides a better description of the phenomenon than the tangent modulus theory, as explained here, experimental testing of a large number of columns, each of which approximates the ideal column, has shown that Eq. 17–13 is *reasonably accurate* in predicting the column's critical stress. Furthermore, the tangent modulus approach to modeling inelastic column behavior is relatively easy to apply.

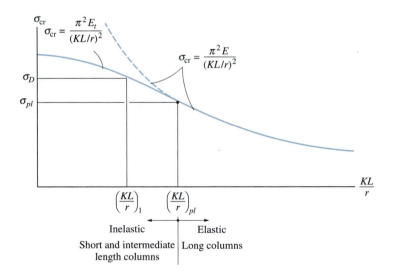

Fig. 17–16

# EXAMPLE 17.5

A solid rod has a diameter of 30 mm and is 600 mm long. It is made of a material that can be modeled by the stress–strain diagram shown in Fig. 17–17. If it is used as a pin-supported column, determine the critical load.

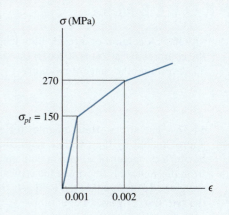

Fig. 17–17

**Solution**

The radius of gyration is

$$r = \sqrt{\frac{I}{A}} = \sqrt{\frac{(\pi/4)(15)^4}{\pi(15)^2}} = 7.5 \text{ mm}$$

and therefore the slenderness ratio is

$$\frac{KL}{r} = \frac{1(600 \text{ mm})}{7.5 \text{ mm}} = 80$$

Applying Eq. 17–13 yields

$$\sigma_{cr} = \frac{\pi^2 E_t}{(KL/r)^2} = \frac{\pi^2 E_t}{(80)^2} = 1.542(10^{-3})E_t \qquad (1)$$

First we will assume that the critical stress is elastic. From Fig. 17–17,

$$E = \frac{150 \text{ MPa}}{0.001} = 150 \text{ GPa}$$

Thus, Eq. 1 becomes

$$\sigma_{cr} = 1.542(10^{-3})[150(10^3)] \text{ MPa} = 231.3 \text{ MPa}$$

Since $\sigma_{cr} > \sigma_{pl} = 150$ MPa, inelastic buckling occurs.
From the second line segment of the $\sigma$–$\epsilon$ diagram, Fig. 17–17, we have

$$E_t = \frac{\Delta\sigma}{\Delta\epsilon} = \frac{270 \text{ MPa} - 150 \text{ MPa}}{0.002 - 0.001} = 120 \text{ GPa}$$

Applying Eq. 1 yields

$$\sigma_{cr} = 1.542(10^{-3})[120(10^3)] \text{ MPa} = 185.1 \text{ MPa}$$

Since this value falls within the limits of 150 MPa and 270 MPa, it is indeed the critical stress.
The critical load on the rod is therefore

$$P_{cr} = \sigma_{cr}A = 185.1 \text{ MPa}[\pi(0.015 \text{ m})^2] = 131 \text{ kN} \qquad \textit{Ans.}$$

# PROBLEMS

**\*17-36.** A column of intermediate length buckles when the compressive stress is 40 ksi. If the slenderness ratio is 60, determine the tangent modulus.

**17-37.** The stress–strain diagram for a material can be approximated by the two line segments shown. If a bar having a diameter of 60 mm and a length of 2 m is made from this material, determine the critical load if both ends are pinned. Assume that the load acts through the axis of the bar. Use Engesser's equation.

**17-38.** The stress–strain diagram for a material can be approximated by the two line segments shown. If a bar having a diameter of 60 mm and a length of 2 m is made from this material, determine the critical load provided the ends are fixed. Assume that the load acts through the axis of the bar. Use Engesser's equation.

**17-39.** The stress–strain diagram for a material can be approximated by the two line segments shown. If a bar having a diameter of 60 mm and length of 2 m is made from this material, determine the critical load provided one end is free and the other is fixed. Assume that the load acts through the axis of the bar. Use Engesser's equation.

**17-40.** Construct the buckling curve, $P/A$ versus $L/r$, for a column that has a bilinear stress–strain curve in compression as shown.

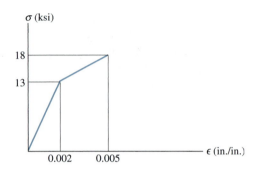

**Prob. 17–40**

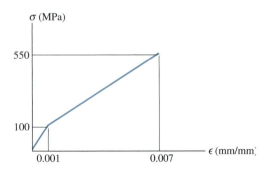

**Probs. 17–37/38/39**

# CHAPTER REVIEW

- *Elastic Buckling.* Buckling is the sudden instability that occurs in columns or members that support an axial load. The maximum axial load that a member can support just before buckling occurs is called the critical load $P_{cr}$. The critical load for an ideal column is determined from the Euler equation, $P_{cr} = \pi^2 EI / (KL)^2$, where $K = 1$ for pin supports, $K = 0.5$ for fixed supports, $K = 0.7$ for pinned and fixed supports, and $K = 2$ for a fixed support and free end.

- *Inelastic Buckling.* When the axial load tends to cause yielding of the column, then the tangent modulus should be used with Euler's equation to determine the buckling load. This is referred to as Engesser's equation.

# REVIEW PROBLEMS

**17-41.** The steel pipe is constrained between the walls at $A$ and $B$. If there is a gap of 3 mm at $B$ when $T_1 = 4°C$, determine the temperature required to cause the pipe to become unstable and begin to buckle. The pipe has an outer diameter of 40 mm and a wall thickness of 10 mm. Assume that the collars at $A$ and $B$ provide fixed connections for the pipe. Neglect their size. $\alpha_{st} = 12(10^{-6})/°C$, $E_{st} = 200$ GPa, $\sigma_Y = 250$ MPa.

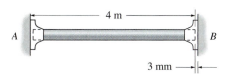

**Prob. 17–41**

**17-42.** Construct the buckling curve, $P/A$ versus $L/r$, for a column that has a bilinear stress-strain curve in compression as shown.

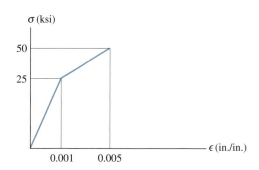

**Prob. 17–42**

# A Mathematical Expressions

## Quadratic Formula

If $ax^2 + bx + c = 0$, then $x = \dfrac{-b \pm \sqrt{b^2 - 4ac}}{2a}$

## Hyperbolic Functions

$\sinh x = \dfrac{e^x - e^{-x}}{2}$, $\cosh x = \dfrac{e^x + e^{-x}}{2}$, $\tanh x = \dfrac{\sinh x}{\cosh x}$

## Trigonometric Identities

$\sin^2 \theta + \cos^2 \theta = 1$

$\sin(\theta \pm \phi) = \sin \theta \cos \phi \pm \cos \theta \sin \phi$

$\sin 2\theta = 2 \sin \theta \cos \theta$

$\cos(\theta \pm \phi) = \cos \theta \cos \phi \mp \sin \theta \sin \phi$

$\cos 2\theta = \cos^2 \theta - \sin^2 \theta$

$\cos \theta = \pm\sqrt{\dfrac{1 + \cos 2\theta}{2}}$, $\sin \theta = \pm\sqrt{\dfrac{1 - \cos 2\theta}{2}}$

$\tan \theta = \dfrac{\sin \theta}{\cos \theta}$

$1 + \tan^2 \theta = \sec^2 \theta \qquad 1 + \cot^2 \theta = \csc^2 \theta$

## Power-Series Expansions

$\sin x = x - \dfrac{x^3}{3!} + \dfrac{x^5}{5!} - \dfrac{x^7}{7!} + \cdots$

$\cos x = 1 - \dfrac{x^2}{2!} + \dfrac{x^4}{4!} - \dfrac{x^6}{6!} + \cdots$

$\sinh x = x + \dfrac{x^3}{3!} + \dfrac{x^5}{5!} + \cdots$

$\cosh x = 1 + \dfrac{x^2}{2!} + \dfrac{x^4}{4!} + \cdots$

## Derivatives

$\dfrac{d}{dx}(u^n) = nu^{n-1}\dfrac{du}{dx}$
$\qquad \dfrac{d}{dx}(\sin u) = \cos u\dfrac{du}{dx}$
$\qquad \dfrac{d}{dx}(\cot u) = -\csc^2 u\dfrac{du}{dx}$
$\qquad \dfrac{d}{dx}(\sinh u) = \cosh u\dfrac{du}{dx}$

$\dfrac{d}{dx}(uv) = u\dfrac{dv}{dx} + v\dfrac{du}{dx}$
$\qquad \dfrac{d}{dx}(\cos u) = -\sin u\dfrac{du}{dx}$
$\qquad \dfrac{d}{dx}(\sec u) = \tan u \sec u\dfrac{du}{dx}$
$\qquad \dfrac{d}{dx}(\cosh u) = \sinh u\dfrac{du}{dx}$

$\dfrac{d}{dx}\left(\dfrac{u}{v}\right) = \dfrac{v\dfrac{du}{dx} - u\dfrac{dv}{dx}}{v^2}$
$\qquad \dfrac{d}{dx}(\tan u) = \sec^2 u\dfrac{du}{dx}$
$\qquad \dfrac{d}{dx}(\csc u) = -\csc u \cot u\dfrac{du}{dx}$

## Integrals

$$\int x^n \, dx = \frac{x^{n+1}}{n+1} + C, n \neq -1$$

$$\int \frac{dx}{a + bx} = \frac{1}{b}\ln(a + bx) + C$$

$$\int \frac{dx}{a + bx^2} = \frac{1}{2\sqrt{-ba}}\ln\left[\frac{\sqrt{a} + 2\sqrt{-b}}{\sqrt{a} - x\sqrt{-b}}\right] + C,$$
$$a > 0, b < 0$$

$$\int \frac{x \, dx}{a + bx^2} = \frac{1}{2b}\ln(bx^2 + a) + C$$

$$\int \frac{x^2 \, dx}{a + bx^2} = \frac{x}{b} - \frac{a}{b\sqrt{ab}}\tan^{-1}\frac{x\sqrt{ab}}{a} + C$$

$$\int \frac{dx}{a^2 - x^2} = \frac{1}{2a}\ln\left[\frac{a + x}{a - x}\right] + C, a^2 > x^2$$

$$\int \sqrt{a + bx} \, dx = \frac{2}{3b}\sqrt{(a + bx)^3} + C$$

$$\int x\sqrt{a + bx} \, dx = \frac{-2(2a - 3bx)\sqrt{(a + bx)^3}}{15b^2} + C$$

$$\int x^2\sqrt{a + bx} \, dx =$$
$$\frac{2(8a^2 - 12abx + 15b^2x^2)\sqrt{(a + bx)^3}}{105b^3} + C$$

$$\int \sqrt{a^2 - x^2} \, dx = \frac{1}{2}\left[x\sqrt{a^2 - x^2} + a^2 \sin^{-1}\frac{x}{a}\right] + C,$$
$$a > 0$$

$$\int x\sqrt{a^2 - x^2} \, dx = -\frac{1}{3}\sqrt{(a^2 - x^2)^3} + C$$

$$\int x^2\sqrt{a^2 - x^2} \, dx = -\frac{x}{4}\sqrt{(a^2 - x^2)^3}$$
$$+ \frac{a^2}{8}\left(x\sqrt{a^2 - x^2} + a^2\sin^{-1}\frac{x}{a}\right) + C, a > 0$$

$$\int \sqrt{x^2 \pm a^2} \, dx =$$
$$\frac{1}{2}\left[x\sqrt{x^2 \pm a^2} \pm a^2 \ln(x + \sqrt{x^2 \pm a^2})\right] + C$$

$$\int x\sqrt{x^2 \pm a^2} \, dx = \frac{1}{3}\sqrt{(x^2 \pm a^2)^3} + C$$

$$\int x^2\sqrt{x^2 \pm a^2} \, dx = \frac{x}{4}\sqrt{(x^2 \pm a^2)^3}$$
$$\pm \frac{a^2}{8}x\sqrt{x^2 \pm a^2} - \frac{a^4}{8}\ln(x + \sqrt{x^2 \pm a^2}) + C$$

$$\int \frac{dx}{\sqrt{a + bx}} = \frac{2\sqrt{a + bx}}{b} + C$$

$$\int \frac{x \, dx}{\sqrt{x^2 \pm a^2}} = \sqrt{x^2 \pm a^2} + C$$

$$\int \frac{dx}{\sqrt{a + bx + cx^2}} = \frac{1}{\sqrt{c}}\ln\left[\sqrt{a + bx + cx^2} + \right.$$
$$\left. x\sqrt{c} + \frac{b}{2\sqrt{c}}\right] + C, c > 0$$
$$= \frac{1}{\sqrt{-c}}\sin^{-1}\left(\frac{-2cx - b}{\sqrt{b^2 - 4ac}}\right) + C, c < 0$$

$$\int \sin x \, dx = -\cos x + C$$

$$\int \cos x \, dx = \sin x + C$$

$$\int x\cos(ax) \, dx = \frac{1}{a^2}\cos(ax) + \frac{x}{a}\sin(ax) + C$$

$$\int x^2\cos(ax) \, dx = \frac{2x}{a^2}\cos(ax) + \frac{a^2x^2 - 2}{a^3}\sin(ax) + C$$

$$\int e^{ax} \, dx = \frac{1}{a}e^{ax} + C$$

$$\int xe^{ax} \, dx = \frac{e^{ax}}{a^2}(ax - 1) + C$$

$$\int \sinh x \, dx = \cosh x + C$$

$$\int \cosh x \, dx = \sinh x + C$$

# B | Average Mechanical Properties of Typical Engineering Materials[a]

## (SI Units)

| Materials | Density (Mg/m³) | Modulus of Elasticity (GPa) | Modulus of Rigidity (GPa) | Yield Strength (MPa) Tens. | Comp.[b] | Shear | Ultimate Strength (MPa) Tens. | Comp.[b] | Shear | % Elongation in 50 mm specimen | Poisson's Ratio | Coef. of Therm. Expansion (10⁻⁶)/°C |
|---|---|---|---|---|---|---|---|---|---|---|---|---|
| **Metallic** | | | | | | | | | | | | |
| Aluminum Wrought Alloys ⌐2014-T6 | 2.79 | 73.1 | 27 | 414 | 414 | 172 | 469 | 469 | 290 | 10 | 0.35 | 23 |
| └6061-T6 | 2.71 | 68.9 | 26 | 255 | 255 | 131 | 290 | 290 | 186 | 12 | 0.35 | 24 |
| Cast Iron Alloys ⌐Gray ASTM 20 | 7.19 | 68.9 | 27 | – | – | – | 179 | 669 | – | 0.6 | 0.28 | 12 |
| └Malleable ASTM A-197 | 7.28 | 172 | 68 | – | – | – | 276 | 372 | – | 5 | 0.28 | 12 |
| Copper Alloys ⌐Red Brass C83400 | 8.74 | 101 | 37 | 68.9 | 68.9 | – | 241 | 241 | – | 35 | 0.35 | 18 |
| └Bronze C86100 | 8.83 | 103 | 38 | 345 | 345 | – | 655 | 655 | – | 20 | 0.34 | 17 |
| Magnesium Alloy [Am 1004-T61] | 1.83 | 44.7 | 18 | 152 | 152 | – | 276 | 276 | 132 | 1 | 0.30 | 26 |
| Steel Alloys ⌐Structural A36 | 7.85 | 200 | 75 | 250 | 250 | – | 400 | 400 | – | 30 | 0.32 | 12 |
| ├Stainless 304 | 7.86 | 193 | 75 | 207 | 207 | – | 517 | 517 | – | 40 | 0.27 | 17 |
| └Tool L2 | 8.16 | 200 | 78 | 703 | 703 | – | 800 | 800 | – | 22 | 0.32 | 12 |
| Titanium Alloy [Ti-6A1-4Y] | 4.43 | 126 | 44 | 924 | 924 | – | 1,000 | 1,000 | – | 16 | 0.36 | 9.4 |
| **Nonmetallic** | | | | | | | | | | | | |
| Concrete ⌐Low Strength | 2.38 | 22.1 | – | – | – | – | – | 12 | – | – | 0.15 | 11 |
| └High Strength | 2.38 | 29.0 | – | – | – | – | – | 38 | – | – | 0.15 | 11 |
| Plastic Reinforced ⌐Kevlar 49 | 1.45 | 131 | – | – | – | – | 717 | 483 | 20.3 | 2.8 | 0.34 | – |
| └30% Glass | 1.45 | 72.4 | – | – | – | – | 90 | 131 | – | – | 0.34 | – |
| Wood Select Structural Grade ⌐Douglas Fir | 0.47 | 13.1 | – | – | – | – | 2.1[c] | 26[d] | 6.2[d] | – | 0.29[e] | – |
| └White Spruce | 3.60 | 9.65 | – | – | – | – | 2.5[c] | 36[d] | 6.7[d] | – | 0.31[e] | – |

[a] Specific values may vary for a particular material due to alloy or mineral composition, mechanical working of the specimen, or heat treatment. For a more exact value reference books for the material should be consulted.

[b] The yield and ultimate strengths for ductile materials can be assumed equal for both tension and compression.

[c] Measured perpendicular to the grain.

[d] Measured parallel to the grain.

[e] Deformation measured perpendicular to the grain when the load is applied along the grain.

**(U.S. Customary Units)**

| Materials | Specific Weight (lb/in³) | Modulus of Elasticity (10³) ksi | Modulus of Rigidity (10³) ksi | Yield Strength (ksi) Tens. | Comp.[b] | Shear | Ultimate Strength (ksi) Tens. | Comp.[b] | Shear | % Elongation in 2 in. specimen | Poisson's Ratio | Coef. of Therm. Expansion $(10^{-6})/°F$ |
|---|---|---|---|---|---|---|---|---|---|---|---|---|
| **Metallic** | | | | | | | | | | | | |
| Aluminum Wrought Alloys — 2014-T6 | 0.101 | 10.6 | 3.9 | 60 | 60 | 25 | 68 | 68 | 42 | 10 | 0.35 | 12.8 |
| Aluminum Wrought Alloys — 6061-T6 | 0.098 | 10.0 | 3.7 | 37 | 37 | 19 | 42 | 42 | 27 | 12 | 0.35 | 13.1 |
| Cast Iron Alloys — Gray ASTM 20 | 0.260 | 10.0 | 3.9 | — | — | — | 26 | 97 | — | 0.6 | 0.28 | 6.70 |
| Cast Iron Alloys — Malleable ASTM A-197 | 0.263 | 25.0 | 9.8 | — | — | — | 40 | 83 | — | 5 | 0.28 | 6.60 |
| Copper Alloys — Red Brass C83400 | 0.316 | 14.6 | 5.4 | 10 | 10 | — | 35 | 35 | — | 35 | 0.35 | 9.80 |
| Copper Alloys — Bronze C86100 | 0.319 | 15.0 | 5.6 | 50 | 50 | — | 95 | 95 | — | 20 | 0.34 | 9.60 |
| Magnesium Alloy [Am 1004-T61] | 0.066 | 6.48 | 2.5 | 22 | 22 | — | 40 | 40 | 22 | 1 | 0.50 | 14.3 |
| Steel Alloys — Structural A36 | 0.284 | 29.0 | 11.0 | 36 | 36 | — | 58 | 58 | — | 30 | 0.32 | 6.60 |
| Steel Alloys — Stainless 304 | 0.284 | 28.0 | 11.0 | 30 | 30 | — | 75 | 75 | — | 40 | 0.27 | 9.60 |
| Steel Alloys — Tool L2 | 0.295 | 29.0 | 11.0 | 102 | 102 | — | 116 | 116 | — | 22 | 0.32 | 6.50 |
| Titanium Alloy [Ti-6A1-4V] | 0.160 | 17.4 | 6.4 | 134 | 134 | — | 145 | 145 | — | 16 | 0.36 | 5.20 |
| **Nonmetallic** | | | | | | | | | | | | |
| Concrete — Low Strength | 0.086 | 3.20 | — | — | — | — | — | 1.8 | — | — | 0.15 | 6.0 |
| Concrete — High Strength | 0.086 | 4.20 | — | — | — | — | — | 5.5 | — | — | 0.15 | 6.0 |
| Plastic Reinforced — Kevlar 49 | 0.0524 | 19.0 | — | — | — | — | 104 | 70 | 10.2 | 2.8 | 0.34 | — |
| Plastic Reinforced — 30% Glass | 0.0524 | 10.5 | — | — | — | — | 13 | 19 | — | — | 0.34 | — |
| Wood Select Structural Grade — Douglas Fir | 0.017 | 1.90 | — | — | — | — | 0.30[c] | 3.78[d] | 0.90[d] | — | 0.29[e] | — |
| Wood Select Structural Grade — White Spruce | 0.130 | 1.40 | — | — | — | — | 0.36[c] | 5.18[d] | 0.97[d] | — | 0.31[e] | — |

[a] Specific values may vary for a particular material due to alloy or mineral composition, mechanical working of the specimen, or heat treatment. For a more exact value reference books for the material should be consulted.

[b] The yield and ultimate strengths for ductile materials can be assumed equal for both tension and compression.

[c] Measured perpendicular to the grain.

[d] Measured parallel to the grain.

[e] Deformation measured perpendicular to the grain when the load is applied along the grain.

# C  Geometric Properties of An Area and Volume

| Centroid Location | Centroid Location | Area Moment of Inertia |
|---|---|---|

**Trapezoidal area**

$$A = \tfrac{1}{2}h(a+b)$$
$$\tfrac{1}{3}\left(\tfrac{2a+b}{a+b}\right)h$$

**Circular sector area**

$$A = \theta r^2$$
$$\tfrac{2}{3}\,\tfrac{r\sin\theta}{\theta}$$

$$I_x = \tfrac{1}{4}r^4\left(\theta - \tfrac{1}{2}\sin 2\theta\right)$$
$$I_y = \tfrac{1}{4}r^4\left(\theta + \tfrac{1}{2}\sin 2\theta\right)$$

**Semiparabolic area**

$$\tfrac{2}{5}a$$
$$A = \tfrac{2}{3}ab$$
$$\tfrac{3}{8}b$$

**Quarter circular area**

$$A = \tfrac{1}{4}\pi r^2$$
$$\tfrac{4r}{3\pi}$$

$$I_x = \tfrac{1}{16}\pi r^4$$
$$I_y = \tfrac{1}{16}\pi r^4$$

**Exparabolic area**

$$A = \tfrac{1}{3}ab$$
$$\tfrac{3}{10}b$$
$$\tfrac{3}{4}a$$

**Semicircular area**

$$A = \tfrac{\pi r^2}{2}$$
$$\tfrac{4r}{3\pi}$$

$$I_x = \tfrac{1}{8}\pi r^4$$
$$I_y = \tfrac{1}{8}\pi r^4$$

**Parabolic area**

$$A = \tfrac{4}{3}ab$$
$$\tfrac{2}{5}a$$

**Circular area**

$$A = \pi r^2$$

$$I_x = \tfrac{1}{4}\pi r^4$$
$$I_y = \tfrac{1}{4}\pi r^4$$

| Centroid Location | Centroid Location | Area Moment of Inertia |
|---|---|---|

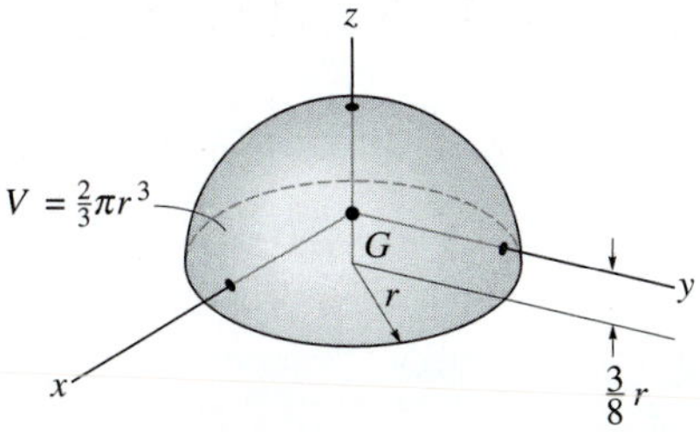

$V = \frac{2}{3}\pi r^3$

$\frac{3}{8}r$

Hemisphere

$A = bh$

$I_x = \frac{1}{12}bh^3$

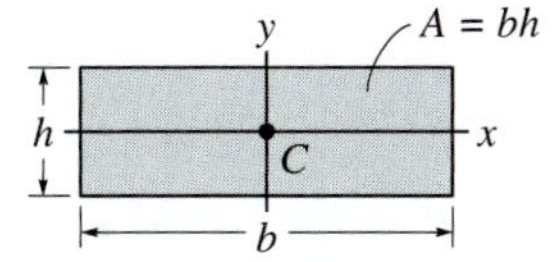

$I_y = \frac{1}{12}hb^3$

Rectangular area

$V = \frac{1}{3}\pi r^2 h$

$\frac{h}{4}$

Cone

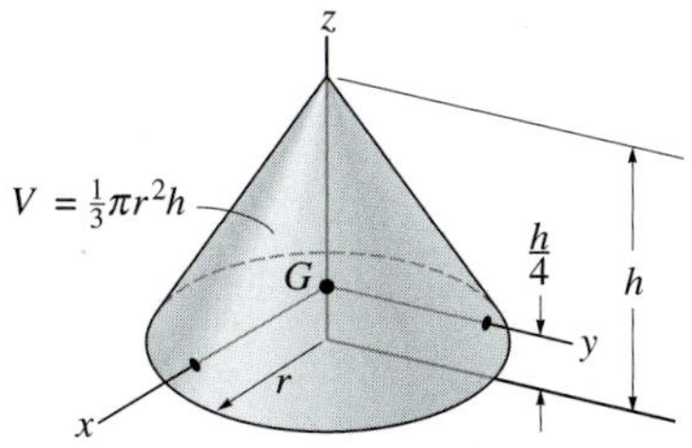

$A = \frac{1}{2}bh$

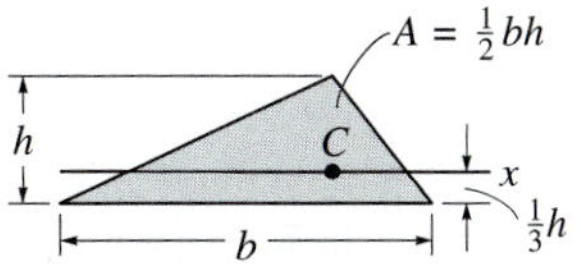

$\frac{1}{3}h$

$I_x = \frac{1}{36}bh^3$

Triangular area

# D Properties of Wide-Flange Sections

| Designation | Area A in² | Depth d in. | Web thickness $t_w$ in. | Flange width b in. | Flange thickness $t_f$ in. | x–x axis I in⁴ | x–x axis S in³ | x–x axis r in. | y–y axis I in⁴ | y–y axis S in³ | y–y axis r in. |
|---|---|---|---|---|---|---|---|---|---|---|---|
| W24 × 104 | 30.6 | 24.06 | 0.500 | 12.750 | 0.750 | 3100 | 258 | 10.1 | 259 | 40.7 | 2.91 |
| W24 × 94 | 27.7 | 24.31 | 0.515 | 9.065 | 0.875 | 2700 | 222 | 9.87 | 109 | 24.0 | 1.98 |
| W24 × 84 | 24.7 | 24.10 | 0.470 | 9.020 | 0.770 | 2370 | 196 | 9.79 | 94.4 | 20.9 | 1.95 |
| W24 × 76 | 22.4 | 23.92 | 0.440 | 8.990 | 0.680 | 2100 | 176 | 9.69 | 82.5 | 18.4 | 1.92 |
| W24 × 68 | 20.1 | 23.73 | 0.415 | 8.965 | 0.585 | 1830 | 154 | 9.55 | 70.4 | 15.7 | 1.87 |
| W24 × 62 | 18.2 | 23.74 | 0.430 | 7.040 | 0.590 | 1550 | 131 | 9.23 | 34.5 | 9.80 | 1.38 |
| W24 × 55 | 16.2 | 23.57 | 0.395 | 7.005 | 0.505 | 1350 | 114 | 9.11 | 29.1 | 8.30 | 1.34 |
| W18 × 65 | 19.1 | 18.35 | 0.450 | 7.590 | 0.750 | 1070 | 117 | 7.49 | 54.8 | 14.4 | 1.69 |
| W18 × 60 | 17.6 | 18.24 | 0.415 | 7.555 | 0.695 | 984 | 108 | 7.47 | 50.1 | 13.3 | 1.69 |
| W18 × 65 | 16.2 | 18.11 | 0.390 | 7.530 | 0.630 | 890 | 98.3 | 7.41 | 44.9 | 11.9 | 1.67 |
| W18 × 50 | 14.7 | 17.99 | 0.355 | 7.495 | 0.570 | 800 | 88.9 | 7.38 | 40.1 | 10.7 | 1.65 |
| W18 × 46 | 13.5 | 18.06 | 0.360 | 6.060 | 0.605 | 712 | 78.8 | 7.25 | 22.5 | 7.43 | 1.29 |
| W18 × 40 | 11.8 | 17.90 | 0.315 | 6.015 | 0.525 | 612 | 68.4 | 7.21 | 19.1 | 6.35 | 1.27 |
| W18 × 35 | 10.3 | 17.70 | 0.300 | 6.000 | 0.425 | 510 | 57.6 | 7.04 | 15.3 | 5.12 | 1.22 |
| W16 × 57 | 16.8 | 16.43 | 0.430 | 7.120 | 0.715 | 758 | 92.2 | 6.72 | 43.1 | 12.1 | 1.60 |
| W16 × 50 | 14.7 | 16.26 | 0.380 | 7.070 | 0.630 | 659 | 81.0 | 6.68 | 37.2 | 10.5 | 1.59 |
| W16 × 45 | 13.3 | 16.13 | 0.345 | 7.035 | 0.565 | 586 | 72.7 | 6.65 | 32.8 | 9.34 | 1.57 |
| W16 × 36 | 10.6 | 15.86 | 0.295 | 6.985 | 0.430 | 448 | 56.5 | 6.51 | 24.5 | 7.00 | 1.52 |
| W16 × 31 | 9.12 | 15.88 | 0.275 | 5.525 | 0.440 | 375 | 47.2 | 6.41 | 12.4 | 4.49 | 1.17 |
| W16 × 26 | 7.68 | 15.69 | 0.250 | 5.500 | 0.345 | 301 | 38.4 | 6.26 | 9.59 | 3.49 | 1.12 |
| W14 × 53 | 15.6 | 13.92 | 0.370 | 8.060 | 0.660 | 541 | 77.8 | 5.89 | 57.7 | 14.3 | 1.92 |
| W14 × 43 | 12.6 | 13.66 | 0.305 | 7.995 | 0.530 | 428 | 62.7 | 5.82 | 45.2 | 11.3 | 1.89 |
| W14 × 38 | 11.2 | 14.10 | 0.310 | 6.770 | 0.515 | 385 | 54.6 | 5.87 | 26.7 | 7.88 | 1.55 |
| W14 × 34 | 10.0 | 13.98 | 0.285 | 6.745 | 0.455 | 340 | 48.6 | 5.83 | 23.3 | 6.91 | 1.53 |
| W14 × 30 | 8.85 | 13.84 | 0.270 | 6.730 | 0.385 | 291 | 42.0 | 5.73 | 19.6 | 5.82 | 1.49 |
| W14 × 26 | 7.69 | 13.91 | 0.255 | 5.025 | 0.420 | 245 | 35.3 | 5.65 | 8.91 | 3.54 | 1.08 |
| W14 × 22 | 6.49 | 13.74 | 0.230 | 5.000 | 0.335 | 199 | 29.0 | 5.54 | 7.00 | 2.80 | 1.04 |

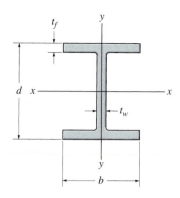

| Designation | Area $A$ in² | Depth $d$ in. | Web thickness $t_w$ in. | Flange width $b$ in. | Flange thickness $t_f$ in. | $x$–$x$ axis $I$ in⁴ | $x$–$x$ axis $S$ in³ | $x$–$x$ axis $r$ in. | $y$–$y$ axis $I$ in⁴ | $y$–$y$ axis $S$ in³ | $y$–$y$ axis $r$ in. |
|---|---|---|---|---|---|---|---|---|---|---|---|
| W12 × 87 | 25.6 | 12.53 | 0.515 | 12.125 | 0.810 | 740 | 118 | 5.38 | 241 | 39.7 | 3.07 |
| W12 × 50 | 14.7 | 12.19 | 0.370 | 8.080 | 0.640 | 394 | 64.7 | 5.18 | 56.3 | 13.9 | 1.96 |
| W12 × 45 | 13.2 | 12.06 | 0.335 | 8.045 | 0.575 | 350 | 58.1 | 5.15 | 50.0 | 12.4 | 1.94 |
| W12 × 26 | 7.65 | 12.22 | 0.230 | 6.490 | 0.380 | 204 | 33.4 | 5.17 | 17.3 | 5.34 | 1.51 |
| W12 × 22 | 6.48 | 12.31 | 0.260 | 4.030 | 0.425 | 156 | 25.4 | 4.91 | 4.66 | 2.31 | 0.847 |
| W12 × 16 | 4.71 | 11.99 | 0.220 | 3.990 | 0.265 | 103 | 17.1 | 4.67 | 2.82 | 1.41 | 0.773 |
| W12 × 14 | 4.16 | 11.91 | 0.200 | 3.970 | 0.225 | 88.6 | 14.9 | 4.62 | 2.36 | 1.19 | 0.753 |
| W10 × 100 | 29.4 | 11.10 | 0.680 | 10.340 | 1.120 | 623 | 112 | 4.60 | 207 | 40.0 | 2.65 |
| W10 × 54 | 15.8 | 10.09 | 0.370 | 10.030 | 0.615 | 303 | 60.0 | 4.37 | 103 | 20.6 | 2.56 |
| W10 × 45 | 13.3 | 10.10 | 0.350 | 8.020 | 0.620 | 248 | 49.1 | 4.32 | 53.4 | 13.3 | 2.01 |
| W10 × 30 | 8.84 | 10.47 | 0.300 | 5.810 | 0.510 | 170 | 32.4 | 4.38 | 16.7 | 5.75 | 1.37 |
| W10 × 39 | 11.5 | 9.92 | 0.315 | 7.985 | 0.530 | 209 | 42.1 | 4.27 | 45.0 | 11.3 | 1.98 |
| W10 × 19 | 5.62 | 10.24 | 0.250 | 4.020 | 0.395 | 96.3 | 18.8 | 4.14 | 4.29 | 2.14 | 0.874 |
| W10 × 15 | 4.41 | 9.99 | 0.230 | 4.000 | 0.270 | 68.9 | 13.8 | 3.95 | 2.89 | 1.45 | 0.810 |
| W10 × 12 | 3.54 | 9.87 | 0.190 | 3.960 | 0.210 | 53.8 | 10.9 | 3.90 | 2.18 | 1.10 | 0.785 |
| W8 × 67 | 19.7 | 9.00 | 0.570 | 8.280 | 0.935 | 272 | 60.4 | 3.72 | 88.6 | 21.4 | 2.12 |
| W8 × 58 | 17.1 | 8.75 | 0.510 | 8.220 | 0.810 | 228 | 52.0 | 3.65 | 75.1 | 18.3 | 2.10 |
| W8 × 48 | 14.1 | 8.50 | 0.400 | 8.110 | 0.685 | 184 | 43.3 | 3.61 | 60.9 | 15.0 | 2.08 |
| W8 × 40 | 11.7 | 8.25 | 0.360 | 8.070 | 0.560 | 146 | 35.5 | 3.53 | 49.1 | 12.2 | 2.04 |
| W8 × 31 | 9.13 | 8.00 | 0.285 | 7.995 | 0.435 | 110 | 27.5 | 3.47 | 37.1 | 9.27 | 2.02 |
| W8 × 24 | 7.08 | 7.93 | 0.245 | 6.495 | 0.400 | 82.8 | 20.9 | 3.42 | 18.3 | 5.63 | 1.61 |
| W8 × 15 | 4.44 | 8.11 | 0.245 | 4.015 | 0.315 | 48.0 | 11.8 | 3.29 | 3.41 | 1.70 | 0.876 |
| W6 × 25 | 7.34 | 6.38 | 0.320 | 6.080 | 0.455 | 53.4 | 16.7 | 2.70 | 17.1 | 5.61 | 1.52 |
| W6 × 20 | 5.87 | 6.20 | 0.260 | 6.020 | 0.365 | 41.4 | 13.4 | 2.66 | 13.3 | 4.41 | 1.50 |
| W6 × 15 | 4.43 | 5.99 | 0.230 | 5.990 | 0.260 | 29.1 | 9.72 | 2.56 | 9.32 | 3.11 | 1.46 |
| W6 × 16 | 4.74 | 6.28 | 0.260 | 4.030 | 0.405 | 32.1 | 10.2 | 2.60 | 4.43 | 2.20 | 0.966 |
| W6 × 12 | 3.55 | 6.03 | 0.230 | 4.000 | 0.280 | 22.1 | 7.31 | 2.49 | 2.99 | 1.50 | 0.918 |
| W6 × 9 | 2.68 | 5.90 | 0.170 | 3.940 | 0.215 | 1.64 | 5.56 | 2.47 | 2.19 | 1.11 | 0.905 |

# Answers to Selected Problems

## Chapter 1

**1–1. a)** 4.66 m, **b)** 55.6 s, **c)** 4.56 kN, **d)** 2.77 Mg
**1–2.** 4.70 slug/ft$^3$ = 2.42 Mg/m$^3$
**1–3. a)** 0.000431 kg = 0.431 g,
    **b)** 35.3(10$^3$) N = 35.3 kN,
    **c)** 0.00532 km = 5.32 m
**1–5.** 55 mi/h = 88.5 km/h, 88.5 km/h = 24.6 m/s
**1–6. a)** (430 kg)$^2$ = 0.185 Mg$^2$,
    **b)** (0.002 mg)$^2$ = 4 $\mu$g$^2$,
    **c)** (230 m)$^3$ = 0.0122 km$^3$
**1–7. a)** 250(10$^3$) slugs = 3.65 Gg,
    **b)** $W_e = mg$ = 35.8 MN,
    **c)** $W_m = mg_m$ = 5.89 MN,
    **d)** $m_m = m_e$ = 3.65 Gg
**1–9.** 1 Pa = 20.9(10$^{-3}$) lb/ft$^2$, 1 ATM = 101 kPa
**1–10. a)** $W$ = 98.1 N, **b)** $W$ = 4.90 mN,
    **c)** $W$ = 44.1 kN
**1–11. a)** (354 mg)(45 km)/0.0356 kN = 0.447 kg·m/N,
    **b)** (0.00453 Mg)(201 ms) = 0.911 kg·s,
    **c)** 435 MN/23.2 mm = 18.8 GN/m
**1–13. a)** 20 lb·ft = 27.1 N·m,
    **b)** 450 lb/ft$^3$ = 70.7 kN/m$^3$,
    **c)** 15 ft/h = 1.27 mm/s
**1–14.** 40 slugs = 584 kg
**1–15.** $\rho_w$ = 1.00 Mg/m$^3$
**1–17. a)** $m$ = 2.04 g, **b)** $m$ = 15.3 Mg,
    **c)** $m$ = 6.12 Gg
**1–18. a)** $m$ = 4.81 slug, **b)** $m$ = 70.2 kg,
    **c)** $W$ = 689 N, **d)** $W$ = 25.5 lb,
    **e)** $m$ = 70.2 kg
**1–19.** $F$ = 7.41(10$^{-6}$) N = 7.41 $\mu$N

## Chapter 2

**2–1.** $F_R$ = 867 N, $\phi$ = 108°
**2–2. a)** $F_R$ = 111 N, **b)** $F_R'$ = 143 N
**2–3.** $F_R$ = 393 lb, $\phi$ = 353°
**2–5.** $F_{1u}$ = 205 N, $F_{1v}$ = 160 N
**2–6.** $F_{2u}$ = 376 N, $F_{2v}$ = 482 N
**2–7.** $F_R$ = 10.8 kN, $\phi$ = 3.16°
**2–9.** $F_{AB}$ = 448 N, $F_{AC}$ = 366 N
**2–10.** $F_{AB}$ = 314 lb, $F_{AC}$ = 256 lb

**2–11.** $F_a$ = 30.6 lb, $F_b$ = 26.9 lb
**2–13.** $F_{AB}$ = 485 lb, $\theta$ = 24.6°
**2–14.** $T$ = 744 lb, $\theta$ = 23.8°
**2–15.** $\theta$ = 53.5°, $F_{AB}$ = 621 lb
**2–17. a)** $F_y$ = 16.3 lb, $F_n$ = −22.3 lb,
    **b)** $F_t$ = 5.98 lb, $F_x$ = 16.3 lb
**2–18.** $\theta$ = 18.6°, $F$ = 319 N
**2–19.** $\phi = \theta$ = 70.5°
**2–21.** $\theta$ = 29.1°, $F_1$ = 275 N
**2–22.** $F_R$ = 1.03 kN, $\theta$ = 87.9°
**2–23.** $\mathbf{F}_1$ = {−15.0**i** − 26.0**j**} kN,
    $\mathbf{F}_2$ = {−10.0**i** + 24.0**j**} kN
**2–25.** $F_R$ = 867 N, $\theta$ = 108°
**2–26.** $F_R$ = 19.2 N, $\theta$ = 2.37° ⭨$_\theta$
**2–27.** $\theta$ = 68.6°, $F_B$ = 960 N
**2–29.** $F_{1x}$ = 141 N, $F_{1y}$ = 141 N, $F_{2x}$ = −130 N,
    $F_{2y}$ = 75 N
**2–30.** $F_R$ = 217 N, $\theta$ = 87.0°
**2–31.** $F_{1x}$ = −200 lb, $F_{1y}$ = 0, $F_{2x}$ = 320 lb,
    $F_{2y}$ = −240 lb, $F_{3x}$ = 180 lb, $F_{3y}$ = 240 lb,
    $F_{4x}$ = −300 lb, $F_{4y}$ = 0
**2–33.** $\theta$ = 54.3°, $F_A$ = 686 N
**2–34.** $F_R$ = 1.23 kN, $\theta$ = 6.08°
**2–35.** $\mathbf{F}_1$ = {90**i** − 120**j**} lb, $\mathbf{F}_2$ = {−275**j**} lb,
    $\mathbf{F}_3$ = {−37.5**i** − 65.0**j**} lb, $F_R$ = 463 lb
**2–37.** $F_R$ = 161 lb, $\theta$ = 38.3°
**2–38.** $F$ = 2.03 kN, $F_R$ = 7.87 kN,
**2–39.** $F_1$ = 87.7 N, $\alpha_1$ = 46.9°, $\beta_1$ = 125°,
    $\gamma_1$ = 62.9°, $F_2$ = 98.6 N, $\alpha_2$ = 114°,
    $\beta_2$ = 150°, $\gamma_2$ = 72.3°
**2–41.** $F$ = 50 N, $\alpha$ = 74.1°, $\beta$ = 41.3°, $\gamma$ = 53.1°
**2–42.** $F_R$ = 39.4 lb, $\alpha$ = 52.8°, $\beta$ = 141°, $\gamma$ = 99.5°
**2–43.** $\beta$ = 90°, $\mathbf{F}$ = {−30**i** − 52.0**k**} N
**2–45.** $\mathbf{F}_1$ = {53.1**i** − 44.5**j** + 40**k**} lb, $\alpha_1$ = 48.4°,
    $\beta_1$ = 124°, $\gamma_1$ = 60°, $\mathbf{F}_2$ = {−130**k**} lb,
    $\alpha_2$ = 90°, $\beta_2$ = 90°, $\gamma_2$ = 180°
**2–46.** $\alpha_1$ = 45.6°, $\beta_1$ = 53.1°, $\gamma_1$ = 66.4°
**2–47.** $\alpha_1$ = 90°, $\beta_1$ = 53.1°, $\gamma_1$ = 66.4°
**2–49.** $\mathbf{F}_1$ = {176**j** − 605**k**} lb,
    $\mathbf{F}_2$ = {125**i** − 177**j** + 125**k**} lb,
    $\mathbf{F}_R$ = {125**i** − 0.377**j** − 480**k**} lb,
    $F_R$ = 496 lb; $\alpha$ = 75.4°, $\beta$ = 90.0°, $\gamma$ = 165°
**2–50.** $F_R$ = 369 N, $\alpha$ = 19.5°, $\beta$ = 78.3°, $\gamma$ = 105°
**2–51.** $F_2$ = 66.4 lb, $\alpha$ = 59.8°, $\beta$ = 107°, $\gamma$ = 144°

**2–53.** $\mathbf{F}_1 = \{86.5\mathbf{i} + 186\mathbf{j} - 143\mathbf{k}\}$ N,
$\mathbf{F}_2 = \{-200\mathbf{i} + 283\mathbf{j} + 200\mathbf{k}\}$ N,
$\mathbf{F}_R = \{-113\mathbf{i} + 468\mathbf{j} + 56.6\mathbf{k}\}$ N,
$F_R = 485$ N, $\alpha = 104°$, $\beta = 15.1°$, $\gamma = 83.3°$

**2–54.** $F_x = 1.28$ kN, $F_y = 2.60$ kN, $F_z = 0.776$ kN

**2–55.** $F = 2.02$ kN, $F_y = 0.523$ kN

**2–57.** $F_3 = 166$ N, $\alpha = 97.5°$, $\beta = 63.7°$, $\gamma = 27.5°$

**2–58.** $\alpha_{F_1} = 36.9°$, $\beta_{F_1} = 90.0°$, $\gamma_{F_1} = 53.1°$,
$\alpha_R = 69.3°$, $\beta_R = 52.2°$, $\gamma_R = 45.0°$

**2–59.** $F_x = 40$ N, $F_y = 40$ N, $F_z = 56.6$ N

**2–61.** $r = 31.5$ m, $\alpha = 69.6°$, $\beta = 116°$, $\gamma = 34.4°$

**2–62.** $\mathbf{r}_{AB} = \{2\mathbf{i} - 7\mathbf{j} - 5\mathbf{k}\}$ ft, $r_{AB} = 8.83$ ft,
$\alpha = 76.9°$, $\beta = 142°$, $\gamma = 124°$

**2–63.** $\alpha = 73.4°$, $\beta = 64.6°$, $\gamma = 31.0°$

**2–65.** $\mathbf{r} = \{-2.35\mathbf{i} + 3.93\mathbf{j} + 3.71\mathbf{k}\}$ ft, $r = 5.89$ ft,
$\alpha = 113°$, $\beta = 48.2°$, $\gamma = 51.0°$

**2–66.** $\mathbf{F} = \{404\mathbf{i} + 276\mathbf{j} - 101\mathbf{k}\}$ lb, $\alpha = 36.0°$,
$\beta = 56.5°$, $\gamma = 102°$

**2–67.** $r_{AB} = 2.11$ m

**2–69.** $r_{AB} = 17.0$ ft, $\mathbf{F} = \{-160\mathbf{i} - 180\mathbf{j} + 240\mathbf{k}\}$ lb

**2–70.** $r_{AB} = 467$ mm

**2–71.** $r_{AD} = 1.50$ m, $r_{BD} = 1.50$ m, $r_{CD} = 1.73$ m

**2–73.** $\mathbf{F} = \{452\mathbf{i} + 370\mathbf{j} - 136\mathbf{k}\}$ lb, $\alpha = 41.1°$,
$\beta = 51.9°$, $\gamma = 103°$

**2–74.** $F_R = 316$ N, $\alpha = 60.1°$, $\beta = 74.6°$, $\gamma = 146°$

**2–75.** $\mathbf{F}_A = \{285\mathbf{j} - 93.0\mathbf{k}\}$ N,
$\mathbf{F}_C = \{159\mathbf{i} + 183\mathbf{j} - 59.7\mathbf{k}\}$ N

**2–77.** $\mathbf{F}_{AB} = \{75.5\mathbf{i} - 43.6\mathbf{j} - 122\mathbf{k}\}$ lb,
$\mathbf{F}_{BC} = \{26.8\mathbf{i} + 33.5\mathbf{j} - 90.4\mathbf{k}\}$ lb,
$F_R = 236$ lb, $\alpha = 64.3°$, $\beta = 92.5°$, $\gamma = 154°$

**2–78.** $\mathbf{F}_A = \{-43.5\mathbf{i} + 174\mathbf{j} - 174\mathbf{k}\}$ N,
$\mathbf{F}_B = \{53.2\mathbf{i} - 79.8\mathbf{j} - 146\mathbf{k}\}$ N

**2–79.** $\mathbf{F}_1 = \{-26.2\mathbf{i} - 41.9\mathbf{j} + 62.9\mathbf{k}\}$ lb,
$\mathbf{F}_2 = \{13.4\mathbf{i} - 26.7\mathbf{j} - 40.1\mathbf{k}\}$ lb, $F_R = 73.5$ lb,
$\alpha = 100°$, $\beta = 159°$, $\gamma = 71.9°$

**2–81.** $\theta = 121°$

**2–82.** $\theta = 109°$

**2–83.** Proj $\mathbf{r}_1 = 2.99$ m, Proj $\mathbf{r}_2 = 1.99$ m

**2–85.** $F_1 = 19.4$ N, $F_2 = 53.4$ N

**2–86.** $\theta = 74.2°$

**2–87.** $r_{BC} = 5.39$ m

**2–89.** $F_\parallel = 82.4$ N, $F_\perp = 594$ N

**2–90.** $F_1 = 333$ N, $F_2 = 373$ N

**2–91.** Proj $\mathbf{F} = 0.667$ kN

**2–93.** $\theta = 70.5°$

**2–94.** $\phi = 65.8°$

**2–95.** $F_1 = 20.0$ N, $F_2 = 31.6$ N

**2–97.** $\theta = 100°$

**2–98.** $F = \{-34.3\mathbf{i} + 22.9\mathbf{j} - 68.6\mathbf{k}\}$ lb

**2–99.** $\mathbf{F}_1 = \{20\mathbf{i} + 34.6\mathbf{j}\}$ lb,
$\mathbf{F}_2 = \{-42.4\mathbf{i} + 42.4\mathbf{j}\}$ lb,
$F_R = 80.3$ lb, $\theta = 106°$

**2–101.** $F_v = 80.4$ lb, $F_u = 75.0$ lb

**2–102.** $x = 6$ ft, $y = 6$ ft, $z = 4.5$ ft

**2–103.** $x = 10.3$ ft, $y = 2.36$ ft

**2–105.** $\mathbf{F}_1 = \{70\mathbf{i} - 121\mathbf{j}\}$ lb, $\mathbf{F}_2 = \{-180\mathbf{j}\}$ lb,
$\mathbf{F}_3 = \{-88.4\mathbf{i} - 88.4\mathbf{j}\}$ lb, $F_R = 390$ lb

**2–106.** $\theta = 50.3°$, $F_{CB} = 369$ N

**2–107.** $90°$

## Chapter 3

**3–3.** If $\mathbf{A} \cdot (\mathbf{B} \times \mathbf{C}) = 0$, then the volume equals zero, so that $\mathbf{A}$, $\mathbf{B}$, and $\mathbf{C}$ are coplanar.

**3–5.** $M_P = 2.37$ kN $\cdot$ m $\curvearrowleft$

**3–6.** $M_O = 2.88$ kN $\cdot$ m $\curvearrowright$

**3–7.** $M_P = 3.15$ kN $\cdot$ m $\curvearrowright$

**3–9.** $M_P = 3.15$ kN $\cdot$ m $\curvearrowleft$

**3–10.** $(M_{F_1})_O = 24.1$ N $\cdot$ m $\curvearrowright$, $(M_{F_2})_O = 14.5$ N $\cdot$ m $\curvearrowright$,

**3–11.** $M_O = 2.42$ kip $\cdot$ ft $\curvearrowleft$

**3–13.** $(M_{F_1})_B = 4.125$ kip $\cdot$ ft $\curvearrowright$,
$(M_{F_2})_B = 2.00$ kip $\cdot$ ft $\curvearrowright$,
$(M_{F_3})_B = 40.0$ lb $\cdot$ ft $\curvearrowright$

**3–14.** $M_B = 90.6$ lb $\cdot$ ft $\curvearrowleft$, $M_C = 141$ lb $\cdot$ ft $\curvearrowleft$

**3–15.** $M_A = 195$ lb $\cdot$ ft $\curvearrowleft$

**3–17.** $M_O = 28.1$ N $\cdot$ m $\curvearrowright$, $\theta = 88.6°$,
$(M_O)_{max} = 32.0$ N $\cdot$ m $\curvearrowright$

**3–18.** a) $(M_A)_{max} = 330$ lb $\cdot$ ft, $\theta = 76.0°$,
b) $(M_A)_{min} = 0$, $\theta = 166°$

**3–19.** $-M_O = 120$ N $\cdot$ m $\curvearrowright$, $+M_O = 520$ N $\cdot$ m $\curvearrowright$

**3–21.** $M_A = 13.0$ N $\cdot$ m $\curvearrowright$, $F = 35.2$ N

**3–22.** $(M_{F_1})_A = 433$ N $\cdot$ m $\curvearrowright$,
$(M_{F_2})_A = 1.30$ kN $\cdot$ m $\curvearrowright$,
$(M_{F_3})_A = 800$ N $\cdot$ m $\curvearrowright$,

**3–23.** $\theta = 7.48°$

**3–25.** $F_A = 28.9$ lb

**3–26.** $\theta = 8.53°$

**3–27.** $F = 1.33$ kip, $F' = 1.63$ kip

**3–29.** $\mathbf{M}_O = \{440\mathbf{i} + 220\mathbf{j} + 990\mathbf{k}\}$ N $\cdot$ m

**3–30.** $\mathbf{M}_O = \{-84\mathbf{i} - 8\mathbf{j} - 39\mathbf{k}\}$ kN $\cdot$ m

**3–31.** $\mathbf{M}_P = \{-116\mathbf{i} + 16\mathbf{j} - 135\mathbf{k}\}$ kN $\cdot$ m

**3–33.** $\mathbf{M}_B = \{-37.6\mathbf{i} + 90.7\mathbf{j} - 155\mathbf{k}\}$ N $\cdot$ m

**3–34.** $\mathbf{M}_A = \{-720\mathbf{i} + 120\mathbf{j} - 660\mathbf{k}\}$ N $\cdot$ m

**3–35.** $\mathbf{M}_C = \{-35.4\mathbf{i} - 128\mathbf{j} - 222\mathbf{k}\}$ lb $\cdot$ ft

**3–37.** $F_{AB} = 18.6$ lb

**3–38.** $\mathbf{M}_A = \{-5.39\mathbf{i} + 13.1\mathbf{j} - 11.4\mathbf{k}\}$ N·m

**3–39.** $\mathbf{M}_B = \{10.6\mathbf{i} + 13.1\mathbf{j} + 29.2\mathbf{k}\}$ N·m

**3–41.** $\mathbf{M}_R = \{-1.90\mathbf{i} + 6.00\mathbf{j}\}$ kN·m

**3–42.** $(\mathbf{M}_{Oa})_P = \{218\mathbf{j} + 163\mathbf{k}\}$ N·m

**3–43.** $\mathbf{M}_{aa} = \{40\mathbf{i} + 40\mathbf{j}\}$ N·m

**3–45.** $(M_{AB})_1 = 72.0$ N·m, $(M_{AB})_2 = (M_{AB})_3 = 0$

**3–46.** $M_x = 44.4$ lb·ft

**3–47.** $M_{aa} = 4.37$ N·m, $x = 33.7°$,
$\beta = 90°$, $\gamma = 56.3°$, $M_{max} = 5.41$ N·m

**3–49.** $\mathbf{M}_y = \{-78.4\mathbf{j}\}$ lb·ft

**3–50.** $M_x = 15.0$ lb·ft, $M_y = 4.00$ lb·ft,
$M_z = 36.0$ lb·ft

**3–51.** $\mathbf{M}_{AC} = \{11.5\mathbf{i} + 8.64\mathbf{j}\}$ lb·ft

**3–53.** $M_z = 109$ lb·in

**3–54.** $M_{CA} = 226$ N·m

**3–55.** $M_C = 18.3$ kN·m ↖

**3–57.** $M_C = 17.6$ kN·m ↖

**3–58.** $F = 108$ lb

**3–59.** $F = 133$ N, $P = 800$ N

**3–61.** $N = 26.0$ N

**3–62.** $F = 625$ N

**3–63.** $F = 167$ lb
Resultant couple can act anywhere.

**3–65.** $\mathbf{M}_C = \{126\mathbf{k}\}$ lb·ft

**3–66.** $M_C = 126$ lb·ft

**3–67.** $\mathbf{M}_C = \{-360\mathbf{i} + 380\mathbf{j} + 320\mathbf{k}\}$ lb·ft

**3–69.** $\mathbf{M}_R = \{11.0\mathbf{i} - 49.0\mathbf{j} - 40.0\mathbf{k}\}$ lb·ft,
$M_R = 64.2$ lb·ft, $\alpha = 80.1°$,
$\beta = 140°$, $\gamma = 129°$

**3–70.** $M_R = 59.9$ N·m, $\alpha = 99.0°$,
$\beta = 106°$, $\gamma = 18.3°$

**3–71.** $M = 18.3$ N·m, $\alpha = 155°$,
$\beta = 115°$, $\gamma = 90°$

**3–73.** $d = 342$ mm

**3–74.** $F_O = 375$ N, $M_O = 100$ N·m ↓

**3–75.** $F_P = 375$ N, $M_P = 737$ N·m ↖

**3–77.** $F_R = 178$ N, $\theta = 73.0°$ ↗, $M_{R_P} = 2.68$ kN·m ↖

**3–78.** $F_R = 274$ lb, $\theta = 5.24°$ ↘$_\theta$, $M_O = 4.61$ kip·ft ↖

**3–79.** $F_R = 274$ lb, $\theta = 5.24°$ ↘$_\theta$,
$M_P = 5.48$ kip·ft ↖

**3–81.** $F_R = 6.57$ kN, $\theta = 57.4°$ ↙,
$M_{R_P} = 31.0$ kN·m ↖

**3–82.** $F_R = 2.10$ kN, $\theta = 81.6°$ ↗$_\theta$,
$M_O = 10.6$ kN·m ↓

**3–83.** $F_R = 2.10$ kN, $\theta = 81.6°$ ↘$_\theta$,
$M_P = 16.8$ kN·m ↓

**3–85.** $F_R = 375$ lb ↑, $x = 2.47$ ft

**3–86.** $F_R = 5.93$ kN, $\theta = 77.8°$ ↗,
$M_{R_A} = 34.8$ kN·m ↓

**3–87.** $F_R = 5.93$ kN, $\theta = 77.8°$ ↗,
$M_{R_B} = 11.6$ kN·m ↖

**3–89.** $F = 798$ lb, $\theta = 67.9°$ $_\theta$↗, $x = 6.57$ ft

**3–90.** $F = 922$ lb, $\theta = 77.5°$ $_\theta$↗, $x = 3.56$ ft

**3–91.** $F = 1302$ N, $\theta = 84.5°$, $_\theta$↗ $x = 7.36$ m

**3–93.** $F_2 = 25.9$ lb, $\theta = 18.1°$, $F_1 = 68.1$ lb

**3–94.** $F_R = 10.75$ kip ↓, $M_{R_A} = 99.5$ kip·ft ↖

**3–95.** $F_R = 10.75$ kip ↓, $d = 9.26$ ft

**3–97.** $\mathbf{F}_R = \{0.232\mathbf{i} + 5.06\mathbf{j} + 12.4\mathbf{k}\}$ kN,
$\mathbf{M}_{R_O} = \{36.0\mathbf{i} - 26.1\mathbf{j} + 12.2\mathbf{k}\}$ kN·m

**3–98.** $\mathbf{F} = \{270\mathbf{k}\}$ N, $\mathbf{M} = \{-2.22\mathbf{i}\}$ N·m

**3–99.** $F_R = 140$ kN ↓, $y = 7.14$ m, $x = 5.71$ m

**3–101.** $(M_1)_A = 1.44$ kip·ft ↓, $(M_2)_A = 7.20$ kip·ft ↓,
$(M_3)_A = 5.20$ kip·ft ↓, $M_{RA} = 13.8$ kip·ft ↓

**3–102.** $(M_1)_B = 3.96$ kip·ft ↓, $(M_2)_B = 4.80$ kip·ft ↓,
$(M_3)_B = 1.30$ kip·ft ↓, $M_{RB} = 10.1$ kip·ft ↓

**3–103.** $\mathbf{M}_{BA} = \{-3.39\mathbf{i} + 2.54\mathbf{j} - 2.54\mathbf{k}\}$ N·m

**3–105.** $(\mathbf{M}_R)_{x'} = 64.8$ kNm, $(\mathbf{M}_R)_{y'} = 40.5$ kNm

**3–106.** $\mathbf{M}_A = \{-59.7\mathbf{i} - 159\mathbf{k}\}$ N·m

**3–107.** $M_{aa} = 59.7$ N·m

**3–109.** $\mathbf{F}_R = \{-45.9\mathbf{j} - 79.5\mathbf{k}\}$ N,
$(\mathbf{M}_R)_A = \{-23.8\mathbf{j} + 13.8\mathbf{k}\}$ N·m

**3–110.** $\mathbf{F}_R = \{-70\mathbf{i} + 140\mathbf{j} - 408\mathbf{k}\}$ N,
$\mathbf{M}_{RP} = \{-26\mathbf{i} + 357\mathbf{j} + 127\mathbf{k}\}$ N·m

## Chapter 4

**4–1.** $B_y = 586$ N, $F_A = 413$ N

**4–2.** $F_A = 30$ lb, $F_B = 36.2$ lb, $F_C = 9.38$ lb

**4–3.** $F_H = 59.4$ lb, $T_B = 67.4$ lb

**4–5.** $F_{BC} = 1.82$ kip, $F_A = 2.06$ kip

**4–6.** $(N_A)_r = 98.6$ lb, $(N_A)_s = 100$ lb

**4–7.** $(N_B)_r = 10.5$ N, $A_x = 4.20$ N, $A_y = 10.5$ N

**4–9.** $T = 74.6$ lb, $A_x = 33.4$ lb, $A_y = 61.3$ lb

**4–10.** $F_B = 6.38$ N, $A_x = 3.19$ N, $A_y = 2.48$ N

**4–11.** $F_{BC} = 574$ lb, $A_x = 1.08$ kip, $A_y = 637$ lb

**4–13.** $F_B = 3.68$ kN, $x_A = 53.2$ mm,
$x_B = 50.5$ mm

**4–14.** $D_x = 0$, $D_y = 1.65$ kip, $M_D = 1.40$ kip·ft
Remove 800-lb line.

**4–15.** $x = 10$ ft, $A_x = 4.17$ kip, $A_y = 5.00$ kip,
$x = 4$ ft, $A_x = 1.67$ kip, $A_y = 5.00$ kip

**4–17.** $h = 15.8$ ft

**4–18.** $N_A = 81.6$ lb, $F_B = 50.2$ lb

**4–19.** $B_x = 989$ N, $A_x = 989$ N, $B_y = 186$ N

**4–21.** $\theta = 26.4°$

**4–22.** $R_A = 40.9$ kip, $R_B = 125$ kip

**4–23.** $C_x = 333$ lb, $C_y = 722$ lb

**4–25.** $F_{CB} = 782$ N, $A_x = 625$ N, $A_y = 681$ N

**4–26.** $F_2 = 724$ lb, $F_1 = 1.45$ kip, $F_A = 1.75$ kip

**4–27.** $d = \dfrac{3a}{4}$

**4–29.** $F_B = 0.3P$, $F_C = 0.6P$, $x_C = \dfrac{0.6P}{k}$

**4–30.** $k = 11.2$ lb/ft

**4–31.** $R_A = 20.6$ lb, $R_B = 11.9$ lb, $R_C = 63.9$ lb

**4–33.** $F_A = 125$ lb, $M_A = 206$ lb · ft

**4–34.** $A_x = -8$ lb, $A_y = 4$ lb, $A_z = 0$, $(M_A)_y = 44$ lb · ft, $(M_A)_z = 32$ lb · ft

**4–35.** $N_C = 375$ lb, $N_A = N_B = 188$ lb

**4–37.** $P = 75$ lb, $B_z = 75$ lb, $B_x = 112$ lb, $A_y = 0$, $A_x = 37.5$ lb, $A_z = 75$ lb

**4–38.** $B_z = 312$ N, $M_{B_y} = 104$ N · m, $B_x = 180$ N, $A_x = 480$ N, $B_y = 0$, $A_z = 831$ N

**4–39.** $A_y = 168$ lb, $A_z = 368$ lb, $B_x = -358$ lb, $B_y = -168$ lb, $C_x = 358$ lb, $C_z = -168$ lb

**4–41.** $F_C = 27.4$ lb, $N_C = 309$ lb

**4–42.** The pole will remain stationary.

**4–43.** $d = 13.4$ ft

**4–45.** $P = 1$ lb

**4–46.** $N_C = 800$ lb, $N_B = 961$ lb

**4–47.** $T = 3.67$ kip

**4–49.** $P = \dfrac{M_0}{\mu_s ra}(b - \mu_s c)$

**4–51.** $P = \dfrac{M_0}{\mu_s ra}(b + \mu_s c)$

**4–53. a)** No, **b)** Yes

**4–54.** It is possible to pull the load without slipping or tipping.

**4–55.** $\theta = 21.8°$

**4–57.** $P = 100$ lb

**4–58.** $m = 54.9$ kg

**4–59.** $b = \dfrac{h}{\mu_s} + \dfrac{d}{2}$

**4–61.** $F = 22.5$ lb, $\mu_m = 0.15$

**4–62.** $F = 30.4$ lb, $\mu_m = 0.195$

**4–63.** $d = 537$ mm

**4–65.** $\theta = 16.7°$, $\phi = 42.6°$

**4–66.** Car $A$ will not move.

**4–67.** $P = 182$ N

**4–69.** Slipping occurs at $A$.

**4–70.** $\theta = 11.0°$

**4–71.** The semicylinder will not slide. $\theta = 24.2°$.

**4–73.** $L = 3.35$ ft

**4–74.** $1.02°$

**4–75.** $k_B = 2.5$ kN/m

**4–76.** $P = 100$ lb, $B_z = 40$ lb, $B_x = -35.7$ lb, $A_x = 136$ lb, $B_y = 0$, $A_z = 40$ lb

**4–77.** $F_{BD} = 171$ N, $F_{BC} = 145$ N

**4–78.** $T = 1.01$ kN, $F_D = 982$ N

**4–79.** $N_{min} = 561$ N

**4–81.** $T_{BA} = 715$ N, $T_{BC} = 104$ kN, $D_x = 490$ N, $D_y = 654$ N, $D_z = 2.29$ kN

## Chapter 5

**5–1.** $F_{BA} = 286$ lb (T), $F_{BC} = 808$ lb (T), $F_{CA} = 571$ lb (C)

**5–2.** $F_{BA} = 286$ lb (T), $F_{BC} = 384$ lb (T), $F_{CA} = 271$ lb (C)

**5–3.** $F_{AD} = 849$ lb (C), $F_{AB} = 600$ lb (T), $F_{BD} = 400$ lb (C), $F_{BC} = 600$ lb (T), $F_{DC} = 1.41$ kip (T), $F_{DE} = 1.60$ kip (C)

**5–5.** $F_{AE} = 8.94$ kN (C), $F_{AB} = 8.00$ kN (T), $F_{BC} = 8.00$ kN (T), $F_{BE} = 8.00$ kN (C), $F_{EC} = 8.94$ kN (T), $F_{ED} = 17.9$ kN (C)

**5–6.** $F_{AE} = 372$ N (C), $F_{AB} = 332$ N (T), $F_{BC} = 332$ N (T), $F_{BE} = 196$ N (C), $F_{EC} = 558$ N (T), $F_{DC} = 582$ N (T)

**5–7.** $F_{BC} = 3$ kN (C), $F_{AC} = 1.46$ kN (C), $F_{AF} = 4.17$ kN (T), $F_{CD} = 4.17$ kN (C), $F_{CF} = 3.12$ kN (C), $F_{EF} = 0$, $F_{ED} = 13.1$ kN (C), $F_{DF} = 5.21$ kN (T)

**5–9.** $F_{CB} = 8.00$ kN (T), $F_{CD} = 6.93$ kN (C), $F_{DE} = 6.93$ kN (C), $F_{DB} = 4.00$ kN (T), $F_{BE} = 4.00$ kN (C), $F_{BA} = 12.0$ kN (T)

**5–10.** $F_{AG} = 471$ lb (C), $F_{AB} = 333$ lb (T), $F_{BG} = 0$, $F_{BC} = 333$ lb (T), $F_{DE} = 943$ lb (C), $F_{DC} = 667$ lb (T), $F_{EC} = 667$ lb (T), $F_{EG} = 667$ lb (C), $F_{CG} = 471$ lb (T)

**5–11.** $F_{AG} = 1179$ lb (C), $F_{AB} = 833$ lb (T), $F_{BC} = 833$ lb (T), $F_{BG} = 500$ lb (T), $F_{DE} = 1650$ lb (C), $F_{DC} = 1167$ lb (T), $F_{EC} = 1167$ lb (T), $F_{EG} = 1167$ lb (C), $F_{CG} = 471$ lb (T)

**5–13.** $F_{GB} = 30$ kN (T), $F_{AF} = 20$ kN (C), $F_{AB} = 22.4$ kN (C), $F_{BF} = 20$ kN (T), $F_{BC} = 20$ kN (T), $F_{FC} = 28.3$ kN (C), $F_{FE} = 0$, $F_{ED} = 0$, $F_{EC} = 20.0$ kN (T), $F_{DC} = 0$

**5–14.** $F_{AB} = 330$ lb (C), $F_{AF} = 79.4$ lb (T),
$F_{BF} = 233$ lb (T), $F_{BC} = 233$ lb (C),
$F_{FC} = 47.1$ lb (C), $F_{FE} = 113$ lb (T),
$F_{EC} = 300$ lb (T), $F_{ED} = 113$ lb (T),
$F_{CD} = 377$ lb (C)

**5–15.** $F_{AB} = 377$ lb (C), $F_{AF} = 190$ lb (T),
$F_{BF} = 267$ lb (T), $F_{BC} = 267$ lb (C),
$F_{FC} = 189$ lb (T), $F_{FE} = 56.7$ lb (T),
$F_{ED} = 56.7$ lb (T), $F_{EC} = 0$, $F_{CD} = 189$ lb (C)

**5–17.** $F_{GF} = 29.0$ kN (C), $F_{CD} = 23.5$ kN (T),
$F_{CF} = 7.78$ kN (T)

**5–18.** $F_{DF} = 2.00$ kip (C), $F_{GF} = 5.70$ kip (C),
$F_{DE} = 6.40$ kip (T)

**5–19.** $F_{KJ} = 13.3$ kN (T), $F_{BC} = 14.9$ kN (C),
$F_{CK} = 0$

**5–21.** $F_{JI} = 7.50$ kip (T), $F_{EI} = 2.50$ kip (C)

**5–22.** $F_{GF} = 8.08$ kN (T), $F_{BC} = 7.70$ kN (C),
$F_{GC} = 0.770$ kN (C)

**5–23.** $F_{FG} = 8.08$ kN (T), $F_{CD} = 8.47$ kN (C),
$F_{CF} = 0.770$ kN (T)

**5–25.** $F_{BG} = -200\sqrt{L^2 + 9}$, $F_{BC} = -200L$,
$F_{HG} = 400L$

**5–26.** $F_{IC} = 5.62$ kN (C), $F_{CG} = 9.00$ kN (T)

**5–27.** *AB*, *BC*, *CD*, *DE*, *HI*, and *GI* are zero-force members.
$F_{JE} = 9.38$ kN (C), $F_{GF} = 5.625$ kN (T)

**5–29.** $F_{CD} = 11.2$ kN (C), $F_{CF} = 3.21$ kN (T),
$F_{CG} = 6.80$ kN (C)

**5–30.** $F_{GF} = 1.80$ kip (C), $F_{FB} = 693$ lb(T),
$F_{BC} = 1.21$ kip (T)

**5–31.** $F_{GJ} = 2.00$ kip (C)

**5–33.** $F_{GF} = 1.78$ kN (T), $F_{CD} = 2.23$ kN (C),
$F_{CF} = 0$

**5–34.** **a)** $P = 25.0$ lb, **b)** $P = 33.3$ lb,
**c)** $P = 11.1$ lb

**5–35.** $F_B = 61.9$ lb, $F_A = 854$ lb

**5–37.** $R_E = 177$ lb, $R_A = 128$ lb

**5–38.** $P = 40.0$ N, $x = 240$ mm

**5–39.** $P = 21.8$ N, $R_A = 43.6$ N, $R_B = 43.6$ N,
$R_C = 131$ N

**5–41.** $A_y = 9.59$ kip, $B_y = 8.54$ kip, $C_y = 2.93$ kip,
$C_x = 9.20$ kip

**5–42.** $P = 743$ N

**5–43.** $A_y = 300$ N, $A_x = 300$ N, $C_x = 300$ N,
$C_y = 300$ N

**5–45.** $B_y = 1.33$ kN, $B_x = 5.00$ kN,
$A_x = C_x = 5.00$ kN, $A_y = C_y = 6.67$ kN
$M_D = 10.0$ kN·m, $D_y = 8.00$ kN, $D_x = 0$

**5–46.** $C_x = 75$ lb, $C_y = 100$ lb

**5–47.** $A_x = 4.20$ kN, $B_x = 4.20$ kN, $A_y = 4.00$ kN,
$B_y = 3.20$ kN, $C_x = 3.40$ kN, $C_y = 4.00$ kN

**5–49.** $T = 100$ lb, $\theta = 14.6°$

**5–50.** $x = 9.43$ ft

**5–51.** $T = 350$ lb, $A_y = 700$ lb, $A_x = 1.88$ kip,
$D_x = 1.70$ kip, $D_y = 1.70$ kip

**5–53.** $A_x = 80$ lb, $A_y = 80$ lb, $B_y = 133$ lb,
$B_x = 333$ lb, $C_x = 413$ lb, $C_y = 53.3$ lb

**5–54.** $F_{AB} = 9.23$ kN, $C_x = 2.17$ kN, $C_y = 7.01$ kN,
$D_x = 0$, $D_y = 1.96$ kN, $M_D = 2.66$ kN·m

**5–55.** $C_y = 34.4$ lb, $C_x = 16.7$ lb, $B_x = 66.7$ lb,
$B_y = 15.6$ lb

**5–57.** $C_x = D_x = 160$ lb, $C_y = D_y = 107$ lb,
$B_y = 26.7$ lb, $B_x = 80.0$ lb, $E_x = 0$,
$E_y = 26.7$ lb, $A_x = 160$ lb

**5–58.** $F_E = 3.64F$

**5–59.** $A_y = 657$ N, $C_y = 229$ N, $C_x = 0$, $B_x = 0$,
$B_y = 429$ N

**5–61.** $m = 366$ kg, $F_A = 2.93$ kN

**5–62.** $M = 314$ lb·ft

**5–63.** $P = 46.9$ lb

**5–65.** $A_y = 34.0$ N, $A_x = 0$, $C_y = 6.54$ N, $C_x = 0$,
$x = 292$ mm, $B_y = 1.06$ N, $B_x = 0$

**5–66.** $F_{DE} = 1.07$ kN

**5–67.** $C_y = 1.33$ kN, $B_y = 549$ N, $C_x = 2.98$ kN,
$A_y = 235$ N, $A_x = 2.98$ kN, $B_x = 2.98$ kN

**5–69.** $F = 9.42$ lb

**5–70.** $F_{AC} = 2.51$ kip, $F_{AB} = 3.08$ kip,
$F_{AD} = 3.43$ kip

**5–71.** $D_y = 2.5$ kip, $B_y = 12$ kip, $A_y = 1.5$ kip,
$A_x = 0$

**5–73.** $F_{AD} = 990$ lb (C), $F_{AB} = 700$ lb (T),
$F_{DB} = 495$ lb (C), $F_{DC} = 1.48$ kip (C),
$F_{CB} = 1.05$ kip (T)

**5–74.** $F_{HD} = 7.07$ kN (C), $F_{CD} = 50$ kN (T),
$F_{GD} = 5$ kN (T)

**5–75.** $F_{HB} = 21.2$ kN (C), $F_{BC} = 50$ kN (T),
$F_{HI} = 35$ kN (C)

**5–77.** $B_y = 449$ lb, $A_x = 92.3$ lb, $A_y = 186$ lb,
$M_A = 359$ lb·ft

**5–78.** $B_x = 84.9$ lb, $B_y = 84.9$ lb, $A_x = 84.9$ lb,
$A_y = 265$ lb, $M_A = 953$ lb·ft

# Chapter 6

**6–1.** $\bar{x} = 0, \bar{y} = \dfrac{2}{5}$ m

**6–2.** $\bar{x} = 3.20$ ft, $\bar{y} = 3.20$ ft, $T_A = 384$ lb,

$T_C = 384$ lb, $T_B = 1.15$ kip

**6–3.** $\bar{x} = \dfrac{n+1}{2(n+2)} a, \bar{y} = \dfrac{n+1}{2(2n+1)} h$

**6–5.** $\bar{y} = \dfrac{4b}{3\pi}, \bar{x} = \dfrac{4a}{3\pi}$

**6–6.** $\bar{x} = \dfrac{\pi}{2} a, \bar{y} = \dfrac{\pi}{8} a$

**6–7.** $\bar{y} = 2.80$ m, $\bar{x} = 6.00$ m

**6–9.** $\bar{x} = \bar{y} = 0, \bar{z} = \dfrac{4}{3}$ m

**6–10.** $\bar{z} = \dfrac{3}{8} a$

**6–11.** $\bar{z} = \dfrac{5}{6} h$

**6–13.** $\bar{x} = 0.4a$

**6–14.** $\bar{z} = 2.50$ ft

**6–15.** $\bar{y} = 2.67$ m

**6–17.** $\bar{y} = 154$ mm

**6–18.** $\bar{y} = 11.9$ in.

**6–19.** $\bar{x} = \dfrac{\frac{2}{3} r \sin^3\alpha}{\alpha - \frac{\sin 2\alpha}{2}}$

**6–21.** $\bar{y} = 272$ mm

**6–22.** $\bar{x} = 77.2$ mm, $\bar{y} = 31.7$ mm

**6–23.** $\bar{y} = 135$ mm

**6–25.** $\bar{y} = 291$ mm

**6–26.** $h = 323$ mm

**6–27.** $\bar{z} = 128$ mm

**6–29.** $\theta = 53.1°$

**6–30.** $\bar{x} = 4.74$ in., $\bar{y} = 2.99$ in.

**6–31.** $\bar{z} = 101$ mm

**6–33.** $\bar{x} = 1.47$ in, $\bar{y} = 2.68$ in, $\bar{z} = 2.84$ in

**6–34.** $\bar{z} = 0.70$ ft

**6–35.** $h = 2.00$ ft

**6–37.** $F_R = 27$ kN, $x = 2.11$ m

**6–38.** $F_R = 18.0$ kip, $x = 11.7$ ft

**6–39.** $F_R = 13.5$ kip, $x = 6.00$ ft

**6–41.** $b = 9.0$ ft, $a = 7.50$ ft

**6–42.** $F_R = 107$ kN ←, $h = 1.60$ m

**6–43.** $F_R = 1.87$ kip, $x = 3.66$ ft

**6–45.** $F_R = 133$ lb, $\bar{y} = 750$ ft, $\bar{x} = 0$

**6–46.** $F_R = 42.7$ kN, $\bar{y} = 2.40$ m, $\bar{x} = 0$,
$C_z = B_z = 12.8$ kN, $A_x = 0, A_y = 0$,
$A_z = 17.1$ kN

**6–47.** $F_R = 7.62$ kN, $\bar{y} = 3.00$ m, $\bar{x} = 2.74$ m

**6–49.** $I_y = 8.53$ m$^4$

**6–50.** $I_x = 23.8$ ft$^4$

**6–51.** $I_x = 19.5$ in$^4$

**6–53.** $I_x = \dfrac{2}{15} bh^3$

**6–54.** $I_x = 1.54$ in$^4$

**6–55.** $I_y = 0.333$ in$^4$

**6–57.** $I_y = \dfrac{2}{15} hb^3$

**6–58.** $I_x = 10.7$ in$^4$

**6–59.** $I_x = 0.533$ m$^4$

**6–61.** $I_x = 0.571$ in$^4$

**6–62.** $I_y = 1.07$ in$^4$

**6–63.** $\bar{y} = 2.00$ in., $\bar{I}_{x'} = 64.0$ in$^4$

**6–65.** $\bar{x} = 68.0$ mm, $I_{y'} = 36.9(10^6)$ mm$^4$

**6–66.** $I_{x'} = 49.5(10^6)$ mm$^4$

**6–67.** $I_x = 1.21(10^3)$ in$^4$, $I_y = 364.8$ in$^4$

**6–69.** $I_x = 154(10^6)$ mm$^4$, $I_y = 91.3(10^6)$ mm$^4$

**6–70.** $\bar{y} = 80.7$ mm, $\bar{I}_{x'} = 67.6(10^6)$ mm$^4$

**6–71.** $\bar{x} = 61.6$ mm, $\bar{I}_{y'} = 41.2(10^6)$ mm$^4$

**6–73.** $I_{x'} = 222(10^6)$ mm$^4$, $I_{y'} = 115(10^6)$ mm$^4$

**6–74.** $\bar{y} = 22.5$ mm, $\bar{I}_{x'} = 34.4(10^6)$ mm$^4$

**6–75.** $I_{y'} = 122(10^6)$ mm$^4$

**6–77.** $I_y = 1971$ in$^4$

**6–78.** $\bar{y} = 2$ in., $\bar{I}_{x'} = 128$ in$^4$

**6–79.** $I_x = I_y = 503$ in$^4$

**6–81.** $\bar{I}_{y'} = \dfrac{ab \sin\theta}{12} (b^2 + a^2 \cos^2\theta)$

**6–82.** $\bar{y} = 53.0$ mm, $I_{x'} = 3.67(10^6)$ mm$^4$

**6–83.** $I_{x'} = 30.2(10^6)$ mm$^4$

**6–85.** $\bar{I}_{x'} = \frac{1}{36} bh^3$, $\bar{I}_{y'} = \frac{1}{36} hb(b^2 - ab + a^2)$

**6–86.** 90.5 mm

**6–87.** 0.4 m

**6–89.** $\bar{z} = 0.422$ m, $\bar{x} = \bar{y} = 0$

**6–90.** $I_x = 246$ in$^4$, $I_y = 61.5$ in$^4$

**6–91.** 10.8 kN, 2.26 m

**6–93.** $\bar{x} = 3.30$ in., $\bar{y} = 3.30$ in.

**6–94.** $\bar{x} = -0.262$ in., $\bar{y} = 0.262$ in.

**6–95.** $0.0954\ d^4$

# Chapter 7

**7-1.** $V_A = 0$, $N_A = 12.0$ kN, $M_A = 0$,
$V_B = 0$, $N_B = 20.0$ kN, $M_B = 1.20$ kN·m

**7-2.** $N_A = 550$ lb,
$N_B = 250$ lb, $N_C = 950$ lb

**7-3.** $N_A = 5.00$ kN, $N_C = 4.00$ kN, $N_B = 3.00$ kN

**7-5.** $M_C = -15.0$ kip·ft, $N_C = 0$, $V_C = 2.01$ kip,
$M_D = 3.77$ kip·ft, $N_D = 0$, $V_D = 1.11$ kip

**7-6.** $N_C = 0$, $V_C = -1.00$ kip, $M_C = 56.0$ kip·ft,
$N_D = 0$, $V_D = -1.00$ kip, $M_D = 48.0$ kip·ft

**7-7.** $N_C = 0$, $V_C = -386$ lb, $M_C = -857$ lb·ft,
$N_D = 0$, $V_D = 300$ lb, $M_D = -600$ lb·ft

**7-9.** $N_D = \mp 800$ N, $V_D = 0$, $M_D = 1.20$ kN·m

**7-10.** $w = 100$ N/m

**7-11.** $M_C = 48$ kip·ft, $V_C = 6$ kip

**7-13.** $N_D = 0$, $V_D = 800$ lb, $M_D = -1.60$ kip·ft,
$N_C = 0$, $V_C = 0$, $M_C = 800$ lb·ft

**7-14. a)** $N_a = 500$ lb, $V_a = 0$,
**b)** $N_b = 433$ lb, $V_b = 250$ lb

**7-15.** $N_E = -1.92$ kN, $V_E = 800$ N,
$M_E = 2.40$ kN·m

**7-17.** $N_C = -406$ lb, $V_C = 903$ lb,
$M_C = 1.35$ kip·ft

**7-18.** $N_D = -464$ lb, $V_D = -203$ lb, $M_D = 2.61$ kip·ft

**7-19.** $N_C = -30$ kN, $V_C = -8$ kN, $M_C = 6$ kN·m

**7-21.** $N_B = 0$, $V_B = 28.8$ kip, $M_B = -115$ kip·ft

**7-22.** $\dfrac{a}{b} = \dfrac{1}{4}$

**7-23.** $N_C = 20.0$ kN, $V_C = 70.6$ kN,
$M_C = -302$ kN·m

**7-25.** $M_C = -17.8$ kip·ft

**7-26.** $N_D = 0$, $V_D = 0.75$ kip, $M_D = 13.5$ kip·ft,
$N_E = 0$, $V_E = -9$ kip, $M_E = -24.0$ kip·ft

**7-27.** $N_D = 0$, $V_D = 0$, $M_D = 9.00$ kN·m,
$N_E = 0$, $V_E = -7.00$ kN, $M_E = -12.0$ kN·m

**7-29.** $V = 0.293rw_0$, $N = 0.707\, rw_o$,
$M = -0.0783\ r^2 w$

**7-30.** $C_x = -170$ lb, $C_y = -50$ lb, $C_z = 500$ lb
$M_{Cx} = 1$ kip·fit, $M_{Cy} = -900$ lb·ft,
$M_{Cz} = -260$ lb·ft

**7-31.** $(V_C)_x = 104$ lb, $N_C = 0$,
$(V_C)_z = 10.0$ lb,
$(M_C)_x = 20.0$ lb·ft, $(M_C)_y = 72.0$ lb·ft,
$(M_C)_z = -178$ lb·ft

**7-33.** For $0 \le x < 5$ ft: $V = 100$, $M = 100x - 1800$
For $5 < x \le 10$ ft: $V = 100$,
$M = 100x - 1000$

**7-34.** For $0 \le x < 1.5$ ft: $V = -300$ lb,
$M = -300x$ lb·ft,
For $1.5$ ft $< x \le 3$ ft: $V = 300$ lb,
$M = \{300x - 900\}$ lb·ft

**7-35.** For $0 \le x < \dfrac{L}{3}$: $V = 0$, $M = 0$

For $\dfrac{L}{3} < x < \dfrac{2L}{3}$: $V = 0$, $M = M_0$

For $\dfrac{2L}{3} < x \le L$: $V = 0$, $M = 0$

For $0 \le x < \dfrac{8}{3}$ m: $V = 0$, $M = 0$

For $\dfrac{8}{3}$ m $< x < \dfrac{16}{3}$ m: $V = 0$, $M = 500$ N·m

For $\dfrac{16}{3}$ m $< x \le 8$ m: $V = 0$, $M = 0$

**7-37.** $w = 400$ lb/ft

**7-38.** For $0 \le x < 2$ m: $V = 0.250$ kN,
$M = 0.250x$ kN·m

For $2$ m $< x \le 3$ m: $V = \{325 - 1.50x\}$ kN

$M = \{-0.750x^2 + 3.25x - 3.00\}$ kN·m

**7-39.** $V = \{250(10-x)\}$ lb,
$M = \{25(100x - 5x^2 - 6)\}$ lb·ft

**7-41. Member AB:**
For $0 \le x < 12$ ft: $V = \{875 - 150x\}$ lb,
$M = \{875x - 75.0x^2\}$ lb·ft,
For $12 < x \le 14$ ft:
$V = \{2100 - 150x\}$ lb,
$M = \{-75.0x^2 + 2100x - 14700\}$ lb·ft,
**Member CD:**
For $0 \le x < 2$ ft: $V = 919$ lb,
$M = 919x$ lb·ft For $2 < x \le 8$ ft:
$V = 306$ lb, $M = \{2450 - 306x\}$ lb·ft

**7-42.** $V = \frac{w}{4}(3L - 4x)$,
$M = \frac{w}{4}(3Lx - 2x^2 - L^2)$

**7-43.** For $0 \le x < L$: $V = \dfrac{w}{18}(7L - 18x)$,
$M = \dfrac{w}{18}(7Lx - 9x^2)$, For $L < x < 2L$:
$V = \dfrac{w}{2}(3L - 2x)$,
$M = \dfrac{w}{18}(27Lx - 20L^2 - 9x^2)$,
For $2L < x \le 3L$: $V = \dfrac{w}{18}(47L - 18x)$,
$M = \dfrac{w}{18}(47Lx - 9x^2 - 60L^2)$

**7–45.** $w = 22.2$ lb/ft

**7–46.** $x = \dfrac{L}{2}, P = \dfrac{4M_{max}}{L}$

**7–47.** For $0 \le x < 3$ m:
$V = \{ - 650 - 50.0x^2\}$N,
$M = \{ - 650x - 16.7x^3\}$ N $\cdot$ m
For $3$ m $< x \le 7$ m:
$V = \{2100 - 300x\}$ N,
$M = \{-150(7 - x^2)\}$ N $\cdot$ m

**7–49.** $V_x = 1.5$ kip, $V_y = 0$, $V_z = 800(4 - y)$ lb,
$M_x = 400(4 - y)^2$ lb $\cdot$ ft, $M_y = -3$ kip $\cdot$ ft,
$M_z = -1500(4 - y)$ lb $\cdot$ ft

**7–50.** $N = \frac{P}{5}(4 \cos \theta + 3 \sin \theta)$,
$V = \frac{P}{5}(4 \sin \theta - 3 \cos \theta)$,
$M = \frac{Pr}{5}(4 - 4 \cos \theta - 3 \sin \theta)$

**7–51.** $V_x = 0$, $V_z = \{24.0 - 4y\}$ lb,
$M_x = \{2y^2 - 24y + 64.0\}$ lb $\cdot$ ft,
$M_y = 8.00$ lb $\cdot$ ft, $M_z = 0$

**7–70.** For $0 \le x < 3$ m: $V = -1.33x^2$ kN,
$M = \{-0.444x^3 + 24\}$ kN $\cdot$ m
For $3$ m $< x \le 6$ m:
$V = \{-1.33x^2 + 16x - 48\}$ kN,
$M = \{-0.444x^3 + 8x^2 - 48x + 96\}$ kN $\cdot$ m

**7–71.** $N_C = 0$, $V_C = 108$ lb, $M_C = 433$ lb $\cdot$ ft

**7–74.** $N_D = 6.08$ kN, $V_D = 2.60$ kN,
$M_D = 13.0$ kN $\cdot$ m

**7–77. a)** For $0 \le x < 2.75$ in.: $M = (7.5x)$ kip in.
For $2.75$ in. $\le x \le 5.5$ in.:
$M = (-7.5x + 41.2)$ kip in.
**b)** For $0 \le x \le 2$ in.: $M = (7.5x)$ kip in.
For $2$ in. $\le x \le 3.5$ in.:
$M = (-5x^2 + 27.5x - 20)$ kip in.
For $3.5$ in. $\le x \le 5.5$ in.:
$M = (-7.5x + 41.2)$ kip in.

**7–78.** For $0 \le x < 5$ m: $V = \{-2x + 2.5\}$ kN,
$M = \{-x^2 + 2.5x\}$ kN $\cdot$ m
For $5$ m $< x \le 10$ m: $V = -7.5$ kN,
$M = \{-7.5x + 25\}$ kN $\cdot$ m

**7–79.** For $0 < x < 5$ ft: $V = \{-5x - 20\}$ lb,
$M = \{-2.5x^2 - 20x\}$ lb $\cdot$ ft
For $5$ ft $< x < 15$ ft:
$V = \{-5x - 40\}$ lb,
$M = \{-2.5x^2 - 40x + 100\}$ lb $\cdot$ ft

## Chapter 8

**8–1.** $\sigma = 1.82$ MPa

**8–2.** $P_{allow} = 6240$ lb $= 6.24$ kip

**8–3.** $\sigma = 15.4$ psi

**8–5.** $\sigma_{max} = 357$ psi

**8–6.** $\theta = 47.4°$, $F_{AB} = 34.66$ lb, $F_{AC} = 44.37$ lb,
$\sigma_{AC} = 353$ psi, $\sigma_{AB} = 177$ psi

**8–7.** $\sigma_B = 151$ kPa, $\sigma_C = 32.5$ kPa,
$\sigma_D = 25.5$ kPa

**8–9.** $F = 36$ kN, $d = 110$ mm

**8–10.** $\tau_{avg} = 119$ kPa

**8–11.** $(\tau_B)_{avg} = 6053$ psi $= 6.05$ ksi

**8–13.** $(\tau_D)_{avg} = 6621$ psi $= 66.2$ ksi,
$(\tau_E)_{avg} = 6217$ psi $= 6.22$ ksi

**8–14.** $(\tau_D)_{avg} = 13.2$ ksi, $(\tau_E)_{avg} = 12.4$ ksi

**8–15.** $\bar{x} = 4$ in., $\bar{y} = 4$ in., $\sigma = 9.26$ psi

**8–17.** $x = 0.4$ m, $\sigma = 306$ MPa

**8–18.** $\tau_A = 0$, $(\tau_B)_{avg} = 11.0$ ksi

**8–19.** $P = 68.3$ kN

**8–21.** $1.59$ ksi

**8–22.** $\sigma_{AB} = 1.63$ ksi,
$\sigma_{BC} = 0.819$ ksi

**8–23.** $\tau_{avg} = 16$ psi

**8–25.** $22.9$ ksi

**8–26.** $\sigma_{AB} = 10.7$ ksi (T),
$\sigma_{AE} = 8.53$ ksi (C),
$\sigma_{ED} = 8.53$ ksi (C), $\sigma_{EB} = 4.80$ ksi (T),
$\sigma_{BC} = 23.5$ ksi (T), $\sigma_{BD} = 18.7$ ksi (C)

**8–27.** $P = 6.82$ kip

**8–29.** $\sigma = \dfrac{w_0}{2aA}(2a - x)^2$

**8–30.** $\sigma_{max} = 21.7$ MPa, $x = 0$

**8–33.** $\sigma_{AB} = 21.2$ MPa, $\sigma_{BC} = 23.6$ MPa

**8–34.** $\sigma_{AB} = 16.7$ MPa, $\sigma_{BC} = 8.64$ MPa

**8–35.** $h = 2\dfrac{3}{4}$ in.

**8–37.** $d = \dfrac{5}{8}$ in., $h = \dfrac{3}{8}$ in.

**8–38.** $h = 25.0$ mm
$d = 5.93$ mm

**8–39.** $a = 0.253$ in.

**8–41.** $P = 55.0$ kN

**8–42.** $t = 5.33$ mm,
$b = 24.0$ mm,
$a = 4.31$ mm

**8–43.** $d_A = 0.441$ in., $d_B = 0.794$ in.

**8–45.** $P = 0.491$ kip

**8–46.** $(\tau_{avg})_A = 1.53$ ksi,
$(\tau_{avg})_B = 1.68$ ksi

**8–47.** $d_A = 0.155$ in., $d_B = 0.162$ in.

**8–49.** $d_1 = 44.6$ mm, $d_3 = 26.4$ mm, $t = 15.8$ mm

**8–50.** For bolt $A$: $\tau = 23.9$ ksi, F.S. $= 1.05$
For rod $AB$: $\sigma = 30.6$ ksi, F.S. $= 1.24$

**8–51.** $w = 0.452$ kip/ft

**8–53.** $d_B = 611$ mm, $d_w = 15.4$ mm

**8–54.** $\varepsilon_{avg} = 0.250$ in./in.

**8–55.** $\varepsilon = 0.885$ in./in.

**8–57.** $\Delta p = 11.2$ mm

**8–58.** $(\varepsilon_{avg})_{approx} = 1.50\,\theta$,
$(\varepsilon_{avg})_{approx} = 52.4(10^{-3})$ m/m

**8–59.** $\varepsilon_{AB} = 6.50(10^{-3})$ mm/mm

**8–61.** $\varepsilon_{AB} = 0.152$ in./in., $\varepsilon_{AC} = 0.0274$ in./in.

**8–62.** $\varepsilon_{AB} = 16.8(10^{-3})$ m/m

**8–63.** $\Delta L = \dfrac{kL^2}{2}$

**8–65.** $\gamma_{xy} = -0.0200$ rad

**8–66.** $\gamma_{x'y'} = -7.27(10^{-3})$ rad

**8–67.** $\varepsilon_x = -1.25(10^{-3})$ in./in.,
$\varepsilon_y = 2.50(10^{-3})$ in./in.,
$\varepsilon_{x'} = \varepsilon_{y'} = 0.627(10^{-3})$ in./in.

**8–69.** $(\gamma_B)_{xy} = 0.0116$ rad $= 11.6(10^{-3})$ rad,
$(\gamma_A)_{xy} = -0.0116$ rad $= -11.6(10^{-3})$ rad

**8–70.** $(\gamma_C)_{xy} = -11.6(10^{-3})$ rad,
$(\gamma_D)_{xy} = 11.6(10^{-3})$ rad

**8–71.** $\varepsilon_{AC} = 1.60(10^{-3})$ mm/mm,
$\varepsilon_{DB} = 12.8(10^{-3})$ mm/mm

**8–73.** $\varepsilon_{AC} = 0.01665$ mm/mm $= 16.7(10^{-3})$ mm/mm,
$\varepsilon_{BD} = 0.01134$ mm/mm $= 11.3(10^{-3})$ mm/mm

**8–74.** $(\gamma_B)_{xy} = 5.24(10^{-3})$ rad

**8–75.** $\varepsilon_{AB} = 1.61(10^{-3})$ mm/mm
$\varepsilon_{CD} = 126(10^{-3})$ mm/mm

**8–77.** $\varepsilon_{AB} = \dfrac{v_B \sin\theta}{L} - \dfrac{u_A \cos\theta}{L}$

**8–78.** $\sigma_{AB} = 417$ psi (C), $\sigma_{BC} = 469$ psi (T),
$\sigma_{AC} = 833$ psi (T)

**8–79.** 267 kPa

**8–81.** $N_D = 17.6$ kN, $V_D = 2.61$ kN,
$M_D = 5.22$ kN·m, $N_E = 10.1$ kN,
$V_E = 4.89$ kN, $M_E = 15.3$ kN·m

**8–82.** $N_C = 0$, $V_C = 3.50$ kip, $M_C = 47.5$ kip·ft,
$N_D = 0$, $V_D = 0.240$ kip, $M_D = 0.360$ kip·ft

**8–83.** $N_B = 0$, $V_{BR} = 22.5$ kN, $M_B = 102$ kN·m,
$V_{BL} = 62.1$ kN

**8–85.** $\tau = 7.50$ MPa

**8–86.** $\varepsilon_{BE} = 3.82(10^{-3})$ in./in.,
$\varepsilon_{CF} = 7.31(10^{-3})$ in./in.,
$\varepsilon_{AD} = 0.333(10^{-3})$ in./in.

**8–87.** $\gamma_{xy} = 0.00125$ rad

# Chapter 9

**9–1.** $E_{approx} = 3.275(10^3)$ ksi

**9–2.** $E = 55.3(10^3)$ ksi, $u_r = 9.96\,\dfrac{\text{in} \cdot \text{lb}}{\text{in}^3}$

**9–3.** $(u_t)_{approx} = 85.0\,\dfrac{\text{in} \cdot \text{lb}}{\text{in}^3}$

**9–5.** $(E)_{approx} = 229$ GPa, $(\sigma_u)_{approx} = 528$ MPa,
$(\sigma_f)_{approx} = 479$ MPa

**9–6.** $(u_t)_{approx} = 117$ MJ/m$^3$

**9–7.** The amount of elastic recovery $= 0.00350$ in.,
Permanent elongation $= 0.1565$ in.

**9–9.** $(u_t)_{approx} = 18.0\,\dfrac{\text{in} \cdot \text{kip}}{\text{in}^3}$, $u_r = 20$ in. · lb/in$^3$

**9–10.** $E_{approx} = 173$ GPa, $\sigma_{pl} = 260$ MPa,
$\sigma_u = 400$ MPa, $u_r = 195$ kJ/m$^3$,
Elastic recovery $= 0.00208$ mm/mm,
Permanent set $= 0.0729$ mm/mm

**9–11.** $L = 50.0123$ in.

**9–13.** $\Delta P = 50.05$ kip

**9–14.** copolymer

**9–15.** $E = 28.6(10^3)$ ksi

**9–17.** $A = 0.209$ in$^2$, $P = 1.62$ kip

**9–18.** $d_{AB} = 3.54$ mm, $d_{AC} = 3.23$ mm,
$L_{AB} = 750.49$ mm

**9–19.** $\delta L = 0.778$ mm, $\delta d = -0.0264$ mm

**9–21.** $d = 1.5034$ in.

**9–22.** $\delta L = -0.0173$ mm, $d = 20.00162$ mm

**9–23.** $\nu = 0.300$

**9–25.** $L = 50.0377$ mm, $d = 12.99608$ mm

**9–26.** $\delta = \dfrac{Pa}{2bhG}$

**9–27.** $\delta = \dfrac{P}{2\pi hG}\ln\dfrac{r_o}{r_i}$

**9–29.** $L = 2.792$ in.

**9–30.** $(\Delta L)_{8mm} = 1.19$ mm, $(\Delta L)_{6mm} = 2.12$ mm

**9–31.** $x = 1.53$ m, $d'_A = 30.00782$ mm

# Chapter 10

**10–1.** $\delta_A = -3.64(10^{-3})$ mm

**10–2.** $\delta_A = -0.0603$ in.

**10–3.** $P_1 = 70.7$ kip, $P_2 = 141$ kip

**10–5.** $\delta_{A/C} = 0.0298$ in.

**10–6.** $\delta_B = -0.00632$ in., $\delta_A = 0.0670$ in.

**10–7.** $P_1 = 16.4$ kip, $P_2 = 35.3$ kip

**10–9.** $\delta_C = 0.00843$ in., $\delta_E = 0.00169$ in.,
$\delta_B = 0.0333$ in.

**10–10.** $\delta_{C_y} = 0.150$ mm

**10–11.** $P = 13.3$ kN

**10–13.** $\alpha = 0.00341°$, $\beta = 0.0295°$

**10–14.** $\delta_B = 0.0311$ in.

**10–15.** $P = 15.4$ kip

**10–17.** $W = 9.69$ kN

**10–18.** $\delta_F = 0.0113$ in.

**10–19.** $0.00878°$

**10–21.** $x = 4.24$ ft, $w = 1.02$ kip/ft

**10–22.** $P = 6.80$ kip

**10–23.** $P = 11.8$ kip

**10–25.** $A_{st} = 18.2$ in$^2$, $\delta = 0.00545$ in.

**10–26.** $\sigma_{al} = 27.5$ MPa, $\sigma_{st} = 79.9$ MPa

**10–27.** $\sigma_{st} = 65.9$ MPa, $\sigma_{con} = 8.24$ MPa

**10–29.** $F_C = \dfrac{P}{3}$, $F_A = \dfrac{2}{3}P$

**10–30.** $F_B = \dfrac{P}{3}$, $F_D = \dfrac{2}{3}P$

**10–31.** $P_s = 35.61$ kN, $P_b = 14.4$ kN

**10–33.** $d_{AC} = 1.79$ mm

**10–34.** $\sigma_{AB} = \dfrac{7P}{12A}$, $\sigma_{CD} = \dfrac{P}{3A}$, $\sigma_{EF} = \dfrac{P}{12A}$

**10–35.** $F_A = F_B = 40.0$ kN

**10–37.** $A'_1 = \left(\dfrac{E_1}{E_2}\right) A_1$

**10–38.** $F = 4.20$ kN

**10–39.** $F = 116$ kip

**10–41.** $\delta = 0.348$ in., $F = 19.5$ kip

**10–42.** $F = 2.54$ kip, $F = 2.54$ kip

**10–43.** $F = 1.68$ kip

**10–45.** $\sigma_{br} = \sigma_{al} = 9.77$ ksi, $\delta_B = 0.611(10^{-3})$ in. $\rightarrow$

**10–46.** $T_2 = 116°F$, $\sigma_{al} = 11.6$ ksi

**10–47.** $\sigma_{al} = 25.4$ ksi, $L'_{al} = 8.00462$ in.

**10–49.** $F = 7.60$ kip

**10–50.** $F_{AC} = 10.0$ lb, $F_{AD} = 136$ lb

**10–51.** $\sigma_s = 40.1$ MPa, $\sigma_b = 29.5$ MPa

**10–53.** $F = 2.39$ kip

**10–54.** $\Delta_{A/B} = 0.491$ mm,

**10–55.** $\Delta = 0.129$ mm, $h' = 49.9988$ mm,
$w' = 59.9986$ mm

**10–57.** $\sigma_B = 8.90$ ksi, $\Delta_{A/B} = 0.307(10^{-3})y$ ft

# Chapter 11

**11–1. a)** $T = 7.95$ kip$\cdot$in., **b)** $T' = 6.38$ kip$\cdot$in.

**11–2. a)** $r' = 0.707\, r$, **b)** $r' = 0.707\, r$

**11–3.** $\tau_{max} = 6.62$ ksi

**11–5.** $\tau_C = 37.7$ MPa, $\tau_D = 75.5$ MPa

**11–6.** $\tau_{AB} = 7.82$ ksi, $\tau_{BC} = 2.36$ ksi

**11–7.** $(\tau_{BC})_{max} = 5.07$ ksi, $(\tau_{DE})_{max} = 3.62$ ksi

**11–9.** $d = 2\frac{1}{2}$ in.

**11–10.** $\tau_A = 1.72$ ksi, $\tau_B = 3.02$ ksi

**11–11.** $\tau_{max}^{abs} = 3.59$ ksi

**11–13.** $\tau_{max}^{abs} = 49.7$ MPa

**11–14.** $d = 1\dfrac{1}{4}$ in.

**11–15.** $\tau_{max} = 216$ psi

**11–17.** $d = \dfrac{1}{2}$ in.

**11–18.** $n = \dfrac{2\,r^3}{Rd^2}$

**11–19.** $\tau_{avg} = 1.17$ MPa

**11–21.** $\omega = 130$ rpm

**11–22.** $\tau_{max} = 6018$ psi $= 6.02$ ksi

**11–23.** $T_0 = 670$ N$\cdot$m, $\tau_{max}^{abs} = 6.66$ MPa

**11–25.** $t = 0.104$ in.

**11–26.** Use $d = 1\frac{3}{8}$ in.

**11–27.** $\tau_{max} = 44.3$ MPa, $\phi = 11.9°$

**11–29.** $\tau_{max} = 2.83$ ksi, $\phi = 4.43°$

**11–30.** $\phi_{B/A} = 0.578°$

**11–31.** $\phi_{A/D} = 0.638°$

**11–33.** $t = 0.00753$ m $= 7.53$ mm

**11–34.** $\omega = 131$ rad/s

**11–35.** $\tau_{max}^{abs} = 65.2$ MPa, $\phi_{C/F} = 11.4°$

**11–37.** $\phi_C = 0.113°$

**11–38.** $\phi_{D/C} = 0.0823$ rad, $\tau_{max}^{abs} = 34.0$ ksi

**11–39.** $\phi_A = 2.70°$

**11–41.** $T = 2.25$ kip$\cdot$in

**11–42.** $\phi_A = 0.432°$

**11–43.** $\phi = \dfrac{7TL}{12\pi r^4 G}$

**11–44.** $\theta_{A/D} = 8.96°$

**11–46.** $\theta = \dfrac{T}{4\pi\, hG}\left[\dfrac{1}{r_i^2} - \dfrac{1}{r_o^2}\right]$

**11–47.** $(\tau_{AC})_{max} = 14.3$ MPa, $(\tau_{CB})_{max} = 9.55$ MPa

**11–49.** $F = 23.4$ lb

**11–50.** $\tau_{max} = 389$ psi

**11–51.** $T_D = 385$ N$\cdot$m, $T_A = 115$ N$\cdot$m

**11–53.** $\tau_{max}^{abs} = 3.67$ ksi

**11–54.** $T_A = 19.4$ lb·ft, $T_B = 481$ lb·ft

**11–55.** $T = 1.29$ kip·ft

**11–57.** $T_B = 222$ N·m, $T_A = 55.6$ N·m

**11–58.** $\phi_E = 1.66°$

**11–59.** $\phi_C = 0.002019$ rad $= 0.116°$,
$(\tau_{st})_{max\ BC} = 394.63$ psi $= 395$ psi (max),
$(\gamma_{st})_{max} = 34.3(10^{-6})$ rad,
$(\tau_{br})_{max} = 96.1$ psi (max),
$(\gamma_{br})_{max} = 17.2(10^{-6})$ rad

**11–61.** $T' = 510$ N·m

**11–62.** $d = 4.39$ mm

**11–63.** 6.67% for both stress and twist

**11–65.** $\phi_{AB} = \dfrac{T}{2\pi aG}(1 - e^{-4ab})$

**11–66.** $\tau_{max} = \dfrac{T}{2\pi r^2 ih}$

## Chapter 12

**12–1.** $\%\left(\dfrac{M'}{M}\right) = 84.6\%$

**12–2.** $M = 36.5$ kN·m, $\sigma_{max} = 40.0$ MPa

**12–3.** $\sigma_{max} = 74.7$ MPa,
% of effectiveness $= 53.0\%$

**12–5.** $(\sigma_t)_{max} = 3.72$ ksi, $(\sigma_c)_{max} = 1.78$ ksi

**12–6.** $F = 3296.5$ lb $= 3.30$ kip

**12–7.** $F = 5494.19$ lb $= 5.49$ kip

**12–9.** $(\sigma_t)_{max} = 6.71$ MPa, $(\sigma_c)_{max} = 3.61$ MPa

**12–10.** $\sigma_{max} = 86.5$ psi

**12–11.** $F = 846$ lb

**12–13.** $\sigma_{max} = 95.0$ MPa in design
(b), 38.4% more effective

**12–14.** $\sigma_{max} = 244$ MPa

**12–15.** $\sigma_{max} = 4.42$ ksi

**12–17.** $d = 31.3$ mm

**12–18.** $P = 1.67$ kN

**12–19.** $\sigma_{max} = 9.00$ MPa

**12–21.** $d = 1.28$ in.

**12–22.** $a = 160$ mm

**12–23.** $P = 5.97$ kip

**12–25.** $\varepsilon_{max} = 0.711(10^{-3})$ mm/mm

**12–26.** $\sigma_{max} = 158$ MPa

**12–27.** $d = 86.3$ mm

**12–29.** $\sigma_A = -119$ kPa, $\sigma_B = 446$ kPa,
$\sigma_D = -446$ kPa, $\sigma_E = 119$ kPa

**12–30.** $\bar{y} = 4.83$ in., $M = 75.7$ kip·ft

**12–31.** $\bar{y} = 4.83$ in., $\sigma_A = 2.35$ ksi (C), $\sigma_B = 2.62$ ksi

**12–33.** $\bar{z} = 36.6$ mm, $\sigma_A = 4.38$ MPa,
$\sigma_B = -1.13$ MPa, $\sigma_D = -1.977$ MPa,
$\sigma_E = 5.23$ MPa

**12–34.** $P = 14.2$ kN

**12–35.** $(\sigma_A)_{max} = 7.60$ MPa (T and C)

**12–37.** $\sigma_{max} = 66.8$ ksi

**12–38.** $\sigma_{max} = 161$ MPa

**12–39.** $d = 28.9$ mm

**12–41.** $d = 62.9$ mm

**12–42.** $\sigma_{max} = 5.0$ ksi, $F_A = 17.7$ kip, $F_B = 13.7$ kip

**12–43.** $M_{D_M} = 22.6\%$

**12–45.** $M = 1.25$ kip·ft

**12–46.** $(\sigma_{max})_t = 18.4$ ksi, $(\sigma_{max})_c = 19.2$ ksi

**12–47.** $\sigma_A = 36.7$ ksi, $\sigma_B = 25.9$ ksi

**12–49.** $F_R = 70.8$ kip

**12–50.** $\sigma_{max} = 12.2$ ksi

**12–51.** $w_2 = 800$ lb/in., $w_1 = 533$ lb/in.,
$\sigma_{max} = 45.1$ ksi

## Chapter 13

**13–1.** $\tau_A = 74.7$ MPa

**13–2.** $\tau_B = 98.7$ MPa

**13–3.** $\tau_{max} = 0.499$ ksi, $(\tau_{AB})_F = 0.166$ ksi,
$(\tau_{AB})_W = 0.498$ ksi

**13–5.** $Q_{max} = 88.23$ in$^4$, $\tau_{max} = 276$ psi,
shear stress jump $= 156$ psi

**13–6.** $V_f = 3.05$ kip

**13–7.** $\tau_B = 4.41$ MPa

**13–9.** $V = 9.96$ kip

**13–10.** $V = 32.1$ kip

**13–11.** $\tau_{max} = 4.22$ MPa

**13–13.** $\tau_{max} = 3.16$ MPa, $(\tau_A)_w = 1.87$ MPa,
$(\tau_A)_f = 1.24$ MPa

**13–14.** $P = 80.1$ kip

**13–15.** $\tau_{max} = 99.8$ psi

**13–17.** $\tau_{max} = 281$ psi

**13–18.** $w = 5.69$ kip/ft

**13–19.** $w = 13.3$ kip/ft, $\tau_{max} = 625$ psi

**13–21.** $V_f = 76.2$ lb

**13–22.** $F_t = 512$ lb

**13–23.** $V = 4.97$ kip, $s_{top} = 1.14$ in, $s_{bot} = 1.36$ in.

**13–25.** $F = 5.31$ kN

**13–26.** $\tau_B = 646$ psi, $\tau_A = 592$ psi

**13–27.** $F = 12.5$ kN

**13–29.** $V = 34.5$ kip

**13–30.** $P = 238$ N

**13–31.** $\tau_{nail} = 159$ MPa

**13–34.** $\tau = \dfrac{4V(c^2 - y^2)}{3\pi c^4}$

**13–35.** $\tau_{\max} = 4.48$ ksi

**13–37.** $F = 1.47$ kip

**13–38.** $V = 28.8$ kip

**13–39.** $\tau_{\text{avg}} = 5.06$ MPa, $\tau_{\max} = 7.38$ ksi

## Chapter 14

**14–1.** $t = 18.75$ mm

**14–2.** $\sigma_1 = 600$ psi, $\sigma_2 = 0$

**14–3.** $\sigma_1 = 600$ psi, $\sigma_2 = 300$ psi

**14–5.** $T = 18.2$ kip·ft, $P = 18.1$ kip, $F = 9.05$ kip

**14–6.** $p = 25$ psi, $\delta = 0.00140$ in.

**14–7.** $\sigma_1 = 5.06$ MPa, $\sigma_2 = 2.53$ MPa

**14–9.** $\delta r_i = \dfrac{pr_i^2}{E(r_o - r_i)}$

**14–10. a)** $\sigma_1 = 127$ MPa, **b)** $\sigma_1' = 63.3$ MPa, **c)** $(\tau_{\text{avg}})_b = 258$ MPa

**14–11.** $\sigma_{\max} = 44.0$ ksi (T)

**14–13.** 2.13 ksi (T), 1.07 ksi (C)

**14–14.** $\sigma_A = 0.533$ ksi (T), $\sigma_B = 1.07$ ksi (C), $\tau_A = 0.600$ ksi, $\tau_B = 0$

**14–15.** $w = 79.7$ mm

**14–17.** $\sigma_{\max} = 26.7$ MPa (C), $\sigma_{\min} = 13.3$ MPa (T)

**14–18.** $\sigma_{\max} = 53.3$ MPa (C), $\sigma_{\min} = 40.0$ MPa (T)

**14–19.** $d = 13.1$ mm

**14–21.** $\sigma_A = 0.318$ MPa, $\tau_A = 0.735$ MPa

**14–22.** $\sigma_B = -21.7$ MPa, $\tau_B = 0$

**14–23.** $(\sigma_C)_{\max} = 11.0$ MPa, $(\sigma_C)_{\min} = 8.33$ MPa

**14–25.** $T = 2.16$ kip

**14–26.** $T = 2.16$ kip

**14–27.** $\sigma_B = 235.1$ psf (C). The chimney is *safe*.

**14–29.** $\sigma_A = 25.0$ psi (C), $\sigma_B = 75.0$ psi (C)

**14–30.** $\sigma_A = 25.0$ psi (C), $\sigma_B = 75.0$ psi (C), $\sigma_C = 25.0$ psi (C), $\sigma_D = 25.0$ psi (T)

**14–31.** $\sigma_D = -88.0$ MPa, $\tau_D = 0$

**14–33.** $\sigma_A = 0.444$ MPa (T), $\tau_A = 0.217$ MPa

**14–34.** $\sigma_B = 0.522$ MPa (C), $\tau_B = 0$

**14–35.** $\sigma_A = 107$ MPa, $\tau_A = 15.3$ MPa, $\sigma_B = 0$, $\tau_B = 14.8$ MPa

**14–37.** $\sigma_A = 17.7$ MPa (T), $(\tau_{xz})_A = 4.72$ MPa, $(\tau_{xy})_A = 0$

**14–38.** $\sigma_B = 81.3$ MPa (C), $(\tau_{xz})_B = 2.36$ MPa, $(\tau_{xy})_B = 0$

**14–39.** $\sigma_C = 103$ MPa (C), $(\tau_{xy})_C = 3.54$ MPa, $(\tau_{xz})_C = 0$

**14–41.** $(\sigma_D)_y = -126$ psi, $(\tau_D)_{yx} = -57.2$ psi, $(\sigma_E)_y = -347$ psi, $(\tau_E)_{yz} = -66.4$ psi

**14–42.** $\sigma_A = 4.37$ MPa (C), $\sigma_B = 0.318$ MPa (C), $\tau_A = 0$, $\tau_B = 0.477$ MPa

**14–43.** $\sigma_A = 10.1$ ksi (T), $\sigma_B = 10.3$ ksi (C), $\tau_A = \tau_B = 0$

**14–45.** $\sigma_A = 27.3$ ksi (T), $\sigma_B = 0.289$ ksi (T), $\tau_A = 0$, $\tau_B = 0.750$ ksi

**14–46.** $(\sigma_A)_y = 16.2$ ksi, $(\tau_A)_{yx} = 2.84$ ksi, $(\tau_A)_{yz} = 0$

**14–47.** $(\sigma_B)_y = 7.80$ ksi, $(\tau_B)_{yz} = 3.40$ ksi, $(\tau_B)_{yx} = 0$

**14–49.** $(\sigma_{\max})_t = 106$ MPa, $(\sigma_{\max})_c = 159$ MPa

**14–50.** $(\sigma_{\max})_t = 228$ MPa, $(\sigma_{\max})_c = 168$ MPa

## Chapter 15

**15–2.** $\sigma_{x'} = -4.05$ ksi, $\tau_{x'y'} = -0.404$ ksi

**15–3.** $\sigma_{x'} = -4.05$ ksi, $\tau_{x'y'} = -0.404$ ksi

**15–5.** $\sigma_x = -2.71$ ksi, $\tau_{x'y'} = 4.17$ ksi

**15–6.** $\sigma_{x'} = -388$ psi, $\tau_{x'y'} = 455$ psi

**15–7.** $\sigma_{x'} = -388$ psi, $\tau_{x'y'} = 455$ psi

**15–9.** $\sigma_{x'} = 329$ psi, $\sigma_{y'} = -28.9$ psi, $\tau_{x'y'} = -69.9$ psi

**15–10. a)** $\sigma_1 = 10.1$ ksi, $\sigma_2 = -4.07$ ksi, $\theta_{p_1} = -40.9°$, $\theta_{p_2} = 49.1°$ **b)** $\tau_{\text{in-plane}}^{\max} = 7.07$ ksi, $\theta_s = 4.07°$ and $-85.9°$, $\sigma_{\text{avg}} = 3.00$ ksi

**15–11.** $\sigma_{x'} = -19.9$ ksi, $\tau_{x'y'} = 7.70$ ksi, $\sigma_{y'} = 9.89$ ksi

**15–13. a)** $\sigma_1 = 474$ psi, $\sigma_2 = -1034$ psi, $\theta_{p1} = -55.9°$, $\theta_{p2} = 34.1°$, **b)** $\tau_{\text{in-plane}}^{\max} = 754$ psi $\theta_s = -10.9°$ and $79.1°$, $\sigma_{\text{avg}} = -280$ psi

**15–14.** $\sigma_1 = 40.0$ psi, $\sigma_2 = -40.0$ psi

**15–15.** $\tau_{\text{in-plane}}^{\max} = 0.500$ MPa, $\sigma_{\text{avg}} = 3.50$ MPa

**15–17.** $\sigma_x = 33.0$ MPa, $\sigma_y = 137$ MPa, $\tau_{xy} = -30$ MPa

**15–18.** $\tau_a = -1.96$ ksi, $\sigma_1 = 80.1$ ksi, $\sigma_2 = 19.9$ ksi

**15–19.** $\sigma_b = 7.46$ ksi, $\sigma_1 = 8.29$ ksi, $\sigma_2 = 2.64$ ksi

**15–21.** $\sigma_1 = 2.29$ MPa, $\sigma_2 = -7.20$ kPa, $\theta_p = -3.21°$

**15–22.** Point A: $\sigma_1 = \sigma_x = 0$, $\sigma_2 = \sigma_y = -192$ MPa, Point B: $\sigma_1 = 24.0$ MPa, $\sigma_2 = -24.0$ MPa, $\theta_{p_1} = -45.0°$, $\theta_{p_2} = 45.0°$

**15–23.** Point C: $\sigma_1 = \sigma_y = 256$ MPa, $\sigma_2 = \sigma_x = 0$ Point D: $\sigma_1 = 0.680$ MPa, $\sigma_2 = -154$ MPa, $\theta_{p_1} = 3.80°$, $\theta_{p_2} = -86.2°$

**15–25.** Point A: $\sigma_1 = \sigma_x = 152$ MPa, $\sigma_2 = \sigma_y = 0$, Point B: $\sigma_1 = 0.229$ MPa, $\sigma_2 = -196$ MPa, $\theta_{p_1} = -88.0°$, $\theta_{p_2} = 1.96°$

**15–26.** Point $A$: $\sigma_1 = \sigma_y = 0$, $\sigma_2 = \sigma_x = -87.1$ MPa,
Point $B$: $\sigma_1 = \sigma_x = 93.9$ MPa, $\sigma_2 = \sigma_y = 0$,
$(\tau_{\text{in-plane}}^{\max})_A = 43.6$ MPa, $(\tau_{\text{in-plane}}^{\max})_B = 47.0$ MPa

**15–27.** Point $A$: $\sigma_1 = \sigma_x = 13.7$ MPa, $\sigma_2 = \sigma_y = 0$,
Point $B$: $\sigma_1 = \sigma_y = 0$, $\sigma_2 = \sigma_x = -13.0$ MPa,
$(\tau_{\text{in-plane}}^{\max})_A = 6.83$ MPa, $(\tau_{\text{in-plane}}^{\max})_B = 6.50$ MPa

**15–29.** Point $A$: $\sigma_1 = 150$ MPa, $\sigma_2 = -1.52$ MPa,
Point $B$: $\sigma_1 = 1.60$ MPa, $\sigma_2 = -143$ MPa

**15–30.** $\sigma_1 = \dfrac{2}{\pi d^2}\left(-F + \sqrt{F^2 + \dfrac{64T_0^2}{d^2}}\right)$,

$\sigma_2 = -\dfrac{2}{\pi d^2}\left(F + \sqrt{F^2 + \dfrac{64T_0^2}{d^2}}\right)$,

$\tau_{\text{in-plane}}^{\max} = \dfrac{2}{\pi d^2}\sqrt{F^2 + \dfrac{64T_0^2}{d^2}}$

**15–31. a)** $\sigma_{\text{avg}} = 100$ psi, $R = 700$ psi
**b)** $\sigma_{\text{avg}} = -1.00$ ksi, $R = 1.00$ ksi
**c)** $\sigma_{\text{avg}} = 0$, $R = 20.0$ MPa

**15–33. a, b)** $\sigma_{\text{avg}} = 0$, $R = 30.0$ ksi

**15–35.** $\sigma_1 = 1.50$ ksi, $\sigma_2 = -0.0235$ ksi

**15–37.** $\sigma_1 = 141$ MPa, $\sigma_2 = 9.05$ MPa,
$\tau_{\text{in-plane}}^{\max} = 66.0$ MPa

**15–38.** $\sigma_{x'} = 500$ MPa, $\tau_{x'y'} = 167$ MPa

**15–39.** $\sigma_1 = 29.4$ ksi, $\sigma_2 = -17.0$ ksi

**15–41. a)** $\sigma_{\max} = 6$ ksi, $\sigma_{\text{int}} = \sigma_{\min} = 0$
**b)** $\sigma_{\max} = 50$ MPa, $\sigma_{\text{int}} = 0$, $\sigma_{\min} = -40$ MPa
**c)** $\sigma_{\max} = 600$ psi, $\sigma_{\text{int}} = 200$ psi,
$\sigma_{\min} = 100$ psi
**d)** $\sigma_{\max} = 0$, $\sigma_{\text{int}} = -7$ ksi, $\sigma_{\min} = -9$ ksi
**e)** $\sigma_{\max} = \sigma_{\text{int}} = \sigma_{\min} = -30$ MPa

**15–42.** For $x$–$y$ plane: $\tau_{\text{in-plane}}^{\max} = 4$ ksi, $\sigma_{\text{avg}} = 11$ ksi
For $x$–$z$ plane: $\tau_{\text{in-plane}}^{\max} = 2$ ksi, $\sigma_{\text{avg}} = 13$ ksi
For $y$–$z$ plane: $\tau_{\text{in-plane}}^{\max} = 2$ ksi, $\sigma_{\text{avg}} = 9$ ksi

**15–43.** $\sigma_{\max} = 7.00$ ksi, $\sigma_{\text{int}} = 5.00$ ksi,
$\sigma_{\min} = -5.00$ ksi, $\tau_{\max}^{\text{abs}} = 6.00$ ksi

**15–45.** For $x$–$y$ plane: $\tau_{\max} = 6.0$ ksi, $\sigma_{\text{avg}} = -2.0$ ksi
For $x$–$z$ plane: $\tau_{\max} = 3.0$ ksi, $\sigma_{\text{avg}} = 7.0$ ksi
For $y$–$z$ plane: $\tau_{\max} = 9.0$ ksi, $\sigma_{\text{avg}} = 1.0$ ksi

**15–46.** $\sigma_1 = \sigma_2 = \sigma_3 = -p$

**15–47.** $\sigma_{\max} = 0.615$ kPa, $\sigma_{\text{int}} = 0$,
$\sigma_{\min} = -4.62$ MPa, $\tau_{\max}^{\text{abs}} = 2.31$ MPa

**15–49.** $\sigma_{\max} = 48.8$ ksi, $\sigma_{\text{int}} = 0$,
$\sigma_{\min} = -25.4$ ksi, $\tau_{\max}^{\text{abs}} = 37.1$ ksi, $\theta_s = 9.22°$

**15–50.** $\sigma_{\max} = 34.7$ ksi, $\sigma_{\text{int}} = 0$
$\sigma_{\min} = -34.7$ ksi, $\tau_{\max}^{\text{abs}} = 34.7$ ksi

**15–53.** $\epsilon_{x'} = -116(10^{-6})$,
$\epsilon_{y'} = 466(10^{-6})$,
$\gamma_{x'y'} = 393(10^{-6})$

**15–54.** $\epsilon_{x'} = 466(10^{-6})$, $\epsilon_{y'} = -116(10^{-6})$,
$\gamma_{x'y'} = -393(10^{-6})$

**15–55.** $\epsilon_{x'} = 649(10^{-6})$, $\gamma_{x'y'} = -85.1(10^{-6})$,
$\epsilon_{y'} = 201(10^{-6})$

**15–57. a)** $\epsilon_1 = 1039(10^{-6})$, $\epsilon_2 = 291(10^{-6})$,
$\theta_{p1} = 30.2°$, $\theta_{p2} = 120°$
**b)** $\gamma_{\text{in-plane}}^{\max} = 748(10^{-6})$, $\epsilon_{\text{avg}} = 665(10^{-6})$,
$\theta_s = -14.8°$ and $75.2°$, $\gamma_{x'y'} = 748(10^{-6})$

**15–58. a)** $\epsilon_1 = 441(10^{-6})$, $\epsilon_2 = -641(10^{-6})$,
$\theta_{p1} = -24.8°$, $\theta_{p2} = 65.2°$,
**b)** $\gamma_{\text{in-plane}}^{\max} = 1082(10^{-6})$, $\epsilon_{\text{avg}} = -100(10^{-6})$,
$\theta_s = 20.2°$ and $-69.8°$, $\gamma_{x'y'} = -1082(10^{-6})$

**15–59. a)** $\epsilon_1 = 385(10^{-6})$, $\epsilon_2 = 195(10^{-6})$
$\theta_{p1} = 54.2°$, $\theta_{p2} = -35.8°$,
**b)** $\gamma_{\text{in-plane}}^{\max} = 190(10^{-6})$, $\theta_s = 9.22°$, $-80.8°$,
$\epsilon_{\text{avg}} = 290(10^{-6})$

**15–62.** $\epsilon_{x'} = -309(10^{-6})$, $\epsilon_{y'} = -541(10^{-6})$,
$\gamma_{x'y'} = -423(10^{-6})$

**15–63.** $\epsilon_{x'} = -116(10^{-6})$, $\gamma_{x'y'} = 393(10^{-6})$,
$\epsilon_{y'} = 466(10^{-6})$

**15–65.** $\epsilon_{x'} = 649(10^{-6})$, $\gamma_{x'y'} = -85.1(10^{-6})$
$\epsilon_{y'} = 201(10^{-6})$

**15–66.** $\epsilon_1 = 441(10^{-6})$, $\epsilon_2 = -641(10^{-6})$,
$\theta_{p1} = -24.8°$, $\gamma_{\text{in-plane}}^{\max} = -1082(10^{-6})$,
$\theta_s = 20.2°$

**15–67.** $\epsilon_{\text{avg}} = 275(10^{-6})$, $\epsilon_1 = 368(10^{-6})$,
$\epsilon_2 = 182(10^{-6})$,
$\theta_{p1} = 52.8°$, $\gamma_{\text{in-plane}}^{\max} = -187(10^{-6})$, $\theta_s = 7.76°$

**15–69.** $\epsilon_{\text{avg}} = 290(10^{-6})$, $\epsilon_1 = 385(10^{-6})$,
$\epsilon_2 = 195(10^{-6})$, $\theta_{p_1} = 54.2°$
$\gamma_{\text{in-plane}}^{\max} = 190(10^{-6})$, $\theta_s = 9.22°$

**15–70.** $\epsilon_1 = 341(10^{-6})$, $\epsilon_2 = -291(10^{-6})$,
$\theta_{p_1} = 30.3°$

**15–71. a)** $\varepsilon_1 = 1434(10^{-6})$, $\varepsilon_2 = -304(10^{-6})$,
**b)** $\gamma_{\text{in-plane}}^{\max} = 1738(10^{-6})$, $\varepsilon_{\text{avg}} = 565(10^{-6})$

**15–77.** $k = 7.78\,(10^3)$ ksi

**15–78.** $\nu_{pvc} = 0.614$

**15–79.** $\varepsilon_{max} = 30.5(10^{-6})$, $\varepsilon_{int} = \varepsilon_{min} = -10.7(10^{-6})$

**15–81.** $\epsilon_{y'} = \epsilon_{x'} = 2.52(10^{-3})$

**15–82.** $\varepsilon_{max} = 0.755(10^{-3})$, $\varepsilon_{int} = 0.621(10^{-3})$,
$\varepsilon_{min} = -1.31(10^{-3})$

**15–83.** $E_p = 769$ ksi, $e = 0.254(10^{-3})$

**15–85.** $\sigma_1 = 60.4$ MPa, $\sigma_2 = -31.9$ MPa

**15–86.** $\nu = 0.309$, $E = 28.1(10^3)$ ksi

**15–87.** $p = 3.43$ MPa, $\tau_{\substack{max \\ in\text{-}plane}} = 0$, $\tau_{max}^{abs} = 85.7$ MPa

**15–89.** $P = 83.0$ lb

**15–90.** $\varepsilon_{max} = 289(10^{-6})$, $\varepsilon_{int} = 0$,
$\varepsilon_{min} = -289(10^{-6})$

**15–91.** $\Delta d = 0.800$ mm, $\sigma_{AB} = 315$ MPa

**15–93.** $\epsilon_x = \epsilon_y = 0$, $T = 41.3$ N · m

**15–94.** $\delta V = \dfrac{1 - 2\nu}{2E}(2PL + w_0L^2)$

**15–97.** $k = 1.35$

**15–99.** $\varepsilon_1 = \dfrac{pr}{2Et}(2 - \nu)$, $\delta d = 2.72$ mm,
$\varepsilon_2 = \dfrac{pr}{2Et}(1 - 2\nu)$, $\delta L = 1.60$ mm

**15–101.** $\sigma_x = \sigma_y = -70.0$ ksi, $\sigma_z = -55.2$ ksi

**15–102.** $\varepsilon_x = \varepsilon_y = 0$, $\varepsilon_z = 5.44(10^{-3})$

**15–103.** $\Delta T = 67.7°F$

**15–105.** $\sigma_1 = 236$ psi, $\sigma_2 = -236$ psi

**15–106. a)** $\sigma_1 = 310$ psi, $\sigma_2 = -1160$ psi,
**b)** $\gamma_{\substack{max \\ in\text{-}plane}} = 735$ psi, $\sigma_{avg} = -425$ psi,
$\theta_{p_1} = 62.7°$, $\theta_{p_2} = -27.3°$,
$\theta_s = 17.7°$ and $-72.3°$

**15–107. a)** $\epsilon_1 = 862(10^{-6})$, $\epsilon_2 = -782(10^{-6})$,
$\theta_{p_2} = 2.01°$ ↲,
**b)** $\epsilon_{avg} = 40(10^{-6})$, $\gamma_{\substack{max \\ in\text{-}plane}} = 1.64(10^{-3})$,
$\theta_s = 43.0°$ ↰

**15–109.** $\sigma_1 = 3.17$ ksi, $\sigma_2 = -3.86$ ksi,
$\tau_{\substack{max \\ in\text{-}plane}} = 3.51$ ksi

**15–110.** $\sigma_1 = 4.00$ ksi, $\sigma_2 = -0.0317$ ksi

**15–111.** $\sigma_{x'} = 736$ MPa, $\sigma_{y'} = -156$ MPa,
$\tau_{x'y'} = 188$ MPa

**15–113.** $\sigma_1 = 32$ psi, $\sigma_2 = -32$ psi

# Chapter 16

**16–1.** $h = 6$ in.

**16–2.** $h = 4\dfrac{3}{4}$ in.

**16–3.** $b = 3.40$ in.

**16–5.** W12 × 16

**16–6.** W12 × 26

**16–7.** Yes

**16–9.** 0.394 in.

**16–10.** $w = 1.66$ kip / ft

**16–14.** W14 × 22

**16–15.** 11.4 mm

**16–17.** $P = 12.5$ kip, $\tau_{req'd} = 466$ psi

**16–18.** $h = 8.0$ in., $P = 3200$ lb

**16–19.** $b = 4.00$ in.

**16–21.** 4.25 in.

**16–22.** W12 × 22

**16–23.** 0.595 in.

**16–25.** $\sigma = 100$ MPa

**16–26.** $\sigma = 582$ MPa

**16–27.** $v_1 = \dfrac{Pb}{6EIL}(x_1^3 - (L^2 - b^2)x_1)$,
$v_2 = \dfrac{Pa}{6EIL}[3x_2^2L - x_2^3 - (2L^2 + a^2)x_2 + a^2L]$

**16–29.** $\theta_A = \dfrac{Pa(a - L)}{2EI}$, $v_1 = \dfrac{Px_1}{6EI}[x_1^2 + 3a(a - L)]$,
$v_2 = \dfrac{Pa}{6EI}[3x(x - L) + a^2]$,
$v_{max} = \dfrac{Pa}{24EI}(4a^2 - 3L^2)$

**16–30.** $v_1 = \dfrac{Px_1}{12EI}(-x_1^2 + L^2)$,
$v_2 = \dfrac{P}{24EI}(-4x_2^3 + 7L^2x_2 - 3L^3)$,
$v_{max} = \dfrac{PL^3}{8EI}$

**16–31.** $v_1 = \dfrac{-Pb}{6aEI}[x_1^3 - a^2x_1]$,
$v_2 = \dfrac{P}{6EI}[-x_2^3 + (-2ab + 3b^2)x_2 - 2(b^3 + ab^2)]$

**16–33.** $\theta_A = \dfrac{Pab}{2EI}$,
$v_1 = \dfrac{P}{6EI}[-x_1^3 + 3a(a + b)x_1 - a^2(2a + 3b)]$,
$v_3 = \dfrac{Pax_3}{2EI}(-x_3 + b)$, $v_C = \dfrac{Pab^2}{8EI}$

**16–34.** $\theta_A = \dfrac{M_0a}{2EI}$, $v_{max} = -\dfrac{5M_0a^2}{8EI}$

**16–35.** $\rho = 336$ ft, $\theta_{max} = \dfrac{M_0 L}{EI}$, $v_{max} = -\dfrac{M_0 L^2}{2EI}$

**16–37.** $\theta_{max} = \dfrac{-M_0 L}{EI}$, $v = -\dfrac{M_0 x^2}{2EI}$, $v_{max} = -\dfrac{M_0 L^2}{2EI}$

**16–38.** $\theta_{max} = \dfrac{M_0 L}{3EI}$, $v_{max} = -\dfrac{\sqrt{3} M_0 L^2}{27EI}$

**16–39.** $|\theta_{max}| = \dfrac{M_0 L}{2EI}$, $v = \dfrac{M_0 x}{2EI}(x - L)$,

$v_{max} = -\dfrac{M_0 L^2}{8EI}$

**16–41.** $\theta_A = -\dfrac{3}{8}\dfrac{PL^2}{EI}$, $v_C = \dfrac{-PL^3}{6EI}$

**16–42.** $v_B = -\dfrac{11 PL^3}{48EI}$

**16–43.** $\theta_A = 0.0611$ rad, $v_A = -3.52$ in.

**16–45.** $\Delta_C = 0.895$ in.

**16–46.** $M_0 = \dfrac{Pa}{6}$

**16–47.** $\Delta_A = 0.781$ in.

**16–49.** W14 × 34

**16–50.** $\Delta = PL^2\left(\dfrac{1}{k} + \dfrac{L}{3EI}\right)$

**16–51.** $\Delta_A = PL^3\left(\dfrac{1}{12EI} + \dfrac{1}{8JG}\right)$

**16–53.** $\Delta_G = 5.82$ in.

**16–54.** $M_A = 0.639$ kip · in., $M_B = 1.76$ kip · in.

**16–55.** $d = 0.708$ in.

**16–57.** $A_x = 0$, $B_y = 35.0$ kip, $A_y = 15.0$ kip, $M_A = 40.0$ kip · ft

**16–58.** $T_{AC} = \dfrac{3w A_2 E_2 L_1^4}{8[3E_1 I_1 L_2 + A_2 E_2 L_1^3]}$

**16–59.** $C_x = 0$, $B_y = \dfrac{2P}{3}$, $C_y = \dfrac{P}{3}$

**16–61.** $A_y = B_y = 2.15$ kN

**16–62.** $A_x = 0$, $B_y = \dfrac{5wL}{4}$, $C_y = \dfrac{3wL}{8}$

**16–63.** $F_{sp} = \dfrac{5wk L^4}{4(6EI + kL^3)}$

**16–65.** $B_y = 634$ lb, $A_y = 243$ lb, $C_y = 76.8$ lb

**16–66.** $F_{sp} = \dfrac{3kw L^4}{24EI + 8kL^3}$

**16–67.** Segment $AB$: W16 × 26, Segment $BC$: W12 × 14

**16–69.** $\theta_A = 6.54°$ ↓

**16–70.** $\theta_A = \dfrac{8}{EI}$ ↱, $\theta_B = \dfrac{16}{EI}$ ↓, $\theta_C = \dfrac{40}{EI}$ ↓, $\Delta_C = \dfrac{84}{EI}$

**16–71.** $V_{max} = \dfrac{0.00652 w_0 L^4}{EI}$

# Chapter 17

**17–1.** $P_{cr} = \dfrac{5kL}{4}$

**17–2.** $d = \dfrac{5}{8}$ in.

**17–3.** $P_{cr} = 35.1$ kip

**17–5.** $d = \dfrac{9}{16}$ in.

**17–6.** $P_{cr} = 22.7$ kN

**17–7.** $P_{cr} = 46.4$ kN

**17–9.** $P_{cr} = 70.4$ kip

**17–10.** $P_{cr} = 1.30$ MN

**17–11.** $P_{cr} = 325$ kN

**17–13.** $d = 1.81$ in.

**17–14.** $P_{cr} = 2.92$ kip

**17–15.** $P_{cr} = 5.97$ kip

**17–17.** F.S. $= 2.19$

**17–18.** $P = 475$ kip

**17–19.** $P = 83.6$ lb

**17–21.** $P_{cr} = 28.4$ kip

**17–22.** $P_{cr} = 58.0$ kip

**17–23.** $P = 4.23$ kip

**17–25.** $P = 5.79$ kip

**17–26.** $d = 1\dfrac{3}{4}$ in.

**17–27.** $P = 4.57$ kip

**17–29.** $P = 29.9$ kN

**17–30.** $P = 207$ lb

**17–31.** $P = 2.42$ kip

**17–33.** No, $AB$ will fail.

**17–34.** F.S. $= 2.12$

**17–35.** $P = 63.6$ kip

**17–37.** $P_{cr} = 157$ kN

**17–38.** $P_{cr} = 471$ kN

**17–39.** $P_{cr} = 39.2$ kN

**17–41.** $T_2 = 92.2°C$

**17–42.** $0 < L/r < 49.7$, $P/A = 61.7(10^3)/(L/r)^2$; $49.7 < L/r < 99.3$, $P/A = 25$ ksi; $99.3 < L/r$, $P/A = 247(10^3)/(L/r)^2$

# Index

# Simply Supported Beam Slopes and Deflections

| Beam | Slope | Deflection | Elastic Curve |
|------|-------|------------|---------------|
| | $\theta_{max} = \dfrac{PL^2}{16EI}$ | $v_{max} = \dfrac{-PL^3}{48EI}$ | $v = \dfrac{-Px}{48EI}(3L^2 - 4x^2)$<br><br>$0 \le x \le L/2$ |
| | $\theta_1 = \dfrac{Pab(L + b)}{6EIL}$<br><br>$\theta_2 = \dfrac{Pab(L + a)}{6EIL}$ | $v \Big|_{x=a} = \dfrac{-Pba}{6EIL}$<br>$\cdot (L^2 - b^2 - a^2)$ | $v = \dfrac{-Pbx}{6EIL}(-x^2 - b^2 + L^2)$<br><br>$0 \le x \le a$ |
| | $\theta_1 = \dfrac{ML}{3EI}$<br><br>$\theta_2 = \dfrac{-ML}{6EI}$ | $v_{max} = \dfrac{-ML^2}{\sqrt{243}\ EI}$ | $v = \dfrac{-Mx}{6LEI}(x^2 - 3Lx + 2L^2)$<br><br>$0 \le x \le L$ |
| | $\theta_{max} = \dfrac{wL^3}{24EI}$ | $v_{max} = \dfrac{-5wL^4}{384EI}$ | $v = \dfrac{-wx}{24EI}(x^3 - 2Lx^2 + L^3)$<br><br>$0 \le x \le L$ |
| | $\theta_1 = \dfrac{3wL^3}{128EI}$<br><br>$\theta_2 = \dfrac{-7wL^3}{384EI}$ | $v \Big|_{x=L/2} = \dfrac{-5wL^4}{768EI}$ | $v = \dfrac{-wx}{384EI}(16x^3 - 24Lx^2 + 9L^3)$<br><br>$0 \le x \le L/2$<br><br>$v = \dfrac{-wL}{384EI}(8x^3 - 24Lx^2$<br>$\qquad\qquad + 17L^2x - L^3)$<br>$L/2 \le x \le L$ |
| | $\theta_1 = \dfrac{7w_0L^3}{360EI}$<br><br>$\theta_2 = \dfrac{-w_0L^3}{45EI}$ | | $v = \dfrac{-w_0x}{360LEI}(3x^4 - 10L^2x^2 + 7L^4)$<br><br>$0 \le x \le L$ |